2nd International Conference on Artificial Intelligence Applications and Technologies (AIAAT 2018)

IOP Conference Series: Materials Science and Engineering Volume 435

Shanghai, China
8 – 10 August 2018

ISBN: 978-1-5108-7445-9
ISSN: 1757-8981

Printed from e-media with permission by:

Curran Associates, Inc.
57 Morehouse Lane
Red Hook, NY 12571

Some format issues inherent in the e-media version may also appear in this print version.

Printed by Curran Associates, Inc. (2019)

For permission requests, please contact the Institute of Physics
at the address below.

Institute of Physics
Dirac House, Temple Back
Bristol BS1 6BE UK

Phone: 44 1 17 929 7481
Fax: 44 1 17 920 0979

techtracking@iop.org

Additional copies of this publication are available from:

Curran Associates, Inc.
57 Morehouse Lane
Red Hook, NY 12571 USA
Phone: 845-758-0400
Fax: 845-758-2633
Email: curran@proceedings.com
Web: www.proceedings.com

TABLE OF CONTENTS

(PHM) PROGNOSTIC AND HEALTH MANAGEMENT

OPTIMIZATION ALGORITHM

CONTROL STRATEGY

IOP Conf. Series: Materials Science and Engineering **435** (2018) 011001 doi:10.1088/1757-899X/435/1/011001

PREFACE

AIAAT 2018 was held in Shanghai, China during Aug.8-10, 2018. AIAAT 2018, which is co-organized by Shanghai University Of Engineering Science and Hong Kong Society of Mechanical Engineers (HKSME), sponsored by Donghua University and technical supported by York Univeristy. The conference provides a useful and wide platform both for displaying the latest research and for exchanging of research results and thoughts in broad Artificial Intelligence Applications and Technologies, Automation, Control and Robotics and other topics. The participants of the conference were from almost every part of the world, with background of either academia or industry, even well-known enterprise. The success and prosperity of the conference is reflected by high level of the papers received.

The proceeding is a compilation of the accepted papers which represents an interesting outcome of the conference. This book covers 5 chapters: Object Recognition; AI(artificial Intelligence) Platform; PHM(Prognostic and Health Management); Optimization Algorithm; Control Strategy

We would like to acknowledge all of those who supported AIAAT 2018. Each individual and institutional help was very important for the success of this conference. Especially we would like to thank the organizing committee for their valuable advices in the organization and helpful peer review of the papers.

We sincerely hope that AIAAT 2018 will be a forum for excellent discussions that will put forward new ideas and promote collaborative researches. We are sure that the proceedings will serve as an important research source of references and the knowledge, which will lead to not only scientific and engineering progress but also other new products and processes.

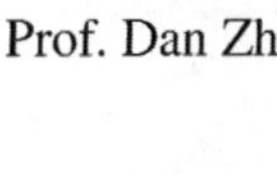

Prof. Dan Zhang, York University, Canada

IOP Conf. Series: Materials Science and Engineering **435** (2018) 011002 doi:10.1088/1757-899X/435/1/011002

COMMITTEES

Honorary Chair

Prof. Jianguo Xia, Shanghai University of Engineering Science , China

Conference Chairman

Prof. Dan Zhang, York University, Canada

Conference Committee Chair

Prof. Yansong Wang, Shanghai University of Engineering Science , China

Prof. Ruxu Du, The Chinese University of Hong Kong, Hong Kong

Program Committee Chair

Prof.Ge Chen, Donghua University, China

Prof. Jie Zhang, Donghua University, China

Publication Chair

Prof. Wilson Q. Wang, Lakehead University, Canada

Local Committee

Prof. Jinsong Bao, Donghua University, China

Technical Program Committee

Dr. Martin Stommel, AUT University, Auckland, New Zealand

Prof. Krzysztof Jemielniak, Warsaw University of Technology,Poland

Prof. Qin Jiahu, University of Science and Technology of China, China

Prof. Shao Cheng, Dalian University of Technology, China

Dr. Tansel Yucelen, University of South Florida, USA

Dr. Irene Fassi, Institute of Industrial Technologies and Automation, National Research Council, Italy

IOP Conf. Series: Materials Science and Engineering **435** (2018) 011002 doi:10.1088/1757-899X/435/1/011002

IOP Conf. Series: Materials Science and Engineering **435** (2018) 011003 doi:10.1088/1757-899X/435/1/011003

Peer review statement

All papers published in this volume of *IOP Conference Series: Materials Science and Engineering* have been peer reviewed through processes administered by the proceedings Editors. Reviews were conducted by expert referees to the professional and scientific standards expected of a proceedings journal published by IOP Publishing.

IOP Conf. Series: Materials Science and Engineering **435** (2018) 012001 doi:10.1088/1757-899X/435/1/012001

Superpixel Segmentation with Adaptive Nonlocal Random Walk

Hui Wang*

Beijing Laboratory of Intelligent Information Technology, School of Computer Science, Beijing Institute of Technology, Beijing, China

*316213375@qq.com

Abstract. In this paper, we propose a novel superpixel segmentation method using adaptive nonlocal random walk (ANRW) algorithm. There are three main steps in our image superpixel algorithm. Our method is based on the random walk model, and the seed points are produced to generate the initial superpixels by a gradient-based method in the first step. In the second step, the ANRW is proposed to get the initial superpixels by adjusting the nonlocal random walk (NRW) to be more suitable for image segmentation. In the last step, we merge these small superpixels to get the final regular and compact superpixels. The experimental results have demonstrated that our method achieves better performance than the state-of-the-art methods.

1. Introduction

Superpixel is an image segmentation method to cluster the pixels together that have the similar features such as colors and texture in an image. The concept of superpixel is first proposed in [1], and it was developed fast during the last decades. Now it has been widely used especially in the general tasks of image processing and computer vision.A lot of superpixel algorithms have been proposed so far, such as normalized cuts [2], Turbo-pixel [3], lazy random walk (LRW) [4], SLIC [5] and so on. Compared to the other superpixel segmentation methods, random walk [6-8] consider the relationship of the adjacent pixels, which can deal with complicated texture information and weak boundaries. However, the performance of these random walk based algorithms are sensitive to the initial seed points. By considering this drawback of the previous random walk based superpixel segmentation methods, we propose a novel superpixel algorithm using ANRW.

2. Our approach

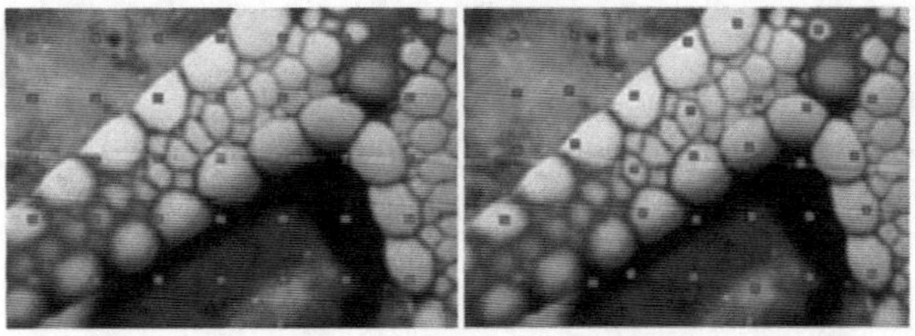

(a) (b)

Figure 1. Illustration of initial seeds and the final seeds. (a) is the initial seed points, and (b) is the final seed points.

In this section, we will introduce our superpixel segmentation method that mainly includes three stages. The first stage is to generate the gradient-based initial seeds, the second step is to obtain the initial superpixels by adaptive none-local random walk(ANRW), and the last step is to refine superpixels by locally merging.

2.1. Initial seeds generation

We first introduce some symbols to be used. N_1 is denoted as the number of desired superpixels. $[h, w]$ represents the size of input image. N=h*w is the number of pixels in the input image. $x_k(k = 1, 2, \ldots, N_1)$ is denoted as the pixels in the image and the corresponding coordinate is $[h_{x_k}, w_{x_k}]$.Firstly, we separate the image into N_1 uniform parts and use $part_i(i = 1, 2, \ldots, N_1)$ to denote each part.Then the points in the center of each part are selected as the initial seed points. We denote $S = \{s_1, s_2, \ldots, s_{N_1}\}$ as the set of initial seed points, where S_i is located in $part_i$ and its coordinate is $[h_{s_i}, w_{s_i}]$. Then we calculate the gradient of every pixels of the image, where $[hg'(x_k), wg'(x_k)]$ represents the absolute value of the gradient of the pixel x_k. We let $[hg(x_k), wg(x_k)]$ to denote the sum of the absolute values of the gradients among x_k and its 8-neighbouring points, which can reflect the smoothness of a small part around x_k. A function f(x) is designed to calculate the score of every pointswhich is used to choose the final seed points S'.

$$
\begin{aligned}
f(x_k) &= gra(x_k) + \alpha * dis(x_k), \\
gra(x_k) &= (hg(x_k) + wg(x_k)) / 2, \\
dis(x_k) &= \sqrt{\frac{N_1 * ((hx_k - hs_i)^2 + (wx_k - ws_i)^2}{N}}, \; if \; x_k \; is \; located \; in \; part_i
\end{aligned}
\tag{1}
$$

We set $\alpha = 0.5$. The $gra(x_k)$ is the constraint to make seed points located in the smooth area, and the $dis(x_k)$ is the constraint to get the regular shape of superpixels.We choose the final seeds based on the following rules: $x_k \in part_i$ is selected as the seed of $part_i$, if $f(x_k)$ is the minimum score among the points in $part_i$.

We show the example of initial seed points and final seed points in Figure 1 (a) and (b).

2.2. Get initial superpixels

For describing ANRW algorithm, we first define a graph $G = (V, E)$ on a given image $I(x)$, which represents a weighted graph containing a set of nodes V and edges $E \subseteq V \times V$. Then every pixel x_i is represented uniquely by a node vertex $v_i \in V$ in our undirected graph, where the degree of each vertex is computed as $d_i = \sum_j \omega_{ij}$ for all the edges connected on v_i. The edge-weight ω_{ij} measures the similarity between two adjacent nodes v_i and v_j. Different to the NRW [9], we make some adjustments to the edge-weight in NRW. We denote the edge-weight in NRW as ω'_{ij}, which is the following Gaussian weighting function:

$$
\omega'_{ij} = \exp(-\frac{||g_i - g_j||^2}{2\sigma^2}), \tag{2}
$$

Where g_i and g_j denote the image intensity values at two nodes v_i and v_j, and σ is the user defined parameter. The value of $2\sigma^2$ is fixed to 1/30 in our experiments.

To reduce the situation where the object boundaries are fall inside the superpixels, we design an adaptive adjusting strategy to improve the boundary adherence.We use N_{kind} to denote the number of different values of ω', and ω'_{sort} to denote these values in an ascending order. We define the adaptive ω as follows:

$$\omega_{ij} = \begin{cases} \lambda * \omega'_{ij}, \; if \; \omega'_{ij} \le \max(\delta, \omega'_{sort}(N_{Kind}/2)), \\ \omega'_{ij}, \; otherwise, \end{cases} \tag{3}$$

Where λ and δ are constants. We set $\delta = 0.15$ and $\lambda = 0.01$ in all experiments. This strategy can not only great reduce the possibility to cross a boundary during the random walk, but also decrease the sensitivity of the seeds.

The ANRW contains another kind of edge-weight which is the same to the KNN-linking edges [10] in NRW. We use ω_{ij}^{KNN} to denote the KNN-linking edge e_{ij}. We propose an adaptive method to choose the number of KNN-linking points according to the size of the superpixel. We choose $\sqrt{N / N_1}$ KNN linking points in our experiments.We apply Fast Library for Approximate Nearest Neighbors (FLANN) in VL-feat [11] to compute KNN in the feature space.The nonlocal weight ω_{ij}^{KNN} is defined as:

$$\begin{aligned} \omega_{ij}^{KNN} &= \exp(-\frac{||g_i - g_j||}{2\sigma^2} - \beta ||y_i - y_j||, \\ y_i &= [\frac{hx_i}{\sqrt{h^2+w^2}}, \frac{wx_i}{\sqrt{h^2+w^2}}], \\ y_j &= [\frac{hx_j}{\sqrt{h^2+w^2}}, \frac{wx_j}{\sqrt{h^2+w^2}}], \end{aligned} \tag{4}$$

We use E^8 denote the 8-neighbouring pixels, and E^{KNN} denote the KNN links. The combinatorial Laplacian matrix L is defined as:

$$L_{ij} = \begin{cases} 10\sum_{e^{ij}\in E^8} \omega_{ij} + \sum_{e^{ij}\in E^{KNN}} \omega_{ij}^{KNN}, \; if \; i = j, \\ -10\omega_{ij}, \; if \; i \ne j \; and \; e_{ij} \in E^8, \\ -\omega_{ij}^{KNN}, \; if \; i \ne j \; and \; e_{ij} \in E^{KNN} \\ 0, \; otherwise. \end{cases} \tag{5}$$

The reason that local weights are multiplied by 10 is to ensure local neighbours with similar feature have stronger influence to than KNN. We define the Dirichlet problem formulation as:

$$D(X) = X^T LX, \tag{6}$$

We re-arrange the vertices into two parts: the marked vertices (seed points) V_M and the unmarked vertices V_U ,$V_M \cup V_U = V$, $V_M \cap V_U = \varnothing$. According to this we can reorder the Laplacian matrix L and X. Followed by unmarked pixels, the marked pixels (seed points) appear first, then: $L = \begin{bmatrix} L^S & B \\ B^T & L^U \end{bmatrix}$, and $X = [(X^s)^T, (X^u)^T]^T$.

And at last we can rewrite the Dirichlet problem as:

$$D(X) = [(X^s)^T, (X^u)^T]^T L \begin{bmatrix} X^s \\ X^u \end{bmatrix}, \tag{7}$$

Differentiating $D(X)$ with respect to X^u, we can get:

$$L^u X^u = -B^T X^s. \tag{8}$$

After solving the above problem, we obtain the probabilities X^u of pixels belonging to the seed points. The probabilities indicate the pixels' similarities to the seed points. We assign different labels to seed points and assign the same label to an unmarked pixel which has largest probability among all seed points. After assigning the labels, we get the initial superpixels.

2.3. Refine the initial superpixels

Adaptive nonlocal random walk doesn't enforce the connectivity of superpixels. We refine the initial superpixels by taking a coarse-to-fine strategy. We first reallocate the small parts according to the color information and the location information. In this stage, but there are still a little very small parts need to be merge. Now we can directly randomly merge these parts into the superpixels which are adjacent to them since such small parts don't make much differences to the results.

In the first merge stage, we define two constant:

$$PA = \sqrt{N / N_1}, areathresh = PA^2 / 10. \tag{9}$$

Where PA^2 is the approximate area of a regular superpixel and areathresh is a threshold to distinguish connected parts. First, we need to relabel the connected parts to ensure every connected parts have unique labels. We separate the relabelled connected parts into two categories $area_r$ and $area_l$. If the number of points in a connected part is less than areathresh, we assign it into a set $area_r$, otherwise we assign it into the other set $area_l$, we need to merge all the parts in $area_r$ into $area_l$. We use the average value of the color and coordinate of the parts in $area_r$ and $area_l$ to help us to compute the combination distance. We merge $area_r(k)$ to the element in $area_l$ which has the smallest combination distance to it. After the first merge stage, we randomly merge the rest of parts to the connected parts adjacent to them, then we can get the final results.

3. Experimental results

We evaluate the performance of the proposed ANRW superpixel algorithm by comparing with state-of-the-art algorithms, including Turbo-pixel [3], SLIC [5], PB [12], LRW [4], fastSCSP[13], USEQ [14]. Their results are generated by running publicly available implementation by the original authors. All the experiments are performed on the Berkeley Segmentation Database (BSD) [15], which contains three hundred test images with human segmentation as the ground-truth data.

3.1. Visual Comparison

In Figure 2, we give the representative superpixel results by SLIC [5], fastSCSP[13], LRW[4], USEQ [14] and our algorithm. It is worth mentioning that our method performs well in terms of the weak boundary and small irregular object. As shown in Figure 2, our method successfully detects the edges between leaf and the green background (weak boundary) and the talon of falcon (small irregular object).

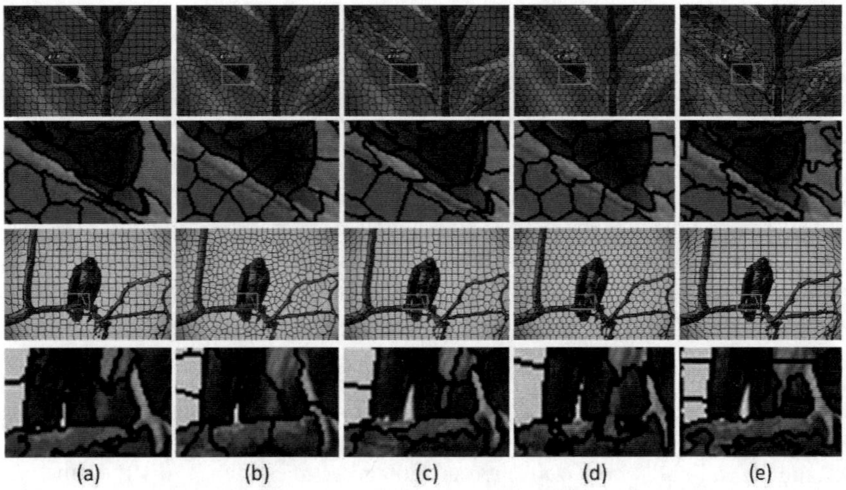

Figure 2. Comparisons of visual details among (a) our algorithm and the other four well-known algorithms: (b) LRW [4], (c) SLIC [5], (d) fastSCSP[13], (e) USEQ [14]. The second and fourth lines show the enlarged images of green rectangles in the first and third lines.

3.2. Quantitative Comparison

The boundary adherence is one of the most important requirements to the superpixel. There are three commonly used evaluation metrics [9] in superpixel segmentation for measuring the boundary adherence: under-segmentation error (UE), boundary recall (BR), and achievable segmentation accuracy (ASA). We show the results of UE, BR, ASA of all the methods for comparison in Figure 3.

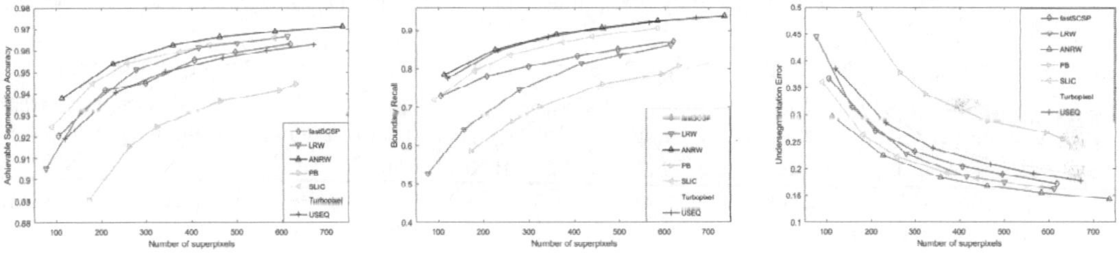

Figure 3. Quantitative evaluation on BSD benchmark: (a)the curves of UE; (b)the curves of BR; (c)the curves of ASA. Our ANRW algorithms has a better performance than the state-of-the-art methods (fastSCSP[13], LRW [4], PB [12], SLIC [5], Turbo-pixel [3], USEQ [14]).

Considering the three evaluation metrics, the BR of USEQ is almost the same as our method, but USEQ has a higher UE and lower ASA. It means that too many superpixels of USEQ have crossed the object boundary. While our method can reduce these errors by using ANRW and adjusting strategy.

Therefore, our method has the best performance among the state-of-art superpixels algorithms in comparison.

3.3. More Discussion

We now further analyze the computation complexity of the algorithms. The computation complexity of Turbo-pixel [3], USEQ [14], SLIC [5], PB [12], fastSCSP [13] is $O(N)$. Because LRW [4] and our method both use the symmetric and highly sparse matrix to solve large sparse linear system, the computation complexity of LRW and our algorithm are higher than the methods above. The computation complexity LRW is $O(n_l N^2)$, where n_l is the number of iterations. While LRW is slower than our algorithm which's complexity is $O(N^2)$, since ours doesn't include a iteration step.

4. Conclusion

We propose a novel superpixel segmentation method using adaptive nonlocal random walk. We use a gradient-based seeds generation method to generate the initial seed points and use ANRW to get the initial superpixels. At last we refine the initial superpixels to get the final results. In the experiment we measure the UE, BR, ASA of our algorithm and some state-of-the-art algorithms on the Berkeley Segmentation Database (BSD), and the experiment results show that our algorithm has better boundary adherence than these methods.

Reference

[1] X. Ren and J. Malik, -Learning a classification model for segmentation, in Proceedings of the IEEE International Conference on Computer Vision (ICCV), pp. 10-17, 2003.
[2] J. Shi and J. Malik, "Normalized cuts and image segmentation," IEEE Trans. Pattern Anal. Mach.Intell., vol. 22, no. 8, pp. 888-905, 2000.
[3] A. Levinshtein, A. Stere, K. N. Kutulakos, D. J. Fleet, S. J. Dickinson, and K. Siddiqi, "TurboPixels: Fast superpixels using geometric flows," IEEE Trans. Pattern Anal. Mach. Intell., vol. 31, no. 12,pp. 2290-2297, Dec. 2009.
[4] J. Shen, Y. Du, W. Wang, and X. Li, "Lazy random walks for superpixel segmentation," IEEETrans. Image Process., vol. 23, no. 4, pp. 1451-1462, 2014.
[5] R. Achanta, A. Shaji, K. Smith, A. Lucchi, P. Fua, and S. S, sstrunk, "SLIC superpixels compared to state-of-the-art superpixel methods," IEEE Trans. Pattern Anal. Mach. Intell., vol. 34, no. 11,pp. 2274-2282, Nov. 2012.
[6] L. Grady, "Random walks for image segmentation," IEEE Trans. Patt. Anal. Mach. Intell., vol. 28,no. 11, pp. 1768-1783, 2006.
[7] T. H. Kim, K. M. Lee, and S. U. Lee, "Generative image segmentation using random walks with restart," in Proc. ECCV, pp. 264-275, 2008.
[8] X. Dong, J. Shen, L. Shao, and L. Van Gool, "Sub-Markov random walk for image segmentation," IEEE Trans. Image Process., vol. 25, no. 2, pp. 516-527, Feb. 2016.
[9] H. Yuan, S. Wu, P. Cheng, P. An, and S. Bao, "Nonlocal random walks algorithm for semi-automatic2d-to-3d image conversion," IEEE Signal Processing Letters, vol. 22, no. 3, pp. 37-374, 2015.
[10] M. Muja and D. G. Lowe, "Scalable Nearest Neighbor Algorithms for High Dimensional Data," in IEEE Trans. Patt. Anal. Mach. Intell., Vol. 36, 2014.
[11] A. Vedaldi and B. Fulkerson, "VLFeat: An open and portable library of computer vision algorithms," [Online]. Available: http://www.vlfeat.org/ 2014
[12] Y. Zhang, R. Hartley, J. Mashford, and S. Burn, "Superpixels via pseudo-Boolean optimization," in Proc. IEEE Int. Conf. Comput. Vis., Nov. 2011, pp. 1387-1394.
[13] Freifeld O,Y. Li, and J. Fisher, "A fast method for inferring high-quality simply-connected superpixels," in Proceedings of IEEE Conf. Image Processing, pp. 2184-2188, 2015.
[14] C. Huang, W. Wang, and S. Lin, "USEQ: Ultra-fast superpixel extraction via quantization," in Proceedings of International Conference on Pattern Recognition, pp. 1965-1970, 2016.
[15] D. Martin, C. Fowlkes, D. Tal, and J. Malik, "A database of human segmented natural images and its application to evaluating segmentation algorithms and measuring ecological statistics," in Proceedings of the IEEE International Conference on Computer Vision (ICCV), pp. 416-423,2001.

IOP Conf. Series: Materials Science and Engineering **435** (2018) 012002 doi:10.1088/1757-899X/435/1/012002

Better Person Re-identification Using ResNet Model and Re-ranking Strategy

Nengchao Pan*

Beijing Laboratory of Intelligent Information Technology, School of Computer Science, Beijing Institute of Technology, Beijing, China

*pannengchao@126.com

Abstract. In this paper, we propose a novel person re-identification algorithm based on deep learning. To start with, deep features are extracted from images by ResNet model. Then a set of related images are selected under the low similarity constraint. In the end, we re-rank the retrieval result with related images. Our algorithm has three advantages. First, the transfer learning method reduces the cost needed to train the model compared with the others. Second, we employ the genetic algorithm under adaptive sparsity constraint to detect related images, which reduces the computation complexity. Third, the detected related images can be applied to promote the initial retrieval result. Experiments on Market1501 have demonstrated the effectiveness of our algorithm and our algorithm is robust against variations in background, pose, illumination and viewpoint.

1. Introduction

Person re-identification(ReID) aims to search an expected pedestrian's image(probe image) in the gallery for images containing the same person and those images were taken under non-overlapping environment. Person re-identification is an important task in the surveillance system and has various applications and re-identification was developed rapidly during the last decades for its great potential.

Although the person ReID problem has been studied for years, there are still several challenges. Pedestrians' images vary largely in background, pose, illuminance and viewpoint because these images were taken under non-overlapped environment. To tackle these problems, traditional methods are mainly studied in two aspects: designing a robust feature descriptor[1] and learning the distance metrics. Some representative methods, like symmetry-driven features[2], probabilistic colour histogram[3] and PSD constrained asymmetric metric learning[4], have promoted the experimental results. Deep learning is adopted for it can reduce the impact of human intervention on models. Ahmed[5] proposed a binary classification network to determine whether two images containing the same pedestrian. Xiao[6] mixed all databases to pre-train the model and finetune on the target database. State-of-the-art mainly studies the distance between probe image and each image in the gallery to obtain an initial ranking list. A good practice consists of adding a re-ranking step can improve the searching result.

We propose a novel re-ranking algorithm using ResNet model and related images. Our main contributions are summarized as follows:1)We finetune the pre-trained ResNet model on target person ReID database, and extract features from images with the trained model to obtain the initial ranking list.2)Use the genetic algorithm under adaptive sparsity constraint to detect related images in which each image is relevant to each other. Aggregate the original distance and the distance between related images and gallery images to generate the final distance. Experimental results on Market1501[7] show the effectiveness of our algorithm.

2. Our Approach

2.1. Problem Definition

Given a probe image and the gallery set with N images $G = \{g_i | i = 1,2, \dots, N\}$, we obtain an initial ranking list $\bar{L}(p, G) = \{g_i^0 | i = 1,2, \dots, N\}$, simplified as $\bar{L}$, by metric the distance between two images p and $g_i . d(p, g_i) = (x_p - x_{g_i})^T M (x_p - x_{g_i})$, where x_p and x_{g_i} respectively represents the feature of probe image p and gallery g_i, M is a positive semidefinite matrix. Our goal is to detect a set of images, called related images, in $\bar{L}$, and use the related images to re-rank $\bar{L}$ with the expectation of improving the results that more positive samples rank higher in the list than previous.

2.2. Initial Ranking List Generation

To ensure the features can represent the images and reduce the cost of computing, we finetune the pre-trained neural network model on target person ReID database. Transfer learning is very common in practical use, with the benefits of saving time and calculation. We choose the ResNet model to extract image features. ResNet model solves the problem of vanishing gradient and exploding gradient by shortcut connection. After finetuning the ResNet model on the train data of the target dataset, we take the outputs of the model as the features extracted from the images. The distance between probe image p and gallery image g_i is defined as follows:

$$d(x_p, x_{g_i}) = x_p \odot x_p + x_{g_i} \odot x_{g_i} - 2 * x_p \odot x_{g_i} \tag{1}$$

The initial ranking list $\bar{L}$ is acquired by sorting the gallery images according to their distance to the probe image in ascending order. Let the images of $\bar{L}$be $\bar{G} = \{\bar{g}_i | i = 1,2, \dots N\}$.

2.3. Detest Related Image

In this subsection, we describe how to detect the related images in the initial ranking list. As mentioned earlier[8], those images relevant to probe image should have high similarity to each other because the expected individual's images share the same visual pattern in different cameras. It means that the total distance of a positive sample to all other images in the initial ranking list should be smaller than the total distance of a negative sample because a negative sample does not share a similar visual pattern with other images in the ranking list. We name the set of images in the initial ranking list which is highly similar to each other related images. To detect related images from top M items in the initial sequence, we name a matrix $\boldsymbol{K} \in R^{M*M}$ in which $\boldsymbol{K}_{ij}$ is the distance between$\boldsymbol{x}_{\bar{g}_i}$ and $\boldsymbol{x}_{\bar{g}_j}$, let $\boldsymbol{s} = \{\boldsymbol{s}_i | i = 1,2, \dots, M\}$ be the sum of gallery image $\boldsymbol{x}_{\bar{g}_i}$ to every image in the top M items.

$$\boldsymbol{s}_i = \sum_{j=1}^{M} d(\bar{g}_i, \bar{g}_j) \tag{2}$$

Let $\boldsymbol{c} \in \{0,1\}^M$ be the index of our detection of related images in initial ranking list, $\boldsymbol{c}_i = 1$ means image $\bar{g}_i$ been chosen as related image, and vice versa for $\boldsymbol{c}_i = 0$. If the total distance of an expected image to others is low, the expected image is possible to be similar to other images and could possibly be a relevant image. In order to find the set of images which are most similar to each other, we aim to maximize the result of ||**s-Kc**||. We add a sparsity constraint to **c** to ensure the precision of selected images.

$$max\ ||\boldsymbol{s} - \boldsymbol{Kc}||_p^q + a||\boldsymbol{c}||_2 \tag{3}$$

$$s.t. \boldsymbol{c} \in \{0,1\}^M$$

Wherea is the sparsity constraint on **c**. To simplify the problem, we solve the dual problem of the original problem, as Equation 3.

$$max\ ||\boldsymbol{s} - \boldsymbol{K}\bar{\boldsymbol{c}}||_p^q + b||\bar{\boldsymbol{c}}||_2 \tag{4}$$

$$s.t.\bar{\boldsymbol{c}} \in \{0,1\}^M$$

Where$\bar{\boldsymbol{c}}$ is the index the unrelated images, **c** equals $\bar{\boldsymbol{c}}$ flipped. b is the constraint to ensure that most of the unrelated images are selected.

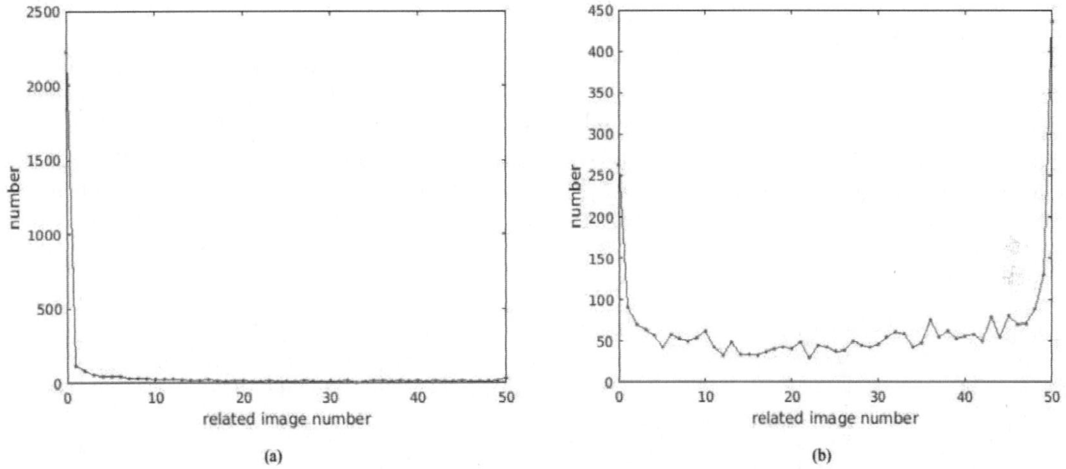

Figure 1. The impact of *b* on detection of related images. We detect related images in top 50 items. The horizontal axis represents the number of related image of an initial ranking list, the longitudinal axis represents the number of initial ranking lists. (a*) is the statistical result of* b=*20*, (b) is the resu*lt of* b = *30*

Both the original problem and the dual problem are NP-hard, and exhaustive search is infeasible to solve as the demand for computation is hard to meet. Instead, we obtain the suboptimal solution by optimizing. It is not doable to detect related images of different initial ranking list under the same b because it is not guaranteed that all the related images detected from the different initial ranking list meet the sparsity constraint, as shown in Figure1. We propose a related image detected algorithm based on the adaptive sparse parameter. Set the upper and lower bounds on the sparse parameter for every initial ranking list, the result of related images detection of a particular initial ranking list is iteratively calculated under current b by binary search the appropriate value of b, until the number of related images conforms to the sparse constraint. We use the genetic algorithm to detect related images because **c** is not differentiable and the number of related images depends on the value of b, the process of detecting related images from the expected initial ranking list at different weight may need to be iterated many times until the final result is obtained. The genetic algorithm has a small computational cost and can solve non-differential problems.

2.4. Final Distance

In this subsection, we focus on how to re-rank the initial ranking list with the related images. The importance of original distance cannot be neglected, so we jointly aggregate the original distance and the distance between related images and the gallery images, the final distance $d(g_i)$ is defined as equation 5.

$$d(g_i) = d(\boldsymbol{x}_p, \boldsymbol{x}_{g_i}) + vd(g_i, \boldsymbol{c}) \tag{5}$$

$$d(g_i, \boldsymbol{c}) = \sum_{j=1}^{N} c_j * d(g_i, \overline{L_j})$$

Where $v \in [0,1]$ denote the weight factor of the distance between related images and the gallery images, $d(g_i, \boldsymbol{c})$ is the total distance of a gallery image to all the related images. The final distance of a relevant image should be lower than the total distance of an uncorrelated image. Only the original distance is considered when v=0.

3. Experiment Result

Experiments are conducted on the Market1501[7]. Market1501 is the largest person re-identification dataset at the moment, consisting of 3368 images with 750 identities for querying, 12936 images with 751 identities for training and 19732 images with 750 identities for testing. We compare the two indices of rank-i and mAP with state-of-the-art algorithms to evaluate the proposed ResNet model and related image based person ReID algorithm. Rank-i measures the proportion of the queries that find the positive samples in top-i items in all the queries. Higher rank-i means the method can find at least one positive sample in top-i items faster. The mAP is the mean average precision, which reflects global performance.

Table 1. The impact of a, detect 2~3 related images in top-5 items

	Rank1	Rank5	mAP
NFST[9]	55.4	-	-
PersonNet[11]	48.2	-	-
S-CNN[12]	65.9	-	-
BoW-best[7]	44.4	63.9	-
WARCA[10]	45.16	-	-
CAN[13]	48.24	-	24.43
ResNet	73.9	88.24	51.52
ResNet+re-rank	74.91	87.35	53.44

We demonstrate the effectiveness of our approach by comparing with the following algorithm. Traditional methods, like NFST[9] is a representative metric learning algorithm which projects a high dimension vector to a low dimension vector by a projection matrix. BoW[7] is a method of encoding feature which aims at reducing intra class distance and increase inter class distance. WARCA[10] is a Mahalanobis distance based metric algorithm. Deep learning has promoted the effect. Both PersonNet[11] and S-CNN[12] are binary classification model. CAN[13] proposed a new soft attention based model. The results are exhibited in Table 1.

It is clear from the results that our algorithm outperforms other existing person ReID algorithms. The initial ranking list generated from ResNet model achieves a better result than the traditional methods and the deep learning methods mentioned above. And the re-ranking stage based on related images improves the ranking result. The main reason is that the feature extracted from images by

ResNet model is robust against variations of background, pose, illuminance and viewpoint, and the features can represent the images well. At the same time, the images detected from the initial ranking list are similar to each other, the related images are more likely to be relevant to the query image, and in the ideal case, the final ranking result is the total distance of the expected individual in multi-cameras, which can mitigate the difference between same class samples caused by taken under non-overlapping environment.

4. Conclusion

We present a novel person ReID method using ResNet model and related images. We use ResNet model to generate the initial ranking list, then detect related images from the list and aggregate the original distance and the distance between gallery images and related images to obtain the final ranking list. The experiments conducted on Market1501 show our algorithm achieves a better ranking result compared with some state-of-the-art algorithms.

Reference

[1] Liu C, Chen C L and Chen C L 2012 *International Conference on Computer Vision*Person re-identification: what features are important?pp.391–401.
[2] Farenzena M, Bazzani L and Perina A 2010 *Computer Vision and Pattern Recognition*Person re-identification by symmetry-driven accumulation of local features pp. 2360–2367.
[3] D'Angelo A and Dugelay J L 2011 *Proceedings of SPIE - The International Society for Optical Engineering*People re-identification in camera networks based on probabilistic color histogramspp.2155–2157.
[4] Liao S and Li S Z 2015 *IEEE International Conference on Computer Vision*Efficient PSD Constrained Asymmetric Metric Learning for Person ReIdentificationpp.3685–3693.
[5] Ahmed E, Jones M and Marks T K 2015 *Computer Vision and Pattern Recognition*An improved deep learning architecture for person re-identificationpp.3908–3916.
[6] Xiao T, Li H andOuyang W 2016 *Computer Vision and Pattern Recognition*Learning Deep Feature Representations with Domain Guided Dropout for Person Re-identificationpp.1249–1258.
[7] Zheng, Liang, ShenLiyue and Tian Lu, Wang Shengjin, Wang Jingdong and Tian Qi 2015 *IEEE International Conference on Computer Vision*Scalable Person Re-identification: A Benchmarkpp.1116-1124.
[8] Morioka N and Wang J 2011 *ACM International Conference on Multimedia*Robust visual reranking via sparsity and ranking constraints pp.533-542, 2011.
[9] Zhang Li, Xiang Tao and Gong Shaogang 2016 *Computer Vision and Pattern Recognition*Learning a Discriminative Null Space for Person Re-identificationpp.1239–1248.
[10] Jose, Cijo and Fleuret François 2016 *European Conference on Computer Vision*Scalable Metric Learning via Weighted Approximate Rank Component Analysispp.875-890.
[11] Wu Lin, ShenChunhua, Hengel and Anton Van Den 2016 *CoRP*PersonNet: Person Re-identification with Deep Convolutional Neural Networksvol.1601.07255.
[12] Varior, Rahul Rama, Haloi, Mrinal and Wang Gang 2016 *European Conference on Computer Vision*Gated Siamese Convolutional Neural Network Architecture for Human Re-identificationpp.791-808.
[13] Liu Hao, FengJiashi, QiMeibin, Jiang Jianguo and YanShuicheng 2016 *IEEE Transactions on Image Processing A Publication of the IEEE Signal Processing Society*End-to-End Comparative Attention Networks for Person Re-identificationvol.PP,number.99,pp.1-1.

IOP Conf. Series: Materials Science and Engineering **435** (2018) 012003 doi:10.1088/1757-899X/435/1/012003

Vibration State Classification and Identification of the Quayside Crane Motor Based on Genetic Algorithm

Chao Hu[1*], Gang Tang[1] and Xiong Hu[1]

[1]Logistics Engineering College, Shanghai Maritime University, 1550 Harbor Avenue, Pudong New Area, Shanghai, China

*1376454586@qq.com

Abstract: After a period of operation, the quayside crane motor will change. It is necessary to classify and identify the vibration state of the quayside crane motor. This paper presents an application of the genetic algorithm in vibration state classification and identification of the quayside crane motor. This article shows an engineering example and classifies and identifies the vibration state of the quayside crane motor in the last day successfully. This paper provides some suggestions on the use and maintenance of the quayside crane based on the vibration state of the quayside crane motor.

1. Introduction

Because of the rapid development of container transportation, the utilization rate of the quayside crane is higher and higher in port logistics transportation. Whether the normal operation of the quayside crane can be guaranteed will directly affect the working efficiency and economic benefit of the port. [1, 2] As a result, more and more crane motor operating state of the quayside crane is detected and evaluated.

Under normal running condition, the vibration of the motor output shaft is regular and the vibration intensity is smaller because of the symmetry of the structure. But with the motor running, the material fatigues and even the internal components damages of motor. For example, the inner and outer rings of the bearing appear to be punctured and rotor section or rotor eccentricity of motor will make the vibration law of motor output shaft become more complex and increase the vibration intensity. The working environment of quayside crane motor is complex, and the vibration law of motor output shaft is also more complicated, but the vibration intensity of motor output shaft has obvious linear relationship with motor damage [3]. By classifying the vibration intensity of the lifting motor, the classification and recognition of the vibration state of the lifting motor on the quayside crane can be realized [4].

Genetic algorithm (GA) is an adaptive global optimization search algorithm for simulating natural selection and genetic mechanism in biological world. It uses simple coding and reproduction mechanisms to solve very complex problems. It was first proposed by Professor Holland of Michigan University in the United States. The algorithm is robust, stochastic and global. In this paper, search five distribution centers of vibration intensity data of output shaft of crane motor for a period of time based on genetic algorithm and count the number of data at each central point [5]. The vibration intensity distribution of the output shaft of the crane motor is obtained during this period because of the fixed interval of each data acquisition time [6]. Some suggestions can be provided on the use and maintenance of the quayside crane based on the distribution because of the distribution can reflect the damage of crane motor [7].

2. Detection method of vibration state of the quayside crane motor

First of all, we collect data through sensors on the shore bridge. Then using genetic algorithm to find the classification center of crane vibration state from the data. Finally, based on the vibration classification center of the quayside crane motor, the collected vibration data are classified and counted.

2.1 Data collection

System of monitoring and assessment for crane based on intranet(Net-CMAS) can automatically obtain the mechanical condition detection data of the shore bridge, which brings advanced detection and monitoring technology for the maintenance and management of port equipment. The experimental data in this paper are derived from the vibration signal on the shaft of the crane in Net-CMAS. Because Net-CMAS collects data every 10 seconds, it can collect about 8,000 data a day. The raw data are shown in figure 1.

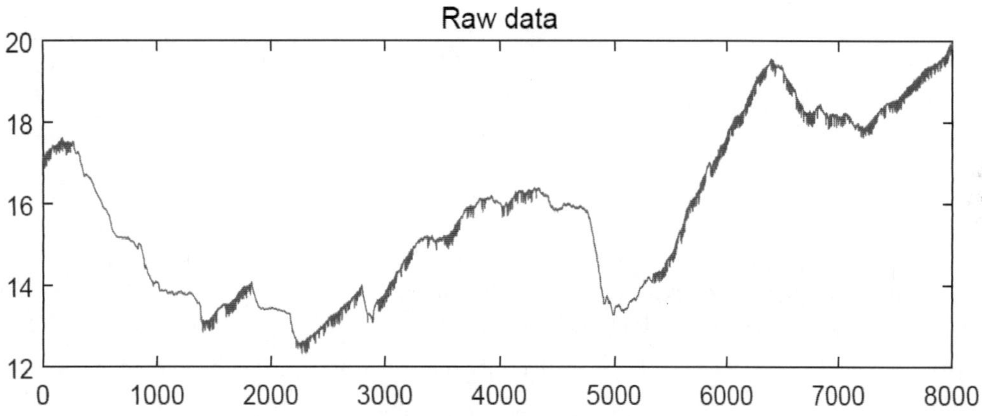

Figure 1. The raw data

2.2 Genetic algorithm

The calculation of fitness is very important in genetic algorithm. Fitness is the criterion for determining whether an individual needs to be retained. The adaptability in this article should include the following factors:

1) The Euclidean distance (dik) of the cluster center i and the other point k.
2) Comparison of Euclidean distance of other points k and cluster center i with cluster center j.
3) The fitness of different individuals should have a certain degree of differentiation, which is convenient for screening.

Therefore, the objective function equation is first established:

$$J = \sum_{k=1}^{n} \sum_{i=1}^{c} u_{ik} d_{ik}^2 \quad \ldots\ldots \tag{1}$$

In the form, uik is the matching function. It plays the role of comparing the rest of point k with cluster center i and its Euclidean distance with cluster center j; dik2 is the square of the Euclidean distance between the cluster center I and the other points k.

$$d_{ik}^2 = (x_i - m_k)^2 \tag{2}$$

$$u_{ik} = \frac{1}{\sum_{j=1}^{c} (d_{ik} / d_{jk})^{10}} \tag{3}$$

The larger the fitness degree, the smaller the objective function value should be, so the fitness function should be established:

$$f = \frac{J'}{J} \tag{4}$$

In the form, J is the minimum value of the objective function; J is the target function value.

The probability of crossover should be as follows:

$$P_c = \begin{cases} \frac{k_1}{2}\left(1-\sin\left(\frac{\pi}{1-\bar{f}}\left(f' - \frac{1+\bar{f}}{2}\right)\right)\right) & f' \geq \bar{f} \\ k_1 & f' < \bar{f} \end{cases} \tag{5}$$

In the form, f is the fitness of the individual; fmax is the highest fitness in the population; $\bar{f}$ is the average fitness in the population; k_1 is the blending coefficient.

The probability of mutation should be as follows:

$$P_m = \begin{cases} \frac{k_2}{2}\left(1-\sin\left(\frac{\pi}{1-\bar{f}}\left(f - \frac{1+\bar{f}}{2}\right)\right)\right) & f \geq \bar{f} \\ k_2 & f < \bar{f} \end{cases} \tag{6}$$

In the form, f is the fitness of the individual; fmax is the highest fitness in the population; $\bar{f}$ is the average fitness in the population; k_2 is the blending coefficient.

$$\begin{cases} N = \left[\frac{n}{500}\right] \\ k_1 = e^{-\frac{N}{20}} \\ k_2 = e^{-\frac{N^2}{10}} \end{cases} \tag{7}$$

In the form, refers to the total number of individuals generating the initial population; The algorithm for the operator "[]" is defined as "[a] denotes the minimum integer greater than a".

In this paper, the vibration state of the quayside crane motor is divided into five categories (They are named A B C D and E.), and the classification center increases in turn.

Based on the above adaptive genetic algorithm (GA), the MATLAB is used to find the five classifying centers of the vibration state of the crane motor in the data. Finally obtain the crane motor vibration status classification center (figure 2).

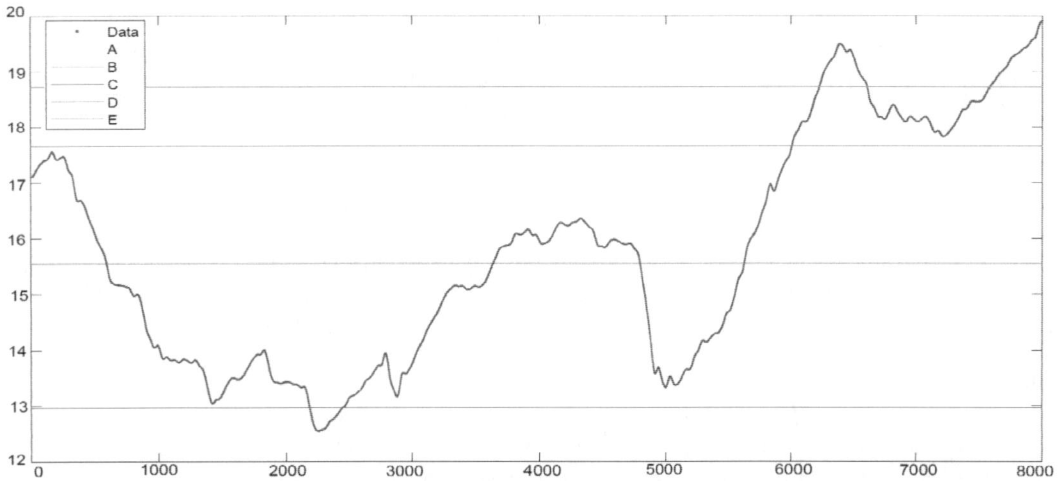

Figure 2. Classification center for vibration state of the quayside crane motor

Because Net-CMAS collects a data every 10 seconds, the proportion of the number of data in each class is equivalent to the proportion of time the motor works in this vibration state. Count the number of data in each crane vibration classification center and calculate the proportion of the data. The specific statistical results are shown in figure 3.

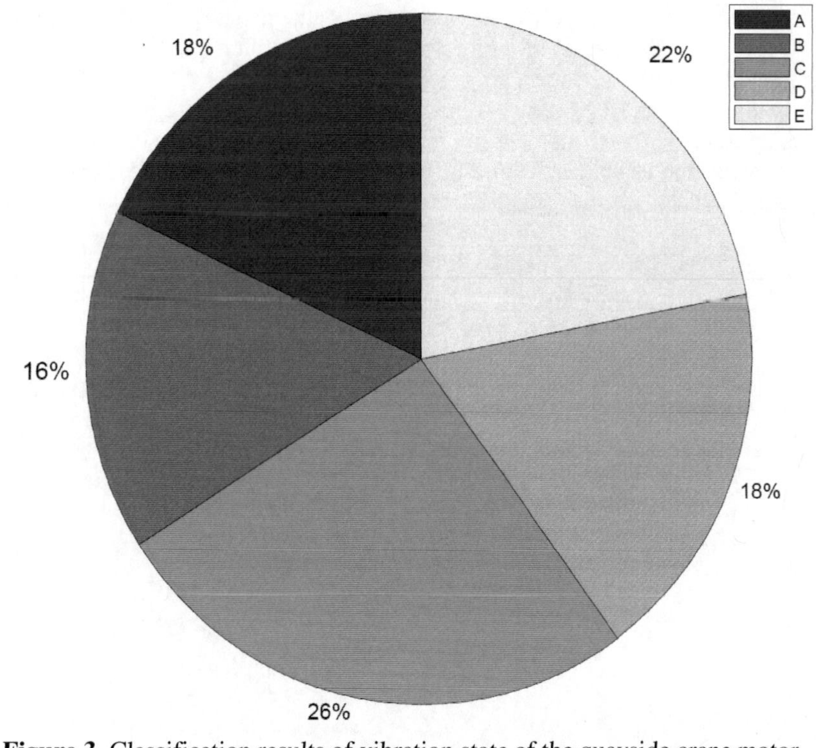

Figure 3. Classification results of vibration state of the quayside crane motor

3. Conclusion

In this paper, the classification center of vibration state of the quayside crane motor is successfully identified by genetic algorithm, and the working time distribution of the classification center of the vibration state of the quayside crane motor is obtained.For example in this article, the vibration state of the quayside crane motor is 26%B and 22%A, and the distribution centers of A and C are 13.49276 and 16.14292, respectively. Therefore, the degree of motor wear is not large and the quayside crane motor can be used normally as long as it needs normal maintenance.

4. Acknowledgements

This work was supported by the National Natural Science Foundation of China (No. 31300783), China Postdoctoral Science Foundation (No. 2014M561458), Doctoral Fund of the Ministry of Education Jointly Funded Project (No. 20123121120004), the Shanghai Maritime University Research Project (No. 20130474), the Shanghai Top Academic Discipline Project- management Science & Engineering, and the High-tech Research and Development Program of China (No. 2013A2041106).

References

[1] Wang, Zhixin, X. Hu, and Z. Chen. "Study of Remote Condition Monitoring and Assessing on Quayside-Container-Cranes." Springer London, pp. 898-903, 2006.
[2] Wen, Bo, et al. "Life cycle assessment of Quayside Crane: A case study in China." Journal of Cleaner Production, pp. 1-11, 2017.
[3] Delgado, M., A. García, and J. A. Ortega. "Evaluation of feature calculation methods for electromechanical system diagnosis." IEEE International Symposium on Diagnostics for Electric Machines, Power Electronics & Drives IEEE, pp. 495-502, 2011.
[4] Kuehne, Stephan. "Model assisted computing of three-phase asynchronous motor torque." 1999.
[5] Xiaofang Guo. "A Genetic Algorithm for a Class of Fractional Bilevel Programming Problems with Interval Coefficients." Advances in Applied Mathematics, pp. 63-69, 2015.
[6] YIN Y, HU X, LI Y. "Heavy lifting motor vibration research and health evaluation standard to determine." Vibration and Shock, pp. 67-71, 2013.
[7] Yin, Yinan, X. Hu, and T. R. Lin. "A practical approach to analyze the non-stationary signals of a quayside container crane motor using a combined empirical mode decomposition and wavelet packet quantization technique." Noise Control Engineering Journal, pp. 126-133, 2016.

A Deep Learning Approach for Face Detection and Location on Highway

Yang Zhang [1,2], Peihua Lv[1,2]and Xiaobo Lu[1,2,*]

[1]School of Automation, Southeast University, Nanjing 210096, China
[2]Key Laboratory of Measurement and Control of Complex Systems of Engineering, Ministry of Education, Southeast University, Nanjing210096, China

[*]E-mail: xblu2013@126.com

Abstract. Face detection and location technique is a hot research direction during recent years. Especially, driver face detection on highway is still a challenging problem in social safty deserving research. This paper proposes a novel algorithm based on the improved Multi-task Cascaded Convolutional Networks (MTCNN) and Support Vector Machine (SVM) to realize accurate face region detection and feature location of driver's face on highway, predicting face and feature location via a coarse-to-fine pattern. The proposed algorithm is verified under various complex highway conditions. Experimental results show that the proposed model shows satisfied performance compared to other state-of-the-art techniques used in driver face detection and alignment, keeping robust to the occlusions, varying pose and extreme illumination on highway.
Keywords. Driver face detection; Face alignment; Convolutional Networks; Support Vector Machine

1. Introduction

Face detection and alignment technology has been widely used in various practical fields, especially in driver face detection and alignment area, which involves public security and traffic order. With the fast development of digital image processing technology, various detection and alignment techniques have been proposed [1]. However, in real applications, there exit varying illumination, occasion as well pose affect the detection and alignment performance.

Face detection is a hot research direction in these years. In 2004, Viola [2] put forward cascade detection method based on AdaBoost with Haar-Like features to perform cascaded classifiers. Unfortuntely, some later researches [3, 4] indicate that it cannot keep continued competitiveness in real applications which affect the visual consistency of faces. Afterwards, deep CNNs are utilized in face detection. Yang et al. [5] proposed the deep neural networks for facial feature recognition. However, this algorithm is time consumely in real condition. Later, Kaipeng Zhang designed the MTCNN [6] model, which is consisted of three layers of precise deep convolutional networks. Nevertheless, due to its fixed training samples and limited network structure, when it comes to real condition, such as driver face detection, it show unsatisfied performance.

At the same time, face alignment also attracts wide study interests. The research fruits in this area can be approximately divided into two parts[7-12], regression-based techniques and template fitting methods. Nevertheless, previous methods about face detection and alignment methods overlooked the inherent relationship among these two issues.

In this paper, we propose a driver face detection and alignment model based on improved MTCNN and SVM[13], called IMT-SVM, forming adaptive detection and alignment model. We train this model on our own traffic driver face database, which is constructed by Public Security Department of Jiangsu Province, verifying the performance on the testing dataset. According to the experimental result, the IMT-SVM model shows high detection rate as well as low error detection rate compared to other state-of-the-art method.

2. Basic Theory

In this paper, we do research on the structural improvement of MTCNN and realize the fusion of SVM, improving the performance of the original algorithm. In this section, the basic theory about face detection and alignment which are utilized in our model are introduced.

2.1. MTCNN

The main purpose of MTCNN is to construct the image pyramid of the corresponding face image. The overall stages of MTCNN which is constituted of three convolutional networks are illustrated below.

2.1.1. Proposal Network (P-Net). The fully convolutional network is utilized to get the rough facial windows as well the corresponding vectors contain bounding box regression information. This can be concluded as a two-class classification issue which can be solved by the cross-entropy loss.

$$L_1^{\det} = -\left(y^{\det}\log(p_i) + \left(1 - y_i^{\det}\right)\left(1 - \log(p_i)\right)\right) \tag{1}$$

$$y_i^{\det} \in \{0,1\} \tag{2}$$

In Eq(1), xi is the input image, pi is the probability represents xi being the face. Eq(2) indicates thelabel of ground-truth.

2.1.2. Refine Network (R-Net). This layer of network play the role of calibration based on bounding box regression and NMS, aiming at rejects major false rough facial windows. This objective can be summed as a regression problem, as well overcomed by Euclidean loss.

$$L_1^{box} = \left\|\hat{y}_i^{box} - y_i^{box}\right\|_2^2 \tag{3}$$

In Eq(3), $\hat{y}_i^{box}$ and y_i^{box} represent the regression target calculated by the network and the corresponding real coordinate, respectively.

2.1.3.Output Network (O-Net). This stage proposes more supervisions to mark face region. Most important of all, this stage marks out five facial features' coordinates. Facial features detection belongs to the regression issue through minimizing the defined Euclidean loss:

$$L_1^{landmark} = \left\|\hat{y}_i^{landmark} - y_i^{landmark}\right\|_2^2 \tag{4}$$

In Eq(4), $\hat{y}_i^{landmark}$ and $y_i^{landmark}$ are the coordinates of facial features correspond to trained network and real condition for the i-th input image, respectively. The facial features are consisted of five feature points, including left eye, right eye, nose, left mouth as well as right mouth.

2.2. SVM

SVM is a classical classification method applied in pattern recognition field. SVM maps the pixels' data into the space consisted of higher dimensional which is contribute to constructing the optimal separating hyperplane, aiming at solving the quadratic programming and local minima issue. In our model, we propose SVM for classification issues of two classes, whether it is face region or not, realizing multiple classification judgment. The combination of SVM and MTCNN can be utilized in complex conditions.

3. The proposed method

3.1. The Whole Procedure of IMT-SVM

Fig.1 performs the procedure chart of the IMT-SVM algorithm. Firstly, the P-Net can get the rough facial windows in the input driver face image. Secondly, the accurate face region is labeled through R-Net. Thirdly, SVM model is utilized to judge whether it is driver face region. If not, the false sample will be deleted. Finally, the five facial features' coordinates are labeled through O-Net.

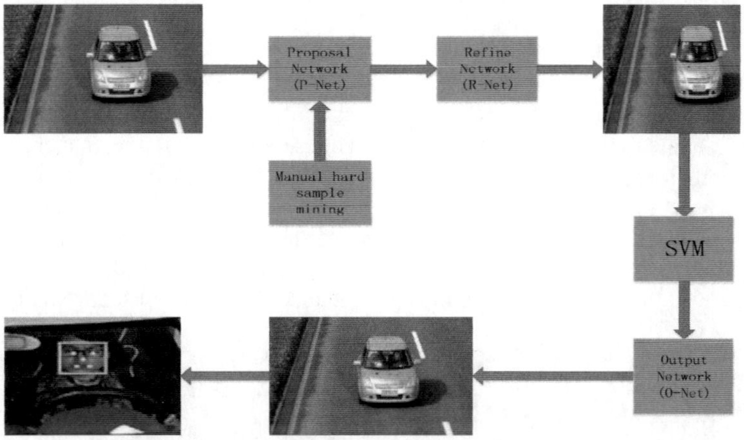

Figure 1. The Procedure of IMT-SVM algorithm

3.2. Architecture of Proposed Convolutional network

In IMT-SVM model, we make improvement on the architectures of P-Net, R-Net as well O-Net, which share similar architecture (Fig.2). The size of input image is 28*28 . C1 and C2 are the first and second convolutional layer, respectively, which consist of six feature maps. They share the same convolutional kernel of 5*5.

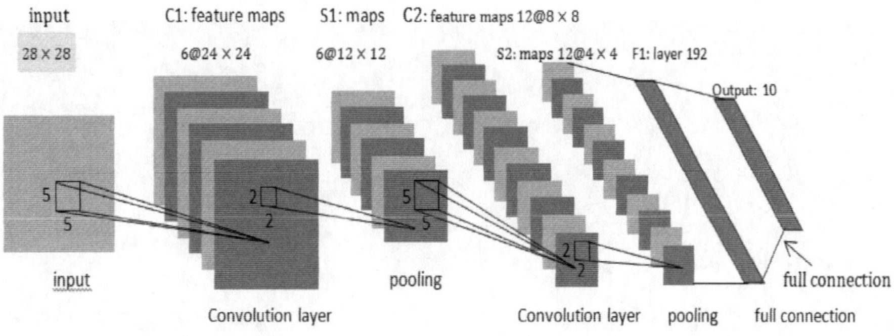

Figure 2. The Architecture of Proposed Convolutional network

3.3. Manual Hard Sample Mining

We use the training set of our own traffic driver face database to train this model. The testing set is utilized to verify the performance. What should be noticed is that, even though the online hard sample mining is conducted in MTCNN. However, in real application, especially in highway condition, as the result of complex environment, it is necessary to manual generate representative negative sample, strengthening the detector in training procedure. Experimental results indicate that this strategy performs better performance with adding manual hard sample, showing in Section IV.

4. Experimental Results

In this section, We use our own traffic driver face database, which is constructed by Public Security Department of Jiangsu Province, as dataset, containing approximately 1500 driver face images in different traffic conditions. We randomly choose 1000 images for training IMT-SVM model, remaining 500 images for verifying the model's performance. Double kernel CPU are used to train the networks. The performances of our method on face detection and alignment are shown below.

4.1. The Performances of Face Detection

Fig.3 and Fig.4 show the testing sample of IMT-SVM model under different environment. The detection and alignment results of IMT-SVM model in the dark environment and day time are performed. We can get the conclusion that the proposed model in this paper can realize good performance in complex environment on highway . Table 1 is the comparing table, indicating the detecting performances of IMT-SVM technique with MTCNN as well as Cascade CNN [14]. From the experimental result, the proposed technique in this paper shows high detection rate as well as low error detection rate compared to other state-of-the-art method [15]. This fully proves the IMT-SVM algorithm own excellent performance on improving the accuracy of driver face detection.

(a) Original image (b) Result picture (c) Magnified labelled face

Figure 3. The detection and alignment result of IMT-SVM model under the dark environment

(a) Original image (b) Result picture (c) Magnified labelled face

Figure 4. The detection and alignment result of IMT-SVM model during the day time

Table 1. Detecting Performances Between our Method and Other comparision Techniques

Detection Method	***IMT-SVM (OUR)***	***MTCNN [6]***	***Cascade CNN [19]***
Detection Rate	83%	75.6%	68.9%
Error Detection Rate	1.35%	11.28%	18.7%

4.2. The Performances of Face Alignment

Then, we compare the alignment performances of our method with MTCNN as well as Cascade CNN. Average alignment errors on left eye(LE), right eye(RE), nose(N), left mouth(LM) as well as right mouth(RM) of the above methods are shown in Fig.5. For nose tip, which is the most difficult in detection, the average error of IMT-SVM method is approximately 2.3% lower than MTCNN and about 4.2% lower than Cascde CNN. When it comes to other face features, including LE, RE, LM and RM, IMT-SVM algorithm' error rates still rank the lowest in the contrast methods. According to the result, we can prove that the IMT-SVM model shows outstanding performance on improving the precision of driver face alignment.

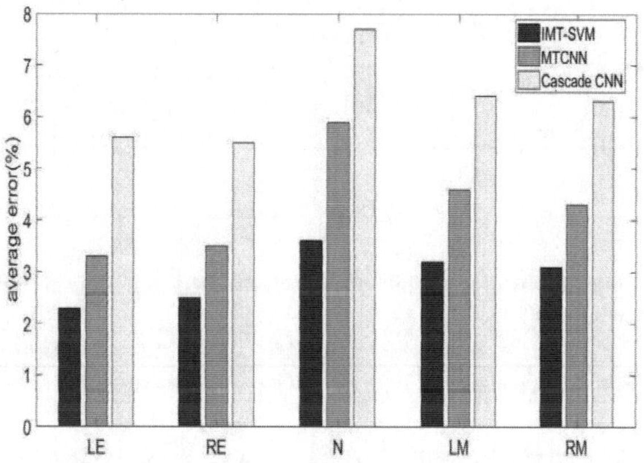

Figure 5. The alignment results of our model and other comparison techniques

5. Conclusions

This paper proposes a driver face detection and alignment model based on improved MTCNN and SVM. As face detection through convolutional network in MTCNN show unsatisfied error detection rate, it is desirable to use SVM for multiple classification judgment. In addition, improved MTCNN is proposed to realize precise facial point detection under complex highway environment. Experimental results show that the IMT-SVM model in this paper is effective, keeping robust to the occlusions, varying pose and extreme illumination on highway.

However, our work has some limitations under complex illumination environment and shadowed face images. Thus, in future work, denoising method [16][17] is a direction worth studying and making improvements[18][19].

Acknowledgments

This work was supported by the National Natural Science Foundation of China (No.61871123), Key Research and Development Program in Jiangsu Province (No.BE2016739) and a Project Funded by the Priority Academic Program Development of Jiangsu Higher Education Institutions.

References

[1] B. Yang, J. Yan, Z. Lei, and S. Z. Li, "Aggregate channel features formulti-view face detection," in IEEE International Joint Conference onBiometrics, 2014, pp. 1-8.
[2] P. Viola and M. J. Jones, "Robust real-time face detection.Internationaljournal of computervision," vol. 57, no. 2, pp. 137-154, 2004
[3] M. T. Pham, Y. Gao, V. D. D. Hoang, and T. J. Cham, "Fast polygonalintegration and its application in extending haar-like features to improveobject detection," in IEEE Conference on Computer Vision and PatternRecognition, 2010, pp. 942-949.
[4] Q. Zhu, M. C. Yeh, K. T. Cheng, and S. Avidan, "Fast human detectionusing a cascade of histograms of oriented gradients," in IEEE ComputerConference on Computer Vision and Pattern Recognition, 2006, pp.1491-1498.
[5] S. Yang, P. Luo, C. C. Loy, and X. Tang, "From facial parts responses toface detection: A deep learning approach," in IEEE International Conferenceon Computer Vision, 2015, pp. 3676-3684.
[6] Zhang K, Zhang Z, Li Z, et al. "Joint Face Detection and Alignment Using Multitask Cascaded Convolutional Networks," IEEE Signal Processing Letters, 2016, 23(10):1499-1503.
[7] Y. Sun, X. Wang, and X. Tang. Deep Convolutional Network Cascade for Facial Point Detection. In Proceedings of IEEE Conference on Computer Vision and Pattern Recognition (CVPR), 2013
[8] B. Amberg ,T. Vetter. Optimal landmark detection using shape models and branch and bound. In Proc. ICCV, 2011.
[9] P. N. Belhumeur, D. W. Jacobs, D. J. Kriegman, and N. Kumar. Localizing parts of faces using a consensus of exemplars. In Proc. CVPR, 2011.
[10] J. Zhang, S. Shan, M. Kan, and X. Chen, "Coarse-to-fine auto-encodernetworks (CFAN) for real-time face alignment," in European Conferenceon Computer Vision, 2014, pp. 1-16.
[11] X. Zhu and D. Ramanan. Face detection, pose estimation,and landmark localization in the wild. In Proc. CVPR, 2012.
[12] X. Cao, Y. Wei, F. Wen, and J. Sun. Face alignment by explicit shape regression. In Proc. CVPR,2012.
[13] Suykens J A K, Vandewalle J. "Least Squares Support Vector Machine Classifiers," Neural Processing Letters, 1999, 9(3):293-300.
[14] H. Li, Z. Lin, X. Shen, J. Brandt, and G. Hua, "A convolutional neuralnetwork cascade for face detection," in IEEE Conference on ComputerVision and Pattern Recognition, 2015, pp. 5325-5334.
[15] K. He, X. Zhang, S. Ren, J. Sun, "Delving deep into rectifiers: Surpassinghuman-level performance on imagenet classification," in IEEE InternationalConference on Computer Vision, 2015, pp. 1026-1034.
[16] D.Shi, "Face image information processing and recognition," Electronic Industry Press, 2010.
[17] Hu Changhui, Lu Xiaobo, Ye Mengjun, Zeng Weili. "Singular value decomposition and local near neighbors for face recognition under varying illumination," Pattern Recognition, 2017,64: 60-83.
[18] W. Zeng and X. Lu, "Region-based nonlocal means algorithm for noise removal," Electronics Letters, vol. 47, pp. 1125-1127, 2011.
[19] H. Li, Z. Lin, X. Shen, J. Brandt, and G. Hua, "A convolutional neuralnetwork cascade for face detection," in IEEE Conference on ComputerVision and Pattern Recognition, 2015, pp. 5325-5334.

IOP Conf. Series: Materials Science and Engineering **435** (2018) 012005 doi:10.1088/1757-899X/435/1/012005

Singer Recognition Based on Convolutional Deep Belief Networks

Yang Li[1] and Chu Li[2]

[1]School of Information Technology, Shanghai Jianqiao University, Shanghai 201306, China
[2]Pudong Foreign Languages School, Shanghai 201203, China

E-mail: 16083@gench.edu.cn

Abstract. Singer recognition is an important branch of music retrieval and classification. This paper focuses on the application of convolutional deep belief networks (CDBN) for singer recognition. First, the system architecture of singer recognition based on CDBN is given, and then the pre-processing of the song signal is described in detail, including sampling, framing, pre-emphasis and windowing. The feature extraction of song signal based on MFCC is described in detail, and the composition and principle of CDBN and its application in singer recognition are introduced. Experiment based on three different feature extraction techniques of LPCC, MFCC and CDBN is carried out and the result is compared and analysed, the experimental results show that CDBN is effective for singer recognition.

1. Introduction

With the rapid development of computer technology, network technology and multimedia technology, the amount of audio data on the Internet is increasing rapidly. How to effectively retrieve and classify audio information is a very important research topic [1].

Singer recognition is an important branch of music retrieval and classification. Its main purpose is to distinguish the singers from the unknown songs and find out which singers they belong to, so that they can automatically classify the songs by different singers [2]. Singer recognition can also be used for copyright protection, and record companies can automatically scan whether the songs provided by suspicious websites contain music that belongs to the company's copyright. In addition, self-service singing operators can use singer recognition technology to get users' favourite music types, so as to recommend some personalized music services.

Deep learning is a deep simulation of the human brain, and it can acquire the characteristics needed by itself. The deep network can analyse the input data by itself, and learn the characteristics of the data classification, which reduces the workload of the artificial extraction of classification features, and makes the classification recognition system more intelligent [3]. Therefore, it is very meaningful and valuable to study deep learning framework model and apply deep learning knowledge to music information retrieval task. This paper focuses on the application of convolutional deep belief networks (CDBN) in singer recognition.

2. Singer recognition system architecture

The singer recognition system is basically divided into two parts: training and testing. As shown in Figure 1, the training stage is mainly composed of selected training samples, pre-processing, feature extraction and CDBN training & modelling. The test stage is mainly composed of test samples input, pre-processing, feature extraction and CDBN testing [4]. Among them, the same pre-processing part of the two stages includes framing, windowing and pre emphasis. In the pre emphasis stage, the frequency spectrum of the signal can be easily analysed and processed by enhancing the high frequency signal in the signal. In the framing and windowing parts, the song signal is divided into a series of song frames based on Hamming windows based on the short time stationary characteristics of song signals. The feature extraction stage is the analysis and extraction of the pre-processed song frames in the time domain and frequency domain, so as to describe the characteristics of these song frames. For example, Mel frequency cepstrumcoefficient (MFCC) and its variants. The extracted feature vectors are compared with each singer's CDBN model to determine the result of recognition.

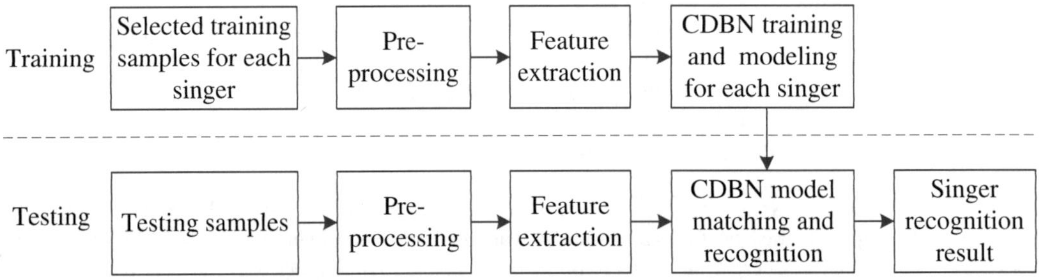

Figure 1. Singer recognition system architecture.

3. Pre-processing of song signals

3.1. Sampling

Both songs on the Internet and songs in CD belong to digital music. They all have a certain sampling frequency and encoding mode. The way of music coding and the frequency of sampling have a great influence on the analysis of music information. Therefore, it is necessary to unify the sampling frequency and coding mode of the music signal before framing. In most applications, digital music does not require too high sampling rate, and the high sampling rate will increase the computational complexity. In this paper, the sampling frequency of 22.05kHz is adopted, thus the efficiency of the algorithm is improved. Under the premise of not losing the basic recognition characteristics of music, the test music signal (44.1kHz) can be sampled, which has little influence on the recognition effect.

The song signals that are downloaded from the Internet can not be directly used to extract feature vectors. Before analysing and extracting the features, they should be handled in a unified way to meet the requirements of extracting features. This section introduces the pre-processing method. This stage mainly includes four steps: sampling, framing, pre-emphasis and windowing.

3.2. Framing

Considering the time continuity and short-time stability of the song signal, song signals should be processed in frames after sampling. The length of the frames will directly affect the extraction and recognition results.In order not to lose the information of the song signal change, the sliding window is used to divide the frame, and the window function smoothly slides on the song signal, and the signal is divided into frames. The framing can be continuous or overlapping, so that there is an overlap between the frames. The overlapping part is called frame shift. The frame shift value is generally 1/2 of the frame length. The framing is usually used with the windowing process described in section 3.4.

3.3. Pre-emphasis

Because of the radiation at the lip, the energy loss of speech or song signal energy is obviously higher than that at high frequency. Therefore, the aim of pre-emphasis is to improve the high frequency part of the speech signal, so that the frequency of the high frequency part becomes gentle, which helps to analyse the spectrum of the whole speech signal.

The influence of the energy loss of the speech or song signal is shown in the relationship between the frequency and the power spectrum: if the frequency of the speech signal is increased by two times, the amplitude of the power spectrum will be reduced by about 6dB. Therefore, theoretically, the pre-emphasis of the high frequency part should also be processed on the 6dB/oct value. In this way, the amplitude of the high-frequency part of the signal is increased by pre-emphasis, and the amplitude is not much different from that of the middle and low frequency parts.

Pre-emphasis can be implemented by hardware or software. This paper adopts software method. The method is to use a digital filter to process the sampled signal. The transfer function of the filter is: $H(z)=1-az^{-1}$, among it, a the pre-emphasis factor, the value is usually close to 1, such as a=0.9375.

3.4. Windowing

Both above framing and pre- emphasis are inseparable from the speech signal "short time analysis technology". The so-called short time analysis technology is based on the theory of short time stationarity of speech signals. Although the speech signal is constantly changing with time, however, in a very short time range, it is generally considered that the short time is 10-30ms, the characteristics of the speech signal are basically unchanged, which can be regarded as a short and stationary procedure, that is, the short time analysis technique can be used to process the speech signal. Correspondingly, in the short period of 10-30ms, the model parameters representing the characteristics of speech signals will not change.

The purpose of windowing is to divide the song signal into frames, and the frame length is N. For each frame, the time window function $\omega(n)$ is multiplied with the original voice signal $s(n)$, which contains the speech signal sequence of N points in the sample, and N is the length of the window.

There are usually two kinds of window functions commonly used in windowing.

(1) Rectangle window function:

$$\omega(n) = \begin{cases} 1, & 0 \le n \le N-1 \\ 0, & otherwise \end{cases} \tag{1}$$

(2) Hamming window function:

$$\omega_H(n) = \begin{cases} 0.54 - 0.46cos\left(\frac{2\pi n}{N-1}\right), & 0 \le n \le N-1 \\ 1, & otherwise \end{cases} \tag{2}$$

From the two window function formula, we can see that the window length of the window function has a great impact on the signal. When the length of the window is large, the window function can be regarded as a narrow low pass filter, the length of which is close to the period of several fundamental tones. The short time information of the signal is almost same, the detail of the waveform can be ignored; the length of the signal may be close to or even less than the cycle of a fundamental tone when N is relatively small, and the change of short time energy of the signal is obvious, and can't get smooth short time information.

The rectangular window is usually used in time domain analysis, and Hamming window is usually used in frequency domain analysis. It can be seen from the formulas that the boundary of the rectangular window is discontinuous, and the leakage will appear in the calculation. But the function boundary of the Hamming window is smooth, which can effectively avoid the leakage phenomenon.

4. Feature extraction of song signals

Feature extraction is crucial for singer identification. the features can present human voice include: short time energy, short time average amplitude, short time pitch period, linear prediction coefficient (LPC), partial correlation coefficient (PARCOR), line spectrum pair (LSP), short time spectrum, cepstrum feature, Mel frequency cepstrum coefficient (MFCC) and so on. MFCC is an auditory feature based on human ear, and it is also the most widely used feature parameter in the audio recognition model [5].

Human ears have different perceptual abilities to sound signals of different frequencies. In general, the human ear perception ability and the sound frequency are basically linear in the frequency region below 1kHz. But in the area above 1kHz, the perceptual ability of human ear is logarithmic with frequency. In order to simulate the special perceptual characteristics of human ears, the concept of Mel frequency was put forward. The conversion formula between the linear frequency of sound and Mel frequency is as follows:

$$f_{mel} = 2595 \times \log_{10}\left(1 + \frac{f}{700}\right) \tag{3}$$

The feature extraction process of MFCC is:

(1) Pre-processing of song signals, as described in section 3.

(2) First, fast Fourier transform (FFT) is used to get the spectrum of each frame, then the spectrum is squared to get the amplitude spectrum of each frame. The discrete Fourier transform (DFT) of the song signal is:

$$X_a(k) = \sum_{n=0}^{N-1} x(n)e^{-\frac{j2\pi kn}{N}}, 0 \le k < N \tag{4}$$

(3) A set of Mel filters is designed to filter the amplitude spectrum of the previous step. The logarithmic energy of the output of each filter group is calculated as:

$$\mathrm{s(m)} = \log_{10}\left(\sum_{k=0}^{N-1} |X_a(k)|^2 H_m(k)\right), 0 \le m \le M \tag{5}$$

(4) Then discrete cosine transform (DCT) to get MFCC:

$$\mathrm{C(n)} = \sum_{m=0}^{N-1} s(m)\cos(\frac{\pi n(m-0.5)}{M}), n = 1,2 \ldots, L \tag{6}$$

In this paper, we first pre-process the song signal and extract the MFCC feature, and then use this feature as the input of the deep belief network.

5. Singer recognition based on CDBN

5.1. The basic idea of deep learning

Deep learning is a new field in machine learning. It is similar to the neural network, which is similar to the layered structure of the biological brain. Its structure is that the system is a multi-layer network consisting of the input layer, the hidden layers (multi layers), the output layer, and each layer of the network can be regarded as a logistic regression model, the adjacent layer nodes of the model are interconnected, and there is no connection between the same layer and the cross layer node.

The basic idea of deep learning is for multiple stacking layers, the input of this layer is the output of the upper layer, that is, the hierarchical expression of input information. By adjusting the parameters in the system, a series of hierarchical features can be obtained automatically [6]. The process of extracting network parameters is widely used in feature extraction and pattern classification.

5.2. Restricted Boltzmann machine

Restricted Boltzmann machine (RBM) is energy based generative structural model, which is usually used to construct deep neural network structure [7]. RBM is an undirected graph, contains two layers of the visual layer v and the hidden layer h, the neuron nodes are not connected in the respective layers of the visual and hidden layers, but the interlayer is fully connected, as shown in Figure 2.

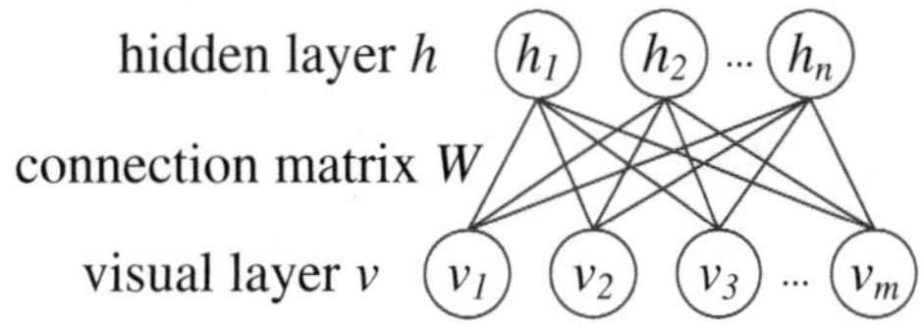

Figure 2. Restricted Boltzmann machine.

From Figure 2, we can see that there are *m* nodes in the visual layer of the RBM, which are used to represent the input data. The hidden layer has *n* nodes, which is used to extract features, the *W* representation layer to layer neuron connection weight matrix, and the calculation formula of the energy function is as follows:

$$E(v,h,\theta) = -\sum_{i=1}^{m}\sum_{j=1}^{n} v_i h_j w_{ij} - \sum_{i=1}^{m} b_i v_i - \sum_{j=1}^{n} a_j h_j \quad (7)$$

Among them, v_i and b_i represent the *i* unit of visual layer and its bias, h_j and a_j represent the *j* unit of hidden layer and its bias, and w_{ij} represents the weight of the connection between the *i* unit of visual layer and the *j* unit of hidden layer, θ is the actual parameter used to limit the RBM.

5.3. Deep belief network

In 2006, Hinton proposed the concept of deep belief network (DBN) and proposed an efficient learning algorithm [8]. It is a greedy layer by layer unsupervised training algorithm. It is a complex directed acyclic graph, consisted of RBM, so the same layer is not connected. The relationship between visual layer and hidden layer of DBN can be expressed by the joint probability distribution:

$$P(v,h^1,h^2,\ldots,h^l) = P(v|h^1)P(h^1|h^2)\ldots P(h^{l-2}|h^{l-1})P(h^{l-1}|h^l) \quad (8)$$

In this formula, *l* is the number of the DBN hidden layers.

DBN adopts the method of training RBM from low to high and layer by layer, and unlabelled data is used during this stage. After this pre-training, the traditional global learning algorithm, such as reverse propagation or wake-sleep algorithm, can be used to further adjust the network by using the labelled data, that is, to use the maximum like function as the target function to make the network reach the best. DBN overcomes the high complexity of DNN global training and converges to the local optimal solution caused by improper training of BP. The training mode of DBN can be summarized as two parts: first, the better initial parameter values are obtained by training RBM layer by layer, and then the network is further optimized through the traditional learning algorithm. Because the training method combining DBN and traditional learning algorithm has many advantages, it has been widely applied in the field of speech recognition.

5.4. Convolutional deep belief network

DBN has good flexibility, it can be easily extended to other networks or combined with other models, and one of the typical DBN extensions is the convolutional deep belief networks (CDBN). CDBN is a stack based on the convolution restricted Boltzmann machine (CRBM). CRBM is also divided into hidden layer and visual layer. Unlike RBM, the connection between the CRBM visual layer and the hidden layer is not a full connection. Each hidden layer node is connected only to a portion of the visual layer nodes, and each visual layer node is connected only to a portion of the hidden layer nodes. All hidden nodes are operated by the same convolution kernel. CRBM is characterized by partial receptive field and weight sharing, that is, the hidden layer and the visual layer are partially connected, and the weights of the models are shared. Weight sharing can reduce the number of parameters to be trained in the neural network. CDBN is a hierarchical probability generating model [9]. CDBN can be understood as a combination of DBN and convolutional neural network (CNN).

The training mode of CDBN is similar to DBN, unsupervised training is first carried out, and CDBN is trained using CRBM layer by layer. After pre-training, a supervised fine tune is adopted, and the weight value updating formula during the reverse propagation process is the same as in CNN [10].

The advantages of CBDN are:

(1) Absorbing the unsupervised training method of DBN, using a large number of unlabelled samples to initialize weights.

(2) Absorbing convolution concept of CNN, can use different convolution kernel to capture different features in music or song.

(3) By increasing the convolution step size, can reduce the number of hidden layer nodes.

(4) By sharing the weight, can reduce the number of parameters to be trained in the neural network.

CDBN is used in this paper for singer recognition because of above advantages.

6. Experiment and result analysis

6.1. Experimental environment

The experiment is done on personal computer. The software and hardware environment includes: CPU is Intel i7-4700, memory size is 16GB, operation system is Linux Ubuntu 16.04 LTS Server, deep learning framework is TensorFlow 1.0.1, programing software is Python2.7.

6.2. Song samples selection

In singer recognition experiment, the selection of singer has a great impact on the recognition result, because different singers may have different identities, and there are no open music databases for singer recognition. In this experiment, 10 singers were selected, with 5 male singers and 5 female singers. Each singer selected five songs as training samples, and the other five songs as test samples.

6.3. Pre-processing

Because songs are downloaded from the Internet, the encoding format is MP3 format, and the sampling rate is 44.1kHz, dual channel. In this paper, a unified sampling of these songs is used, the sampling rate is 22.05kHz. And cut the song into samples of 3s length. Because the song contains pure musical segments and silent segments, these segments have big impact on singer recognition. In this paper, samples of the 3s length segments are screened. If the singer's voice length is less than 1s, the sample is regarded as a pure musical segment and is discarded.

6.4. Feature extraction

For the CDBN, the number of convolution kernel used in this paper is 300, the size of convolution kernel is 8, the number of iterative rounds is 100, and the number of training times in each round is 100. The initial value of the connection weight used in the first tier model is the Gauss distribution random value with a mean value of 0 and a variance of 0.01. The bias of the visual layer takes is 0, and the bias of the hidden layer takes -0.1.

In order to compare the effect of feature extraction, the traditional features are also extracted. The traditional features mentioned here refer to LPCC feature and MFCC feature.

6.5. Comparison and analysis of experimental results

Because the focus of this experiment is to compare the feature extracted by deep learning with the traditional single feature, instead of comparing the model of the classifier, this paper uses the support vector machine (SVM) model as the classifier [11]. In the experiment, the three features and their corresponding classification labels are input into the SVM model respectively to train the classifier. After obtaining the classifier model, song segments of the 10 singers are classified and recognized. The recognition results are shown in Table 1.

IOP Conf. Series: Materials Science and Engineering **435** (2018) 012005 doi:10.1088/1757-899X/435/1/012005

Table 1. Recognition result comparison.

	LPCC	MFCC	CDBN
Recognition rate	53.6%	58.4%	77.8%

According to the results in Table 1, the recognition rate of the three features is not very high, mainly due to the interference of background music. But the recognition rate of CDBN is much higher than the other two features, that shows CDBN has better recognition effect on singer recognition than traditional LPCC and MFCC features. It proves that CDBN is effective in singer recognition.

7. Conclusions

On the basis of reading a large number of literatures about singer recognition, music information retrieval and speech processing, this paper studies the singer recognition from several aspects, such as the pre-processing of song signal, the extraction of auditory feature parameters, singing recognition and the establishment of the singer model. As a new feature extraction technology, deep learning has achieved success in the field of speech signal processing. This paper draws on the research results of deep learning in speech signal processing, based on the combination of music classification and deep learning theory, analyses the characteristics of CDBN in detail, and uses CDBN for singer recognition. In the experiment, for the music songs with background accompaniment on the Internet, the feature extraction based on LPCC, MFCC and CDBN are compared. The experimental results show that the CDBN learning features have better effects than traditional LPCC and MFCC features in singer recognition. The experimental results show that CDBN has great research value in singer recognition.

References

[1] Li T and Ogihara M 2006 Toward intelligent music information retrieval *J. IEEE Transactions on Multimedia***8(3)**564-574

[2] Yuan T T and Cao M M 2015 Voice-recognition-based music retrieval system *J. Bulletin of Science and Technology***31(7)** 156-159

[3] Liu F Y, Wang S H and Zhang Y D 2017 Survey on deep belief network model and its applications *J. Computer Engineering and Applications***54(1)** 11-18

[4] Gong A, Jing M B and Dou F 2017 Music mood classification method based on deep belief network and multi feature fusion *J. Computer Systems and Applications***26(9)** 158-164

[5] Lv L L 2016 Singing voice detection in songs based on clustering of MFCC *J. Computer Knowledge and Technology***12(31)** 170-171

[6] Xu X X 2017 Study on gesture recognition based on PCA and DBN *J. Artificial Intelligence***36(13)** 55-58

[7] Salakhutdinov R and Hinton G 2012 An efficient learning procedure for deep Boltzmann machines *J. Neural Computation***24(8)**1967-2006

[8] Hinton G E and Salakhutdinov R R 2006 Reducing the dimensionality of data with neural networks *J. Science***313(5786)** 504-507

[9] Yi L and Ya E 2017 Based on Gabor feature and deep belief network face recognition methods *J. Computer Simulation* **34(11)** 417-421

[10] Brosch T, Tam R 2015 Efficient training of convolutional deep belief networks in the frequency domain for application to high resolution 3D images *J. Neural Computation***27(1)** 211-227

[11] Chen W H 2016 Emotion classification of music based on support vector machine *J. Software Engineering***19(12)** 20-23

Forest Fire Smoke Recognition Based on Multiple Feature Fusion

Chang Lu[1,a], Mingqi Lu[2,3], Xiaobo Lu[2,3], Min Cai[2,3] and Xiaoqiang Feng[4]

[1] School of Software Engineering, University of Science and Technology of China, Hefei, Anhui 230026, China
[2]School of Automation, Southeast University, Nanjing 210096, China,
[3]Key Laboratory of Measurement and Control of CSE, Ministry of Education, Southeast University, Nanjing 210096, China
[4] Enbo Technology Company Limited, Nanjing 210022, China

[a] Corresponding author's e-mail address: 763397378@qq.com

Abstract. In order to discover the forest fire at the early stage, the video based fire smoke detection system should be developed. Three static features and three dynamic features are selected to recognize the fire smoke through the analysis of forest monitoring video. To solve the fusion problem of different features, a cascade classifier based on BP neutral network is designed. Each feature vector would be assigned a classifier as the first stage of the cascaded classifier. The output of the first level of the classifier used as the input of the second level of the classifier. The weight for each feature that given by the trainer is more scientific. The result of the experiments show that the cascade classifier has better performance compared to the single BP neural network.
Keywords: Fire smoke, Feature extraction, Cascade classifier

1. Introduction

Forest smoke recognition algorithm based on video analysis determines whether there is fire smoke in video through a series of processing of forest monitoring video[1][2]. The feature based smoke recognition method is a tradition method in fire smoke recognition[3]. It extracts appropriate dynamic and static features and designs effective classifiers to fuse and classify features for smoke recognition.

Töreyin proposes to use spatial wavelet transform to monitor the change of the high frequency component of the image, especially the change of the edge of the image[4]. He thinks that the edge of the image can provide the local extremum in the wavelet domain, which suggest the emergence of the fire smoke. Lin Wang et al. extract color features from RGB and HSI space and use two-dimensional wavelet transform to extract the background fuzzy features[5]. From Kolesov's point of view, the motion vector can be extracted as the motion feature by the optical flow method and SVM can be used as classifier which has a good performance[6]. Tiantian Tang introduces a local binary model to extract the texture features of the image and the experiment shows that the smoke texture features are effective in smoke recognition[7]. Chen T-H et al. determine whether the suspected region is a smog area through the static and dynamic features[8]. The static feature mainly refers to the color features extracted from the RGB space, while the dynamic feature refers to the diffusion characteristic of the smoke. Maruta H identifies the fire smoke by comparing the Hurst index of suspected smoke area and standard smoke image[9]. Ha C employs motion information for detecting the correct smoke block by using the characteristics that smoke goes almost upward[10].

This paper extracts 6 kinds of static and dynamic features to recognize the fire smoke region. The static features include the color feature, the texture feature and the edge contour feature, while the dynamic features include the motion direction feature, the area growth feature and the periodic flutter intensity feature. In addition, a cascaded classifier based on the BP neural network is designed to realize multi-feature fusion. Each feature vector would be assigned a classifier as the first stage of the cascaded classifier. The output of the first level of the classifier used as the input of the second level of the classifier, which is more efficient and solves the fusion problem of different features.

2. Feature Extraction

2.1 Static Features

We mainly consider 3 static features, including color feature, the texture feature and the edge contour feature to identify the fire smoke region.

2.1.1 Colour Feature. Color moment is a simple, effective and low dimension color feature developed from the concept “moment” in linear algebra. The pixel values in the image are regarded as random variables. The color information is mainly distributed in the low order moment. So the first order original moment which shows in (1), the second order center moment which showed in (2) and the third order center moment which showed in (3) are the most common used to describe the characteristic of a colored image. $p_{i,j}$ refers to the probability of occurrence of pixels in the channel i of the color image with the value of j, and N represents the number of pixels in the image.

$$\mu_i = \frac{1}{N}\sum_{j=1}^{N} p_{i,j} \tag{1}$$

$$\sigma_i = \left(\frac{1}{N}\sum_{j=1}^{N}(p_{i,j} - \mu_i)^2\right)^{\frac{1}{2}} \tag{2}$$

$$s_i = \left(\frac{1}{N}\sum_{j=1}^{N}(p_{i,j} - \mu_i)\right)^{\frac{1}{3}} \tag{3}$$

2.1.2Texture Feature. The fire smoke is composed of fine particles. When the smoke density reaches a certain level, the smoke completely blocks the background area in the camera monitoring scene. The uneven diffusion of smoke can lead to the diversity of the surface texture of the smoke area and have a certain effect on the high-frequency components, which can help us to identify the smoke area. Texture features can be quantified by gray level co-occurrence matrix. The gray level co-occurrence matrix contains the information of the direction, distance and amplitude of the image by calculating the correlation between the 2 pixels separated by a certain distance and a certain angle in the image.

2.1.3 Edge Contour Feature. Smoke spreads to the surrounding area with the air flow, which leads to the change of contour. Compared to the objects with fixed shapes such as vehicles or human beings, the contour of smoke has the irregular shapes. We can see that the length of the contours of the smoke are always longer than human beings and vehicles when they have same area. Therefore, the contour of the smoke can be measured by the correlation between the circumference and the area, which shows in (4). P refers to the circumference of the motion area while s represents the area of the motion area.

$$\gamma = \frac{p^2}{s} \tag{4}$$

2.2 Dynamic Features
We take the motion direction feature, the area growth feature and the periodic flutter intensity feature into consideration as dynamic features of the fire smokes.

2.2.1 The motion Direction Feature. There will be a lot of heat at the place of forest fire. The fire smoke presents a general trend of rising due to the effect of the heat and the air flow. The moving direction of people and vehicles is obviously distinct with the smoke. The analysis of moving direction of the moving area can also be one of the features to recognize the fire smoke along with other moving object.

We can use the overlap rate to trace the motion area. As it shows in Figure 1, we extract the moving area A and moving area B from frame k and frame k+1 and then calculate the overlap rate R_s following (5). In (5), S_A refers to the area of region A, while S_B refers to the area of the region B and $S_{A\cap B}$represents the overlap area of region A and region B.

$$R_s = \frac{S_{A\cap B}}{max\{S_A, S_B\}} \tag{5}$$

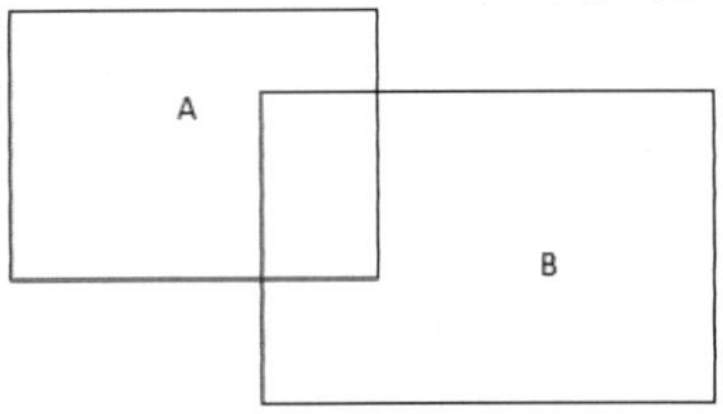

Figure 1. Trace of the moving area

Threshold T_s will be set and if $R_s > T_s$, area A and area B are regarded as the same moving area. The centroid of the region A and the region will be calculated respectively. The state of the moving area is fitted with the motion state of the centroid.

2.2.2 Area Growth Feature. As it shows in Figure 2, we extract the moving area in frame k and frame k+50. With the increasing of fire and the diffusion of smoke, the area of smoke region gradually grows. Detecting the relative growth rate of the area in the continuous frame can measure the diffusion characteristics of the smoke, which help us in fire smoke recognition.

(a) (b)

Figure 2. Change of the area of the smoke

We can assume that the area of region A in frame k equals to S_{A_k} and area of region A in frame k-Δ equals to$S_{A_k-\Delta}$. So the relative growth rate of the area $V_{_S_k}$can be calculate in (6).

$$V_{\_S_k} = \frac{S_{A_k} - S_{A\_k-\Delta}}{S_{A\_k-\Delta}} \tag{6}$$

2.2.3 Periodic Flutter Intensity Feature. In the case of forest fires, smoke has certain speed characteristics besides the movement direction characteristics under the influence of heat and air flow. Suppose that a series of centroid $C_t(t = 1,2,3, \ldots)$of the motion region A is a two-dimensional vector. $C_t[1]$ and $C_t[2]$ respectively represent the transverse and vertical coordinates of the centroid. The position of the centroid of the smoke area is shown in Figure 3.

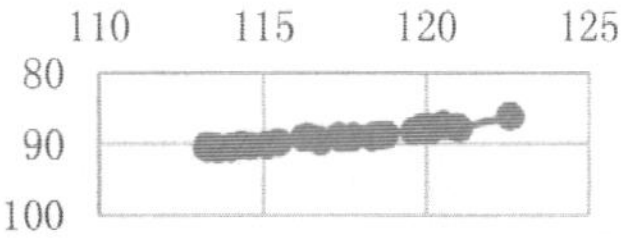

Figure 3. Position of the centroid of the smoke area

It can be seen that the moving direction of the centroid moves upward generally and the intensity of movement is regular. However, human beings have uncertain moving direction and irregular intensity of movement, as it shows in Figure 4.

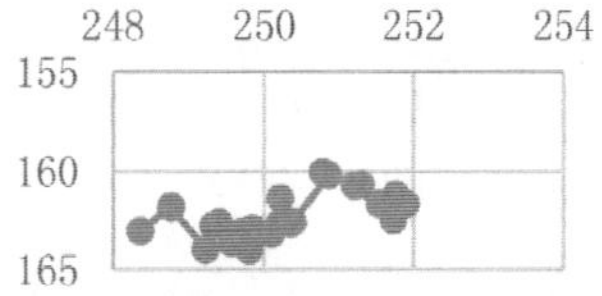

Figure 4. Position of the centroid of human beings

We definite the intensity of the movement in (7) to distinguish the fire smoke and other moving area. W_T refers to the number of frames in the video.

$$FI^* = \sum_{t=2}^{W_T} \sqrt{(C_{t-1}[1] - C_t[1])^2 + (C_{t-1}[2] - C_t[2])^2} \tag{7}$$

3. Classifier Design

Considering that the selected features have different orders and the optimal parameters of the classifiers for each features are not uniform, it is not rational to combine the feature vectors directly. The feature like texture feature who has small order can hardly affect the final result and we can't know the contribution of each feature to allocate the weight in the process of recognition. A cascaded classifier based on the BP neural network is designed in this paper. The specific structure is shown in Figure5, where M is the number of selected features.

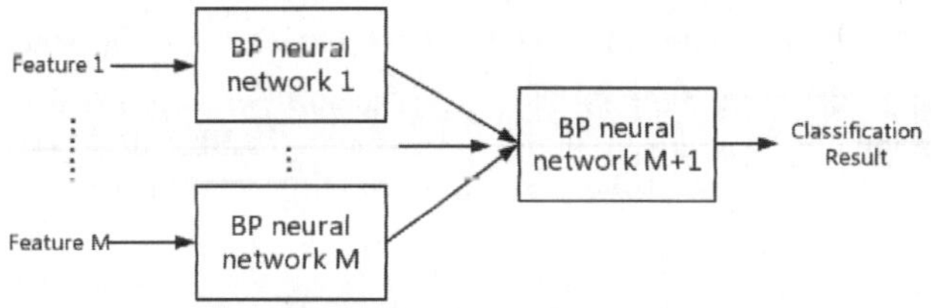

Figure 5. Structure of the classifier

The cascade classifier of double level BP neural network takes into account the self-learning and self-adaptive ability of the BP neural network. We use an independent BP neural network to train each type of feature as the first lever of the classifier. The second level of the cascade classifier uses a single BP neural network with M dimension inputs and one dimension output. The parameters for the classifier of each feature would not affect each other and the weights that given by the trainer are more scientific.

4. Experiment

In this paper, the recognition results are tested by 2 indicators, the recognition rate and efficiency rate, which show in (8) and (9), where N_{sc} refers to the number of smoke areas recognized correctly and N_{sf} refers to the number of smoke areas that are in fact exist. N_{dt} represents the number of smoke areas that are detected in total. The experiments are done from 2 aspects.

$$\text{Recognition Rate} = \frac{N_{sc}}{N_{sf}} \tag{8}$$

$$\text{Efficiency Rate} = \frac{N_{sc}}{N_{dt}} \qquad (9)$$

We would compare the recognition rate and the efficiency rate for the single BP neural network and our cascaded classifier. The results are shown in Table 1.

Table 1. Comparison between the single BP neural network and the cascade classifier

	Single BP Neural Network	Cascade Classifier
Recognition Rate	0.7009	0.9094
Efficiency Rate	0.4033	0.4098

We can see in Table 1 that the cascade classifier can get the recognition rate even higher than 0.9 while the single BP neural network only get about 0.7. Our cascade classifier has better performance. As it shows in Figure 6, the smoke area can be accurately recognized.

(a) (b)

Figure 6. Recognition results

5. Conclusion

This paper extracts 3 static features and 3 dynamic features that suitable for fire smoke recognition. In order to solve the problem of the fusion of different features, a cascade classifier based on BP neutral network is designed. The experiments show that the performance of the cascade classifier is good.

Acknowledgment

This work was supported by the National Natural Science Foundation of China (No.61374194), Key Research and Development Program in Jiangsu Province (No.BE2016739) and a Project Funded by the Priority Academic Program Development of Jiangsu Higher Education Institutions.

References

[1] Cai M, Lu X, Wu X, et al. Intelligent video analysis-based forest fires smoke detection algorithms[C]// International Conference on Natural Computation, Fuzzy Systems and Knowledge Discovery. IEEE, 2016:1504-1508.

[2] Song L, Wang B, Zhou Z, et al. The Research of Real-Time Forest Fire Alarm Algorithm Based on Video[C]// Sixth International Conference on Intelligent Human-Machine Systems and Cybernetics. IEEE, 2014:106-109.

[3] Yu C, Mei Z, Zhang X. A Real-time Video Fire Flame and Smoke Detection Algorithm [J]. Procedia Engineering, 2013, 62:891-898.

[4] Toreyin B U, Dedeoglu Y, Cetin A E. Wavelet based real-time smoke detection in video[C]// Signal Processing Conference, 2005, European. IEEE, 2005:1-4.

[5] Wang L, Li A. Early Fire Recognition Based on Multi-Feature Fusion of Video Smoke[C]// Chinese Control Conference. 2017:5318-5323.

[6] Kolesov I, Karasev P, Tannenbaum A, et al. Fire and smoke detection in video with optimal mass transport based optical flow and neural networks[C]// IEEE International Conference on Image Processing. IEEE, 2010:761-764.

[7] Tang T, Dai L, Yin Z. Smoke Image Recognition Based on Local Binary pattern[C]//

International Conference on Mechatronics, Materials, Chemistry and Computer Engineering. 2017:6-10.
[8] Chen T H, Yin Y H, Huang S F, et al. The Smoke Detection for Early Fire-Alarming System Base on Video Processing[C]// International Conference on Intelligent Information Hiding and Multimedia Signal Processing, 2006. Iih-Msp. IEEE, 2006:427-430.
[9] Maruta H, Nakamura A, Yamamichi T, et al. Image based smoke detection with local Hurst exponent[C]// IEEE International Conference on Image Processing. IEEE, 2010:4653-4656
[10] Ha C, Jeon G, Jeong J. Vision-Based Smoke Detection Algorithm for Early Fire Recognition in Digital Video Recording System[C]// Seventh International Conference on Signal-Image Technology and Internet-Based Systems. IEEE, 2012:209-212.

IOP Conf. Series: Materials Science and Engineering **435** (2018) 012007 doi:10.1088/1757-899X/435/1/012007

Improving Named Entity Recognition using Bilingual Constraints and Word Alignment

An T Dao[1], Thinh H Truong[1] , Long Nguyen , Dien Dinh
Faculty of Information Technology, University of Science, Hochiminh city, Vietnam
[1] These authors contributed equally to this work.
E-mail: dtan@apcs.vn, ththinh@apcs.vn, long.hb.nguyen@gmail.com, ddien@fit.hcmus.edu.vn

Abstract. Named entities carry essential meanings and information in natural language. Therefore, Named Entity Recognition has many applications in different Natural Language Processing tasks such as Information Retrieval, Information Extraction, Machine Translation, and Question Answering. State-of-the-art Named Entity Recognition systems are based on supervised machine learning algorithms which require huge amounts of training data. The main problem, however, is that constructing named entity annotated corpora is an expensive, labor-intensive, and time-consuming task. Therefore, in this paper, we propose an approach to improve monolingual Named Entity Recognition systems by exploiting an existing unannotated English-Chinese bilingual corpus. The system jointly recognizes named entities in both English and Chinese sentences through the use of bilingual constraints. Experimental results show an improvement in Named Entity Recognition of both Chinese and English compared to the strong baseline StanfordNER. In particular, Chinese Named Entity Recognition improves significantly by 20.81% in term of F_1-score. As for the English language, Named Entity Recognition F_1-score increases slightly from 75.75% to 76.08%. When comparing to the state-of-the-art system in improving Named Entity Recognition based on bilingual resources, we manage to outperform in Chinese Named Entity Recognition task by 5.99% and achieve comparable results for the English side. Our proposed method can also be generalized to apply to resource-limited languages.

1. Introduction

With the development of the Internet, the amount of data is growing exponentially [1]. Consequently, the demand for managing and interpreting data has become urgent [2]. In order to interpret information effectively, names of objects like persons, organizations, locations, and numeric expressions need to be recognized. The task of identifying references of these entities is called Named Entity Recognition (NER) which plays an important role in Information Extraction [3]. Named entities are also fundamental to other tasks such as Machine Translation [4], Question Answering [5], Information Retrieval[6], Co-reference Resolution [7], and so on.

Most state-of-the-art NER systems are based on supervised machine learning algorithms which typically require large amounts of high-quality annotated training data in order to produce accurate results [8, 9]. Because of the performance limit, most NER systems make these common types of errors [10] (example in Figure 1):

- NE Boundary Errors: Only some parts of an NE are recognized. For example, the entity "提姆·库克(Tim Cook)" in Chinese sentence is only recognized as "提姆(Tim)" having type PERSON (PER).

- NE Type Errors: NEs are tagged with incorrect types. For example, "Apple Inc." is tagged with label PERSON (PER) although the correct label is ORGANIZATION (ORG)
- Unrecognized NE Errors: NEs are not recognized at all. For instance, the word "美国(United State)" in the Chinese sentence is not recognized as an entity.

Nevertheless, annotating NE manually would take a lot of time and human resources. This problem restricts the availability of annotated corpus, especially for resource-limited language.

Instead of annotating data manually, another way to improve NER is to make use of information from a resourceful language like English. Different languages contain complementary clues which help in disambiguating each other. For example, in Figure 1 the word "库克(Cook)" in the Chinese sentence is not recognized as an entity. However, it is aligned with the entity [Cook, PER] in English, providing information that it might also be an entity of type PERSON.

Thanks to the use of cross-lingual corpora from Machine Translation task which has been extensively studied and improved, previous works showed improvements of NER performance. In 2012, Li et al. [11] proposed a novel model based on linear chain Conditional Random Field which projects features between English and Chinese through word alignment. The information is transferred on feature-level instead of label-level. The model combined both monolingual and bilingual features and performed decoding on two languages simultaneously to help improve the tagging process. However, this model relied on a bilingual corpus in which NEs are manually annotated. A method that formulates the problem of exploring complementary cues about entities on an unannotated parallel corpus between English and Chinese was described by Che et al. [12]. This inspiring idea was based on encoding bilingual features into objective functions which are solved by Integer Linear Programming technique [13]. Nonetheless, this work required initial Named Entity Recognition systems trained on corpora with BIO annotation scheme which are not widely available. According to these two works, using additional information from different languages could help correct the errors made by monolingual taggers. In particular, the high performance of English NER can help improve Chinese NER through word alignment.

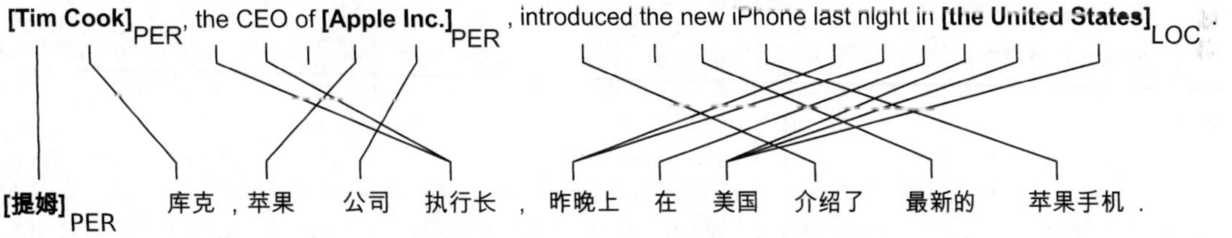

Figure 1. Example of NE errors in a parallel English-Chinese sentence pair.

To fix the mentioned Named Entity Recognition errors, our initial intuition is to make use of additional information in a bilingual corpus. These observations motivate us to make the following contributions:

- We explore a projecting method that takes advantage of an unannotated English-Chinese bilingual corpus aligned at sentence level to boost the performance of monolingual NER systems in this language pair (Section 2.3). This simple but effective method does not require any further annotations such as named entities annotation or gazetteers.
- We introduce an additional bilingual feature which penalize words tagged with none-type (Section 2.7) to improve the proposed method.

The experimental results (Section 3) agree with our intuition, in which the proposed method could improve the strong English and Chinese NER baselines. The related work is discussed in Section 4. Finally, Section 5 is the conclusion and future work.

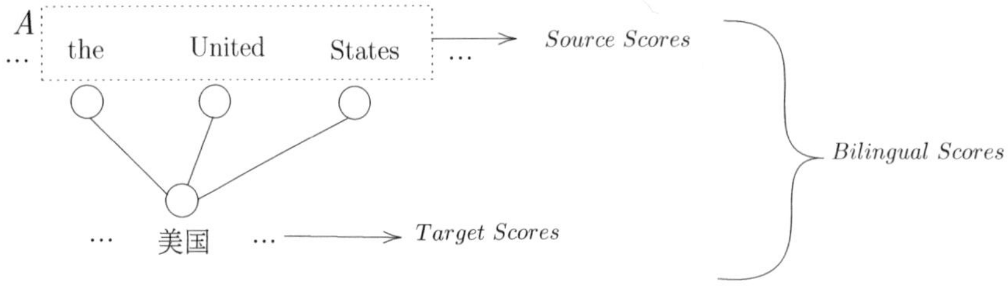

Figure 2. System Overview

2. Method

In this section, we describe in details the proposed system where NEs are jointly tagged in both English and Chinese sentences. Our main idea is to combine bilingual information obtained from both monolingual taggers and word alignment results to project NE tags. In this work, we focus on recognizing the four most common types of NE, which are geopolitical entities (GPE), location names (LOC), organization names (ORG), and person names (PER)

2.1. Overview

This section describes the general idea of our proposed system. Following the work of Che et al. (2013) [12], we re-use the Stanford Word Segmenter with LDC standard [14] to segment Chinese text into words. Given a pair of parallel sentences in English and Chinese, we first use StanfordNER toolkit [15] to recognize NEs in these sentences. Then, we project the NE tags from English to Chinese and vice-versa. In order to project NEs, for each word in the target sentence, we get the Constraint-based Monolingual NER scores P_{target} (Section 2.2). These scores show how the target sentence contributes in defining the tags of its own words, which describes its "self-context" information. This is when most monolingual Named Entity Recognition systems get their results.

Nonetheless, to take advantages of bilingual text, we also get the "cross-lingual context" of each word in the target sentence from the source sentence. In other words, we want to use additional information from the aligned words in the source sentence (P_{source}) to help annotate the corresponding target words with the correct labels. As a result, by combining these source and target context information, we calculate Bilingual Type Reassignment Score P_{bi}(Section 2.4) for each target word. However, not all words in the source sentence should contribute to finding the correct label of the target word. The main problem here is that we do not know which word in the source sentence should affect recognizing the type of the target word. Therefore, we need to find the alignment set A which contains suitable source words contributing to the probability of assigning type for the corresponding target word. This set A is constructed using two separate strategies: TopK (Section 2.3.1) and Threshold (Section 2.3.2). Finally, based on this Bilingual Type Reassignment Score, we assign the appropriate type for each word (Section 2.5).

From the example in Figure 1, suppose that we want to find the appropriate tag for the word "美国(United States)" in the Chinese sentence. The whole process of our system goes through these steps which are illustrated in Figure 2 :

- First, we calculate the Constraint-based Chinese Monolingual NER score $P_{\text{mono}}(T_R \mid$ 美国(United States)) for $T_R \in \{GPE, LOC, ORG, PER, O\}$ respectively.
- Next, we find the alignment set A of that word "美国". Assume that $A = \{\text{the}, \text{United}, \text{States}\}$.
- We calculate the scores of the alignment set A , which denote the contribution of source sentence in defining the tag of target word "美国". In detail, the probability that word

"美国" has tag T_R $P(T_R \mid$ 美国(United States)) is calculated based on $P(T_R \mid \text{the})$, $P(T_R \mid \text{United})$ and $P(T_R \mid \text{States})$ in the English sentence.

- Finally, we calculate the Bilingual Type Reassignment Score $P_{\text{bi}}(T_R \mid$ 美国(United States)) and tag it with type T_R - LOC which has the highest score.

The core of our projection method is Bilingual Type Reassignment Score, which describes the probability that a word w_T has type T_R in the target sentence is introduced in equation (1).

$$P_{\text{bi}}\left(T_R \mid w_T\right) = \lambda P_{\text{target}}\left(T_R \mid w_T\right) + (1-\lambda)\, P_{\text{source}}(T_R \mid A) \quad (1)$$

where the interpolation parameter λ (*mono_weight*), which indicates the weight of the target word monolingual NER score, will be tuned on the development set. This weight shows the contribution of "self-context" in defining the tag of w_T. $P_{\text{target}}\left(T_R \mid w_T\right)$ is the Constraint-based Monolingual NER probability that a target word w_T has type T_R. $P_{\text{source}}\left(T_R \mid A\right)$ is the probability that a target word w_T has type T_R given the alignment set A.

2.2. Constraint-based Monolingual NER

The Constraint-based Monolingual NER probability is obtained from StanfordNER toolkit. StanfordNER implements a linear chain Conditional Random Field (CRF) model [16], which is a popular approach in sequence-tagging tasks. At first, using StanfordNER, we get the conditional monolingual probability for each class as in equation (2).

$$P_{CRF}(y \mid x) = \frac{1}{Z(x)} \prod_i M_i(y_i, y_{i-1} \mid x) \quad (2)$$

where x is the input sequence, y is the output sequence, $Z(x)$ is the partition function, M_i is the clique potential for edge clique i.

In particular, we calculate $P_{\text{target}}(T_R \mid w_T)$ where $T_R \in \{GPE, LOC, ORG, PER, O\}$ which is the probability that the target word w_T has type T_R as in equation (3).

$$P_{\text{target}}(T_R \mid w_T) = P_{CRF}(y_i \mid x_i) \quad (3)$$

Where x_i word is recognized with tag y_i.

Previous work showed that word clustering could improve CRF tagger performance [12]. Therefore, we also use word clustering for both English NER and Chinese NER. In particular, we re-use the word cluster results obtained from the work of Che et al. (2013) [12] to train monolingual CRF taggers.

2.3. Projecting Method

In this section, our main projection mechanism is described. Theoretically, only word alignment is needed to project named entities from source language to target language. Thus, we can just simply assume two words aligned with each other would have the same type. Nonetheless, in practice, word alignment systems can have errors and require large bilingual corpora to generate reliable results. Specifically, word alignment for named entities is less reliable than other words since NEs are usually proper names and one proper name is not frequently repeated in the corpus (i.e., NEs are sparsely distributed) [10]. Therefore, instead of relying solely on the word alignment results, we consider the alignment probability to build the Alignment Set A. Corresponding to one target word, this set A contains source words that contribute to the probability of assigning the NE type of that word. The larger the alignment probability of the target word with the source word is, the more contribution the source word make to the tagging process of that target word. The source score for each word w_T is calculated as in equation (4)

$$P_{\text{source}}(T_R \mid A) = \frac{\sum_{w_S \in A} [1 + P_{\text{mono}}(T_R \mid w_S)]^{P_{\text{align}}(w_T \mid w_S)} - 1}{|A|} \tag{4}$$

where $P_{\text{align}}(w_T \mid w_S)$ is the word alignment probability that the target word w_T is aligned with the source word w_S which can be obtained from IBM models [17]. Similar to P_{target}, we can estimate $P_{\text{mono}}(T_R \mid w_S)$ by $P_{\text{mono}}(T_R \mid w_S) = P_{CRF}(T_R \mid w_S)$.

For example in Figure 1, assume that due to errors in word alignment, the alignment set A of "苹果(Apple)" contains ["Apple", ORG], ["Tim", PER], ["Cook", PER]. Nevertheless, the correct alignment "苹果(Apple)" - "Apple" should have far higher alignment probability compared to "苹果(Apple)" - "Tim" and "苹果(Apple)" - "Cook". Therefore, the word "Apple" with tag ORG contribute more in choosing the final tag for "苹果(Apple)" than the other two words. Thus, we are able to correctly tag "苹果(Apple)" as an entity of type ORG.

In order to find the alignment set A, we proposed two ranking strategies based on beam search: TopK and Threshold. Beam search has been proven to be effective in many Natural Language Processing tasks [18, 19].

2.3.1. TopK Beam Search In order to find the most appropriate source words, we need to rank all possible alignments according to some heuristics. Previous work showed the importance of word translation in Named Entity Alignment task [10]. Therefore, we use the word alignment probability between a target word and a source word obtained from IBM models [17] as the heuristic to compare. We rank all the alignments in descending order and get top K source words which have the highest alignment probability with the considered target word to obtain the alignment set A. Heuristically, the alignment set A is considered as the most appropriate source words set describing the "cross-lingual context". To find the optimal value for K, a development set is used.

2.3.2. Threshold Beam Search Similar to *TopK* approach, we also sort the possible alignments set in descending order based on word alignment probability. We get all source words w_S with word alignment probabilities greater than or equal to some threshold value θ as in equation (5).

$$A = \{w_S \mid P_{\text{align}}(w_T \mid w_S) \geq \theta\} \tag{5}$$

where $P_{\text{align}}(w_T \mid w_S)$ is the word alignment probability and the threshold θ would be chosen using a development set.

2.4. Bilingual Type Reassignment Score

Finally, we combine all the mentioned scores by using interpolation as in equation (6) to calculate the Bilingual Score, which describes the probability that a word w_T has type T_R in the target sentence.

$$\begin{aligned} P_{\text{bi}}(T_R \mid w_T) &= \lambda P_{\text{target}}(T_R \mid w_T) + (1-\lambda) P_{\text{source}}(T_R \mid A) \\ &= \lambda P_{\text{target}}(T_R \mid w_T) + (1-\lambda) \frac{\sum_{w_S \in A} [1 + P_{\text{mono}}(T_R \mid w_S)]^{P_{\text{align}}(w_T \mid w_S)} - 1}{|A|} \end{aligned} \tag{6}$$

where the interpolation parameter λ will be tuned on the development set

2.5. Type Assignment
For each word, we select the highest Bilingual Type Reassignment Score and assign it with the corresponding label, as in equation (7).

$$T_R(w_T) = \underset{T_R \in \{GPE, PER, LOC, ORG, O\}}{\text{argmax}} P_{\text{bi}}(T_R \mid w_T) \tag{7}$$

2.6. Distortion Feature
To further refine the Bilingual Type Reassignment Score, we apply the distortion feature. The intuition for this feature is that the positions of words in sentences are also clues to align target words and source words between languages. If a word appears at the ith position in English sentence, the aligned Chinese word may locate close to the ith position in Chinese sentence. The larger the difference in position is, the less possibility they are aligned together. Assume that a target word w_T appears at position pos_t in the target sentence S_T, and its aligned word w_S appears at position pos_S in the source sentence S_S. Following the work of Nguyen et al [10], the distortion score for w_T and w_S is calculated as in equation (8)

$$Dist(w_T, w_S) = 1 - f\left(\frac{pos_T}{len(S_T)} - \frac{pos_S}{len(S_S)}\right) \tag{8}$$

where $f(x)$ is the *distance* function. In this work, we try f with modulus function. Thus, the source score is redefined as in equation (9)

$$P_{\text{source}}(T_R \mid A) = \frac{\sum_{w_S \in A} [1 + P_{\text{mono}}(T_R \mid w_S)]^{P_{\text{align}}(w_T \mid w_S) * Dist(w_T, w_S)} - 1}{|A|} \tag{9}$$

2.7. O-type Penalty
Typically, the O label (not entity) appears far more frequently than other labels in a corpus. This leads to the unbalanced class problem of training data in which the number of instances for each class is not distributed equally. Therefore, we only tag a word with label O when the score corresponding to class O is considerably larger than the score of other classes, as in equation (10). We introduce the O-type penalty, which is the amount of difference in score between O and other classes needed to tag the word with type O. To find the optimal O-type penalty θ_O, we use a development set.

$$\forall T'_R \in \{PER, LOC, ORG, GPE\}, P(O \mid w) > P(T'_R \mid w) \Leftrightarrow P(O \mid w) - P(T'_R \mid w) \geq \theta_O \tag{10}$$

3. Experiment
3.1. Experimental Settings
3.1.1. Data We conducted experiments with the data used in the work of Che et al. (2013) [12]. In particular, this dataset is extracted from named entity annotated corpus of OntoNotes 4.0 and contains 4 classes: GPE, LOC, ORG, PER.

Training Set:

- We trained word alignment using BerkeleyAligner [20] with 8249 English-Chinese sentence pairs aligned at sentence level.
- For English NER, we trained StanfordNER using a training set of 10,000 sentences. The number of instances for each class is: GPE - 1891, LOC - 211, ORG - 1005, PER - 2044.
- For Chinese NER, we trained StanfordNER using a training set of 10,000 sentences. The number of instances for each class is: GPE - 1326, LOC - 500, ORG - 720, PER - 1440.

Test Set: The *TestSet* consists of 8249 English-Chinese sentence pairs. The detail distribution of each class is: English GPE - 6721, LOC - 596, ORG - 3525, PER - 3845; Chinese GPE - 6055, LOC - 932, ORG - 3387, PER - 3402.

3.1.2. Evaluation Metrics We evaluate NE recognition with 3 metrics: Recall (R), Precision (P), and F_1-score (F_1). The same *TestSet* is used for all experiments.

$$P = \frac{\text{number of correctly predict NE}}{\text{number of predict NE}}, R = \frac{\text{number of correctly predict NE}}{\text{number of reference NE}}, F_1 = \frac{2 \times P \times R}{P + R}$$

3.1.3. Comparison System The StanfordNER toolkit is used as a baseline to compare with our proposed system. We also compare with the work of Che et al. (2013) [12] (*Che2013*) which is considered the state-of-the-art method in improving Named Entity Recognition performance in English-Chinese.

3.2. Results

3.2.1. Monolingual NER Table 1 indicates the monolingual NER results of English and Chinese. From this table, we could see that the performance of class PER is the highest for both languages because it is the most common type in the training set. Overall, the difference in NER performance between English and Chinese is rather significant (32.92 % in term of F_1-score). The average results (*AVG*) of monolingual NER systems is then used as a baseline for comparison (*StanfordNER*) in further experiments.

Table 1. The monolingual NER Results of English / Chinese

Class	P(%)	R(%)	F_1(%)
LOC	57.86 / 31.13	39.06 / 21.70	46.64 / 25.57
ORG	75.68 / 72.53	62.84 / 23.89	68.67 / 35.94
PER	**90.63** / **86.03**	76.94 / **42.28**	**83.22** / **56.70**
GPE	88.40 / 81.72	**77.25** / 31.57	82.45 / 45.54
AVG	82.75 / 75.90	69.90 / 30.13	75.75 / 42.83

3.2.2. Bilingual NER Table 2 shows the results of improving NER without using Distortion feature and O-type Penalty for our system. Two beam search strategies *TopK* and *Threshold* are investigated. In both approaches, the mono weight of Chinese λ is 0.1125 which means that the "cross-lingual context" contributes greatly to fixing the NER errors. By projecting from English to Chinese to recover the unrecognized NEs in Chinese side, we are able to improve Recall by a large amount compared to previous methods (21.49% and 5.71% respectively). Therefore, there is an improvement in Chinese Named Entity Recognition from 42.83% to 59.84% comparing to the *StanfordNER* and by 2.19% comparing to the work of Che at al. (2013). For English side, *Threshold* yields an F_1-score of 75.80%, which is comparable to *StanfordNER* (75.75%) but slightly lower than *Che2013* (77.08%).

3.2.3. Bilingual NER with Distortion In this section, we illustrate the effectiveness of using Distortion Feature to further improve our proposed method as in Table 3. For Chinese side, the mono weight λ in *TopK* and *Threshold* approaches are relatively small (0.05 and 0.0625 respectively), which showed again that the "cross-lingual context" plays an important role in tagging unrecognized NEs. In details, F_1-score improves from 42.83% (*StanfordNER*) and 57.65

Table 2. The Bilingual NER Results of English / Chinese

Method	P(%)	R(%)	F_1(%)
StanfordNER	82.75 / 75.90	69.90 / 30.13	75.75 / 42.83
Che2013	**83.44** / **79.99**	**71.66** / 45.91	**77.08** / 57.65
TopK	83.14 / 72.32	69.71 / **51.62**	75.79 / **59.84**
Threshold	83.16 / 72.44	69.71 / 51.20	75.80 / 59.57

% (*Che2013*) to 60.22%. As for the English side, the NER performance is almost the same as Bilingual NER without Distortion feature (Table 2).

Table 3. The Bilingual NER Results with Distortion of English / Chinese

Method	P(%)	R(%)	F_1(%)
StanfordNER	82.75 / 75.90	69.90 / 30.13	75.75 / 42.83
Che2013	**83.44** / **79.99**	**71.66** / 45.91	**77.08** / 57.65
TopK	83.14 / 70.95	69.72 / 52.73	75.80 / 60.12
Threshold	83.14 / 70.22	69.71 / **53.28**	75.80 / **60.22**

3.2.4. Bilingual NER with O-type Penalty In this section, we demonstrate the effectiveness of using O-Type Penalty to further improve our proposed method as in Table 4. Similar to Bilingual NER with Distortion, the mono weight λ of Chinese is small (0.1625) which showed the importance of using information from English side. Chinese NER improve significantly by 20.17% compared to *StanfordNER* and 5.35 % compared to *Che2013* in term of F_1-score. On the other hand, English NER improved slightly from 75.75% (*StanfordNER*) to 76.03% but this result is still lower than *Che2013* (77.08%).

Table 4. The Bilingual NER Results with O-type Penalty of English / Chinese

Method	P(%)	R(%)	F_1(%)
StanfordNER	82.75 / 75.90	69.90 / 30.13	75.75 / 42.83
Che2013	**83.44** / **79.99**	**71.66** / 45.91	**77.08** / 57.65
TopK	82.83 / 71.14	70.32 / **57.00**	76.03 / 63.00
Threshold	82.84 / 71.35	70.40 / 56.95	76.08 / **63.04**

3.2.5. Bilingual NER with Distortion and O-type Penalty We illustrate the effectiveness of using both Distortion feature and O-type Penalty to further improve our proposed method as in Table 5. As shown in the table, using Distortion feature and O-type Penalty with *Threshold* strategy yields the best results, achieving F_1-score of 63.64% for Chinese NER. These results agree with our intuition that combining these two additional bilingual features could refine the performance of our initial Bilingual NER system. As for English NER, the F_1-score is 76.08%, which shows no further improvement when compared with Bilingual NER with O-type Penalty (Table 4). In general, our system achieved the best performance with Distortion and O-Penalty features, which is 76.08% for English NER and 63.64% for Chinese NER in terms of F_1-score.

4. Related Work

Bilingual corpora have long been exploited in improving different monolingual as well as cross-lingual Natural Language Processing tasks. In 2001, Yarowsky et al. [21] presented a detailed

Table 5. The Bilingual NER Results with O-type Penalty and Distortion of English / Chinese

Method	P(%)	R(%)	F_1(%)
StanfordNER	82.75 / 75.90	69.90 / 30.13	75.75 / 42.83
Che2013	**83.44** / **79.99**	**71.66** / 45.91	**77.08** / 57.65
TopK	82.90 / 70.41	70.28 / 58.24	76.03 / 63.47
Threshold	82.80 / 69.69	70.42 / **58.99**	76.08 / **63.64**

survey for cross-language annotation projection for four different NLP tasks: POS-tagging, noun phrase parsing, named entity recognition, and morphological analysis. The work showed that high-quality results of English could be transferred across languages through word alignment. In 2009, Huang et al. [22] incorporated bilingual constraints into a shift-reduce parser to improve parsing performance in dependency parsing task. Using only three simple alignment features, the authors managed to resolve shift-reduce conflicts with negligible overhead.

Specifically for Named Entity Recognition task, there were various works which made use of existing named entity annotated bilingual resources. Li et al. (2012) [11] proposed a bilingual CRFs framework which incorporates bilingual constraints to jointly tag named entities in a parallel English-Chinese corpus. Besides the use of bilingual corpora, other bilingual resources have proven to be useful in bilingual NER task. The work of Huang and Vogel (2002) [23] used bilingual named entity dictionaries to support the process of extracting named entity pairs from an unannotated corpus. Kim et al. (2012) [24] utilized Wikipedia metadata in combined with bilingual corpora through CRF model to label named entities in sentences of different foreign languages. Despite high performance, these methods required annotated bilingual corpora which are costly to obtain.

Other approaches explored the use of unannotated bilingual corpora which are aligned at sentence level. In 2014, Fu et al. [25] presented intuitive and effective heuristics to project English named entities into Chinese ones. Results showed that the generated corpus achieved comparable results to a manually annotated corpus in Named Entity Recognition task. This method could be expanded to different domains to solve the common domain over-fitting problem. Che et al. (2013) [12] used integer linear programming to enforce entities to agree through bilingual constraints. This method could jointly tag named entities in both languages without any annotated data.

5. Conclusion

In this paper, we explore the use of bilingual resources to improve monolingual Named Entity Recognition systems of English and Chinese. In general, the main idea of our system is incorporating projecting methods with additional bilingual features to fix existing NER errors. Our method is performed on an existing unannotated bilingual corpus without any further annotations like named entity annotations or gazetteers. Experimental results show that the proposed system managed to make an improvement in Chinese NER performance. In particular, F_1-score of Chinese NER increase significantly from 42.83% (*StanfordNER*) and 57.65% (*Che2013*) to 63.64%. Regarding the English side, we managed to outperform *StanfordNER*, in which F_1-score increase from 75.75% to 76.08%. The proposed system could be applied to broader domains to recognize more entity types such as MONEY, TIME, PRODUCT or PROTEIN and GENE in Bio-informatics. Moreover, a modified version of our method can be applied to other tasks like Part-of-speech Tagging, Co-reference Resolution, Semantic Role Labeling, as well as other resource-limited languages.

Acknowledgements
This research is supported by research funding from Advanced Program in Computer Science, University of Science, Vietnam National University - Ho Chi Minh City.

6. References

[1] Chen M, Mao S and Liu Y 2014 Big data: A survey *Mobile Networks and Applications* ISBN 1383-469X ISSN 1383469X
[2] Labrinidis A and Jagadish H V 2012 *Proceedings of the VLDB Endowment* **5** 2032–2033
[3] Grishman R and Sundheim B 1995 Design of the muc-6 evaluation *Proceedings of the 6th conference on Message understanding* (Association for Computational Linguistics) pp 1–11
[4] Babych B and Hartley A 2003 Improving machine translation quality with automatic named entity recognition *Proceedings of the 7th International EAMT workshop on MT and other Language Technology Tools, Improving MT through other Language Technology Tools: Resources and Tools for Building MT* (Association for Computational Linguistics) pp 1–8
[5] Mollá D, Van Zaanen M, Smith D *et al.* 2006
[6] Chen H H, Ding Y W and Tsai S C 1998 *Computer Processing of Oriental Languages* **12** 75–85
[7] Bagga A and Baldwin B 1998 Entity-based cross-document coreferencing using the vector space model *Proceedings of the 36th Annual Meeting of the Association for Computational Linguistics and 17th International Conference on Computational Linguistics-Volume 1* (Association for Computational Linguistics) pp 79–85
[8] Santos C N d and Guimaraes V 2015 *arXiv preprint arXiv:1505.05008*
[9] Chiu J P and Nichols E 2015 *arXiv preprint arXiv:1511.08308*
[10] Nguyen L H, Dinh D and Tran P 2016 *ACM Transactions on Asian and Low-Resource Language Information Processing (TALLIP)* **16** 9
[11] Li Q, Li H, Ji H, Wang W, Zheng J and Huang F 2012 Joint bilingual name tagging for parallel corpora *Proceedings of the 21st ACM international conference on Information and knowledge management* (ACM) pp 1727–1731
[12] Che W, Wang M, Manning C D and Liu T 2013 Named entity recognition with bilingual constraints *Proceedings of the 2013 Conference of the North American Chapter of the Association for Computational Linguistics: Human Language Technologies* pp 52–62
[13] Nemhauser G L and Wolsey L A 1988 *Wiley, Chichester. GL Nemhauser, MWP Savelsbergh, GS Sigismondi (1992). Constraint Classification for Mixed Integer Programming Formulations. COAL Bulletin* **20** 8–12
[14] Tseng H, Chang P, Andrew G, Jurafsky D and Manning C 2005 A conditional random field word segmenter for sighan bakeoff 2005 *Proceedings of the fourth SIGHAN workshop on Chinese language Processing*
[15] Manning C D, Surdeanu M, Bauer J, Finkel J, Bethard S J and McClosky D 2014 The Stanford CoreNLP natural language processing toolkit *Association for Computational Linguistics (ACL) System Demonstrations* pp 55–60 URL `http://www.aclweb.org/anthology/P/P14/P14-5010`
[16] Lafferty J, McCallum A and Pereira F C 2001
[17] Brown P F, Pietra V J D, Pietra S A D and Mercer R L 1993 *Computational linguistics* **19** 263–311
[18] Och F J 2003 Minimum error rate training in statistical machine translation *Proceedings of the 41st Annual Meeting on Association for Computational Linguistics-Volume 1* (Association for Computational Linguistics) pp 160–167
[19] Zhang Y and Clark S 2011 *Computational linguistics* **37** 105–151
[20] BerkeleyAligner Google Code Archive URL `https://code.google.com/archive/p/berkeleyaligner/`
[21] Yarowsky D, Ngai G and Wicentowski R 2001 Inducing multilingual text analysis tools via robust projection across aligned corpora *Proceedings of the first international conference on Human language technology research* (Association for Computational Linguistics) pp 1–8
[22] Huang L, Jiang W and Liu Q 2009 Bilingually-constrained (monolingual) shift-reduce parsing *Proceedings of the 2009 Conference on Empirical Methods in Natural Language Processing: Volume 3-Volume 3* (Association for Computational Linguistics) pp 1222–1231
[23] Huang F and Vogel S 2002 Improved named entity translation and bilingual named entity extraction *Multimodal Interfaces, 2002. Proceedings. Fourth IEEE International Conference on* (IEEE) pp 253–258
[24] Kim S, Toutanova K and Yu H 2012 Multilingual named entity recognition using parallel data and metadata from wikipedia *Proceedings of the 50th Annual Meeting of the Association for Computational Linguistics: Long Papers-Volume 1* (Association for Computational Linguistics) pp 694–702
[25] Fu R, Qin B and Liu T 2014 *Frontiers of Computer Science* **8** 629–641

A New Approach for Low-resolution Image Registration Based on Weighted MI and Powell

Shuping Zhang

College of Engineering, Shanghai Polytechnic University, Shanghai 201209

E-mail: spzhang@sspu.edu.cn

Abstract: Mutual information (MI) has been successfully applied to the registration of medical imagery and SR reconstruction because of its robustness. However, it also has obvious drawbacks that it is susceptible to local minima. For this purpose, an image registration scheme based on weighted mutual information is proposed in this paper. By introducing a reasonable weight coefficient, the gray-levels can exert greater impact on computation of entropy. And then, the weighted mutual information (WMI) which is designed as a new similarity measure is exploited by the Powell optimizer to refine the preregistration results. The experimental results suggest that the proposed algorithm has strong robustness and adaptive capacity for noise, and is convergent and efficient in resolving the given test problems.
Keywords: image registration; low-resolution; weighted MI; Powell optimizer

1.Introduction

Image registration is a vital process which determines the most precise match between two images of the same scene. Many feature-based[1-4] and intensity-based[5-6] image registration methods have been proposed. Mikolajczyk et al. [7] pointed that scale-invariant feature transform (SIFT) has good performances in scale change, rotation, and image blur. Three years later, Herbert Bay[8] presented SURF which has lower computation time than SIFT. Despite the attractive advantages of SURF, there exist some problems when it is directly applied to the LR images, e.g. feature vector is not unique, similarity measurement accuracy is not high enough, and the discrimination between the local features is small. The emergence of MI, which represents a measure of statistical dependence between two images, provides a new research direction for the study of the image registration. It has been demonstrated by Cole-Rhodes[9] that MI is robust to noise and nonlinear intensity relationship between images. It has been successfully applied to the registration of medical imagery [10-11], and SR reconstruction[12-13]. However, what should not be ignored is the mutual information alone is susceptible to local minima. It cannot be employed in SR problems as the Gaussian noise may lead to large mis-registrations.

Considering the high accuracy of MI and the influence of noise on it, a weighted MI-based registration scheme is proposed in this paper. By introducing a reasonable weight coefficient, the gray-levels, which play a significant role to improve the entropy, can exert greater impact on computation of entropy. So as to make the most promising registration area obtains more information. Finally, the Powell algorithm is exploited as an optimization strategy to achieve the most precise registration results, by searching rigid body transformation parameters when the mutual information calculated in the weighted MI reaches the maximum value.

2. Review of the registration model

Given a pair of 2-D gray-level image between which there exist some geometric and radiometric differences, let $f_R(x,y)$ and $f_S(x,y)$ represent the reference and sensed image, respectively, where coordinates $(x,y)\in\Delta\subset\mathbb{R}^2$ and Δ is a region of interest. To register these two images is to find the optimal geometric transformation $T_\mu(\bullet)$ by which $f_S(T_\mu(x,y))$ best matches $f_R(x,y)$ for all (x,y), where μ is a set of transformation parameters.

As being widely used in the image registration, the rigid body transformation model can be written as

$$T_\mu(x,y)=\begin{bmatrix}\cos\theta & -\sin\theta & \delta_x\\ \sin\theta & \cos\theta & \delta_y\\ 0 & 0 & 1\end{bmatrix}\begin{bmatrix}x\\ y\\ 1\end{bmatrix} \tag{1}$$

Where the transformation origin is considered to be the upper left corner of the reference image; θ represents the rotation angle and (δ_x,δ_y) are the shifts between the two images.

The main purpose of image registration is to find the optimal geometric transformation, which maximize (or minimize) some similarity metric between the two images. Therefore, the problem of image registration is mapped as a problem of searching the optimal transformation parameters μ^*, which can be expressed as:

$$\mu^*=\arg opt(M(f_R(x,y),T_\mu(f_S(x,y)))) \tag{2}$$

Where T_μ is the transformation model; μ is the transformation parameter and M represents the similarity metric.

3. Image registration framework based on a weighted MI

3.1. The standard MI

As a basic concept of information theory, mutual information (MI) represents a measure of statistical dependence between two random variables, or represents the amount of information one variable containing another. For random variables A and B, the mutual information can be defined as equation (3). Where, $P_A(a)$ and $P_B(b)$ represent marginal probability distribution; $P_{A,B}(a,b)$ represents joint probability distribution.

$$MI(A,B)=\sum_{a,b}P_{A,B}(a,b)\log\frac{P_{A,B}(a,b)}{P_A(a)P_B(b)} \tag{3}$$

In statistics, if the random variables A and B are independent of each other, $P_{A,B}(a,b)=P_A(a)\bullet P_B(b)$ and $MI(A,B)=0$; if they depend on each other completely, $P_{A,B}(a,b)=P_A(a)=P_B(b)$ and $MI(A,B)$ reaches the maximum value. Based on this theory, MI, as a similarity measure, is introduced into the field of image registration.

3.2. A weighted MI

According to the equation (3), the mutual information of the reference image $f_R(x,y)$ and the sensed image $f_S(x,y)$,can be defined as equation (4):

$$MI(f_R,f_S)=\sum_{i=0}^{L-1}\sum_{j=0}^{L-1}P_{R,S}(i,j)\log\frac{P_{R,S}(i,j)}{P_R(i)P_S(j)} \tag{4}$$

Where, L ($L=256$) is the gray-level contained by $f_R(x,y)$ and $f_S(x,y)$; $P_{R,S}(i,j)$ is their joint probability distribution; $P_R(i)$ and $P_S(j)$ are their marginal probability distribution. And they are defined as :

$$P_{R,S}(i,j)=\frac{h(i,j)}{\sum_{i=0}^{L-1}\sum_{j=0}^{L-1}h(i,j)} \tag{5}$$

$$P_R(i)=\sum_{j=0}^{L-1}P_{R,S}(i,j) \tag{6}$$

$$P_S(j)=\sum_{i=0}^{L-1}P_{R,S}(i,j) \tag{7}$$

$h(i,j)$ is the joint histogram of $f_R(x,y)$ and $f_S(x,y)$, and it is a 2-D metric as follows:

$$h=\begin{bmatrix} h(0,0) & h(0,1) & \cdots & h(0,L-1) \\ \vdots & & & \\ \vdots & & & \\ h(L-1,0) & \cdots & \cdots & h(L-1,L-1) \end{bmatrix} \tag{8}$$

It's obvious that the correlation between $f_R(x,y)$ and $f_S(x,y)$ can be well reflected by the MI-based registration. However, this method is also insufficient in some aspects: 1) it is easy to appear mismatching when the noise of the image is larger, the correlation between the images is weaker or large deformation in the image occurs; 2) the convergence precision is damaged due to image pixels are treated indistinguishably during the computation of MI. Therefore, a weighted MI-based registration scheme is designed in this section to overcome this problem. By introducing a reasonable weight coefficient, the gray-levels, which play a significant role to improve the entropy, can exert greater impact on computation of entropy. So as to make the most promising registration area obtains more information.

The entropy of an image $I(x,y)$ after introducing the weight coefficient can be defined as:

$$H(I)=-\sum_{a\in I}\omega(a)\times P(a)\times\log(P(a)) \tag{9}$$

Where, $\omega(a)$ represents the extent of the impact on computation of entropy of the pixel with a gray-level. In this definition, $\omega(a)$ also should satisfy the following constraints:

$$\sum\omega(a)=1 \quad (0<\omega(a)<1) \tag{10}$$

In order to meet the needs of two aspects, i.e. calculate conveniently and can effectively highlight the difference between extreme values, the proportion of each gray-level in the total number of pixels is employed to define the weighted coefficient of the mutual information: $\omega(a)=P(a)$. Under this condition, a normalized measure of mutual information can be rewritten as:

$$NMI_w(f_R,f_S)=\frac{H_w(f_R)+H_w(f_S)}{H_w(f_R,f_S)} \tag{11}$$

$$H_\omega(f_R)=-\sum_{a\in f_R}P^2(a)\times\log P(a) \tag{12}$$

$$H_\omega(f_S)=-\sum_{b\in f_S}P^2(b)\times\log P(b) \tag{13}$$

$$H_\omega(f_R,f_S)=-\sum_{a\in f_R}\sum_{b\in f_S}P^2(a,b)\times\log P(a,b) \tag{14}$$

The novel similarity measure $NMI_w(f_R,f_S)$ is adopted as the objective function in the fine-tuning registration. It corresponds to registration of the images when maximization of their $NMI_w(f_R,f_S)$ is achieved. Therefore, the process of registration is mapped as a problem of searching the optimal transformation parameters μ^*, and the equation(2) is revised as follows:

$$\mu^*=\arg\max NMI_w(f_R(x,y),T_\mu(f_S(x,y))) \tag{15}$$

4. Optimal search strategy based on the Powell

Powell's optimization method is a reasonably fast yet generally successful optimization method. Given an *n*-dimensional search space, Powell's method performs one-dimensional optimizations for each of the *n* dimensions in turn. For each one-dimensional optimization, the position of the optimum found in one direction is the starting point for the optimization in the next direction. In one iteration, the registration function will be optimized once in all *n* dimensions. The method ceases iterating when the difference in the registration measure at the beginning and the end of each iteration falls below a user-defined tolerance. At the end of each iteration, the average direction of movement is determined by calculating the direction vector between the starting point and the end point of the iteration. This new direction will replace the direction in which the largest increase in function value was found. This direction is chosen because it is most similar to the new direction and this choice prevents convergence of the directions. In some cases the new direction is rejected, for example when an additional optimization in the new direction leads to a decrease in function value.

The one-dimensional optimizations are performed with Brent's method. Beginning with a small interval around the starting position of optimization, this method changes the position and size of the interval until it incorporates a maximum. The position of the maximum is then determined within the interval found. Note that this method does not necessarily return the global optimum in a certain direction. The detailed pseudo-code of registration parameters optimization based on the Powell algorithm is presented as Table1 shown.

Table1 Registration parameters optimization based on the Powell algorithm

Initialize the starting transform parameter vector $\overline{\mu_0}$, independent search direction vectors $\overline{d_i} = \overline{e_i}$, $(i = 1, 2, 3)$, the tolerance for stop criteria ε, set $\overline{\mu_0} = \overline{\mu}$ and $\overline{\mu}$ is the transform parameter vector obtained by the coarse registration, $f_0 = f(\mu_0)$, $k = 1$

While (stopping criterion is not met, namely $|\Delta f| > \varepsilon$) **do**

  for $i = 1:3$ **do**

    If $k \ge 2 \& |\eta| \ge temp$ **then,** $\overline{d_i} = \overline{d_{i+1}}$, **End If**

    $\eta_i = \arg\max_{d_i}(NMI_w)$, $\mu_i = \mu_{i-1} + \eta_i d_i$, $f_i = f(\mu_i)$, $i = i+1$

  End for

  $d_4 = \mu_3 - \mu_0$, $\eta = \arg\max_{d_4}(NMI_w)$, $\mu_4 = \mu_3 + \eta d_4$, $f = f(\mu_4)$ $diff = [f - f_1, f_1 - f_2, f_2 - f_3]$

  $\text{temp} = sqrt((f_3 - f) / \max(diff))$ $\Delta f = \|\mu_3 - \mu_0\|$ $\mu_0 = \mu_4$ $k = k+1$

End While

5.Experiments and analysis

In order to validate the performance of the proposed algorithm, a pair of outdoor car images with LR resolution 360×470 has been employed as the reference image and sensed image respectively(shown as in Fig.1(a) and Fig.1(b)). All the experiments were implemented in Matlab and run on a PC with Intel(R) Core(TM)2 2.33 Gz CPU, and 2 GB RAM.

5.1 Convergence and robustness of the weighted MI

The convergence speed and accuracy of a matching algorithm can be evaluated according to several properties, e.g. the sharpness of peak and smoothness, of the criterion function curve between the reference image and sensed image. The statistical curve of an outstanding criterion function for matching should be smooth. The sharper the curve waves, the easier the optimizing course is trapped the local extreme value and the poorer robustness the algorithm has.

The peak of curve which corresponds to the correct match in an algorithm should be sharp and stand out clearly. The extent of sharpness and standing out of the curve directly determines the convergence speed and accuracy of a matching algorithm. The clearer the peak of curve stands out from the other extreme values and the sharper the peak of curve is, the easier the optimizing course converges to the maximum of the criterion function accurately. Meanwhile, faster convergence speed and higher convergence accuracy are achieved simultaneously.

Therefore, two experiments are performed in this section to evaluate the weighted MI similarity measure (WMI) and the standard MI similarity measure (SMI).

Experiment-A the sensed image $f_s(x, y)$ which contains no noise, as shown in Fig.1(b), is transformed by altering just one element of the transformation parameter vector while keeping other transformation parameters constant. For example, $f_s(x, y)$ is moved in the horizontal direction on a scale of [-30 30] while the transformation parameters in the vertical and rotation direction retain the same. The criterion function curve of WMI and SMI in the horizontal direction has been shown as Fig 1(c) and Fig 1(d). In the same way that the criterion function curves of WMI and SMI in vertical direction on a scale of [-30 30] (Fig 1(e)-Fig 1(f)) and rotation direction on a scale of [-5 5] (Fig 1(g)-Fig 1(h)) have been performed.

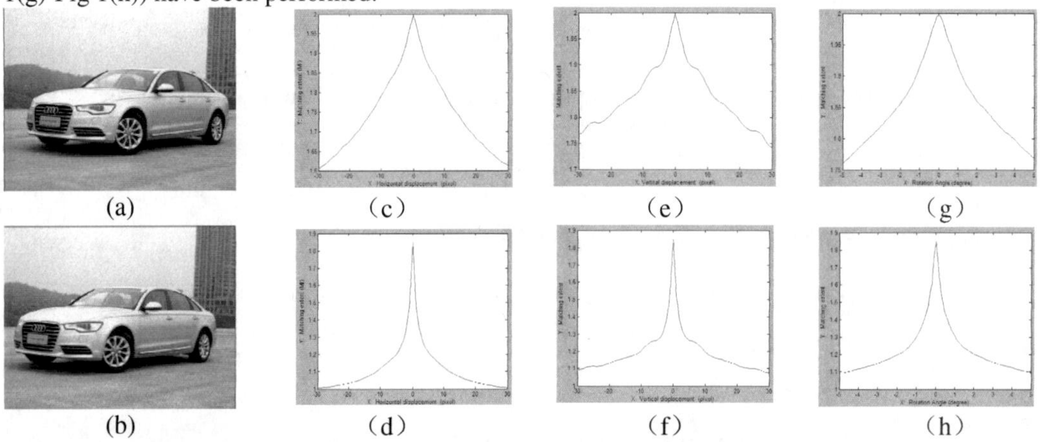

Figure 1 The criterion function curve of WMI and SMI in three directions
(a) the reference image; (b) the sensed image; (c),(e), (g) are the criterion function curves of SMI in horizontal, vertical and rotation direction respectively; (d),(f) ,(h)are the criterion function curves of WMI in horizontal, vertical and rotation direction respectively

Experiment-B WMI and SMI as similarity measures to register the reference image and the sensed image $f_s(x, y)$ which is polluted by Gaussian noise and shown in Fig.2 (b). With the same test method as that of the experiment-A, the criterion function curves in three directions have been drawn and shown in Fig.2(c)-(h).

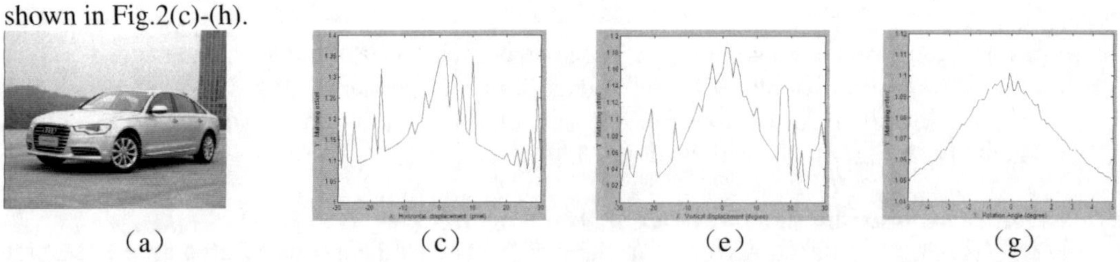

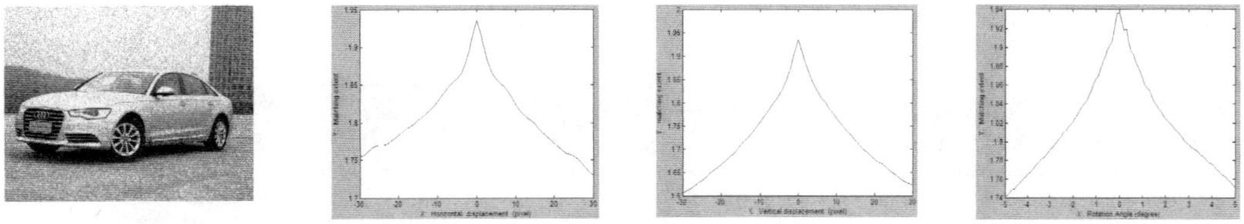

Figure.2 the criterion function curves of WMI and SMI in three directions
(c),(e),(g) are the criterion function curves of SMI in horizontal, vertical and rotation direction respectively; (d),(f),(h) are the criterion function curves of WMI in horizontal, vertical and rotation direction respectively

Results in experiment-A have shown that the criterion function curves of WMI has obtained higher matching degree as proved by the mutual information on vertical axis. Meanwhile the WMI similarity measure is smoother and sharper than that of SMI, which indicates the WMI similarity measure is easier to accurately convergence to excellent approximate result. This tendency shows more evidence in the situation that the sensed image is polluted by more noise, as the results of experiment-B shown.

In Fig.2 the criterion function curves of SMI fluctuate sharply, even have little or no access to convergence the maximum value in horizontal (Fig.2(c)) and vertical direction (Fig.2(e)). And yet this phenomenon is significantly improved in the proposed method. The criterion function curves of WMI are smoother and have more outstanding peak than that of SMI. It means it's less likely trapped local extreme value in the search process for optimal transformation parameters. The results of comparing two groups of experimental data show that the method presented in this paper has stronger robustness.

5.2 Efficiency and accuracy of the Powell optimization strategy

The Powell optimization strategy is utilized to search the optimal transformation parameters with WMI as a similarity measure. In order to verify the efficiency and accuracy of this process, three simulation images $S_i(x,y)(i=1,2,3)$, as shown in Fig.3(a-c), are obtained by transforming the reference image $f_R(x,y)$ as known rigid transformation model $P_i(i=1,2,3)$ whose parameters have been shown in Table 2.

The accuracy of the presented method in this context was evaluated through visual inspection by generating the checkerboard mosaic images using the corresponding registration parameters between simulation images and the reference image, as shown in Fig.3 (d-f).

Table 2 Parameters of three known transformation models

rigid transformation model	parameters
P_1	$\delta_x = 36(pixel), \delta_y = -14(pixel), theta = 5(\deg ree)$
P_2	$\delta_x = 17(pixel), \delta_y = -13(pixel), theta = 3(\deg ree)$
P_3	$\delta_x = -21(pixel), \delta_y = 18(pixel), theta = -3(\deg ree)$

The checkerboard mosaic image [Fig.3 (a-c)] validates that the image features of the reference image and the simulated images such as the body part, other than several erroneously matched positions marked in red ovals, are precisely overlapped, which demonstrates the accuracy of our obtained registration parameters.

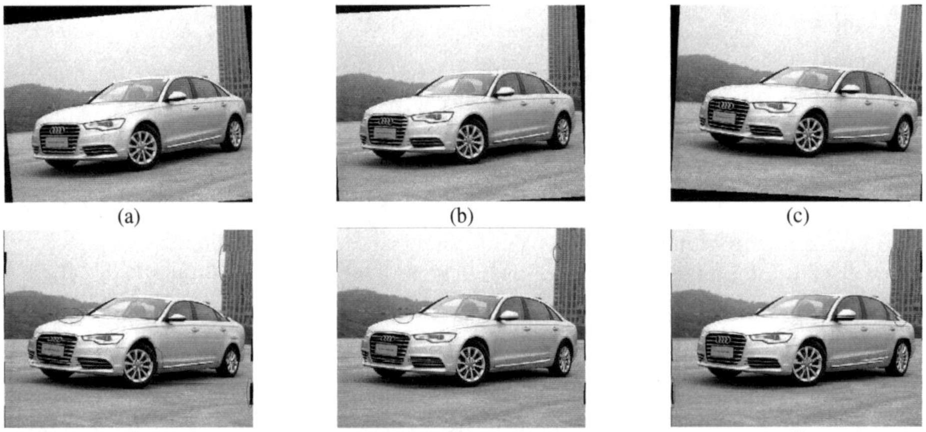

(a) (b) (c)

(d) (e) (f)

Figure.3 The checkerboard mosaic images

(a), (b), (c) are the simulated images with rigid transformation model P_1, P_2, P_3 respectively; (d), (e), (f) are the generated checkerboard mosaic images using (a), (b), (c) and the reference image

6 .Conclusion and Discussion

In this paper, an image registration algorithm based on WMI and the Powell's optimization is proposed. Based on the standard MI , a weighted coefficient of the mutual information between the reference image and the sensed image is defined. A normalized novel similarity measure which is less sensitive to changes in overlap is adopted as the objective function. And then the Powell algorithm is exploited as an optimization strategy to refine the preregistration results to search the most precise registration results. In this process, the weighted mutual information reaches the maximum value, the optimal rigid body transformation parameters are obtained.

Results of experiments have demonstrated two aspects of the proposed method. First, it is convergent and effective because it is easier to accurately convergence to excellent approximate result and prevent optimizing course from trapping the local minimum. Moreover, it is robust and highly accurate, especially in the case of containing noise. However, the proposed method is not perfect. When the angles between LR images vary greatly, the rigid body transformation model is not suitable. Therefore, both the affine transformation model and thin-plate-spline are alternative models according to research objects in the future study.

References

[1] Dong Zhao, Yan Yang, Zhihang Ji,et al. Rapid multimodality registration based on MM-SURF[J], *Neurocomputing*, 2014, 131:87-97.

[2] Yang Gui, Ang Su, Jing Du. Point-pattern matching method using SURF and Shape Context[J], *Optik*, 2013,124:1869-1873.

[3] Maoguo Gong, Shengmeng Zhao, Licheng Jiao,et al. A Novel Coarse-to-Fine Scheme for Automatic Image Registration Based on SIFT and Mutual Information[J], *IEEE transactions on geoscience and remote sensing*, 2014, 52(7):4328-4338.

[4] Xiangzeng Liu, Zheng Tian, Qiang Lu, et al. A new affine invariant descriptor framework in shearlets domain for SAR image multi-scale registration [J], *International Journal of Electronics and Communications*, 2013, **67**(9):743-753.

[5] Marius de Groot, Meike W. Vernooij, Stefan Klein, et al. Improving alignment in Tract-based spatial statistics: Evaluation and optimization of image registration. *Neuro Image*, 2013, 76: 400- 411.

[6] Xiangzeng Liu, Zheng Tian, Qiang Lu, et al. A new affine invariant descriptor framework in

shearlets domain for SAR image multi-scale registration [J], *International Journal of Electronics and Communications*, 2013, **67**(9):743-753.

[7] Krystian Mikolajczyk, Cordelia Schmid. A Performance Evaluation of Local Descriptors[J], *IEEE Transaction on pattern analysis and machine intelligence*, 2005, **27**(10): 1615-1630.

[8] Herbert Bay, Andreas Ess, Tinne Tuytelaars, et al. Speeded-Up Robust Features (SURF) [J], *Computer Vision and Image Understanding*, 2008, 110:346-359.

[9] A. A. Cole-Rhodes, K. L. Johnson, J. LeMoigne et al. Multi-resolution registration of remote sensing imagery by optimization of mutual information using a stochastic gradient [J], *IEEE Transactions on Image Process*, 2003, **12**(12): 1495-1511.

[10] Lian, J., Zheng, Y., Jiao, W., Yan, F., Zhao, B. Deblurring sequential ocular images from multi-spectral imaging (MSI) via mutual information[J], *Medical and Biological Engineering and Computing*,2018, 56(6):1107-1113.

[11] Lin, L.-H., Chen, T.-J., Mutual information correlation with human vision in medical image compression[J], *Current Medical Imaging Reviews*, 2018, 14(1) : 64-70.

[12] Michalis Vrigkas, Christophoros Nikou, Lisimachos P. Kondi, Accurate image registration for MAP image super-resolution [J], *Signal Processing: Image Communication*, 2013,28: 494-508.

[13] Boyang Zhang, Ju Liu, Jinyu Chu, etal. A Mutual Information Based Sub-Pixel Registration Method for Image Super Resolution[C], Fifth International Conference on Intelligent Information Hiding and Multimedia Signal Processing, Kyoto, 2009, 422-425.

IOP Conf. Series: Materials Science and Engineering **435** (2018) 012009 doi:10.1088/1757-899X/435/1/012009

Method for Recognizing Mechanical Status of Container Crane Motor Based on SOM Neural Network

X Q Yao[1,a], G Tang[1,b] and X Hu[1,c]

[1]Logistics Engineering College, Shanghai Maritime University, Shanghai 201306, China

[a]xiaoqiangshmtu@163.com, [b]gangtang@shmtu.edu.cn , [c]huxiong@shmtu.edu.cn

Abstract. The recognition of mechanical state of container crane motor is a complex nonlinear mode classification problem.In this paper, a deep learning method based on self-organizing feature map (SOM) neural network was developed for the cluster analysis of motor vibration signals to characterize the mechanical state of the lifting motor. Through Python software simulation, the vibration intensity signal of the lifting motor can be divided into five different categories, and the interval range of each type of vibration intensity signal is statistically obtained, which corresponds to the five mechanical states of the motor. The SOM neural network can realize the effective and rapid self-adaptive classification of the motor vibration signal, realize the recognition of the mechanical status of the crane motor, and provide a certain basis for the maintenance of the motor.

1. Introduction

The Container crane is huge and complex in structure, and the Container crane fault diagnosis is a complex nonlinear classification problem, The nonlinear classification of SOM neural network is very suitable for the research of the Container crane complex system[1]. The author in [2] based on SOM neural network bearing fault detection and diagnosis, the results show that using SOM neural network clustering to improve the classification performance of fault detection process has a good effect; The author in [3] use improved Hidden Markov model and self-organizing map data drive to identify bearing faults, it has a very significant effect that the recognition rate of up to 99.58%. In addition, the SOM neural network has the advantages of simple structure, high self-organization and self-learning ability [4]. In this paper applies the SOM neural network to the state cluster analysis of the Container crane lifting motor.

2. Signal acquisition and processing

2.1. Vibration signal acquisition

Acceleration sensors are installed on the output terminals of the right and left lifting motors of the Container crane. The sampling period is set to collect a vibration signal every 10 seconds and the signals are collected by the signal conditioning collector. The installation position of motor sensor as shown in Figure 1. The sensor collects about 8000 vibration signals every day. In this paper, select the vibration signal from 0 o'clock on January 10 to 12 o'clock on January 16 for research.

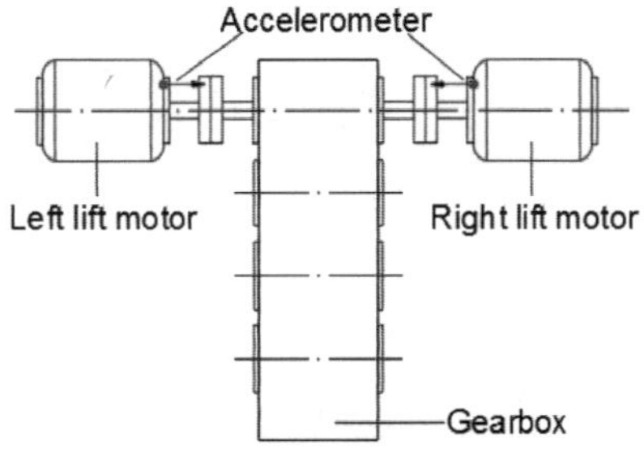

Figure 1. Lift motor sensor installation diagram.

2.2 . Vibration signal time-domain diagram

In order to observe the trend of the vibration signal, We can use MATLAB to draw a time domain diagram of the motor vibration signal for a week, as is shown in Figure 2. The amplitude of the vibration can be seen directly from the figure 2. When the lifting motor works for a long time with a large vibration amplitude, the motor or even the Container crane may be damaged. Therefore, it is necessary to identify the state of the motor through the collected vibration signal to ensure that the motor operates in a relatively safe state.

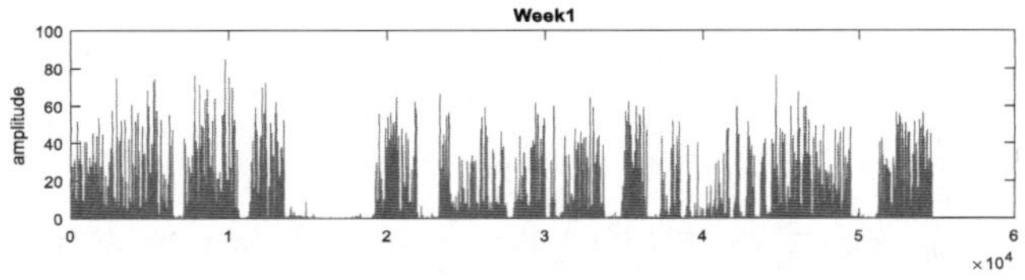

Figure 2. Lift motor vibration signal time domain diagram.

3. Vibration signal feature extraction

Statistical feature extraction is an important basis for judging the operating status of Container crane motor. The characteristic parameters of the motor reflect its mechanical properties and the abnormal data contains information on the degree of vibration hazard. The degree of danger is proportional to the degree of deviation from the average value of the data. Then we can formulate reference standards for analysis. When analyzing the status of the lift motor, it can be divided into five states: good, normal, early warning, warning and danger [5]. The physical concept of vibration of the motor is used in the diagnosis process, The clustering analysis was performed by extracting the effective values X_{RMS} of the vibration intensity of the motor as the eigenvalues [6]-[8].

$$X_{RMS} = \sqrt{\frac{1}{N}\sum_{i=1}^{N} x_i^2} \tag{1}$$

4. SOM neural network clustering principle and steps

4.1. SOM neural network clustering principle

The SOM neural network consists of an input layer and a competition layer, the input layer corresponds to a high-dimensional input vector and the competition layer consists of ordered nodes on a two-dimensional grid, they are connected by weight vectors[9]. The basic structure as shown in Figure 3. Through network training established a network layout, which can make each weight vector of the competition layer is located in the center of the input vector cluster [10][11].

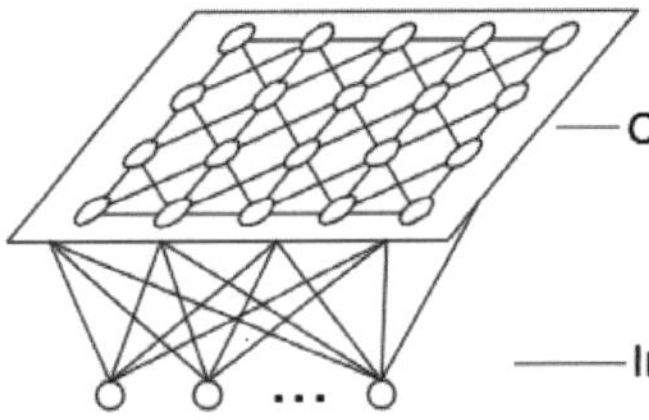

Figure3. SOM neural network topology.

4.2.SOM neural network clustering steps

The SOM network can selectively respond to input data and automatically configure similar input data on the network according to similarity between input vectors[12]. The Network training steps are as follows.

4.2.1.Step1 (*Initialization and normalization of weights*) Initialization of weights is such that the initial position of the weights and the distribution area of the input samples sufficiently overlap. By selecting part of the data from the vibration intensity data of the motor as a training sample for the SOM neural network, then randomly selected m input sample from all training sample sets as initial weights and normalized.

$$\hat{w}_{ji} = \frac{w_{ji}}{\|w_{ji}\|}, (1 \le i \le m) \tag{2}$$

Note: w_{ji} is the initial weight, $\|w_{ji}\|$ is the Euclidean norm of the weight vector and $\hat{w}_{ji}$ is the normalized weight.

4.2.2. Step2 (*Training sample normalization*)

$$X = \frac{\hat{X}}{\|\hat{X}\|} = \left(\frac{x_1}{\sqrt{\sum_{j=1}^{n} x_j^2}}, \cdots, \frac{x_n}{\sqrt{\sum_{j=1}^{n} x_j^2}}\right)^T \tag{3}$$

Note: $\hat{X}$ is the training sample, $\|\hat{X}\|$ is the Euclidean norm of the training sample vector and x_j is the j characteristic component value of the training sample vector.

4.2.3. Step3 (*Get winning neurons*) The normalized vibration intensity sample data of the motor is put into the input layer of the SOM network, the Euclidean distance of the training sample data and the weight vector is calculated, and the neuron with the smallest distance is defined as the winning neuron d_j.

$$d_j = \|X - W_j\| = \sqrt{\sum_{i=1}^{m} (x_i(t) - \omega_{ji}(t))^2} \tag{4}$$

4.2.4. Step4 (*Defining a winning neighborhood*) The winning neighborhood determines the weight adjustment domain at time t with the winning neuron j^* as the center. The initial neighborhood $N_{j*}(0)$ is larger, and $N_{j*}(t)$ gradually shrinks with the training time. The size of the winning neighborhood is expressed by the radius of the neighborhood.

$$r(t) = C_1 e^{-B_1 t/T} \tag{5}$$

Note: C_1 is a constant greater than zero related to the number of nodes in the output layer, B_1 is a constant greater than one and T is the preset maximum number of times of training.

4.2.5. Step5 (*Adjustment weight*) Adjust the weights of all neurons in the winning neighborhood.

$$w_{ij}(t+1) = w_{ij}(t) + \eta(t,N)[x_i^p - w_{ij}(t)], (i = 1,2,\cdots,n, j \in N_{j^*}(t)) \quad (6)$$

Note: $W_{ij}(t+1)$ is the adjusted weight, $\eta(t,N)$ is a function of the training time t and the topological distance N between neurons in the neighborhood.

4.2.6. Step6 (*training requirements*) The $\eta(t)$ is the functional value of the topological distance between the j neuron and the winning neuron after training, $\eta_{\min}$ is the set minimum topology distance function value. If $\eta(t) < \eta_{\min}$, the network training is ended. Otherwise, come back 4.2.3 to continue.

5. Python simulation

The system adopts Python 3.6 as the simulation test platform. The SOM network training sample set is obtained after the experimental data was normalized. the cluster center of the vibration intensity data of the lifting motor is obtained After training, as is shown in Table 1.

Table1. Lifting motor vibration intensity data clustering center.

date	Center 1	Center 2	Center 3	Center 4	Center 5
January 10	0.73132647	0.52094046	0.88128821	0.20799421	0.98397678
	0.6819084	0.85353541	0.47250587	0.97799975	0.17786049
January 11	0.70053833	0.54252202	0.83403607	0.23099253	0.96693977
	0.7135338	0.83996019	0.55152768	0.97295295	0.25481011
January 12	0.70336614	0.54701719	0.83178712	0.23983624	0.96237557
	0.71072541	0.83709021	0.5549968	0.97080239	0.27162745
January 13	0.70474831	0.53329382	0.83521524	0.218813	0.96852065
	0.70938256	0.84585213	0.54987844	0.97570655	0.24867736
January 14	0.70835896	0.53834956	0.84790267	0.22074443	0.97738899
	0.70574111	0.8426558	0.53011293	0.97529536	0.21113709
January 15	0.72004292	0.5386122	0.8448222	0.22185959	0.97447465
	0.69385631	0.84248636	0.53483934	0.97502712	0.22415564
January 16	0.68699258	0.48156783	0.85929699	0.22554832	0.97644026
	0.72657318	0.87636944	0.51139136	0.97413152	0.21563879

According to Table 1, we can get the cluster center of the vibration intensity of the Container crane lifting motor and draw the change trend of the cluster center through SPSS, as is shown in Figure 4.

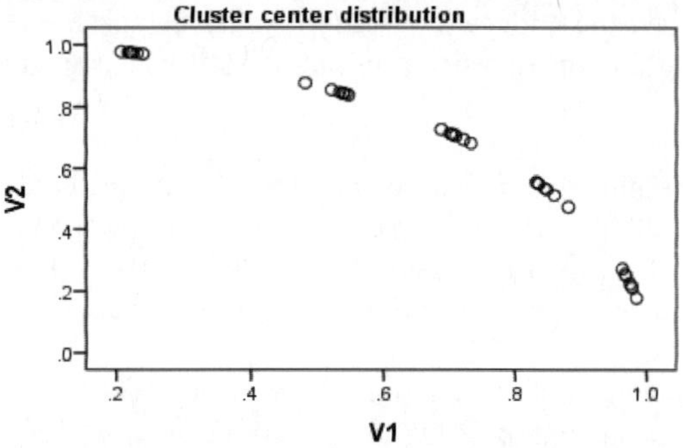

Figure 4. Cluster center distribution.

From January 10th to January 16th, the cluster center of the Container crane lifting motor vibration data is approximately equal. Therefore, the average value of the cluster center at each day can be equivalent to the data of all the vibration intensities within a week. We can obtains the classification intervals for each cluster center by MATLAB, as shown in Table 2.

Table2. Lifting motor vibration intensity data classification interval.

date	Interval 1	Interval 2	Interval 3	Interval 4	Interval 5
January 10	(0,1.719803106)	(1.719803106,6.604036273)	(6.604036273,11.48826944)	(11.48826944,16.37250261)	(16.37250261,∞)
January 11	(0,1.783360843)	(1.783360843,6.813725508)	(6.813725508,11.84409017)	(11.84409017,16.87445484)	(16.87445484, ∞)
January 12	(0,1.337818312)	(1.337818312,5.196238951)	(5.196238951,9.054659591)	(9.054659591,12.91308023)	(12.91308023, ∞)
January 13	(0,1.133435932)	(1.133435932,4.937436915)	(4.937436915,8.741437897)	(8.741437897,12.54543888)	(12.54543888,∞)
January 14	(0,1.475390459)	(1.475390459,5.683604877)	(5.683604877,9.89181929)	(9.89181929,14.10003371)	(14.10003371,∞)
January 15	(0,1.34825868)	(1.34825868,5.360417335)	(5.360417335,9.372575981)	(9.372575981,13.38473463)	(13.38473463,∞)
January 16	(0,1.57578106)	(1.57578106,6.03307690)	(6.03307690,10.49037275)	(10.49037275,14.9476686)	(14.9476686,∞)

Table 2 shows the interval critical value and classification interval of the vibration intensity data. the classification threshold of each day is very close and the existence of a slight difference may be due to measurement error and other factors. Therefore, we can reduce the error by calculating the average of the critical values within a week. The calculated critical values for the intervals are: 1.51298707, 5.67584588, 9.83673070, and 13.9986025. The classification intervals 1 to 5 are (0, 1.51298707), (1.51298707, 5.67584588), (5.67584588, 9.83673070), (9.83673070, 13.9986025), (13.9986025,∞).

According to the lifting motor vibration severity data classification interval, the motor status is divided into five categories: good, normal, early warning, warning and danger. Interval 1 represents the motor's in good state, interval 2 represents the motor's in normal state, interval 3 represents the motor's in warning state, interval 4 Indicates motor in warning status and interval 5 represents motor in dangerous status.

In this paper, SOM neural network is used to excavate the corresponding relationship between the vibration intensity data of lifting motor and the working state of the motor. By comparing the collected vibration intensity data with the classification interval, the working status of the lifting motor can be judged.

6. Conclusion

This method can accurately identify the working status of the Container crane lifting motor, so as to feedback to the Container crane operator to respond in time and reduce the economic loss. The SOM neural network has the advantages of simple structure, high degree of self-organization and self-learning ability, high efficiency and strong anti-interference ability. It also can be used in the other Container crane motor state recognition. In addition, because there are few sample data in this training, the classification thresholds and classification intervals obtained through SOM neural network training are not accurate enough. When the collected vibration signal database is large enough, the comparison between motor state and vibration intensity data can be established accurate relationship.

References

[1] Wang Zhixin, "Data mining and state recognition of dynamic characteristic information of shore bridge machinery" 2008. *Shanghai Jiaotong University,*pp.79-81.

[2] Li Zefang, Fang Huajing and Huang Ming, "Data-driven bearing fault identification using improved hidden Markov model and self-organizing map" 2018. *Computers and Industrial Engineering*, pp.37-46.

[3] Zhang Bide, Ou Jian and Sun Caixin, "SOM Neural Network Method for Multiple Fault Diagnosis of Turbine generator" 2005. *Journal of Chongqing University*, pp.36-38.

[4] Guo Wei, "Statistical Characteristics and State Cluster Analysis of Mechanical Performance Parameters of Bridge Crane Motors" 2006. *Journal of Shanghai Maritime University,* pp.25-39.

[5] Mohanty Satish, Gupta Karunesh K and Raju Kota Solomon, "Vibration Feature Extraction and Analysis of Industrial Ball Mill Using MEMS Accelerometer Sensor and Synchronized Data Analysis Technique" 2015. *Program Computer Science*, pp.217-224.

[6] Tong Peisheng, "Study on feature extraction and fault degree identification of static monitoring for rolling bearings under variable conditions"2014.*Journal of Nanjing University of Aeronautics*,pp.22-24.
[7] Huo Hua, "Research on oil monitoring information entropy and clustering multiple features extraction theory and method"2005.*Shanghai Jiaotong University*, pp.24-35.
[8] Song Qiang and Wang Ai Min, "Fault diagnosis of marine diesel engines based on SOM neural network"2011. *Advanced Materials Research*, pp.809-813.
[9] Meng Linghui,Wang Peizhen and Liu Zhigang, "Safety assessment for electrical motor drive system based on SOM neural network"*2016.Mathematical Problems in Engineering*,pp.221-229.
[10] YI RONG QING, LI WENHUI and WANG DUO, "Feature recognition based on self-organized neural network"2009. *Journal of Jilin University*, pp.148-153.
[11] Tao Gang,Yan Yonggang and Liu jun, "Improved algorithm for discretion of continuous attributes based on SOM clustering"2015.*computer application*,pp.89-92.
[12] GAN MINLIANG and LIU HONGXIN, "On five amplitude demodulating approaches for vibration acceleration signals"2000. *Vibration and Shock*, pp.32-33.

IOP Conf. Series: Materials Science and Engineering **435** (2018) 012010 doi:10.1088/1757-899X/435/1/012010

The Prediction of the Motor State of the Shore Bridge Based on the Generative Resistance Network GAN

C Shi [1,*], G Tang [1] ,Y Li [1]and X Hu [1]

[1]Logistics Engineering College, Shanghai Maritime University, 1550 Harbor Avenue, Pudong New Area, Shanghai, China

*1565455809@qq.com

Abstract: In order to forecast the trend and the vibration characteristic parameters of the shore bridge driving system and find the internal faults. This paper analyze the trend by using a large number of bridge lifting motor historical data based on the theory of generative information against network GAN, introduce the base theory of the generative theory against network in detail, establish a prediction model based on GAN and use the model to predict crane driving system performance parameters. The experimental results show that the prediction model based on GAN can achieve satisfactory results in the prediction and realize the accurate prediction of the change trend of the bridge driving system.

1. Introduction

With the development of port and shipping, the monitoring of the state of the bridge is very necessary, especially the evaluation and trend analysis of the driving system. The traditional data analysis tools were based on the analysis and verification, however the data mining method were based on the use of discovery, applying algorithm to discover the important relationship between the data [1]. From the perspective of machine understanding data, the generative model usually needs Markov chain to train model, the computation complexity is high and the efficiency is low, which limits its application to some extent. In this paper, a lot of historical operation data of the shore crane are analyzed by the generation of counter network GAN to predict the trend of future vibration of the drive system of the shore crane. GAN (Generative adversarial networks) is a generative model proposed by Goodfellow et al. [2] in 2014, its basic idea is derived from the two game theory of zero sum game, consisting of a generator and a classifier, by the way of learning to fight the iterative training, approaching Nash equilibrium finally [3]. GAN as a generative model, it does not directly estimate the distribution of the sample data, but estimate the distribution of new sample data samples and generate the same distribution optimization model through training generator. The training process don't have to use the Markov chain methods [4] and do all kinds of approximate reasoning, no complex variational lower bound, reduce the difficulty of training greatly and improve training efficiency.

2. The basic theory and overview of GAN

2.1. The basic principle of GAN

The core idea of GAN is derived from Nash equilibrium of the game theory [5], it includes two neural networks: a neural network generator, a neural network discriminator [6]. The two neural networks optimize in the process of training, improve their ability to generate and discriminate, eventually reach a Nash equilibrium. Any differentiable function can be used to express the GAN's generator and discriminator, as shown in Figure 1 we can use differential function represent discriminator, its input is the real data *x*; use the differential function represent generator, its input is a random variable *z*, G (z) represent samples generated by the generator G as far as possible obey to real data.

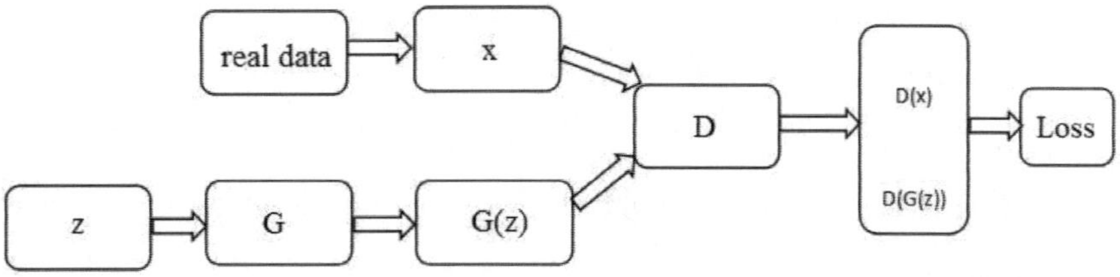

Figure1. The basic schematic diagram of the GAN

2.2. The learning method of GAN

For the learning process of the GAN [7], as shown in Figure 2, it adopts the alternating optimization method:

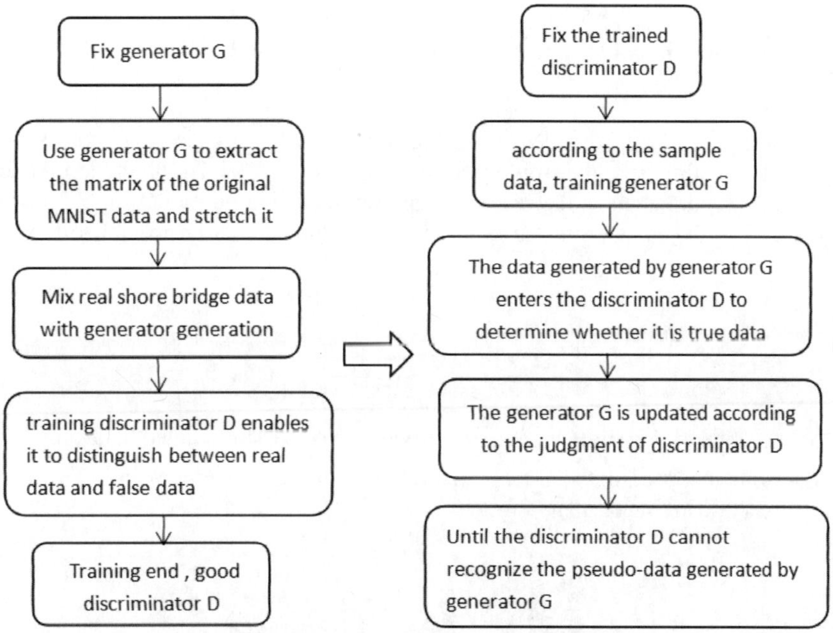

Figure2. The learning method of the GAN

Like two classification model training based on Sigmoid generally, training discriminator D is also a process of minimizing cross entropy, x sampling in real data distribution $P_{data}(x)$, z sampling in prior distribution $P_z(z)$, and its loss function is:

$$Obj^D(\theta_D, \theta_G) = -\frac{1}{2}E_{x\sim Pdata(x)}[\log D(x)] - \frac{1}{2}E_{z\sim Pz(z)}[\log(1 - D(g(z)))] \tag{1}$$

The actual training is different from the conventional two valued classification model [8]. The training data set of the discriminator comes from two parts, the real data set distribution $P_{data}(x)$ and the generator data distribution $p_g(x)$. For given the generator G, we need to minimize (17) to get the optimal solution. In the continuous space, the formula (1) can be written as follows:

$$Obj^D(\theta_D+\theta_G)=-\frac{1}{2}\int_x p_{data}(x)\log(D(x))dx-\frac{1}{2}\int_z p_z(z)\log(1-D(g(z)))dz$$
$$=-\frac{1}{2}\int_x\left[p_{data}(x)\log(D(x))+p_g(x)\log(1-D(x))\right]dx \tag{2}$$

For any nonzero real number a and b, and the real value $y\in[0,1]$, the expression

$$-a\log(y)-b\log(1-y) \tag{3}$$

Get the minimum at the place $\frac{a}{a+b}$. Therefore, in the case of giving the generator G, the target function (2) obtain minimum value at the place

$$D_G^*(x)=\frac{p_{data}(x)}{p_{data}(x)+p_g(x)} \tag{4}$$

which is the optimal solution of the discriminator D. We can see that GAN estimates the ratio of two probability distributions density, which is also the key difference from other lower bound optimization or Markoff chain method [9].

3. State prediction of hoisting motor of shore crane based on GAN

3.1. Data preprocessing

Data this paper analyzed come from NetCMAS (network condition monitoring and assessment system) in 28 days of data sampling, take the front three weeks of data to establish prediction model, other data were used to compare and analyze the results according to prediction model. In order to improve the convergence speed of the model and the accuracy of the model, firstly normalize the data by using formula.

$$\hat{x}_i=\frac{x_i-\frac{1}{n}\sum_{i=1}^{n}x_i}{\max\limits_{1\le i\le n}\{x_i\}-\min\limits_{1\le i\le n}\{x_i\}} \tag{5}$$

Among: x_i is the value of the original data, $\hat{x}_i$ is the value of the normalized data

All the data are mapped to the [-1,1] interval for subsequent processing. The results of data preprocessing are show in Figure 3 and Figure 4.

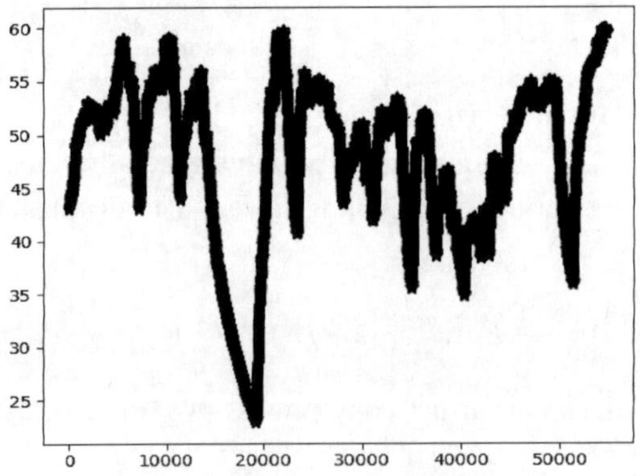

Figure.3 The original data map

IOP Conf. Series: Materials Science and Engineering **435** (2018) 012010 doi:10.1088/1757-899X/435/1/012010

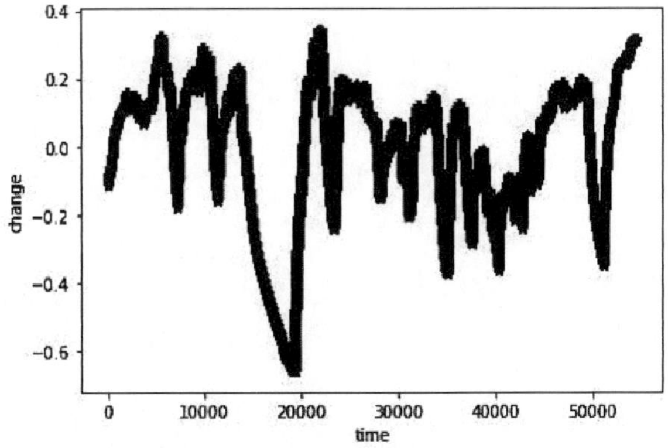

Figure.4 The processed data graphs

3.2. Establishment of data state prediction model for hoisting motor of shore bridge

The first three weeks of data are taken as training samples to establish a prediction model based on GAN, the flow chart of establishing model is shown in Figure 5.

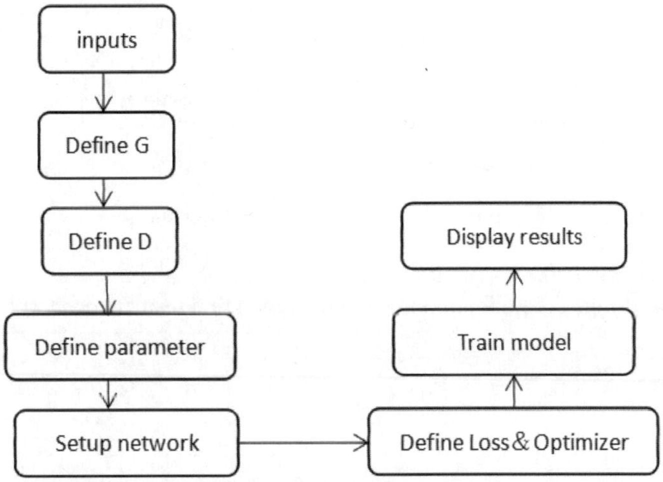

Figure 5 The flowchart based on GAN prediction model

Training model flow chart is show in Figure 6.

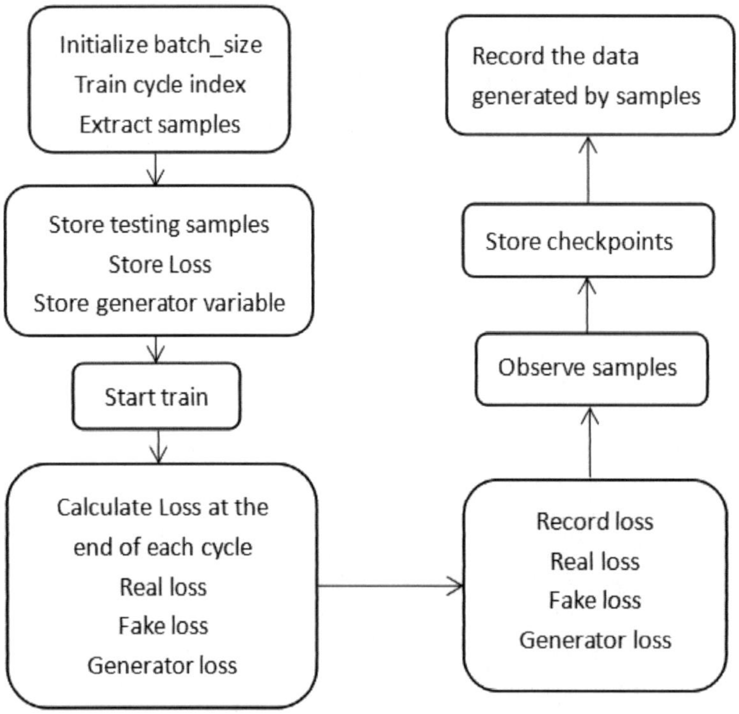

Figure 6 The training model flow chart

3.3. Test for data state prediction model for hoisting motor of shore bridge

The data are brought into the prediction model, and the prediction results are shown in Figure 7.

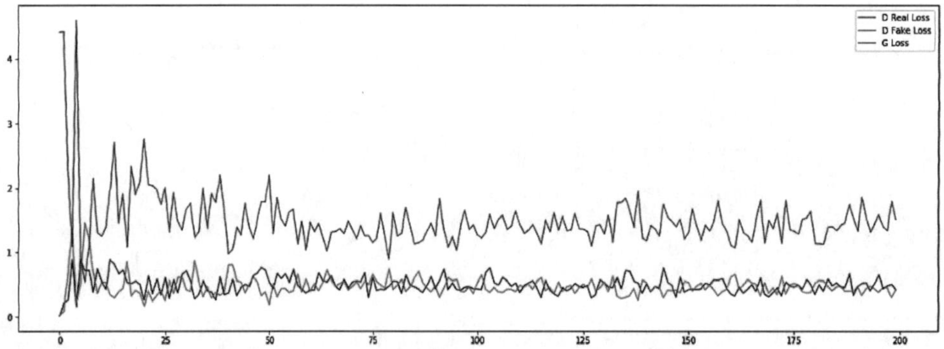

Figure.7 the test diagram of the data state prediction model of shore bridge hoisting motor

From Figure 7, we can see that the final discriminator overall loss fluctuates around 1, while real loss and fake loss fluctuate almost on a horizontal line. This indicates that the discriminator has no ability to distinguish true or false data at last, but to make judgement randomly. That is, the prediction results of the prediction model based on GAN are all satisfactory. Therefore, the prediction model based on the GAN model has high accuracy and reliability in the long term prediction of the running performance characteristic parameters of the hoisting motor.

4. Conclusion

IOP Conf. Series: Materials Science and Engineering **435** (2018) 012010 doi:10.1088/1757-899X/435/1/012010

Generative adversarial networks (GAN) is introduced to predict the performance parameters of motor of shore crane, first introduce the generative adversarial networks (GAN) theory, and then establish the prediction model of the shore bridge drive system based on GAN, forecast the crane hoisting motor trend by using this model. The experimental results show that the prediction model based on GAN can achieve satisfactory results in prediction, and can provide scientific prediction results for condition monitoring, maintenance and maintenance of shore crane.

Reference

[1] Z X Wang, "Data mining and state recognition of dynamic characteristic information of Quayside Bridge crane," 2008. Shanghai Jiaotong University, pp.79-90.
[2] Goodfellow I, Pouget-Abadie J, Mirza M, Xu B, WardeFarley D, Ozair S, Courville A, and Bengio Y, "Generative adversarial nets," in Proceedings of the 2014 Conference on Advances in Neural Information Processing Systems 27, 2014, pp. 2672-2680.
[3] Ratliff L J, Burden S A, and Sastry S S, "Characterization and computation of local Nash equilibria in continuous games," in Proceedings of the 51st Annual Allerton Conference on Communication, Control, and Computing (Allerton), 2013. pp.917-924.
[4] Hinton G E, Sejnowski T J, and Ackley D H, "Boltzmann Machines: Constraint Satisfaction Networks that Learn," in Technical Report No. CMU-CS-84-119, 1984.
[5] Hinton G E and Salakhutdinov R R, "Reducing the dimensionality of data with neural networks," in Science, vol. 313, no.5786, pp. 504-507, 2006.
[6] BRILLOUNI, "Science and information theory," in New York Academic Press Inc, 1956.
[7] Coodfellow I, Bengio Y, and Courville A, "Deep Learning," in MIT Press, 2016.
[8] K F Wang,C Xun,Y J Duan,P L Ling,X H Zheng,and F Y Wang, "Research progress and Prospect of generative antagonism network," in Journal of Automation, 2017,pp. 322-332.
[9] Szegedy C, Zaremba W, Sutskever I, Bruna J, Erhan D, Goodfellow I, and Fergus R, "Intriguing properties of neural networks," arXiv preprint arXiv: 1312.6199, 2013.

IOP Conf. Series: Materials Science and Engineering **435** (2018) 012011 doi:10.1088/1757-899X/435/1/012011

Quayside Crane Hoist Motor State Recognition Based on Hierarchical Clustering Algorithm

Y J Li[1,a] ,G Tang[1,b] and X Hu[1,c]

[1]Logistics Engineering College, Shanghai Maritime University, Shanghai 201306, China

E-mail: gangtang@shmtu.edu.cn,

Abstract: To analyze the dynamic performance of quayside crane, Hierarchical Clustering Algorithm is applied to recognise the different states of Quayside Crane Hoist Motor. The experiment is based on the data collected by NetCMAS (Condition Monitoring & Assessing System on Network). Provided that without knowing the dynamic performance of quayside crane, the vibratory intensity of hoist motor output to the left of quayside crane (sensor location: L1V) is divided into four different states: starting status, lightly vibrating status, moderate vibrating status and severe vibrating status, meanwhile, the floating upper limit and lower limit could be observed.

1. Introduction

With the development of international trading, freight transportation plays a very important part in promoting trading economic development. As necessary large equipment during freight transportation, quayside crane should be monitored in real time to prevent unnecessary loss during work, which is regarded as an urgent work recently.

Before data mining, the priori knowledge is unknown. State recognition, namely, cluster analysis, is an effectual method to monitoring dynamic performance of quayside crane. Unsupervised learning method could be used to solve the above problem[1].

For the problem listed above, this paper presents a cluster analysis method: Hierarchical Clustering Algorithm to classify the data form hoist motor output to the left of quayside crane for recognizing the different status of dynamic performance of quayside crane.

2. Position of Sensors on Quayside Crane

The data from NetCMAS(Condition Monitoring & Assessing System on Network) collected from quayside crane makes sure the authenticity and reliability of this analysis.

In this analysis, using Hierarchical Clustering Algorithm to analyze the vibratory intensity of motor output to the left of quayside crane (sensor location: L1V). Figure 1 displays the positions of sensors on hoist motors & gearbox. As shown in figure 1.

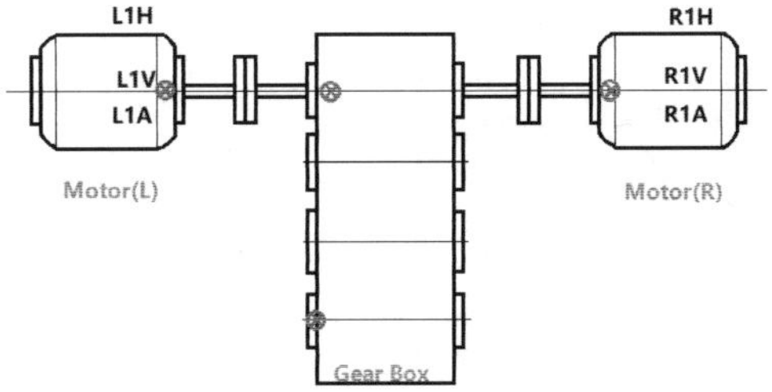

Figure 1. The positions of sensors on hoist motors & gearbox

3. Concept of Hierarchical Clustering Algorithm

Use cluster analysis to cluster something with the same property, which makes the elements in different clusters are less similar, and the elements in the same clusters are more similar. Hierarchical Clustering Algorithm is widely applied to many different fields, such as statistics, machine learning, space database, biology and marketing[2]. For example, in management and engineering, Hierarchical Clustering Algorithm is used to deal with the relationship of customers of bank[3]; for clustering stock samples, and define the investment portfolio on the basis of the result[4] etc. in geography, Hierarchical Clustering Algorithm is applied to regionalism[5]; applied to analyze the structure of consumption[6] etc. in information and communication engineering, Hierarchical Clustering Algorithm is used to traffic communication in real time[7]; used to filter junk e-mail[8] and test mendacious ID[9]; recognition and classification of digital pictures and videos[10] etc. in biology, Hierarchical Clustering Algorithm is used to analyze the system and dynamics of biological networks[11]; used for multi-clustering analysis for different kinds of animals[12]; used to classify the protein clusters, and rank the effectors of the protein for the fungi[13].

Clustering is an unsupervised learning process; the difference between classification and clustering is that it is necessary to know the data characteristics in advance for classification. Thus, in many applications, as a kind of data pre-processing, clustering analysis is the basis of further data analysis and processing[3]. For quayside crane, the machine status of related data collected by NetCMAS is unknown, so that clustering analysis is used to handle this kind of problem. Comparing to K-mean Clustering Algorithm, it is unnecessary for Hierarchical Clustering Algorithm to define the initial value during the data processing, which simplified the data processing; comparing to SOM(Self-organizing Maps), the calculation of Hierarchical Clustering Algorithm is easier and better-understanding, which saves the computer memory.

Hierarchical Clustering Algorithm and Partitional Clustering Algorithm are two mainstays for clustering analysis. Both of them are simple and could be used for handle large size of database.

Hierarchical clustering method is carried out according to distance measure criterion among clusters. Structure and maintain clustering tree made up of clusters and subsidiary clusters, and put it to an end until the specific condition is met. Hierarchical clustering method is divided into agglomerative and divisive according to whether the process is from the bottom to the up or from the up to the bottom. The quality of a simple Hierarchical clustering method is limited by the following condition: once a process of combining or splitting is carried out, it could not be modified.

AGNES(Agglomerative NESting) of which principle is from the bottom to the up, first consider every element as a cluster, and then, combine the clusters into larger clusters until all elements are included to one cluster, or until the specific condition is met. Most Hierarchical clustering methods belong to the above content; they differ from the definition of similarity[14].

Using Hierarchical Clustering Algorithm to cluster the data, choose one of the “bottom-up” AGNES (Agglomerative NESting) method:

(1) Euclidean distance calculation method is used to calculate the distance between two elements; Ward's linkage is used to calculate the distance between two clusters:

$$d(r,s) = \sqrt{\frac{2n_r n_s}{(n_r+n_s)}}\|\overline{x_r} - \overline{x_s}\|_2 \qquad (1)$$

where: n_r is the number of elements in cluster r
n_s is the number of elements in cluster s
$\overline{x_r}$ is the average value of elements in cluster r
$\overline{x_s}$ is the average value of elements in cluster s

(2) (initialized) regard every element as one cluster, and calculate the distance between them. The distance also called the similarity between two samples.
   a) search for the nearest distance, and consider them as one cluster;
   b) repeat to calculate the distance between the new cluster and the old clusters(use the above equations), and repeat a) and b);

(3) condition of convergence: repeat a) and b), until all elements are clustered to one cluster, and then end the process. As shown in Figure 2.

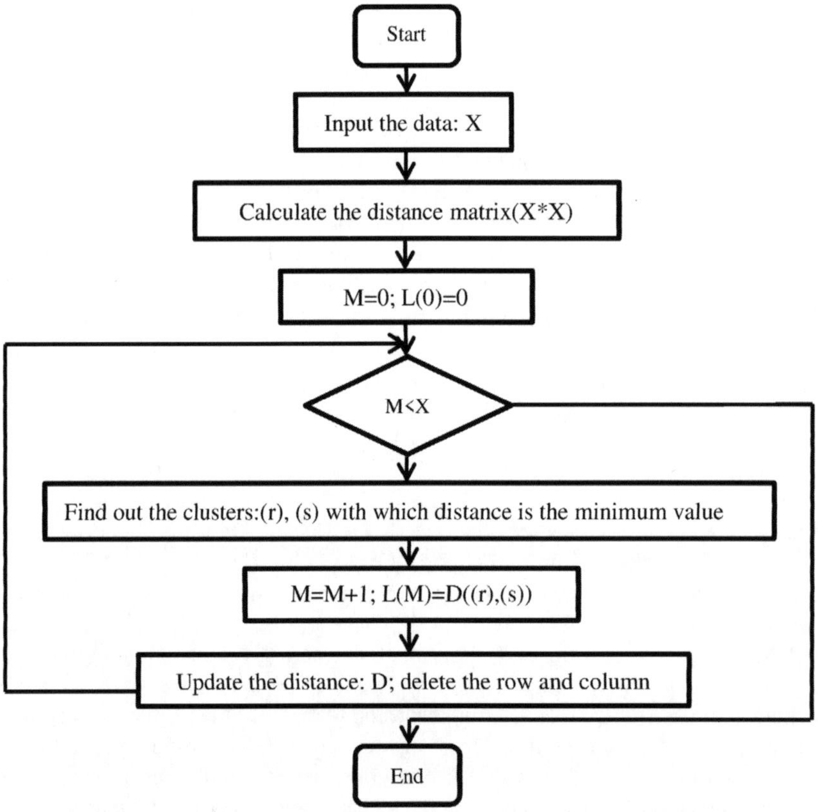

Figure 2. The flow diagram of AGNES (Agglomerative NESting)

4. Data Analysis

Choosing the vibratory intensity from 25th, January, 2010 to 7th, February, 2010, from the motor output to the left of quayside crane (sensor location: L1V), use Hierarchical Clustering Algorithm to cluster the data. The following table is the clustering result, which presents the vibratory intensity interval, classification, and the proportion of every classification. As shown in table1, table 2, table 3, table 4, table 5, table 6, and table 7.

Table 1 Hoist motor vibratory intensity intervals

Date	1-25	Point	%	1-26	Point	%
I	[0.043 1.806]	5692	74.62	[0.035 0.788]	3332	42.00
II	[1.809 4.150]	1536	20.14	[0.789 2.326]	3548	44.73
III	[4.171 8.443]	341	4.47	[2.329 4.179]	685	8.64
IV	[8.543 13.981]	59	0.77	[4.202 14.325]	367	4.62

Table 2 Hoist motor vibratory intensity intervals

Date	1-27	Point	%	1-28	Point	%
I	[0.034 0.898]	5927	70.22	[0.036 0.645]	4529	54.59
II	[0.900 0.971]	1798	21.30	[0.647 1.898]	2667	32.15
III	[1.975 5.482]	617	7.33	[1.900 3.721]	841	10.14
IV	[5.793 13.413]	97	1.15	[3.750 14.011]	258	3.11

Table 3 Hoist motor vibratory intensity intervals

Date	1-29	Point	%	1-30	Point	%
I	[0.037 1.198]	6028	73.66	[0.041 0.603]	3138	38.63
II	[1.200 2.197]	1457	17.80	[0.605 1.629]	3399	41.84
III	[2.202 5.537]	554	6.77	[1.630 5.310]	1440	17.73
IV	[5.594 13.603]	144	1.76	[5.366 13.142]	146	1.80

Table 4 Hoist motor vibratory intensity intervals

Date	1-31	Point	%	2-1	Point	%
I	[0.025 0.354]	3677	44.60	[0.050 0.868]	3089	39.12
II	[0.355 1.910]	3648	44.25	[0.869 2.592]	4042	51.19
III	[1.912 4.132]	739	8.77	[2.601 5.340]	538	6.81
IV	[4.167 16.501]	179	2.13	[5.372 15.681]	227	2.87

Table 5 Hoist motor vibratory intensity intervals

Date	2-2	Point	%	2-3	Point	%
I	[0.031 0.800]	4257	54.22	[0.034 0.643]	3712	45.74
II	[0.804 3.557]	3184	40.55	[0.644 1.843]	3182	39.21
III	[3.571 7.334]	318	4.04	[1.844 4.340]	959	11.82
IV	[7.476 14.535]	93	1.18	[4.363 13.246]	263	3.24

Table 6 Hoist motor vibratory intensity intervals

Date	2-4	Point	%	2-5	Point	%
I	[0.049 0.790]	4274	57.11	[0.037 0.630]	2763	37.77
II	[0.792 3.160]	2929	39.14	[0.631 2.830]	4080	55.77
III	[3.172 7.086]	236	3.15	[2.835 6.690]	405	5.54
IV	[7.178 12.762]	45	0.60	[6.820 13.776]	68	0.93

Table 7 Hoist motor vibratory intensity intervals

Date	2-6	Point	%	2-7	Point	%
I	[0.046 1.137]	5283	62.39	[0.060 1.453]	5867	70.52
II	[1.138 2.164]	2301	27.17	[1.454 3.014]	1958	23.53
III	[2.168 5.284]	696	8.22	[3.024 7.159]	418	5.02
IV	[5.335 14.282]	188	2.22	[7.254 14.717]	77	0.93

According to the above analysis, Hierarchical Clustering Algorithm divides the vibratory intensity into four parts, the first part I present the hoist motor stays in starting status, the second part II stands for the hoist motor stays in lightly vibrating status, the third part III means the hoist motor stays in moderate vibrating status and, the fourth part IV indicates the hoist motor stays in severe vibrating status.
Take the average value of every interval. As shown in table 8:

Table 8 Hoist motor vibratory intensity intervals

classes	Intervals (vibratory intensity)
I	[0.040 0.900]
II	[0.902 2.446]
III	[0.254 5.717]
IV	[5.800 14.141]

According to analysis, 80% points are included to part I and part II, which means the hoist motor stay in starting status and lightly vibrating status.

5. Conclusion

Using Hierarchical Clustering Algorithm to analyze the vibratory intensity of motor output to the left of quayside crane (sensor location: L1V), from 25th, January, 2010 to 7th, February, 2010. The data is divided into four parts(I,II,III,IV), which separately stand for the hoist motor staying in starting status, lightly vibrating status, moderate vibrating status and severe vibrating status. 80% points are included to part I and part II, which means the hoist motor stay in starting status and lightly vibrating status. The final vibratory intensity interval is defined by taking the average value of every data interval of every day.

Reference

[1] Tang gang, Yang zhiqi, Hu xiong. "SOFM-based quay-bridge lift motor status identification and visualization" 2017. *Journal of Donghua University*.pp.559-564.l
[2] Anil K.Jain, Richard C.Dubes. "Algorithms for Clustering Data"1988. *Prentice Hall*.pp.1~334
[3] Qin xiujie. "Research on Application of Data Mining in Bank Customer Relationship Management".2014.
[4] Zhao Fengyun, Wang zongwei, Wang shun, Zhang shitao. "An Empirical Study on the Value of Stock Investment Based on Hierarchical Clustering" 2015. *Journal of Chaohu University*.17(6):pp14-19
[5] Jin xingyu." Hierarchical Cluster Analysis of Regional Division in China"2004. *City planning*.(2):pp23-28.
[6] Wang zhongfu, Wang erda, Li zuozhi, Yu yang. "Cluster Analysis of Domestic Tourism Consumption Structure of Urban Tourism Destinations in Dalian"2009. *Journal of Dalian University of Technology (Social Sciences Edition)*.30(1):pp72-78.
[7] E Dror, A Chen, Z Lotker. "Fast randomized algorithm for hierarchical clustering in Vehicular Ad-Hoc Networks "2011. *Ad Hoc Networks*.11(7):pp1-8
[8] Liao mingtao, Zhang deyun, Li jinku. "Two-stage spam filtering method based on naive Bayes and hierarchical clustering" 2007.Microelectronics & Computers.24(8):pp1-3

[9] Fang yong, Liu daosheng, Huang chen. "False User Detection Based on Hierarchical Clustering"2017. *Journal of Tsinghua University (Natural Science Edition).* (6):pp620-624

[10] S Krishnamachari, M Abdel-Mottaleb. "Image Browsing using Hierarchical Clustering" 2002. *IEEE international Symposium on Computer & Communications*.2000:301

[11] Lu hongchao. "Gene Function Research Based on Protein Network Clustering". *Graduate University of Chinese Academy of Sciences (Institute of Computing Technology)*.2006

[12] Gong jun, Zhang shiping, Xiong bangxi, Liu dingzhu, Li Jinzhong, Cao jun. "Multivariate cluster analysis of summer plankton in different types of lakes" 2010. *Aquatic Life Report*.34(1):pp43-50

[13] Diane G. O. Saunders1, Joe Win1, Liliana M. Cano1, Les J. Szabo2, Sophien Kamoun1, Sylvain Raffaele1. "Using Hierarchical Clustering of Secreted Protein Families to Classify and Rank Candidate effectors of rust fungi" 2012. *Plos One*.7(1):e29847

[14] Duan mingxiu. "Research and Application of Hierarchical Clustering Algorithm" 2009. *Computer Application Technology, Central South University.*

Cluster-based Alpha-Beta Search for Real-Time Strategy Games

X Han[1], H P Yan[1, *], J G Zhang[2] and L F Wang[3]

[1] Department of Information Engineering, China University of Geosciences, No. 29 College Road, Haidian District, Beijing, China
[2] CRIPAC & NLPR, Institute of Automation, Chinese Academy of Sciences, Beijing, China, No. 95 Zhongguancun East Road, Haidian District, Beijing, China
[3] NLPR, Institute of Automation, Chinese Academy of Sciences, Beijing, China, No. 95 Zhongguancun East Road, Haidian District, Beijing, China

[*] Corresponding Author: yanhp@cugb.edu.cn

Abstract. From an artificial intelligence point of view, Real-Time Strategy (RTS) game has been proven to be one of the most challenging areas. Due to the huge action state space, partial observability and real time property, the previous AI solutions are sill low in terms of speed. In this paper, we propose a Cluster-based Alpha-Beta Considering Durations (CABCD) algorithm that searches the optimal action for each unit based on the cluster. When the improved algorithm is applied to large RTS games, the enormous branching factors are significantly reduced. The approach is evaluated in StarCraft I, and outperforms competing methods promisingly.

1. Introduction

Real-Time Strategy (RTS) game belongs to a kind of strategy game that progresses immediately instead of incrementally in turns. In RTS game, we can accomplish activities including building, producing soldiers, developing technology or fighting with the enemy [1]. Due to the partial observability and the huge action state space, real-time strategy games still face plenty of challenges [2, 3]. It is difficult to build an integrated artificial intelligence to solve RST games, therefore, amounts of work split the tasks into collecting resources, producing soldiers, and fighting, etc.

War is one of the core components of the RTS game. The most important operation of a war is to choose the action of each unit. Nowadays, researchers mainly use Alpha-Beta search [4, 5], UCT [6, 7], Portfolio [8] or other algorithms to solve the problem mentioned above. However, RTS game requires huge resources to possess the accumulated units (more than 10 units) for actions and states, which hinders the applicability of the game tree search method such as Alpha-Beta search, UCT algorithm, etc. In order to ensure that the optimal action is obtained within a limited time, most work adopts the strategy of exchanging time with accuracy. For example, Churchill D et al. used Portfolio to search strategy, then adopted the strategy to determine the action. However, using these methods leads to less accurate action, lower intelligence, and less adaptability to the environment. So the war scene in the RTS game poses a great challenge to artificial intelligence.

Churchill D et al. have demonstrated that the Alpha-Beta search algorithm searches better actions when the number of units is small [4]. As shown in Figure 1, with the increase of the number of the units and actions, the number of branches searched by Alpha-Beta search algorithm increases exponentially.

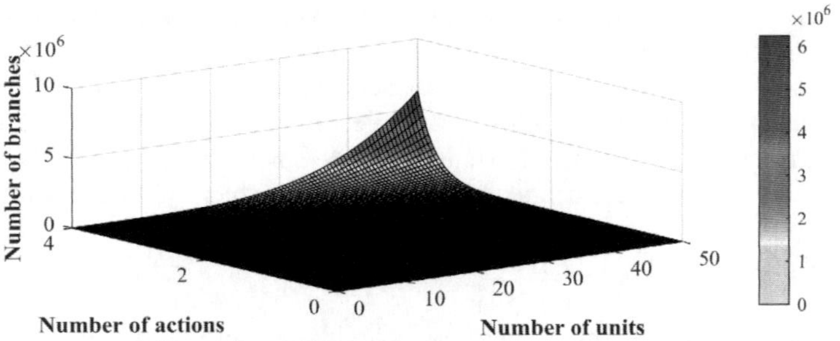

Figure 1. The relationships of the number of units, actions and the branches searched in the Alpha-Beta search algorithm.

This paper proposes an improved Alpha-Beta search algorithm to reduce the action state space, that is, the Cluster-based Alpha-Beta Considering Durations (CABCD) algorithm. CABCD can perform better when the action state space is large. The main idea of the CABCD is to aggregate units to reduce the number of branches and ultimately reduce the action state space. The specific steps of CABCD are: Firstly, CABCD groups units according to their locations and types; Secondly, CABCD performs the Alpha-Beta search algorithm for each group to find the current optimal action of each unit; Finally, CABCD integrates the best actions of units in all groups and implements them simultaneously.

The main contributions of this paper are listed as follows:

1). This paper proposes the CABCD algorithm to achieve the purpose of determining the optimal action of each unit within a limited time, meanwhile, CABCD ensures the accuracy of the searched action.

2). The CABCD algorithm is built to enable the agent to construct the Alpha-Beta tree according to the group to achieve the effect of reducing the action state space and shortening the search time.

The experimental results show that the CABCD algorithm can search for the better action in a limited time. The winning rate is above 0.5 comparing with other search algorithms. In addition, the influence of cluster on decreasing the number of branches is qualitatively analysed in this paper.

2. Related work

One of the earliest and most classic methods in the tactical decision-making of RTS games uses tree search to determine the action for each unit. Stanescu et al. proposed a hierarchical implementation of the search to reduce the action state space [9]. Ontanón S presented a new Monte Carlo Tree Search (MCTS) algorithm based on Naive Sampling [10]. Uriarte A and Ontanón S put forward the game state abstraction [11]. The map is decomposed into regions, and then the action based on the region is selected. However, the grouping error based on the region is large, and the units at the two regional boundaries might be divided into different groups. Ontanón S brought forward an improvement of the MCTS algorithm combined with the Bayesian probability distribution estimation model [12].

Notably, the RTS game is the real-time game, when the number of units is large, there is a huge space of states. Therefore, it is impossible to use a tree search algorithm to traverse the action state space to select suitable actions within a limited time. So most work proposes to find suitable strategies for the unit, such as attacking the nearest unit, attacking the lowest hit point of the unit, and then select the action according to the strategy. Churchill D et al. proposed the Portfolio Greedy Search algorithm to select actions via searching the predefined strategies for each unit [8]. Wang C et al. presented the improvement of the Portfolio Greedy Search algorithm that combined with online evolution [13]. Nicolas A. B et al. put forward a new tactical search framework based on strategy [14]. Although searching strategy can reduce the number of searched branches, it also reduces the accuracy of the searched action. Therefore, this paper first reduces the action state space by clustering, and then determines the action of each unit in a more precise decision-making way.

The work of Justesen N et al. is similar to our work, which groups units and then applies Upper Confidence Bound Apply to Tree (UCT) algorithm to select the appropriate strategy for each cluster [15]. The main difference is that we use the Alpha-Beta search algorithm to determine the action for each unit after grouping. The advantage of our work lies in that the Alpha-Beta search algorithm can achieve higher accuracy when the action state space is smaller. The more appropriate action is determined for each unit, the higher accuracy can be obtained, which will result in the higher winning rate.

3. Background

In this section, we first introduce the Alpha-Beta search algorithm, and then provide the improved method which is clustering.

3.1. Alpha-Beta tree search algorithm and Alpha-Beta Considering Durations algorithm

Minimax search algorithm is commonly used in game search, such as Go, chess, etc., as shown in Figure 2. The Minimax search algorithm traverses the entire game tree to select the optimal route as far as possible. As the search depth increases, the tree grows exponentially. Therefore, Knuth D E et al. proposed the Alpha-Beta search, which is an improved algorithm of Minimax search algorithm, reducing the number of searched branches by pruning to make the result more accurate [16].

Alpha-Beta search based on the theory that when you have a choice, as long as the other options will not be better than the current situation, we can not explore other options and abandon them. The idea is to transfer two parameters in the search. The first parameter is Alpha, which is the minimum lower bound for the best score; the second parameter is Beta, which is the maximum upper bound for the best score.

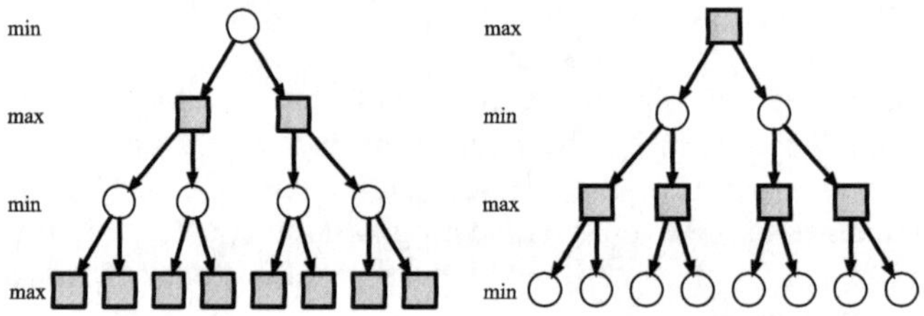

(a) Max-Min Algorithm (b) Min-Max Algorithm

Figure 2. Minimax search algorithm.

Alpha-Beta search is widely used for game problems, and it is suitable for round strategy games. However, the RTS game is a real-time game. The player can control multiple units, which is equivalent to multi-agent problem. So more than one instruction will be assigned to multiple units at each frame. Considering the weapon's cooldown and spatial granularity, Churchill D et al. proposed the Alpha-Beta Considering Durations (ABCD) algorithm for combat games which need continuous movements [4]. If the state does not correspond to a simultaneous decision, the action set m_0 remains $\emptyset$. Otherwise, we apply the effect of m_0 after the action set m is generated, since m_0 should not affect the generation of complementary actions.

ABCD has been proven to be the best search algorithm when the number of units is small, but it is less effective in complex situation. If there exists 20 units in a war and each unit can move in four directions, the branches of move can up to 20^4. Hence when the number of branches becomes enormous and search time is holding steady, the ABCD algorithm will perform poorly.

3.2. Cluster

Applying cluster is the extension of ABCD. Cluster refers to the process of grouping multiple classes of similar objects in a collection of physical or abstract objects.

Given the input $\{u_i\}_{i=1}^n \in U_j \in P_j$, the K-means algorithm is applied to CABCD,

$$u_i \to \tau_c(u) \quad (1)$$

Where P_j is the player j, and U_j is all n units owned by P_j. $u_i = \{(x_i, y_i), t_i\}$ indicates the unit i. (x_i, y_i) represents the coordinate of u_i, and t_i is the type of u_i, such as the Zealot or the Dragoon in StarCraft I.

As shown in equation (1), the unit u_i is assigned to the group c, $1 \le c \le k$. k is the total number of current groups.

Units with similar positions and more consistent types would be assigned to one group, and their movements would be similar. However, it is not certain that the same type of units will be assigned to one group. This paper considers equalizing firepower when grouping.

4. Cluster-based Alpha-Beta Considering Durations algorithm

This section first introduces the combat modal. Then the details of implementation are described and we propose the Cluster-based Alpha-Beta Considering Durations (CABCD) algorithm. Finally, we theoretically analyse the influence of the grouping number k on time complexity.

4.1. The combat model

The CABCD algorithm is evaluated in SparCraft and modifies the combat model of SparCraft, which can be used to construct Alpha-Beta trees in groups. The SparCraft combat model consists of three data components and two logical components.

State $s = <t, U_1, U_2>$
- Current game time t
- Sets of units U_i under control of player i

Unit $u = <p, hp, t_a, t_m, c, k, type>$
- Position $p = <x, y>$ in R^2
- Current hit points hp
- Time step when unit can next attack t_a, or move t_m
- Current group number c.
- The total number of current groups k.
- StarCraft unit type, defining all static unit properties such as damage, maximum hp, armor, speed, etc.

Move $m = <a_1, ..., a_k>$ which is a combination of unit actions a_i = <*u, type, target, t*>, with
- Unit u to perform this action
- The type of action to be performed: Attack unit target, Move u to position target, or Wait until time t

Player function p $[m = p(s, U)]$
- Input state s and units U under player's control
- Perform Move decision logic
- Return move m generated by p

Game function g $[r = g(s, p_1, p_2)]$
- Initial state s and players p_1, p_2
- Perform game simulation logic
- Return game result r (win, lose or draw)

4.2. CABCD algorithm

CABCD algorithm is as shown in Algorithm 1. The core idea is to use the K-means to reduce the branching factors, then build the Alpha-Beta tree for each group. In this paper, the experimental environment is the classic RTS game StarCraft I. When human players play StarCraft I, they tend to divide units of similar types and positions into one group. Consequently, we consider the location and type of units when grouping, then find the optimal action for each unit. Figure 3 shows an example of grouping in StarCraft I.

Algorithm 1 Cluster-based Alpha-Beta Considering Durations

Initialize d, m_0, α, β
Obtain s
if $s.num < 10$ **then**
    $ABCD(s, d, m_0, \alpha, \beta)$
else
    $data \leftarrow s.x, s.y, s.type$
    $K-means(data)$ //Use $K-means$ grouping and store the group number c .
    for $i = k$ **to** 1 **do**
        for $j = n$ **to** 1 **do**
            if $c_j = i$ **then**
                $s_i = [s_i; s_j]$
        end for
        $ABCD(s_i, d, m, \alpha, \beta)$
end for

The enemy can also be grouped, but this experiment considers the best situation where the enemy is not grouped, which is equal to all the units of the enemy is one group.

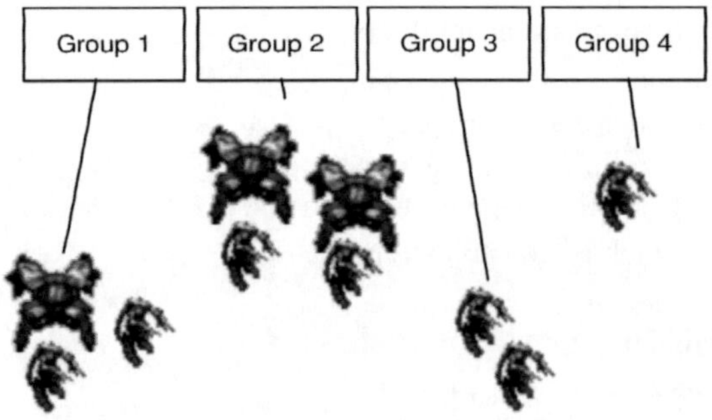

Figure 3. The example of grouping in StarCraft I.

There are two main problems when grouping. The first is whether the units should be the same type. In this paper, we just take the weighting type into account, and select the action for each unit in the same group, instead of selecting a strategy for the group. As a result, we do not need to stipulate the same type of the unit must be in the same group.

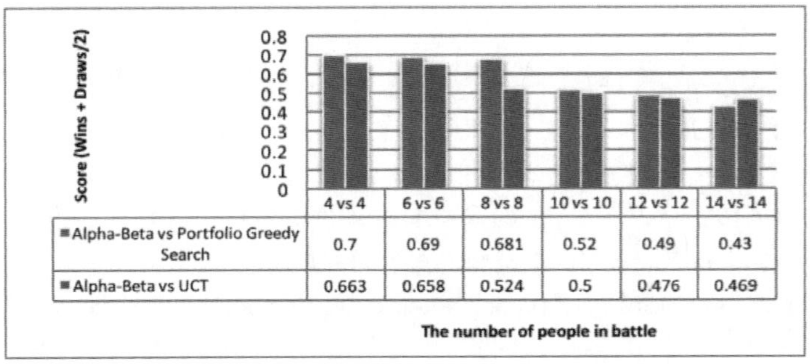

Figure 4. The winning rate between Alpha-Beta and Portfolio Greedy Search.

The second problem is that the total number of current groups k should be selected dynamically. In RTS games it is one of the core elements of the game to save time, whereas dynamic mechanism takes a lot of time. Therefore, we set k manually which uses less time and is more efficient. The number of the group should balance the higher number of units in the group and the performance requirements of the computer. Over-dividing groups must require a relatively high performance for computer. The results of ABCD in different numbers of units are shown in Figure 4. The maximum number of 10 units in each group is superior. The number of units in the group should not be divided equally.

When the player has units fewer than 10, the action state space has been greatly reduced. And no further grouping is required. So all units belong to one group that directly builds an Alpha-Beta tree to find the best action for each unit.

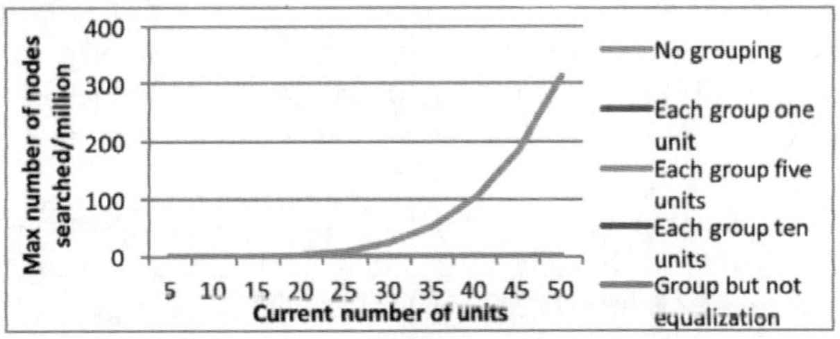

(a)

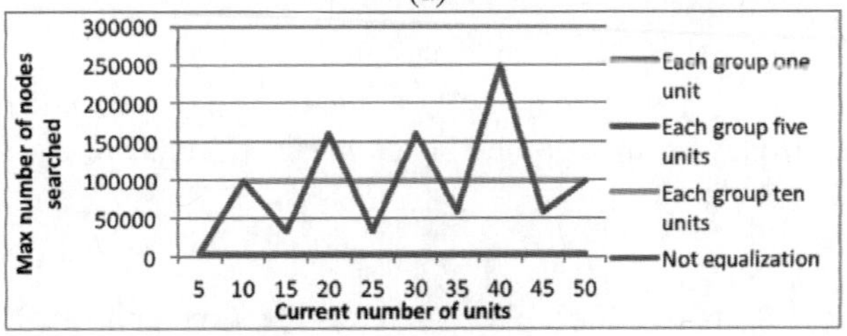

(b)

Figure 5. The relationship between the number of units and max searched nodes. (b) only shows the situation of grouping.

4.3. The influence of k on the time complexity of CABCD

Theorem 1. For $b \in \{1,2,\cdots,n\}$, the number of total nodes searched by Minimax algorithm is:

$$N_d \approx b^d \tag{2}$$

In equation (2), b is the average number of branching factors in the search tree, d is the search depth. The complexity of algorithm is $O(b^d)$. Theorem 1 is proved in the appendices.

Theorem 2. For $b \in \{1,2,\cdots,n\}$, the total number of nodes searched by Alpha-Beta search algorithm is:

$$\begin{cases} N_d = 2b^{d/2} - 1 \\ N_d^{'} = b^{(d+1)/2} + b^{(d-1)/2} - 1 \end{cases} \tag{3}$$

In equation (3), N_d is the total number of the searched nodes when the depth of search is even, $N_d^{'}$ is the total number of the searched nodes when the depth of search is odd. If the nodes are optimally arranged, the algorithm complexity is $O(b^{d/2})$; If the nodes are arranged randomly, the algorithm complexity is $O(b^{3*d/4})$. This theorem has been proved by Knuth D E et al [16].

Consequently, according to Theorem 1, the number of branches increases drastically with the depth, in other words, the number of search points increases exponentially with the number of branches. Hence, this paper proposes to group the units to significantly reduce the number of search points, and the quantitative analysis is shown in Theorem 3.

Theorem 3. The max number of nodes searched using CABCD algorithm is:

$$\begin{cases} N_d^G = k*[2*(b_{\max})^{d/2} - 1] \\ N_d^{'} = k*[(b_{\max})^{(d+1)/2} + (b_{\max})^{(d-1)/2} - 1] \end{cases} \tag{4}$$

$b_{\max} = m^{n_{\max}}$ is the maximum number of branches after grouping, $n_{\max} = \max_k n_c$.

In equation (4), m is the number of actions of each units. k is the total number of groups. n_c is the number of units in group c. Theorem 3 is proved in the appendices.

From Theorem 3, we know that $N_d \gg N_d^G$ (or $N_d \gg N_d^{'}$). So after grouping it can exponentially decrease the number of the branches to reduce the search points. In the case of limited time and depth, it is possible to search all possibilities on the first floor to make the results more accurate.

Figure 5 (a) represents the difference of the maximum number of nodes between grouping and no grouping. It can be seen that there is an exponential increase in the number of nodes compared grouping with no grouping.

The difference between grouping and no grouping is large. So in figure 5 (b), only a few cases of grouping are shown. The number of branches per group 1 (or 5) unit is very small, but it constructs plenty of the Alpha-Beta trees. With the reasonable allocation arrangement, the effect of uneven distribution is better.

5. Experiment

In this section, we evaluate the algorithm CABCD in the war platform SparCraft of the classic RTS game-StarCraft I, which mainly set up a group of experiments to compare CABCD, ABCD, UCTCD, and Portfolio Greedy Search algorithm.

5.1. Test scenarios

The algorithm is tested in SparCraft. Each player controls the same units and sets the parameter n in the scene to determine the number of units each player owns. There are five scenarios. The experimental result uses Equation (5) to calculate the winning rate.

$$Winningrate = \frac{(Wins + Draws/2)}{total} \tag{5}$$

Wins indicates the number of wins. *Draws* is the number of draws. *total* represents the total number of tests, which is 100 there. When the number of units is less than 10, no grouping is performed.

5.2. Environment configuration

All experiments run on Windows 8.1 Intel (R) Core (TM) i5-5287U CPU @ 2.90GHz, and a total of 8GB of RAM is available. All experiments are single-threaded.

Search algorithm settings

Configuration of all Alpha-Beta search algorithms:

-Time Limit: 40 ms
-Max Children: 20
-Evaluation: NOK-AV vs. NOK-AV Playout
-Transposition Table Size: 100000 (13.2 MB)

CABCD:

-Max Clusters： 5
-The Minimum Number of Units Divide into Groups:10

UCT search:

-Time Limit: 40 ms
-Max Children: 20
-Evaluation: NOK-AV vs. NOK-AV Playout
-Final Move Selection: Most Visited
-Exploration Constant: 1.6
-Child Generation: One-at-leaf
-Tree Size: No Limit (6 MB largest seen in 40 ms)

Portfolio Greedy search:

-Time Limit: 40 ms
-Improvement Iterations I: 1
-Response Iterations R: 0
-Initial Enemy Script: NOK-AV
-Evaluation: Improved Playout
-Portfolio Used: (NOK-AV, Kiter)

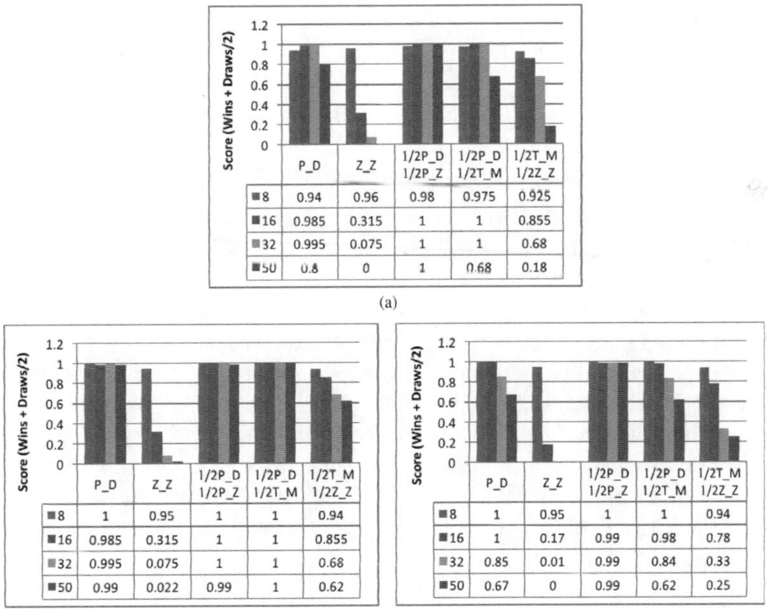

	P_D	Z_Z	1/2P_D 1/2P_Z	1/2P_D 1/2T_M	1/2T_M 1/2Z_Z
8	0.94	0.96	0.98	0.975	0.925
16	0.985	0.315	1	1	0.855
32	0.995	0.075	1	1	0.68
50	0.8	0	1	0.68	0.18

(a)

	P_D	Z_Z	1/2P_D 1/2P_Z	1/2P_D 1/2T_M	1/2T_M 1/2Z_Z
8	1	0.95	1	1	0.94
16	0.985	0.315	1	1	0.855
32	0.995	0.075	1	1	0.68
50	0.99	0.022	0.99	1	0.62

(b)

	P_D	Z_Z	1/2P_D 1/2P_Z	1/2P_D 1/2T_M	1/2T_M 1/2Z_Z
8	1	0.95	1	1	0.94
16	1	0.17	0.99	0.98	0.78
32	0.85	0.01	0.99	0.84	0.33
50	0.67	0	0.99	0.62	0.25

(c)

Figure 6. Results of CABCD vs other algorithms. (a) Results of CABCD vs Alpha-Beta; (b) Results of CABCD vs UCTCD; (c) Results of CABCD vs Portfolio. The results show the war for n vs n units in five different scenarios, n=8, 16, 32, 50.

5.3. Experimental results

Figure 6 shows the experimental results of CABCD vs ABCD, UCTCD and Portfolio Greedy Search algorithm in different scenes. Abscissa corresponds to the scene, P_D means that each player only controls Protoss_Dragoon; Z_Z means that each player only controls Zerg_Zergling; $n/2$ P_D $n/2$

P_Z means that each player controls $n/2$ Protoss_Dragoon and $n/2$ Protoss_Zealot; $n/2$ P_D $n/2$ T_M means that each player controls $n/2$ Protoss_Dragoon and $n/2$ Terran_Marine; $n/2$ T_M $n/2$ Z_Z means that each player controls $n/2$ Terran_Marine and $n/2$ Zerg_Zergling. Run 100 innings in each scenario.

It can be seen from Figure 6 that the winning rate of CABCD vs ABCD, UCTCD can be as high as 0.9 when the unit is the Protoss. Because of no use of parallelism, the winning rate of CABCD vs Portfolio Greedy Search algorithm decreases with the increase of the unit number, but it can also reach about 0.7. When the number of units is large, the more Alpha-Beta trees are constructed. The time of constructing each tree in a limited time is less leading that it can not search for better action.

When the unit is Zerg, the winning rate of CABCD is very low. CABCD needs to be grouped first, which takes time. However, the Zerg has low blood volume. Therefore, it may be defeated while grouping.

CABCD vs ABCD still has a higher rate of success in less units, namely 8 vs 8. Since the CABCD is divided into a group when less than 10 units, it is more likely to concentrate on fire to win.

Figure 7 shows the overall test results of CABCD vs ABCD, UCTCD and Portfolio Greedy Search algorithm in the comprehensive five scenarios, which indicates that the winning rate of CABCD is above 0.5. Figure 7 fully illustrates the effectiveness of CABCD algorithm.

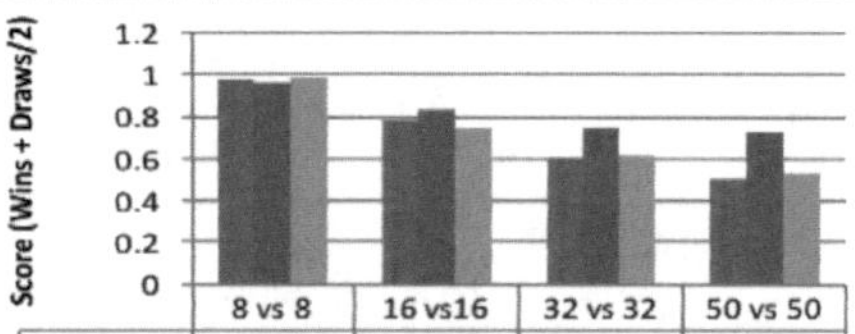

	8 vs 8	16 vs16	32 vs 32	50 vs 50
Portfolio	0.978	0.784	0.604	0.506
ABCD	0.96	0.828	0.75	0.726
UCTCD	0.982	0.744	0.614	0.532

Figure 7. The total winning rate of CABCD vs ABCD, UCTCD and Portfolio in five scenarios

6. Conclusions and discussion

Although search algorithms have made progress in recent years, choosing appropriate actions in large RTS games is still a great challenge. The difficulty is that the RTS game has a large amount of action state space, is partially observability and real-time. In this work, we propose CABCD algorithm, which is an extension of the ABCD algorithm for searching better actions for each unit. CABCD is a search algorithm for RTS games with a large number of units.

Our experimental evaluation shows that the CABCD algorithm is superior to the ABCD, UCTCD and Portfolio Greedy Search algorithms in the RTS game. The winning rate is above 0.5. CABCD algorithm can search for better actions for each unit when the number of units is large, and it can also be applied to the search problems with the large action state space, not only in RTS games.

However, the CABCD algorithm has a low winning rate when using the Zerg, because the grouping takes time and the Zerg blood volume is lower, which may be eliminated when grouping. This should be considered in the future work.

7. Appendices

Proof of Theorem 1

For $b \in \{1, 2, \cdots, n\}$, the total number of nodes N_d searched by the minimal maximum algorithm is:

$$\begin{cases} N_d = 1 + b + b^2 + \cdots + b_d \\ \quad = b^d \dfrac{1-(1/b^d)}{1-(1/b)} \\ \quad \approx b^d \end{cases} \tag{6}$$

b is the average branch number of the search tree, d is the search depth (Approximate value is obtained when b is large enough).

Proof of Theorem 3

The total number of nodes searched for Alpha-Beta search algorithm is:

$$\begin{cases} N_d = 2b^{d/2} - 1 \\ N_d^{'} = b^{(d+1)/2} + b^{(d-1)/2} - 1 \end{cases} \tag{7}$$

After grouping:

$$m^{\sum_{i=1}^{k} n_c} = m^N = b \tag{8}$$

m is the number of actions of the unit, N is the total number of units currently owned by the player.

$$\begin{cases} n_{\max} = \max_k n_c \\ b_{\max} = m^{n_{\max}} \end{cases} \tag{9}$$

Therefore $b_{\max}$ is the maximum number of branches after grouping. The max searched points after grouping are:

$$\begin{cases} N_d^G = k*[2*(b_{\max})^{d/2} - 1] \\ N_d^{'} = k*[(b_{\max})^{(d+1)/2} + (b_{\max})^{(d-1)/2} - 1] \end{cases} \tag{10}$$

References

[1] Ontanón S , Synnaeve G, Uriarte A, Richoux F, Churchill D, Preuss M et al. 2013 A survey of real-time strategy game AI research and competition in starcraft J. IEEE *Transactions on Computational Intelligence and AI in games* **5** 293-311
[2] Robertson G and Watson I 2014 A review of real-time strategy game AI *J. AI Magazine* **35** 75-104
[3] Šustr M, Malý J and Čertický M 2018 Multi-platform Version of StarCraft: Brood War in a Docker Container. Technical Report *J. Preprint* arXiv/1801. 02193
[4] Churchill D, Saffidine A and Buro M 2012 Conf. *Artificial Intelligence and Interactive Digital Entertainment Conference* (USA: Stanford University) pp 112-7
[5] Zhang C P, Liu C H and Xu J C 2008 Research on Pruning Technique in Game Tree Heuristic Search *J. Computer Engineering and Applications* **44** 54-6
[6] Silver D and Schrittwieser J *et al.* 2017 Mastering the game of Go without human knowledge *J. Nature* **550** 354-9.
[7] Silver D and Huang A *et al.* 2016 Mastering the game of Go with deep neural networks and tree search *J. Nature* **529** 484-9
[8] Churchill D and Buro M 2013 *Conf.* Computational Intelligence in Games (CIG) 2013 IEEE (Canada: Niagara Falls) pp 1-8
[9] Stanescu M, Barriga N A and Buro M 2014 *Conf. Artificial Intelligence and Interactive Digital Entertainment Conference* (USA: Raleigh)
[10] Ontanón S 2013 *Conf.* Ninth *Artificial Intelligence and Interactive Digital Entertainment* Conference (USA: Northeastern University)
[11] Uriarte A and Ontañón S 2014 *Conf. Tenth Artificial Intelligence and Interactive Digital Ent*ertainment Conference (USA: Raleigh)
[12] Ontañón S 2016 *Conf. Computational Intelligence and Games (CIG) 2016 IEEE* (Greece: Santorini) pp 1-8
[13] Wang C, Chen P, Li Y, Holmgård C and Togelius J 2016 *Conf. Twelfth Artificial Intelligence and Interactive Digital Entertainment Conference* (USA: San Francisco)
[14] Barriga N A, Stanescu M and Buro M 2015 *Conf. Eleventh Artificial Intelligence and Interactive Digital Entertainment Conference* (USA: California University)
[15] Justesen N, Tillman B, Togelius B and Risi S 2014 *Conf. Computational Intelligence and Games (CIG) 2014 IEEE* (Germany: Dortmund) pp 1-8
[16] Knuth D E and Moore R W 1975 An analysis of alpha-beta pruning *J. Artificial intelligence* **6** 293-326

IOP Conf. Series: Materials Science and Engineering **435** (2018) 012013 doi:10.1088/1757-899X/435/1/012013

Nodemcu-based Low-cost Smart Home Node Design

Tong Qiang[1], Gao Guangling[2], Liu hai*[1],Cao Lina[1] and Wang Han[1]

[1]School of Mechanical, Electrical and Information Engineering, Shandong University (Weihai), Weihai 264209, China
[2]State Grid of China Technology College, Jinan 250002, China

liangLikai@sdu.edu.cn

Abstract. This design proposes a low-cost node solution based on the Wi-Fi wireless sensor network. It aims to solve the problem of the popularity of smart homes due to high costs in real life and is suitable for applications in the smart home field. This solution saves cost, uses SOC solution, and has few peripheral circuits. Nodemcu is used to collect sensor data directly, which saves a separate main control chip, thereby further reducing costs. Various hardware interfaces are provided to facilitate the expansion of different types of sensor functions. The hardware of different functional nodes is designed, and the data is uploaded to a back-end server. Based on the software processing, the data between different nodes is aggregated and uploaded to the cloud platform for display to achieve remote access and control.

1. Introduction

In recent years, the vocabulary closely related to smart homes has been understood by most families, such as smart home local area networks, home gateways and smart devices. Some domestic universities have also begun research and development of smart home appliances systems.S3C2410 series high-performance single-chip microcomputers are used as central control cores to extend the Bluetooth module interface and home gateway solutions for GSM module interfaces, but they do not relate to home node networking solutions.In addition, in a smart home network where a large number of nodes are deployed, the use of the Bluetooth networking technology is complicated and the cost is high. Most domestic smart homes use ZigBee as the inter-node communication module. ZigBee nodes have the advantages of flexible networking and reliable operation. However, the ZigBee protocol stack with high network efficiency needs to be charged and the cost of a single node is high. The gateway is required to convert the network during operation. You can upload data to the Internet. In view of the above situation, we use the Nodemcu chip as the main control and use GSM to upload data to form a low-cost smart home solution.

2. Overall system design

2.1. Overall system framework design

*2.1.1.Overview of the overall plan example.*The smart home system in this design is based on the ESP8266 chip as a hardware platform, secondary development based on Nodemcu firmware and a gateway control platform through a public cloud server.Through the TCP protocol, the cloud server gateway is connected to various home Wi-Fi nodes and accesses to the Internet network. A smart home network system that remotely monitors and controls hand-held terminals is implemented [1].

2.1.2.Networking scheme. Each Wi-Fi module connects to the cloud server gateway through a router. The nodes are not affected and each connection remains independent.

2.1.3.Overall functional design. The smart home system designed in this paper includes intelligent lighting control system, intelligent temperature control system, etc[2].The overall functional design of the smart home is shown in Figure1.

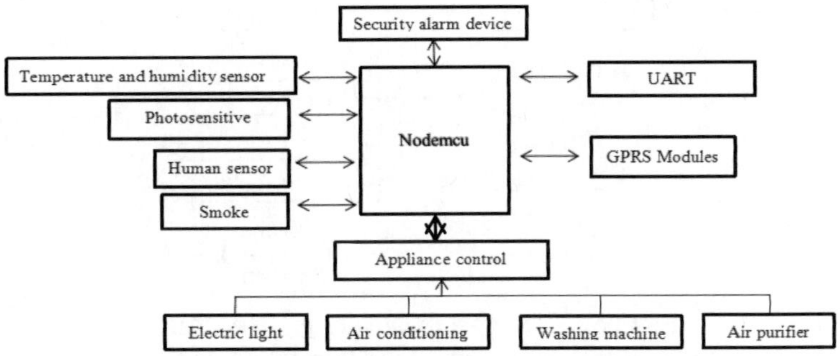

Figure 1. Overall function design of smart home

2.2. System main module hardware and software components

*2.2.1.System main module hardware components.*The main modules of the system consist of smoke sensors, human sensors, temperature and humidity sensors, lighting controllers, cloud server gateways, and GPRS modules [3].The specific role of each part is as follows:

- Smoke sensor: It is mainly installed in places such as home kitchens where main electricity is used. It is used to detect the presence of smoke in the surrounding environment. The fire is discovered in time and a signal is sent to the main control.
- Human Sensor: It carries out security auxiliary inspection and automatically controls the light.
- Temperature and humidity sensor: The temperature and humidity of the ambient air are detected every ten seconds, and the air temperature and humidity data are returned to the main control chip.
- Lighting controller: According to the main control chip to control the light brightness.
- Cloud Server Gateway: The gateway is used to implement the data processing aggregation function of each node. Users can access the cloud platform through the Internet to obtain various data and perform effective monitoring. The discovery data exception immediately informs the gsm module to alert the user.
- GPRS module: Through the Internet without Wi-Fi, the user is notified of the real-time situation by SMS and phone.

2.2.2. Main module software components. This design adopts a cloud server as the gateway of the smart home system. Because the node used in this design is a Wi-Fi connection directly to the Internet, no local other device is required to forward data across the network. Therefore, cloud servers can be directly used for data processing, storage, and control commands. Our operating system uses Win Server 2016. This system is used because it has the following advantages:

- Development, the system can install many development environments.
- Low cost, operators have lower cost control on cloud servers.

- Multiple users, with public network IP, can directly send and receive data from many nodes without forwarding.
- Safe and reliable, the cloud server has special maintenance and high security.

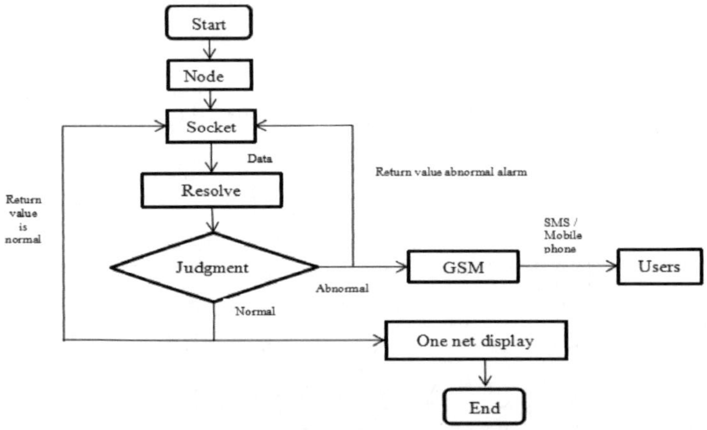

Figure 2. Server software design flow chart

The JDK is installed on the cloud server, and the temperature and humidity data of each Wi-Fi node is processed through a Java script, uploaded to the one net cloud platform, and displayed on the free cloud platform. Users can monitor data through the platform, and can also send commands to the cloud server to control the smart home system. When the data is abnormal, the program connects the GPRS module to enable the platform to have the ability to notify the user of text messages and telephones. The server software design flow is shown in Figure 2.

This design is based on Nodemcu's low-cost smart home node, which realizes the construction of a smart home system at a lower cost, completes the user's remote monitoring of home environment parameters, and controls some of the controlled home devices[4].By connecting home control equipment and cloud servers, the node realizes the exchange of information data between home equipment and nodes, nodes and owners, and finally realizes the regulation and remote real-time monitoring of indoor equipment.

3. System design and implementation

3.1. Overall system framework design

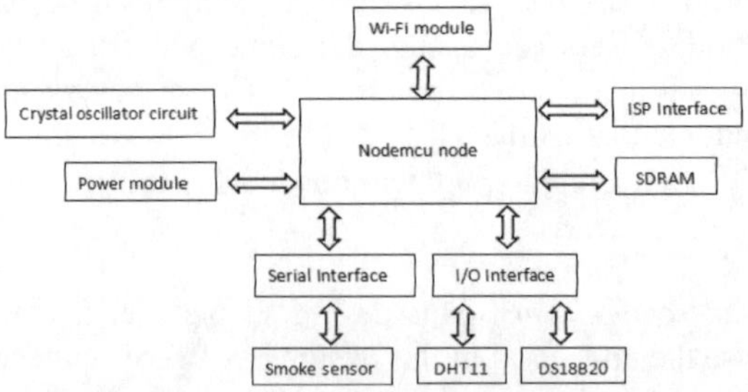

Figure 3. Smart home system structure

The Nodemcu-based low-cost smart home node design mainly uses the Nodemcu as the master. Figure 3 shows the composition of the associated hardware platform. Mainly by the MCU within the ESP8266 to control and process, which can be through the various communications port sensor data acquisition and processing, and data through the Wi-Fi module connected to the Internet to upload data to the cloud server gateway. In this way, effective information transmission can be achieved and control information actions can be completed in real time. The node chip designed in this paper is ESP8266. The implementation of all functions is related to the master control. Instead of using other masters to control the ESP8266, it is directly controlled by the ESP8266. The biggest advantage of using this scheme is its low cost. The node design includes the ESP8266 and CP2102 USB-to-serial modules.

3.1.1.Functions of ESP8266 module. The work of ESP8266 needs 5V power supply and crystal oscillator circuit to ensure its normal operation, and it needs to design the ISP download port to facilitate the software debugging. In this design, the ESP8266 is programmed directly to control the peripheral modules. This not only improves resource utilization but also reduces node hardware costs.

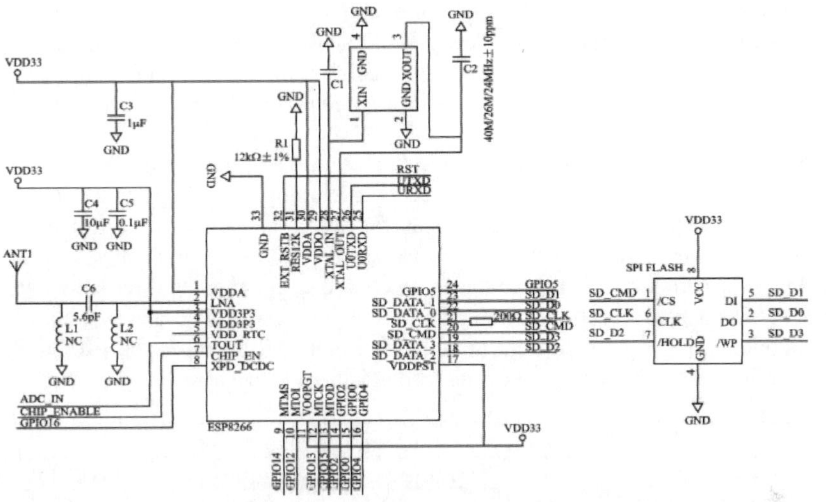

Figure 4. ESP8266 module circuit diagram

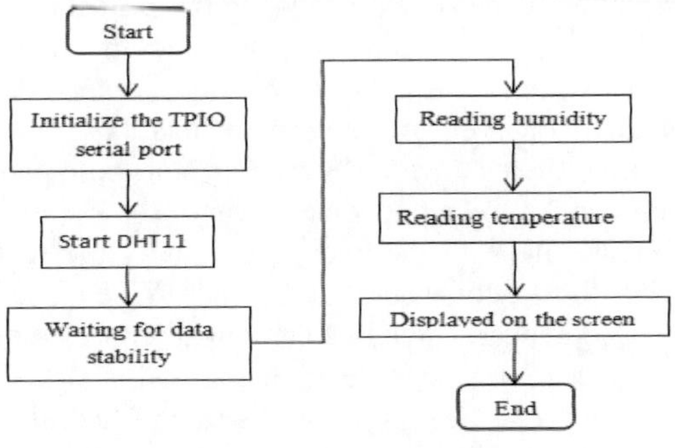

Figure 5. DHT11 data collection process

3.1.2.ESP8266 module circuit design. Because of the high degree of integration, the peripheral design of the ESP8266 module is very simple. In addition to the main chip, only one passive crystal, one SPI Flash, and several resistors, capacitors and inductors are required.ESP8266 module circuit diagram shown in Figure 4.

3.2. System software design and implementation

3.2.1.Smart home node software design. Based on the ESP8266 IoT, IoT SDK provides users with a simple, fast and efficient software platform for the development of IoT products. The SDK provides users with a set of function interfaces that can receive and send data. Users can only focus on the development of the upper application of the Internet of Things, without having to care about the structure of the underlying network. All network functions are implemented in the library, and being opaque to users is a major feature of the ESP8266.Users can do this in user_main.c. Void user_init(void) is the entry function of the upper program, which provides the user with an initialization interface. The user can increase the hardware initialization and other functions. The specific software implementation flow is shown in Figure 5.

3.2.2.Introduction to Nodemcu. Nodemcu is a programmable firmware based on the ESP8266 chip. It allows developers to implement programming logic through simple upper-level code to operate the underlying hardware to achieve software requirements. At the same time Nodemcu also provides a network API, through the event driver can greatly facilitate the developers to achieve their own design features. The Arduino IDE that this design uses carries on the development to the chip, the programming style is succinct, facilitates the developer to realize the rapid development.

3.2.3.Smart home background server development. This design adopts the currently popular cloud server as the smart home gateway server designed by the company. It aggregates all data, monitors each sensor data in real time and creates messages for users. The initialization of the GSM module registrants the network first, waits for the module to register successfully, connects to the background server port after successfully connecting to the Internet, and completes the three-way handshake in the TCP/IP protocol and then successfully connects to the cloud server. The master obtains data such as temperature, humidity, and smoke concentration every second. It averages the data and packages the data that needs to be sent to the server. The data format is shown in Table 1.

Table 1. Data frame format

Send direction	Baotou	ID	Command code	Data length	Data pack	Terminator
Server → Node	0x68	ADDR	0x10	0x01	0x10	0x16
Node → Server	0x69	ADDR	0x10	LEN	DATA	0x16

A communication module is started every ten seconds to upload data to the server. After the server receives a normal return value, the connection is closed and the module is set to sleep mode. Configure the operating mode so that the master chip also enters the standby state.Java as a background programming language, Server Socket class in Java has a variety of methods to provide calls to achieve socket connection. Create a Server Socket on a specific port. After creation, the server starts listening for requests on port 8899.The node connects to the server via TCP/IP, and the server creates a new socket to communicate with the node. After the server calls the accept() method, it enters a blocking state and waits for the connection of the node. When receiving a request from a node, the server creates a thread pool to allocate threads. Each thread creates a new socket to communicate with the node. After the communication is completed, the thread is closed. This allows the server to connect to multiple nodes[5]. After a socket connection is established, the receiving node sends data and analyzes the contents of the packet. Table 1shows the contents of the packet resolution. Each node is defined with a unique address. When the data check is normal, it will be sent to the one net cloud platform for data display. If the data is abnormal, the gateway will not only notify the cloud platform of the abnormality, thus notifying the user through the platform APP, and will also notify the GSM node to notify the user via SMS or by telephone.

4. System test results

An important part of this design is software testing, which is also the key to ensuring the normal operation of the program. The smart home access network is connected to Wi-Fi and IoT connections[6]. Users can check the relevant test data in real time by logging in to the mobile phone or browser to monitor the relevant home appliances in the house.

4.1. System function test and analysis instructions

4.1.1.Hardware function test. According to the hardware requirements, design the hardware scheme and connect the circuit according to the hardware function interface. The appearance of the node is shown in Figure 6.Connect the node to the 5V circuit, test the voltage of the power supply port is normal, the brightness of the power indicator on the Nodemcu motherboard is normal, and the power supply of the DHT11 sensor module is normal. The chip can be found by the debugger, and both the chip and the peripherals can operate normally.

4.1.2. Software function test. First, the user logs in to the mobile phone client app. The smart home platform system builds it on the basis of the login command so that it can come to the login display interface. Based on the user name and password entered to obtain the system verification match, the corresponding security login can be completed. After entering the system, the smart home that has accessed the cloud server can be discovered. The connected smart home device can be operated in real time. The data observed by the user in real time through the mobile phone is shown in Figure 7.

4.2. System test results

According to the above related content, the relevant functions of this system are tested, and the test results are as follows:

- The GSM module networking is not affected by the communication distance, geographical location, etc., to meet the general family needs and other special needs;
- The main control chip has a rich interface and is compatible with common data acquisition modules. After the expansion of the connection module, the system can work normally.
- The system is running normally, and the temperature and humidity data can be uploaded to the server, and the monitoring data can be reflected in the monitoring interface in real time;
- There is no mutual interference between each module to maintain stable operation, which improves system stability.

Figure 6. Node appearance

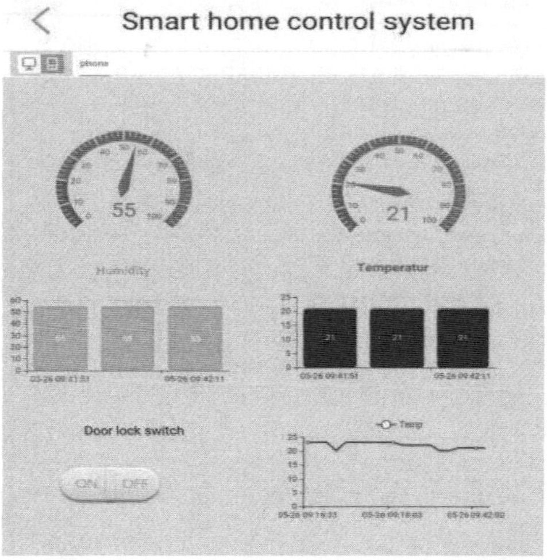

Figure 7. Users observe relevant data in real time through mobile phones

Acknowledgment

This work was supported by the National Natural Science Foundation of China under Grant No. 61473174 and Weihai Science and Technology Project (2018). Corresponding author: Liu hai.

Reference

[1] Meng Kun. Smart home system design based on MSP430 [D]. Master's degree thesis, Suzhou: Suzhou University, 2014.05, 9-12.
[1] Liu Yan, Zhao Xumin, Jiang Jiawei. Development Status and Development Trend of Smart Home[J]. Household Appliances, 2016.07, (07): 17-19.
[3]Liu Miao. Research and Implementation of Intelligent Home Control System Based on Internet of Things Technology [D]. Master thesis, Lanzhou: Lanzhou Jiaotong University, 2015.10, 18-25.
[4] Wei Qingmei. Design and implementation of smart home system based on wireless sensor network [D]. Master's thesis, Xi'an Xidian University, 2015.09, 5-6.
[5] Chen Guojun, Chen Lei, Chen Xixi. Java 2 Programming Design Foundation [M]. Beijing: Tsinghua University Press, 2006.01, 206-210.
[6] Xu Longjie, Ying Wenda. Analysis and Application Practice of Smart Home Application Development[J]. Telecommunication Science, 2015.11, 31(11): 116-122.

IOP Conf. Series: Materials Science and Engineering **435** (2018) 012014 doi:10.1088/1757-899X/435/1/012014

The Design of Game Platform Based on Java

Tong Qiang[1],Zhou Xiaofeng[2],Liang Likai*[1],Yao Sheng[1] and Wang Han[1]

[1]School of Mechanical, Electrical and Information Engineering, Shandong University (Weihai), Weihai 264209, China
[2]Department of Mechanical, Electrical Engineering, Weihai Vocational College, Weihai 264210, China

liangLikai@sdu.edu.cn

Abstract. In the face of rich game resources, it is necessary to have a good tool to manage games and improve the user's game experience. This is the meaning of the game platform. The game platform described in this paper adopts Java technology, and uses C/S mode and My SQL database to complete the design of the platform. This game platform has the functions of login module, management module and so on. Through the platform, it can realize jigsaw puzzle, Snake game and real-time battle of LAN tank war game. The paper describes in detail the early preparation, overall design, specific implementation and functional testing of the game platform. All functions of the platform can be used normally, the game can also run normally under the rules, which has completed the requirements.

1. Introduction

In 1952, the first video game to be recorded by humans, the Tic Tac Toe game was born. In this year, humans invented the first generation of computers based on vacuum tube technology, which is the carrier of this video game. Today, all kinds of excellent games are also blooming. Collecting and classifying games through the game platform facilitates the functions of game management, protection of games, acceleration of games, downloading of updated games, and password-free login games after login to the platform. It greatly facilitates the operation and maintenance of game companies and gives players a better game experience[1].

Completing the design of the platform can also provide valuable reference value for the development of such projects in the future. Based on this framework template, the required management functions are perfected.

2. The overall design of the platform

2.1. The overall architecture design of the platform

The functions and attributes of the integrated platform can be divided into three layers: the login layer, platform management layer, and game operation layer. The platform structure is shown in Figure 1.

Platform development environment	Computer system: windows 10
JDK version:JDK1.8.0_171	Development tools: Eclipse IDE for Java
Database: My SQL-5.7.21	Computer: lenovo laptop

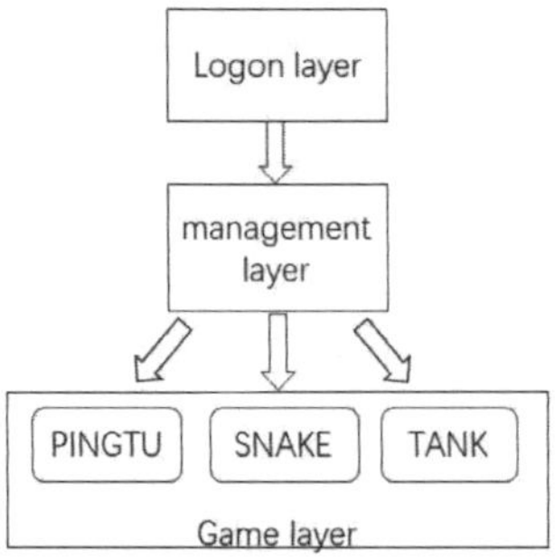

Figure 1. Platform structure

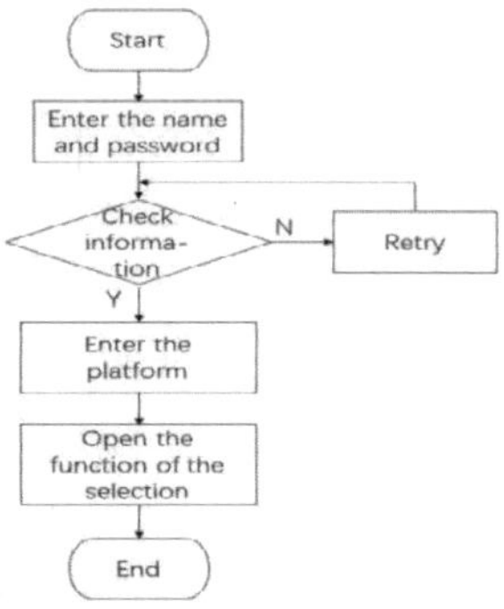

Figure 2. Platform use process

2.2. Platform process design

When users using this game platform, they first need to open the login interface to enter their account name and matching password to login. The platform compares the obtained account name and password with the information in the user table in the database. When the account name and password match at the same time, the user is allowed to log in and open the platform management interface.

After successful login, the user selects the services provided in the platform according to his own needs. The user can view the number of games he owns, the name, etc. through the platform, or open the game directly through the platform. Users can also view game data such as personal game scores and rankings stored in the database on the platform. Platform use process is shown in Figure 2.

2.3. Applied technology

- The Java language. The program written in the Java language has a concise and rigorous structure. The syntax used is concise and clear. The encapsulation call of the core class provides great convenience for the program's future development, maintenance, and modification. Because it supports network applications and communication protocols, it will be very powerful in promoting the development of Internet and Web-related software[2].
- Object-Oriented programming. The three major characteristics of object-oriented languages in the definition of history are encapsulation, inheritance and polymorphism. The most convenient place for programming is to create the project and encapsulate the operating code of the data and the data in the developed project together in the same object[3].
- Socket. Socket is mainly based on two communication protocols. One is based on TCP protocol to provide reliable connection-oriented two-way communication, and the other is based on UDP protocol to provide unreliable unconnected two-way communication[4].
- MySQL database. This game platform uses My SQL database to store management information. My SQL is currently the most popular open source cross-platform relational data management system developed by Swedish MySQL AB Company. It can be regarded as the fastest SQL database management system in the world today, and many domestic companies use it to store data[5].

3. The specific realization of the platform function

3.1. Login module

In this design, it is planned to use a small window to display it. The window contains an input text box for user name and password, a login button and a clear button. The login screen is shown in Figure 3.

3.2. Platform module

When designing the platform, the interface of this layer is expected to be divided into upper and lower parts. The upper half is used to display the functions of personalized information, display platform logo and information query. In the lower half of the game, the game is available in the display game library.

3.3. One of the game puzzle modules

When you start the game, you first show the player a picture. Then divide the picture into nine pieces. Remove the piece in the bottom right corner, which is the ninth piece. Then randomly disturb the picture sequence. Players use the free area to move the rest of the picture blocks, restoring the picture using as few moves as possible.

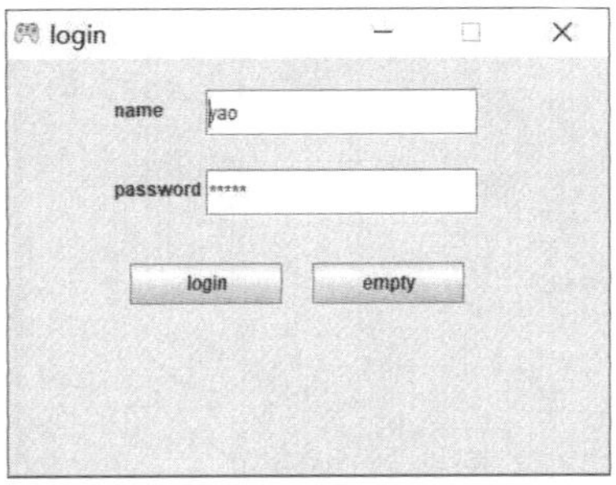

Figure 3. Login screen

Figure 4. Puzzle interface

Figure 5. Snake game interface

3.3.1. Interface implementation. In the design of this game, the complete picture is first displayed in the operation area, and when the game is started, the line of the well type is added to divide the picture into nine pieces and display the blank in the position of the ninth piece. Add three function buttons at the bottom of the interface to control the game. Players can select different pictures to play by selecting buttons. The puzzle interface is shown in Figure 4.

3.3.2. Function implementation. First import the game picture already stored in the project folder. The system generates a random number, moves the eight picture blocks at random, and updates the coordinates of the blank block each time it moves. When the listener receives the mouse click information, the clicked image is moved to a blank area, and the effect of the movement is achieved by the continuous redrawing of the window. In this process, the sequence of pictures is continuously judged. When the sequence is the same as the beginning, the game is successful. Use the JOptionPane method to pop up the puzzle box.

3.4. The game two greed snake module

Players use the keyboard to control the snake's direction so that snakes can eat more food. The system randomly generates a food on the map. As the player helps the Snake grow up, the Snake's body will become longer and its movement speed will increase slightly. The game ends when a snake eats a snake across the map's borders or ties itself.

3.4.1. Interface implementation. The interface design of this game is to create a game map area first, and the snake's movement and food location are limited to this area. Bottom add control game tips and current score display. Snake game interface shown in Figure 5.

3.4.2. Function implementation. In this game, the background of the game is drawn as a white rectangle. Since there is no artwork, the red rectangle is used to represent the food, and the black rectangle is used to represent the snake. Using the Linked List doubly linked list to store the snake's

IOP Conf. Series: Materials Science and Engineering **435** (2018) 012014 doi:10.1088/1757-899X/435/1/012014

body, the Linked-list's remove-last() method is ideal for changing the snake's physical state. The keyboard monitor is used during the game to get the player to change the direction of the snake, and continue to determine whether the rectangle where the snake's head is located intersects with the obstacle.

3.5. Game 3 LAN tank wars module

Tank Wars is a famous game that requires players to kill all enemy computer tanks while protecting themselves. This game is the most complex and involved module in the platform.

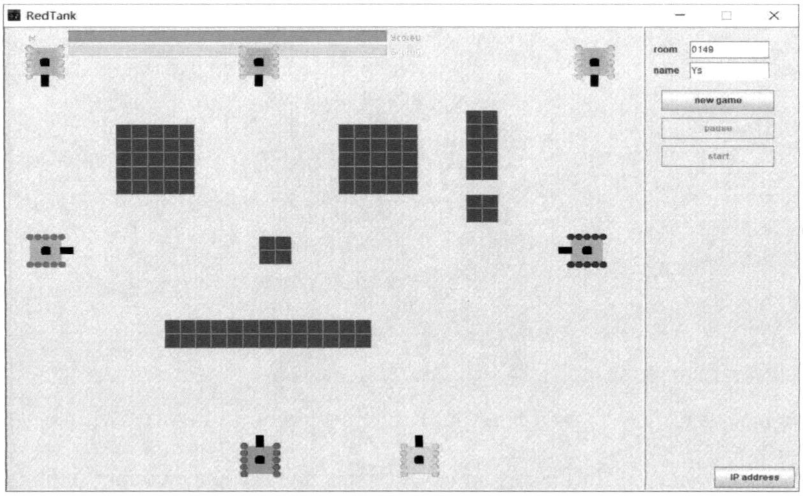

Figure 6. The Interface

3.5.1.Interface implementation. The server-side interface is the host interface and the right-hand panel provides control functions. The game needs to be played on the same LAN, so the method of viewing the IP address of the local machine is provided in the lower right corner of the panel. The host player enters the room number and nickname to create a room waiting for the slave to join. The interface is shown in Figure 6.The interface of the slave machine is the client interface. The right panel provides the user with the function of inputting the host IP address, room number and their own nickname. In order to ensure reliable communication at both ends, only the client enters the same information as the server to play online.

3.5.2. Function implementation. The steps for writing this game can be divided into five steps: The first step, when building a server-side framework, is to use some simple statements to assume that subsequent functions have been implemented, or to use a blank structure to add to the design of the program. This not only makes the framework more three-dimensional, but also reduces the work behind the amount.

In the second step, because the interfaces of the host and the slave are very similar, the client framework is very simple to build. It only needs to copy the server-side framework to rewrite the function and layout of the operation panel. The framework for both steps is to use the J Frame class to construct the form and add text boxes and buttons.

The third step is to write a server thread class that references Socket sockets to connect two clients. Create server proxy thread class and client proxy thread class to send and receive specific communication information at both ends. Then, the method of sending and receiving information is perfected into the framework of both ends, and the server and the client use information that has been agreed upon to represent the specific meaning. Next, design a complete session for both ends and perform a simple test.

The fourth step, call the java.awt package to draw geometric figures to represent the corresponding object. Draw with different colored brushes to distinguish between master and slave computers and computer tanks. In the development of the map class, it is necessary to create a two-dimensional array that is as large as the game interface. The shape of the map is designed by changing the values of the members in the array.

In the fifth step, first use the created public class to design the game board faces on both sides in detail, and add logic rules such as game rules. Secondly, write the server-side drawing thread and draw the client's thread. The functions of these two threads are very similar. The difference is that the forms they repaint are not the same. Then add the game message handling mechanism to the two communication agent threads that have been created. Finally, add the main board of the game to the server and the client to replace the blank part of the structure, and perfect the functions reserved for the construction of the two ends of the framework.

3.6. Database design

3.6.1.Analysis and design of tables. As shown in Table 1. User information table mainly used to store the user's account name and password. Player score sheet: It mainly records players' scores in each game. As shown in table 2

Table 1. User Information Table

Field Name	type of data	Field size	Is the primary key	Instructions
Mg-no	Var-char	50	yes	account name
Password	Var-char	50	no	password

Table 2. Player Score Sheet

Field Name	type of data	Field size	Is the primary key	Instructions
Name	Var-char	50	yes	nickname
Score-one	int	50	no	Score
Score-two	int	50	no	Score

3.6.2.Database operation method. Through the interface provided by JDBC connect to the My SQL database. The use of SQL statements can achieve the specific operations of the database functions[6].

4. Platform functional test

The main purpose of the software testing process is not to show the correctness of the software product, but to explore possible hidden defects in the software[7].According to the source of information used in the test design, the test is mainly divided into two concepts: white box test and black box test[8].The structural test technology and the programming of the program were performed simultaneously. The test was generally successful. The functional test items and results are presented in Table 3.

5. Summary

This platform is a Java-based game management platform. In the early stages of development, the overall framework was designed. Then the tasks were completed in sequence according to planned routes. The entire life cycle of development was accompanied by corresponding tests until the game platform was completed according to plan,among which the major shortcomings are as follows:

- Due to technical issues, copyright issues, etc., this game only supports small games that you write and cannot add other outside games. In the future development, the author will advance towards more complex and extensible design goals.

- All elements of the game platform, such as windows, buttons, text boxes, and game models, are implemented in the Java toolkit. They belong to the most original design and do not carry out aesthetic design, which makes the overall image of the platform somewhat rigid.

Table 3. Test item list

No	Test items	Test content	Results	No	Test items	Test content	Results
1	Login test	Enter an exact account name and password that matches the database	Normal login	5	Snake game test	During the running of the game, check whether the functions are normal, and test the game rules by way of wall crash, self-mutilation, etc.	The game is running normally
		Enter three incorrect account names and passwords	Login failed with appropriate prompt				
2	View score information	Click on the link button of the viewing information to see if the scoring panel pops up	Ability to see score information	6	LAN Tank Battle Test	Detect whether the master and slave can establish a connection. After the connection is successful, the two ends are prompted. Whether the party will prompt the other party after exiting.	Communication function is normal
3	Platform opens the game	Open the game one by one through the platform	Can open the game			After the host sets up the game room, the slave machine enters an unmatched IP address or room number.	Give an error after a short delay
4	Jigsaw puzzle testing	Perform multiple game tests by selecting different pictures to see if they are operating normally	The game is running normally			Test whether the tank can move on the map; if it encounters other tanks or obstacles, it will not restrict movement; whether the two bullets can hurt	The functions of the game are operating normally

Acknowledgment
This work was supported by the National Natural Science Foundation of China under Grant No. 61473174 and Weihai Science and Technology Project (2018). Corresponding author: Liang Likai.

References
[1]Liang Jincai. Design and implementation of high-performance and highly flexible game distribution platform [D]. South China University of Technology, 2016.
[2]Bruce Eckel, Translated by Chen Yupeng. Java programming ideas (4th edition) [M]. Beijing: Mechanical Industry Press, 2007.6
[3](United States) Matt Weissfeld, Huang Bowen translation. Object-oriented thinking process (Fourth Edition) [M]. Beijing: Mechanical Industry Press, 2016.11
[4]Hu Jingjing, Lu Manhong. Design and Implementation of Chat Room System Based on Java Socket[J]. Microprocessors, 2010, 31(05): 66-67+70
[5]Du Lijuan. Performance comparison between relational database and NoSQL database[J]. Intelligent Computers & Applications, 2017, 7(03): 132-134.
[6]Lin Yuanyuan. Analysis of the method of JDBC connection to MySQL database[J]. Journal of Changsha Communication Polytechnic, 2009,8(01):27-30.
[7]M G Limaye, Huang Xiaolei, Zeng Qiong. Software testing principles, techniques and tools [M]. Beijing: Tsinghua University Press, 2011.03
[8](plus) Nike, (United States) Trippati, Yu Lian et al. Software Testing and Quality Assurance: Theory and Practice [M]. Beijing: Electronic Industry Press, 2013.01

IOP Conf. Series: Materials Science and Engineering **435** (2018) 012015 doi:10.1088/1757-899X/435/1/012015

The Internet of Things-based Rehabilitation Equipment Monitoring System

Qiaoling Meng[1,2,3], Hui Zhang[1,2,3] and Hongliu Yu[1,2,3,*]

[1] Institute of rehabilitation engineering and technology, University of Shanghai for Science and Technology, Shanghai, P. R. China
[2] Shanghai Engineering Research Center of Assistive Devices, Shanghai, P. R. China
[3] Key Laboratory of Neural-functional Information and Rehabilitation Engineering of the Ministry of Civil Affairs, Shanghai, P. R. China

*Yhl98@hotmail.com

Abstract. With the rapid development of the internet and embedded sensor technology, the Internet of things has been gradually becoming part of life from once a generalized concept. Especially, the application of the Internet of things in the field of rehabilitation is very helpful for the patients who have to do their rehabilitation therapy at hospital and home. The design and implement a rehabilitation equipment monitoring system based on the Internet of things are presented, for the purpose of providing more convenient equipment management for rehabilitation hospitals and institutions. This system adopts B/S architecture to implement, realizing information fusion from the perception layer, network layer and application layer this three parts, let the doctor can monitor related state information of equipment in real-time, and make rehabilitation training plan for patients and control equipment remotely.

1. Introduction

The Internet of Things (IoT) is a concept reflecting a connected set of anyone, anything, anytime, anyplace, any service, and any network [1]. The Internet of things can somehow reconstruct the healthcare systems so all people from anywhere in the world can access the desired medical facilities [2]. The IoT has the potential to give rise to many medical applications such as remote health monitoring [3,4], fitness programs [5], chronic diseases [6], and elderly care [7].

A Rehabilitation Equipment becomes an essential part of a remote rehabilitation monitoring system when medical applications in IoT systems are considered. Rehabilitation equipment can sense and record user-related data during rehabilitation training. Moreover, frequent examinations and capturing the moment-to-moment rehabilitation indices have an incredible impact on the best chance of recovery and the development of the best training program. What is important is to implement the demanded infrastructure for new technologies in healthcare systems [2]. In view of the existing management and operation of rehabilitation equipment, there are still some problems. The existing rehabilitation equipment is generally operated by buttons. A few are equipped with matching clients. Physicians need to install corresponding clients [8,9] before they can use rehabilitation equipment. Data records are also stored on the local client, and it is difficult to access remotely. The existing management and operation methods of rehabilitation equipment can no longer meet people's growing demands. From the perspective of the user experience, there is still a lot of room to rise. In the new concept of remote rehabilitation, the system we want to design is to provide doctors and patients with digital, remote, cross-platform, and cross-system operation, so that doctors can monitor the rehabilitation equipment at any time and place, make the rehabilitation prescription for patients remotely, and can view the rehabilitation training results of patients remotely. So is the patient. At the same time, it provides channels for information exchange between doctors and patients.

Therefore, we propose an IoT-based rehabilitation equipment monitoring system that aims to connect rehabilitation equipment to the network through communication technologies. The rehabilitation equipment can collect relevant status information in real time and upload it to the server for remote operation by a doctor or therapist. And it allows the doctor to operate on the browser-equipped terminal device, realize the networking management of the rehabilitation equipment in a new form, and include the rehabilitation device incorporated in the monitoring system, so that information has a more comprehensive and intuitive grasp and control.

The remainder of the paper is organized as follows. Section II describes the details of the Proposed System, section III explains about the Experiment Results, section IV concludes the paper and gives the details about the Future work.

2. Proposed System

The Internet of Things-based rehabilitation equipment monitoring system proposed in this paper is implemented using the B/S architecture model. In many B/S architecture development languages in recent years, the Java language has been widely respected by software developers for its cross-platform, security, multithreading, ease of use, and object-oriented thinking. Java programming using Java Web technology (JSP + J2EE + MySQL + Tomcat) [10] programming, local development is carried out in the Windows environment, the project deployment using the Linux environment CentOS system. Linux system security is higher and more lightweight and simple.

2.1. Overall Framework

The overall architecture of the system is divided into three parts: the view access layer, the business logic layer, and the data persistence layer. Each part corresponds to its own functions and services. Because the dependency between layers is down, the bottom layer is relatively closed for the upper layer. This is a weakly dependent relationship, namely the idea of "high cohesion, low coupling" [11]. The overall architecture of the system is shown in Figure 1.

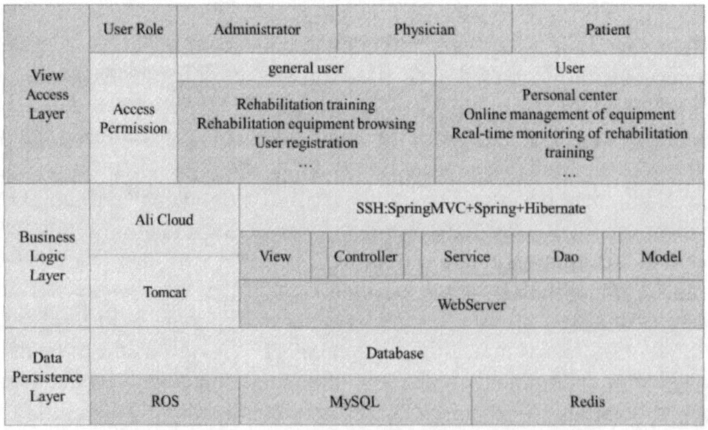

Figure1. The overall architecture of the system

The view access layer receives the user's request in the form of Web or WinForm and returns the background data to the user. Its main function is to provide application access operations. The view access layer is at the outermost layer in the overall three-layer structure of the system and is closest to the user. It is a part that the user can visually and intuitively experience. It is used to display data and receive data input by the user, providing an interactive operation interface for the user. The view access layer of this platform is presented to users in a JSP interface.

The business logic layer is generally the overall program part of the background server. It is the bridge of business communication between the view access layer and the data persistence layer. It is responsible for receiving the view access layer request and assigning it to the corresponding Controller and then assigning it to it. The corresponding Service and Dao get the response data from the data to be displayed on the interface. Simply speaking, the business logic layer is mainly the operation of the data persistence layer. It can filter and logically process the data in the database according to the requirements of the view access layer. It abstracts some logical data from the original data to interface/API. The Application Programming Interface (Application Programming Interface) layer provides related functions, and its purpose is to integrate the underlying logic of the data layer to form a business rule.

The data persistence layer is mainly the operation layer of the original data. In the system, the MySQL database is used to store the data. The backend implements the operation of the original data through the Hibernate framework. It is the operation of the data rather than the operation of the database. The purpose is to provide data services for the application service layer or the view access layer.

2.2. Function Design

The monitoring system is mainly divided into two functional modules, namely, online management of equipment and real-time training monitoring. The system functional block diagram is shown in Figure 2.

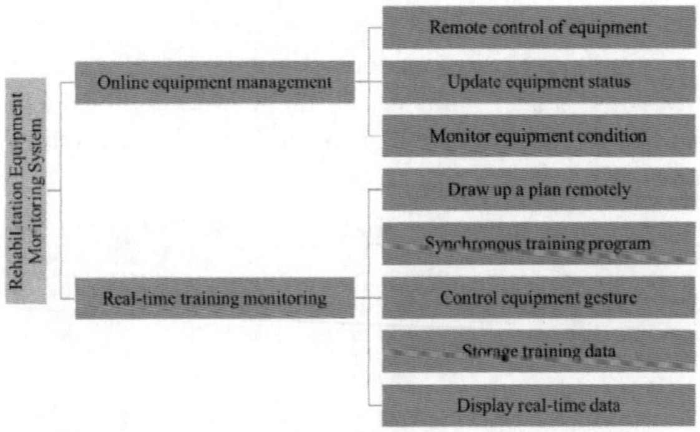

Figure 2. The system functional block diagram

Online management of equipment is divided into a remote control of equipment, equipment status update, and equipment condition monitoring. A doctor can remotely control online equipment through the network, such as controlling the posture of the equipment to lock it. After the doctor sent the control instruction, the device parses the instruction message after receiving the request through the monitoring server port, and obtains the instruction content and sends the instruction content to the control module, and then completes the control of the equipment through the control module. When the current state of the equipment changes, the smart gateway uploads its status information to the server via the HTTP protocol [12] and updates the corresponding equipment information records in the database. In addition, the doctor can view at the personal system center on the web page, such as the geographic location of the equipment, whether it is online, current posture and other information. That is equipment condition monitoring.

Real-time training monitoring is divided into drawing up a plan remotely, synchronized training plan, real-time gesture control, training data storage and real-time data display: Doctors can remotely log in to the personal center through the network to designate rehabilitation equipment and draw up a training plan for the patient. Synchronous training plan refers to click on the "synchronous plan" button in the real-time training interface to send the corresponding plan to the equipment and then let the equipment undergo rehabilitation training. Real-time gesture control means that doctor can click different gesture buttons in the real-time training interface to remote control equipment posture. During the training process of the patient, the equipment uploads the relevant parameters of the patient training to the cloud server for data storage, that is training data storage. Real-time data display means that the data is displayed in the form of a graph in the implementation of the training interface during the process of training so that the doctor can monitor the condition of the patient's rehabilitation training in real time.

The system provides users with three kinds of role operation authority, namely system administrator, doctor and patient, each with different operation authority. The user can apply for a platform account through registration. The administrator has the highest operating authority in the system and can perform operations on all data in the system, including doctor management, patient management, and equipment management. The doctor has the ability to view the equipment status online, remotely monitor the equipment, develop remote rehabilitation training plans for patients online, real-time monitor the patients' training and other operating privileges; patients have access to their own related rehabilitation training program and historical rehabilitation training results.

2.3. Communication Protocol

The system needs to start a Server Socket [13] service when the server is started and always monitor a special port in order to receive the command sent by the Client Socket. A Server Socket simultaneously supports multiple Socket Client requests. Rehabilitation equipment with communication modules like Wi-Fi, GPRS, etc., as a Client Socket, establish a Socket connection with the Server Socket of the system. After the Socket is connected, the robot can upload the data of the sensors to the system's database so that the system can render the data graphically and intuitively. Moreover, the user who owned the privilege of controlling the robot can send an HTTP request to the server by clicking the button on the user interface of the browser, and then the server will forward the corresponding command to the rehabilitation equipment's Socket to achieve remote control of the equipment. The process is shown in Figure 3.

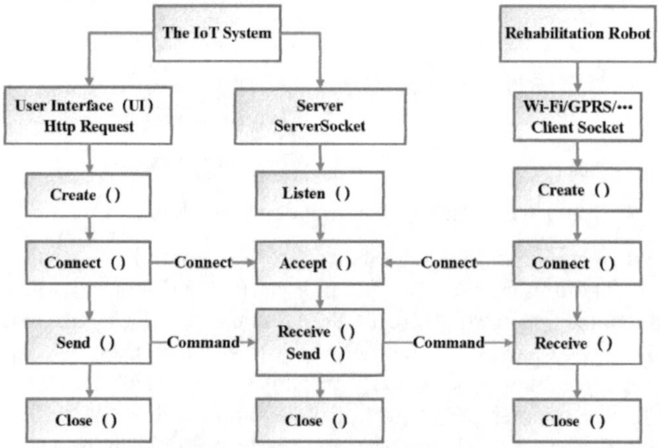

Figure 3. Socket communication block diagram

Here is an example of a rehabilitation wheelchair for explaining specific message formats. The rehabilitation wheelchair is based on an electric wheelchair, integrate with rehabilitation engineering techniques and become a kind of rehabilitation equipment that can both assist patients in training and use instead of walk. The wheelchair uploads data to the server and the server sends remote control commands to the wheelchair via HTTP protocol and Socket communication. For the sake of safety and accuracy of the instruction, the message is divided into three parts: the equipment number (equID), the control command (command) and the data content (data), and is packaged in the form of Json [14]. Equipment number, the unique number used by the server to identify the wheelchair; command is

IOP Conf. Series: Materials Science and Engineering **435** (2018) 012015 doi:10.1088/1757-899X/435/1/012015

specific control instruction so that the wheelchair according to the instruction table to make the corresponding action or posture conversion; data is the specific data packet collected by the sensor and uploaded to the server. Table 1 is a comparison table of control instruction codes and actual actions.

Table 1. Comparison table of control instruction codes and actual actions

Instruction encoding	Actual actions
1	Sit up
2	Standing
3	Flat lay
4	Synchronization plan
5	In place
6	Start training
7	Suspension training
8	Continue training
9	Complete training
10	Lock the wheelchair and stop the power supply
11	Unlock the wheelchair and turn on the power

3. Experiment results

After the doctors successfully log in to the system, they can click on "Online Equipment" to view the online equipment's status information, such as the equipment's operating status, current status, battery level, and geographic location, as shown in Figure 4. They can also monitor the geographical position of the equipment in real time through the map. At the same time, remote control of the device can be achieved through the interface buttons, as shown in Figure 5.

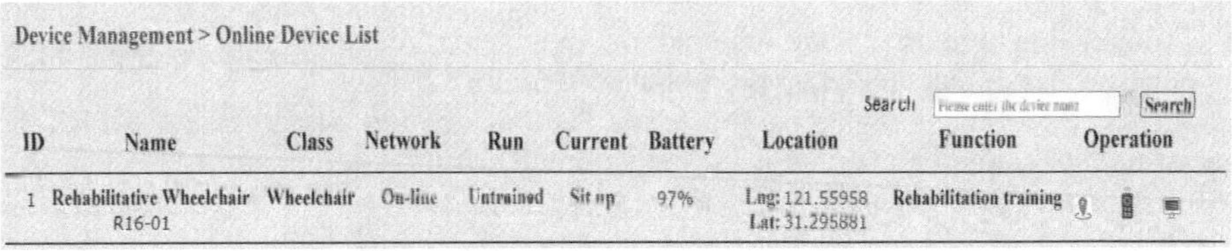

Figure 4. Equipment condition monitoring

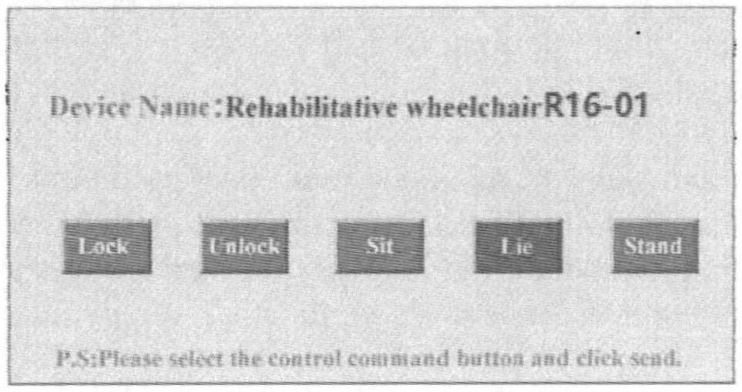

Figure 5. Remote control of equipment

After the doctor logs on to the platform, then click the link to the rehabilitation training plan. After the rehabilitation training plan is established, they open the rehabilitation training interface and synchronizes the training plan with the equipment. During the training process, the equipment uploads the data collected by the sensor and the power supply amount once a second to the server. While the interface takes a record from the database every second for display, and at the same time, the interface clicks the corresponding button on the interface to realize real-time control of the equipment.

4. Conclusion and future work

In this paper, the main idea of the Internet of Things -based rehabilitation equipment monitoring system is to provide better and effective rehabilitation services to patients by remote operation of doctors. Doctors as the main user role can perform remote guidance and monitoring for patients; administrators can manage all equipment and viewing equipment usage in the system, patients can view their own rehabilitation training records and rehabilitation assessment results. On the basis of the completion of the study of this article, the experiment was conducted. Through the analysis of the experimental data, it can be seen that the system realizes the remote monitoring of the equipment and proves the effectiveness and reliability of the system. In the future, this work can be extended by adding more rehabilitation equipment to the existing system. The proposed model can also be deployed on the cloud server so that the system becomes more mobile and easy to access anywhere across the globe.

Acknowledgments

This research was supported by Research Programs (Projects Numbers: 16441905202, 15DZ2251700, 16060502500).

References

[1] G. K. Saha and S. Kumar, "Security Issues in IoT-Based Healthcare," vol. 8, no. 3, p. 385, 2017.
[2] Z. Alansari, N. B. Anuar, A. Kamsin, S. Soomro, and M. R. Belgaum, "The Internet of Things Adoption in Healthcare Applications," in The IEEE International Conference on Engineering, Technologies and Social Sciences, 2017.
[3] E. Aguirre et al., "Design and performance analysis of wireless body area networks in complex indoor e-Health hospital environments for patient remote monitoring," ,12,9(2016-9-01), vol. 12, no. 9, 2016.
[4] C. Doss, M. Anwar, and N. Manjaro, "A Context-Aware Remote Health Monitoring Service for Improved Patient Care," 2016.
[5] X. U. Xiao-Ying and A. I. Shuang-Chun, "Application Situation and Prospect of Wearable Equipment in Rehabilitation Field," World Latest Med. Inf., 2018.
[6] Nguyen, H. H., Mirza, F., Naeem, M. A., & Nguyen, M. (2017). A review on IoT healthcare monitoring applications and a vision for transforming sensor data into real-time clinical feedback. IEEE, International Conference on Computer Supported Cooperative Work in Design (pp.257-262). IEEE.
[7] S. M. R. Islam, D. Kwak, M. H. Kabir, M. Hossain, and K. S. Kwak, "The Internet of Things for Health Care: A Comprehensive Survey," IEEE Access, vol. 3, pp. 678–708, 2017.
[8] Li Junqiang, Wang Juan, Zhao Haiwen, and Liu Jinyue, "An exoskeleton-based long-distance rehabilitation system based on virtual reality," Mechanical Design and Research, vol. 27, no. 4, pp. 35–38, 2011.
[9] Liu Xiuquan, Ma Yan, and Song Aiguo, "Study of an Upper Limb Rehabilitation Robot System Based on IoT Remote Control," High-Tech Communication, vol. 25, no. 1, pp. 82–88, 2015.
[10] T. Qiu, "Design and Implementation of Online Registration System Based on J2EE and JSP Technology——A Case Study on Students in Xinjiang University," J. Xiangyang Vocat. Tech. Coll., 2013.
[11] L. Cao and L. I. Shenluo, "An Application-independent Database Classification Method Based on High Cohesion and Low Coupling," Comput. Digit. Eng., 2016.
[12] B. Aziz, "A Formal Model and Analysis of an IoT Protocol," Ad Hoc Networks, vol. 36, no. P1, pp. 49–57, 2015.
[13] W. Ma, "Socket processing method and device and web server," 2018.
[14] N. Nurseitov, M. Paulson, R. Reynolds, and C. Izurieta, "Comparison of JSON and XML data interchange formats: A case study," in ISCA International Conference on Computer Applications in Industry and Engineering, Caine 2009, November 4-6, 2009, Hilton San Francisco Fisherman's Wharf, San Francisco, California, Usa, 2009, pp. 157–162.

IOP Conf. Series: Materials Science and Engineering **435** (2018) 012016 doi:10.1088/1757-899X/435/1/012016

Predict the Service Life of Rotary Lip Seals by Machine Learning Methods

Yiyi Xu, Wenjie Zhang, Erping Zhou and Peter Myler

School of Engineering, University of Bolton, Bolton, BL3 5AB, UK

E.Zhou@bolton.ac.uk

Abstract This paper aims to use machine learning methods to predict the service life of rotary lip seals to aid manufacturers and users improving the current maintenance procedures. Seals are widely used in most engineering applications. The knowledge of condition of seals throughout their working life is important due to the fact that they are often used on high value engineering products. As the current material properties of the seal and the working environment various, it is difficult to predict useful life of the rotary lip seal. In this paper, the factors relating to life of rotary lip seals are investigated and discussed. The application of machine learning methods using actual testing data in order to estimate the useful life of the seals has been presented. The early results show good agreement between actual and predicted values.

1. Introduction

Rotary lip seals were developed to work for a wide range of mechanical and equipment applications, such as power plant machines, automobiles, turbines, compressors and industrial pumps [1]. Typical rotary shaft seal components, which are shown in Figure 1, include a flexible inner lip and a rigid outer component. The seal lip can be spring-loaded or spring less. The factors relating to life of rotary lip seals have been investigated, they are friction, compression, fatigue, wear, aging, thermal hardening and oxidation, etc. First of all, different modes could coexisted between the rotary lip and the shaft in which they are defined as dry friction, boundary lubrication and fluid lubrication, and these three states may be continuously altered during working cycles. Dry friction leads to wear and tear, fluid lubrication is a cause of leakage, but the boundary lubrication is an oil film layer interfaced between the lip and the axis which generated a stable fluid dynamics. This oil film layer, except to produce lubrication function, has also provided the sealing effect. While the oil film layer is too thick, the fluid will leak; the oil film layer is too thin, lubrication function is weak then the lip will wear. Secondly, the excessive temperatures may result in hardening of the elastomer, which the sealing function could be loss, the deformation or permanently damage of the seals would happen. During the working cycles rubber seals, under the action of mechanical forces, will repeatedly generate fatigue which is the third failure mode of the seals.

The knowledge of condition of seals throughout their working life is important due to the fact that they are often used on high value engineering products. As the current material properties of the seal and the working environment various, it is difficult to predict useful life of the rotary lip seal. Many research works have been conducted to predict the life of the seals. Experimental data analysis combined with experiences is one of the most commonly used methods to predict the life of the rotary lip seal. The advantage of this method could be cost effective; however, it would potentially introduced errors because of the subjective, variable results of manual vision inspection method. It also needs to spend a lot of time to collect data for different kinds of seals.

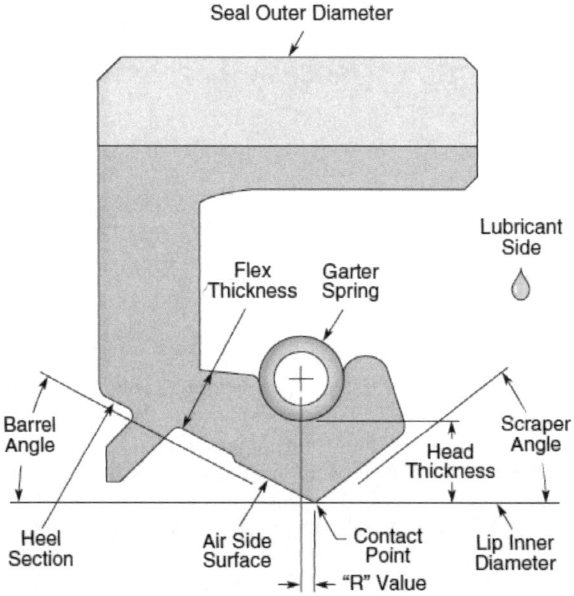

Figure 1. Structure of Rotary shaft seal [2]

To avoid conducting huge experimental and human resources, condition monitoring systems by using various sensors have been applied into many industrial equipment and machines, as well as seals in material handling, construction, waste handling, forestry and other extreme performance applications. Sensors normally have been undergone exhaustive laboratory and field testing to validate its ability to maintain signal fidelity in extremely challenging environments, e.g. exposed to vibration, dust, gravel, corrosives, chemicals, axial load, side load, and immersion. However, these sensor systems could be very expensive, and required to be regularly updated and maintained.

J. Sun et al [3] applied fault tree analysis method to quantitatively investigate the failure causes by oil leakage on mechanical seals. They noticed that the change of surface topography is the main cause of leakage of the mechanical seal under the constant preload. Based on the relationship among surface morphology, working cycles and allowable leakage, a prediction model for mechanical seal life was established. A variety of seals used for predicting the sealing life under different working environment have been investigated. The results have shown that it is possible to achieve the low-cost operation and high-reliability under the scheduled maintenance plan for industrial devices, as well as to predict the life of mechanical seals. ARDL [4] provides service life prediction by using analytical and innovative testing techniques to predict the life of a product or component in actual service conditions. Predicting service life enables users to understand and benchmark product life cycles in the field. By applying various and combination techniques as Statistical Life Prediction Approach, Dynamic and Vibration Studies, Field and Laboratory Data Collection, Finite Element Analysis etc. to predict failure and fatigue resistance on: Belts, Engine mounts, Gaskets, Hoses, Seals, etc.

In this paper, the factors relating to life of rotary lip seals are investigated and discussed. The application of machine learning methods using actual testing data in order to estimate the useful life of the seals has been presented.

2. Maintenance methodologies

2.1. Types of maintenance

Maintenance can be divided into the following three types. Reactive maintenance is to perform maintenance when there is a problem with the equipment machine. The problem with this type of maintenance is that unexpected failures can be expensive and potentially dangerous. Scheduled maintenance or preventive maintenance means that maintenance is performed from time to time regardless of whether the equipment is damaged or has problems. This kind of maintenance will result in the waste of resources. The most significant problem is that it doesn't eliminate all possible faults - they may still exist. The predictive maintenance method is through a model to predict the time of equipment problems occurs to avoid greater losses. The purpose of predictive maintenance is to predict when equipment failure will occur, as well as to prevent failures happen by performing maintenance. Ideally, predictive maintenance allows the maintenance frequency to be as low as possible to prevent accidental reactive maintenance without incurring the costs associated with performing excessive preventive maintenance. However, this type of maintenance method is difficult to make accurate predictions for complex devices [5].

2.2. Modern preventive maintenance

With the advent of machine learning and big data technologies, Condition-based maintenance (CBM) has once again attracted many manufacturers and users. However, CBM can only monitor the machine equipment at a certain time interval to determine whether the system condition is normal. The collected information is real-time information of the monitoring device status without predicted function. Therefore, the emergence of predictive maintenance to CBM is a great turning point. Based on continuous measurement and analysis, predictive maintenance can predict critical indicators such as remain useful life of machine parts. The monitored information can be used to determine the operating status of machinery and equipment, optimise its maintenance time, and thus reduce certain maintenance costs. Remain useful life (RUL) is a random variable which defined as "the useful life on an asset at a particular time of operation" [6].

At present, equipment applications require high performance, leak-free seals to ensure that they do not prematurely fail over their useful life. Early failure of the equipment will not only bring downtime, increase costs, and may be potentially dangerous to humans [7]. Therefore, failure modes analysis and predictive maintenance are essential to eliminate or reduce the potential risk of the premature equipment failure.

3. Case study

3.1. Database and assumptions

Due to the multiple factors of the rotary lip seal and the working environment, such as wear and leakage mechanisms, it is difficult to accurately predict its performance. In addition, the components in the rotating lip seal assembly are manufactured by different industries, which further increase to the complexity. The main challenges when dealing with input data are as follows:

- There is an unknown relationship between the different parameters;
- The collected data could be inaccurate or with noisy;
- Lack of suitable data processing techniques.

In this paper, it is assumed that the measured leakage quantity index is related to the following parameters as shown in the table 1. According to Plath, S. et al [8], the friction torques various with temperature and rotation speed. For the relationship between the rotation speed and the friction torque, the friction torque changes in a small range at different rotation speeds. A further analysis found that the friction torque decreases approximately linear with an increase of the temperature from 60°C to 100°C. In this case, five features have been considered. There are four input variables: speed, temperature, friction torque, coefficient of friction and one output (response) variable: the amount of leakage. In this paper, the range of the temperature is assumed from 45 °C to 100 °C; the range of rotation speed from 500 rpm (run per minute) to 2000 rpm and the range of friction torque between 0.2 Nm and 0.75 Nm.

IOP Conf. Series: Materials Science and Engineering **435** (2018) 012016 doi:10.1088/1757-899X/435/1/012016

Table 1. Seals parameters used in this case study

Variables	Unit
Leakage	ml
Speed	1/min
Temperature	°C
Friction torque	Nm
Coefficient of Friction	/

3.2. The application of machine learning

For machine learning applications, it can be divided into two situations. For scenario 1, there is no failure data available, which means that only operating cycle data before failure are entered into the model. This is unsupervised learning to group and interpret input data. As for scenario 2, historical failure data is available in which supervised learning can be applied to predict the RUL of the seals.

In this case study, supervised machine learning is performed based on testing data with failures. The basic workflow is shown in Figure 2. The whole testing data has been separated into two sets: two-third of the data are used for training and built up prediction model and one-third of the data are used to verify the accuracy of the prediction model. In model training and building stage, the first step is to pre-process the data from real testing data. Then machine learning methods are used to train model based on processed data. In order to obtain model with good performance, the improvement of the model techniques applied. After the improvement achieved, train for getting the best model has been conducted and the prediction model for seals determined. In verifying stage, the remaining one-third new testing data which never been input into the modelling stage have been applied into prediction model to verify the performances of the prediction model.

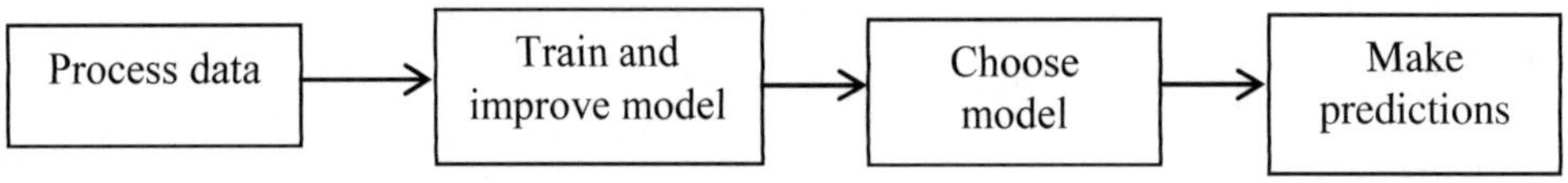

Figure 2. Basic workflow for case study

4. Results

The Figure 3 shows the prediction results with the new verify testing data in selected model and Figure 4 is the actual data. Compare to these two figures, the new data prediction results, in this case, present a good agreement between the actual value and the predicted value. In other words, the machine learning model developed could be used as a prediction model for the seals involved in this case study.

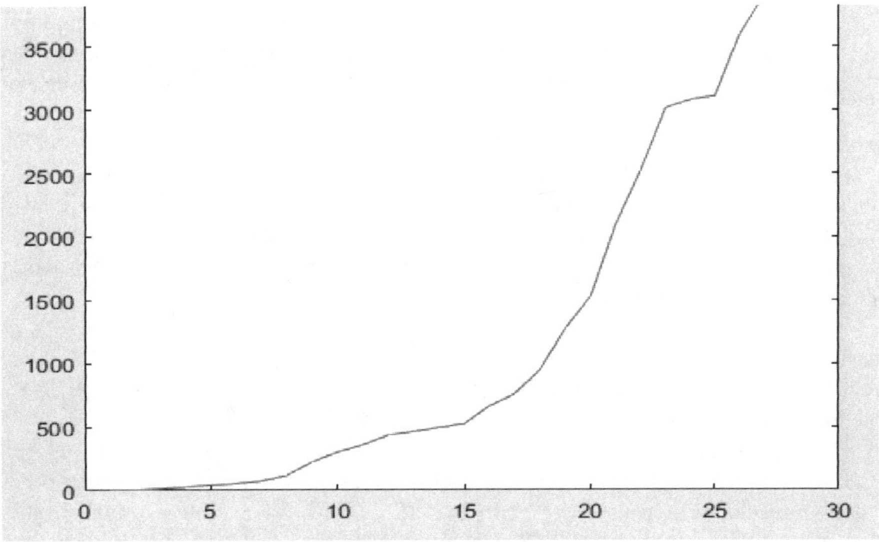

Figure 3. Prediction results with verifying data

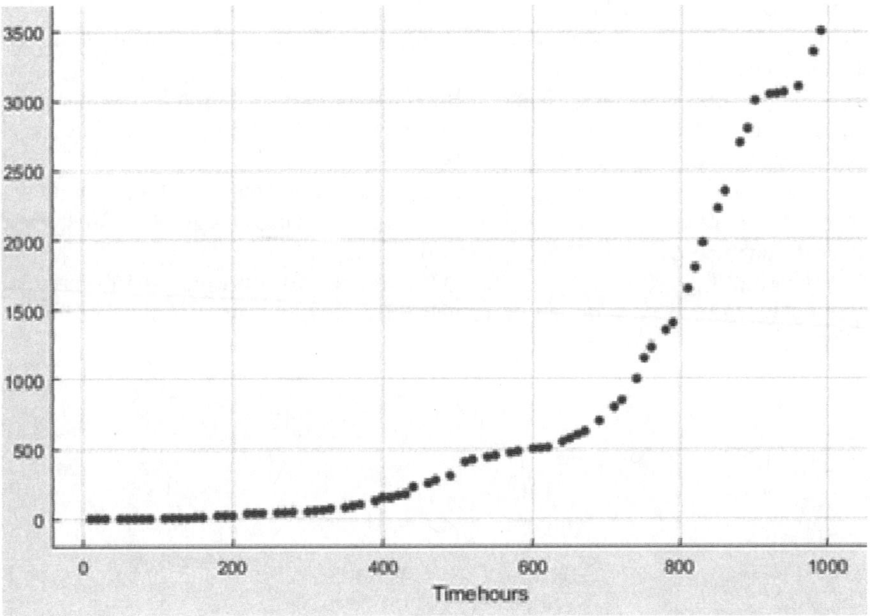

Figure 4. Actual data model

5. Conclusions and future work

5.1. Conclusions

The application of machine learning methods using actual testing data in order to estimate the useful life of the rotary lip seal has been presented. The early results show good agreement between actual and predicted values. Machine learning can be used as predictive maintenance methods to minimise machine failure and maximise equipment efficiency. The life of the seal is estimated and predicted by using machine learning training and selecting a model with good performance.

However, due to the material properties of the seal and the influence of the working environment, obtaining an accurate prediction model requires accurate sensor data, noise removal, and feature data selection. Inaccuracy of any one of these factors will affect the performance of final prediction model.

5.2. Recommendations and future work
In this paper, the model is only for one type of rotary lip seals and mainly trained based on actual leakage data and four input variables measured by sensors. The characteristics considered are relatively small compared to the parameters involved in real systems. Moreover, the amount of data used in this case is not large enough to meet the big data training requirement. Further work could be focused on building up a much bigger database for different seals.

References
[1] Jimoh, I. A., 2015. Numerical Modelling and In-Process Expert System for the Assessment of Large Scale Seals. PhD Thesis ed. Bolton: University of Bolton
[2] Rotary Seal Design Guide, 2006.1st ed. Parker Hannifin Co., Cleveland., USA
[3] Sun, J. J., He, X. Y., Wei, L., Feng, X., 2009. Failure analysis and seal life prediction for contacting mechanical seals. Proc. SPIE 7375, ICEM 2008: International Conference on Experimental Mechanics 2008, 73753L (25 August 2009); doi: 10.1117/12.839260
[4] ARDL, 2007. Shelf life prediction and service life prediction. [online] Available from: http://www.ardl.com/sites/default/files/docs/LifePrediction_SalesSheet_Revised_4-4-14_ForWeb.pdf/ [Accessed 11 May 2017]
[5] Willingham, D., Predictive maintenance using simulation & machine learning. ©2016 The MathWorks Inc.
[6] Chen, X.Z., Yu, J.S., Tang D.Y., Wang,Y.X., 2011. Remaining useful life prognostic estimation for aircraft subsystems or components: A review. International Conference on Electronic Measurement Instruments. Vol. 2. pp. 94–98
[7] Systemseals, 2014. Failure analysis guide: eliminate potential and future problems [online] Available from: http://www.systemseals.com/wp-content/uploads/2014/12/3050_System_Seals _Failure_Handbook .pdf [Accessed 11 May 2017]
[8] Plath, S., Meyer, S., Wollesen, V.M., 2005. "Friction torque of a rotary shaft lip type seal – a comparison between test results and finite element simulation," MECHANIKA, vol. 54 no. 4, pp: 55-59

IOP Conf. Series: Materials Science and Engineering **435** (2018) 012017 doi:10.1088/1757-899X/435/1/012017

Design and Implementation of Wind Resources Web Platform

Yuan Jiang[1], Likai Liang[1], Qiang Tong[2*], Ruitong Yuan[3] and Ruilin Li[1]

[1]School of Mechanical, Electrical&Information Engineering, Shandong University, Weihai, China
[2]General Affairs Management Office, Shandong University, Weihai, China
[3]School of Electrical Engineering, Northeast Electric Power University, Jilin, China

809672965@qq.com; lianglikai@sdu.edu.cn; tronq@sdu.edu.cn; 473129971@qq.com; 531802421@qq.com

Abstract. In the context of China's rapid industrialization and rapid growth of energy demand, wind power development has become an inevitable choice for China's sustainable development. Therefore, it is of great practical significance to study the development status of wind power in China. This paper demonstrates a web platform designed by Dreamweaver CS5 software and HTML5. It clearly and intuitively shows the situation of the development and utilization of wind power in China by four main parts: wind power status, wind power distribution, wind power prediction and wind power application, the development and implementation of this platform can provide convenience for wind power information query and research in China.

1. Introduction

The aggravation of environmental pollution and the shortage of traditional energy are two major problems facing the energy industry. China has vast territory and abundant wind energy resources. According to incomplete statistics, China's available wind energy storage capacity is 3.023 billion kW, with great potential for wind energy application [1]. Therefore, China has taken the development and utilization of renewable energy as the key content of the national medium and long term program of science and technology, and at the same time, how to improve the power grid's capacity to absorb wind power as the main research direction[2-5].

2. Overall system design

2.1. A.Design language -- HTML5

HTML5 is a language used to describe web pages. The language is a markup language. It is mainly used to describe how content in hypertext is displayed. The biggest advantage of HTML5 is that it has a simple grammar structure and is easy to write web pages. So we chose Dreamweaver CC produced by Adobe, which has a very friendly user interface and is an excellent web development tool.

2.2. Design flow chart

The overall design flow chart of wind resources network platform is shown in Fig.1.

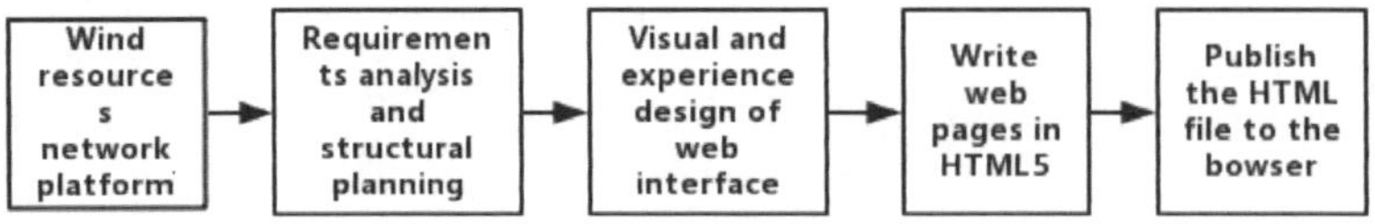

Figure 1. Design flow chart

3. Design and implementation of wind resources web platform

3.1. Home page

The platform is mainly divided into four modules: wind power status, wind power distribution, wind power prediction and wind power application. The system block diagram is shown in Fig.2.

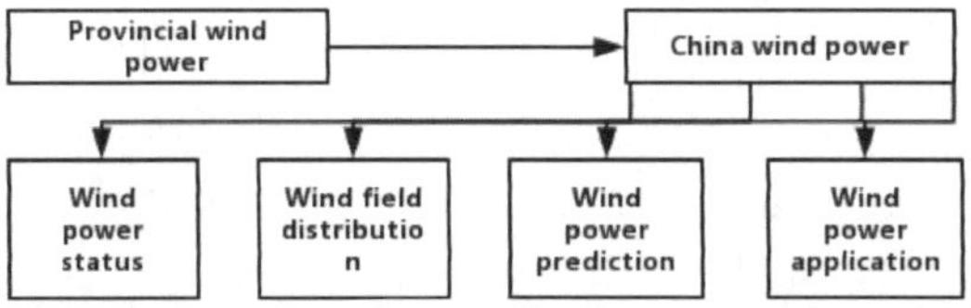

Figure 2. Platform design system block diagram

A web page with a map of China as the main picture was written in HTML5. The division of the webpage is also written in HTML, which successfully realizes the current situation of wind power, the distribution of wind power, the prediction of wind power and the application of wind power. Through these four tags, it is convenient to jump to the corresponding module and query the home page of relevant information about wind power development in China you want to know, as shown in Fig.3.

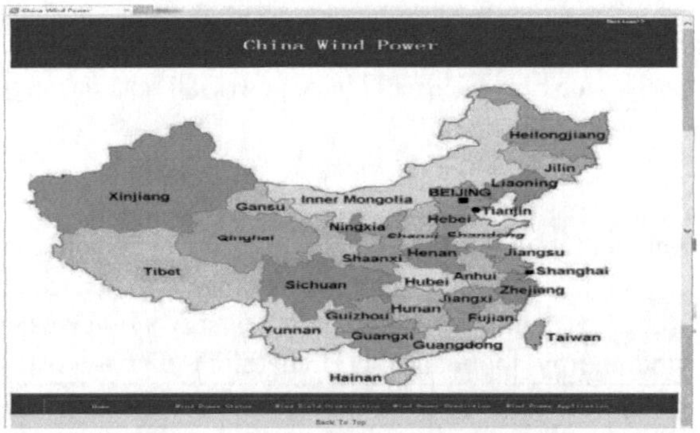

Figure 3. China wind power home page

The main code is as follows:

```
        <li style="border-left:1px #cccccc dashed;"><a href="China Wind Power.html">
Home</a></li>
        <li><a href="Wind Power Status.html">Wind Power Status</a></li>
        <li><a href="Distribution Map of Wind Farm in China.html">Wind Field
Distribution</a></li>
        <li><a href="Wind Power Prediction.html">Wind Power Prediction</a></li>
        <li><a href="Wind Power Application.html"> Wind Power Application</a></li>
</ul></div>
```

Web platform also implements the provinces wind power module, the module did not design a separate module label "provincial power", but will it into the web platform home page of the map of China, provinces of the map of China with obvious click on any one of the provinces of China map, and can be easily open on the provinces of wind power development is introduced. Fig.4 is the website of introducing the development of wind power in xinjiang. Through searching this website, we can get to know the storage and distribution of wind energy resources in Xinjiang, the status quo of wind power distribution and utilization, the impact of wind power development on the environment. It is of great help to the state to solve the development problem of wind power in Xinjiang.

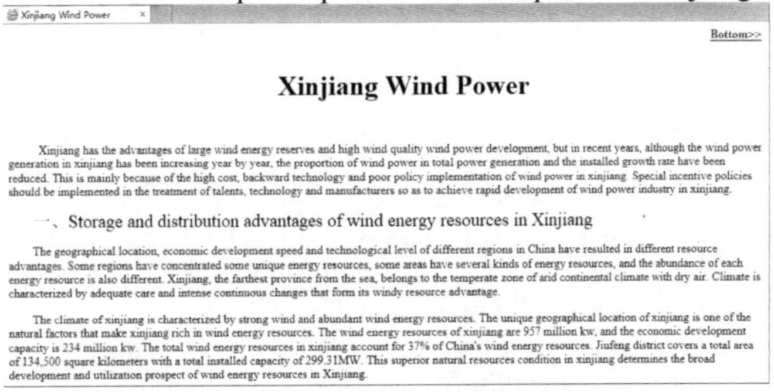

Figure 4. Xinjiang wind power website

3.2. Wind power status

The system block diagram of wind power status module design is shown in Fig.5. This module shows China's energy situation, wind power development situation, wind power industry situation and wind power application situation.

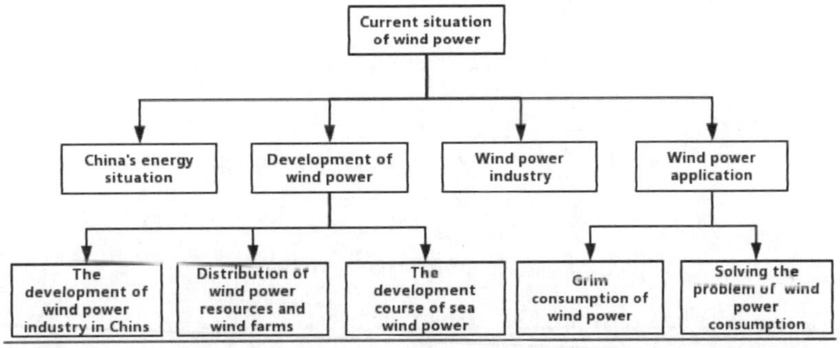

Figure 5. Wind power status module design system block diagram

As shown in Fig.6, web page shows China's energy situation, it can be seen in China's energy resources are relatively rich, but with low per capita energy resources capacity, along with our country economy continues to develop and continue to promote industrialization, urbanization, while the energy production and consumption is growing, but now there is a large energy gap and the trend of present gradually become bigger.

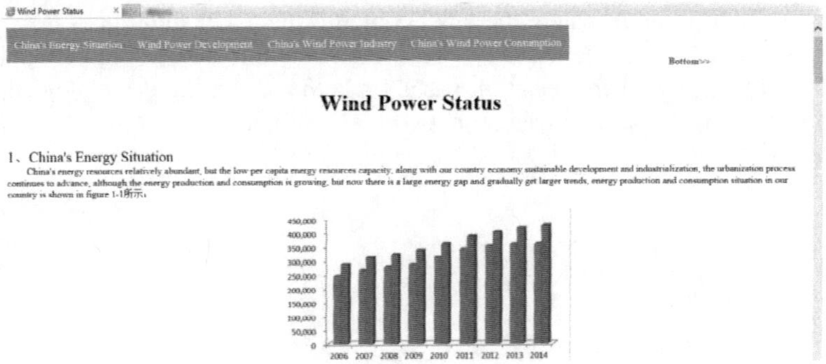

Figure 6. China's energy situation

Wind power has its own characteristics, such as intermittence, fluctuation and randomness. Large-scale grid-connection of wind power system will bring great impact on the stability and reliability of power system, which seriously restricts the grid-connection and absorption of wind power. Developed countries, such as Europe and the United States, cooperate with each other in pumped storage and wind power generation. However, water resources are scarce in China's wind-rich inland areas of northwest China, and the construction and promotion of pumped storage power stations are too limited, so China needs to study energy storage technology suitable for China [6].

3.3. Wind field distribution

Through this website, the distribution of wind power plants in China can be clearly understood. As can be seen from Fig.7, China's wind energy resources are mainly distributed in xinjiang, Inner Mongolia, jilin, hebei and other inland areas in the northwest and north China.

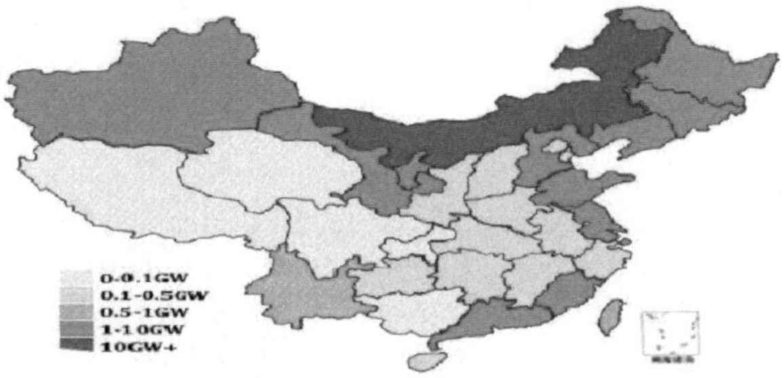

Figure 7. Distribution of wind resources in China

3.4. Wind power prediction

The overall design flow chart of the wind power application section is shown in Fig.8.

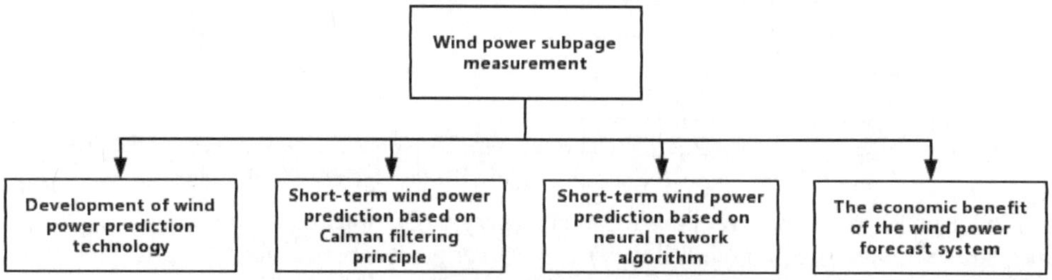

Figure 8. The system block diagram

Wind power prediction web page as shown in Fig.9, through the upper left corner of the wind power prediction technology, the economic benefits of prediction system of four labels, can be convenient and quick to jump to where the label content.This paper introduces the related content of wind power prediction.And in the website two kinds of wind power prediction technologies are presented: short-term wind power prediction based on kalman filtering principle and short-term wind power prediction technology based on neural network algorithm.

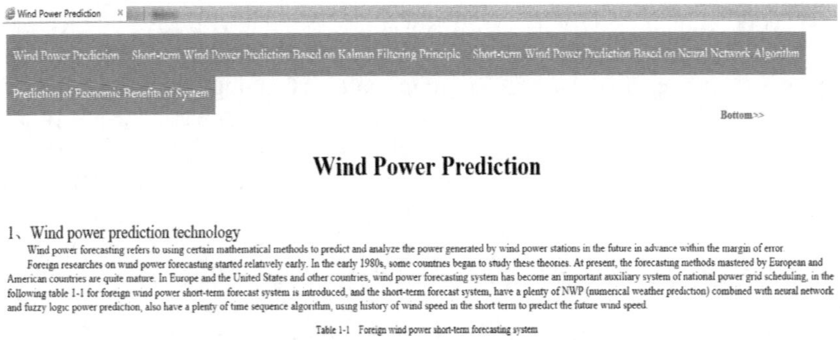

Figure 9. Wind power prediction web design

3.5. Wind power application

Wind power application module design of the system block diagram as shown in Fig.10, this module contains a given wind power development situation, the power grid wind power capacity analysis and the method of improve grid given wind power capacity of three parts, through the analysis of wind power development and application situation in our country, given wind power capacity is to restrict China's wind power to power grid main problem of large scale, improve the network given wind power capacity for the application and development of China's wind power industry is of great significance.

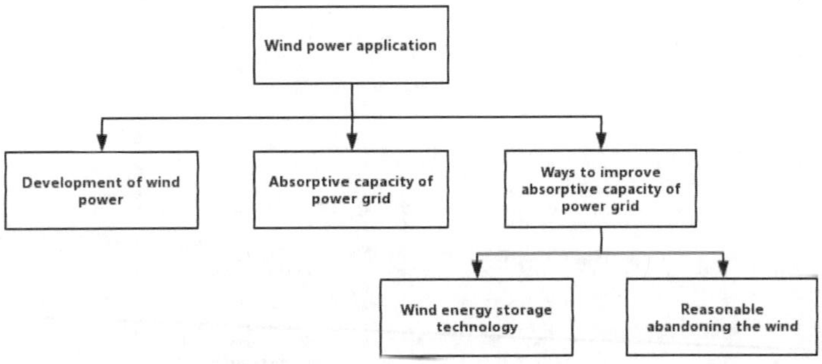

Figure 10. Module design system block diagram

Wind power application page as shown in Fig.11, the interface is divided into wind power development situation, the power grid given wind power capacity analysis and improve the method of grid given wind power capacity three child module, wind power application situation in our country are introduced, and according to limit China's wind power of large scale wind power given problem, put forward in accordance with China's national conditions to increase the ability of given wind power grid method.

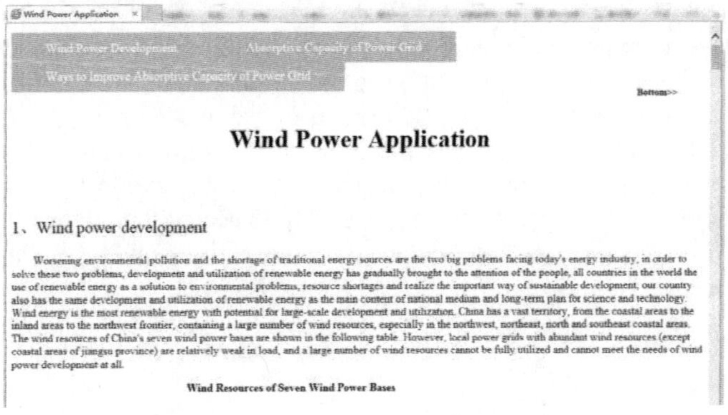

Figure 11. Wind power applications

The wind power consumption capacity of the power grid is mainly affected by the system load size, load characteristics, power structure, regulatory capacity and wind power output characteristics. Among them, many influencing factors are constantly changing, such as the power output characteristics of wind power in different seasons and periods, the regulation capacity of hydropower in rainy season and dry season, and the system load characteristics of different seasons. Therefore, the wind power consumption capacity of the system must take into account the peak capacity, transmission space and safety margin of the power grid.

3.6. Ways to improve absorptive capacity of power grid

3.6.1. The utilization of wind - electric energy storage technology to improve wind - power absorption capacity. The basic principle of the thermal energy storage and power generation technology of wind power plant is to convert the electrical energy generated by the surplus wind energy into heat energy and store it in the high temperature storage medium. When power is needed, the high temperature storage medium is used to heat the water to generate high temperature and high pressure steam and drive the steam turbine generator to generate electricity, so as to complete the energy storage function of wind power. In order to improve the efficiency of energy storage and circulation and realize the "cascade utilization" of heat energy, the combined heat and power production method is generally adopted, and the efficiency can reach 70% to 85%, which is of great significance to areas in north China where there is high demand for winter wind and high demand for heating and severe power limitation.The web interface design of this part is shown in Fig.12.

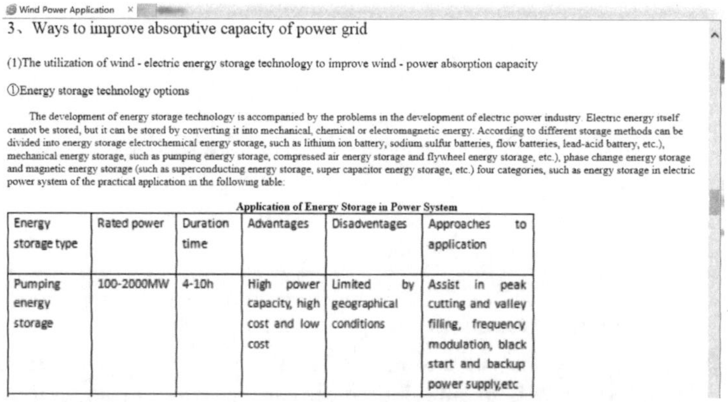
Wind Power Application

3、Ways to improve absorptive capacity of power grid

(1)The utilization of wind - electric energy storage technology to improve wind - power absorption capacity

①Energy storage technology options

The development of energy storage technology is accompanied by the problems in the development of electric power industry. Electric energy itself cannot be stored, but it can be stored by converting it into mechanical, chemical or electromagnetic energy. According to different storage methods can be divided into energy storage electrochemical energy storage, such as lithium ion battery, sodium sulfur batteries, flow batteries, lead-acid battery, etc.), mechanical energy storage, such as pumping energy storage, compressed air energy storage and flywheel energy storage, etc.), phase change energy storage and magnetic energy storage (such as superconducting energy storage, super capacitor energy storage, etc.) four categories, such as energy storage in electric power system of the practical application in the following table:

Application of Energy Storage in Power System

Energy storage type	Rated power	Duration time	Advantages	Disadventages	Approaches to application
Pumping energy storage	100-2000MW	4-10h	High power capacity, high cost and low cost	Limited by geographical conditions	Assist in peak cutting and valley filling, frequency modulation, black start and backup power supply,etc

Figure 12. Wind electric energy storage technology

3.6.2. Reasonable wind abandoning to improve wind power consumption capacity on a large scale. Wind power is intermittent and random, and its output cannot be arranged and controlled like other conventional power sources. When the system load is at the minimum value, the power output of wind power may reach the maximum, while other power sources need to adjust the power output, so that the load is supplied by wind power. However, due to the limitation of the conventional thermal power micro-increment, reducing the output operation or closing down will increase its power generation cost. If the wind can be properly abandoned during the low load period of the power grid, it can not only reduce the pressure of the conventional thermal power generation to reduce the output power operation, but also improve the capacity of the power grid to accept wind power to be connected to the grid, so as to improve the security and economy of the entire power system operation. The web interface design of this part is shown in Fig.13.

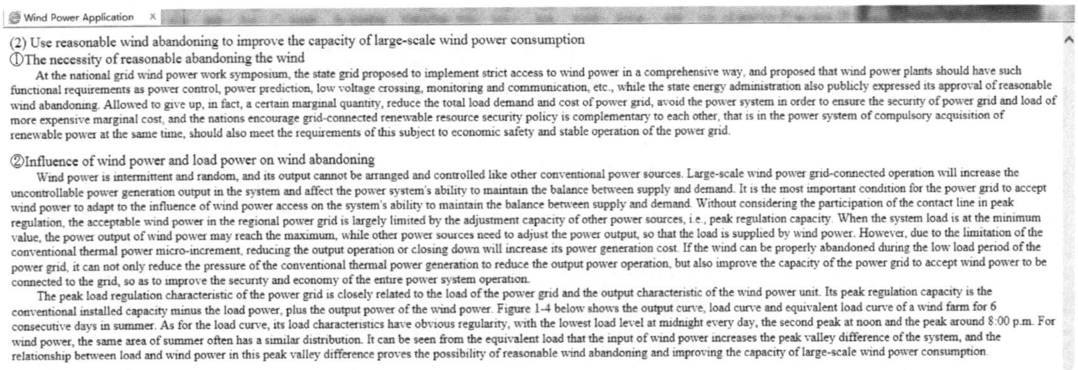

Figure 13. Reasonable abandoning the wind

4. Conclusion

The realization of the wind resources website platform centralizes the fragmented wind power information in various regions of the country. Through the web design, the development status of China's wind power is introduced simply and clearly and intuitively. It is convenient for the inquiry of wind power research data and provides theoretical support for the formulation of wind power policy. At the same time, the phenomenon of wind abandoning and the prediction method of wind power are introduced, and the methods to improve the power consumption capacity of the power grid are put forward: wind power storage technology and reasonable wind abandoning method, which provide the solutions for the severe situation of wind power consumption in China.

Acknowledgments

This work was supported by the National Natural Science Foundation of China (61473174) and Weihai Science and Technology Project (2018).Corresponding author: Qiang Tong.

References

[1] Wang Changlu, Wang Weigong, Zhang liyong, Qiao Xuetao. Analysis on the development of China's wind power industry [J]. Journal of Chongqing University,2015,1:148-154.
[2] Outline of the national medium - and long-term plan for scientific and technological development (2006-2020). Website of the development planning division of the national development and reform commission of the People's Republic of China.
[3] Medium - and long-term development plans for renewable energy. Website of the development planning division of the national development and reform commission of the People's Republic of China.
[4] Chi Yongning, Liu Yanhua, Wang Weisheng. Effects of wind power access on power system [J]. Grid technology,2007,31(3):76-81.
[5] Zhu Lingzhi, Chen Ning, Han Hualing. Analysis on key problems and countermeasures of wind power consumption [J], automation of power system,2011,35(22):90-96.
[6] Jia Hongxin, Zhang Yu, Wang Yufei, etc. Application of energy storage technology in wind power system [J], renewable energy,2009,27(6):11-15.Yang Zhongkai, patent measurement and patent system [M], Dalian University of technology press, 2008, first edition.

Design and Implementation of Software Planning System for Dental Robot

Lihua Yang[1,a], Chuanhong Zhou[1,b] and Tao Zhou[1,c]

[1]School of Mechatronic Engineering and Automation, Shanghai Key Laboratory of Intelligent Manufacturing and Robotics Shanghai University, Shanghai, China

Email: [a]871218424@qq.com; [b]13817297415@139.com;[c]1358533088@qq.com

Abstract: With the improvement of people's living standards and the aging of the population, the demand for the dental restoration market continues to increase. However, due to the low efficiency and high prices of traditional technologies, it cannot meet the needs of most people. The dental robot can make up for the above deficiencies. The dental robot is mainly divided into two systems, namely the software planning system and the robot body and control system. This article mainly uses the secondary development function of three-dimensional animation software Blender and the knowledge of prosthodontics to research and develop a CAD software system (a software planning system for dental robots) for dental implants.
Keywords: dental implant, dental robot, secondary development

1. Introduction

With the improvement of people's living standards, changes in eating habits and structure, and aging population, the demand for dental restoration market continues to increase. However, due to traditional technologies and backward consumption concepts, China's oral remediation and cultivation market is still relatively small. Traditional restorations cannot meet people's increasing needs in terms of technology, materials, aesthetics, and price. CAD/CAM technology has penetrated almost all areas of engineering technology and human life. One of the most typical examples of this technology in the field of oral medicine is the pioneering introduction of the design and manufacture of oral fixed prostheses [1].

Modern dental prosthetic technology is combined with the secondary development technology of three-dimensional animation software Blender. By constructing a three-dimensional CAD system for dental restoration, the dental implant process is modeled in three-dimensional simulation, and the simulated data is transmitted to in the robot system. The robot implants new teeth into the patient's mouth based on the acquired data and pre-programmed procedures. This software system can precisely plan the planting plan according to the patient's oral structure, and achieve full navigation and autonomous control during the operation.

2. The Overall Structure of the Dental Robot Software System

The software system of the dental robot researched and developed in the article adopts standard dental data and standard implant data methods. That is based on standard dental implants and standard implants. The CT scan model of the patient's oral cavity is introduced, according to its specific basics. The characteristic structure is to quantify the standard teeth and standard implants and then re-simulate the implant process on the patient's oral model until the implants and implants that match the patient's mouth are generated. The shape of dental implants was modified using Blender's Hook modifier and

Laplacian modifier, combined with various methods of surface creation to reconstruct the dental implant's data structure model.

This software system is composed of four modules: the oral implant user interface module, the dental implant and implant module developed using the Python language, the auxiliary tool module, and the communication module with the robot part.

The software system controlling panel user interface developed is shown in Figure 1. For the system function of oral dentistry, the control panel is designed as five main menus: Teeth, Implants, Bridges, Splints, Ortho. The five main functions are set as buttons on the control panel. When the button is pressed, the corresponding drop-down bar below will show the content. The operation flow of the software system is:

The first step: A CT scan is performed on the oral cavity of a patient who needs to implant a tooth to obtain a patient oral data model, and then the resulting data model is imported into the software system.

The second step: According to the imported patient oral data model, select the appropriate dental data model and implant data model, then perform three-dimensional simulation of the entire planting process, and modify the initial selected implant and tooth shape data to achieve the best position and depth of planting.

The third step: The three-dimensional simulation data will be saved and transmitted to the robot system. After the simulation the patient oral data model is saved as an STL format file.

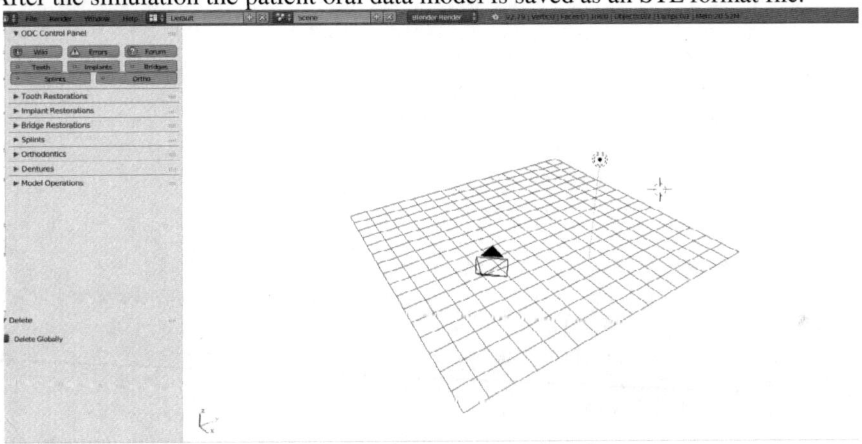

Figure 1. software system interface

Due to the different design of dental implants and the design of industrial products, the design of industrial products is based on a uniform design standard, and this software system is based on the patient's oral form. The software system structure is shown in Figure 2.

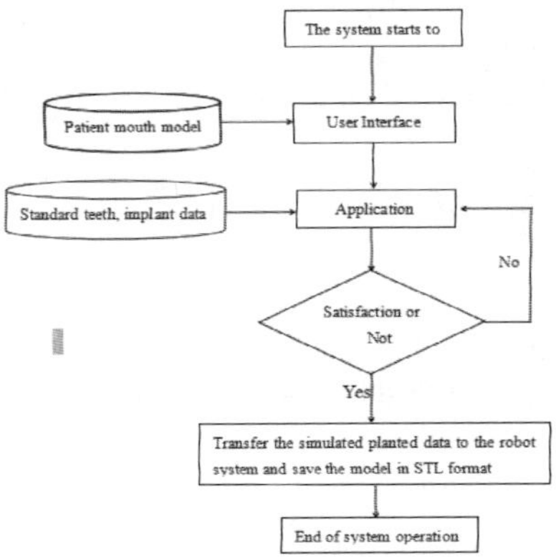

Figure 2. System Structure

```
import bpy
class SimpleOperator(bpy.types.Operator):
    bl_idname = "object.simple_operator"
    bl_label = "Tool Name"
    def execute(self, context):
        print("Hello World")
        return {'FINISHED'}
def register():
    bpy.utils.register_class(SimpleOperator)
def unregister():
    bpy.utils.unregister_class(SimpleOperator)
if __name__ == "__main__":
    register()
```

Figure 3. Python script

3. Key Technologies for Implementing Dental Robot Software System

3.1.Blender Secondary Development Technology

Blender provides very powerful API support. Because Blender is built on the Python language. Its API is fully compatible with the Python language module, so it has a very strong multimedia application scalability [2]. Blender embeds a Python interpreter that is started and active by Blender. This interpreter runs scripts to draw the user interface and is used for some of Blender's internal tools. Blender provides a bpy module for the Python interpreter that can import this module into a script and provide access to Blender data, classes and functions. In order to interact with Blender, scripts need to use tightly integrated APIs. Blender provides a lot of interface functions for secondary development of the Python language.

Blender provides a Python API, which provides a platform for deep interaction between Blender and Python. Users can use standard or custom Python modules to invoke Blender's default Python library, which will give Python its power [3]. The Blender Python API provides the following major modules for manipulating objects:

(1) bpy.ops.object: Selected operation object.

(2) bpy.data: Operation of the currently open Blender file data.

(3) bpy.context: Operates the text content of the selected object.

The secondary development Python script based on Blender is mainly composed as follows Figure 3: First, defining a subclass of bpy.types, both class attributes begin with the bl_ prefix. This is a convention used to distinguish Blender-defined variables from your own. The Execute function receives an operator and an instance of the current context. Finally, the registration function is called and it loads the class into Blender. The Blender module loaded at startup needs the register() and unregister() functions. These are the only functions that can be called in your code, otherwise it is a regular Python module.

3.2 Design of Dental Model Library

Since the standard tooth data determines the basic morphology of the outer surface of the implanted tooth, the acquisition of standard dental data is also crucial. The use of conventional methods, such as mechanical scanning or laser scanning, cannot obtain complete and accurate surface data for individual teeth. The reason is that it is difficult to obtain surface data for areas below the high point of the crown profile in the previous scanning methods. With the successful development of a new generation of 3D laser scanners, we have a new means to obtain data [4]. Therefore, we used a 3D dental model laser scanner [5]. to scan a superhard plaster model of 28 standard crowns/bridges provided by the hospital as 1:1. Through the scanner scan, we obtained the full dentition of surface 3D point cloud data of 28 standard dental crowns [6].The completed standard dental model library is shown in Figure 4.

Figure 4. Standard Dental Model Library

The module code of the dental system to retrieve dental data from the standard dental model library is shown in Figure 5:

```
class CBGetCrownForm(bpy.types.Operator):
    '''Inserts a crown form from the tooth library at the cursor location'''
    bl_idname = 'opendental.get_crown_form'
    bl_label = "Get Crown Form"
    bl_options = {'REGISTER','UNDO'}
    bl_property = "ob_list"

    def item_cb(self, context):
        return [(obj.name, obj.name, '') for obj in self.objs]

    objs = bpy.props.CollectionProperty(type=bpy.types.PropertyGroup)
    ob_list = bpy.props.EnumProperty(name="Tooth Library Objects", description="A List of the tooth library", items=item_cb)

    def invoke(self, context, event):
        self.objs.clear()
        settings = get_settings()
        #here we grab the asset library from the addon prefs
        libpath = settings.tooth_lib

        #a list of all objects in the asset library
        assets = odcutils.obj_list_from_lib(libpath, exclude = '_')
        for asset_object_name in assets:
            self.objs.add().name = asset_object_name
        #context.window_manager.invoke_search_popup(self.ob_list)
        context.window_manager.invoke_search_popup(self)
        return {'FINISHED'}
```

Figure 5. To retrieve the tooth code from the standard tooth library

Figure 6. Standard implant model library

3.3 Design of the Implant Model Library

Implants are generally made of artificial materials (such as metal, ceramics, etc.), and threaded columnar and root-shaped implants have become widely accepted implant forms worldwide. At present, implants with threaded roots and parallel-wall designs are the mainstream of clinical application [7]. Therefore, we also used a three-dimensional laser scanner to scan the standard implants provided by the hospital to obtain three-dimensional point cloud data to form a standard implant model library. The completed standard implant model library is shown in Figure 6:

The module code for calling the implant from the standard implant library in the software system is shown in Figure 7.

4. Software System Operation and Results

After a series of compiled Python code packages are compressed and installed on the Blender in the form of plug-ins, the entire software system interface will be generated. Firstly, after scanning the patient's oral three-dimensional model into the software system (Figure 8a), and then import the implant and import standard dental implants (Figure 8b and Figure 8c), for oral dental implant simulation (Figure 8d). Finally obtain the desired data and pass the data to the robot system.

```
class OPENDENTAL_OT_place_implant(bpy.types.Operator):
    '''Places Implant or swaps existing implant with new implant of your choice'''
    bl_idname = "opendental.place_implant"
    bl_label = "Place Implant"
    bl_options = {'REGISTER','UNDO'}
    bl_property = "imp"

    def item_cb(self, context):
        return [(obj.name, obj.name, '') for obj in self.objs]

    objs = bpy.props.CollectionProperty(type=bpy.types.PropertyGroup)

    imp = bpy.props.EnumProperty(name="Implant Library Objects",
                                 description="A List of the tooth library",
                                 items=item_cb)
    hardware = bpy.props.BoolProperty(name="Include Hardware", default=False)

    #classmethod
    def poll=(cls, context):
        return len(context.scene.odc_implants) > 0

    def invoke(self, context, event):
        self.objs.clear()
        settings = get_settings()
        libpath = settings.imp_lib
        assets = odcutils.obj_list_from_lib(libpath, exclude = '_')

        for asset_object_name in assets:
            self.objs.add().name = asset_object_name

        context.window_manager.invoke_search_popup(self)
        return {'FINISHED'}
```

Figure 7. Calling the implant module code diagram

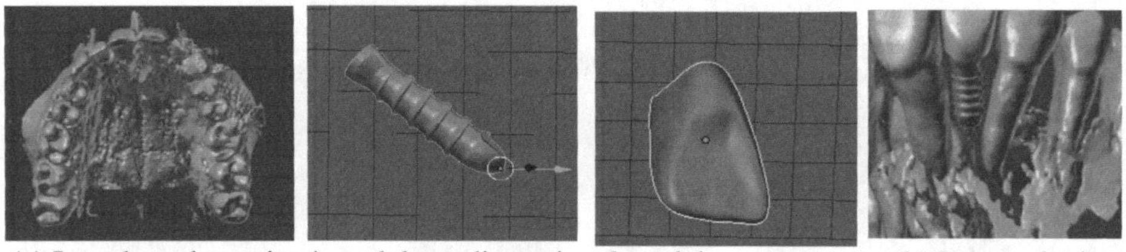

(a) Introduce the patient's oral three-dimensional model (b) Import the implant
(c) Import standard dental implants (d) Conduct dental implant simulation

Figure 8. Three-dimensional simulation and simulation process diagram of dental robot software system

5. Communication Between Software System and Robot System

The robot system uses ABB's robots--model IRB 140TW. IRB series general-purpose robots are the main products of ABB robots. This series of robots is a very mature 6-axis industrial robot. This series of robots is an open-architecture design to adapt to different occasions, and can easily communicate with external systems [8]. Because the software system communicates with the robot system, the simulated data is transmitted to the robot system, and ABB robot provides a rich I/O communication mechanism, which can easily communicate with the PC and surrounding equipment [9]. table 1 shows the communication method of ABB robots. Therefore, the Socket message communication method is adopted between the software system and the robot system, and the main mode of connection-based Socket communication is the client/server method [10]. Among them, the PC acts as a server and the ABB robot acts as a client.

Table 1. ABB robot communication

PC	Field bus	ABB standard
RS232	Device	Standard I/O edition
OPC server	Profibus	PLC
Socket Message	Profibus-DP	.....
	Profinet	.....
	EtherNet IP	.....

The code module using Socket message communication between the software system and the robot system is shown in Figure 9.

```
TCP_IP = '127.0.0.1'
TCP_PORT = 5005
SOCKET = None
RAYPUMP_PATH = None
class MessageRenderOperator(bpy.types.Operator):
    bl_idname = "object.raypump_message_operator"
    bl_label = "Save & Send To data"
    bl_description = "Saves, sends and schedules current scene to Robot system"

    def connect(self):
        global SOCKET, RAYPUMP_PATH, TCP_IP, TCP_PORT
        print("New connection!")
        if SOCKET != None:
            SOCKET.close
        # Create a connection:
        try:
            SOCKET = socket.socket(socket.AF_INET, socket.SOCK_STREAM)
        except socket.error as msg:
            self.report({'ERROR'}, "Cannot create a socket")
            SOCKET = None
            return {'CANCELLED'}
```

Figure 9. Code module for communication between software system and robot system

6. Conclusions

This article adopts Blender's secondary development technology to develop a software system for dental robots. This system is mainly aimed at oral dental implants. The main target is dentists. The software system has a very good user interface and can easily import models required for various dental implantation procedures. It is very easy to learn and at a glance. The development of this software system has solved the domestic traditional oral dental implant methods in the past, greatly improving the precision and speeding of dental implant surgery, and making the whole dental implant surgery process quick and accurate.

Acknowledgment

We thank Shanghai Key Laboratory of Intelligent Manufacturing and Robotics for assistance with our work.

Reference

[1] Wang Shujie, Lin Ying, Song Lina. Orthopaedics fixing CAD system based on surfacer secondary development[J]. Computer Engineering and Applications, 2004(28):191-194.

[2] Hao Zhenhua. Research on dynamic simulation application of python in Blender engine[J]. Software Guide, 2012, 11(11):66-68.

[3] Hao Zhenhua. Research on dynamic simulation application of python in Blender engine[J]. Software Guide, 2012, 11(11):66-68.

[4] Zou Bo, Lu Peijun, Wang Yong. Acquisition of standard crown data [J]. Journal of Practical Stomatology, 2002(06): 550-552.

[5] Zou Bo, Lv Peijun, Wang Yong. A comparative study on the reliability of a new three-dimensional dental model laser scanner. Journal of Practical Stomatology, 2002,18 (6): 546

[6] Cheng Yusheng, Dai Ning, Liao Wenhe, Yu Qing. Establishment of standard crown and bridge database[J]. Journal of Southeast University(Medical Science Edition), 2007(01):1-3.

[7] Huang Yuanliang. Development trend of dental implant materials and morphology design[J]. Journal of Dental Materials and Devices, 2015, 24(03): 113-117.

[8] Deng Sihao. Research and implementation of real-time robot monitoring and communication system[D]. Wuhan University of Technology, 2003.

[9] Sun Zhiwei. Research on ABB industrial robot network integrated control system[D]. Yanshan University, 2017.

[10] Zhao Y. Research and implementation of message queue middleware based on Socket [D]. Inner Mongolia University, 2007.

IOP Conf. Series: Materials Science and Engineering **435** (2018) 012019 doi:10.1088/1757-899X/435/1/012019

Improved Set Covering Location Model for Charging Facility Deployments

Ji Haorui[1,*], Liu Manhao[1], Zhang Bowen[2] and Zhang Xin[1]

[1]School of Information and Communication Engineering, Beijing University of Posts and Telecommunication, 10 Xitucheng Rd., Beijing. 100876, People's Republic of China

[2] Department of Mathematical Sciences, Tsinghua University, Hai Dian, Beijing. 100084, People's Republic of China

[*]Corresponding author: Ji Haorui(email: jerryjhr@163.com)

Abstract. While electric vehicles (EVs) can assist in resolving environmental pollution and resource scarcity problems, they require specific infrastructure developments. Destination chargers, which are based on current charging technologies and EV characteristics, have been the main charging facilities employed. Using an Improved Set Covering Location Model (ISCLM), the primary objective of this paper is to determine how destination chargers could be deployed in urban areas. As the model basis, the purpose, time, and parking conditions of urban car-use behaviors are first analyzed, based on which the ISCLM is constructed. As the basic Set Covering Location Model (SCLM) is an NP-hard problem, improvements in the assumptions and solution algorithms are made so that the ISCLM treats all demand points as equivalent regardless of the internal sizes. As it is also assumed that the service range for all charging stations is circular and have the same radius, the ISCLM is a polynomial problem; therefore, the Minimum Covering Circle algorithm and the Voronoi Graph are used to derive the destination charger deployment solution. To verify the effectiveness of the ISCLM, the model is used to analyze a specific area in Beijing. In this case study, the simulated demand points were selected based on the location of actual parking lots in residential communities and office buildings. The results validated the enhancement in the ISCLM network architecture.
Key words: Car-Use Behavior; Charging Network Architecture; Electric Vehicle; Improved Set Covering Location Model; Minimum Covering Circle; Voronoi Graph.Introduction

1. Introduction

As electric vehicles (EVs) use clean energy, they can reduce greenhouse gas emissions, alleviate urban pollution, and assist in resolving environmental problems. However, to encourage people to switch gasoline and diesel cars to EVs, there needs to be a commensurate rise in the construction and deployment of charging facilities. Currently, there are two types of charging facilities available to EV users; destination chargers and superchargers. What markets are currently assessing is the minimum investment needed to meet the EV charging requirements in a specific area and the charging facility construction that can cover as large an area as possible.

Research has therefore been ongoing on the development of site selection algorithms[1-3], for which two mainstream models have been developed; point-based demand models and path-based demand models[4].

Point-based demand models assume that the demand is generated at certain nodes in the road network, and therefore the minimum distance between the demand points and the charging stations has been the optimization goal. Major models currently available are the P-Center Model[5], the P-Median Model[6] and the Covering Location Model[7-8]. Of these models, the Covering Location Model has been found to have the best ability to improve network architecture[9]. For path-based demand models, the energy demand is not limited to the nodes but is based on road section traffic volumes, with the optimization goal being to achieve the maximum traffic volume that can be served by each charging station[10]; therefore, adjacent road sections with high traffic have been found to be the preferred charging station sites.

Existing research has tended to focus on the rapid charging and refueling typical of the supercharging mode, for which road section traffic volume and path-based demand model implementation and improvements have been the main focus[11-13]. Therefore, road conditions, dynamic vehicle driving area changes, and market-oriented services are the focus of this model.

However, the current EV charging technologies have long-term charging characteristics, which is the main difference between conventional fuel vehicles and EVs. Therefore, it has become more important to first meet the basic requirements necessary for long-term EV charging, which means that the infrastructure should be aimed at fixed demand points. Therefore, in this paper, we focus on point-based demand models, or destination charging, rather than path-based demand models.

Based on above research, network layout optimization and the associated algorithms for deploying destination chargers within a certain boundary can still be improved; therefore, an Improved Set Covering Location Model is proposed to improve the efficiency and accuracy of the existing models.

2. Preparation of the Model

Charging facility deployments are a technical economics optimization problem. As the optimization model should be constructed and evaluated based on real world situations, the following background settings are proposed before the establishment of the models.

2.1 Charging Facilities

Existing EV charging technologies take about 6-10 hours to fully charge a battery, after which the EV can travel from 300-500 kilometers[14]. Depending on the environment and the charging requirements, Tesla, the leading corporation in this field, offers two types of charging facilities; destination chargers and superchargers.

As destination chargers are designed to fully charge the battery, they have a relatively low charging rate and require several hours to reach a full charge. Because of these characteristics, destination chargers are usually installed in mall parking lots, hotels, residential communities or other public places where people can park and charge their cars.

Superchargers, however, can charge the battery in a relatively short period of time, and are designed to meet urgent charging needs; therefore, these charging facilities are generally installed on city streets in the same way as gas stations.

2.2 Car-use Behavior in Urban Areas

As the premise for the later analysis, car use patterns and rules are briefly discussed in the following sub-section.

2.2.1 Purpose. City traffic can be divided into commuting and non-commuting depending on the purpose and the destination[15-17].Commuting refers to people moving between their homes and work places, and non-commuting traffic is when people travel to other destinations. Commuting traffic has relatively fixed routes and travel times, and non-commuting traffic has random routes and travel times. Both car use trends tend to be the most important for people in urban areas.

2.2.2 Time. Through an analysis of user travel time distributions, most people use their cars between 5 a.m. and 10 p.m [18]. As these car use data can give some indication as to the possible EV charging demand during the day, they provide guidance for the installation of effective charging networks.

2.2.3 Parking Conditions. Generally, the best time to charge an EV is when it is parked and idle. Most people, therefore, would choose to charge their cars at night from 10 p.m. to 6 a.m. As most people park in their own neighborhood or in other public parking lots, installing charging stations in these places would be the most effective plan.

2.3 Strategies for the Deployment of Charging Stations

Based on the above analysis, the typical car-use patterns for urban residents are as follows:

1. Relatively fixed routes and travel times.
2. Car-use distribution time is centralized and has a short time span.
3. People tend to park in places that can accommodate many cars.
4. Travel distances are relatively short.

Because of the short time-span, EVs owners have plenty of time to use destination chargers. Therefore, when deciding where to deploy charging networks, destination rather than superchargers should a priority to satisfy the daily charging demands of EV owners, which is a key difference between gasoline and diesel cars and EVs.

Further, all mass parking environments such as residential communities, malls, public parking lots, and work places should be fully considered when demand points are selected. To minimize the number of charging stations within a certain degree of need, an Improved Set Covering Location Model (ISCLM) is proposed, which has the following assumptions:

1. There is a strong correlation between the charging station site distribution and the number of EVs in the area. However, site configuration is an infrastructure construction problem and requires forward-looking approaches. Therefore, this model assumes that the EV ownership in a certain area is linearly related to the local population density or the working place areas, and that this ownership is uniformly distributed in a spatial dimension.
2. Sufficient charging equipment is installed at each charging station to fully meet the requirements of all EVs within the service coverage area.

3. The Establishment of the Optimal Charging Facilities Distribution Model

3.1 Description

Fig. 1 below is an abstract graph of a city map (without roads), in which the smaller black dots are the actual locations with a certain EV charging demand, and the red dots are the charging stations. However, as can be seen, as the charging stations are randomly distributed, they are unable to service all demand points; therefore, in the following, we discuss how these stations can be distributed.

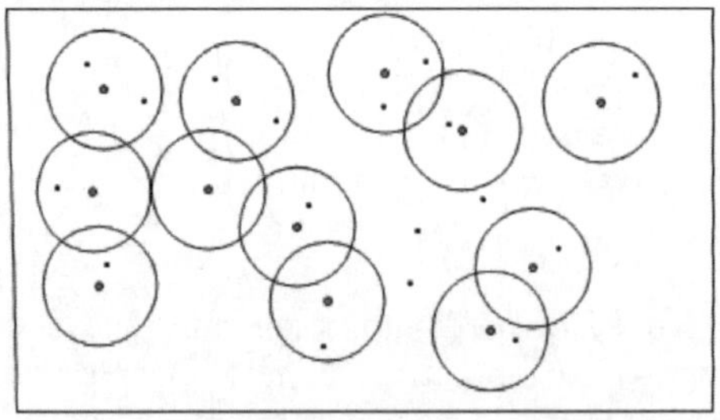

Figure 1. Abstract graph of a city without roads

A reduced-form version of the problem can be described as follows:

Let A be a set of points in an area $\Omega \subset \mathbb{R}^2$. We want to find another set of points B in which the number of elements is minimal s.t. for each $p \in A$; there is $q \in B$ and $d(p,q) < r$. Here, r is a positive constant.

3.2 Improved Set Covering Location Model

Church and Revell proposed a Maximum Covering Location Model (MCLM) for a sites selection problem[19], which was mainly applied to deploy a certain number of facilities to cover the maximum demand. While the Set Covering Location Model (SCLM) is a better choice for determining the minimum number of sites in a certain district.

However, the basic SCLM is an NP-hard problem[20], which makes it difficult to solve. So, an Improved Set Covering Location Model (ISCLM) is proposed, the goal of which is to minimize the number of charging stations to cover certain demand.

The term cover is used here to describe a demand point that is within the service range of at least one charging station.

The ISCLM model is formulated as follows:

Given a set of elements $E = \{e_1, e_2, \ldots, e_n\}$ and a set of m subsets of E, $S = \{S_1, S_2, \ldots, S_m\}$. If $J \subseteq \{1, 2, \ldots, m\}$ and $\bigcup_{j \in J} S_j = E$, then define $\{S_j\}_{j \in J}$ as a covering set for E. The goal is to find a covering set for the given set E with the least number of elements.

For every subset S_j, the following variable is introduced:

$$x_j = \begin{cases} 1, j \in J \\ 0, else \end{cases} \tag{1}$$

Based on these notations, the SCLM is constructed:

$$\text{Min} \sum_{j=1}^{m} x_j \tag{2}$$

$$\text{s.t} \sum_{j:e_i \in S_j} x_j \geq 1, i = 1, 2, \ldots, n \tag{3}$$

$$x_j = 0, 1, j = 1, 2, \ldots, m \tag{4}$$

that is, to make sure every element e_i in set E is covered by at least one subset S_j in the set of subsets S.

Here, the model makes two important assumptions based on the basic SCLM; the service range of all charging stations is a circle with the same radius, and every charging station can provide unlimited service and can satisfy all demand point requirements once it is within the station's service range. It also means that the model treats all demand points in the same way, ignores the number of facilities inside the charging station, focuses only on the network topology.

To conduct research on how a charging station can cover the surrounding demand points, the Minimum Covering Circle algorithm is introduced.

3.3 Minimum Covering Circle

In a location planning problem such as this, the main effect of the Minimum Covering Circle theory is to assist in determining the least cost for a charging station to serve the furthest demand point in its coverage.

The model can be abstracted into the following problem:

There are n points in $\mathbb{R}^2$; therefore, the goal is to determine the circle O s.t that contains all n points and has a minimum radius. Given this definition, the Minimum Covering Circle Model is constructed as:

$$Min\{Max(d(s_i, u_j), j = 1, 2, \ldots, n_i), i = 1, 2, \ldots, m\} \tag{5}$$

Notations:

$U=\{u_1,u_2,\ldots,u_n\}$: all demand points in the area

s_i: location of the i th charging station

n_i: number of demand points that has the shortest distance to the i th charging station

$d(s_i,u_j)$: Euclidean distance between s_i and u_j

The solution to this model is indicated in Fig.2 which shows the flow chart for the Minimum Covering Circle algorithm.

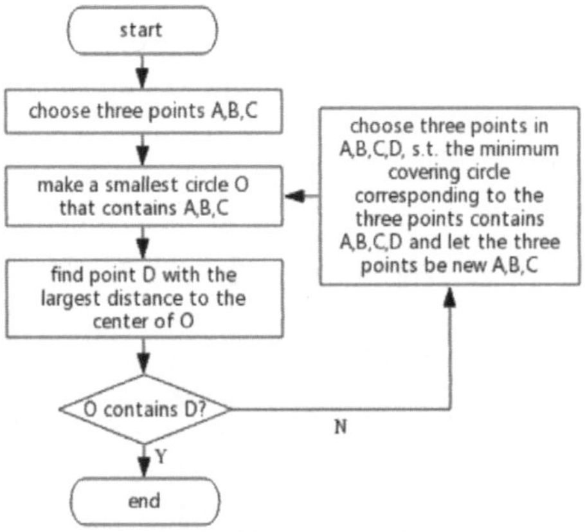

Figure 2. Flow chart for the Minimum Covering Circle algorithm

3.4 Voronoi Graph

Obviously, as users select the nearest charging station, it is necessary to determine the corresponding relationships between ***n*** demand points and the ***m*** charging stations to be implemented. For this purpose, the Voronoi Graph algorithm is introduced[21];

Let $x_i, i=1,2,\ldots,n$ be n different points in an area D, let

$$V_i=\{p\in\mathbb{R}^2|\ d(p,x_j)\geq d(p,x_i)\forall j\} \tag{6}$$

be the V-polygon of x_i. The boundary of V_i and D determined the Voronoi graph for these n points.

The Voronoi Graph for the following five red points is constructed in Fig.3.

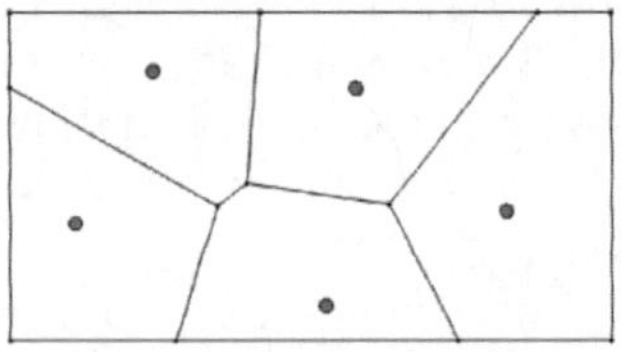

Figure 3. Example of a Voronoi Graph

3.5 Solution

To effectively deploy the destination chargers, the ISCLM is applied and the Minimum Covering Circle algorithm and the Voronoi Graph are employed to solve the model.

Step 1: Initialize N suitable points to make up set B with an associated Voronoi graph and V-polygon for V_x for each $x\in B$

Step 2: For each V_x, consider V_x to be the area, and $V_x\cap A$ be the set of points. Let O_x be the associated minimum covering circle; if there is y s.t. y and O_y are different, then replace x with the center of O_x for all $x \in B$, and then repeat Step 2.

Step 3: If the radius of all O_x is smaller than r, the B is what is desired; then, use $N+1$ to replace N, and repeat Step 1.

Here, the N in Step 1 for the first time is determined from the area of Ω and r; if the distance between the two points is close enough, they are considered one point.

In this way, the elements in B are the locations at which to establish the charging stations. Fig.4 shows the flow chart for the Improved Set Covering Location Model algorithm.

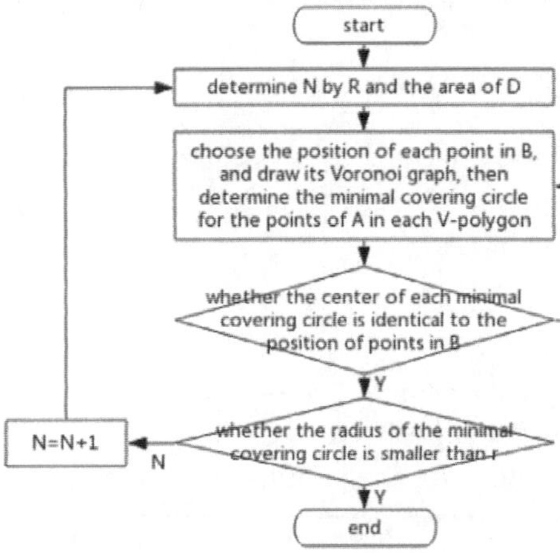

Figure 4. Flow chart for Improved Set Covering Location Model algorithm

4. Case Study

To verify the effectiveness of the ISCLM, a simulation was conducted based on geographical information from Beijing, the capital of China.

Fig. 5 shows the Beijing city map with the simulated demand points extracted from Google Maps. As discussed in 2.2, an area in the city center was first selected, after which the demand points were chosen based on the people's parking behaviors. In this case, these selections were mainly centralized at parking places, such as communities, office buildings and public parking lots.

IOP Conf. Series: Materials Science and Engineering **435** (2018) 012019 doi:10.1088/1757-899X/435/1/012019

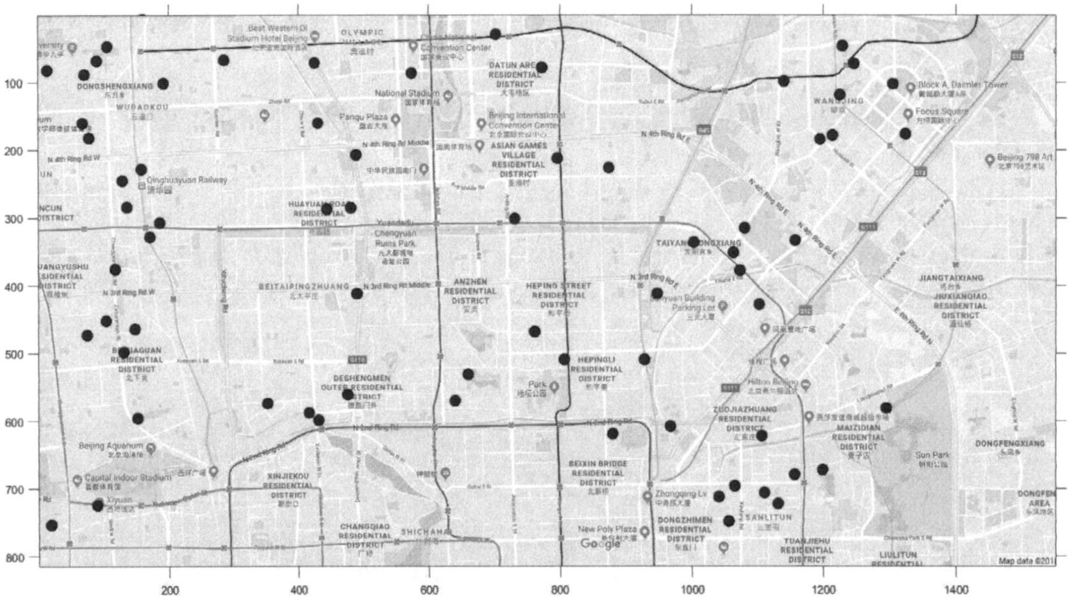

Figure 5. Beijing City map with simulated demand points

Coordinates for each demand point listed in Table 1

Table 1. Coordinates for the simulated demand points

No.	x	y	No.	x	y	No.	x	y	No.	X	y
1	1295	581	**18**	90	68	**35**	874	226	**52**	796	212
2	104	452	**19**	432	599	**36**	1102	428	**53**	1131	722
3	351	574	**20**	68	160	**37**	762	468	**54**	703	29
4	1199	672	**21**	732	301	**38**	1056	748	**55**	1106	622
5	78	182	**22**	1214	178	**39**	1324	176	**56**	1157	333
6	640	570	**23**	131	498	**40**	417	588	**57**	1063	351
7	20	753	**24**	118	376	**41**	1003	336	**58**	75	473
8	1041	712	**25**	773	78	**42**	806	509	**59**	106	47
9	1110	706	**26**	1156	679	**43**	574	86	**60**	129	245
10	1065	696	**27**	158	228	**44**	90	724	**61**	1140	98
11	285	67	**28**	1195	184	**45**	1229	46	**62**	192	101
12	136	284	**29**	490	412	**46**	1225	118	**63**	1246	72
13	445	287	**30**	426	71	**47**	489	207	**64**	1305	102

14	660	531	**31**	1080	315	**48**	880	619	**65**	171	328
15	148	464	**32**	15	82	**49**	1073	378	**66**	928	509
16	481	285	**33**	71	88	**50**	152	596	**67**	476	561
17	431	160	**34**	967	608	**51**	186	307	**68**	947	412

The statistics show that Beijing has an area of 16,410 square kilometers with a total vehicle ownership of 5.94 million. There are currently 1963 gas stations in Beijing, which means that if distributed uniformly, the distance between two neighboring gas station is about 3.2 kilometers.

The area selected from a specific Beijing district in Fig.5 is 142.34 square kilometers, which was abstracted to a graph of the size 1552*815. Within this area, the existing vehicle ownership was approximately 8000. Once the conversion from conventional fuel vehicles to EVs is complete, it was assumed that the EV ownership would remain the same.

From the statistics discussed above, it was assumed that the service range for each charging station was 3.2 kilometers. The simulation based on the above ISCLM was run on MATLAB and the results are shown in Fig. 6. From the output, it can be seen that the model converged to the optimal solution when the number of charging stations reached 12; the coordinates for each are shown in Table 2.

In contrast, if these charging stations were distributed uniformly, the result was $142.34\left[\pi^*(1.6)^2\right]^{-1} \approx 18$. Therefore, it was concluded that the ISCLM was effective in deploying the destination chargers.

Deploying 12 destination charging stations across the 142.34 square kilometer area would meet the charging requirements of approximately 680 EVs, which is equivalent to a medium-sized residential community in Beijing. Charging for 6 hours continuously overnight would allow an EV to travel about 300 kilometers, which would be adequate for most driver's everyday needs; therefore, this assumption is reasonable. Of course, EV charging station construction is a complex problem and also needs to take account of the impact of the vehicles, road networks, existing power grids, and the provision of a viable charging station business model. In short, the research in this paper provides a model and algorithmic support for municipal construction decisions and possible charging station investments.

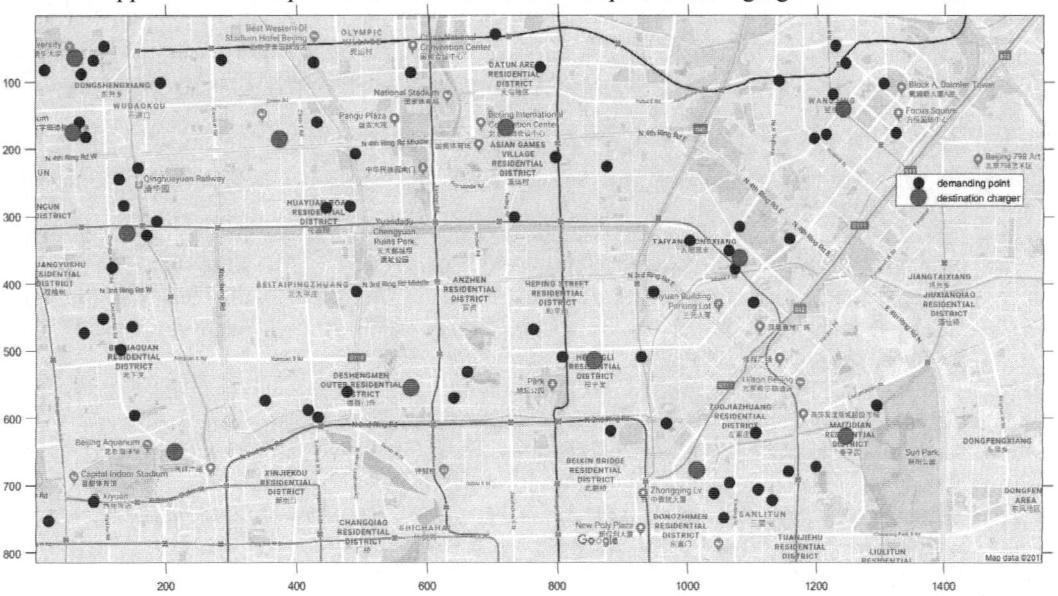

Figure 6. Planning results for the charging stations

Table 2. Coordinates for the charging stations

No.	x	y	No.	x	y	No.	x	Y
1	1013.8	676.5	**5**	212.8	649.6	**9**	1242.1	140.1
2	60.5	64.5	**6**	573.9	555.1	**10**	718.7	168.5
3	141.0	325.0	**7**	58.0	175.0	**11**	1247.0	626.5
4	1079.7	362.2	**8**	853.5	513.7	**12**	373.2	184.7

5. Conclusion

Based on current charging technologies and EV behavioral characteristics, destination chargers are currently being used at charging facilities. In this paper, an Improvement Set Covering Location Model (ISCLM) was presented to solve the destination charger deployment problem in urban areas. Two extra assumptions were appended to the ISCLM and the model improved to become a polynomial problem. The Minimum Covering Circle algorithm and the Voronoi Graph were then employed to find the solution to the ISCLM. The ISCLM was then demonstrated in a practical example for the deployment of destination charging stations in a specific Beijing district of 142.34 square kilometers, with the assumptions that the service range for each charging station was 3.2 kilometers and the charging time was 8 hours. Based on the calculation results, the optimal solution was found to be 12 charging stations. If a uniform distribution model were used, 18 charging stations would be needed; therefore, the ISCLM was proven to give a more effective solution than uniform distribution models.

Further research, however, is still necessary. First, while the ISCLM assumes that all demand points are the same, in practice, different places have charging different needs. Therefore, another demand point weight variable is needed to account for these differences. Second, this paper only considered one of the two types of charging facilities; therefore, it is necessary to develop models that also consider supercharger deployments.

References

[1] Rahman I, Vasant P M, Singh S M, et al. Review of recent trends in optimization techniques for plug-in hybrid, and electric vehicle charging infrastructures[J]. Renewable & Sustainable Energy Reviews, 2016, 58:1039-1047.

[2] Xi X, Sioshansi R, Marano V. Simulation–optimization model for location of a public electric vehicle charging infrastructure[J]. Transportation Research Part D Transport & Environment, 2013, 22(4):60-69.

[3] Cao Y, Song H, Kaiwartya O, et al. Mobile Edge Computing for Big Data-Enabled Electric Vehicle Charging[J]. IEEE Communications Magazine, 2017.

[4] Liu Kai, Sun Xiaohui, Zuo Zhi. Review on Location Optimization of Recharging Stations for Electic Vehicles[J]. Journal of Wuhan University of Technology(Transportaion Science & Engineering), 2015(3):523-528.

[5] Lu C C, Sheu J B. Robust vertex p-center model for locating urgent relief distribution centers[J]. Computers & Operations Research, 2013, 40(8):2128-2137.

[6] Berman O, Krass D, Menezes M B C. Facility Reliability Issues in Network p-Median Problems: Strategic Centralization and Co-Location Effects[M]. INFORMS, 2007.

[7] Berman O, Drezner Z, Krass D. Generalized coverage: New developments in covering location models[J]. Computers & Operations Research, 2010, 37(10):1675-1687.

[8] Edson Luiz Franca Senne, Pereira M A, Lorena L A N. A Decomposition Heuristic for the

Maximal Covering Location Problem[J]. Advances in Operations Research, 2010, 2010.
[9] Church R L, Revelle C S. Theoretical and Computational Links between the p-Median, Location Set-covering, and the Maximal Covering Location Problem[J]. Geographical Analysis, 1976, 8(4):406-415.
[10] Hodgson M J. A Flow-Capturing Location-Allocation Model[J]. Geographical Analysis, 1990, 22(3):270-279.
[11] Riemann R, Wang D Z W, Busch F. Optimal location of wireless charging facilities for electric vehicles: Flow-capturing location model with stochastic user equilibrium[J]. Transportation Research Part C Emerging Technologies, 2015, 58:1-12.
[12] Ismail Capar, Michael Kuby. An efficient formulation of the flow refueling location model for alternative-fuel stations[J]. Iise Transactions, 2012, 44(8):622-636.
[13] Lim S, Kuby M. Heuristic algorithms for siting alternative-fuel stations using the Flow-Refueling Location Model[J]. European Journal of Operational Research, 2010, 204(1):51-61.
[14] Liu G, Ouyang M, Lu L, et al. Driving Range Estimation for Electric Vehicles Based on Battery Energy State Estimation and Vehicle Energy Consumption Prediction[J]. Automotive Engineering, 2014, 36(11):1302-1309 and 1301.
[15] Inata K, Raksincharoensak P, Nagai M. Driver behavior modeling based on database of personal mobility driving in urban area[C]// International Conference on Control, Automation and Systems. IEEE, 2008:2902-2907.
[16] Zeng L. Research on Tidal Traffic Road Extraction Based on Real-time Traffic Data[J]. Geospatial Information, 2017.
[17] Sun Gengjie. Study on Urban Transportation Accessibility Based on Trip Purpose[D]. Beijing Jiaotong University, 2016.
[18] Ou Zhou. Research on City Residents Trip Behavior and Characteristics[D]. Changsha University of Science & Technology, 2012
[19] Church R,ReVelle C. The maximal covering location problem[J]. Papers of the Regional Science Association,1974,32:101
[20] Garey M R, Johnson D S. Computers and Intractability: A Guide to the Theory of NP-Completeness[M]. W.H. Freeman and company, 1979.
[21] Choset H, Burdick J. Sensor based planning. I. The generalized Voronoi graph[C]// IEEE International Conference on Robotics and Automation, 1995. Proceedings. IEEE, 1995:1649-1655 vol.2.

IOP Conf. Series: Materials Science and Engineering **435** (2018) 012020 doi:10.1088/1757-899X/435/1/012020

Design and Implementation of a Handling Robot Based on Target Recognition

Chenyang Ma*

Shanghai World Foreign Language Academy, Shanghai 200233, China

*dannymcy@sina.cn

Abstract: In this paper, aiming at robot target recognition and path planning, an algorithm of target recognition based on clustering analysis, which eliminates the traditional boundary extraction algorithm, decreases CPU time, and improves the real-time performance, is proposed, and a related hardware and software module design are implemented, including a claw grab mechanism, a drive control unit, visual unit, Mecanum four-wheeled mobile platform, universal driver, robot control app, etc. The robot can realize automatic operation and manual control by target recognition training calibration procedure, automatic operation program, and manual procedure. Test results show that the robot is working properly and achieves the desired goal.

1. Introduction

In recent years, robots for competition have developed rapidly and various games have grown significantly. Participating in challenging robotics competition can greatly stimulate the scientific interest of young people. The handling robot is a common type of robot in various types of competition robots. It usually follows a preset procedure to transfer a designated target to a predetermined position. This involves target acquisition and identification, path planning and various types of control servos. It is a complex and comprehensive system. Each part needs to coordinate work to ensure the completion of the task.

In [1], a mobile robot based on embedded system is studied. By using a CCD camera, the target can be detected and captured, and image processing techniques such as histogram, spatial resolution of the image, and adjacent pixels can be used. In [2], aiming at target recognition and tracking, an efficient algorithm is introduced to overcome the contradiction between the complexity of the algorithm, tracking accuracy, and real-time performance. Literature [3] proposes a high-speed visual system of table tennis robot design algorithm and a ball detection algorithm to improve the image processing speed up to 200 frames per second. Literature [4] studies the visual system for RoboCup to improve the accuracy and stability of the visual system. There is also research on the visual system of robots used for harvesting wild plants [5].

This paper proposes a target recognition method based on cluster analysis in point of the specific target recognition. The algorithm can greatly simplify the complex steps of the traditional target recognition one, and effectively improve the efficiency of target recognition. A robot device is achieved through the relevant software and hardware module design in the paper. The paper is organized as follows: The section II proposes a target recognition algorithm based on clustering analysis; the section III gives the overall framework of the robot and its hardware and software design; and finally the conclusion is given.

2. Target Recognition Based on Cluster Analysis

At present, there are many methods to identify the shape of a target, such as boundary feature method, Fourier shape descriptor method, geometric parameter method, shape invariant moment method, Finite Element Method (FEM) and so on. The traditional target boundary extraction methods use the shape feature to identify the target, so the boundary of the target needs to be extracted first, which belongs to the field of image segmentation. In general, the boundaries of the target and the background are intertwined. This can consume a lot of CPU time and affect the real-time performance of target recognition. This paper proposes a target recognition algorithm based on cluster analysis to identify the center position and shape of the workpiece. The algorithm eliminates the traditional boundary extraction algorithm and directly performs target recognition.

The robot vision recognition system designed in this paper is the target shape, color and distance recognition, so as to complete the robot's recognition and tracking for the shape and color of target. The target identification process is shown in Figure 1.

The target in this design is a blue ball. According to the characteristics of the target, there are two tasks to be performed: one is to determine the color is blue, and another is to determine the shape as a sphere.

We can divide the entire area of the captured image into 16 sub-blocks, each of which presets a motion path for the robot. We first use "cluster analysis" to detect the initial target. Fig. 2(a) is the first frame of the collected image containing the blue ball. According to the collected RGB original image, it can be determined which pixel belongs to the blue based on the pixel value feature. Therefore, for the pixels of the same color, cluster can perform as formula (1):

$$z = \frac{1}{N}\sum x_i \tag{1}$$

Where z is the cluster center, x_i is the sample, and N is the number of samples of the cluster. Then, the distances between the most peripheral pixels and the cluster center are calculated. If these distance errors are smaller than a certain threshold, the cluster may be determined to be a circle.

The actual distance between the ball and the robot must be calibrated during the training phase so that the distance between the robot and the ball can be calculated based on the calculated cluster radius.

According to the analysis of Figure 2(a), the corresponding servo drive mechanism is adjusted to obtain the next sampled frame image, as shown in Figure 2(b). It can be seen that the orientation of the robot has been calibrated at this time. According to formula (1), cluster analysis is performed to obtain the corresponding cluster center and cluster radius, and then the robot's travel path is adjusted. By analogy in turn, the target is tracked continuously. Figure 2(c) is the third frame of the collected image.

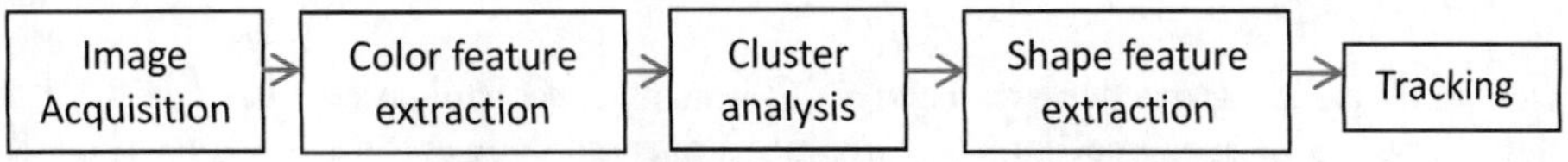

Figure 1. Identification based on cluster analysis

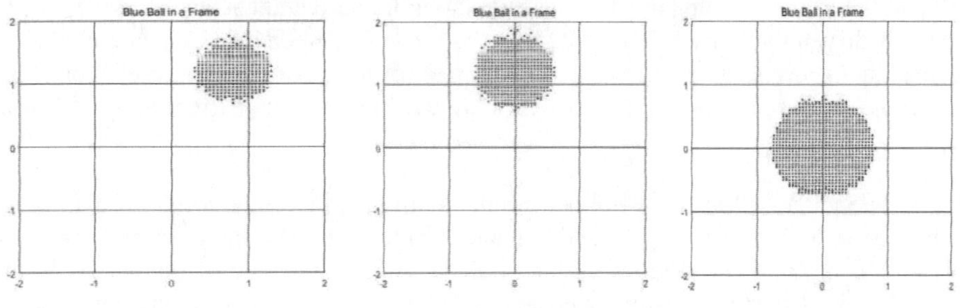

(a)Frame 1 (b) Frame 2 (c) Frame 3

Figure 2. Image frame containing blue balls

IOP Conf. Series: Materials Science and Engineering **435** (2018) 012020 doi:10.1088/1757-899X/435/1/012020

3. Robot System Design

3.1 System Hardware

In order to complete the target recognition and corresponding drive, the overall block diagram design of this robot system is shown in Figure 3:

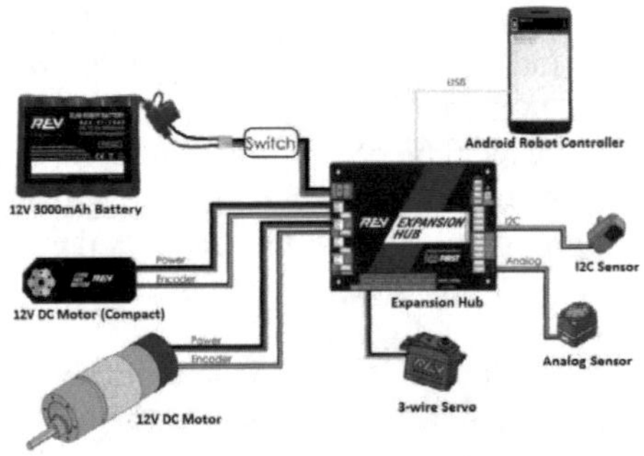

Figure 3. System structure

The robot is mainly composed of mechanical claw gripping mechanism, drive control unit, Mecanum four-wheeled mobile platform, visual unit, sensor unit and robot control app. The robot can shift back and forth. With the mechanical arm of lifting mechanism, it can grasp and release the specified objects with different heights flexibly. At the same time, the robot has accurate positioning capability. Using a gyro sensor and a rotary encoder, a rotational error less than 1 degree and a horizontal movement error less than 1 cm are achieved through the PID control algorithm. The functions of main module are as follows:

3.1.1 Drive control unit: Drive control unit of the robot system consists of two EXPANSION HUBs provided by REV Robotics – a company. EXPANSION HUB is powered by 12Vdc power supply. Two EXPANSION HUBs exchange control data with each other through RS485, and communicate with Xiaomi 4 on the robot side via Mini USB. Each EXPANSON HUB supports 4 motor control interfaces, 4 rotary encoder input interfaces, 6 servo control interfaces, 8 digital input interfaces and 4 IIC bus interfaces.

3.1.2 Mecanum four-wheeled mobile platform: The robot mobile platform adopts four Mecanum wheels. All four wheels are driving wheels and driven by AndyMark's 12V dc motor (NeveRest 40). The maximum RPM is 160 RPM. The motor comes with a rotary encoder for simple closed-loop control.

3.1.3 Mechanical claw grasping mechanism: The robot's grasping mechanism consists of lifting slide rail, lifting plate, lifting arm seat, mechanical claw and upper and lower limit switch. The lifting mechanism is driven by 12V dc motor (NeveRest 60) with maximum speed of 110 RPM. The motor controls uplift and fall of the grasping mechanism by rotating forward and backward, respectively. The mechanical claw is driven by two CYS-BLS9825 servos made by Supersonic, which includes helical gears with a maximum torque of 15kg·cm. and owns the advantages of overcurrent protection and long service life. The upper and lower limit switches are used to limit the extreme position of lifting and falling.

3.1.4 Visual unit: The robot's visual acquisition is obtained through the camera of the Xiaomi 4 mobile phone, which is used to monitor robot handling and identify moving target. The mobile phone placement platform can rotate 180 degrees left and right, and 60 degrees up and down, using motors and servos, respectively. The mobile platform is mounted on the robot frame structure above the lifting slide, which can provide the stereo image and close-up image of the environment and moving objects.

3.1.5 Sensor unit: The rotating center of the robot is equipped with a gyroscope sensor, which is used to precisely control the robot's rotation angular displacement, so that the lifting plate and mechanical claw of the robot can be used for the designated transport.

3.2 System Software

The system software includes the remote App running on the remote mobile phone and the robot control App running on the robot side phone. Both of these apps are programmed with Java and are compiled and debugged in Android Studio 3.01 development environment.

3.2.1 Target recognition training calibration procedure: The target recognition training calibration procedure can automatically establish a quantitative array of the actual distance between the ball and the robot and the cluster radius of the recognized ball. The program controls the mobile phone camera of the robot platform to collect the target image. The main method is to control the mobile platform rotation through accessed image area containing recognized target to locate recognized target at the center of the image area (Figure 4), calculate the radius of the cluster and the actual distance between the robot and the ball at this time, and record the result in the corresponding global array variable as the basis for automatic program execution. The actual distance between the robot and ball can be achieved by factors of robots and the initial position of ball, the rotary encoder values from the drive motors of robots Mecanum wheels and Mecanum wheel radius. In the execution of the identification training, the initial position of the robot and the ball is determined as 5m according to site conditions, and a set of calibration data is determined every 0.2m. The flow chart of calibration program for the target identification training is shown in Figure 5.

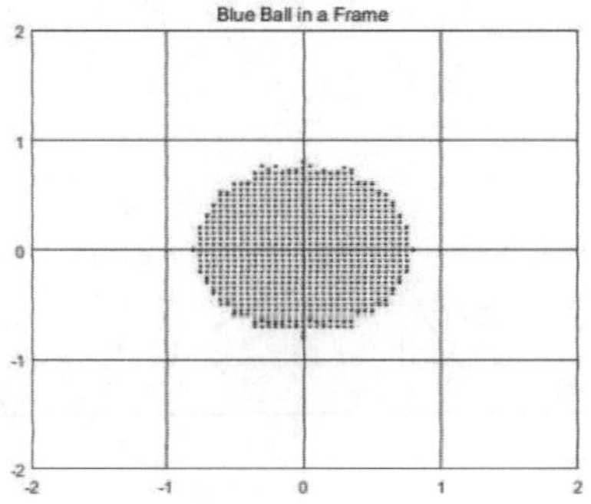

Figure 4. Target in the center of the image area.

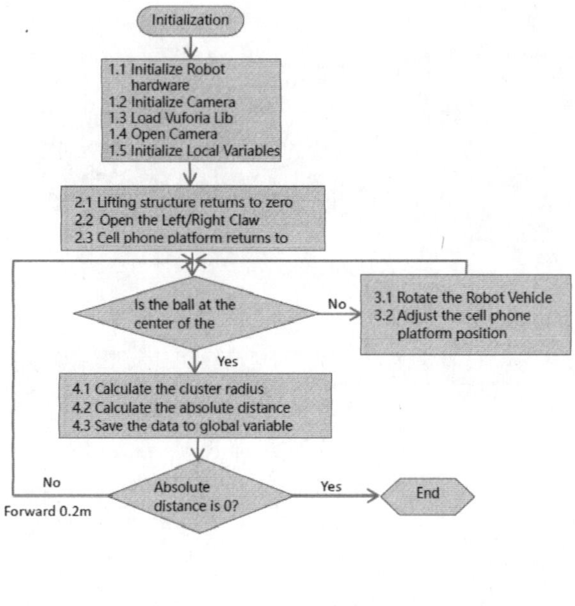

Figure 5. Flow chart of calibration program

3.2.2 Automatic operation program: The robot can complete scheduled action automatically according to the program set in figure 6.

3.2.3 Manual procedure: The current popular augmented reality platform Vuforia 7 is directly used when considering the camera monitoring function in the process of manual programming. Manual control of robot movement can be realized when necessary.

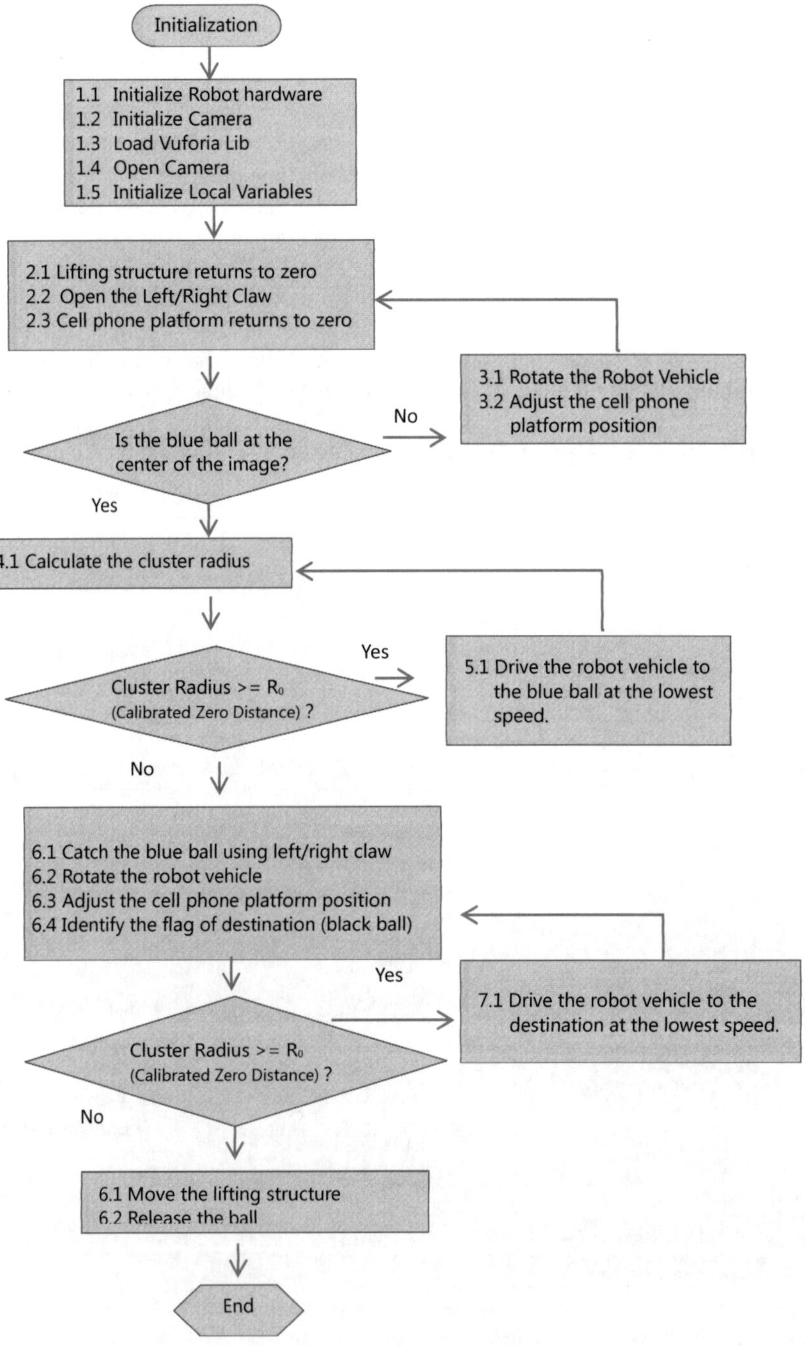

Figure 6. Flow chart of automatic mode

IOP Conf. Series: Materials Science and Engineering **435** (2018) 012020 doi:10.1088/1757-899X/435/1/012020

3.2.4 Universal driver: Universal driver includes hardware initialization modules and Mecanum wheel control modules. When performing manual program, the operator sets platform forward (backward) speed, to the right (left) speed, rotating angular velocity through game handle, calculates driver voltage of the dc motor used in the four-wheel driving platform by calling the universal driven Mecanum wheel algorithm, and controls the corresponding motor angular velocity by voltage, as shown in Figure 7.

3.2.5 Robot control App: The robot side mobile phone establishes wireless communication with remote control phone through Wifi Direct connection. The robot control App mainly supports the following functions: receiving the command from the remote control mobile phone, sending the robot system running status and monitoring video data to the remote control mobile phone.

Remote App obtains robot system running status information, gives instructions to the robot to perform manual program, automatic program or target recognition training calibration procedure, sends game handle code and acquires mobile phone camera surveillance video data on the robot side.

Figure 8 shows the completed robot system. After testing, the robot is working properly to achieve the desired goal.

```
fl_value = (forward + clockwise + right);//left front wheel
fr_value = (forward - clockwise - right);//right front wheel
rl_value = (forward + clockwise - right);//left rear wheel
rr_value = (forward - clockwise + right);//right rear wheel
max_value = getMaxValue(fl_value,fr_value,rl_value,rr_value);
if (max_value > 1.0 ) {
    fl_value = fl_value / max_value;
    fr_value = fr_value / max_value;
    rl_value = rl_value / max_value;
    rr_value = rr_value / max_value;
}
```

Figure 7. Output voltage of each drive motor

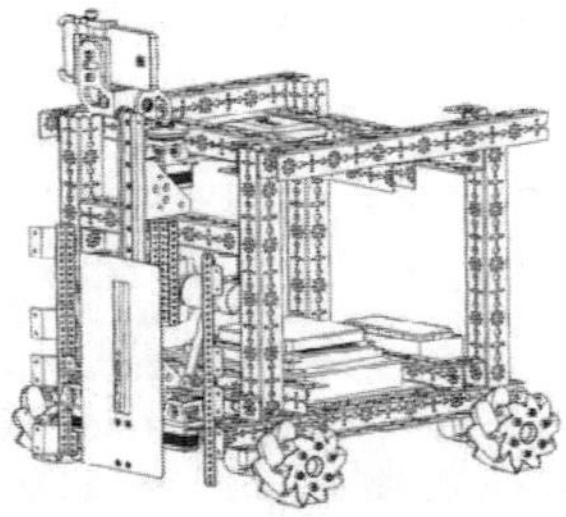

Figure 8.Outline of robot system Conclusion

With the continuous increase of labor costs in modern society, there are more and more replacement of people with robots to reduce costs and increase production efficiency. The recognition-based robot proposed in this paper can achieve the specified path planning and object moving. Future work can consider more complex target recognition algorithms to achieve more complex applications such as fruit picking and qualified product classification, so as to truly realize the replacement from human to machine to complete general repetitive tasks, thus allowing more people to engage in creative work.

References

[1] Prajin Palungsuntikul and Wichian Premchaiswadi. Object Detection and Keep on a Mobile Robot by Using a Low Cost Embedded Color Vision System, 2010 Eighth International Conference on ICT and Knowledge Engineering,pp.70-76

[2] Guihua Xia, Zhuoyi Xing. A New Algorithm for Target Recognition and Tracking for Robot Vision System.2007 IEEE International Conference on Control and Automation ThA9-1,Guangzhou, CHINA,2007,pp.1004-1008

[3] Jianran Liu , Zaojun Fang, Kun Zhang, and Min Tan. Improved high-speed Vision System for Table Tennis Robot. Proceedings of 2014 IEEE International Conference on Mechatronics and Automation, August, China,pp.652-657

[4] Guo Hui, Yu Qiushui, Fan Min. A Study of Camera Calibration in the Vision System of Soccer Robots. 2014 IEEE Workshop on Advanced Research and Technology in Industry Applications (WARTIA),pp.866-869

[5] Tyryshkin A.V. and Kuznetsov A.I. The Technical Vision System for Diagnosis of the Harvesting Unit of the Robot for Gathering Wild Plants. CSIT 2017, September, 2017, Lviv, Ukraine, pp.405-408

IOP Conf. Series: Materials Science and Engineering **435** (2018) 012021 doi:10.1088/1757-899X/435/1/012021

Design and Simulation of Trajectory Tracking Guidance Law Based on LQR for Target Missile

X Zheng[1,*], S C Yang[1] and K Q Zhang[1]

[1]Army Engineering University, No.97 Heping West Road, Shijiazhuang, Hebei, 050003, China

*Corresponding e-mail: 810935784@qq.com

Abstract. Aiming at the trajectory tracking problem of the target projectile, a ballistic tracking guidance law of the target projectile is designed based on the linear quadratic regulator (LQR) theory. Using time as an independent variable, the motion equation of the center of mass of the target bomb was linearized using the small disturbance method to establish the center of mass motion of the Space expressions; The LQR theory was used to design the trajectory tracking guidance law of the target projectile. Finally, the semi-physical simulation test was conducted to verify the trajectory tracking guidance law of the designed target projectile.

1. Introduction

The target missile is a commonly used target for evaluating the performance of air defense weapons in the process of development, type setting, batch sampling and identification. It is used to simulate the external dimensions, ballistic characteristics, and confrontation characteristics of the incoming missile target [1,2].

The accuracy of the target's ballistic tracking program is one of the important indexes that affect the target's elastic energy. During the actual flight of the target bomb, the deviation of the actual flight trajectory of the target missile from the project trajectory will be greatly affected by factors such as aerodynamic deviation, position deviation of the start and control, and wind. Therefore, it is necessary to design an appropriate trajectory tracking guidance law to ensure its accuracy. Trajectory flight along the program can still be performed with high accuracy under various disturbance conditions [3]. The early program trajectory is mainly achieved through program control, and its control accuracy is greatly affected by disturbance factors.

In the design of today's target guidance control method, the control method of position control and feedback of position deviation is often used [4,5]. With the development and application of modern control theory such as synovial control, optimal control, and intelligent control, many new methods have emerged for the trajectory tracking of guided weapons. Literature [3] applied the method of dynamic inversion and synovial control in the ballistic tracking of missiles. Tang Shantong designed a new air defense missile trajectory tracking guidance law based on Lyapunov stability principle in Literature [6]. Greg Dukeman introduced the LQR theory in the design of ballistic tracking control methods for reentry vehicles, effectively improving the ballistic tracking accuracy of reentry vehicle programs [7]. In this paper, a certain type of rocket missile reconstructed target missile is taken as the research object. The LQR theory is used to design the trajectory tracking guidance law of this project. The designed guidance law is verified by semi-physical simulation.

2. Target Centroid Movement Model

In the design of the trajectory tracking guidance law of the target project, the main consideration is to change the spatial position of its center of mass, and the rotation process of the target around the center of mass can be ignored. The target bomb is regarded as a manipulable particle during the entire flight. With the assumption of instantaneous equilibrium, it is assumed that the target control system is an ideal control system with no error and no time delay. Ignoring the lateral plane motion of the target bomb, the equation of motion for the center of mass of the target bomb in the longitudinal plane is [8]:

$$\begin{cases} m\dfrac{\mathrm{d}V}{\mathrm{d}t} = -X - mg\sin\theta \\ mV\dfrac{\mathrm{d}\theta}{\mathrm{d}t} = Y - mg\cos\theta \\ \dfrac{\mathrm{d}x}{\mathrm{d}t} = V\cos\theta \\ \dfrac{\mathrm{d}y}{dt} = V\sin\theta \\ \delta_z = -\dfrac{m_z^{\alpha}\alpha}{m_z^{\delta_z}} \\ m = m_0 \\ \varepsilon = 0 \end{cases} \tag{1}$$

Among them, m is the target projectile mass; V is the target projectile velocity; α is the angle of attack; θ is the ballistic inclination; X and Y is the air resistance and lift force.

The calculation formula for the lift and resistance of the target is:

$$\begin{cases} X = C_x qS = \dfrac{1}{2}\rho V^2 SC_x \\ Y = C_y qS = \dfrac{1}{2}\rho V^2 SC_y \end{cases} \tag{2}$$

In the formula, ρ is the air density; S is the reference area of the airfoil; C_x and C_y is the drag coefficient and the lift coefficient. Both the drag coefficient and the lift coefficient are functions of the angle of attack and the Mach number. The corresponding aerodynamic parameters can be obtained from the wind tunnel test.

The approximate formula for air density and gravitational acceleration:

$$\begin{cases} \rho = \rho_0 \exp(-\dfrac{h}{h_p}) \\ g = g_0 \dfrac{R_e^2}{(R_e + h)^2} \end{cases} \tag{3}$$

Among them, ρ_0 is the air density at the surface; g_0 is the acceleration of gravity at sea level; R_e is the average radius of the earth.

3. Linear Quadratic Regulator (LQR) Theory

For a linear time-varying system with the following formal state space equations:

$$\dot{x} = A(t)x + B(t)u \tag{4}$$

Where, x is the state variables, u is the control vectors, $A(t)$ and $B(t)$ are $n \times n$ matrices that satisfy unique conditions for solutions.

Given quadratic performance index functions for state variables and control variables:

$$J = \frac{1}{2}x^T(t_f)Fx(t_f) + \frac{1}{2}\int_{t_0}^{t_f}[x^T Q(t)x + u^T R(t)u]dt \tag{5}$$

According to the LQR theory, there exists a comprehensive optimal control variable, and the system state is adjusted from the initial state to the equilibrium state while ensuring that the above performance index takes a minimum value.

The optimal control volume calculation method is:

$$\begin{cases} u^* = -K(t)x \\ K(t) = -R^{-1}(t)B^T(t)P(t) \end{cases} \tag{6}$$

Among them, $P(t)$ is a $n \times n$ positive semidefinite matrix satisfies the following matrix Riccati differential equation:

$$\begin{aligned} &\dot{P}(t) + P(t)A(t) + A^T(t)P(t) - \\ &P(t)B(t)R^{-1}(t)B^T(t)P(t) + Q(t) = 0 \end{aligned} \tag{7}$$

4. LQR program ballistic tracking guidance law design

4.1 Linearization of the motion equation of the center of mass of the target bomb

Using the small perturbation assumption, linearize the equation of motion of the center of mass in the longitudinal plane of the target bomb. After linearization, the mass disturbance perturbation equations of motion are:

$$\begin{cases} \dfrac{\mathrm{d}\Delta V}{\mathrm{d}t} = -\dfrac{X^V}{m}\Delta V - \dfrac{X^\alpha}{m}\Delta\alpha - g\cos\theta\Delta\theta - \\ \qquad \dfrac{X^y + G^y}{m}\Delta y \\ \dfrac{\mathrm{d}\Delta\theta}{\mathrm{d}t} = \dfrac{Y^V}{mV}\Delta V + \dfrac{Y^\alpha}{mV}\Delta\alpha + \dfrac{g\sin\theta}{V}\Delta\theta + \\ \qquad \dfrac{Y^y - G^y\cos\theta}{mV}\Delta y + \dfrac{Y^{\delta_z}}{mV}\Delta\delta_z \\ \dfrac{\mathrm{d}\Delta x}{\mathrm{d}t} = \cos\theta\Delta V - V\sin\theta\Delta\theta \\ \dfrac{\mathrm{d}\Delta y}{\mathrm{d}t} = \sin\theta\Delta V + V\cos\theta\Delta\theta \end{cases} \tag{8}$$

Where, X^V, X^α and X^y are the partial derivative values of the undisturbed movement resistance at the corresponding moments for speed, angle of attack, and the target missile's flight altitude. Y^V, Y^α

and Y^y are the partial derivative values of the undisturbed movement lift for speed, angle of attack, and altitude at the corresponding time.

$$\begin{cases} Y^{\alpha} = \dfrac{\rho V^2 S C_y^{\alpha}}{2} \\ Y^{\delta_z} = \dfrac{\rho V^2 S C_y^{\delta_z}}{2} \\ Y^{V} = \dfrac{1}{2c}\rho V^2 S\alpha \dfrac{\partial C_y^{\alpha}}{\partial Ma} + \rho V S C_y \end{cases} \tag{9}$$

Approximate calculation formula of gravitational acceleration can obtain the partial derivative of gravity to height in the course of target missile flight:

$$G^y = -2mg_0 \frac{R_e^2}{(R_e + y)^3} \tag{10}$$

The partial lift of the target missile to the height of flight:

$$Y^y = -\frac{1}{2}V^2 S C_y \frac{\rho_0}{h_p}\exp(-\frac{y}{h_p}) \tag{11}$$

In the ballistic coordinate system, the longitudinal overload of the target during flight:

$$\Delta n_y = \frac{Y^{\alpha}\Delta\alpha + Y^{\delta_z}\Delta\delta_z}{G} \tag{12}$$

Since the target's velocity characteristics have been designed during the ballistic design phase of the target project, the speed of the target projectile is almost uncontrollable after it is launched at a certain angle. When designing the trajectory tracking guidance law of the target bomb, the control of the flying speed is not considered, so the equation concerning the speed can be saved.

Substituting equation (15) into equation (9), taking the disturbance state variables:

$$x = [\Delta\theta \quad \Delta y] \tag{13}$$

Control variables: $u = \Delta n_y$, The linearized state equation of the center of mass movement in the longitudinal plane of the target can be obtained:

$$\dot{x} = A(t)x + B(t)u \tag{14}$$

Where:

$$A(t) = \begin{bmatrix} \dfrac{g\sin\theta}{V} & \dfrac{Y^y - G^y\cos\theta}{mV} \\ V\cos\theta & 0 \end{bmatrix} \tag{15}$$

$$B(t) = \begin{bmatrix} \dfrac{g}{V} \\ 0 \end{bmatrix} \tag{16}$$

4.2. Determining Weight Matrix

In practical applications, the weight matrix Q, R is generally selected as a diagonal matrix. Therefore, the matrix can be written as:

$$Q=\begin{bmatrix} Q_1 & 0 \\ 0 & Q_2 \end{bmatrix} \tag{17}$$

$$R=\left[R_1\right] \tag{18}$$

The corresponding quadratic performance index can be written as:

$$J=\frac{1}{2}\int_{t_0}^{t_f}[Q_1(\Delta\theta)^2+Q_2(\Delta y)^2+R_1(\Delta n_y)^2]dt \tag{19}$$

In order to simplify the calculation, it is also possible to ensure that the actual flight trajectory height y of the target missile tracks the program ballistic altitude in real time. According to the Bryson principle, we have:

$$Q_2(\Delta y_{max})^2=R_1(\Delta u_{max})^2 \tag{20}$$

Where, the subscript "max" represents the maximum allowable deviation of the corresponding variable. Since the selection of any fixed-weighted weight matrix does not affect the final result,so can let $Q_2=1$,then:

$$R_1=\frac{(\Delta y_{max})^2}{(\Delta u_{max})^2} \tag{21}$$

4.3. Design of ballistic tracking guidance law for target project

During the process of the ballistic flight of the target projectile, the dynamic characteristics of the target projectile with the flight altitude, speed, and other conditions vary greatly, and it is difficult to accurately calculate the dynamic characteristic parameters at any time during the flight process. In order to simplify the calculation considering the linearization accuracy of the target's center-of-mass equation of motion, several trajectory feature points that can fully represent the full trajectory dynamics are selected in the trajectory of the target projectile, and the linearized state equation of the target missile at the trajectory feature point is calculated.

The disturbance motion of the target centroid near the ballistic feature point can be approximated as a linear time-invariant system. Its state equation is:

$$\dot{x}=A_0x+B_0u \tag{22}$$

Among them, A_0, B_0 is determined by the dynamic coefficient of the target bomb at the feature point of the undisturbed motion trajectory.

According to the infinite-time LQR theory, for the linear time-invariant system shown in Equation (22), in Equation (5) $t_f\to\infty$, the terminal constraint $x^T(t_f)Fx(t_f)/2$ loses its meaning. Therefore, the quadratic performance index function can be simplified as:

$$J=\frac{1}{2}\int_{t_0}^{\infty}[x^TQ(t)x+u^TR(t)u]dt \tag{23}$$

From the weight matrix determined by 3.2, the optimal state feedback gain K at each trajectory feature point is calculated according to Equations (7) and (8).

The feature points on the program trajectory and the optimal state feedback gain K are bound to the on board computer. During the flight of the target bomb, the integrated state navigation system is used to acquire the state information such as the position and velocity of the target bo555mb, and the real-time state deviation of the actual flight trajectory and the project trajectory of the target bomb is obtained:

$$\begin{cases} \Delta\theta = \theta - \theta_{bz} \\ \Delta y = y - y_{bz} \end{cases} \tag{24}$$

Using flight time as an independent variable, the optimal state feedback gain at this moment is obtained from the computer on board by interpolation method. The corresponding optimal feedback guidance law is:

$$u^* = K[\Delta\theta \quad \Delta y]^T \tag{25}$$

5. Simulation

In the simulation process, the numerical integration method of the fourth-order Runge-Kutta was used to solve differential equations. The ballistic solution step length was chosen to be 0.01 s, the target missile was launched with the initial emission angle $\theta_0 = 60°$, and the engine thrust was 6×10^4 N, work Time $t = 6.4$ s, 10 s after the target launches, the corresponding control law is applied. Before the control law is inserted, only the target missile attitude is controlled stably. The range of target height for target bombs is 2.5~20 km in the descending phase, and the target height error should not exceed 100 m.

Under the non-interference conditions, the results of semi-physical simulation are shown in the figure 1 to figure 3. It can be seen from figure 1 that the LQR scheme's trajectory tracking guidance law and classical control method can effectively track the trajectory of the program under no-jamming conditions. From figure 2 and figure 3, it can be seen that the trajectory accuracy of the classic control method tracking scheme is greatly influenced by the aerodynamic parameter change during the flight of the target bomb. The maximum height deviation is up to 120 m, the maximum ballistic angle deviation is -0.803°, and the target segment height is The deviation range is -52.8 to 80.5 m, and the deviation range of ballistic inclination is -0.803 to -0.115°. The LQR guidance law keeps the trajectory height and trajectory inclination error of the tracking program within a relatively small range during the entire flight of the target bomb. The maximum height deviation is 17.5 m, for the target segment height deviation range -1.7 to 12.5 m, ballistic inclination deviation range -0.069 to -0.057°. Therefore, LQR guidance law, which is available under no interference conditions, provides higher accuracy than the classical control method.

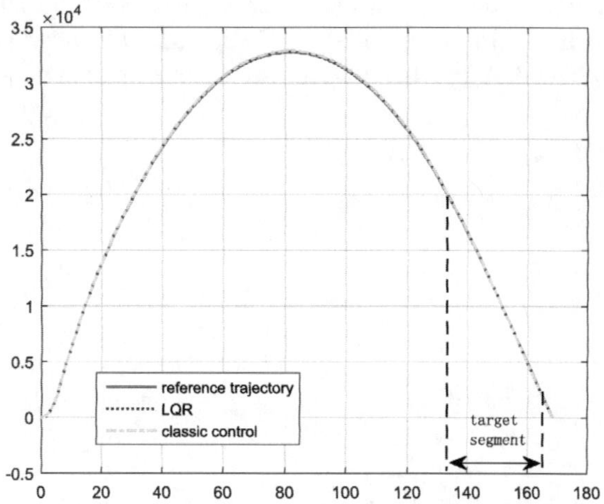

Figure 1. Interference free conditional time-height curve

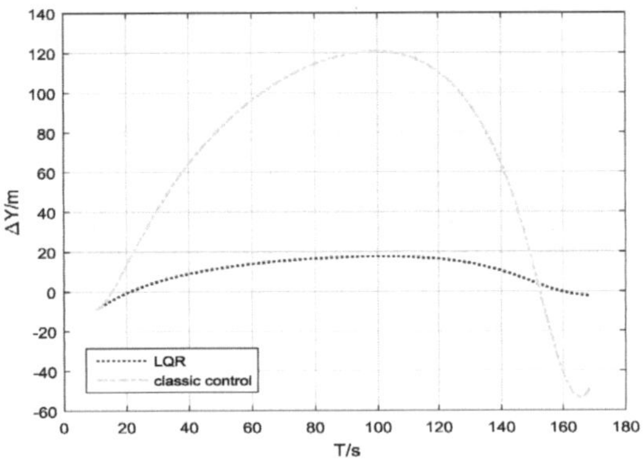

Figure 2. Interference free conditional time-height deviation curve

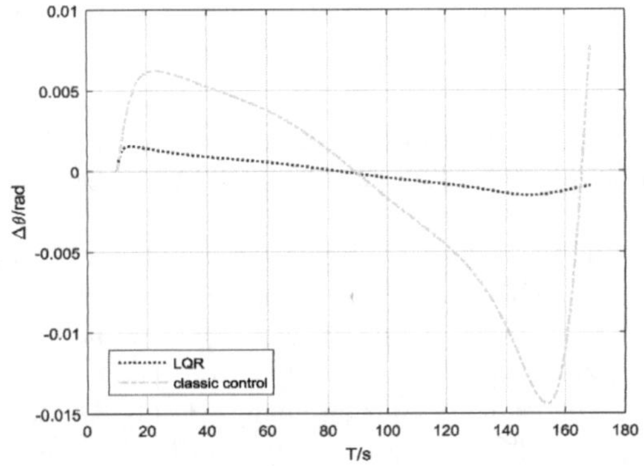

Figure 3. Interference free conditional time-ballistic inclination deviation curve

The results of the lower physical simulation with the starting and controlling position deviation and the 10% aerodynamic deviation are shown in figure 4 to figure 6. It can be seen from figure 4 that under the disturbance conditions, the trajectory of the classical control method tracking scheme produces large errors, and the accuracy of the LQR guidance law tracking scheme is higher than the classical control method. Figure 5 and figure 6 show that the classical control method provides a target range height deviation range of -642.9 to 294.1 m, a ballistic range deviation range of -0.752 to 5.205°, and the height deviation exceeding the allowable target height tolerance range; LQR guidance law The height deviation range of the target segment is -98.5 to 41.78 m, and the deviation range of the ballistic inclination is 0 to 0.434°. Compared with the classical control method, the target accuracy is higher, and the target height error is controlled within the allowable range.

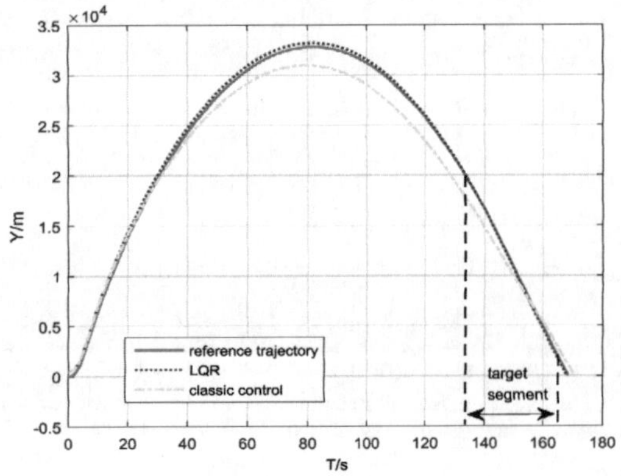

Figure 4. Start to control position deviation conditional time-height curve

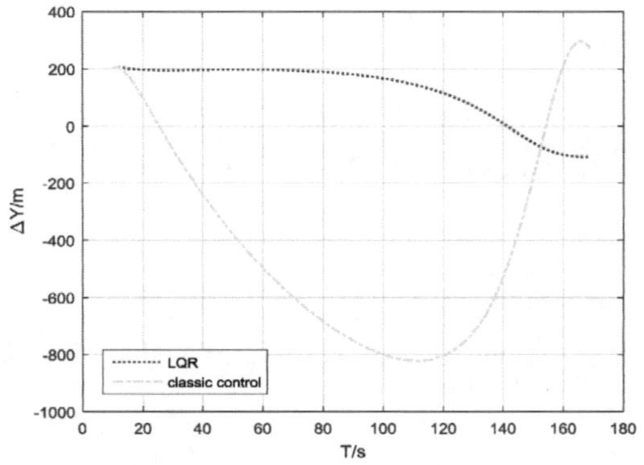

Figure 5. Start to control position deviation time-height curve

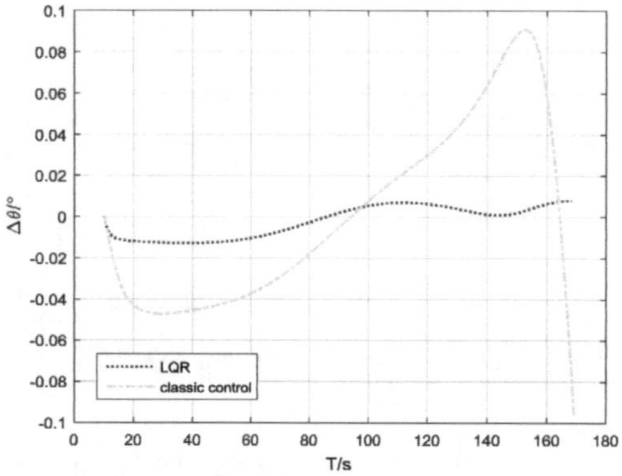

Figure 6. Start to control position deviation time-ballistic inclination deviation curve

6. Conclusion

Targeting high-performance target missiles requires the actual flight trajectory to track the flight trajectory of the program with a high degree of accuracy. After applying the small disturbance method to linearize the linear center of mass motion equation of the target bomb, the LQR theory is used to design the target bomb in the longitudinal plane. The program's trajectory tracking guidance law, and the use of semi-physical simulation method for simulation verification. The simulation results show that the LQR-based trajectory tracking guidance law of the target projectile can effectively track the trajectory of the project, and the performance of suppressing the position deviation of the start-control and aerodynamic parameter deviation is better than that of the classical control method. It provides a target trajectory tracking for the target missile. The control problem provides a new solution to improve the target ballistic accuracy of the target bomb.

References

[1] Liu C J, Zong Y D, Zhang Z H 1997 *Target for Air-defence Weapons* (Beijing: Aviation Industry Press).
[2] Zhou D H, Li P, and Du J 2007 New Characteristic of MissileTarget System Development *Modern Defence Technology*, vol. 35, no. 2, pp. 27-30.
[3] Yang R J, Ye Y, and Yan D H, *et al* 2015 Application of Sliding Mode Control on Trajectory Following for Guided Munitions *Journal of Ballistics*, vol. 27, no. 4, pp. 24-29.
[4] Shen G J, Feng S S, and Cao H S 2015 Research on a Program Angle Optimization Method for Supersonnic Rochet Target *Acta Armamentarii*, vol. 36, no. 4, pp. 644-651.
[5] Yang X C, Sun L, and Lu H L 2009 Design of Multipurpose Target Control System *Journal of Projectiles, Rockets, Missiles and Guidance*, vol. 29, no. 6, pp. 85-89.
[6] Tang S T 2003 A Guidance Law for Air Defence Missile Trajectory Tracking *Modern Defence Technology*, vol. 31, no. 4, pp. 13-16.
[7] Greg A D. 2002 Profile-following enter guidance using linear quadratic regulator theory *AIAA guidance, navigation, and control conference and exhibit*, Monterey, California, August 5-8.
[8] Qian X F and Fang Q 2015 *Missile Flight Mechanics* (Beijing: Beijing Institute of Technology Press).

IOP Conf. Series: Materials Science and Engineering **435** (2018) 012022 doi:10.1088/1757-899X/435/1/012022

Bayesian Network Based Netica for Respiratory Diseases

Chu Jing, Tang Gang*, Li Yong and Hu Xiong

Logistics Engineering College, Shanghai Maritime University, Shanghai 201306, China

*E-mail: 345796099@qq.com

Abstract. The disease of the influenza, whose early symptoms are similar to other diseases, it is very difficult to diagnose a patients. Such as viral colds, bacterial colds and early symptoms of pneumonia. Based on the structural characteristics of Bayesian network in statistical probability, we do a diagnostic analysis of human respiratory diseases, so that doctors can provide basis for diagnosis of patients' diseases. General principle is the use of Bayesian network based on probabilistic reasoning directed acyclic graph model, the complex relationship between variables in the specific problems are expressed in a network structure, and through the network model to reflect the dependencies between the variables in the field of applied research, expression and reasoning on uncertainty knowledge. This is an example of applying the Bias network to the diagnosis of disease.

1. Introduction

Bayesian theory originated in the paper published by Reverend Thomas Bayes "comments on the solution of probability problems". In 1950s, the empirical Bayes method was combined with the classical method with Robbins as the representative. In 1980s, Pearl proposed the Bayesian network, and Bayesian network are applied to the field of artificial intelligence for probabilistic reasoning[1], based on this and Bayesian network successfully applied in the field of expert system, the Bayesian network has become one of the important methods of uncertainty reasoning and expert knowledge. With the continuous deepening of study theory and application level unceasing enhancement, a method of Bayesian network through the graphical representation and the use of probability knowledge, it has overcome the rule system based on many concepts and computational difficulties, it has been widely recognized in academic circles and attention. It mainly has the following characteristics: solid theoretical basis, flexible learning mechanism and strong knowledge expression, and the diversity of reasoning ability and application model [2,3].

First, several concepts and theorems are introduced.

- A priori probability: it refers to the probability obtained according to the previous experience and analysis.
- A posteriori probability: things have happened, and the reason for this thing to happen is the size of the possibility caused by a certain factor.
- Conditional probability: the occurrence probability of event A under the condition that another event B has already occurred, and the conditional probability is expressed as P (A|B).

$$P(A|B)=\frac{P(A\cap B)}{P(B)} \tag{1}$$

- The whole probability formula: if the event is B1, B2,... In the form of a complete set of events and there is a positive probability, the following formula is established for any event A.

$$P(B)=\sum_{i=1}^{n}P(A_i)P(B|A_i) \tag{2}$$

- Bayes formula:

$$P(B_i|A)=\frac{P(AB_i)}{P(A)}=\frac{P(B_i)P(A|B_i)}{\sum_{i=1}^{n}P(B_i)P(A|B_i)} \tag{3}$$

- Bayesian networks: composed of a directed acyclic graph (DAG) and conditional probability table (CPT). The Bias network uses a directed acyclic graph to express the conditional dependence of a set of random variables with them. It is parameterized by conditional probability distribution. Each node is parameterized by P (node|Pa (node)), and Pa (node) represents the parent node in the network. For example, figure 1 is a simple Bayesian network, and its corresponding full probability formula is[3]:

$$P(A, B, C)=P(C|A, B)P(B|A)P(A) \tag{4}$$

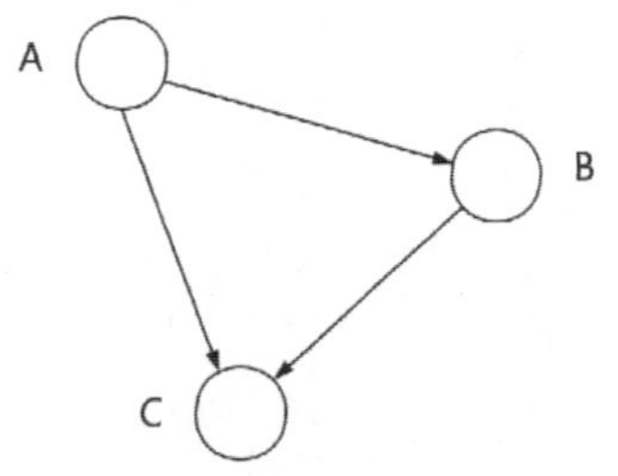

Figure 1. Bayesian network schematic diagram.

The structure of Bias network is divided into two steps: to determine the network topology and to determine the network parameters.

Determine the network topology:

1) Select a set of random variables {X1, X2,... Xn}, which depict the problem.

2) Select a variable sequence a={X1, X2,... Xn};

3) Starting from an empty graph, the variable is added to the network diagram in accordance with the order a.

4) When the variable Xi is added, the variables in the network diagram include X1, X2,..., X (i-1);

4.1) Using background knowledge, select a subset of the C as small as possible in these variables (Xi), makes the assumption that given the subset of C (Xi), Xi and other variables in the network diagram are independent and reasonable.

4.2 from a subset of C (Xi) in each node to add a pointer to a Xi directed edge.

Determining network parameters:

The Bias network parameter is the probability distribution of each variable, which is generally obtained by the training sample statistics.

Considering the basic structure that two variables A and B are indirectly connected by third variables C, the Bayesian network can be divided into 3 substructure forms:

- Head-to-tail:

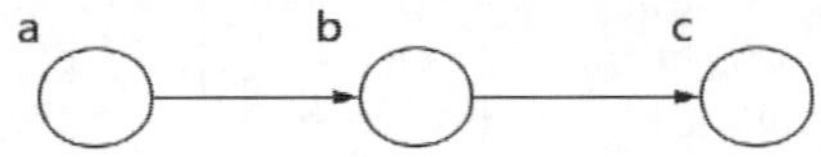

Figure 2. The structure of head-to-tail

$$P(a, b, c)=P(a)P(c|a)P(b|c) \tag{5}$$

As shown in Figure 2. In the case of known C, understanding of a will not affect the reliability of C, and it will not affect the reliability of B. The information channel between a and B is blocked, a and B are independent. But in the case of C, the understanding of a affects the reliability of the C and then affects the reliability of the B.

- Tail-to-tail:

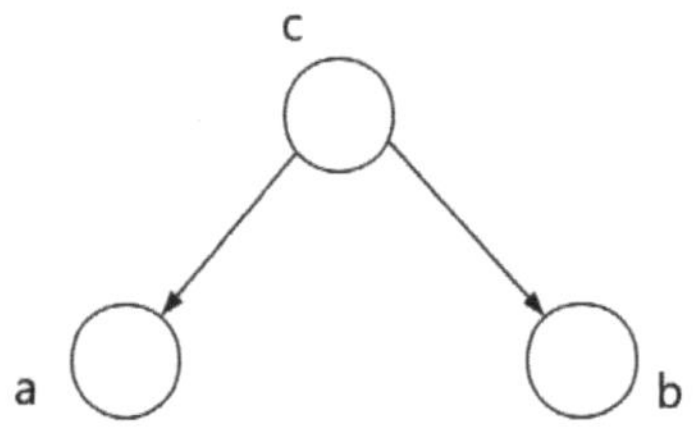

Figure 3. The structure of Tail-to-tail

$$P(a, b, c) = P(c)\,P(a|c)\,P(b|c) \tag{6}$$

As shown in Figure 3. In the case of unknown C, information can be passed between a and B. They are associated. When C is known, information can not be transferred between a and B. They are blocked and blocked. A and B are independent.

- Head-to-head:

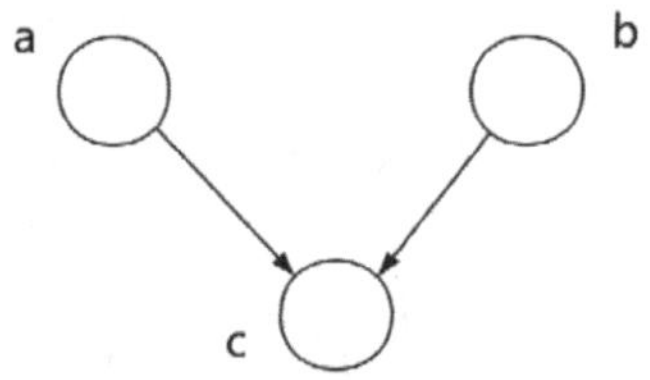

Figure 4. The structure of head-to-head

$$P(a, b, c) = P(a)\,P(b)\,P(c|a, b) \tag{7}$$

As shown in Figure 4. Completely opposite to the two previous cases. In the case of known C, a and B are interrelated. In the case of unknown C, a and B are blocked and are independent[4,5].

2. Preliminary

The clinical symptoms of bacterial and viral respiratory infections are similar. The following are some of the differences between bacterial and viral respiratory infections. 1, the viral respiratory infection has the characteristic of the obvious group disease, but the bacterial respiratory infection is mostly sporadic. 2, viral respiratory infection has no purulent secretion, and pyogenic sputum is an important evidence of bacterial infection. 3. The total number of leukocytes in the virus infected people is low or normal. In the case of bacterial infection, the total number of white blood cells is high. Therefore, it is difficult to determine whether it is a virus or a bacterium if the clinical blood is not tested. At the same time, there are some similarities between the symptoms of pneumonia at the early stage, so now we use Bayesian network to build a network topology based on Netica software to quickly identify the respiratory diseases of patients.

3. Model Architecture

According to the existing data, the incidence of some respiratory diseases and the typical clinical symptoms of these diseases are obtained, such as viral influenza, bacterial colds and pneumonia. Therefore, the Bayesian network is established according to the theoretical knowledge. As in Figure 5, a Bayesian network model like this does not use some of the patient's data and does not reflect the patient's actual situation. Even if there is not enough data, the Bayesian network can play a role, which is the power of it [6].

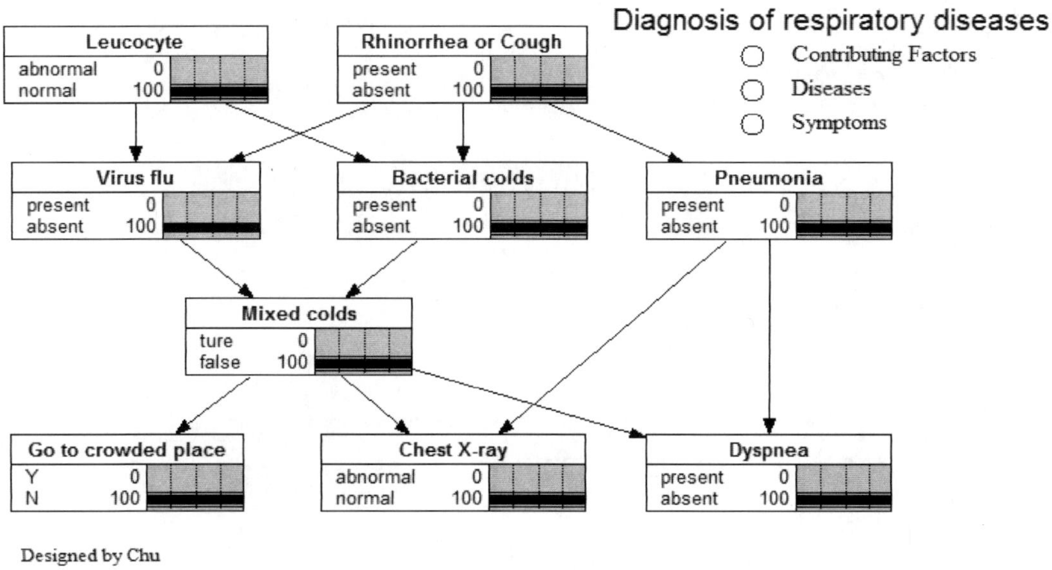

Figure 5. Initial state model

Figure 6 reflects a new patient to seek medical advice from a clinic. We inquired about the patient's condition. We learned that we had been to the crowds recently, and there were some runny nose or cough. Inputting information into the network, the probability of the three diseases increased because of the risk of infection.

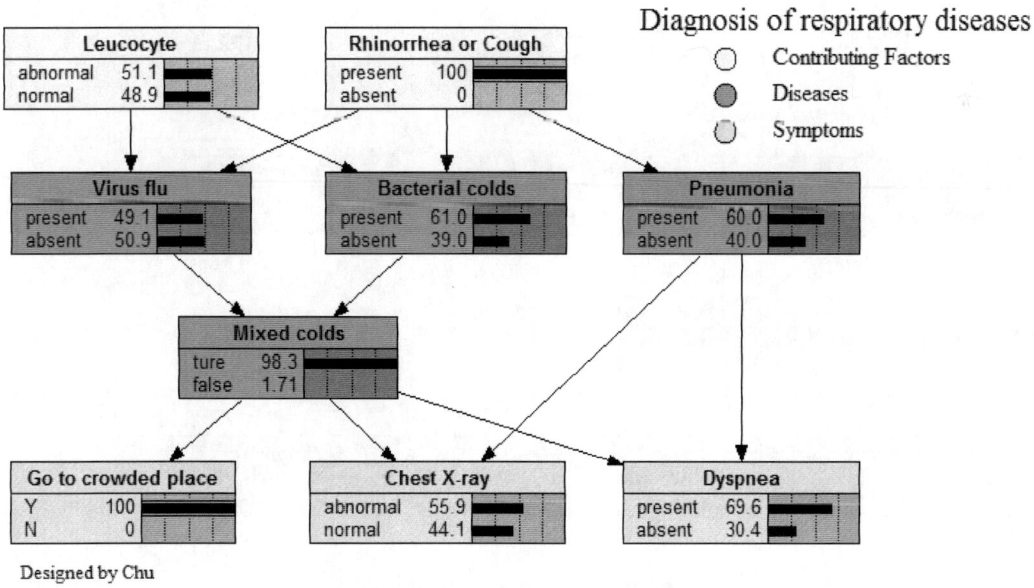

Figure 6. Primary diagnostic model

Look at the detailed inference process: the probability of bacterial colds increased to 61%, and the probability of abnormal breathing and X rays increased correspondingly. At present, the information is too small, and we can not conclude the patient's disease. Continue to ask whether the patient has symptoms of dyspnea. The patient answered "no". The information obtained now affects the Bayesian model, such as Figure 7.

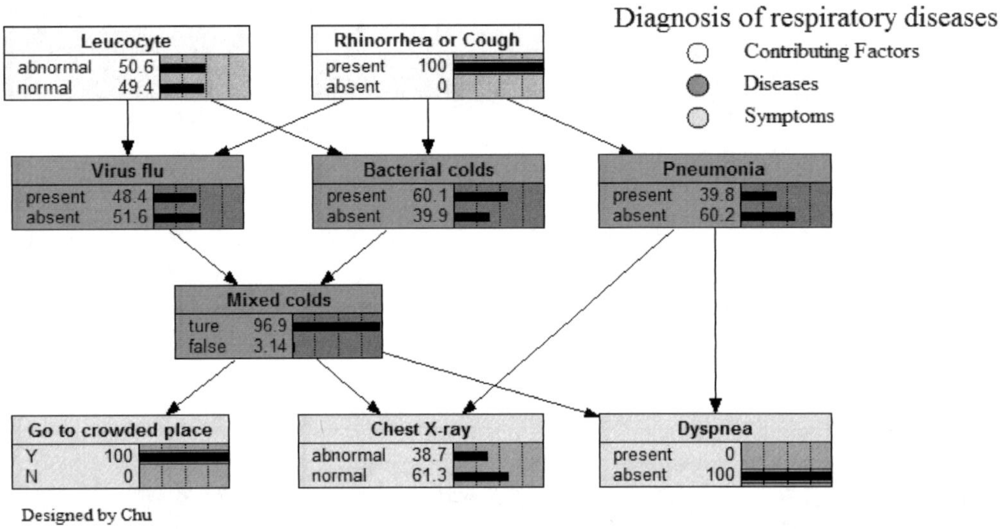

Figure 7. Second diagnosis model

At this time the probability of the three diseases is reduced, especially the pneumonia to 20.2%, which is because the cause of dyspnea is more prone to pneumonia. In order to exclude the possibility of pneumonia, we asked the patient to do a chest X-ray. The results showed normal. The network changed to figure 8. From the chart, we could see that the influence factor of pneumonia dropped to 6.56%. If the patient did not have obvious symptoms, he could basically exclude pneumonia.

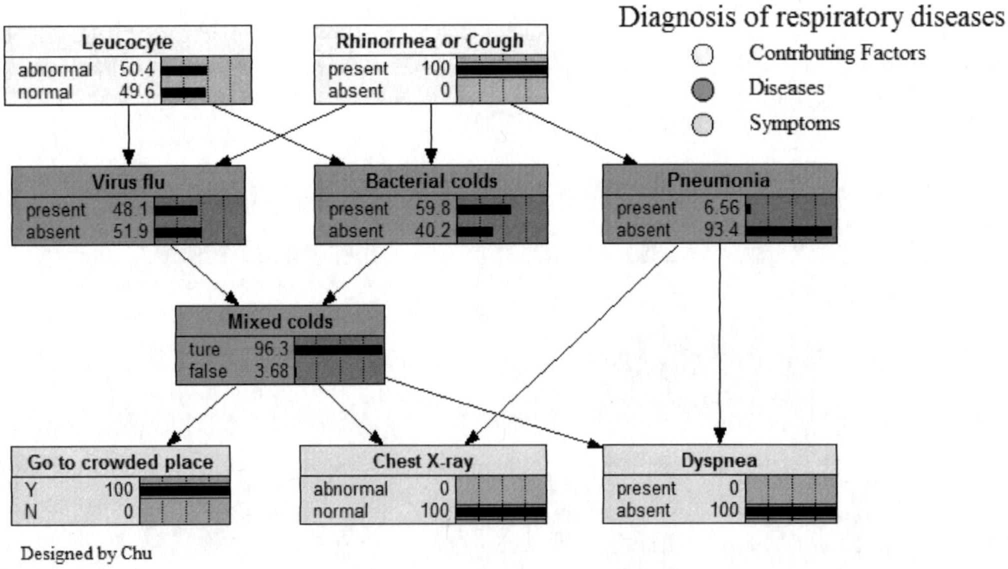

Figure 8. Third diagnosis model

Now we can roughly determine that the patient has a cold, but it is not sure whether it is viral or bacterial. For further confirmation, we asked the patient to test the blood. The results of the blood test showed an increase in the leukocytes of the patient, as shown in Figure 9.

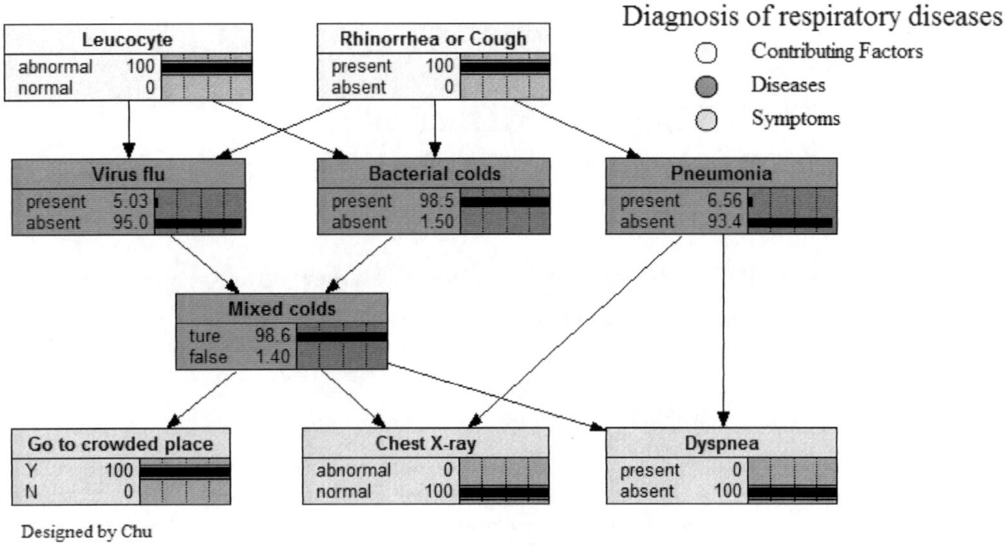

Figure 9. Final diagnosis model

This can basically determine the patient is suffering from bacterial cold, can be the right medicine. If white blood cells are shown to be normal, the probability of a cold is the highest.

4. Conclusions

When we consult the patient information, the probability of the Bayesian network will be automatically adjusted, which is the most perfect and powerful of the Bayesian inference. The most powerful feature of Bayesian network is that the probability obtained from each stage is reflected by mathematics and science. In other words, if we know enough information of patients, we can get statistical knowledge based on these information, and Bayesian network can draw reasonable inference[7]. When the current Bayesian network is unable to give clear results, we can add other inspection items to further test. Even though the current Bayesian network does not include these tests, it is easy to expand, just add extra nodes as the diagnostic program from the new statistical data. We don't have to throw away any part of the past. This is another powerful function of the Bayesian network. They are easy to expand to adapt to changing needs and changes in knowledge.

References

[1] Shuangcheng W 2010 *Bayesian network Learning,Reasoning and Application* (Shanghai: Lixin Accounting press)
[2] Laisheng W 2016 *Bayesian Statistics* (Beijing: Higher Education press)
[3] Haijun L,Dengwu M,Xiao L and Tao J 2009 *The application of Bayesian network theory in fault diagnosis of equipment* (Beijing:National Defend Industry Press)
[4] Laisheng W and Weipin Z 2013 *Bayesian Analysis* (Anhui： University of Science & Technology China press)
[5] Hang L 2012 *Statistical learning method* (Beijing: Tsinghua university press)
[6] Qinkun X,Song G and Xiaoguang G 2007 *Theory and application of dynamic Bayesian Network Reasoning Learning* (Beijing: National Defend Industry Press)
[7] Qinkun X and Song G 2012 *Application of Bayesian Network in Intelligent Information Processing* (Beijing: National Defend Industry Press)

Research on Fault Diagnosis Method of Electric Vehicle Battery System Based on Wavelet-RBF Neural Network

Jing-Bo ZHAO [1], Zhong WANG[1,*], Han-Wen Shen[1] and Peng-Hao Liao[1]

[1]College of Information and Control Engineering, Qingdao University of Technology, Qingdao, 266520, China

[*]1181695718@qq.com

Abstract. Combining the good pattern recognition performance of neural network and the data processing performance of wavelet decomposition, this paper combines wavelet analysis with neural network and uses particle swarm optimization radial basis function neural network to diagnose the fault of electric vehicle battery. In this paper, for large data volume and redundant data, wavelet decomposition is used to process signal processing, including data noise reduction and feature vector extraction. In the fault diagnosis of electric vehicle battery system, a radial basis function RBF neural network based on particle swarm optimization (PSO) optimization is proposed. Finally, the reliability and feasibility of the proposed algorithm are proved by simulation.

1. Introduction

With the continuous enhancement of social environmental awareness and the rapid development of modern battery technology, emerging energy vehicles represented by electric vehicles are booming. As an important part of the electric vehicle system, the battery system plays a decisive role in the normal operation of electric vehicles. Therefore, research on the battery management system of electric vehicles has become a key topic for many research institutes. The monitoring and diagnosis structure of electric vehicle battery systems is as shown in Figure 1.

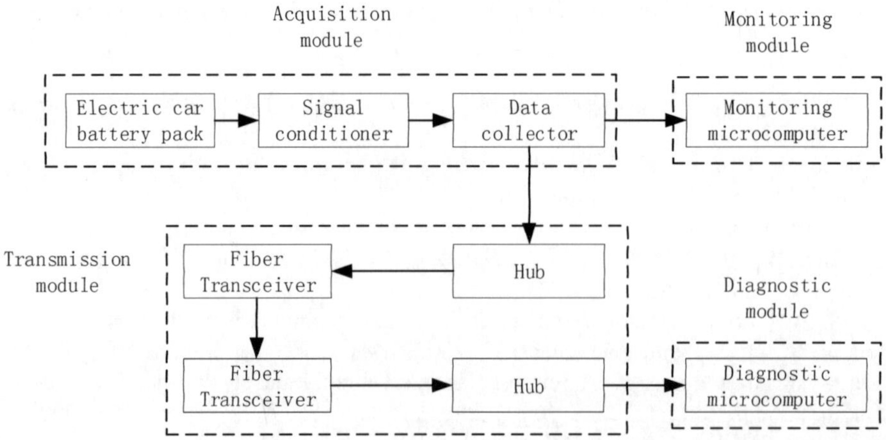

Figure 1. Electric vehicle battery system monitoring and diagnosis structure block diagram

In order to ensure the normal operation of electric vehicles, domestic and foreign experts have used DS evidence theory, Petri net, BP neural network algorithm, SVM fault diagnosis algorithm and other theoretical algorithms for the fault diagnosis of battery management system [1-4], and optimized the fault diagnosis model of electric vehicle battery. Based on the analysis of previous algorithms, this paper proposes a fault diagnosis method based on wavelet-RBF neural network for electric vehicle battery system.

2. Battery pack status signal acquisition

The basis for battery system state analysis and detection is the acquisition of raw data. In this paper, the disturbance signal of the battery pack is collected to analyze the operating state of the battery system. If the measurement system is in disrepair or the maintenance is not timely, the signal accuracy may not be high. In addition, the measurement position and sampling frequency may also affect the signal quality. Battery pack status signal acquisition system is shown in Figure 2.

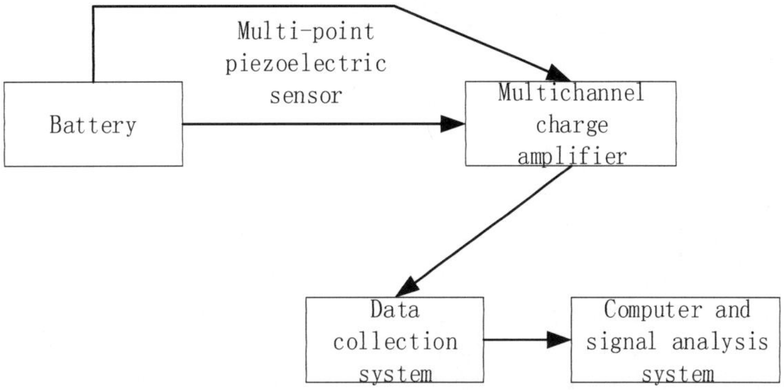

Figure 2. Battery pack status signal acquisition system block diagram

3. Wavelet Denoising of Signals and Extraction of Feature Vectors

3.1. Experimental data denoising

In the experiment, we set the sampling frequency of the battery disturbance signal to be, the number of sample points is, so the frequency resolution of the signal is. As shown in Figure 3 as its time domain diagram, it can be seen from the time domain diagram of the original signal that the signal is subject to severe noise interference. Therefore, noise is eliminated by signal preprocessing, including high-pass filtering and averaging, and db10 compactly supported wavelet base and 4-layer wavelet are selected for noise reduction.

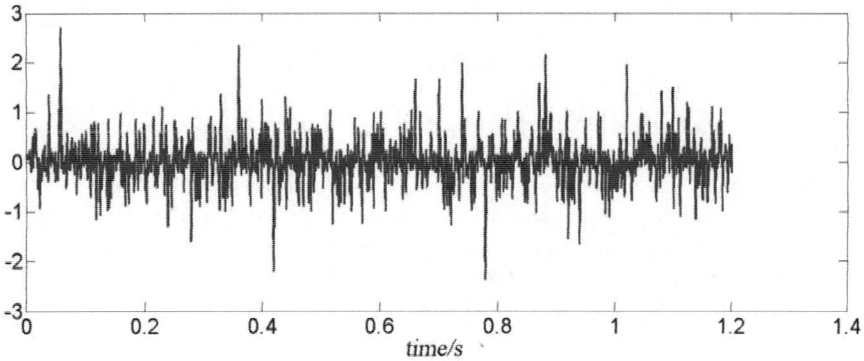

Figure 3. The original vibration signal

In matlab, the signal is decomposed by wavelet, and the four layers of wavelet coefficients are respectively decomposed into a layer and the wavelet is approximated to a4. As shown in Figure 4, the decomposition signals of each layer are shown.

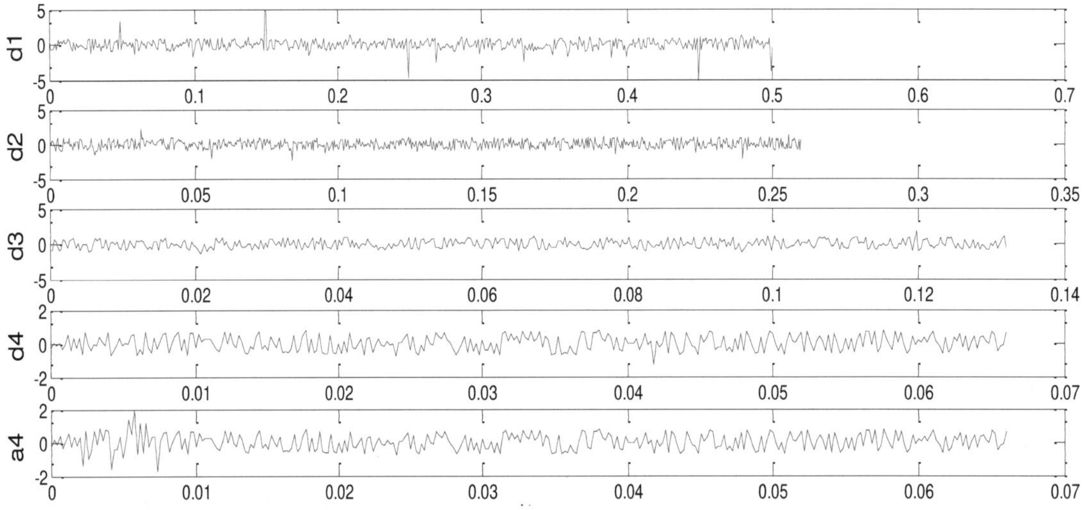

Figure 4. Wavelet denoising of battery pack disturbance signal

When the data is denoised, the noise reduction threshold is obtained for each layer of the wavelet signal according to the default threshold algorithm, thereby obtaining the corrected noise reduction threshold, and finally, the corrected threshold is substituted into the wavelet coefficient, and the reconstruction is performed. The signal is already denoised, as shown in Figure 5.

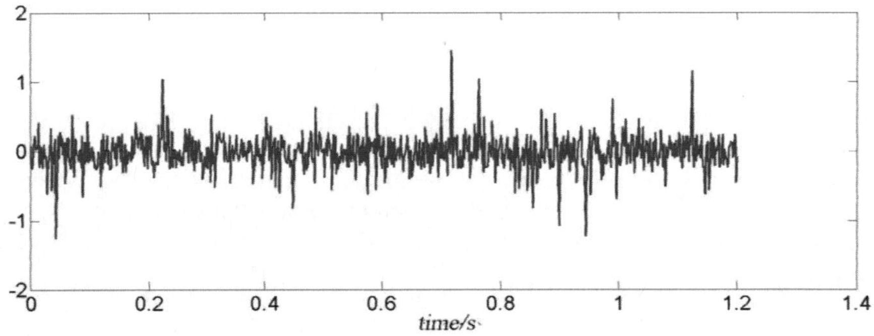

Figure 5. Signal after noise reduction

3.2. Time domain fault feature extraction

For signal analysis, time domain waveforms are usually used when the requirements are not high, because the time domain waveform can visually display the characteristics of the signal, which is convenient for the observer to analyze, especially the disturbance waveform when the fault occurs, and different faults. The resulting waveform has a different shape, such as when the external vibration or voltage is unstable, an impact signal is generated in the waveform.

The range parameter values include dimension range parameters and dimensionless range parameters. The former includes peak value, root mean square value, and kurtosis. The latter includes waveform indicators and pulse indicators. The time domain analysis is based on these range parameters, and they perform certain mathematical processing to analyze the data [5].

3.3. Frequency domain fault feature extraction

Although it is more intuitive to observe the operating state of the device with the signal, the natural frequency of the device is usually affected when the fault occurs, so the dynamics of the device can be observed through frequency domain analysis. The failure of the battery pack usually causes a change in the frequency component of the system signal, so frequency domain analysis is indispensable in the fault diagnosis of the battery system. There are many methods for frequency domain analysis, and the common methods used in engineering are detailed spectrum analysis and power spectrum analysis [6].

The random distribution of signal energy can be reflected by the signal power spectrum. The center position of the power spectrum changes with the energy ratio of the frequency components in the signal. Moreover, the power spectrum energy distribution will be more as the signal frequency component increases. concentrated. That is to say, the change of the position of the center of gravity of the spectrum and the degree of dispersion of the spectral energy distribution can be used as the basis for analyzing the signal products and features.

4. Fault Diagnosis of RBF Neural Network Based on PSO Optimization

The general idea of neural network applied to fault diagnosis is to obtain the process parameters of the device under set fault and no fault, respectively, and normalize it into network input; then use the known data to train the weight parameters of the neural network [7].The next step is the performance test of the neural network. The input symptom vector is output, and the output data is processed to obtain the fault result. Figure 6 is a troubleshooting flowchart.

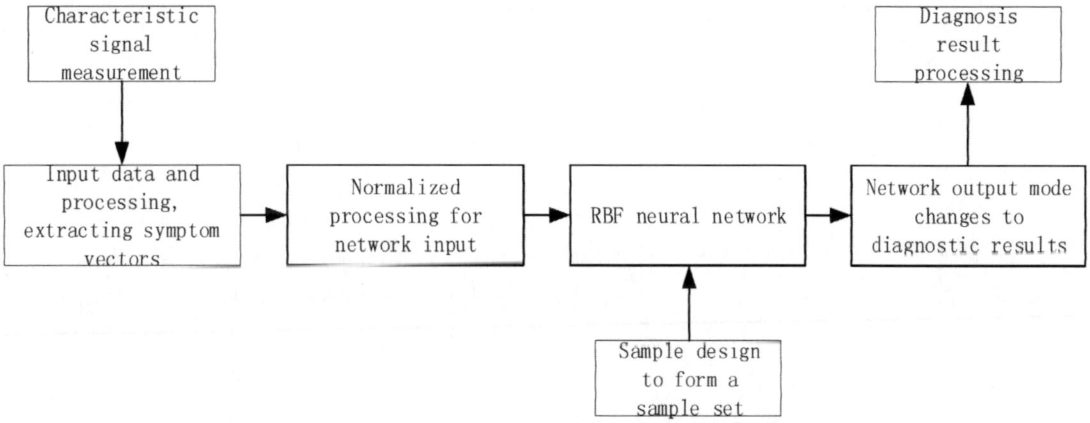

Figure 6. Neural network troubleshooting flowchart

4.1. The coding and fitness function of PSO optimized neural network algorithm

In the PSO algorithm, the particles are in one-to-one correspondence with the feasible solutions. Therefore, the center value and width of the basis function and the fitness of the particle velocity should be included in the particle coding. Suppose there is a center, each center is dimensioned, the position of the particle is dimension, and the corresponding particle velocity is also dimensioned, plus a fitness. The coding structure of the particles is as follows:

$$\begin{matrix} Z_{11}Z_{12}\cdots Z_{lk}\sigma_i\cdots Z_{21}Z_{22}\cdots \\ Z_{1k}\sigma_1\cdots Z_{m1}Z_{m2}\cdots Z_{mk}\sigma_m\cdots \end{matrix} \quad V_1V_2\cdots V_{m\times(k+1)} \tag{1}$$

The training of neural networks is mainly to minimize the parameter errors in the network, so the fitness function selects the average squared error. Then the fitness of the first individual is:

$$f_i = R_i$$
$$R_i = \frac{1}{N}\sum_{k=1}^{N}\left(y_k - \hat{y}_k\right)^2 \quad (2)$$

4.2. Determination of neural network sample set

In order to facilitate the observer to understand the fault of the equipment, we divide the fault of the equipment into four levels, namely 0, 1, 2, and 3, and the fault degree from 0 to 3 is getting higher and higher, 0 The level is no fault and the third level is a serious fault. Then the network output is set to the following:

If $0.85<S<1.50$,then $S=3$ (serious failure);
If $0.50<S<0.85$,then $S=2$ (medium failure);
If $0.15<S<0.50$,then $S=1$ (slight failure);
If $S<0.15$ or $S>1.50$,then $S=0$ (no fault);

4.3. Comparison of diagnostic results of neural networks

In order to reflect the superiority of neural network based on particle swarm optimization, we use the same training sample set to train the original neural network and particle swarm optimization neural network,and input the test data into two neural networks respectively. comparing. In the particle swarm optimization neural network, the number of particles is set to 40, and the maximum number of iteration steps is 300. The fitness function curve at this time is shown in Figure 7.

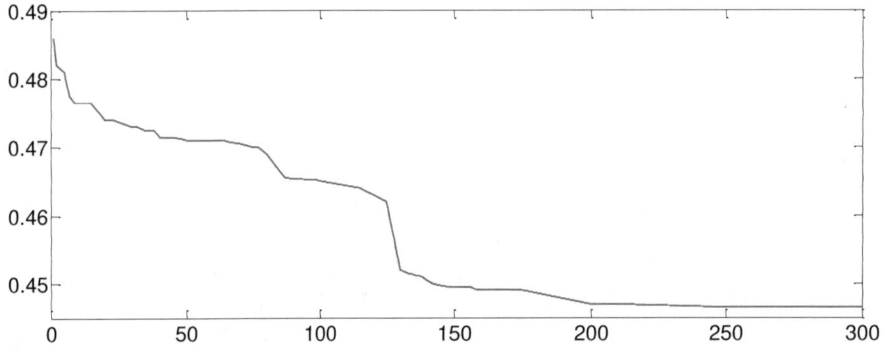

Figure 7. The curve of optimal fitness function

It can be seen from the diagnostic simulation that both neural networks have achieved rapid training and accurate diagnosis of the fault. However, comparing the two neural networks, it can be seen that the neural network diagnosis by particle swarm optimization is more accurate. The original neural network will have errors when the diagnostic data is large, such as the fault diagnosis error or fault type of the same fault type. Diagnostic errors, etc.,[8] and the particle network-optimized neural network has greatly improved the accuracy of diagnosis.

5. Conclusion

Through the fault diagnosis analysis of the battery system. Firstly, the battery pack fault diagnosis system is designed. [9] It is a local area network consisting of four parts: acquisition module, transmission module, monitoring module and diagnostic module. Then the wavelet analysis is performed on the collected data to denoise the experimental data, and the collected neural network is optimized by the particle swarm optimization algorithm, and some data is used for performance verification of the neural network. It is concluded that the radial basis function neural network optimized by the particle swarm optimization algorithm shows good performance in fault diagnosis.

IOP Conf. Series: Materials Science and Engineering **435** (2018) 012023 doi:10.1088/1757-899X/435/1/012023

References

[1] Gao Dizhen, Lan Xi, Shen Aidi. A Lithium Battery Fault Diagnosis System Based on Petri Net[J]. Journal of System Simulation, 2018, 30(02): 614-621.

[2] Xia Fei, Ma Wei, Zhang Hao, Peng Daogang, Sun Peng, Luo Zhijiang.Application of Improved DS Evidence Theory in Fault Diagnosis of Lithium Battery for Electric Vehicles[J].Journal of Intelligent Systems,2017,12(04):526- 537.

[3] Meng Qingwu, Zhao Hongwei, Sun Runcheng. Research on Battery Pack Fault Diagnosis Based on BP Neural Network Algorithm[J]. Automation and Instrumentation, 2017(11): 45-47.

[4] Zhang Wei, Guo Wei, Xie Wenlong. Research on Fault Diagnosis of Electric Vehicle Power Battery System[J]. Modern Business and Industry, 2017(01): 179-180.

[5] Meng Xiangmin, Song Ping, Tan Jiwen. Fault Diagnosis of Ball Screw Based on Wavelet Packet and SVM[J]. Machine Tool & Hydraulics, 2014, 42(19): 181-184.

[6] Zhang J, Walter G G, Miao Y B, et al. Wavelet Neural Networks for Function Learning[J].IEEE Trans on Signal Processing, 1995,43(6):1485-1497.

[7] Souza J. C. S.,Meza E. M., Schilling M. T., et al. Alarm processing in electrical power systems through a neuro-fuzzy approach. IEEE Transactions on Power Delivery[J], 2004,19(2):537-544.

[8] Slachevsky A, Villalpando J M, Sarazin M, et al. Frontal assessment battery and differential diagnosis of frontotemporal dementia and Alzheimer disease.[J]. Archives of Neurology, 2004, 61(7):1104.

[9] Choi W, Choi M. Charger having battery diagnosis function and method of driving the same[J]. 2018.

The Parameter Identification of Least Absolute Deviation Based on an Improved Gravitational Search Algorithm

Baochang Xu[1,*], Hua Zhang[1], Likun Yuan[1] and Jinshan Wang[2]

[1]Department of Automation, China University of Petroleum (Beijing), Beijing, 102400, China
[2]PetroChina Tarim Oilfield Company, Korla ,841000, China

[*]E-mail:xbcyl@163.com

Abstract. The least squares (LS) identification algorithm is vulnerable to outliers and has large residual square when the measured data is mixed with impulse noise which obeys symmetrical alpha stable (*SaS*) distribution, so the least absolute deviation (LAD) is selected as the objective function to get better identification performance when impulse noise exists. And taking the non-difference of least absolute deviation into consideration, we adopt an improved gravitational search algorithm as optimal algorithm to search for optimal solution globally. Then the parameter identification method based on LAD objective function using an improved gravitational search algorithm (LAD-IGSA) is put forward creatively. Simulation results show that the LAD-IGSA method can restrain the influence of impulse noise effectively and achieve higher identification accuracy. Moreover, LAD-IGSA method presents better robustness and bettidentification accuracy than LP method with small data sets.

1. Introduction

LS estimation is relatively mature and perfect in theory and methods, which has been widely applied to time series system identification, state estimation, functional approximation, and many other research fields [1]. LS also has some shortcomings: when measurement noise exists, especially, when impulse noise exists, LS shows poorer optimization precision [2]. In order to minimize the sum of squares, it's inevitable to submit to these outliers [3]. The studies show that the LS method has better identification effect when the stochastic noise is normally distributed. But the LS recursive method may come along with ill-posed solutions because of matrix inversion. Therefore, LAD is selected as the objective function to solve the problem of large residual when the measured data is disturbed by the impulse noise [4]. However, LAD does not gain enough attention in estimation and research compared with LS until Charnes applied LAD to a specific management question that describing the deviation with the difference between two nonnegative numbers [5]. This outstanding research laid the foundation of LAD. LAD avoids the complexities of matrix inversion and decreases the effect of impulse noise interference. Many experiments have proved that LAD demonstrates a strong robustness compared with LS [6]. But the LAD criterion function is non-differentiable, so the traditional optimization method based on gradient information can not be adopted. So far many effective algorithms for optimizing the LAD criterion function had been proposed such as linear programming [7], maximum entropy method [8] and intelligent optimization algorithm [9].Intelligent optimization algorithm for the model parameters identification is essentially a probabilistic search, which does not require the gradient information of the objective function, and optimizes the objective function without requirement of continuity.

In this paper, the parameter identification based on the least absolute deviation and an improved gravitational search algorithm (IGSA) is firstly put forward. Gravitational search algorithm (GSA) is a novel meta-heuristic stochastic optimization algorithm inspired by the law of gravity and mass interactions [10]. The GSA method has been employed to filter modelling [11] and unit commitment of renewable energy sources [12] and slope stability analysis [13]. The course of the parameter identification is also to seek the minimum of an objective function. It is important to carry out the precise neighbourhood search in the late period of optimization for enhancing the identification precision.

2. The improved algorithm of GSA

2.1. GSA

In GSA, considering a system with *N* agents which are a collection of masses. Each agent is composed of four characteristics: position X_i, inertial mass M_{ii}, active gravitational mass M_{ai} and passive gravitational mass M_{pi}, where *i* represents the dth agent and $X_i = (x_i^1, \cdots, x_i^d, \cdots, x_i^n)$, $i = 1,2,\cdots,N$, x_i^d is the dth dimension value of agent i.

According to Newton's gravitational force that the force F_{ij} acts on agent i by agent *j* at time *t* is

$$F_{ij}^d(t) = G(t)\frac{M_{pi}(t) \times M_{aj}(t)}{R_{ij}(t) + \varepsilon}(x_j^d(t) - x_i^d(t)) \tag{1}$$

Where ε is a small constant and $G(t)$ is gravitational constant at time *t* which is decreased with time gradually so as to control the search accuracy.

$$G(t) = G_0 e^{-\tau \frac{t}{t_{\max}}} \tag{2}$$

Where $G = 100$, $\tau = 20$ and $t_{\max}$ is the maximum number of iterations.

Assuming that the total force acting on object i in the dth dimension is

$$F_i^d(t) = \sum_{j \in Kbest, j \neq i} rand_j F_{ij}^d(t) \tag{3}$$

Where *Kbest* means the *Kbest* agents with the best fitness value and biggest mass, $rand_j$ is a random constant in the interval [0,1] as well. With the initial value *Kbest=N* at the beginning, *Kbest* is a function that decreases with time. Under this circumstance, all agents apply the force mutually at the beginning, and with time going on, *Kbest* is decreased linearly. Eventually, only one object is left to apply force to other agents at the end [14].

As a result, the acceleration on the basis of Newton Second Law is

$$a_i^d(t) = \frac{F_i^d(t)}{M_{ii}(t)} \tag{4}$$

Gravitational and inertial masses are updated by the following equations:

$$M_{ai} = M_{pi} = M_{ii} = M_i, i = 1,2,...N, m_i(t) = \frac{fit_i(t) - worst(t)}{best - worst(t)}, M_i(t) = \frac{m_i(t)}{\sum_{j=1}^{N} m_j(t)}, \tag{5}$$

Where $fit_i(t)$ represent the fitness value of the agent i at time *t*. Then the speed and position of the agent i will be updated as follows in GSA:

$$\begin{aligned} v_i^d(t+1) &= rand_i * v_i^d(t) + a_i^d(t), \\ x_i^d(t+1) &= x_i^d(t) + v_i^d(t+1). \end{aligned} \tag{6}$$

Where $rand_i$ is a uniform random variable between 0 and 1.

2.2. Main improved strategies of IGSA

2.2.1. Decrease the movement velocity of agents. The next velocity of an agent is considered as a fraction of its current velocity added to its acceleration. Then in GSA, the speed and position of the agent i will be updated as (6). But the current velocity is so large that the current position is far from the next position. Some optimal position may be located between the current position and the next position that leads to the local minimum. Therefore, in IGSA the velocity of agents is decreased in order to improve the iterative search capability:

$$\begin{aligned} v_i^d(t+1) &= rand_i * v_i^d(t) + a_i^d(t) \\ x_i^d(t+1) &= x_i^d(t) + \eta * v_i^d(t+1) \end{aligned} \tag{7}$$

Where η is the number between 0 and 1.

2.2.2. Orbital change of poor agents' positions. During the course of GSA searching, all agents gradually converge to a small local zone, which results in a low searching efficiency in the late period, so orbital change operation should be established in order to jump out of the local minimum. And orbital change operation is to enlarge or contract the position of poor agents at a certain probability (named Jump rate). $rands * x_i$ is called orbital change radius. That is, if agents' positions converge to a smaller value, the orbital change radius will be smaller. The orbital change operation is good for jumping out of the local minimum and improving the convergence speed, and not making a big disturbance upon the global.

$$xn_i = x_i + rands * x_i \text{, i=1,2...N} \tag{8}$$

Where $rands$ is the random number between -1 and 1.

2.2.3. Updating optimal agent using trial-and-error method. In IGSA, the optimal object x_{best} is updated using the trial-and-error method. In GSA, all current agents change at each step, if the optimal agent's fitness becomes bad, the next search will begin from a worse position. The optimal position of those historical search steps, *Lbest*, and its fitness *Fbest*, only play a role for comparison, rather than participate into each step of iterative search. In order to utilize the information of those historical search steps *Lbest*, after one iteration is implemented for the optimal agents, the search will continue to the next step in case the fitness turns better. On the contrary, the position of optimal object's position and fitness will be replaced by *Lbest* and *Fbest*.

2.2.4. Further search of optimal agent position. The GSA algorithm generally converges quickly in the early 70% iterations, and then the convergence speed becomes slow. In order to further intensify the optimal searching ability of the algorithm in the late period, the optimal agent is further optimized by coordinate descent method and turns the multivariable optimization problem into some single-variable sub-problems.

3. Simulation and analysis for algorithm performance

3.1. Model description

Considering the time series as below:

$$\begin{aligned} A(z^{-1})y(k) &= B(z^{-1})x(k) + v(k) \\ A(z^{-1}) &= 1 + 1.4z^{-1} + 0.45z^{-2} \\ B(z^{-1}) &= z^{-1} + 0.7z^{-2}. \end{aligned} \tag{9}$$

Where the input sequence $u(k)$ is white Gaussian noise with variance $\sigma_u^2 = 1$. The noise sequence $v(k)$ is white Gaussian noise sequence with variance $\sigma_u^2 = 0.2$. The objective function can be constructed as $J(\theta)=\sum_{k=1}^{m}|e(k)|$, where $e(k)=y(k)-\varphi^T(k)\theta$. LAD-IGSA is adopted to find the optimal solution of the LAD objective function. The evaluation standard in the experiments is the relative error δ of the parameters estimation.

The impulse noise in the experiment subjects to *SaS* distribution, the probability density function of the standard *SaS* distribution is

$$f_\alpha(x)=\begin{cases}\dfrac{1}{\pi x}\sum_{k=1}^{\infty}\dfrac{(-1)^{k-1}}{k!}\Gamma(\alpha k+1)|x|^{-\alpha k}\sin(k\alpha\pi/2),0<\alpha<1\\ \dfrac{1}{\pi x}\sum_{k=1}^{\infty}\dfrac{(-1)^{k}}{2k!}\Gamma[(2k+1)/\alpha]x^{2k},1<\alpha\le 2\end{cases} \tag{10}$$

Where $\Gamma(*)$ is gamma function, α is characteristics exponent. The less α is, the more probability the large amplitude sample of the random variable occurs, and the stronger the pulse strength is [15,16].

3.2. Simulation results under different situations

In order to guarantee the reliability of the simulation results, 100 times repeated independent experiments are carried out and the data length is $L=500$.

Case 1: measured data with white noise only

LAD-IGSA method, LAD-IGSA method and LS method are adopted for parameter identification separately and the compared results are shown as Table 1, Figure 1 and Figure 2 are the curves of the optimal fitness *Fbest* by LAD-IGSA method and LAD-GSA method.

Table 1. The identification results with white noise only

algorithm	a_1	a_2	b_1	b_2	δ (%)	iterations	time (s)
LAD-IGSA	-1.3883	-0.4403	1.0094	0.7001	0.94	60	2.5085
LAD-GSA	-1.4856	-0.5000	1.0314	0.8486	9.18	73	3.3522
LS	-1.4343	-0.4786	1.0097	0.7512	3.59	-	0.4742
true value	-1.4	-0.45	1.0	0.7	0	-	-

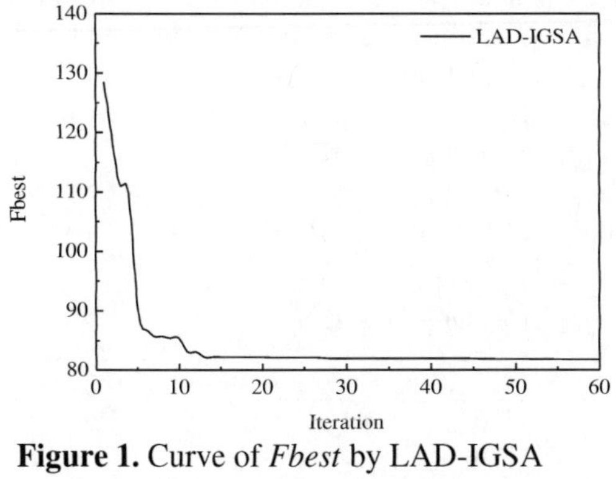

Figure 1. Curve of *Fbest* by LAD-IGSA

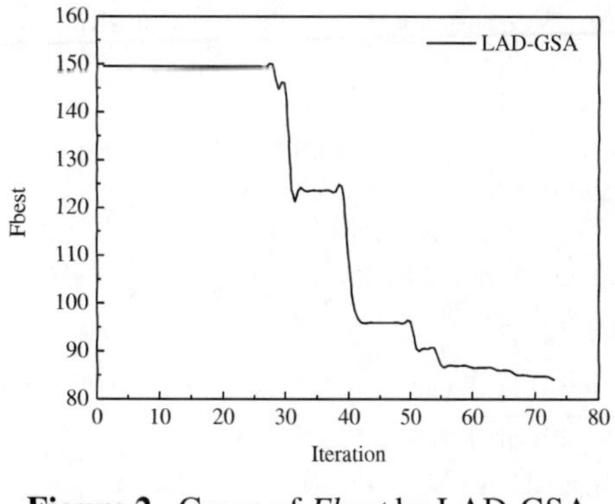

Figure 2. Curve of *Fbest* by LAD-GSA

a_1, a_2, b_1, b_2 are estimated parameters. Table 1 indicates that the relative error of LAD-IGSA method is far less than that of LAD-GSA method and LS method. But LP method runs 0.4742s which presents higher estimation speed. Compared with Figure 1, the curve of *Fbest* by LAD-GSA in Figure 2 has several parallel segments. That is to say, the objective function value does not always keep falling and falls into local optimal solution. Figure 1 presents that the improved strategies in LAD-IGSA method can improve the optimization accuracy and estimation speed greatly and ensures that the objective function value is optimized in the direction of minimization which contributes to jump out local optimal solution.

Case 2: measured data with white noise and impulse noise which subjects to SaS distribution

The measured data are contaminated with both white noise and impulse noise. Let the characteristics exponent α of the impulse noise be 1.5, 1.2 and 0.9. Then the measured data are estimated by LAD-IGSA method, LAD-GSA method and LS method. The parameter identification results and the optimal fitness *Fbest* of LAD-IGSA and LAD-GSA are shown in Table 2.

Table 2. The identification results when $\alpha = 1.5$

algorithm	a_1	a_2	b_1	b_2	δ (%)	iterations	time (s)
LAD-IGSA	-1.3987	-0.4442	1.0041	0.6998	0.87	50	2.2535
LAD-GSA	-1.2957	-0.3711	1.0077	0.5641	9.83	98	4.8557
LS	-1.3656	-0.4272	0.9888	0.6607	3.04	-	0.0352
true value	-1.4	-0.45	1.0	0.7	0	-	-

Table 3. The identification results when $\alpha = 1.2$

algorithm	a_1	a_2	b_1	b_2	δ (%)	iterations	time (s)
LAD-IGSA	-1.3976	-0.4486	1.0007	0.7186	0.67	40	1.9011
LAD-GSA	-1.2930	-0.3482	1.0049	0.5890	9.90	115	5.2288
LS	-1.4005	-0.4503	1.1630	0.8056	10.16	-	0.0351
true value	-1.4	-0.45	1.0	0.7	0	-	-

Table 4. The identification results when $\alpha = 0.9$

algorithm	a_1	a_2	b_1	b_2	δ (%)	iterations	time (s)
LAD-IGSA	-1.3881	-0.4385	0.9975	0.7088	0.98	77	4.4196
LAD-GSA	-1.3265	-0.3783	1.0579	0.5497	9.97	116	11.4576
LS	-1.3960	-0.4434	0.5287	0.7966	25.17	-	0.0355
true value	-1.4	-0.45	1.0	0.7	0	-	-

From Table 2 ~ 4, the relative errors of the LAD-IGSA method are less than 1% which is much better than LAD-GSA method and LS method. With the characteristics exponent α decreasing, the identification accuracy of LAD-IGSA method and LS method become worse. The relative error of LS method changes greatly and the identification accuracy is unstable. LS method even can not identify the parameters successfully when $\alpha = 0.9$. But LAD-IGSA method maintains high precision and the relative error converges to a small value under the interference of impulse with different intensities. So it can be summarized that LAD-IGSA method overcomes the impact of impulse noise successfully.

Case 3: The influence of length of the sample data on identification accuracy

We carry on simulation tests with LP method and LAD-IGSA method in the context of the measured data which are of different length. The compared results under different circumstances are as following in Table 5 ~Table 8.

Table 5. The identification results when L=10

algorithm	a_1	a_2	b_1	b_2	δ (%)	iterations	time (s)
LAD-IGSA	-1.4077	-0.4433	1.1094	0.6726	5.92	10	0.1439
LP	-1.0432	-0.2391	0.9688	0.3179	29.54	-	0.1724
true value	-1.4	-0.45	1.0	0.7	0	-	-

Table 6. The identification results when L=50

algorithm	a_1	a_2	b_1	b_2	δ (%)	iterations	time (s)
LAD-IGSA	-1.3340	-0.3678	0.9944	0.7048	5.53	17	0.2756
LP	-1.5751	-0.6168	1.0343	0.9341	17.7	-	0.1717
true value	-1.4	-0.45	1.0	0.7	0	-	-

Table 7. The identification results when L=100

algorithm	a_1	a_2	b_1	b_2	δ (%)	iterations	time (s)
LAD-IGSA	-1.3824	-0.4399	1.0506	0.6961	2.86	67	1.1820
LP	-1.5918	-0.6133	1.0378	0.9187	17.57	-	0.1862
true value	-1.4	-0.45	1.0	0.7	0	-	-

Table 8. The identification results when L=300

algorithm	a_1	a_2	b_1	b_2	δ (%)	iterations	time (s)
LAD-IGSA	-1.3834	-0.4273	1.0159	0.6879	1.81	42	1.2637
LP	-1.4890	-0.5176	1.0201	0.8129	8.38	-	0.2659
true value	-1.4	-0.45	1.0	0.7	0	-	-

Table 5. shows that the identification accuracy of LAD-IGSA method still can get an acceptable identification result and is higher than that of the LP method with L=10. The accuracy of the LP method indicates that LP can not successfully identify the estimation parameter effectively. From Table 6. and Table 7., we can conclude that the accuracy of LP method is improved with the length of sample data increasing such as L=50 and L=100, but the relative error is still unaccepted. On the contrary, the relative error of LAD-IGSA method is still maintained at 5% or less. Table 8 shows that the LP method can identify the parameters effectively when L=300. Obviously, LAD-IGSA method can identify the parameters effectively when the sample data is insufficient or the length is small. LAD-IGSA method shows better robustness than LP method in this situation.

4. Conclusions

In this paper, we firstly expand the application of the LAD based on an improved gravitational search algorithm to the field of time series system identification. The improved strategies in LAD-IGSA enhance the optimization capabilities and decrease the convergence time. LAD-IGSA method can inhibit the influence of impulse noise which obeys $S\alpha S$ distribution on the identification results and improve the robustness to a great extent due to the application of the LAD criterion. The improved gravitational search algorithm contributes to get a better optimization results when the length of measured data is small, which make the available of identification of time series system when the measured dada are not sufficient. The proposed algorithm is carried out easily and should be very valuable for those applications such as identification of complex chemical process.

References

[1] Xie K G, Song Q K and Zhou J Q A Linear Regress Model Based on Least-Absolute Criteria *Journal of System Simulation* vol 14 no 2 pp 189-192 2002
[2] Cao H R System Identification Based on the Least-Absolute Criteria *Control Theory and Applications* vol 28 no 7 pp 8-10 2009
[3] Chen X R and Wu Y H A Problem about the Relationship between the Minimum L1-Norm Estimate and Linear Programming *Journal of Nanjing Institute of Technology* vol 18 no 5 pp 16-21 1988
[4] Xu B C, Lin Z H, Zhang Y D and Xiao Y Y Extended Stochastic Gradient Identification Method for Hammerstein Model based on Approximate Least Absolute Deviation *Mathematical Problems in Engineering* vol 2016 Article ID 9548428 2016
[5] Charnes A, Cooper W W and Ferguson R O Optimal estimation of executive compensation by linear programming *Management Science* vol 1 no 2 pp 138-151 1955
[6] Wan S P Multi-sensor Information Fusion Method Based on Least Absolute Deviation Estimation *Computer Engineering* vol 36 no 2 pp 257-259 2010
[7] Xiao J H and Yang W L Realization of Least-Absolute Regression Using Linear Programming Based on Matlab *Journal of Wuyi University(Natural Science Edition)* vol 20 no 1 pp 5-8 2006

[8] Yao J K and Cui H Y Maximum Entropy Method for Nonlinear l1 Problem *Communication On Applied Mathematics and Computation* vol 15 no 1 pp 79-86 2001
[9] Jing J and Wang J Application of genetic algorithm to nonlinear LAD regression *Hydro-Science and Engineering* no 2 pp 43-47 2008
[10] Rashedi E, Nezamabadi-pour E and Saryazdi S GSA: a gravitational search algorithm *Information Sciences* vol 179 no 13 pp 2232-2248 2009
[11] Rashedi E, Nezamabadi-pour E and Saryazdi S Filter modeling using gravitational search algorithm *Engineering Applications of Artificial Intelligence* vol 24 no 1 pp 117-122 2011
[12] Shukla A and Singh S N Multi-Objective Unit Commitment With Renewable Energy Using GSA Algorithm *INAE Letters* vol 1 no 1 pp 21-27 2016
[13] Jiang J G, Tan Y, Dong L M and Wang C Application of modified gravitational search algorithm in slope stability analysis *Chinese Journal of Geotechnical Engineering* vol 38 no 3 pp 419-425 2016
[14] Bazaraa M S and Shetty C M Nonlinear Programming: Theory and Algorithms *John Wiley & Sons* pp 779-841 2005
[15] Xu B C and Zhang Y Y An improved gravitational search algorithm for dynamic neural network identification *International Journal of Automation and Computing* vol 11 no 4 pp 434-440 2014
[16] Coates M J and Kuruoglu E E Time-frequency based detection in impulsive noise environments using alpha-stable noise model *Digital Signal Processing* vol 82 no 3 pp 1917-1925 2002

Water Quality Prediction in the Waste Water Treatment Process Based on Ridge Regression Echo State Network

Jing Zhao[1,2,*], Chao Zhao[1,2], Fan Zhang[1,2], Gang Wu[1,2] and Haitao Wang[1,2]

[1] China National Institute of Standardization, Beijing 100191, China
[2] AQSIQ Key Laboratory of Human Factors and Ergonomics(CNIS), Beijing 100191, China

[*]E-mail: nlzhaojing0506@163.com

Abstract. Echo state network (ESN), a novel recurrent neural network, has a randomly and sparsely connected reservoir. Since the output weights are computed by Moore-Penrose inverse, the ill-posed problem may exist in the ESN. To overcome this problem, ridge regression echo state network (RESN) is proposed, in which the ridge regression algorithm is used to calculate the output weights instead of linear regression. Simulation results show that the RESN has better performance than some other existing methods, thus can deal with the ill-posed problem.

1. Introduction

Water quality prediction in the wastewater treatment process (WWTP) can provide strong support for water treatment plant management decisions [1-2]. However, it is very difficult to predict the water quality, since WWTP is a complex system including a variety of physical and biochemical reactions. Moreover, due to the nonlinear characteristics, delay-time and uncertainty, it is difficult to measure effluent qualities parameters in the WWTP. The quality of treated wastewater is measured by some parameters, such as biochemical oxygen demand (BOD), total phosphorous (TP), ammonia nitrogen (NH_4-N), and so on, which must meet the national standard. The traditional measurement approaches usually depend on the laboratory analysis which takes a long time [3-4]. For example, it takes about five days to get BOD values according to the conventional chemical measurement approaches, which cannot have a real time monitoring process of the water pollution situation. This lack of real- time process variable information limits the effective operation of effluent water quality prediction. Furthermore, the online monitoring instrument needs high economic costs and is difficult to be conducted in the WWTP. Therefore, water quality prediction model is essential to support water quality parameters.

To solve the above-mentioned problems, soft computing method has been widely used in complex system modelling, where the hard-to-measure process variables are estimated by the other easy-to-measure variables. The artificial neural networks (ANNs) are significant nonlinear approaches for nonlinear system modelling and have attracted a lot of attention. In [5], Takens embedding theorem based phase space reconstruction is used to extract more information from the limited datasets of the chaotic system, where principal component analysis and artificial neural network is then adopted to estimate the effluent BOD value. In [6], an improved T–S fuzzy neural network (TSFNN) is used to predict BOD values based on the knowledge representation ability and learning capability, where a gradient descent method with the momentum item is used to adjust antecedent parameters and consequent parameters. In [2], a flexible structure radial basis function neural network (FS-RBFNN) is proposed for water quality prediction, where the hidden neurons in the RBF neural network can be added or removed online based on the neuron activity and mutual information. In [1], a self-organizing cascade neural network (SCNN) with random weights is proposed for water quality prediction on effluent BOD and TP. The SCNN is constructed via simultaneous structure and parameter learning processes. The simulation results show that the proposed SCNN has better performance.

Recurrent neural network (RNN) is a significant nonlinear approach for nonlinear system modelling and has attracted increasingly more and more attention. However, the traditional RNN based on the gradient method has gradient disappearance and gradient explosion problem. As an effective alternative for RNN training, echo state network (ESN) [7] can overcome the local minima and gradient vanishing problems. A typical ESN has a large and sparsely connected reservoir. During the training process, only the output weights are computed by simple Moore-Penrose inverse method, while the input weights and reservoir weights remain fixed once they are generated. This simple and effective training approach makes ESN have numerous successful applications such as time-series prediction [8], speech recognition [9], and nonlinear signal processing [10].

Although ESN has better performance than the traditional neural network, there are still some problems in the ESN. Since the output weights are trained by the simple regression, the ill-posed problem may exist in the training process. If it happens, the large output weights might occur and the generalization ability will be degraded. Additionally, when the number of training samples is less than the number of reservoir neurons, the ill-posed problem must happen. To solve the ill-posed problem, noise injection is used to improve the network performance, however, there are no satisfactory methods to determine the amplitude of noise. The intrinsic plasticity rule is used in the reservoir computing setting [11]. However, parameter adaption process based on the gradient method is complicated and some hyper-parameters are obtained by experience. In this paper, ridge regression method is used to address the ill-posed problem, where the regularization term is introduced into the loss function to penalize larger readout weights and the ridge parameter is determined by the cross-validation method.

The remainder of this paper is organized as follows: In section 2, a brief review of the original ESN model is given. In section 3, the ridge regression echo state network model is proposed. In section 4, the stability analysis is given. In section 5, the simulations are carried out to illustrate the performance of the proposed algorithm. Finally, some conclusions are drawn.

2. Echo state network

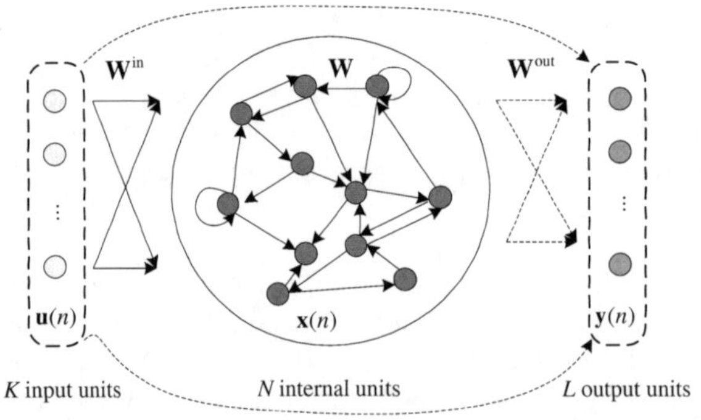

Figure 1. The basic architecture of OESN without feedback connections.

An original ESN (OESN) [7] without feedback connections consists of three layers, input layer, reservoir layer and readout layer. The structure of OESN is presented in figure 1. It is assumed that the OESN has K input neurons, N reservoir neurons and L readout neurons, respectively. $\mathbf{u}(n)$, $\mathbf{x}(n)$, $\mathbf{y}(n)$ denote the input vector, the reservoir state vector and the readout vector, respectively. The updating formula of OESN is given as follows:

$$\mathbf{x}(n) = \mathbf{tanh}(\mathbf{W}^{in}\mathbf{u}(n) + \mathbf{W}\mathbf{x}(n-1)) , \quad (1)$$

$$\mathbf{y}(n) = \mathbf{W}^{out}(\mathbf{u}(n), \mathbf{x}(n)) , \quad (2)$$

Where $\mathbf{W}^{in}$, $\mathbf{W}$, $\mathbf{W}^{out}$ are the input weight matrix, reservoir weight matrix and output weight matrix, respectively. **f** is the reservoir active function, which is usually chosen as hyperbolic tangent function. The output weights are computed by using the linear regression method as follows

$$\mathbf{W}^{out} = ((\mathbf{X}^T\mathbf{X})^{-1}\mathbf{X}^T\mathbf{Z})^T \quad (3)$$

Where $\mathbf{X}=[\mathbf{X}(1),\mathbf{X}(2),\cdots,\mathbf{X}(P)]^T$ (P is the number of training samples) denotes the internal state matrix, $\mathbf{X}(n)= [\mathbf{u}(n)^T, \mathbf{x}(n)^T]^T$, $\mathbf{Z}=[\mathbf{z}(1),\mathbf{z}(2),\cdots,\mathbf{z}(P)]^T$ is the desired output.

3. Ridge regression echo state network

From formula (3), if $\mathbf{X}^T\mathbf{X}$ is singular, the ill-posed problem will happen. To obtain a stabilized solution, the ridge regression, a widely used regularization method, is applied to introduce a penalty term into the loss function as follows.

$$J(\mathbf{W}^{out}) = \arg\min_{\mathbf{W}^{out}} \left\| \mathbf{Z} - \mathbf{X}\mathbf{W}^{out} \right\|_2^2 + \lambda \left\| \mathbf{W}^{out} \right\|_2^2 \quad (4)$$

Where $\lambda > 0$ denotes the ridge parameter. Differentiating with respect to $\mathbf{W}^{out}$, it can be got as follows

$$\frac{\partial J(\mathbf{W}^{out})}{\partial \mathbf{W}^{out}} = -2\mathbf{X}^T(\mathbf{Z} - \mathbf{X}\mathbf{W}^{out}) + 2\lambda \mathbf{W}^{out} \quad (5)$$

Let $\frac{\partial J(\mathbf{W}^{out})}{\partial \mathbf{W}^{out}} = 0$, we have $\mathbf{X}^T\mathbf{X}\mathbf{W}^{out} - \mathbf{X}^T\mathbf{Z} + \lambda \mathbf{W}^{out} = 0$, the following equation can be obtained

$$(\mathbf{X}^T\mathbf{X} + \lambda \mathbf{I})\mathbf{W}^{out} = \mathbf{X}^T\mathbf{Z} \quad (6)$$

Since the matrix $\mathbf{X}^T\mathbf{X} + \lambda \mathbf{I}$ must be invertible, the solution of (4) is obtained as follows

$$\mathbf{W}^{out} = (\mathbf{X}^T\mathbf{X} + \lambda \mathbf{I})^{-1}\mathbf{X}^T\mathbf{Z} . \quad (7)$$

Where ridge parameter is obtained by cross-validation method. From (7), it is known that the ill-posed problem can be solved.

The detailed steps are summarized in Algorithm 1.

Algorithm 1

Step 1: Randomly generate an input weight matrix $\mathbf{W}^{in}$ according to uniform distribution and initialize the reservoir states $\mathbf{X}$ (0).

Step 2: Randomly generate a reservoir weight matrix $\mathbf{W}_0$ with the predefined sparsity and reservoir size. Scale $\mathbf{W}_0$ to $\mathbf{W}= (\alpha_w / \rho(\mathbf{W}_0))\, \mathbf{W}_0$, where $0<\alpha_w<1$ and $\rho(\mathbf{W}_0)$ is the spectral radius of $\mathbf{W}_0$.

Step 3: Drive the reservoir by the input signals as (1), discarded a certain number of initial steps and collect the internal states at an initial transient n_{min}.

Step 4: Compute the output weight matrix $\mathbf{W}^{out}$ using (7).

Step 5: Test the trained RESN.

4. Stability analysis

The key of the ESN is that the reservoir should have the echo state property (ESP). ESP means that the internal states should be uniquely depend on external input and it is related to the input samples and reservoir weight. To describe the ESP, the local dynamics of the system by linearizing the RESN is considered. The nonlinear system (1) can be approximated as follows

$$\mathbf{x}(n) = \mathbf{g}'\,\mathbf{W}\mathbf{x}(n-1) + \mathbf{g}'\,\mathbf{W}^{in}\mathbf{u}(n) \triangleq \mathbf{A}\mathbf{x}(n-1) + \mathbf{B}\mathbf{u}(n) \quad (8)$$

where $\mathbf{g}' = \mathbf{tanh}'$, $\|\mathbf{g}'\| \le 1$, denote $\mathbf{A} = \mathbf{g}'\,\mathbf{W}, \mathbf{B} = \mathbf{g}'\,\mathbf{W}^{in}$.

The sufficient condition of the ESP is that the maximal singular value of reservoir weight matrix $\mathbf{W}$ is less than 1 ($\sigma(\mathbf{W}) < 1$). Since $\|\mathbf{W}\| = \sigma(\mathbf{W})$, the sufficient condition is equivalent to $c \triangleq \|\mathbf{W}\| < 1$.

Suppose $\mathbf{x}(n)$ and $\mathbf{x}'(n)$ are different internal state vectors.

$$\begin{aligned}\|\mathbf{x}(n)-\mathbf{x}'(n)\| &= \|\mathbf{A}\mathbf{x}(n-1)+\mathbf{B}\mathbf{u}(n)-\mathbf{A}\mathbf{x}'(n-1)-\mathbf{B}\mathbf{u}(n)\| = \|\mathbf{A}\mathbf{x}(n-1)-\mathbf{A}\mathbf{x}'(n-1)\| \\ &\le \|\mathbf{A}\|\|\mathbf{x}(n-1)-\mathbf{x}'(n-1)\| \le \|\mathbf{g}'\|\cdot\|\mathbf{W}\|\cdot\|\mathbf{x}(n-1)-\mathbf{x}'(n-1)\| \le c\|\mathbf{x}(n-1)-\mathbf{x}'(n-1)\| \\ &\le \cdots \le c^n\|\mathbf{x}(0)-\mathbf{x}'(0)\|.\end{aligned} \tag{9}$$

This shows that the current reservoir state is determined by its past external input history, which guarantees the ESP.

5. Simulation

In this section, the performance of the RESN is evaluated on Lorenz time series prediction, effluent BOD and TP prediction in the WWTP. To show the effectiveness of the RESN, the simulations are compared with the following models: OESN [7], TSFNN [6], and FS-RBFNN [2]. The reservoir size, spectral radius, and sparsity are selected as 200, 0.85 and 0.05, respectively. All simulations are tested in MATLAB 2013b environment and run on i7-4790 with CPU 3.60GHz and 8.0GB RAM.

The normalized root mean square error (NRMSE) is used as the evaluation criteria of model performance, which is defined as below:

$$\text{NRMSE} = \sqrt{\sum_{t=1}^{S}\frac{(d_i(t)-y_i(t))^2}{S\sigma^2}} \tag{10}$$

Where $d_i(t)$ denotes the desired output, $y_i(t)$ is the corresponding prediction output, σ^2 is the variance of the desired outputs, and S is the total number of $d_i(t)$.

5.1. Lorenz time series prediction

The Lorenz system is a mathematical model for atmospheric convection, which is widely used as a benchmark in many applications [12]. It can be described as follows

$$\begin{cases}\dot{x} = a_1(y-x) \\ \dot{y} = -xz + a_2x - y. \\ \dot{z} = xy - a_3z\end{cases} \tag{11}$$

Where a_1=10, a_2=28 and a_3=8/3 are the system parameter. Figure 2 gives the the trajectory of the Lorenz system.

In this experiment, the data set are generated by the fourth-order Runge-Kutta method with a step size 0.01, and only the *Y*-dimension samples *y*(t) are used for time series prediction. For *y*(t), 5000 data samples are generated, the first 3000 values are used for training, and the last 2000 values are used to test the proposed model. In the training set, the first 1000 samples are discarded to washout the initial transient. The initial value is selected as *x*(0)=1, *y*(0)=1, *z*(0)=0.

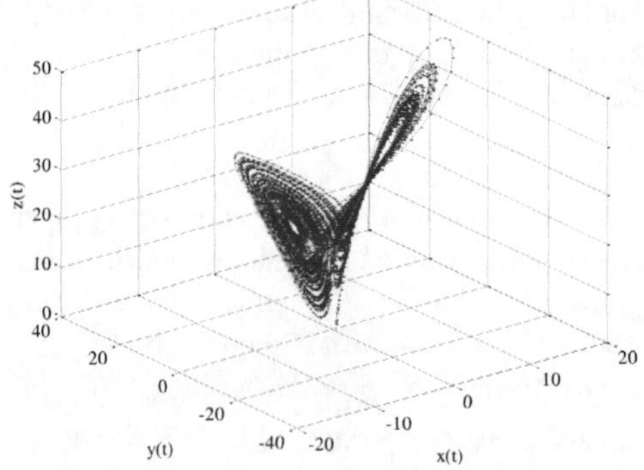

Figure 2. Lorenz attractor

Figure 3 and figure 4 give the testing output and testing error, respectively. Obviously, the proposed

RESN has better prediction performance than OESN. The testing error of RESN is limited in [-0.3×10^{-4}, 0.3×10^{-4}], while the testing error of OESN is limited in [-0.9×10^{-4}, 0.9×10^{-4}].

The detailed comparison results are listed in Table 1 based on the average of 50 independent simulations. As listed in Table 1, although RESN takes more training time than OESN, it has less training NRMSE and testing NRMSE compared with other models.

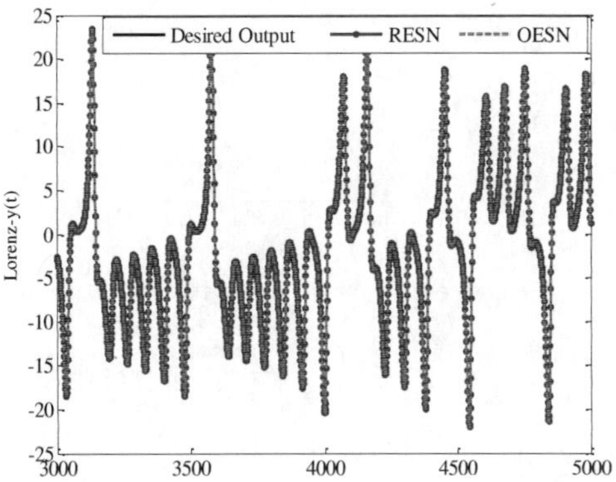

Figure 3. Testing output for Lorenz time series

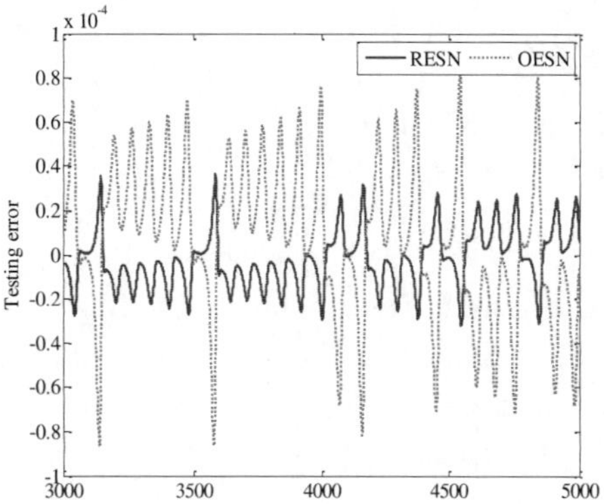

Figure 4. Testing error for Lorenz time series

Table 1. Comparison of different models for Lorenz time series

Models	Training time (s)	Training NRMSE	Testing NRMSE
OESN[7]	**35.18**	4.18×10^{-5}	8.36×10^{-4}
TSFNN[6]	48.65	6.21×10^{-4}	7.68×10^{-3}
FS-RBFNN[2]	65.32	8.15×10^{-4}	8.96×10^{-3}
RESN	41.18	$\mathbf{3.28\times10^{-5}}$	$\mathbf{5.28\times10^{-4}}$

5.2. Water quality prediction in the WWTP

In recent years, wastewater has become one of the major environment problems in the national governments worldwide. Treating wastewater and predicting water quality have become very important [1]. However, the municipal sewage treatment system is a typical nonlinear system, which is difficult to be modeled for complex biochemical reaction. In this section, the main purpose is to provide accurate predictions of effluent BOD and TP, which are one of the most important effluent quality indices and can reflect the water pollution situation.

Table 2. Variables used in the prediction model

Output	Inputs
Effluent BOD	COD; TSS; pH; DO
Effluent TP	Influent TP; T; ORP; DO; TSS; pH

Table 3. List of acronyms used in simulations

Acronyms	Description
BOD	Biochemical oxygen demand
COD	Chemical oxygen demand
TP	Total phosphorus
TSS	Total suspended solids
DO	Dissolved oxygen
ORP	Oxidation reduction potential
T	Temperature

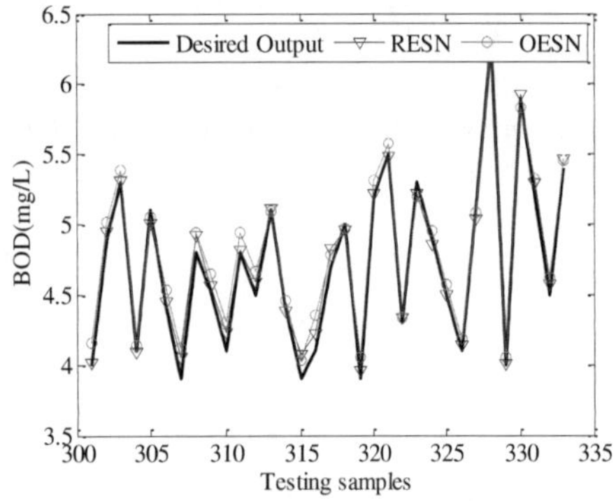

Figure 5. Testing outputs for effluent BOD

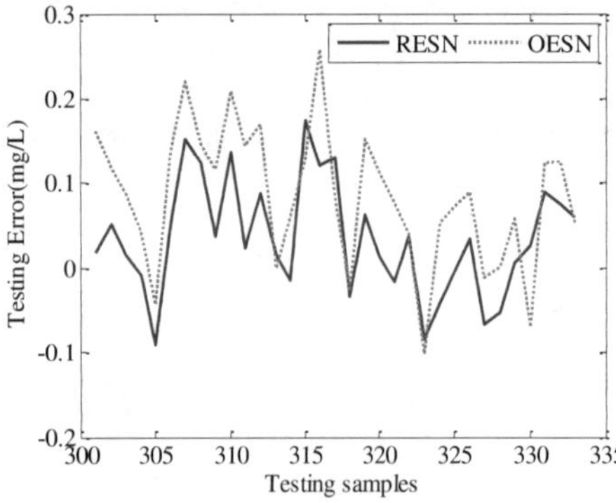

Figure 6. Testing error for effluent BOD

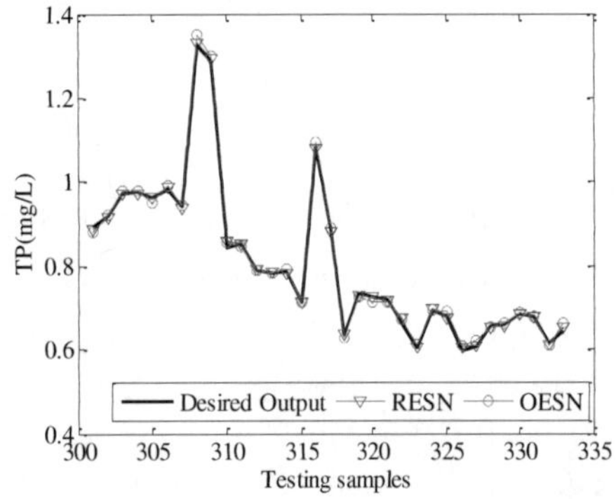

Figure 7. Testing outputs for effluent TP

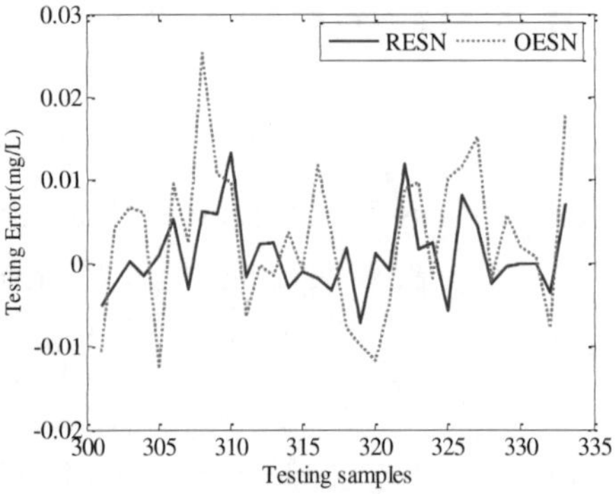

Figure 8. Testing error for effluent TP

Table 4. Comparison of different models for effluent BOD prediction

Models	Training time (s)	Training NRMSE	Testing NRMSE
OESN[7]	**25.13**	0.1742	0.1956
TSFNN[6]	79.65	0.3328	0.4376
FS-RBFNN[2]	93.32	0.3129	0.4267
RESN	32.18	**0.1182**	**0.1329**

Table 5. Comparison of different models for effluent TP prediction

Models	Training time (s)	Training NRMSE	Testing NRMSE
OESN[7]	**24.36**	0.0032	0.0156
TSFNN[6]	78.63	0.0098	0.0398
FS-RBFNN[2]	88.35	0.0069	0.0278
RESN	31.19	**0.0018**	**0.0089**

After deleting the abnormal data, 333 samples are obtained from a sewage treatment plant in Beijing, China. The RESN selects some important subsidiary variables for effluent BOD and TP prediction, respectively. The detailed water quality parameters listed in Table 2 are selected as the input variables of the prediction model, and the description of each variable can be found in Table 3.

For the samples data, the first 200 samples are used for training, 50 samples in training set are discarded to washout initial transient, and the last 133 values are used to test the network performance. Before the simulation, the inputs and the target outputs are normalized into [-1, 1]. After simulations, the outputs are converted.

The simulation results for effluent BOD are shown in figure 5 and figure 6, respectively, which illustrate that the RESN has more accurate prediction than OESN for actual time-series. Likewise, the simulation results for effluent TP are shown in figure 7 and figure 8, respectively. Based on 50 independent simulations, the detailed results for effluent BOD and TP are listed in Table 4 and Table 5, respectively. From the comparison results of different models, it is seen that the training time of RESN is more than OESN, however, RESN is faster than TSFNN and FS-RBFNN. The training NRMSE and the testing NRMSE of RESN are less than other models, which means that the RESN has high accuracy than other models.

6. Conclusion

In this paper, an RESN model based on ridge regression echo state network is proposed. Since the output weights are computed by Moore-Penrose inverse, the ill-posed problem may exist in the ESN. The proposed RESN can solve the ill-posed problem by adding a penalty term to the loss function. The simulation results show that the proposed RESN model can get good performance and has a smaller testing error than other models.

Acknowledgments
This research was supported by China National Institute of Standardization (522018Y-5941, 522018Y-5948).

References

[1] Li F J, Qiao J F, Han H G, Yang C L 2016 A self-organizing cascade neural network with random weights for nonlinear system modeling *Appl. Soft Comput.* **42** 184
[2] Han H G, Chen Q L, Qiao J F 2011 An efficient self-organizing RBF neural network for water quality prediction *Neural Netw.* **24**(7) 717
[3] Han H G, Qiao J F 2010 A self-organizing fuzzy neural network based on a growing-and-pruning algorithm *IEEE Trans. Fuzzy Syst.* **18**(6) 1129
[4] Han H G, Qiao J F 2013 A structure optimisation algorithm for feedforward neural network construction *Neurocomputing* **99** 347
[5] Qiao J F, Hu Z Q, Li W J 2016 Soft measurement modeling based on chaos theory for biochemical oxygen demand (BOD) *Water* **8** 581
[6] Qiao J F, Li W, Han H G 2014 Soft computing of biochemical oxygen demand using an improved T–S fuzzy neural network *Chin. J. Chem. Engin.* **22**(11-12) 1254
[7] Jaeger H, Haas H 2004 Harnessing nonlinearity: predicting chaotic systems and saving energy in wireless communication *Science* **304**(5667) 78
[8] Qiao J F, Li F J, Han H G, Li W J 2017 Growing echo-state network with multiple subreservoirs *IEEE Trans. Neural Netw. Learn. Syst.* **28**(2) 391
[9] Skowronski M D, Harris J G 2007 Automatic speech recognition using a predictive echo state network classifier *Neural Netw.* **20**(3) 414
[10] Xia Y, Jelfs B, Hulle M M V 2011 An augmented echo state network for nonlinear adaptive filtering of complex noncircular signals *IEEE Trans. Neural Netw.* **22**(1) 74
[11] Schrauwen B, Wardermann M, Verstraeten D 2008 Improving reservoirs using intrinsic plasticity *Neurocomputing* **71**(7–9) 1159
[12] Xu M, Han M 2016 Adaptive elastic echo state network for multivariate time series prediction *IEEE Trans. Cybern.* **46**(10) 2173

IOP Conf. Series: Materials Science and Engineering **435** (2018) 012026 doi:10.1088/1757-899X/435/1/012026

Stock Prediction Using Convolutional Neural Network

Sheng Chen[1,*] and Hongxiang He[1]

[1]School of Optical-Electrical and Computer Engineering, University of Shanghai for Science and Technology, Shanghai, People's Republic of China

[*]Email: chnshn@hotmail.com

Abstract. Stock prediction is a very hot topic in our life. However, in the early time, because of some reasons and the limitation of the device, only a few people had the access to the study. Thanks to the rapid development of science and technology, in recent years more and more people are devoted to the study of the prediction and it becomes easier and easier for us to make stock prediction by using different ways now, including machine learning, deep learning and so on. In this paper, we proposed a deep learning method based on Convolutional Neural Network to predict the stock price movement of Chinese stock market. We set the opening price, high price, low price, closing price and volume of stock deriving from the internet as input of the architecture and then run and test the program. The result has shown that it is a bit reliable to use deep learning method based on Convolutional Neural Network to predict the stockprice movement of China.

1. Introduction

It is recorded that there are about 200 billion stock investors in China. Although the number is very large, the real number of the people who make profit from the stock market is very small. If you want to get the win from the stock market, the analysis of the data is necessary. For example, you can study the movement tendency of the stock price in last few weeks and then determine whether it will be up or down in next few days according to some relations between each day's stock data. Of course, you also can use some mathematical models to help you do the analysis. In fact, stock prediction is really a very nice idea in one way, but because of its complexity and dynamic features, the result of the prediction is not satisfied. Some scientists still doubt whether it is possible for us to make the prediction.

In these years, the main method to predict the stock price movement can be concluded in 2 aspects: using machine learning ways or deep learning ways.

Machine learning is a very useful tool to do some work in financial areas, which can be divided into two parts: supervised machine learning and unsupervised machine learning. The significant difference between the former and the latter is whether we have the label of the training data. The core ideology of machine learning is that we should find a supervised mathematical model to fit the data, and then you should train the model to lower the error or improve the fitting degree. Traditional machine learning way or concept that people use to help them with the job in daily life includes support vector machines, decision trees, random forest and so on. For example, we can use regression analysis to predict the price of the house. We also can use logistic regression to detect credit card fraud. Of course, there is no doubt that it can be used to predict the stock price movement, too.

Although machine learning has a long history and very useful, in recent few years, deep learning enjoys popularity among people all over the world. [1] has shown that the Google team used a special machine learning way --deep learning to win the contest of image classification. Besides, when the famous robots named Alphago2.0 who was composed of several deep learning models beat the world Weiqi Champion KeJie with 3:0, it attracts more people to make a study on deep learning.

One we should pay attention to is that deep learning is developed on the basis of machine learning, which shows that both concepts are very similar. Traditional deep learning way also includes two parts: Convolutional Neural Network and Recurrent Neural Network.

Traditionally, the input of a Convolutional Neural Network is often a 2D image, but that does not mean that we cannot use the model to help us make predictions. There are two ways to preprocess the data, one we can convert the 1D-input data into a 2D matrix and the other way is to take advantage of 1D function to help us do the computation of convolution. We chose the latter to help us run and test the experiment.

Compared with Convolutional Neural Network, Recurrent Neural Network may be more useful to deal the prediction problem in financial, for the input of the neural network is time series data. But when people use RNN model to help them, they find that the gradient may be vanished. So, they have improved model, naming it as LSTM (Long Short-Term Memory), which are a useful model in Natural Language Process task.

In this paper, a deep learning model based on Convolutional Neural Network is proposed to predict the stock price movement of Chinese stock market. We use TensorFlow to help us design the model. At first, we will preprocess data and normalize them because each feature has a big difference, otherwise it will affect the result. Then, because the stock data belongs to 1D time series data, we have used a 1D function to do the convolution and set five features including volume, high price, low price, closing price and volume as input.

The rest of this paper is organized as follows: Section 2 will have a literature review. Section 3 will introduce how to preprocess the data from the internet and detail in our model. Section 4 will describe the result of our experiment. Section 5 will conclude the paper and propose some suggestion for further study.

2. Literature review

2.1 Literature review in machine learning method

Before deep learning gains its popularity among people, machine learning is a very good way to make prediction in financial area. [2] has used a SVM wayto forecast financial time series data, and the experiment result of which has indicated that the use of support vector machine (SVM) to make prediction can be a good idea. Besides, SVM also can do the work of classification. It is used in [3] to predict whether the stock price movement will be up in a short term.

In addition to SVM, the other machine learning methods also can make sense in financial area. [4] has used an artificial neural network to predict the stock values and analyze the result when using more or less hidden layers and different activation function. Besides, the hybrid model, which are used in [5] and [6], has absorbed both data mining techniques and traditional methods to make the prediction of stock price movement. [7] has also used a hybrid model, which combines genetic algorithm with neural network to make prediction, gaining a satisfied result.

2.2 Literature review in deep learning method

In recent years, using deep learning methods to deal problem has shown its superiority when compared with previous machine learning methods, especially in image caption [8] and natural language processing tasks. [9] has compared the efficiency of machine learning and deep learning, which shows that the result of the latter is a bit better than the former.

Traditionally, if we want to solve the problem by deep learning, first we should do is to design an architecture similar with neural network, but it includes more hidden layers and needs a large number of data sets. The more layers and data sets you have, the more features it will be extracted for classification.

The generally used model in deep learning includes two parts: Convolutional neural network [10-13] and Recurrent neural networks [14-17]. Generally speaking, as the data of stock usually can be seen as sequential data, the frequency of using Recurrent Neural Network to process the data related with time series are higher compared with using Convolutional Neural Network to deal them. [18] and [19] has used LSTM ways to make predictions, which both can get a good result. But that does not mean that LSTM is the first model you should consider when dealing with the problem in financial area. Because the stock data can be seen as a large 2D matrix, [3] has used ANN model to make prediction and gain a satisfied result, both of which have proved that CNN also can be used to do the same thing. Thus, [1] and [9] have tried to use CNN to predict stock price movement. Of course, the result is not inferior to the people who used LSTM to make prediction.

2.3 Convolutional Neural Network

Convolutional Neural Network is a feed-forward neural network. Like the traditional architecture of a neural network including input layers, hidden layers and output layers, convolutional neural network also contains these features and the input of the layer of convolution are the output of the previous layer of convolution or pooling. Of course, they still have some unique features such as pooling layers, full connection layers, etc. The number of hidden layers in a convolutional neutral network is more than that in a traditional neural network, which, to some extent, shows that the capability of the neural network. The more the hidden layers are, the higher feature it can extract and recognize from the input. People always use convolutional neural network in computer vision, such as face recognition, image classification [20-21].

3. Method

3.1 Data preprocessing

Data preprocessing is a very important step when you want to get some information from data sets to help you make the prediction. As the initial data may have a lot of noise, it is necessary to reduce them so that they will not interfere with you result. Besides, because some features of data may make no sense, in order to improve efficiency, you should neglect them when you train the data.

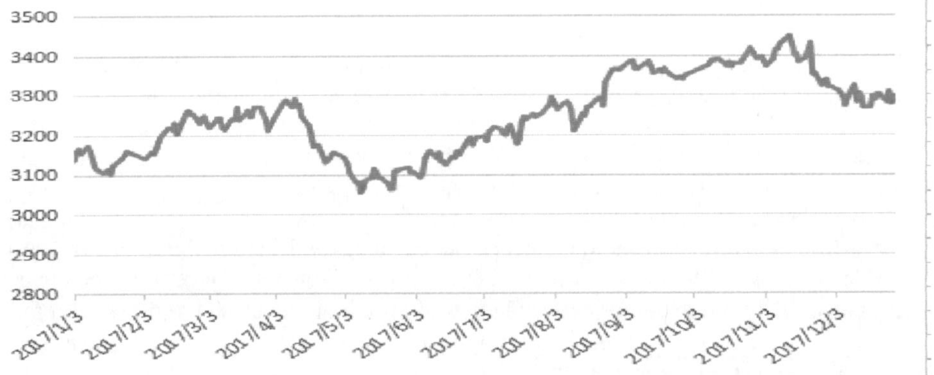

Figure 1. Part of the closing price chart from China stock exchange

We got the historical stock data sets of China which covered from January 5h, 2015 to December 29th, 2017 every trading day from the Internet. The data initially contains not only several features, including opening price, closing price, high price, low price, volume, turnover, change rate, etc., but also the name and the code of the stock. The part of closing price chart is shown in Figure 1. Most of them will change once every five minutes. To accelerate the speed of running a program and improve its efficiency, the first thing we should do is to clean the data. [22] has made some experiments with different features for prediction to show that the experiment will achieve the satisfying result when using open, close, high, low and volume as the input features of model. So, we chose the first five features as our input and neglect irrelevant data such as their names and stock codes. Then, we have removed the data with missing values. Next, as the stock might be suspended one day, it is meaningless to use those data for prediction and they also should be removed from the data sets. One more thing we should note is that there will be a phenomenon in theory that some stock data may not change once every five minutes, so on no account should we regard it as repeated data and remove it from the data sets.

Our task it to do the binary classification instead of regression, so the output of our model only consists of two values--one or zero to show whether the stock price movement will be up or down. Of course, since we adopt supervised way to design the model, labels are of great importance.Unfortunately, the data sets do not contain the labels and we should make the labels ourselves. We have defined a function which will compare the change of closing price between one day and ten days after that day and then get the label. For example, if the closing price would be up after 10 days, the label of that day would be set as 1 to show the stock price would be move up after 10 days and vice versa. People could consider buying the stock when the result of prediction is 1 and selling it after 10 days to get the stock returns. The full data set is divided into two parts: 20% for testing the data and 80% for training the data.

To shorten the computing time, we have vectorized the input data, which means that when we train the model at one time, several numbers of the data will be input into the model. The input form can be defined as following matrix:

$$X = \left(\mathrm{x}^{(1)}, x^{(2)}, \ldots, x^{(m)}\right) \tag{1}$$

Here m indicates the number of samples as inputs. Each sample consist of five features mentioned above, so the input of the neural network is a $\mathrm{m} \times 5$ matrix. The output form is basically similar as the input form:

$$Y = \left(y^{(1)}, y^{(2)}, \ldots, y^{(m)}\right) \tag{2}$$

Here each y is a Boolean value: 1 and 0. Noting that one training sample only has one label.

In order to accelerate the convergence speed when we use gradient descent to find the optimal values, the input data should be normalized because there is too much difference between each feature. We have adopted min-max normalization to keep the value in the range of 0 to 1. The formula is as following:

$$\mathrm{x}^{*} = \frac{x - \min}{\max - \min} \tag{3}$$

Here $\max$ indicates the maximum value of the features and $\min$ indicates the minimum value of the features.

3.2 Convolution function for 1-Dimensional data

Usually, by means of Google's TensorFlow, we can directly use the function "conv2d" to do the computation of convolution in a CNN model which usually accepts a 2D image as its input. But, as our dataof the stock belongs to1D time series data, it is necessary to change the function to help us do the computation. The new function we adopt in the model accepts a number of stock data, which has mentioned above, and other properties such as the number of the filters, width of the filters and stride as its input. Note that the height of the kernel here is not applicable. Then we use a Gaussian distribution to initiate the value of the filters and zero to initiate the biases. The outputs of our function are a number of matrices, and the specific number is decided by the number of the filters. CNN model will extract some features from them and they also will be the input of the next pooling layer after the computation of activation function.

3.3 Convolutional neural network for the stock prediction

By means ofconv1d function, we can builda CNN model to help us make the prediction. For getting the final classification values, which will tell us the stock will be up or down in future, we will use "teaching" number to train the model. Firstly, the vector of stock data will be input into the stock prediction model and we will use the conv1d function to do the computation of convolution. Because our stock prediction model maybe linear and it will be very poor to solve the nonlinear problem, it is necessary to introduce the activation function. Traditionally, the activation function that people always use can be concluded as followings: Tanh, Relu, Sigmoid and Lrelu. To solve the problem of gradient vanishing, we choose the Relu and Lrelu function to do the computation of activation. The Relu function and Lrelu function can be defined as followings:

$$\mathrm{Relu}\ (x) = \begin{cases} 0 & (x \le 0) \\ x & (x > 0) \end{cases} \tag{4}$$

$$\mathrm{Lrelu}(x_i) = \begin{cases} a_i x_i & (x \le 0) \\ x_i & (x > 0) \end{cases} \tag{5}$$

Where x and x_i both are the results of the computation of convolution in the convolutional layer a_i is a hyper-parameter with a very small value and usually we choose 0.01 as its initial value (it also can be fine-tuned to get the best result while training).

Note that Relu function is a very useful activation function to solve the problem of gradient vanishing, but the problem will still exist when the input vectors include negative value [23]. Thus, in the first convolutional layer we choose Relu as our activation function and in the next several convolutional layers it will be replaced by the Lrelu function.

Secondly, the feature maps of the convolutional layer will be the input of pooling layer and they will be pooled to reduce the feature dimension and accelerate the computation. We choose max pooling function to do the work. For example, A pooling layer using a 2×2 filter will subsample its input by the maximum function, that is to say, it will find the maximum in each 2×2 area of the feature maps.

After the computation of several convolutional layers and pooling layers, the output of the last layer will be the input of a fully connected layer, which will help to do the classification tasks. Because our task is to do binary classification, the output of the fully connected layer will be computed by the activation function of SoftMax to get the probability of the stock's movement of up and down and we will choose the bigger to be the final prediction value. Of course, we will also adopt dropout strategy to avoid over-fitting in the fully connected layer and set 0.8 as its value.

The final value of the prediction and the ground-truth will be compared, and we will compute its average error. Our target during the training period is to minimize its error in order to improve its accuracy, so a loss function must be designed and optimized, which can be divided into following two steps:

Firstly, we should compute the result of the SoftMax function used in fully connected layer, which can be represented by the following formula:

$$\mathrm{y_i} = \frac{e^{a_i}}{\sum_{k=1}^{C} e^{a_k}}, i = 2 \tag{6}$$

Where i indicates the number of the classification, a is the output of the fully connected layer and C is the number of a and e is a natural logarithm.

Secondly, the error function will be computed as followings:

$$J(W) = -\sum_{\mathrm{k}=1}^{n} \sum_{i=1}^{C} t_{ki} \log(\mathrm{y}_{ki}) \tag{7}$$

Where n is the number of the samples, C is the number of output of the fully connected layer, t_{ki}

indicates the probability that samples k belongs to class i, y_{ki} indicates the probability that models predict sample k belongs to class I and W is the weightcoefficient of filters. Noting that the loss in each batch will be summed and averaged to get the final loss.

To minimize the loss, the most common way that people use is to take advantage of backpropagation algorithm, which is based on gradient descent [16]. We adopt ADAM algorithm proposed in [17] to do the computation of optimization and use the Equation defined in (8) to update the parameter W.

$$W^{'} = W - \eta \cdot \frac{\partial J(W)}{\partial W} \tag{8}$$

Where $W^{'}$ is the value which will be updated after the computation of each gradient descent step and η is the learning rate set to 0.0001 in the experiment. Figure 2 has visualized the architecture of our model.

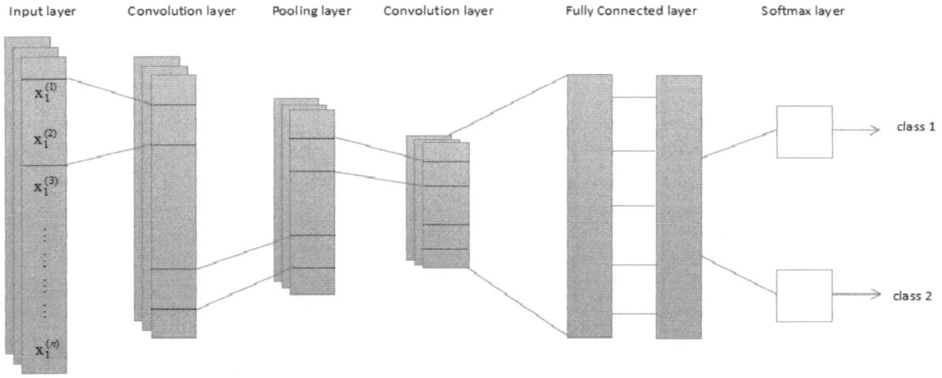

Figure 2. The architecture of our model

4. Result

Some related parameters in our model can be concluded as followings:

(1) Number of layers:6.

Note that too many convolutional layers may lead to complex computation and gradient vanishing or diffusion, and a small number of convolutional layer may make the result unreliable. So, to balance the speed of computation and the efficiency of our model, we have make several tests and decide to choose 6 as our layer number.

(2) Number of fully connection layer:2

We have set 2 fully connection layers and both of them will help to convert the output data into a matrix with different shapes, which will make it easy for us to do the classification by means of SoftMax function.

(3) The number of each filters in convolutional layers and the size of kernel in pooling layers:

32,64 and 128 for the first three convolutional layers and 256 for the others. The size of kernel in pooling layer is 1.

Beside what has been mentioned above, we use batches of 1000 samples and each sample contains 1×4 eigenvector. We have split the data-sets and 80% of that is for training and the rest is for testing. To evaluate the performance, accuracy is far from enough, so we have also proposed another 3 indexes called precision, recall the F1-score to help us make the evaluations.

Accuracy: Given test data-sets, the accuracy is the number of samples correctly classified by the classifier divided by the total number of samples.

Precision and recall: The precision and recall can be calculated by the following formulas from Table 1:

Table 1. Confusion Matrix

Actual class	Prediction class	
	Up	Down
Up	True-positive	False-negative
Down	False-positive	True-negative

$$P = \frac{TP}{TP + FP} \tag{9}$$

$$R = \frac{TP}{FP + FN} \tag{10}$$

Note that it is hard to find a satisfying result where the value of precision and recall are both high. Generally speaking, when the value of recall is very high, the value of precision is often low, and vice versa. Because of that, to judge which the algorithm is better for the samples easily, we also introduce F1-score, which contains an argument β showing the weight of precision and recall according to the demand. The formula likes the following equations:

$$F_{\omega} = \frac{(1+\omega^2) \times P \times R}{(\omega^2 \times P) + R} \tag{11}$$

In this paper, we choose $\omega = 1$ to measure the performance.

We have visualized the result of the experiment in Figure 3, including the change between training sets and testing sets in accuracy and precision-recall curve of testing sets.

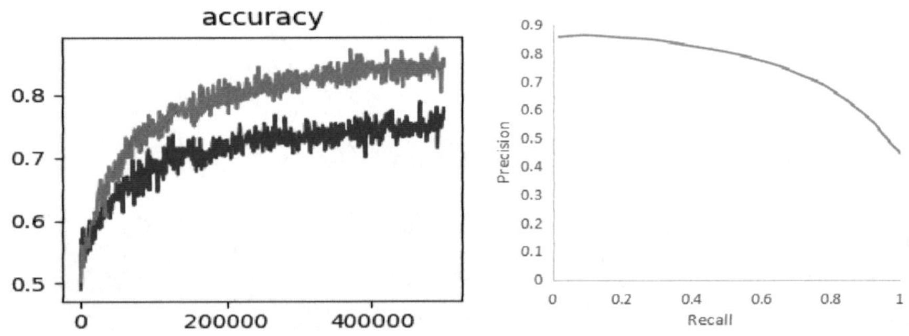

(a) (b)

Figure 3. The related evaluation indexes of out model

Where the red or the higher curve indicates the training data and the blue or the lower curve indicates the testing data in picture (a). The curve in picture (b) indicates precision-recall images of testing sets. X-axis is the number of iterations in picture a and recall in picture b while y-axis is the accuracy in picture a and precision in picture b.

Besides, we have used a table to record the number of accuracy in different number of iterations and F1-score using different thresholds to compare whether the model can fit the stock data of different types.

Note the threshold in Table 2 means that firstly the prediction score of test sets should be sorted in descending order and then we choose the biggest number, median and the smallest number of the prediction score named a, b and c as our model's thresholds for classification and calculation.

By comparison of the result, we can conclude that our model is feasible for different stock data.

Besides, it has shown that in addition to Recurrent Neural Network, the convolutional neural network can also solve the problem like stock movement prediction or other predictions in financial areas thanks to its ability of capturing micro-change of data at different time.

Table2. The result of the experiment using different data sources: (a) is the result of accuracy and (b) is the result of F1-score

Data source	1	2	3
Iteration=25000			
Accuracy	73.2%	73.9%	73%
Iteration=500000			
Accuracy	73.8%	75.2%	74.4%

(a)

Data source	1	2	3
Threshold=a			
F1-score	4.5%	3.7%	4.0%
Threshold=b			
F1-score	70.4%	73.0%	72.3%
Threshold=c			
F1-score	60.0%	61.8%	61.6%

(b)

5. Conclusion

In this paper we have introduced a CNN model to make the stock prediction and used a conv1d function to process the 1D data in the convolutional layer. We have also preprocessed stock data which will be input into the model to improve the result of the model. The result has been evaluated by different stock data, and finally indicates that our CNN model is robust and can also be used to make the predictions even if the source data is 1D sequential.

There are some interesting directions for further study:

(1) We can use the stock data with more features to make predictions, for the movement of stock may not only influenced by the features of open, close, high and low prices. Moreover, we can make a comparison about which features are key to the result.

(2) We can use more data sets to judge whether the result will be better, for the number of the data-sets is of great importance in deep learning.

(3) Using other convolutional neural network architectures can also be considered as a good idea, including Google net, Alex net and so on for they have achieved good performance in ILSVRC. We can finetune them to take advantage of them to better solve the problems in financial areas.

References

[1] J. F. Chen, W. L. Chen, C. P. Huang, S. H. Huang and A. P. Chen, "Financial Time-Series Data Analysis Using Deep Convolutional Neural Networks," *2016 7th International Conference on Cloud Computing and Big Data (CCBD)*, Macau, 2016, pp. 87-92.

[2] L. J. Cao and F. E. H. Tay, "Support vector machine with adaptive parameters in financial time series forecasting," in *IEEE Transactions on Neural Networks*, vol. 14, no. 6, pp. 1506-1518, Nov. 2003.

[3] Kercheval, Alec N., and Y. Zhang. "Modelling high-frequency limit order book dynamics with support vector machines." *Quantitative Finance*15.8(2015):1-15.

[4] K. Abhishek, A. Khairwa, T. Pratap and S. Prakash, "A stock market prediction model using Artificial Neural Network," *Computing Communication &Networking Technologies (ICCCNT), 2012 Third International Conference on*, Coimbatore, 2012, pp. 1-5

[5] Zhai, Yuzheng, A. Hsu, and S. K. Halgamuge. "Combining News and Technical Indicators in Daily Stock Price Trends Prediction." *International Symposium on Neural Networks: Advances in Neural Networks* Springer-Verlag, 2007:1087-1096.

[6] Ruiz, Eduardo J., et al. "Correlating financial time series with micro-blogging activity." ACM, 2012:513-522.

[7] Armano, G., M. Marchesi, and A. Murru. "A hybrid genetic-neural architecture for stock indexes forecasting." *Information Sciences*170.1(2005):3-33.

[8] Xu, Kelvin, et al. "Show, Attend and Tell: Neural Image Caption Generation with Visual Attention." *Computer Science* (2015):2048-2057.

[9] A. Tsantekidis, N. Passalis, A. Tefas, J. Kanniainen, M. Gabbouj and A. Iosifidis, "Forecasting Stock Prices from the Limit Order Book Using Convolutional Neural Networks," *2017 IEEE 19th Conference on Business Informatics (CBI)*, Thessaloniki, 2017, pp. 7-12.

[10] Kim, Yoon. "Convolutional Neural Networks for Sentence Classification." *EprintArxiv* (2014).

[11] Zhang, Xiang, and Y. Lecun. "Text Understanding from Scratch." *Computer Science* (2015).

[12] Kalchbrenner, Nal, E. Grefenstette, and P. Blunsom. "A Convolutional Neural Network for Modelling Sentences." *EprintArxiv* 1(2014).

[13] Collobert, Ronan, and J. Weston. "A unified architecture for natural language processing: deep neural networks with multitask learning." *International Conference on Machine Learning* ACM, 2008:160-167.

[14] Mikolov, Tomas, et al. "Recurrent neural network based language model." *INTERSPEECH 2010, Conference of the International Speech Communication Association, Makuhari, Chiba, Japan, September* DBLP, 2010:1045-1048.

[15] Irsoy, Ozan, and C. Cardie. "Opinion Mining with Deep Recurrent Neural Networks." *Conference on Empirical Methods in Natural Language Processing* 2014:720-728.

[16] Saad, E. W, D. V. Prokhorov, and D. C. Wunsch. "Comparative study of stock trend prediction using time delay, recurrent and probabilistic neural networks." *IEEE Transactions on Neural Networks* 9.6(1998):1456-70.

[17] Rather, Akhter Mohiuddin, A. Agarwal, and V. N. Sastry. "Recurrent neural network and a hybrid model for prediction of stock returns." *Expert Systems with Applications* 42.6(2015):3234-3241.

[18] Chen, Kai, Y. Zhou, and F. Dai. "A LSTM-based method for stock returns prediction: A case study of China stock market." *IEEE International Conference on Big Data* IEEE, 2015:2823-2824.

[19] Nelson, David M. Q., A. C. M. Pereira, and R. A. D. Oliveira. "Stock market's price movement prediction with LSTM neural networks." *International Joint Conference on Neural Networks* IEEE, 2017:1419-1426.

[20] A. Raj, S. Gupta and N. K. Verma, "Face detection and recognition based on skin segmentation and CNN," *2016 11th International Conference on Industrial and Information Systems (ICIIS)*, Roorkee, India, 2016, pp. 54-59.

[21] A. Farooq, S. Anwar, M. Awais and S. Rehman, "A deep CNN based multi-class classification of Alzheimer's disease using MRI," *2017 IEEE International Conference on Imaging Systems and Techniques (IST)*, Beijing, 2017, pp. 1-6.

[22] K. Chen, Y. Zhou and F. Dai, "A LSTM-based method for stock returns prediction: A case study of China stock market," *2015 IEEE International Conference on Big Data (Big Data)*, Santa Clara, CA, 2015, pp. 2823-2824.

[23] A. L. Maas, A. Y. Hannun, and A. Y. Ng. Rectifier nonlinearities improve neural network acoustic models. In *ICML*, 2013.

IOP Conf. Series: Materials Science and Engineering **435** (2018) 012027 doi:10.1088/1757-899X/435/1/012027

Global Optimization of Thermal Fatigue Resistance of Brake Disc Materials Based on Genetic Algorithm and Neural Network

Xuemin Sun [1], Jinsong Bao[1,*], Shiyong Yin[1] and Zhiqiang Li[1]

[1]College of Mechanical Engineering, Donghua University, Shanghai 201620, China

*corresponding author's e-mail: bao@dhu.edu.cn

Abstract. Brake disc is an important component of braking system, its thermal fatigue resistance directly affects the braking effect. In order to avoid the thermal fatigue phenomenon of brake disc, an orthogonal experimental method is presented to optimize thermal fatigue performance, which optimizes the horizontal combination of the four elements, and determines the order of the four elements that affect the thermal fatigue resistance of the material by means of range analysis and variance analysis. A global optimization model based on genetic algorithm and neural network is proposed to solve the local optimal problem in orthogonal experiments. Finally, an comparative experiment is taken, and the results show the efficiency of the orthogonal experimental method combined with the global optimization model on improving the thermal fatigue resistance of the brake disc material.

1. Introduction

Brake system principle is that the brake disc and the hub are assembled together [1-2]. Through dry friction between the brake disc and the brake linings, the kinetic energy of high speed vehicle is turned into heat energy, so the vehicle can be stopped gradually. When braking, the vehicle from high speed to being still only lasts a very short time. The braking system bears huge load. In the braking process, the brake disc has been bearing the effect of friction and wear. If the pressure is too large, the brake disc will deform or crack. The braking is a continuous cycle process. The brake disc has been in alternating hot and cold, so the disc brake in the braking process produces large thermal stress. Under the effect of the thermal stress, the brake disc surface will crack and deformation, namely, thermal fatigue phenomenon. It will make the brake disc invalid [3-5]. Therefore, the brake disc is required to have a good anti fatigue performance.

Zhang Junqing, Zhou Suxia, et al [6] obtained the thermal fatigue experimental of SiCp/A356 particle reinforced composite notched specimens at 20° C~300° C cycle, and obtained the relation between the thermal fatigue crack formation life and geometric dimensions such as notch radius and thickness of specimens. Shi Xiaoling [7] analyzed the variation law of the stress intensity factor of cracks on the friction surfaces of the brake discs. Through a combination of 1:1 gantry experimental and finite element calculation, the mechanism of action between multiple cracks on the friction surface was further studied and further based on fatigue. The damage mechanism theory and fracture mechanics method evaluated the service life of forged steel brake discs. Feng Chenhui [8] studied the effect of carbon-silicon content on the thermal fatigue properties of grey cast iron. The experimental results show that within a certain range of components, the higher the ratio of carbon to silicon, the higher the carbon equivalent of gray cast iron, the greater the thermal fatigue resistance of gray cast iron.

In this paper, the brake disc of ductile iron is studied, the factors of Si, Mn, Mo and Cu are added in ductile iron, and the thermal fatigue performance of ductile iron is optimized by orthogonal experimental design, and the Si, Mn, Mo and the order of influence of Cu factor on the thermal fatigue performance of brake disc; In order to make up for the insufficiency of the orthogonal experimental, the genetic algorithm and neural network algorithm are added, and the global optimization model is set up to optimize it, and finally the level combination of Si, Mn, Mo and Cu factors is obtained. In order to improve the thermal fatigue performance of the brake disc, prolong the service life of the brake disc and increase the braking safety factor of the vehicle, the overall frame of this article is shown in Figure 1.

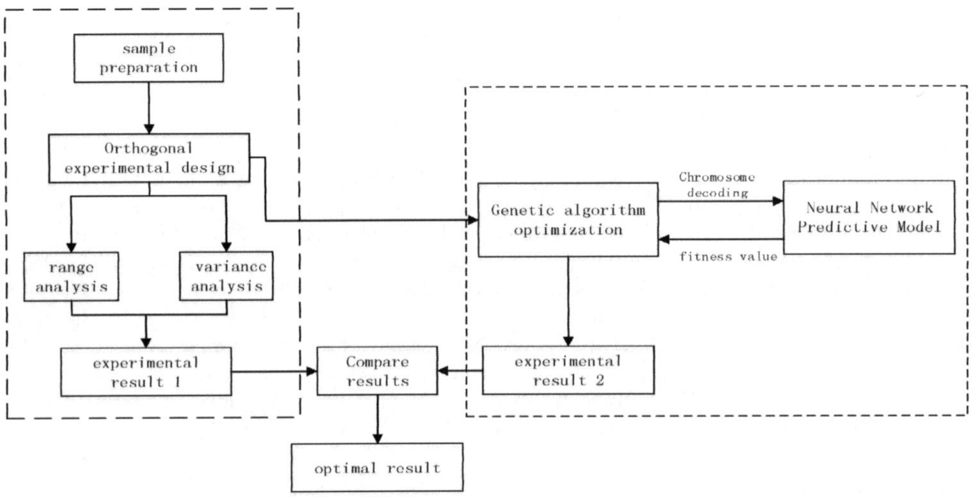

Figure 1. Overall framework

2. Experimental materials and methods

2.1. Preparation of raw materials and samples

The ductile iron is used as raw material in the experimental. Its chemical composition of mass proportion is that of 3.74%C, 1.00%Si, 0.20%Mn, S≤0.02%, P≤0.05%. The spheroidizing agent is rare earth magnesium alloy. The nucleating agent is ferrosilicon 75. After the pig iron in the cupola is melted and desulfurized, the temperature is kept at 1600℃. Then 1.5% of the spheroidizing agent and 1% of the nucleating agent are Added. The treated molten iron is casted into a brake disc. The size of the brake disc for experimental is shown in Figure 2.

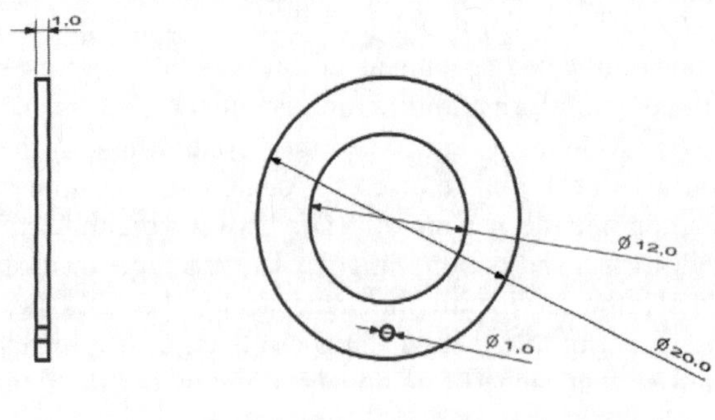

Figure 2. The size of the brake disc for experimental

2.2. Program and method of thermal fatigue experimental

The thermal fatigue experimental is done by metal thermal fatigue experimentaling machine. The specimen is suspended on the transmission device through the hole of φ1. The specimen is made to move vertically to complete the cycle process of heating and cooling. The cycle time is 15s. The electromagnetic relay is used to heat the specimen to 800℃ at 5s. And the cooling water is used to cool it to 20℃ at 10s. From 20℃ to 800℃ by heating and cooled to 20℃ is a cycle. The counter records the time of cycles. The crack occurrence is observed by laser confocal scanning microscope each 50 time cycles. The cycle time is recorded when the crack in the specimen occurs for the first time.

The orthogonal table L9 (3^4) is used to analyze the levels and factors of the experimental. It is shown in Table 1.

Table 1. Levels and factors of the experimental

Levels	Factors			
	Si	Mn	Mo	Cu
1	4.0	0.6	0.1	0.6
2	4.5	0.8	0.3	0.8
3	5.0	1.0	0.5	1.0

2.3. Orthogonal experimental results

Nine groups of specimens are experimentaled. The cycle times of all the groups are recorded when the crack in the specimen occurs. All the cycle times are shown in Table 2. In the experiment, the more the time of cycles in the metal thermal fatigue experimentaling machine is, the stronger the thermal fatigue resistance of the specimen is.

Table 2. Specimen composition and cycle times

Number	Composition（%）				cycles
	Si	Mn	Mo	Cu	
1	4.0	0.6	0.1	0.6	500
2	4.0	0.8	0.3	0.8	750
3	4.0	1.0	0.5	1.0	900
4	4.5	0.6	0.3	1.0	1100
5	4.5	0.8	0.5	0.6	1350
6	4.5	1.0	0.1	0.8	850
7	5.0	0.6	0.5	0.8	1600
8	5.0	0.8	0.1	1.0	1050
9	5.0	1.0	0.3	0.6	1200

2.3.1. Range analysis of orthogonal experiment. Range analysis is used mainly to distinguish the primary and secondary order of each factor in the fight against thermal fatigue, so as to determine what factors can be used to make the thermal fatigue resistance of nodular cast iron better.

①Determine the optimal level combination of experimental factors

Taking Si as an example, the sum of experimental indexes corresponding to the 1 level of Si is:

$$K_1 = y_1 + y_2 + y_3 \tag{1}$$

$$k_1 = K_1 / 3 \tag{2}$$

The sum of the experimental indexes corresponding to the 2 level of Si is:

$$K_2 = y_1 + y_2 + y_3 \tag{3}$$

$$k_2 = K_2 / 3 \tag{4}$$

The sum of the experimental indexes corresponding to the 3 level of Si is:

$$K_3 = y_1 + y_2 + y_3 \tag{5}$$

$$k_3 = K_3 / 3 \tag{6}$$

Range:

$$R = k_{max} - k_{min} \tag{7}$$

The experimental index and the range of the Si, Mn, Mo, Cu can be calculated, as shown in Table 3. According to the data in Table 3, it can be concluded that the optimum level of Si factor is K_3,Mn factor is K_1, Mo factor is K_3, Cu factor is K_2. Therefore, the level combination of the factors with better thermal fatigue resistance is as follows: $Si_3Mn_1Mo_3Cu_2$.

Table 3. experimental index and the extreme difference

	Si	Mn	Mo	Cu
K_1	2150	3200	2400	3050
K_2	3300	3150	3050	3200
K_3	3850	2950	3850	3050
K_1	7166.667	1066.667	800.000	1016.667
K_2	1100.000	1050.000	1016.667	1066.667
K_3	1283.333	983.333	1283.333	1016.667
R	566.666	83.334	483.333	50.000

②Determine the order of the factors

The exponential change trend of factors are shown in Figure 3. The influence of various factors on the thermal fatigue resistance of nodular cast iron is in the order of Si > Mo > Mn > Cu.

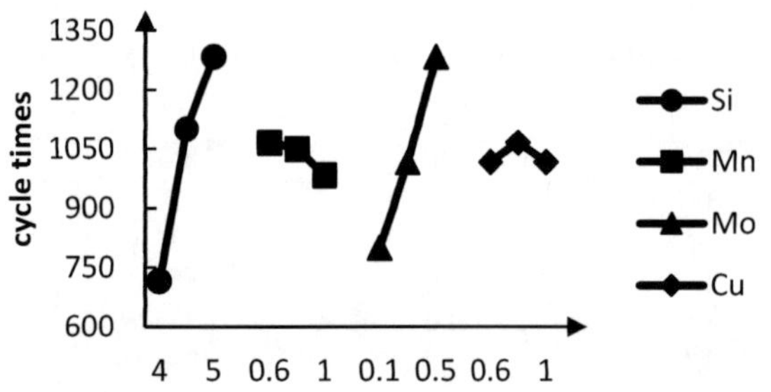

Figure 3. The exponential change trend of factors

2.3.2. Variance analysis of orthogonal experiment. By means of variance analysis, the influence degree of each factor on the thermal fatigue resistance of nodular graphite cast iron is further judged, which makes up the deficiency of extreme difference analysis. The variance analysis table can be go.

①Calculate the sum of squared deviations and degrees of freedom for each factor

Total deviation square sum:

$$SS_T = \sum_{i=1}^{n} X^2{}_i - \frac{\left(\sum_{i=1}^{n} x_i\right)^2}{n} \tag{8}$$

Column deviation square sum:

$$SS_j = \frac{1}{r}\sum_{i=1}^{m} K^2{}_{ij} - \frac{\left(\sum_{i=1}^{n} X_i\right)^2}{n} \tag{9}$$

Total degree of freedom:

$$df_T = n - 1 \tag{10}$$

Factor degree of freedom:

$$\mathrm{df}_{\mathrm{j}} = m - 1 \tag{11}$$

There:j=1,2,3,4.m=3.n=9.r=n/m.

Because Mn, Cu has little effect on the thermal fatigue resistance of ductile iron, it can be seen that Mn,Cu can be taken as the error term.

Error deviation square sum:

$$SS_{\mathrm{e}} = SS_2 + SS_4 \tag{12}$$

Error degree of freedom:

$$\mathrm{df}_{\mathrm{e}} = 2df_j \tag{13}$$

② significance experimental

The analysis of variance table of orthogonal experimental can be obtained by consulting F distribution table, as shown in Table 4. According to the analysis in Table 4, Si and Mo are the significant factors, Mn and Cu are not significant factors.

Table 4. Variance analysis

Factors	Sum of Squared deviation	Degree of freedom	Ratio	Critical value	Significance
Si	490555.556	2	55.187	F0.01=18.00	*
Mn	7222.222	2	0.812		
Mo	357222.222	2	40.187		*
Cu	10555.556	2	1.188		
SS_e	17777.778	4			
SS_T	865555.56	8			

③Determination of optimal process

The more cycles of the specimen in the thermal fatigue experimentaling machine, the better the thermal fatigue resistance of the specimen, and the factors Si and Mo have great influence on the experimental and are significant factors. The influence of Mn and Cu on the experimental is relatively small and not significant. The smaller the mass fraction of Mn and Cu is, the better the mass fraction is. However, as a brake disc material, the strength and hardness are also involved. The factors Cu can promote the formation of pearlite, reduce or completely inhibit the formation of ferrite, so the mass fraction of Cu can be increased.

3. Global optimization model for thermal fatigue resistance of brake discs using neural networks and genetic algorithms

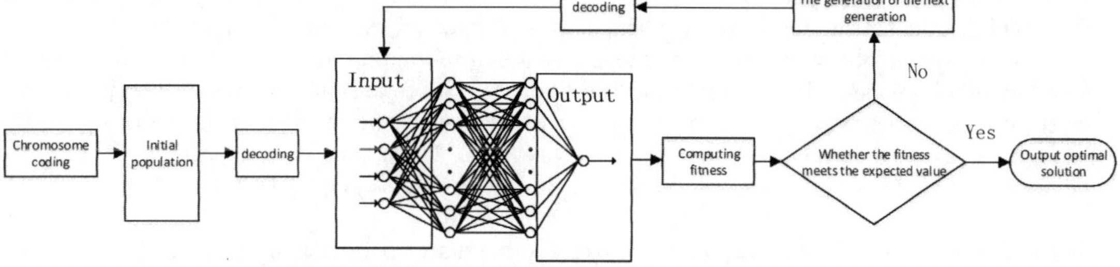

Figure 4. Flow chart of global optimization model

The results of 9 experimentals were obtained by orthogonal experimental in the above section, however, in order to solve the problem, genetic algorithm and neural network algorithm are introduced in this paper. The global optimization model is established by using the global optimizing ability of genetic algorithm and the nonlinear fitting ability of neural network, and its flowchart is shown in Figure 4. As can be seen from Figure 4, the structure framework of global optimization model is built by genetic algorithm in the process, and a neural network predictive model based on neural network is nested.

3.1. Structural framework based on genetic algorithm

In order to obtain the optimum ratio of Si, Mn, Mo and Cu for the thermal fatigue performance of the brake disc, the cycle times are used as the optimization objective, and the optimization model of the parameters in the range of parameters selected by the genetic algorithm is as follows:

$$\begin{cases} Maximum\ numberof\ cycles \\ 4.0\% \le Si \le 5.0\% \\ 0.6\% \le Mn \le 1.0\% \\ 0.1\% \le Mo \le 0.5\% \\ 0.6\% \le Cu \le 1.0\% \end{cases} \tag{14}$$

The initial population number is set to 100, The crossover rate is 0.6, the mutation rate is 0.05, the maximum iteration number is 400, and the genetic iterative counter is gen=0. The initial population was selected by the roulette method. The greater the degree of adaptability, the greater the probability of the individual being selected, the smaller the probability of being selected. After each chromosome is decoded and entered into the neural network predictive model, the output result is used as the input value of the fitness function, so the fitness function is defined as:

$$y = \max\ (S - \text{cycles}) \tag{15}$$

S is a large enough input parameter. The crossover rate randomly decides to participate in the cross chromosome, and to pair it with a single point crossover operation, with the new chromosome to replace the original chromosome, to get new groups. The individuals in the new population were randomly selected to mutate with mutation rate, and the new chromosomes were substituted for the new ones [9]. When the fitness value reaches the expectation, it terminates the algorithm, outputs the individual with maximum fitness, then decodes it and obtains the optimal parameter combination.

3.2. Prediction Model of fatigue cycle number based on Neural Network

In order to obtain the thermal fatigue cycle times of arbitrary Si, Mn, Mo and Cu factors, the prediction model of the relationship between the thermal fatigue performance of the brake disc and the ratio of Si, Mn, Mo and Cu factors can be established by neural network. The key factors that influence the accuracy of neural network prediction model are the kinds of neural networks, the number of training samples, the number of neurons in the hidden layer, the transfer function and the kinds of training functions.

The method used in this paper is the reverse propagation algorithm (BP algorithm), using MATLAB to train the neural network. The 9 sets of data obtained from the orthogonal experimental were used as training samples, and 4 elements of the orthogonal experimental table, Si, Mn, Mo and Cu as input layer nodes, were normalized to the data before training to improve the precision of the neural network model. The more cycle times of the specimen in the metal thermal fatigue experimentaler, the stronger the thermal fatigue resistance of the specimen, the more the cyclic frequency is used as the output node of the neural network. There is no definite method to determine the number of hidden layer nodes, and the theoretical calculation is complicated:

$$J = (n + m)/2 + a \tag{16}$$

There: J is the number of hidden layer nodes, n is the number of nodes in the input layer, m is the number of nodes in the output layer, a is the range of values for1-10.The number of neurons in the hidden layer can be estimated by formula (15) in the range of 4-13.The extensible range is 4-26.

The input and output relationships between each neuron in the hidden layer are as follows:

$$a_k = f(\sum_{i=1}^{r} w_{ki} p_i + b_k)/n \tag{17}$$

IOP Conf. Series: Materials Science and Engineering **435** (2018) 012027 doi:10.1088/1757-899X/435/1/012027

There: p_i Is the output value of the i neuron on the output layer, w_{ki} is the weight of the i th input to the k output, f is an activation function.

Activation function using ReLU function:

$$f(z) = \max(0, z) \tag{18}$$

The advantage of the Relu function compared to the sigmoid function and the Tanh function is that the convergence of SGD has a great acceleration effect [10].

It takes only one threshold to get the activation value, instead of counting the more complex exponential operations.

To solve the problem that the structure of neural network can not be clearly determined, this paper uses the method of trial and error to determine 11 kinds of neural network models of different structures: 4-4-1, 4-6-1, 4-8-1, 4-10-1, 4-4-4-1, 4-5-5-1, 4-6-6-1, 4-7-7-1, 4-8-8-1, 4-9-9, 4-10-10-1. In the training of neural network, the experimental data in Table 2 are used as training samples, but the data samples are too few, and it is easy to fit the phenomenon, so this paper uses the method of cross-validation to take any 8 groups as training sets, and the remaining group as the experimental set, and repeats the cycle. Set the maximum training Times 2000 times, the target error is 0.1.

Aiming at the problem of low training efficiency of traditional BP algorithm, this paper uses additional momentum method and adaptive learning rate algorithm to improve the convergence efficiency of model training.

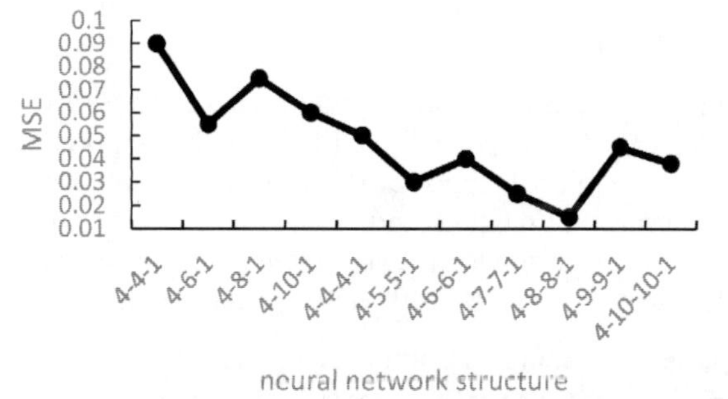

Figure 5. Mean square error of neural network models with different structures

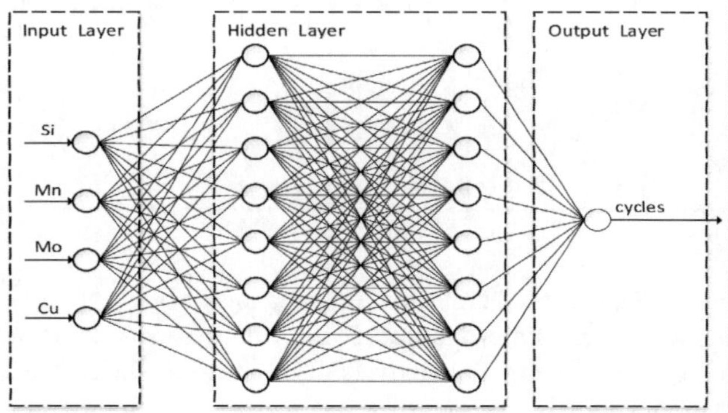

Figure 6. Neural network prediction model

In order to determine the number of neurons in the hidden layer of neural networks, to analyze the errors generated in the training of neural network models with different structures, and the method of mean square error (MSE) is used to evaluate the results, as shown in Figure 5. From Figure 5, we can see that the neural network structure of the double hidden layer is generally good, and the network model of 4-8-8-1 is the best, and the network model is shown in Figure 6.

4. Result contrast analysis

The global optimization model is implemented by MATLAB. After 60 iterations, the number of cycles reaches the maximum, and no change occurs. The genetic algorithm converges, as shown in figure 7.Finally, the optimum alloy factors for improving the thermal fatigue resistance of brake disc materials are obtained as follows: ω (Si) 5.2%, ω (Mn) 0.65%, ω (Mo) 0.5 %, ω (Cu) 0.84%.The number of cycles is 1780.

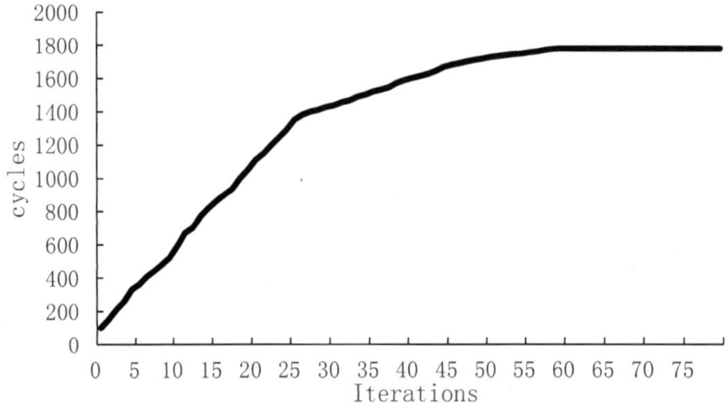

Figure 7. Global optimization process

Finally, the optimal combination obtained by the global optimization model is compared with the optimal combination obtained by the orthogonal experimental. It can be seen from Table 5 that the optimal horizontal combination obtained by the global optimization model is combined with the optimal level obtained by the orthogonal experimental method, and the number of cycles is increased by 11.25 %. Therefore, the effect obtained by the orthogonal experimental method combining the global optimal model is obviously better than that of the single orthogonal experimental method.

Table 5. Comparison of optimal results of optimization methods

Optimization Method	**Si/%**	**Mn/%**	**Mo/%**	**Cu/%**	**cycles**	**relative difference/%**
Orthogonal Experiment	5.0	0.6	0.5	0.8	1600	11.25
GA+BP	5.2	0.65	0.5	0.84	1780	

5. Conclusions

(1) In this paper, the influence of Si, Mn, Mo and Cu on the thermal fatigue resistance of ductile iron is studied by orthogonal experimental by adding a certain proportion of Si, Mn, Mo and Cu factors to the brake disc material. The influence order is obtained by the analysis of the extreme difference and variance of the experimental results: Si>Mo>Mn>Cu.

(2) In view of the problem that the orthogonal experimental can not achieve global optimization, a genetic algorithm and a BP neural network algorithm are proposed for global optimization. The global optimization process is constructed by using the nonlinear fitting ability of neural network and the global optimization ability of genetic algorithm. Finally, the optimal level combination of Si, Mn, Mo and Cu factors for improving the thermal fatigue resistance of the brake disc is obtained, and the optimal combination ratio is ω (Si) 5.2%, ω (Mn) 0.65%, ω (Mo) 0.5 %, ω (Cu) 0.84%.

Acknowledgments
This research was supported by the Natural Science Foundation of Shanghai (17ZR1400600) & National Natural Science Foundation of China (No.51475301).

References

[1] Liu W X. Structural analysis and design calculation of automobile brake system [M]. *Tsinghua university press*, 2014.
[2] Sun X M. Study on lightweight design of heavy duty vehicle disc brake [D]. *Liaocheng University*, 2017.
[3] Yang X, Zhang Z, Wang J, et al. Investigation of nanomechanical properties and thermal fatigue resistance of gray cast iron processed by laser alloying [J]. *Journal of Alloys and Compounds*, 2015, 626: 260-263.
[4] Mackin TJ, Noe SC, Ball KJ, et al. Thermal cracking in disc brakes [J]. *Engineering Failure Anasys*, 2002, 9(1):63-76.
[5] Moonesan M, Madah F. Effect of alloying elements on thermal shock resistance of gray castiron [J], *J.Alloys Comp*, 2012, (520): 226-231.
[6] Zhang J Q, Zhou S X, Yang Y, et al. Research on the thermal fatigue behaviors of the composites material SiCp/A356 used on the high-speed train[J]. *Engineering Mechanics*, 2011, 28(8):252-256.
[7] Shi X L. Research Thermal Fatigue Life Evaluation of Forging Brake Disc used for High Speed Trains [D]. *Beijing jiaotong university*, 2016.
[8] Feng C H. Application of high carbon low sillicon gray cast iron with Nb in the break disc [D]. *Shandong University*, 2016.
[9] Boris Hanin, Mark Sellke. Approximating Continuous Functions by ReLU Nets of Minimal Width [J]. *arXiv preprint arXiv*: 1710-11278,2017.
[10] Chen G L, Wang X F, et al. Genetic algorithm and its application [M]. *Posts & Telecom Press*, 1996.

IOP Conf. Series: Materials Science and Engineering **435** (2018) 012028 doi:10.1088/1757-899X/435/1/012028

Multiple Improvements to the Particle Swarm Optimization Algorithm

T Li [1,2] and Y Chen [1]

[1]School of Transportation Management, Dalian Maritime University, Dalian 116026, P.R.China
[2]China Waterborne Transport Research Institute, Beijing 100088, P.R.China

litao@wti.ac.cn

Abstract. To address the tendency of the particle swarm optimization algorithm to fall into local minima as well as to speed up its final stages of convergence, this paper proposes a combination of several improvements to the algorithm. First, the convergence factor is introduced as a special case of the inertia weight and transboundary particles are reset. Second, to increase the diversity of the particles and break the stagnation states of the particle swarm, a random number is introduced into the speed and position of the particles. The variation of the particles' position and velocity is hence optimized, and the search space for the optimal value is expanded. We use the single-mode sphere and Rosenbrock functions as well as the multimodal Rastrigin and Griewank functions to verify the algorithm. The maximum, mean, variance, stability, and convergence accuracy of the proposed algorithm are improved.

1. Introduction

The particle swarm optimization (PSO) algorithm is a type of bionic intelligence optimization algorithm proposed by Kennedy and Eberhart [1]. It is based on the idea of iterative search for an optimal value and solves the optimization problem by simulating a group of foraging activities and individual foraging action laws. In the foraging behavior of birds, although the target is not clear, each bird knows how far it is away from a target, so the best foraging program is to fly to the food closest to the location of the bird. This is why flying birds will suddenly synchronize their direction of change or aggregation or, alternatively, scatter. In the PSO algorithm, each bird is abstracted as a random search particle, and the distance to the target is measured by the fitness function. Each particle corresponding to a solution of the optimization problem is solved in the d-dimensional search space, and there is a speed vector and position vector that determine the search direction and distance in the search space of each particle. There is also a fitness value based on the objective function. Each particle knows the best position and current position it has experienced so far. This represents the flight experience accumulated by the particle itself. Moreover, each particle also knows the best position experienced by all the particles in the current population, which is called the accumulated flight experience of the population. Using these two pieces of information, the particles continually update their own speed and position. Simultaneously, the global best position and fitness of the local particle and global population are constantly changing until the condition for terminating the iterations is met. The optimal solution is the final global optimal position.

This method records the particle's position vector $X_i = (x_{i1,} x_{i2,} \cdots, x_{id})$ and speed vector $V_i = (v_{i1,} v_{i2,} \cdots, v_{id})$. The individual extrema of the particles $P_i = (p_{i1,} p_{i2,} \cdots, p_{id})$ and the global extrema of the population $P_g = (p_{g1,} p_{g2,} \cdots, p_{gd})$ are also recorded, where $I = 1, 2, \ldots, m$. A population composed of m particles searches in d-dimensional space. For the (t + 1)th iteration, the particles update the particle position and velocity using the following equation.

$$\begin{aligned} v_{ik}^{t+1} &= v_{ik}^{t} + c_1 r_1 (p_{ik}^{t} - x_{ik}^{t}) + c_2 r_2 (p_{gk}^{t} - x_{ik}^{t}) \\ x_{ik}^{t+1} &= x_{ik}^{t} + v_{ik}^{t+1} \end{aligned} \tag{1}$$

Here, c1 and c2 are the learning factors, where generally c1 = c2 = 2. In addition, r1 and r2 are used to increase the particle flight randomness. In Equation (1), the first term is the speed of the last iteration of the particle, followed by the term that specifies the effect of the best position of the particle on the next iteration, which comes from the flight experience of the particle itself. Finally, the last term is the social experience, which indicates the effect of the global extreme point on the current particle position, that is, the experience of all the particles benefits the search process of the particle. In each dimension, the velocity and position of the particles are limited to avoid searching beyond the scope of the search, for example, if the particle speed is too high, it could fly out of the range of the optimal solution, and if the speed is too low, it could perform an invalid or partially optimal search.

Kennedy and Eberhart proposed PSO in 1995. PSO uses a set of particles that represent potential solutions to an optimization problem. The particle moves toward an optimal solution based on its present velocity and the individual best position. It has found in previous iterations while also incorporating the globally best solution found by its companion particles [2]. To gain deeper insight into the mechanism of PSO, many theoretical analyses have been conducted on the algorithm, and most of these studies have focused on the behavior of a single particle in PSO, analyzing the particle's trajectory or stability using deterministic or stochastic methods [3–8]. To view PSO from a new perspective, we constructed a relationship between the dynamic PSO process and the transition process of a control system to identify factors that influence the convergence speed and stability of the basic PSO without increasing the algorithm's complexity [3]. In recent years, PSO has captured a great deal of attention and has been successfully applied in many applications because of its simplicity of implementation [9-12]. However, many experiments have indicated that the canonical PSO suffers from premature convergence when solving complex problems [13-14]. For this reason, a number of methods have been proposed to improve the efficiency and/or effectiveness of the PSO algorithm through adjustments to the code pattern, inertia weight, and maximum speed of the particles, mutation operators, neighborhood operations, and boundary conditions [15-18]. Some of the studies used hybrid (mixture and parallel) algorithms, for instance PSO with immune algorithms, simulated annealing algorithms, and chaos algorithms, to improve search effectiveness [19-21].

Overall, PSO algorithm research includes the following main aspects. For theoretical research, PSO convergence theory mainly focuses on a simplified PSO system. The convergence of the PSO can be analyzed using dynamic system stability theory to obtain some convergence conditions [22–25] by dynamically expressing the behavior of the particles as a nonlinear feedback control system. Furthermore, it obtains a statistical law of the PSO when approaching an attraction center [26]. To solve dynamic optimization problems, a PSO with a dynamic and hierarchical domain structure has been used, and it was proven that this structure can help keep the particles dispersed in a dynamic environment [24].

Using the above techniques, when an initial random solution is used such that the iterations of the particle position bring it close to the optimal solution, which is easy to implement, the algorithm converges faster. Furthermore, there are no complicated crossover and mutation operations as in a genetic algorithm [27]. However, it is also easy for a PSO to fall into a local extremum [28] and the final phase of the convergence is slow, among other problems [29]. To address these shortcomings, this paper presents three improvements to the PSO algorithm as described in detail in the next section.

2. Improvement Strategies for PSO

2.1 Convergence factor

Shi [30] proposed the inertia weight method to guarantee the convergence of the PSO algorithm and changed the speed update formula. This equation is now regarded as the standard PSO model, and is calculated as follows.

$$v_{ik}^{t+1} = wv_{ik}^{t} + c_1 r_1 (p_{ik}^{t} - x_{ik}^{t}) + c_2 r_2 (p_{gk}^{t} - x_{ik}^{t}) \quad (2)$$

In Equation (2), w is the inertia weight, and its value reflects the effect of the historical state of the particle on the current state during particle iteration. The algorithm is optimized using fixed or linearly decreasing inertia weights. Usually, weights in the range [0.1, 0.9] obtain better convergence performance. When the value is greater than 1.2, the particle group will not converge [31].

To improve the convergence speed and performance of the PSO algorithm, Clerc [32] added a convergence factor, as follows.

$$v_{ik}^{t+1} = \chi[v_{ik}^{t} + c_1 r_1 (p_{ik}^{t} - x_{ik}^{t}) + c_2 r_2 (p_{gk}^{t} - x_{ik}^{t})] \quad (3)$$

Here, $\chi = \frac{2}{\left|2-\varphi-\sqrt{\varphi^2-4\phi}\right|}$ is the convergence factor, where $\varphi = c_1 + c_2 > 4$, φ is usually set to 4.1, giving a convergence factor of 0.7298. The convergence factor can be regarded as a special case of the inertia weight. In this paper, the convergence factor is used to optimize PSO.

2.2 Handling transboundary particles.

During particle parameter optimization, if a particle's parameters exceed the range of allowable values, they are usually reset. The usual approach is to set the velocity and position of the transboundary particle parameters to the boundary values. That is, when the particle speed exceeds the upper bound, i.e., $v_{ik} > v_{k,\max}$, it is set to the maximum speed $v_{ik}=v_{k,\max}$, and when the velocity of the particle is below the lower bound, i.e., $v_{ik} < v_{k,\min}$, it is set to the minimum speed, $v_{ik} = v_{k,\min}$. The position of a particle beyond the border of the solution space is treated similarly. This method was suggested by Fu [33]. If there is a locally optimal solution in the vicinity of the boundary value, it will cause a large number of particles to move closer to the boundary value to form a particle aggregation, which may cause the particle group to fall into a locally optimal extremum and result in a combination of parameters that are not optimal.

Hence, this paper proposes an improved way to set the speed and location of transboundary particles. When $v_{ik} > v_{k,\max}$, the velocity of the particle is combined with a random number, $v_{ik} = v_{k,\max} - (v_{k,\max} - v_{k,\min}) * \text{rand}()$. When the velocity of the particle is below the minimum boundary value, $v_{ik} < v_{k,\min}$, it is changed using $v_{ik} = v_{k,\min} + (v_{k,\max} - v_{k,\min}) * \text{rand}()$. Using the same improvement strategy, the position of transboundary particles is also treated with random values; that is, when the particle position exceeds the highest boundary value, i.e., $x_{ik} > x_{k,\max}$, the location of the particles is corrected as $x_{ik} = x_{k,\max} - (x_{k,\max} - x_{k,\min}) * \text{rand}()$. Likewise, when $x_{ik} < x_{k,\min}$, the position of the particle is corrected using $x_{ik} = x_{k,\min} + (x_{k,\max} - x_{k,\min}) * \text{rand}()$. The rand() function generates a random value in [0, 1].

2.3 Adaptive particle mutation.

In the PSO algorithm, the position of the particle is determined by both the optimal merits of the particle itself and the optimal merits of the particles in the whole population. Hence, if there is a particle in the population that is close to the optimal solution, the other particles in the population will move rapidly in the direction of this particle. If there is no better position than the current global extremum, the algorithm is more likely to fall into a search for a locally optimal solution, which may lead to the convergence of the algorithm. Most of the particles are clustered into one place, so there is no way for particles to search the space again. Hence, because the final stage of the convergence is too slow, the lack of particle diversity caused by premature convergence, and other shortcomings, the diversity of the particles should be increased to break the stagnation of particle groups. It is possible to consider increasing the position or velocity of the particles so that the particle swarm can continue to find the optimal value, which strengthens the convergence speed and accuracy of the particle group. Because the key to avoiding locally optimal solutions is to use the optimal position experienced by each individual particle, this paper introduces an adaptive mutation based on the best position experienced by the particle itself.

Consider the example of looking for a minimum value. After the *k*-th search is performed, using $f_{avg}(k) = \frac{1}{n}\sum_{i=1}^{n} f_i(k)$, we can obtain the average and variance of the particle fitness of the particle group:

$$\sigma^2 = \sum_{i=1}^{n} \left(\frac{f_i - f_{avg}}{f}\right)^2$$

$$f = \begin{cases} max\{\,|f_i - f_{avg}|\},max\{\,|f_i - f_{avg}|\} > 1 \\ 1, others \end{cases} \tag{4}$$

Here, f is a normalization factor that controls the population fitness variance of the data range. When the variance of the overall fitness of the particle group is small, the particle group will be more closely aggregated. If the algorithm begins to prematurely converge, to reduce the possibility of local convergence of the current particles, this paper proposes randomly selecting a dimension for the position mutation operation according to a certain probability. This probability is the probability of occurrence of particle variation, which is related to the variance of the population fitness.

$$P{=}P_{max} - (P_{max} - P_{min})\frac{\sigma^2}{N} \tag{5}$$

$$x_{ik} = x_{k,min} + (x_{k,max} - x_{k,min}) * rand \tag{6}$$

In each iteration, a random number r is generated, and if it is smaller than the probability P of the mutation operation, the position of the particle can be updated, and the dimension of the particle position is arbitrarily selected.

When the variance of the overall fitness of the particle group is large, the diversity of the current particle swarm is relatively high and it is still in a state of random search. Here, it is not necessary to increase the diversity of the particle swarm. The probability of mutation should also be relatively small. When the variance of the population fitness is small, the current particle group is relatively dense, and the states of the particles need to be more diverse. Therefore, according to Equation (5), P is large and indicating that mutation of the particles is more likely. Hence, this method of calculating the mutation probability increases the diversity of particles when the algorithm begins to converge.

2.4 Pseudo Code of the Improved PSO Algorithm.

In this paper, the PSO algorithm is improved by introducing the convergence factor, combining it with the random reset for the velocity and position of transboundary particles, and adding particle position mutation to increase the diversity of the particle swarm. The pseudo code of the algorithm after these improvements is as follows.

```
Begin
        %Initial population
```

```
    For i = 1 to N
      For d = 1 to Dimension D
            Randomly initialize the population and speed
          End
    Calculate the initial fitness value fitness (i)
    End
    % Find individual extremes and global extremes
    [global_fit bestindex] = min(fitness)
    Indiv_fit = fitness
    % Iterative optimization
    For i = 1 to MaxIter
            For j = 1 to N
                    For d = 1 to Dimension D
                     Vj = w*Vj + c1*rand*(indiv_xj − xj) + c2*rand*(global_x − xj)  % Update Speed
                    if Vj > Vmax then
  Vj = Vmax − (Vmax − Vmin)*rand    % Cross speed processing
     end if
    if Vj < Vmin then
  Vj  = Vmin + (Vmax − Vmin)*rand
end if
            Xj = Xj + Vj   % Update location
            if          Xj > Xmax then
    Xj = Xmax − (Xmax − Xmin)*rand  %Transboundary position processing
    end if
    if          Xj < Xmin then
    Xj = Xmin + (Xmax − Xmin)*rand
end if
            End
            if rand < p
                    k = ceil(D*rand)
                    Xj, k = Xk, min + (Xk, max − Xk, min)*rand  % Particle position mutation operation
            end if
            update fitness(j), global_fit, indiv_fit(j)
            End
            For k = 1 to N
                    f(k) = abs(fitness(k) − avgfit_gen(i)) % Fitness variance
            End
            For k = 1 to N
                    sigma(k) = (fitness(k) − avgfit (i))/fmax;
            End
            square = sum(sigma^2)
            p = pmax − (pmax − pmin)*square/N         % Mutation probability
    End
End
```

3. Experimental analysis

3.1 Standard test functions.

To_verify whether the improved PSO algorithm proposed in this paper improves its performance, four benchmark functions widely used to evaluate the performance of the PSO algorithm are used. The four benchmark functions are divided into two categories: single-mode functions and multimodal functions. That is, they are divided by their number of extrema. Each category contains two test functions.

A. Single-mode functions

The sphere function is a simple single-peak continuous function, usually used to detect the optimization accuracy of an algorithm. Its value is (0, ..., 0) at the globally optimal solution, and its three-dimensional map shown in Figure 1. The function is expressed as formula(7).

$$f_1 = \sum_{i=1}^{n} x_i^2 \qquad (7)$$

$$f_2 = \sum_{i=1}^{n} 100(x_{i+1}^2 - x_i) + (1 - x_i)^2 \qquad (8)$$

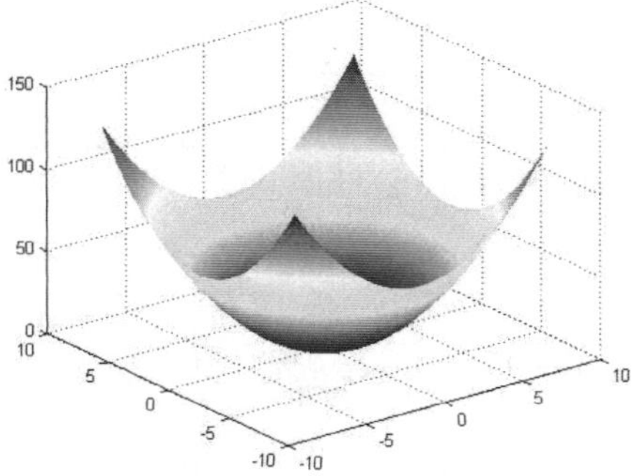

Figure 1 Sphere function

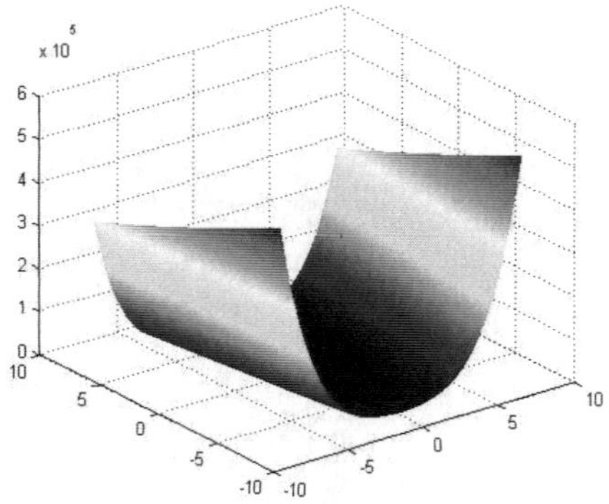

Figure 2 Rosenbrock function

The Rosenbrock function is a nonconvex function with a globally optimal solution at (1, ..., 1). Its three-dimensional graph is a parabolic valley, as shown in Figure 2, and the function is expressed as formula(8) .

B. Multimodal functions

The Rastrigrin function is a multi-peak function of sinusoidal ripple. There is a globally optimal solution at (0, ..., 0). To see the position of its global optimal solution, we present the three-dimensional graph of the negative of the function in Figure 3. There are several local extrema in the range (−5.12, 5.12), the number of which is related to the dimension, and the function expression is as formula(9), where the two coefficients are 10 here, but can be other values.

$$f_3 = \sum_{i=1}^{n} (x_{i+1}^2 - 10\cos(2\pi x_i) + 10) \qquad (9)$$

$$f_4 = \frac{1}{4000}\sum_{i=1}^{n} x_{i+1}^2 - \prod_{i=1}^{n} \cos(\frac{x_i}{\sqrt{i}}) + 1 \qquad (10)$$

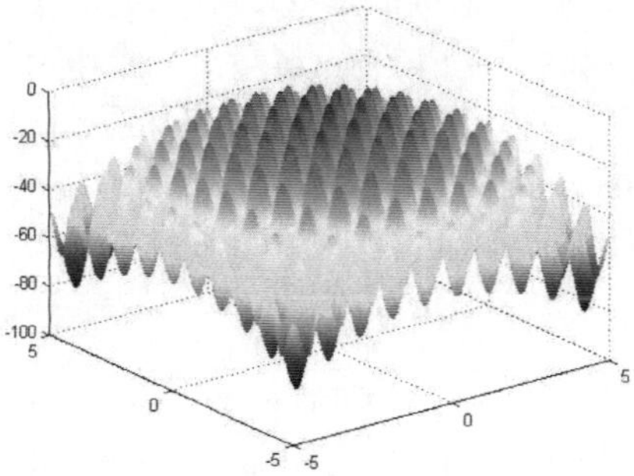

Figure 3 Rastrigrin function

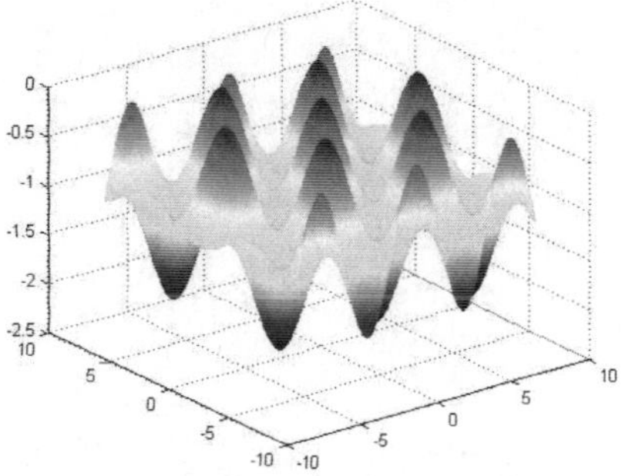

Figure 4 Griewank function

The Griewank function is a typical nonlinear multi-peak function, and it has many local extrema. Their number is related to the dimension of the search space, and the globally optimal solution is obtained at (0, ..., 0). We show the three-dimensional graph of the negative function in Figure 4. The function is expressed as formula(10).

The search ranges of the four test functions are shown in Table 1, and the search space dimension is 10 for all functions.

Table 1 Search range of the benchmark functions

functions	range	optimum	optimum points
Sphere	(-100, 100)	0	(0, …, 0)
Rosenbrock	(-30, 30)	0	(1, …, 1)
Rastrigin	(-5.12, 5.12)	0	(0, …, 0)
Griewank	(-300, 300)	0	(0, …, 0)

3.2 Results

The performance of the improved PSO algorithm is compared with those of the standard PSO, LDWPSO [34] and CLSPSO [35] algorithms on the four benchmark functions. Table 2 shows the parameters of each algorithm in this experiment.

Table 2 Parameter settings for each algorithm

algorithms	w	c_1	c_2	Population size
PSO	0.9	2	2	20
LDWPSO	[0.5, 0.9]	2	2	20
CLSPSO	0.9	1.496	1.496	20
Proposed	0.729	1.496	1.496	20

The four algorithms were run 10 times on each benchmark function and the maximum number of iterations was 1,000. The minimum, maximum, mean, and variance of each PSO algorithm are shown in Table 3. Figures 5–Figures 8 shows the convergence of the four algorithms in different benchmark functions.

Table 3 Comparison of each PSO algorithm on the benchmark functions

functions	algorithms	min	max	average	variance
Sphere	PSO	3.57316E-05	0.995077471	0.17262312	0.096997182
	LinWPSO	1.13012E-07	2.622217715	0.294078248	0.675193305
	CLSPSO	0	183313.1499	18331.31499	3360371093
	Proposed algorithm	0.001006009	0.004771275	0.002622363	1.29E-06
Rosenbrock	PSO	8.612494913	336.0685119	81.45418547	10114.34847
	LinWPSO	9.444753583	1273.833536	201.2536662	152751.5962
	CLSPSO	0	1.46577E+11	14657663272	2.14847E+21
	Proposed algorithm	8.743861127	81.93926906	21.21815365	6.61E+02
Rastrigin	PSO	7.825758006	32.72522317	16.58103419	49.48017738
	LinWPSO	11.94200845	29.85006006	18.41880806	34.80968821
	CLSPSO	0	113737.0288	11373.70288	1293611172
	Proposed algorithm	5.97915844	20.90221235	10.75755236	28.78808246
Griewank	PSO	0.054095474	0.175401891	0.110346599	0.00241495
	LinWPSO	0.118265615	1.552401333	0.781081812	0.259327117
	CLSPSO	0	42.40817951	4.240817951	179.8453689
	Proposed algorithm	0.029118976	0.363126298	0.187540239	0.008228349

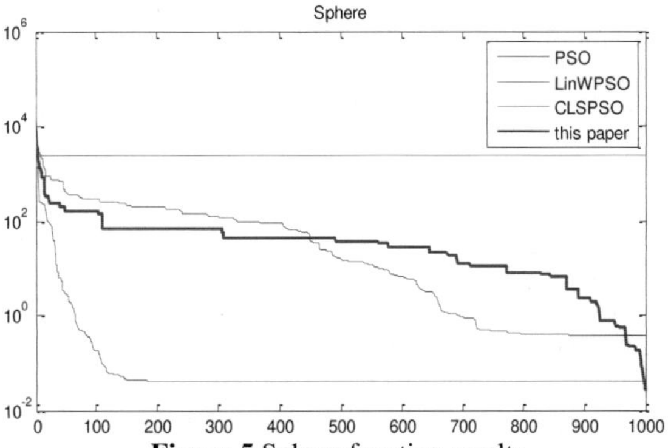

Figure 5 Sphere function results

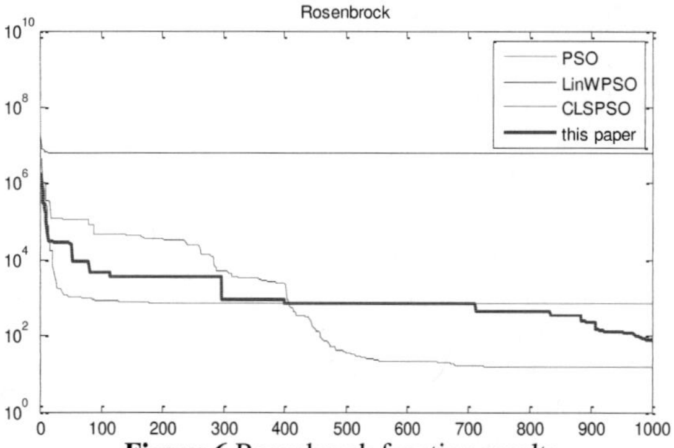

Figure 6 Rosenbrock function results

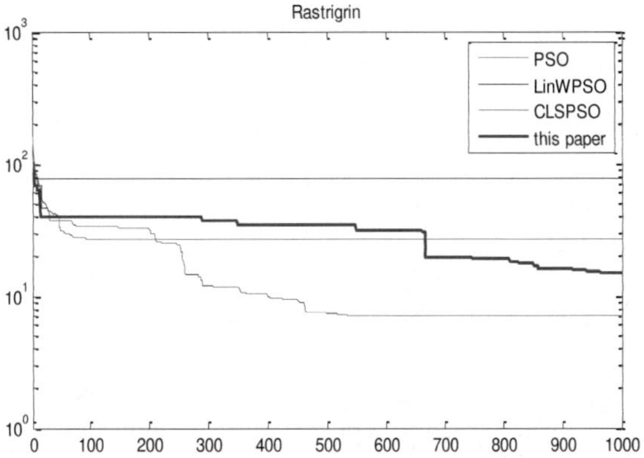

Figure 7 Rastrigrin function results

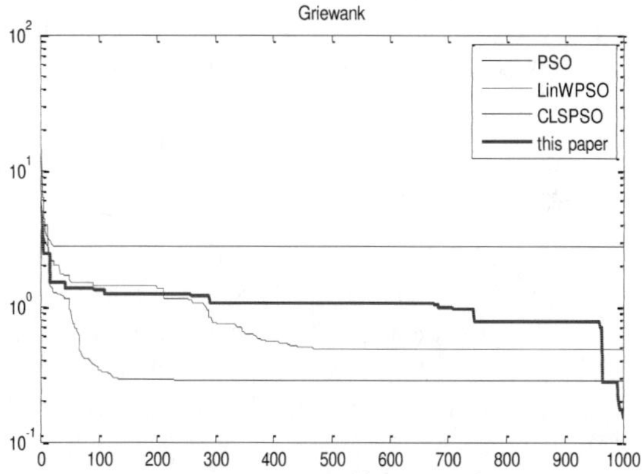

Figure 8 Griewank function results

The results show that the improved PSO algorithm has better local search ability than the other PSO algorithms. Later increases in the diversity of particles to boost the algorithm out of a local search obtains a better convergence. For the sphere function, the proposed algorithm has better stability than the other algorithms with respect to mean and variance, and the convergence precision of the algorithm is better than those of the other algorithms. The complicated Rosenbrock function optimization problem provides more detailed information. Each algorithm has large differences in the single-experiment results, but the proposed algorithm still has better stability and convergence accuracy. For the multimode Rastrigrin and Rosenbrock functions, several obvious inflection points can be seen in Figures-8, which is the embodiment of the particle self-mutation. A comparison of the experimental data obtained on the four test functions shows that the algorithm proposed in this paper performed better with respect to the maximum, mean, and variance than the other three algorithms.

4. Conclusions

In PSO, the individual particles make full use of their own experience and the group's experience to adjust their own state to obtain the optimal solution. However, it has the problem of premature convergence (especially when dealing with complicated, multi-extrema search problems) and poor global optimization ability. In this paper, we increased the convergence factor and used a random number with the velocity and position to reset transboundary particles. To avoid becoming trapped in a locally optimal solution, that is, the optimal position of an individual, the particle's location was modified by adaptive mutation to increase the diversity of the particles in a group. The improved algorithm was compared with the standard PSO, LDWPSO, and CLSPSO algorithms, and the maximum, minimum, mean, and variance are improved on standard single-mode and multimodal test functions. The performance results prove the effectiveness of the improved PSO algorithm.

Acknowledgments
This work was supported by the National Science and Technology Major Project of China (Grant No. 2017YFC1404602).

References

[1] Kennedy J, Eberhart R C. (1995)Particle swarm optimization[C]. *In Proc. of IEEE conference on neural networks*. 1942-1948.
[2] Zhang W, Liu J, Fan L B, Liu Y H, Ma D. (2016) Control strategy PSO[J]. *Applied Soft Computing*, 38(C):75-86.
[3] Clerc M., Kennedy J., (2002) The particle swarm-explosion, stability, and convergence in a multidimensional complex space [J], *IEEE Trans. Evolut. Comput*. 6(1): 58-73.
[4] Kadirkamanathan V., Selvarajah K., Fleming P.J. (2006) Stability analysis of the particle dynamics in particle swarm optimizer [J], *IEEE Trans. Evolut. Comput*. 10(3):245-255.
[5] Eberhart R.C., Shi Y. (2000) Comparing intertia weights and constriction factors in particle swarm optimization [C]. *in: Proceedings of the Congress on Evolutionary Computation*, 1(5):84-88.
[6] Trelea I.C. (2003) The particle swarm optimization algorithm: convergence analysis and parameter selection [J], *Inf. Process. Lett*. 85(6): 317-325.
[7] Jiang M., Luo Y., Yang S. (2007) Particle Swarm Optimization-Stochastic Trajectory Analysis and Parameter Selection., *I-TECH Education and Publishing*.
[8] Jiang M., Luo Y.P., Yang S.Y. (2007) Stochastic convergence analysis and parameter selection of the standard particle swarm optimization algorithm [J], *Inf. Process.Lett.* 102 (1): 8-16.
[9] Ciuprina G, Ioan D, Munteanu I. (2002) Use of intelligent-particle swarm optimization in electromagnetics [J], *IEEE Trans. Magn*. 38(2): 1037-1040.
[10] Ling S.H, Iu H.H.C, Chan K.Y. (2008) Hybrid particle swarm optimization with wavelet mutation and its industrial applications [J], *IEEE Trans. Syst. Man Cybern. Part B: Cybern.*, 38(3): 743-763 .
[11] Soh H., Y.S., Q.C. Nguyen, Nguyen Q.H., (2010) Discovering unique, low-energy pure water isomers: meme Ong tic exploration, optimization and landscape analysis [J], *IEEE Trans. Evol. Comput.*, 14(3): 419-437.
[12] Wachowiak M.P., Smolikova R, Zheng Y.F, (2004) An approach to multimodal biomedical image registration utilizing particle swarm optimization [J], *IEEE Trans. Evol. Comput.* 8(3): 280-288.
[13] Zhan Z.H., Zhang J., L i Y., (2009) Adaptive particle swarm optimization [J], *IEEE Trans. Syst. Man Cybern. Part B: Cybern*. 39(6): 1362-1381.
[14] Xia X W, Xie C W, Wei B, et al. (2017) Particle swarm optimization using multi-level adaptation and purposeful detection operators [J], *Information Sciences*, 385-386:174–195.
[15] Feng Y.M., Li C.J., Zhang M., (2008) Research on the function of reservoir long term operation based on improved particle swarm optimization (IPSO)[J], *Water Power*. 34(2): 94–97.
[16] Li G.Y. (2006) Regulation of water and sediment for the Yellow River based on joint operation of reservoirs and artificial intervention [J], *Journal of Hydraulic Engineering*, 37(12):1439-1446.
[17] Marler R.T., Arora J.S. (2004) Survey of multi-objective optimization methods for engineering [J], *Structural & Multidisciplinary Optimization*, 26(6):369-395.
[18] Massimiliano K., (2013) A multi-start opposition-based particle swarm optimization algorithm with adaptive velocity for bound constrained global optimization [J], *Journal of Global Optimization*, 55(1):165-188.
[19] Wang Y., Zhou J.Z., Zhou C., et al, (2012) An improved self-adaptive PSO technique for short-term hydrothermal scheduling [J], *Expert Systems with Applications*, 39(3):2288-2295.
[20] Yin X.A., Yang Z.F., Yang W., et al. (2010) Optimized reservoir operation to balance human and riverine ecosystem needs: model development, and a case study for the Tanghe reservoir, Tang

river basin, China [J], *Hydrological Processes*, 24(4):461-471.

[21] Yu X.B., Cao J., Shan H.Y., et al, (2014) An adaptive hybrid algorithm based on particle swarm optimization and differential evolution for global optimization [J], *The Scientific World Journal*. (6): 25-40.

[22] Trelea I C. (2003) The particle swarm optimization algorithm: convergence analysis and parameter selection[J]. *Information processing letters*, Elsevier, 85(6): 317-325.

[23] Van Den Bergh F. (2002) Analysis of Particle Swarm Optimizers [D], Doctoral Dissertation, Department of Computer Science, University of Pretoria, Pretoria, South Africa.

[24] Van Den Bergh F, Engelbrecht A P. (2006) A study of particle swarm optimization on particle trajectories[J]. *Information Sciences*, 176(8): 937-971.

[25] Kadirkamanathan V, Selvarajah R, Fleming P. (2006) Stability analysis of the particle dynamics in particle swarm optimizer [J]. *IEEE Transactions on Evolutionary Computation*, 10(3): 245-255.

[26] Poli R. (2009) Mean and Variance of the Sampling Distribution of Particle Swarm Optimizers During Stagnation [J]. *IEEE Transactions on Evolutionary Computation*, 13(4): 712-721.

[27] Zhang J B, Wang X L. (2010) A Novel Algorithm Based on K-Means Clustering and Binary Particle Swarm Optimization [J]. *COMPUTER TECHNOLOGY AND DEVELOPMENT*, 20(6):25-28.

[28] Jiang T, Zhang Y F, Wang Y H. (2006) A Study of Application of an Improved PSO Algorithm in BP Network [J]. *COMPUTER SCIENCE*, 33(9):164-165.

[29] Yu H S. (2015) Research of adaptive mutation of Particle Swarm Optimization [D]. Master Dissertation, Department of Applied Mathematics, Nanjing University of Information Science & Technology, Nanjing, China.

[30] Shi Y E, Berhart R C. (1998) A Modified Particle Swarms Optimizer [C]. *in Proc of IEEE Congress on Evolutionary Computation*. 6:69-73.

[31] Sun X, Zhou D W, Zhang X W. (2010) Convergence analysis and parameter selection of PSO model with inertia weight [J]. *COMPUTER ENGINEERING AND DESIGN*, 31(18):4068-4071.

[32] Clerc M. (1999) The Swarm and the Queen: Towards a Deterministic and Adaptive Particle Swarm Optimization [C], *in Proc of the Congress on Evolutionary Computation*, III: 1951-1957.

[33] Fu G J, Wang S M, Liu S Y, et al. (2005) A PSO with Bounded Mutation Operator [J]. *JOURNAL OF WUHAN UNIVERSITY OF TECHNOLOGY*, 27(9): 101-103.

[34] Lin W M, Zhou N N. (2014) A Particle Swarm Optimization Algorithm of Linear Decreasing [J]. *Computer Technology and Development*, 24(10):67-70.

[35] Meng F, Pan P P. (2011) Neural Network Trained by Hybrid Algorithms Based on Chaos Particle Swarm Optimization and Back-Propagation Algorithm [J]. *COMPUTER SIMULATION*, 28(2):196-199.

Dunhuang Decorative Pattern Digital Intelligent Enhancement Algorithm

Keyan Liu[1], Kaifa Deng[1,*] and Yuantao Jiang[2]

[1] School of Art and Design, Shanghai University of Engineering Science, Shanghai, China

[2] School of Optical-Electrical and Computer Engineering, University of Shanghai for Science and Technology, Shanghai, China

Tel.:+86-139-1764-5709,E-mail: liukeyan888@126.com, *dengkaifa@126.com, j17621607346@163.com

Abstract: Dunhuang Grottoes Art attracts attention from the world. The famous Mogao Grottoes, Yulin Grottoes, and Thousand-Buddha Grottoes not only have valuable artistic treasures such as murals, colorful sculptures, and architecture, but also have rich decorative pattern arts to be studied. However, due to the protection policy for Dunhuang murals and the limited research technology, the study of Dunhuang decorative designs has been greatly hindered. Therefore, this paper proposes a digital intelligent enhancement algorithm for Dunhuang decorative patterns study. Through image restoration and color enhancement processing of Dunhuang decorative pattern photographs, the decorative patterns are restored to the greatest degree, and scientific research is conducted on this. This not only can fully study the value of decorative patterns, but also can protect the Millennium murals in Dunhuang Caves well.

Keyword: Dunhuang Grottoes, decorative patterns, digitization, image restoration, color enhancement

1.Introduction

Dunhuang decorative patterns are interspersed among murals, colorful plastics and architecture, and play an auxiliary role in Dunhuang art. Only a small number of scholars at home and abroad who study Dunhuang or Dunhuang art pay attention to the study of Dunhuang patterns[1]. Through modern people's constant repair and consolidation, it has been found that these decorative patterns not only reflect the artistic characteristics of different historical periods, but also have very strong applicability in the present. The content of Dunhuang decorative patterns is very rich. After thousand years of development, each decorative pattern has different styling characteristics, color applications, patterns, and combinations of patterns that all reflect the era characteristic. After the millennium-developed Dunhuang decorative designs are rich in content, the humanities, history, religion, science and technology, and artistic expression methods contained in the patterns are worth studying[2]. Therefore, the work of sorting, researching, and re-designing Dunhuang decorative patterns needs to be more researchers involved. The current research issues are also faced with various aspects. For example, the Dunhuang Grottoes as a world cultural heritage receives visitors from all over the world all the year round. The gas they breathe and the activities in the caves all cause irreparable damage to the murals of the millennium, and thus they are facing the grotto art. Considering the protection, some of the special caves will be protected by taking turns to be open or not open to the outside world, and the time for the viewers to enter the cave will be limited. Only 15 minutes will make it difficult to study in depth.

After more than a thousand years of experience in Dunhuang patterns, a large number of exquisite murals have fallen off and swelled, showing problems such as crisp alkalis, empty drums, and armor. Therefore, how to use modern digital methods to restore a large number of vague and indistinct decorative patterns. Organizing patterns for more scholars to research and re-design applications, and more effective protection and dissemination of Dunhuang decorative patterns is a modern proposition that is worthy of study.

2.Image enhancement algorithm

The image enhancement algorithm improves the visibility of the scene by enhancing the contrast of the image. The most commonly used methods for image enhancement are histogram equalization, wavelet transform, and methods based on the retinal cortex theory.

The image enhancement method based on the Retinex theory was first proposed by E.Land et al. in 1971[3]. Retinex is an abbreviation of Retinex and Cortex, which indicates that the theory is based on the human visual system.

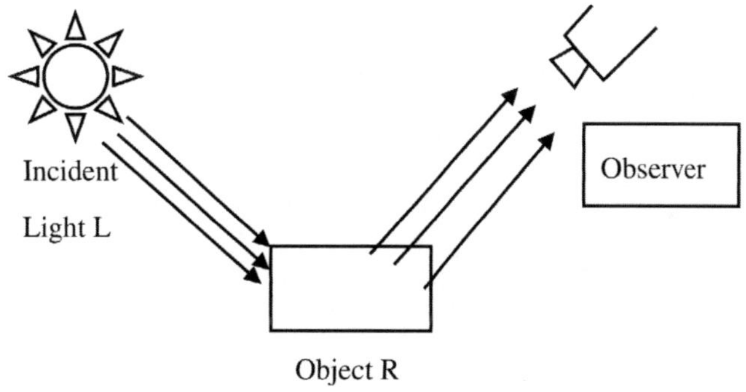

Figure 1. The retinex schematic.

Figure 1 shows that the scene image seen by the observer is actually a combination of incident and reflected light images. Therefore, each point in the image can be expressed as this formula.

$$S(x,y) = R(x,y)L(x,y) \tag{1}$$

In the formula, $S(x,y)$ is the original image, that is ,the fuzzy image that needs to be processed; $R(x,y)$ is the reflection image, which is controlled by the reflection condition of the surface of the object and contains the real details of the image; $L(x,y)$ is the luminance image, which is the incident light, the size of the incident light. Reflects the dynamic range that the image pixel can represent.

At present,the most widely used is the Single Scale Retinex (SSR) algorithm[4], Multiple Scale Retinex(MSR) algorithm[5], Multiple Scale Retinex color restore(MSRCR) algorithm[6]. The Retinex-based approach uses a logarithmic function and a Gaussian wrap function to enhance the visually weaker region. Compared to other commonly used image enhancement algorithms, the processed image has higher local contrast and less color distortion. However, because this algorithm involves the value of standard deviation of the Gauss function, if the value is too small, the visual effect of the enhanced image is not true. Even though MSR and MSRCR solve this problem by using different standard deviations, the problem still exists. Among them, the MSRCR algorithm recovers the color by using the non-linear function of the chromaticity of the original image, so as to compensate for the missing color information in the MSR algorithm, instead of restoring the true color of the image. Therefore, the color compensation of the MSRCR algorithm violates the theory of color constancy in a sense[7].

Meylan [8] used the center/neighborhood Retinex model as the core of the adaptive filtering method, according to the edge of the image adaptively changing the shape of the filter, thereby effectively eliminating the image of the halo phenomenon.

Petro et al. [9] proposed the Multiscale Retinex with Chromaticity preservation (MSRCP) algorithm with color retention, which is an improved MSRCR algorithm. MSRCP only executes on the luma channel, instead of performing an algorithm on each channel of R, G, and B, and then calculates the scaling factor to recover the true color of the image. In most cases, the image processed by the MSRCP algorithm has better enhancement.

Liu et al. [10] proposed an improved MSRCR low-visibility aeronautical image enhancement method to design an MSRCR with three or more scales. Based on mathematical analysis and derivation, an accurate balance of image contrast and color plausibility was proposed. In addition, the histogram truncation technique was introduced as a post-processing to remap the multi-scale Retinex to the dynamic range of the display. The image enhancement effect after processing is good.

Tang et al.[11] proposed an improved Retinex image enhancement algorithm based on steered filtering. In the HSI color space, the algorithm replaces the Gaussian filter with steerable filtering to obtain detailed information at different fine scales. Enhances high frequency information in different frequency bands. Then, guided filtering is used to extract certain low-frequency information in the logarithmic domain, and then the enhanced high-frequency information and low-frequency information are combined to obtain a joint image. Finally, the joint image is stretched to enhance the contrast of the image. Compared with the existing image enhancement methods, the images obtained in this way can effectively avoid the loss of halo effect and details.

This paper proposes a joint algorithm based on HSI color space model and color enhancement and image restoration based on fuzzy theory. The joint algorithm converts the image from RGB model to HSI model. In the fuzzy domain, WLS filtering is applied to the components of HSI space. Instead of the Soft Matting method, the transmission rate is refined, and then image enhancement is performed through the improved adaptive fuzzy logic to obtain an image of the enhanced image of Dunhuang decorative pattern.

3. Pal-King algorithm

This paper will implement the basic content of fuzzy set theory through Pal-King algorithm [12]. The algorithm proposed by Pal and King transforms an image into a fuzzy matrix through a transformation function. Then, it uses the fuzzy set theory to perform enhancement processing on the degradation in the fuzzy domain. Finally, the inverse transformation of the fuzzy domain yields the result. The process of the algorithm is as figure 2:

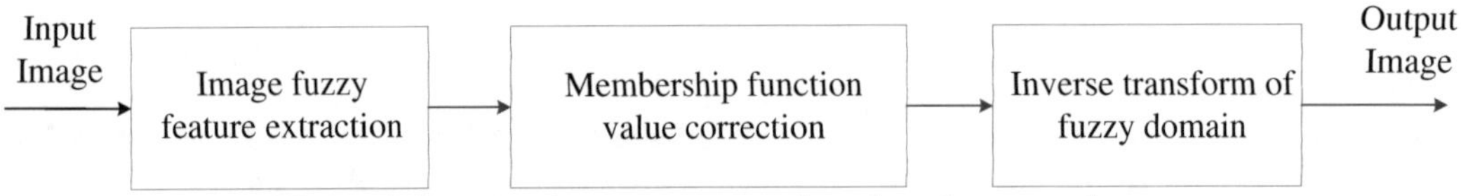

Figure 2. Image enhancement steps based on fuzzy set theory.

(1). Image fuzzy feature extraction.

The degraded image needs to be transformed from the spatial domain to the fuzzy domain. The expression is as follows:

$$\mu_{mn} = G(g_{mn}) = [1 + \frac{g_{\max} - g_{mn}}{F_d}]^{-F_e} \tag{2}$$

In the formula F_e and F_d are exponential and reciprocal transform coefficients respectively. The size of F_e and F_d determines the degree of fuzziness on the fuzzy feature plane.The scope of F_e is [1,3] and the scope of F_d is $[0.5(g_{max}-g_{min}),2(g_{max}-g_{min})]$. g_{max} is the largest gray value in the image; g_{min} is the gray value of the current pixel. When $\mu_{mn} = \mu_c = G(g_c) = 0.5$, and g_c is called the crossing point.And the selection of transition points is usually acquired by a priori knowledge at present.

(2). Membership function value correction.

The regression function of the fuzzy enhancement operator is used to correct the membership function. The expression is as follows:

$$T(\mu_{mn})=\begin{cases}2(\mu_{mn})^2 & 0\le\mu_{mn}\le 0.5\\ 1-2(1-\mu_{mn})^2 & 0.5\le\mu_{mn}\le 1\end{cases} \tag{3}$$

Fuzzy domain enhancement mainly depends on the value of μ_{mn} which is increased or reduced fuzzy enhancement operator, reducing the fuzziness of G. The fuzzy enhancement operator produces another fuzzy set on the fuzzy set G.

$$\mu'_{mn}=T^{(r)}(\mu_{mn})=T(T^{(r-1)}(\mu_{mn})) \qquad r=1,2\ldots,\infty \tag{4}$$

In the formula, $T^{(r)}$ is a multiple call to T.And When $r\to\infty$, $T^{(r)}$ produces a non 0 or 1 two value image. In order to ensure the integrity of image details and fuzzy image enhancement, according to different enhancement targets and image features, the average r value is 1, 2 or 3.

(3). Inverse transform of fuzzy domain.

The new gray level g_{mn} is generated by inverse transformation, and then the data is transformed from the fuzzy domain to the spatial domain of the image. The following expressions are as follows:

$$g'_{mn}=G^{-1}(\mu'_{mn})=g_{mn}-F_d[(\mu'_{mn})^{\frac{-1}{F_e}}-1] \tag{5}$$

The following figure shows the fuzzy enhanced image:

Figure 3. The fuzzy enhanced image.

4. The color space

The color space is also called a color model. The commonly used color models are: RGB model, CMY model, CMYK model, HSI model, NTSC model, and YCbCr model. There are still many current color models because the fields involved in color science are very extensive [13].

4.1 The RGB model

The RGB model is based on the Cartesian coordinate system mainly composed of the three primary colors of red, green and blue. This model as shown in figure 4.

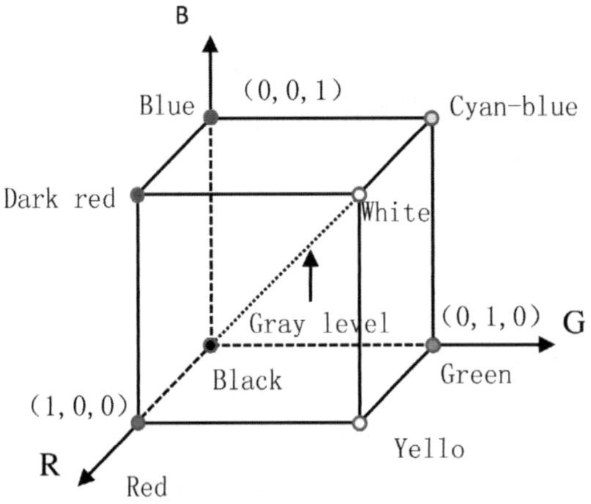

Figure 4. The RGB model.

R, *G*, and *B* are located on three corners of a rectangular coordinate system. In figure 3, the diagonal from black to white in the cube model is the gray level. The image represented by the RGB color model is composed of three component images, each of which represents a component image. The RGB model focuses on hardware and is commonly used in color monitors and a large part of color video cameras.

4.2 The HSI model

The RGB color model works very well for hardware processing. In addition, the RGB model meets the human eye's keen recognition of the red, green, and blue primary colors. However, the RGB model cannot explain human visually understood colors well. The proportion of each primary color given the three primary colors cannot describe the color of a chair. In general, people use hue, saturation and brightness to describe a colored object. Therefore, the HSI color model is produced by American colorist *H. A. Munseu* proposed in 1915. It describes color by hue (*H*), saturation (*S*), and brightness (*I*). Hue represents a color attribute, usually a solid color; saturation shows the intensity of white light diluting a solid color; brightness is subjective and in fact it is unmeasurable. It portrays the concept of colorless strength, and it is one of the important factors that characterize color perception[13].

4.3 RGB and HSI space conversion

An RGB space image is converted into an HSI model. The expression of each RGB pixel component is as follows:

$$H(R,G,B)=\begin{cases}\theta, & B\leq G\\ 360-\theta, & B>G\end{cases} \tag{6}$$

In the formula,

$$\theta=\arccos\left\{\frac{\frac{1}{2}[(R-G)+(R-B)]}{[(R-G)^2+(R-B)(G-B)]^{1/2}}\right\} \tag{7}$$

And the component *S* can be obtained by the following formula:

$$S(R,G,B)=1-\frac{3}{(R+G+B)}[\min(R,G,B)] \tag{8}$$

The component *I* can be obtained by the following formula:

$$I(R,G,B)=\frac{1}{3}(R+G+B) \tag{9}$$

In the formula, the RGB and HSI components are normalized to the interval [0,1].

From HSI model to RGB model conversion, the formula that can be used depends on the value of H. The formula is as follows:

When $H \in [0°,120°)$:

$$R=I[1+\frac{S\cos H}{\cos(60^{\circ}-H)}] \tag{10}$$

$$G=3I-(R+B) \tag{11}$$

$$B=I(1-S) \tag{12}$$

When $H \in [120^{o}, 240^{o}]$:

$$R=I(1\text{-}S) \tag{13}$$

$$G=I[1+\frac{S\cos(H-120°)}{\cos(180^{\circ}-H)}] \tag{14}$$

$$B=3I-(R+\mathrm{G}) \tag{15}$$

When $\mathrm{H} \in [240^{o}, 360^{o}]$:

$$R=3I-(G+B) \tag{16}$$

$$G=I(1-S) \tag{17}$$

$$B=I[1+\frac{S\cos(H-240°)}{\cos(360^{\circ}-H)}] \tag{18}$$

5. Joint algorithm for color enhancement and image restoration

Combining dark-color prior and WLS filtering theory in the fuzzy domain to perform color enhancement on images in the HSI space model. The color image enhances the essence of the enhancement of its luminance component, while the hue and saturation do not change. Therefore, this article will mainly operate on components *I*. The algorithm is as follows:

(1). First,exctracting the component R,G,B from the input original image.And according to the formula (6),(8),(9) transfer to the HIS space to obtain the cpmponent *H,S,I*.

(2). The dark color image is extracted from the component *I* and refined by WLS filtering.

(3). Transform images from spatial domain to fuzzy domain for feature extraction.

(4). The selection of transit points in the Pal-King algorithm is based on prior knowledge and is random, because different thresholds have a greater influence on the effect of image enhancement. Therefore, this paper adopts the OTSU (Maximum Class Variance Method) algorithm to search adaptively. The optimal threshold value of the image to be processed is used as the value of the transition point in the blur enhancement process. Next, the membership function value is corrected by the regression call of the fuzzy enhancement operator.

(5). Perform inverse transformation to convert the HSI space image to RGB space image to obtain the enhanced image of decorative pattern.

6. Experiment and result analysis

The algorithm in this paper is simulated on matlab platform. In this paper, the image enhancement algorithm used in this paper is compared with the original image according to these evaluation indexes include the image contrast, information entropy, average brightness, histogram correlation coefficient and algorithm running time. And the results are shown in table 1.

Table 1. Comparison of original and enhanced image parameters

Image	Contrast	Information Entropy	Average Brightness	Histogram Correlation Coefficient	Running Time /s
Original	260.69	7.54	140.81	—	—
Enhanced	1049.50	7.65	133.08	0.8037	2.79

Figure 5 and figure 6 show the experimental results of this paper's algorithm as well as the channel histograms for the two sets of images. In the histogram, red indicates *R* channels, and green indicates *G* channels,blue indicates the *B* channel.

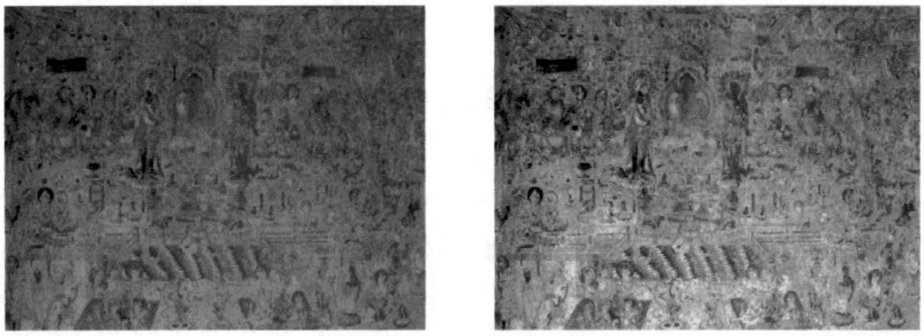

Figure 5. Image enhancement.

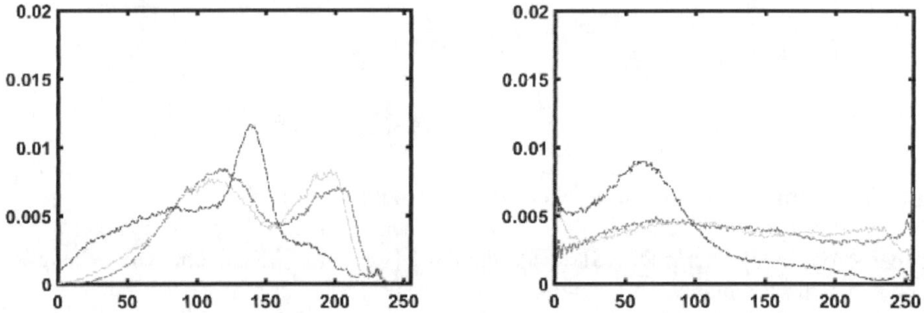

Figure 6. Comparing graphs and channel histograms after the originalgraph and algorithm in this chapter are processed.

The local contrast of the image processed by this algorithm is moderately enhanced, the restored image is clearer, the color reproduction is higher, the image is natural, and it is more in line with the visual perception of human eyes.As can be seen from the table, this algorithm enhances the contrast of the original image, improves the information entropy of the image, which is suitable for human eyes observation, and the correlation coefficient between the processed image and the original image histogram reaches 0.8037. It shows a high degree of similarity. And it reduces the time complexity and has strong robustness.

7. Conclusion

This paper mainly introduces the joint algorithm of image restoration and color enhancement based on Pal-King algorithm and HSI color space model. In the fuzzy domain, WLS filtering is used to replace the component of HSI space. The Soft Matting method refines the transmittance and then enhances the image through fuzzy logic. The OTSU algorithm is used to adaptively obtain the optimal threshold of the image to be processed. Finally, the Dunhuang decorative pattern experiment results obtained after the enhancement show that the algorithm complexity of this chapter is more complex. Low, brightly restored image, better detailing, better visual experience, and stronger robustness. Therefore, an intelligent enhancement algorithm based on image enhancement has provided us with a good solution for the study and investigation of Dunhuang decorative patterns.

Reference

[1] Zhao, J., Department, F. A., & Academy, D. (2017). The heart of a craftsman passed down——mr.duan wenjie's copying,research and inheritance of dunhuang mural art. Dunhuang Research.

[2] Yao, J. (2009). Secular characteristic of the dunhuang decorative pattern art. Art & Design

[3] Land E H, Mccann J J. Lightness and Retinex Theory[J]. Josa, 1971, 61(1): 1–11.

[4] Jobson D J, Rahman Z U. Properties and performance of a center/surround retinex[J]. IEEE Transactions on Image Processing, 1997, 6(3): 451–462.

[5] Rahman Z, Jobson D J. Multi-scale retinex for color image enhancement[J]. Proceedings of 3rd IEEE International Conference on Image Processing, 1996, 3:

[6] Jobson D J, Rahman Z U. A multiscale retinex for bridging the gap between color images and the human observation of scenes[J]. IEEE Transactions on Image Processing, 1997, 6(7): 965–976.

[7] Barnard K, Funt B. Investigations into Multi-Scale Retinex[J]. Colour Imaging: Vision and Technology, 1999: 9–17.

[8] Meylan L, Süsstrunk S. High dynamic range image rendering with a Retinex-based adaptive filter.[J]. IEEE Transactions on image processing, 2006, 15(9): 2820–2830.

[9] Petro A B, Sbert C. Multiscale Retinex[J]. Image Processing On Line, 2014, 4: 71–88.

[10] Liu C, Cheng I. Enhancement of low visibility aerial images using histogram truncation and an explicit Retinex representation for balancing contrast and color consistency[J]. ISPRS Journal of Photogrammetry and Remote Sensing, International Society for Photogrammetry and Remote Sensing, Inc. (ISPRS), 2017, 128: 16–26.

[11] Tang S, Dong M. Color image enhancement based on retinex theory with guided filter[C]//Proceedings of the 29th Chinese Control and Decision Conference, CCDC 2017. 2017: 5676–5680.

[12] Pal S K, King R A. On Edge Detection of X-Ray Images Using Fuzzy Sets[J]. IEEE Transactions on Pattern Analysis and Machine Intelligence, IEEE Computer Society, 1983, 5(1): 69–77.

[13] Gonzalez, Rafael C, & Woods, Richard E. (1987). Digital image processing. Prentice Hall International, 28(4), 484 - 486.

IOP Conf. Series: Materials Science and Engineering **435** (2018) 012030 doi:10.1088/1757-899X/435/1/012030

Multi-objective Optimization of Supply Chain Problem Based NSGA-II-Cuckoo Search Algorithm

Ali Metiaf [1,*], Fathy Elkazzaz [2], Wu Qian Hong [1] and Mohammed Abozied [3]

[1]Beihang University, School of Electronics and Information Engineering100191 China
[2]Benha University, Computer Science Department, Benha, Egypt
[3]MTC, Electrical Engineering Department, Cairo Egypt

E-mail: *alimetiaf23@buaa.edu.cn

Abstract. A hybrid Cuckoo search algorithm and NSGA-II algorithm called NSGA-II-Cuckoo is proposed and applied to find a set of non-dominated solutions for multi-objective supply chain problem. In NSGA-II-Cuckoo, the cuckoo search algorithm is applied to generate the initial population that considers as a good start for another algorithm to reduce the time, avoids the local optima entrapment so its effect on the accuracy of the final solutions obtained by the NSGA-II algorithm. The presented results clarify the enhanced performance for the NSGA-II-Cuckoo algorithm, which provides a sufficient stability and robustness for solving multi-objective supply chain problem.

1. Introduction

The optimization is a critical requirement in the world with a wide range of applications, the problem in these requirements is how to achieve many objectives under complex constraints in short time, by maximizing or minimizing desired and the undesired parameters that involved in the problem. However, with the increase of complexity in the economic world, due to several factors such as the expansion of the market, increased competition and customers' demands on the performance of a companies that requires to continuously evaluate, configure their Supply Chains (SCs) and provide customers with high-quality products/ services at the lowest cost within the shortest time, the multi-objective optimization has become a wide area of research interest.

Meta-heuristic algorithms can be defined as a useful tool that can be used to solve specific optimization problems [1]. The developed metaheuristic algorithms have three main purposes: solving problems faster, solving large problems, and obtaining robust algorithms [2]. There are many metaheuristic algorithms has proposed for solving multi-objective optimization problems like multi-objective genetic algorithm (MOGA), strength Pareto evolutionary algorithm (SPEA), Pareto-archived evolution strategy (PAES), etc. NSGA-II is a well-known, fast sorting and elite multi-objective genetic algorithm which proposed by[3] in 2002. However, the new metaheuristic search algorithm is cuckoo search (CS) which proposed by Yang and Deb[4] in 2009.

Over the past a few years various researchers have used meta-heuristic techniques to solve multi-objective optimization for supply chain cases. Bhattacharya and Bandyopadhyay (2010) [5] presented NSGA II [3] to choose facility location for achieving two contradictory objectives. Also, there are several researches applied NSGA II to solve multi-objective optimization for supply chain [6, 7]. Mastrocinque et al. [8] used bees algorithm for supply chain network with multi-delivery destinations and multi-products to decrease lead-time and total cost. Yuce et al.[9] improved Bees Algorithm to reduce the total lead-time and total cost and find the optimal solution for the given supply chain problem.

There are many studies concerning on multi-objective optimization applied Ant Colony Optimization (ACO) algorithm for SC problem with different configurations and frameworks. Zhao et al. [10] applied ACO algorithm for optimization SC design with a different business environment, and different customer demands to reduce total cost and total lead-times. Monkayo Martinez and Zhang [11] developed SC for a family of the product containing a complex hierarchy of sub-assemblies and components to satisfy two objectives (cost and time). Monkayo Martinez and Chang [12] designed SC to provide a satisfactory level of service to customers with minimizing total cost of SC. Xin-She Yang and Suash Deb[13] developed Cuckoo search algorithm (CS) which enhanced by Lévy flights [14] instead of simple random walks. They got the idea from brood parasitism of some cuckoos. Cuckoos have a beautiful sound, however, their aggressive reproduction strategy through removing others' eggs in the nest of other host birds to lay their eggs so increase the hatching probability.

In this article, the power of the proposed algorithm is increased through integrating the Cuckoo search algorithm and NSGAII algorithm to develop a hybrid algorithm for a multi-objective supply chain to achieve speed up convergence of the heuristic algorithms. We tested this algorithm by comparing it with the other algorithms like Bee (bee colony algorithm) and GA(Genetic Algorithms), and the solutions obtained for the proposed algorithm is better than the other algorithms like Bee and GA. These algorithms are applied to Engineering Optimization.

The remainder of this article is organized as follows. Section 2 presents a case study of Multi-objective optimization for supply chain design. After that, Section 3 presents Hybrid NSGA-II-Cuckoo search algorithm followed by Section 4, which describes Results and discussion. Section 5 provides the summary and conclusions.

2. Multi-objective optimization of supply chain design

In this paper, the multi-objective supply chain problem, obtained from[11] that interests the optimal choice of resource options to minimize TC and TLT.

In this study, the given supply chain is formed of M activities. Each activity i can be achieved by a various number of resource options (M_i), and each resource option j has its cost (C_{ij}) and processing lead-time (T_{ij}).

The cumulative lead-time (LT_i) at activity i denoted by Equation (1), is the sum of the processing lead-time of the activity and the maximum delivery lead-time for the set of activities S_i that input to activity i.

$$LT_i = \sum_{j=1}^{M_i} T_{ij}X_{ij} + \max_{a_q \in S_i}(LT_q) \tag{1}$$

For the sourcing activities, the second term of Equation (1) will be zero because of there is no preceding input.

$$X_{ij} = \begin{cases} 1 & if\ j\ is\ selected \\ 0 & otherwise \end{cases} \quad for\ i \in M \tag{2}$$

$$\sum_{j=1}^{M_i} X_{ij} = 1 \quad for\ i \in M \tag{3}$$

The cumulative lead-time at the activity a_d will be the LT_{a_d} for delivery of a product to its destination. The maximum lead-time amongst delivery nodes represented by Equation (4) which is used in this work:

$$TLT = \max_{a_d \in D} LT_{a_d} \tag{4}$$

The total supply chain cost can be expressed by Equation (5):

$$TC = \omega \sum_{i=1}^{M} \left(\mu_i \sum_{j=1}^{M_i} C_{ij} X_{ij} \right) \tag{5}$$

where μ_i is the average demand per unit time at the activity i, and ω denotes the period of interest depending on the unit time.

3. Hybrid NSGA-II-Cuckoo search algorithm

Heuristic and meta-heuristic algorithms is widely used for solving multi-objective supply chain problem because of their significant advantages, but long computational time is the common drawback of these methods, especially when the solution space is hard to discover. The newest meta-heuristic algorithms is a Cuckoo search (CS) algorithm proposed in 2009 by Yang and Deb[4], it is very easy to understand, implement, apply and requires simple mathematical pre-processing Comparing with other population based algorithms such as PSO, GA and other popular algorithms[15], which can solve general N-dimensional, linear and nonlinear optimization problems. Cuckoo search algorithm introduced to solve multi-objective supply chain problem[16], compared it with genetic algorithm and bee colony algorithm, and obtain better Pareto solutions (non-dominated set) for the supply chain problem.

Many efforts have been already done to speed up convergence of the heuristic algorithms. Most of these works are concentrated on improving selection, mutation and crossover operators and adaptive controlling of parameter settings. However, the initial population has a vital role in evolutionary algorithms, for example, NSGA-II algorithm, where affects the quality and the accuracy of the final result and also the convergence speed, because it helps to reduce the time of those algorithms needs to reach an optimal result. Therefore, we create a hybrid NSGA-II-Cuckoo search algorithm for solving the multi-objective supply chain problem to generate the set of non-dominated solutions. So in this work, we utilized cuckoo search algorithm described in [13,15] to find fast, high-quality initial solutions that used as initial population for the NSGA-II algorithm in order to provide fast convergence to the set of non-dominated solutions with less number of iterations which reduce the algorithm overall execution time. In the following, we present a pseudo-code, which presented that solving multi-objective supply chain problem by the NSGA-II algorithm with initial population created by the cuckoo search algorithm. See all details in Figure 1.

4. Results and discussion

The non-dominated solutions for multi-objective supply chain problem which generated by cuckoo search algorithm, bee colony algorithm, genetic algorithm and NSGA-II with maximum iterations equal to 100, the used parameters are duplicated in table 1, are presented in the figures 2-4. The black circle, Black Square, red star, and blue points respectively represent the set of non-dominated solutions created by cuckoo search algorithm, bee colony algorithm, NSGA-II algorithm and genetic algorithm.

From the Figure 2, we observe that the set of solutions which generated by cuckoo search algorithm dominates the three sets of solutions generated by other algorithms. Also, the set of solutions which generated by the NSGA-II algorithm is dominated by the three sets of solutions generated by other algorithms because of the NSGA-II algorithm, in this case, start with random initial population. So that, the initial population has a worse effect on the quality of the final result and the convergence speed.

Procedure of hybrid NSGA-II-Cuckoo search algorithm is presented as follow:	
1:	Create an initial population P0 of size N by Cuckoo search algorithm
2:	Set t =0.
3:	Apply genetic operators on P0 to create new population Q_0 of size N
4:	While stop criteria is not satisfied
5:	Apply genetic operators on $P_t$to create new population Q_t of size N
6:	Create $R_t = P_t \cup Q_t$.
7:	Rank all individuals in R_t and identify the non-dominated fronts $F_1, F_2, \dots, F_R$

by fast non-dominated sorting algorithm.
For each objective function k, sort the solutions in F_j in the ascending order.
Let $l = |F_j|$, Assign $cd_k(x_{[l,k]}) = \infty$ for $i = 1,2,\ldots,l-1$.
Calculate crowding distance for each objective function $cd_k(x)$
Calculate the total crowding distance $cd(x) = \sum_k cd_k(x)$ of a solution x.
End
Stop and return Pt.

Figure 1. Pseudo code of NSGA-II -Cuckoo search algorithm.

Table 1. Used parameters for GA, NSGA-II, Bee colony, Cuckoo search algorithms

GA/NSGA-II		Bee colony				Cuckoo
Crossover rate	Mutation rate	N of scout Bee	N of selected positions	N of selected elite positions	N of recruited Bees	Probability (pa)
0.64	0.1	20	10	4	4	0.25

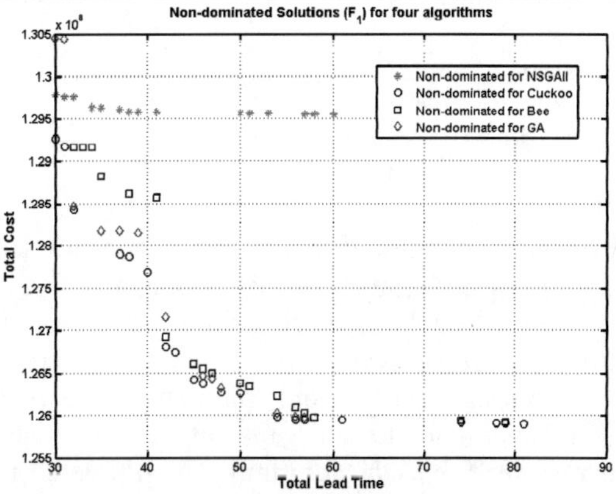

Figure 2. Non-dominated solutions by three algorithms and NSGA-II with the random initial population.

From the Figure 3, the set of solutions which generated by cuckoo search algorithm dominates the three sets of solutions generated by other algorithms. The initial population which contains the ideal solution for each objective function has a significant effect on improving the set of solutions which generated by the NSGA-II algorithm. Therefore, NSGA-II algorithm results better than bee colony algorithm.

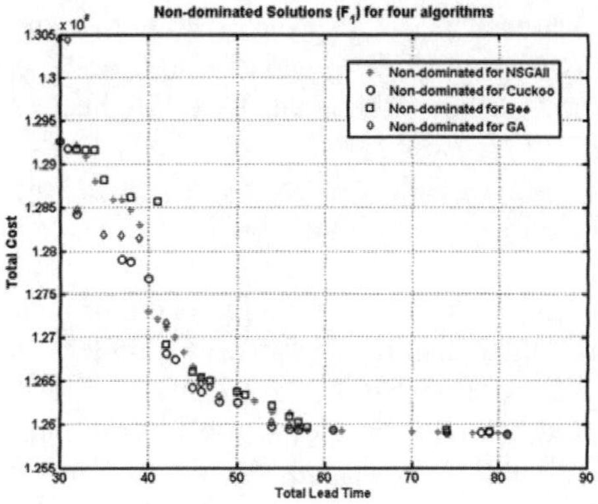

Figure 3. Non-dominated solutions by three algorithms and NSGA-II with improve initial population.

From Figure 4, the set of solutions which generated by NSGA-II-Cuckoo search algorithm dominates the three sets of solutions generated by other algorithms. The initial population which created by cuckoo search algorithm has the best effect on the quality of the final result and the convergence speed. Therefore, NSGA-II-Cuckoo search algorithm results better than other algorithms.

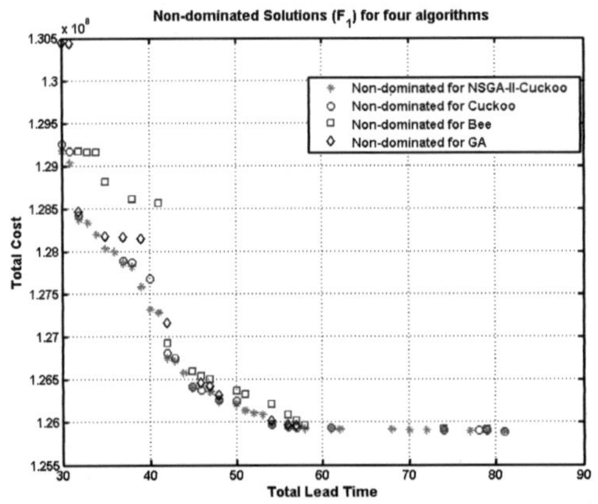

Figure 4. Non-dominated solutions by three algorithms and NSGA-II-Cuckoo search algorithm.

5. Conclusion

This paper presented the integration of the Cuckoo search algorithm and NSGA-II, which called NSGA-II-Cuckoo to solve multi-objective supply chain problem and comparing it with a genetic algorithm, bee colony algorithm, and cuckoo search algorithm. We applied it to find set of non-dominated solutions for multi-objective supply chain problem. In NSGA-II-Cuckoo, we applied the cuckoo search algorithm to generate the initial population that considers as a good start for another algorithm to reduce the time and avoids the local optima entrapment. Also, its effect on the accuracy of the final solutions obtained by the NSGA-II algorithm. The presented results clarify the enhanced performance for the NSGA-II-Cuckoo algorithm, which introduced sufficient stability, and robustness for solving multi-objective supply chain problem.

References

[1] A. H. Gandomi, X. S. Yang, and A. H. Alavi, "Cuckoo search algorithm: A metaheuristic approach to solve structural optimization problems," *Eng. Comput.*, vol. **29**, no. 1, pp. 17–35, 2013.

[2] E. G. Talbi, *Metaheuristics: From Design to Implementation*. John Wiley, New jersey, 2009.

[3] K. Deb, A. Pratap, S. Agarwal, and T. Meyarivan, "A fast and elitist multiobjective genetic algorithm: NSGA-II," *IEEE Trans. Evol. Comput.*, vol. **6**, no. 2, pp. 182–197, 2002.

[4] X. S. Yang and S. Deb, "Cuckoo search via L'evy flights," *2009 World Congr. Nat. Biol. Inspired Comput. NABIC 2009 - Proc.*, pp. 210–214, 2009.

[5] R. Bhattacharya and S. Bandyopadhyay, "Solving conflicting bi-objective facility location problem by NSGA II evolutionary algorithm," *Int. J. Adv. Manuf. Technol.*, vol. **51**, no. 1–4, pp. 397–414, 2010.

[6] S. Liao, C. Hsieh, and P. Lai, "An evolutionary approach for multi-objective optimization of the integrated location – inventory distribution network problem in vendor-managed inventory," *Expert Syst. Appl.*, vol. **38**, no. 6, pp. 6768–6776, 2011.

[7] L. Jia, X. Feng, and G. Zou, "Solving multiobjective bilevel transportation-distribution planning problem by modified NSGA II," *Proc. - 9th Int. Conf. Comput. Intell. Secur. CIS 2013*, pp. 303–307, 2013.

[8] E. Mastrocinque, B. Yuce, A. Lambiase, and M. S. Packianather, "A multi-objective

optimization for supply chain network using the bees algorithm," *Int. J. Eng. Bus. Manag.*, vol. **5**, no. 1, pp. 1–11, 2013.

[9] B. Yuce, E. Mastrocinque, A. Lambiase, M. S. Packianather, and D. T. Pham, "A multi-objective supply chain optimisation using enhanced Bees Algorithm with adaptive neighbourhood search and site abandonment strategy," *Swarm Evol. Comput.*, vol. **18**, pp. 71–82, 2014.

[10] F. Q. Zhao, J. X. Tang, and Y. H. Yang, "A new approach based on Ant Colony Optimization (ACO) to determine the supply chain (SC) design for a product mix," *J. Comput.*, vol. **7**, no. 3, pp. 736–742, 2012.

[11] L. A. Moncayo-Martínez and D. Z. Zhang, "Multi-objective ant colony optimisation: A meta-heuristic approach to supply chain design," *Int. J. Prod. Econ.*, vol. **131**, no. 1, pp. 407–420, 2011.

[12] L. A. Moncayo-Martínez and D. Z. Zhang, "Optimising safety stock placement and lead time in an assembly supply chain using bi-objective MAX-MIN ant system," *Int. J. Prod. Econ.*, vol. **145**, no. 1, pp. 18–28, 2013.

[13] X.-S. Yang and S. Deb, "Engineering Optimisation by Cuckoo Search," vol. **1**, no. 4, pp. 330–343, 2010.

[14] I. Pavlyukevich, "Lévy flights, non-local search and simulated annealing," *J. Comput. Phys.*, vol. **226**, no. 2, pp. 1830–1844, 2007.

[15] P. Civicioglu and E. Besdok, *A conceptual comparison of the Cuckoo-search , particle swarm optimization , differential evolution and artificial bee colony algorithms*. 2013.

[16] F. Elkazzaz, A. Mahmoud, and A. Maher, "Cuckoo search algorithm for multi-objective supply chain problem," *2nd Int. Conf. Mater. Eng. Adv. Manuf. Technol. (MEAMT 2018) , Beijing, China May 25-27.*, 2018.

IOP Conf. Series: Materials Science and Engineering **435** (2018) 012031 doi:10.1088/1757-899X/435/1/012031

An Alert Aggregation Algorithm Based on K-means and Genetic Algorithm

Xianguang Lu[1,*], Xuehui Du[1] and Wenjuan Wang[1]

[1]Zhengzhou Information Science and Technology Institute, Zhengzhou, 450001, China

*Corresponding Author: hzcxlxg@163.com

Abstract. The Intrusion Detection System often produces a large number of alerts, in which 90% are useless. This makes it difficult for security administrator to identify real attack alerts. Using clustering algorithm, such as K-means, DBSCAN, can efficiently cluster the similar alerts, thus greatly reducing the number of alerts that need to be processed. However, the original clustering algorithm has some shortcomings, for example, the K-means has great dependence on the initial value selection, and it is easy to fall into local optimum. Therefore, this paper proposed a new alert aggregation method based on the genetic algorithm and K-means algorithm. We use the Darpa99 dataset to test the performance of our algorithm. And the experimental results show that the algorithm can obtain good aggregation results.

1. Introduction

Intrusion Detection System (IDS) is an important tool to protect the security of computer networks. In order to protect important resources better, people usually deploy multiple IDSs. However, these IDSs will produce a large number of false alert, which will make it difficult for security managers to identify useful information.

The purpose of alert aggregation is to aggregate similar alerts in the original alert. For a large number of alert produced by low level IDSs, alert aggregation can reduce the number of alert effectively. The aggregated alert is more representative, which can help the security administrator to analyse the causality between the alert. But there are some defects in the existing alert aggregation method, such as overdependence on expert knowledge and low degree of aggregation. Therefore, it is a hot topic to design an efficient method which is less dependent on expert knowledge.

K-means algorithm is a common clustering algorithm, which has the advantages of simple, easy implementation, linear and so on. But there are some shortcomings: (1) users need to specify the number of clusters K in advance [1]; (2) the clustering effect depends on the choice of the initial center heavily, and the clustering results of different initial centers are often different; (3) it might fall into local optimum. Aiming at these shortcomings, this paper proposed a new clustering algorithm based on the Genclust algorithm proposed by Rahman [2] firstly, and applied this algorithm to alarm aggregation.

The remaining part of this paper is arranged as follows: The second chapter mainly introduces the research status of alert aggregation. The third chapter mainly introduces the improved clustering algorithm. The fourth chapter analyses the algorithm through vertical and horizontal comparative experiments. Finally, the fifth part briefly summarizes the work and points out the direction of future work.

2. Related Work

Valdes et al. [3] proposed a method of aggregation by calculating the similarity of alert. They defined the degree of similarity between each attack class in advance, and defined the method of similarity calculation for each alert attribute (such as time and IP address, etc.), then calculated the overall similarity by the formula. If the similarity exceeds a certain threshold, they are aggregated. Debar and Wespi [4] proposed an aggregation and correlation components (ACC) alert correlation component, and used predefined rules to do correlation analysis. The definition of the ACC rule took into account the two relationships of alerts: duplicate and consequence, that is, they considered both the similarity and the causality. Lin et al. [5] proposed an improved multisource heterogeneous alert aggregation method, which divided the alert aggregation features into attack mode constraints and time interval constraints, and different attack modes had different time interval constraints. The author used the relative mean variance of time interval proposed in [6] as the dynamic updating coefficient of time interval, and set the average value of the time interval to the threshold of the initial time interval, thus making the time interval threshold adaptive network attack the environment change. Siraj [7] proposed a method that uses Fuzzy Cognitive Modelling for alert aggregation, and used a multilevel clustering method to adapt to inaccurate matching.

In recent years, researchers have begun to use artificial intelligence (AI) technology to solve the problems of intrusion detection technology. In [8], Wang et al. put forward a framework called FCANN. First, they used the fuzzy C mean clustering algorithm to cluster the data, and then classified each cluster by ANN. The framework integrates unsupervised learning and supervised learning, shortens modelling time and improves detection rate. Gogoi et al. [9] put forward a MLH-IDS framework, which included three levels: Supervision layer, unsupervised layer and outlier layer. The supervision layer is used to detect DoS and Probe attacks. The unsupervised layer is used to detect normal data, while outliers are used to distinguish R2L and U2 attacks. Hybrid multilevel frameworks make full use of different machine learning algorithms, which are more flexible and have better performance. In [10], Zhu proposed a method that automatically extracts attack strategies from a large number of Intrusion Alerts without relying on a specific prior knowledge, based on multilayer perceptron and support vector machine, and used an alert correlation matrix to store the correlation between any two types of alerts.

3. IGenClust: an Improved Clustering Algorithm

In view of the shortcomings of the K-means algorithm, there are many works that have improved the original method [11] [12] [13] [14]. Combining genetic algorithm and K-means algorithm, on the one hand, do not require the user to enter the exact number of cluster [15], and can better solve the local optimum problem due to the improper selection of the initial center [16]; on the other hand, can speed up the convergence of genetic algorithm. Rahman proposed a clustering technique called GenClust in [2]. In GenClust, the authors introduced new fitness function and gene rearrangement technology, and achieved good clustering results. This method requires the user to enter multiple clustering radius in advance, but in the actual operation, the good clustering radius may not appear in the initial set provided by the user, so there may not be genes in the initial population that represent the best cluster center. Moreover, as an improved genetic algorithm, this algorithm doesn't solve the inherent premature problem of genetic algorithm, which has a certain influence on the result of clustering. This paper proposed an improved GenClust algorithm, IGenClust, which overcomes the premature problem of the GenClust algorithm by introducing the population entropy, and avoids the poor clustering result caused by the improper selection of the radius by improving the selection of the initial clustering center.

3.1. Main components

The IGenClust algorithm mainly consists of 8 main components:

- Component 1: Standardized data set and distance calculation;
- Component 2: Candidate population;

- Component 3: Chromosome selection;
- Component 4: Crossover;
- Component 5: Twin Removal;
- Component 6: Elitism;
- Component 7: Mutation;
- Component 8: K-means algorithm.

3.2. Detailed description

3.2.1. Standardized data set and distance calculation

3.2.1.1. Standardized data sets. Assume that the original data set has n records, that is $R=\{R_1,R_2,R_3....R_n\}$, the record R_i has m attributes, that is $R_{i,1},R_{i,2},R_{i,3}....R_{i,m}$. That is to say, $R_{i,j}$ represents the *j*th property of the *i*th record, it can be numeric or categorized. We normalize the numeric attributes of the record in order to avoid the scale or measurement unit problem of distance calculation in subsequent calculation, for example, the range of the attribute p is [l , u], $R_{i,p}$ should be normalized to $R_{i,p}$' =($R_{i,p}$ - l)/(u - l).

3.2.1.2. Distance calculation. The calculation of distance between records is usually based on Euclidean distance between attributes, but this method can only calculate the distance between numerical attributes. Giggins designed a method to calculate the similarity of categorical attributes in [17], and Rahman applied it to GenClust. For the record R_a and R_b, the distance of categorical attribute j can be expressed as $\text{Dist}(R_{a,j},R_{b,j})=1-\text{Sim}(R_{a,j},R_{b,j})$.The total distance between record R_a and R_b is:

$$\text{Dist}(R_a,R_b)=\frac{\sum_{i=1}^{N}\text{abs}(R_{a,i},R_{b,i})+\text{dist}_{j=1}^{C}(R_{a,j},R_{b,j})}{N+C} \tag{1}$$

User should sort the attributes of records firstly, and the first N attributes are numerical attributes after normalization, the latter C attributes are categorized.

The above method can be used as a general purpose method for dealing with general problems, but they should be adjusted appropriately when dealing with specific problems. In this paper, our method is mainly used to solve the problem of alert aggregation, so the above method can be adjusted as follows:

(1) Source IP, Destination IP: The similarity of the IP address is calculated by converting the IP address to the 32-bit binary string and comparing them. Assuming that the number of two strings from the left to the right is n, the similarity is $\text{Sim}_{\text{ip}}(ip_a,ip_b)=n/32$. For example, for two addresses: ip_1 = 172.16.112.100, ip_2 = 172.16.112.192, First, convert the two addresses 10101100000100000111000001100100 and 10101100000100000111000011000000, it is easy to see that n = 24, so $\text{Sim}_{\text{ip}}(ip_1,ip_2)=24/32=0.75$.The range of similarity between IP is [0,1].

(2) Source port, Destination port: The maximum value of port is 65535, so the similarity of ports can be normalized $\text{Sim}_{\text{port}}(port_a,port_b)=1-\frac{|port_a-port_b|}{65535}$. For example, for ports 23 and 25625, $\text{Sim}_{\text{port}}(port_a,port_b)=1-\frac{|23-25625|}{65535}$=0.61. The range of similarity between port is [0,1].

(3) Time: The similarity of time is mainly calculated by the time interval and threshold. We set a maximum threshold $T_{\max}$ and a minimum threshold $T_{\min}$, then the similarity of time, t_1,t_2 ,can be expressed as:

- If $|t_1 - t_2| < T_{min}$,then $Sim_t = 1$;
- If $|t_1 - t_2| > T_{max}$,then $Sim_t = 0$;
- If $T_{min} < |t_1 - t_2| < T_{max}$, then $Sim_t = (T_{max} - |t_1 - t_2|)/(T_{max} - T_{min})$.

The range of similarity between time is [0,1].

(4) Type: If the type of the two alert is the same, then $Sim_{type} = 1$,otherwise, $Sim_{type} = 0$.

To sum up, the overall similarity of the alert is:

$$Sim(R_a, R_b) = \frac{Sim_{sip} + Sim_{dip} + Sim_{sport} + Sim_{dport} + Sim_t + Sim_{type}}{6} \tag{2}$$

Distance is: $Dist(R_a,R_b)=1 - Sim(R_a,R_b)$.

Because the ranges of similarity of each attribute is [0,1], the range of $Sim(R_a,R_b)$ and $Dist(R_a,R_b)$ are [0,1].

3.2.2. Candidate population. The function of this component is to generate the candidate population of the genetic algorithm. The chromosome CR_j is composed of K genes, that is, $CR_j=(G_{j1},G_{j2},G_{j3}...G_{jk})$, and the K is the number of clusters in the K-means algorithm. Each of the genes in the chromosome represents the cluster center in the K-means algorithm, which means that the genes with the highest fitness in the final output of the genetic algorithm correspond to the optimal initial cluster center of the K-means algorithm.

In our method, the candidate population consists of three parts, each of which generates W chromosomes in different ways: (1) given W radius r;(2) given W cluster number k; (3) randomly select W cluster numbers K. Finally, the candidate population contains 3W chromosomes.

(1) In the first part, users should provide W radius r, $r_1,r_2,r_3...r_w$. In the original data set, we first calculate the distance between any two points: $Dist(R_i,R_j)$, and then use the given radius r_i to calculate the density of each point: $Density(R_i)=|\{R_j : Dist(R_i,R_j) \le r_x; \forall j\}|$ .The record R_i ,with the highest density ($Density(R_i) > Density(R_j); \forall j$),is set as the first gene G_{j1} , and all the records within the distance r_i ($\{R_j : Dist(R_i,R_j) \le r_x; \forall j\}$) were delete from the data set. As long as our record has a density greater than the user defined threshold T, we continue to select the genes. For a given r_i , we get multiple genes G_{jx} to form chromosomes CR_j . As we often can't obtain the radius of optimal cluster in real life, provide multiple r can adapt to the change of the radius, but also increased the genetic diversity, which can help us to find the optimal solution finally.

(2) The number of alert is very large, so it's not easy to determine the radius. That is to say, the radius provided by users may not contain the optimal radius, and the convergence rate is slower. As pointed out above, when calculating the distance between alert, the similarity of type is calculated simply according to same or not, so the effect of difference of the type is larger than that of other attributes, therefore, the K may be close to the total number of alert data types. For the alert, it is easier to determine the K based on experience than to determine the radius of each cluster. So in the second part, we use the W K values provided by users, and use the radius $Dist(R_i,R_j)$ which calculated in (1) to get the tightness of each point,

$$Tight(R_i)=\frac{1}{\sum_{j \in n, j \ne i} Dist(R_i,R_j)} \tag{3}$$

and the probability of selecting of each point is:

$$P_{\text{Tight}}(R_j)=\frac{\text{Tight}(R_j)}{\sum_{i\in n}\text{Tight}(R_i)} \tag{4}$$

and select K points as genes to form chromosomes. The advantage of such deterministic process is that the selected chromosomes are basically optimal and will speed up the speed of convergence.

(3) After deterministic selection of the two sets of chromosomes, we have obtained most of the useful individual genes, but may loss some individuals in remote clusters. So, we use the method of random selection in the third part. We randomly generate an integer in the range of $[2,\sqrt{n}]$ as the number of genes in the chromosome, and repeat W times to get W chromosomes. Although the overall fitness of chromosomes selected by random process is relatively low, chromosomes still have some useful genes.

3.2.3. Chromosome selection. Before introducing selection, we first introduce the fitness function used in GenClust algorithm. The fitness function used in GenClust is not based on the sum of squares of errors in the general K-means algorithm, but is based on the tightness and separation of clusters. The degree of closeness is the average distance from other points to cluster centers in clusters,

$$\text{Comp}_j=\frac{\sum_{R_a\in C_j}\text{Dist}(R_a,m_j)}{|C_j|} \tag{5}$$

Where $|C_j|$ is the total number of records in C_j; the degree of separation is the minimum value of the distance between the cluster centers and other clusters, $\text{Seq}_j=\min_{\forall k\neq j}\text{Dist}(m_j,m_k)$ .Thus, the fitness function is constructed as follows:

$$\text{Fitness}=\sum_{\forall j}(\text{Seq}_j-\text{Comp}_j) \tag{6}$$

In order to use highly adaptable chromosomes in subsequent work, we need to select W chromosomes from the candidate population by roulette, that is, the probability of each chromosome being selected is $P(CR_j)=\frac{\text{fitness}(CR_j)}{\sum_{i\in 3W}\text{fitness}(CR_i)}$. Finally, we get the initial population containing W chromosomes.

3.2.4. Crossover. Our cross operation basically follows the idea of GenClust algorithm. When selecting the parent chromosomes, we sort the chromosomes according to the fitness firstly, then select the most adaptable chromosome CR_p (delete it after selection from the cluster), finally select another chromosome CR_q in the chromosomes from the remaining chromosomes by roulette technique, and CR_q is also deleted after being selected.

The offspring generated after single point crossover of chromosomes get some genes on the left part of one parent and the right part of another parent, but these genes may be similar to each other and can't be extended to the whole space, so we should rearrange the gene before single point crossover on the chromosome. Set CR_p as a reference chromosome, CR_q as a target chromosome, the general idea of gene rearrangement is as follows:

In all genes of the target chromosome, the gene with the smallest distance from the first gene of the reference chromosome is placed in the first place of the target chromosome rearrangement version. In the remaining genes of the target chromosome, the gene with the smallest distance from the second genes of the reference chromosome is placed in the second place of the rearrangement version. If the length of the target chromosome (that is, the number of genes) is equal to or less than the length of the reference chromosome, calculate the best gene in each position in sequence according to the above method, until all the genes are rearranged on the target chromosome. If the length of the target chromosome is larger than that of the reference chromosome, the rearrangement is divided into two stages. In the first stage, the first X gene of the target chromosome (x is the length of the reference chromosome) is rearranged as above. In the second stage, calculate the distances between the remaining genes of the target chromosome and the genes of the rearranged chromosomes and insert them into the back of their nearest genes.

After the recombination, we perform the traditional single point crossover, that is, the left part of one parent chromosome is connected to the right part of another parent of the parent chromosome to form an offspring chromosome, and the remaining parts are combined together to form another offspring chromosome.

3.2.5. Twin removal. After generating new offspring chromosomes, we need to remove duplicate genes from chromosomes. If the distance of gene G_i and G_j, $\text{dist}(G_i,G_j) < \delta$, δ is a very small positive number, we think that these two genes are duplicate. When the number of genes in the chromosome is more than 2, we regard these genes as duplicate, when the number of genes is equal to 2, one of these genes is randomly mutated until these genes are no longer duplicate.

3.2.6. Elitism. The elitism operation is to store the highest fitness chromosomes of each iteration as CR_b, and compare it with the worst (the best) chromosomes of the offspring. If the fitness of CR_b is larger than that of the worst chromosomes of offspring, the worst chromosome is replaced by CR_b. If the fitness of CR_b is smaller than that of the best chromosomes of offspring CR_b' ,then $CR_b = CR_b'$.

3.2.7. Mutation. Mutation is an operation to change one attribute of an individual into other values within the range of the value of the attribute. Its purpose is to increase the diversity of chromosomes in the population. If the mutation probability is too small, it is difficult to produce a new individual. If the mutation probability is too large, the genetic algorithm will become a pure random search algorithm.

Srinivas and Patnaik proposed an adaptive genetic algorithm (AGA) [14]. When the fitness is high, the mutation probability is low; when the fitness is low, the mutation probability is high, that is to say, the probability is adjusted adaptively according to the fitness. Although the adaptive genetic algorithm has better probability of the individual in each generation, it does not solve the inherent precocious problem of the genetic algorithm. The premature problem is that when the algorithm has not yet searched the global optimal solution, the fitness of the offspring can no longer exceed the parent generation, that is, there are almost no differences between individuals. The problem of precocity is essentially due to the lack of species diversity, which leads to the failure of genetic algorithm to jump out of the local optimal solution. In other words, when two parents are exactly the same or very similar (similar to inbreeding), crossover will be difficult to produce new individuals, which will prevent the algorithm from searching for new solution space, resulting in the loss of population diversity.

In order to solve the premature problem, we introduce the population entropy in this components.

(1) The definition of population entropy: the total number of chromosomes in the population is N, and the similar chromosomes in the population can be divided into M class (M<N). The number of individuals in each class is n_i, then the entropy of the population is defined as:

$$S=-\sum_{i=1}^{M}\frac{n_i}{N}\log_2\frac{n_i}{N} \tag{7}$$

(2) The calculation method for population entropy: first, determine which chromosomes belong to the same class, the number of the genes in the same class of chromosome is the same (the same species), and the chromosomes with small difference of genes belong to the same class. On the other hand, if the number of genes is different or the difference between genes is large, the chromosomes do not belong to the same class.

The method of judging whether genes are similar follows the method of attribute similarity mentioned earlier. After finding similar points, we can get the chromosomes belonging to the same class and then get the population entropy.

(3) The relationship between population entropy and population diversity: if the population has a high diversity, the probability of each individual appears is similar, the population is in the active evolutionary selection stage, and the information entropy of the population is large.

Therefore, when the diversity is low and the population entropy is small, it is necessary to increase mutation probability and increase individual diversity. Suppose the value of entropy is S, and the mutation probability is

$$P = \begin{cases} k_1 e^{-S} \dfrac{f_{max} - f}{f_{max} - f_{avg}} & f > f_{avg} \\ k_1 e^{-S} & f \le f_{avg} \end{cases} \quad (8)$$

Where f_{avg} represents the average fitness of the current population, f_{max} represents the highest fitness of the current population, and f represents the fitness of the current chromosome. Because the range of S is $[0, \log_2 n]$, $0 \le k_1 \le \frac{1}{n}$.

Increasing the mutation probability can increase the diversity of the population, but it is not conducive to the convergence of the population. It will add additional work if the mutation is performed every iteration, so the mutation is performed every 10 generations in this article.

3.2.8. K-means algorithm. We use the genes in the best chromosome as the initial seed of the K-means algorithm, which will produce a high quality cluster solution. In this component, we use the improved distance calculation method mentioned earlier to calculate distance.

4. Experiment

Weka (Waikato Environment for Knowledge Analysis) is an open source machine learning and data mining software based on the Java environment. It is one of the most complete data mining tools in the present day. In order to show the experimental results better, this paper writes the algorithm based on the interface document of Weka, and adds the IGenClust algorithm to the cluster module of Weka, then carries out clustering analysis on the Darpa1999 dataset.

We compare the improved algorithm with the SimpleKMeans, GenClust and other algorithms in the same operating environment according to two aspects, error and execution time. The experimental operating environment is Inter CPU 2.40 GHz and the memory is 8 GB, the software environment is as follows: the operating system is Windows10 Microsoft, the experimental program is written in Java and the development environment is Visual Studio Microsoft 2015. The development tools used are Myeclipse10 and Weka3.6.15.

4.1. Preparation

Firstly, we convert the original alert information for the Darpa data set generated by IDS into *.arff* format files that can be identified by Weka, and then import Weka3.6 engineering file into Myeclipse10 for the compilation of the algorithm.

4.2. Implementation

The experiment is divided into two stages. The first stage is divided into 4 parts to compare different algorithms:

- The first part uses SimpleKMeans algorithm that comes with Weka.
- The second part composes the algorithm which is suitable for distance comparison of alert information, and add it to SimpleKMeans algorithm to form ImproveKMeans algorithm.
- The third part uses the original GenClust algorithm.
- The fourth part uses our IGenClust algorithm.

The second stage is divided into two parts:

- The first part is repeated experiments to measure the stability of the algorithm.
- The second part is to test the performance trend of the IGenClust algorithm under different data sets.

4.2.1. Vertical comparison. The test data used by different algorithms are 2534 records extracted from the alert collected from the Darpa99 dataset for week 5, day 3.

(1) In the first part, the maximum iteration is set to 500 times, and the generated clusters are set to 10. The result is shown as figure 1 (a).

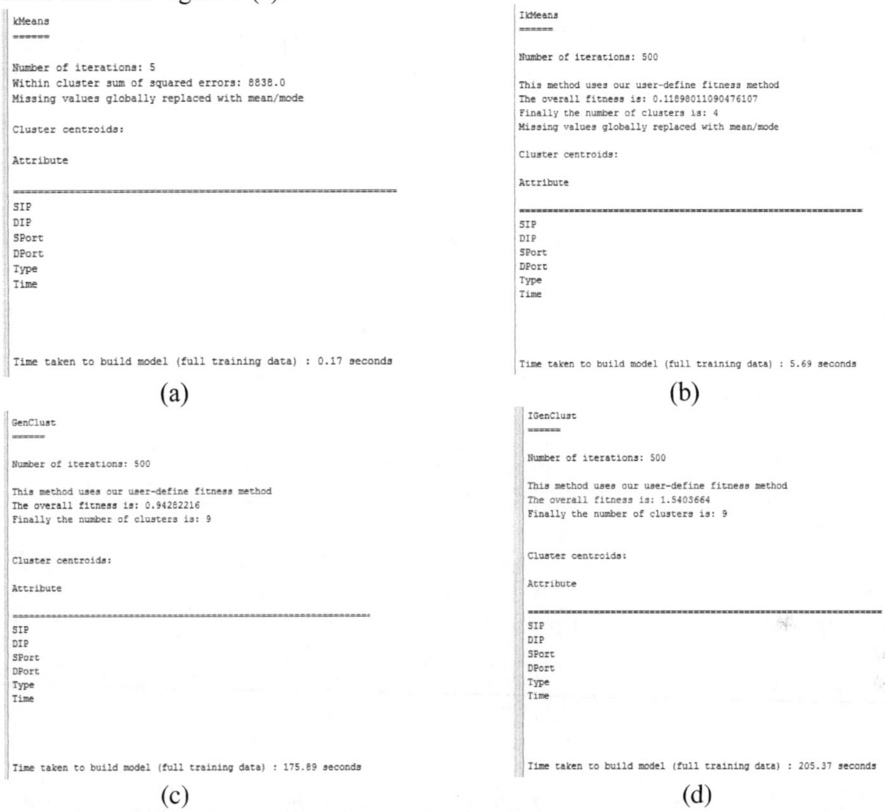

Figure 1. The results calculated using different algorithms.

SimpleKMeans constantly judged whether the polymerization was complete, so the experiment ended up only 5 times. The speed of SimpleKMeans operation is very fast, but this algorithm uses Euclidean distance as fitness function. The reference value as a longitudinal comparison is smaller.

(2) In the second part, the ImproveKMeans algorithm does not make any changes except the use of our own distance calculation function to replace the Euclidean function. The maximum iteration is set to 500 times, and the generated clusters are set to 10. The result is shown as figure 1 (b).

The smaller fitness shows that the degree of aggregation is poor.

(3) In the third part, we use the GenClust algorithm with our own distance function and the maximum iteration 500 times. The result is shown as figure 1 (c).

The fitness indicates that the algorithm can get better clustering results than SimpleKMeans.

(4) In the fourth part, we use the IGenClust algorithm and set the maximum iteration 500 times. The result is shown as figure 1 (d).

We can see that although the IGenClust algorithm is slightly worse than the GenClust algorithm in the execution time for the reason of the population entropy, the overall fitness has been improved.

The results are summarized as shown in Table 1. From this table, we can compare the differences between algorithms more intuitively.

Table 1. Comparison of execution time and fitness of different algorithms.

	SimpleKMeans	ImproveKMeans	GenClust	IGenClust
Fitness	\	0.119	0.943	1.540
Time(ms)	0.17	5.69	175.89	205.37

4.2.2. Horizontal comparison. (1) Part one: In the first part, we have done many experiments on 2534 records to verify the stability of the algorithm. The results of the experiment are as follows:

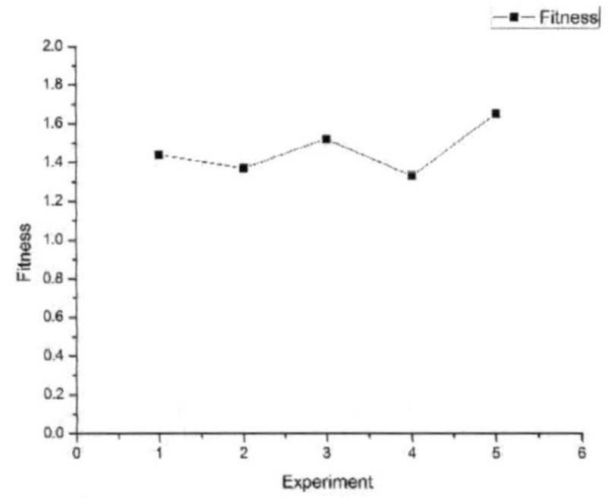

Figure 2. Different fitness in multiple experiments.

From the figure 2, we can see that there is little change in the fitness of the algorithm under repeated tests, which shows that the algorithm is relatively stable.

(2) Part two: In the second part, we used five sets of different datasets, the numbers were 2534, 3813, 4857, 6861, 8192, respectively. The transformation curves of fitness and time are as follows:

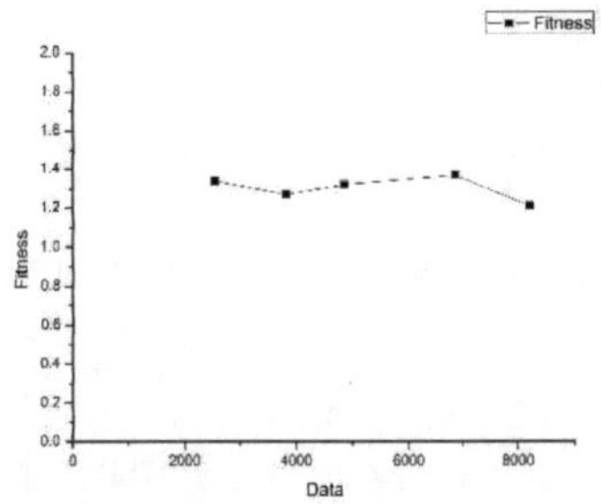

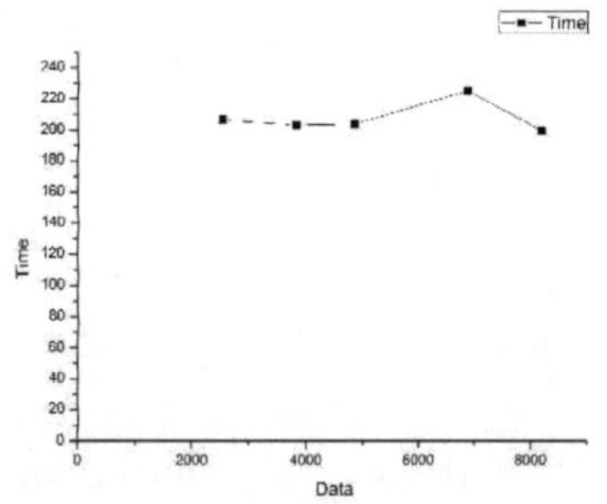

Figure 3. Different fitness under different data sets.

Figure 4. Different time under different data sets.

As shown in the figure 3 and figure 4, the number of experimental data has little effect on fitness and time, and higher fitness often means more time.

4.3. Conclusion

Through the above two stage experiments, we can see that our IGenClust algorithm sacrificed a certain amount of execution time, but improved a certain aggregation effect. In theory, the algorithm can solve the problem of precocious algorithm, and it is relatively stable.

5. Summary and future work

For the large number of similar alerts generated by Intrusion Detection Systems, more and more experts apply the clustering algorithm to alert aggregation to enhance the aggregation effect. The

GenClust algorithm proposed by Rahman has a high degree of aggregation and good scalability. Based on this algorithm, this paper proposes an improved algorithm, IGenClust, and applies it to alert aggregation. Experiments show that the IGenClust algorithm can not only improve the premature problem of GenClust algorithm theoretically, but also improve the degree of aggregation to a certain extent. However, although the number of alerts after aggregation has become smaller, the causality between alerts is not well represented. Therefore, in order to provide more intuitive alert information for the security managers, further association analysis of the aggregated alert should be carried out to reconstruct the real attack scene.

Acknowledgments

This research is supported by the National High-tech R&D Program of China under grant No.2015AA016006, and the Natural National Key Basic Research Program of China under grant No.2016YFB0501901.

References

[1] Lewis R, Krawiec M, Confer E, et al. 2014 Data clustering: 50 years beyond K-means, *Pattern Recognition Letters*, Vol. **31**, No. 8, pp.651-666

[2] Rahman MA, Islam MZ 2014 A hybrid clustering technique combining a novel genetic algorithm with K-Means, *Knowledge-Based Systems*, Vol. **71**, No. 71, pp. 345-365

[3] Valdes, A, & Skinner, K 2000 Adaptive, Model-Based Monitoring for Cyber Attack Detection, *International Workshop on Recent Advances in Intrusion Detection*, Vol.**1907**, pp.80-92

[4] Wespi, A 2001 Aggregation and Correlation of Intrusion-Detection Alerts, *International Symposium on Recent Advances in Intrusion Detection*, Vol.**2212**, pp.85-103

[5] Lin H, Zhijie W, Xiaofang H, et al. 2014 An improved multisource heterogeneous alarm aggregation scheme, *Application Research of Computers*, Vol.**31**, No. 2, pp.579-582

[6] Gong J, Haibin M, Yong D, et al. 2005 Redundant elimination of intrusion events with multi feature association, *Journal of Southeast University (Natural Science)*, Vol.**35**, No. 3, pp. 366-371

[7] Siraj A, Vaughn R 2005 Multi-level alert clustering for intrusion detection sensor data, *Fuzzy Information Processing Society*, pp.748-753

[8] Wang, G, Hao J, Ma J, Huang L 2010 A new approach to intrusion detection using artificial neural networks and fuzzy clustering, *Expert Systems with Applications*, Vol.**37**, No.9, pp. 6225–6232

[9] Gogoi, P, Bhattacharyya DK, Borah B, Kalita JK 2014 MLH-IDS: a multi-level hybrid intrusion detection method, *Computer Journal*, Vol.**57**, No.4, pp. 602-623

[10] Zhu B, Ghorbani A 2006 Alert correlation for extracting attack strategies, *International Journal of Network Security*, Vol.**3**, No.3, pp. 244-258

[11] Rahman M, Islam M 2012 CRUDAW: a novel fuzzy technique for clustering records following user defined attribute weights, *Tenth Australasian Data Mining Conference*, pp.27-41

[12] Yu S, Chu, SW, Wang CM, Chan YK, Chang TC 2017 Two improved k-means algorithms, *Applied Soft Computing*

[13] Srinivas M, Patnaik LM 2002 Adaptive probabilities of crossover and mutation in genetic algorithms, *IEEE Transactions on Systems Man & Cybernetics*, Vol.**24**, No.4, pp.656-667.

[14] Jie L, Liu W, Sun Z, Teng S 2017 Hybrid fuzzy clustering methods based on improved self-adaptive cellular genetic algorithm and optimal-selection-based fuzzy c-means, *Neurocomputing*, No.249, pp.140-156.

[15] Hruschka ER, Campello RJGB, Freitas AA 2009 A survey of evolutionary algorithms for clustering, *IEEE Transactions on Systems Man & Cybernetics Part C Applications & Reviews*, Vol.**39**, No.2, pp.133-155

[16] Pakhira MK, Bandyopadhyay S, Maulik U 2005 A study of some fuzzy cluster validity indices, genetic clustering and application to pixel classification, *Fuzzy Sets & Systems*, Vol.**155**, No. 2, pp.191-214

[17] Giggins H, Brankovic L 2012 VICUS: a noise addition technique for categorical data, *Tenth Australasian Data Mining Conference*, Vol.**134**, pp.139-148

Improving Neural Chinese Word Segmentation Using Unlabeled Data

Yanna Zhang[1], Jinan Xu[2], Guoyi Miao[3], Yufeng Chen[4] and Yujie Zhang[5]

[1, 2, 3, 4, 5]School of Computer and Information Technology, Beijing Jiaotong University Shangyuancun No.3, Beijing, P.R.China

{[1]yannazhang, [2]jaxu, [3]gymiao, [4]chenyf, [5]yjzhang}@bjtu.edu.cn

Abstract. Supervised word segmentation heavily relies on large-scale and high quality labeled data. However, building such a corpus is difficult, especially with respect to domain specific data. In this paper, we propose a novel semi-supervised Chinese word segmentation (CWS) method. Specifically, we seek to select more useful sample sentences from the large-scale unlabeled sentences to extend the training data, by means of a sampling strategy that uses character-based semantic similarity. The presented similarity algorithm is used to calculate the similarity between unlabeled sentences and the training data, which can help select helpful sample sentences from unlabeled data. In addition, we integrate an attention mechanism into our word segmentation model to focus on available contextual information. Experiments on PKU, MSR and Weibo benchmark data sets show that our method outperforms the previous neural network models and state-of-the-art methods.

1. Introduction

Chinese word segmentation (CWS) is a preliminary and important task for many Chinese natural language processing (NLP) tasks, such as machine translations (MT), question answering systems (QA) and named entity recognition (NER). Neural word segmentation has shown promising progress in recent years [1] [2] [3]. However, neural word segmentation, mainly trained by supervised learning, relies on a large-scale labeled data which is usually expensive and tends to be of a limited amount. Lack of labeled data makes it very difficult to capture useful semantic information from contexts to improve the performance of word segmentation model.

For solving the mentioned problem above, researchers have tried to use large-scale unlabeled data to improve the supervised word segmentation. One intuitive idea is to adopt semi-supervised self-training method. It exploits a small amount of labeled data to train an initial word segmentation model, which is used to annotate unlabeled data. The annotation results are added to training data set to update parameters of word segmentation model in the next iteration. However, there still remains some challenges, for example, some existing semi-supervised word segmentation methods cannot pick out samples containing rich features. The above issues have greatly limited the effect of word segmentation.

In this paper, we propose a novel semi-supervised word segmentation method by leveraging semantic similarity. Our work is mainly about of two aspects to select samples: The uncertainty of annotation and the semantic similarity between unlabeled sentences and training data. Moreover, we adopt an attention-based natural word segmentation to dynamically focus on the character-based n-gram contextual information. Experiments show that our approach can select samples containing more useful features and achieve significant improvements on Weibo, PKU and MSR datasets.

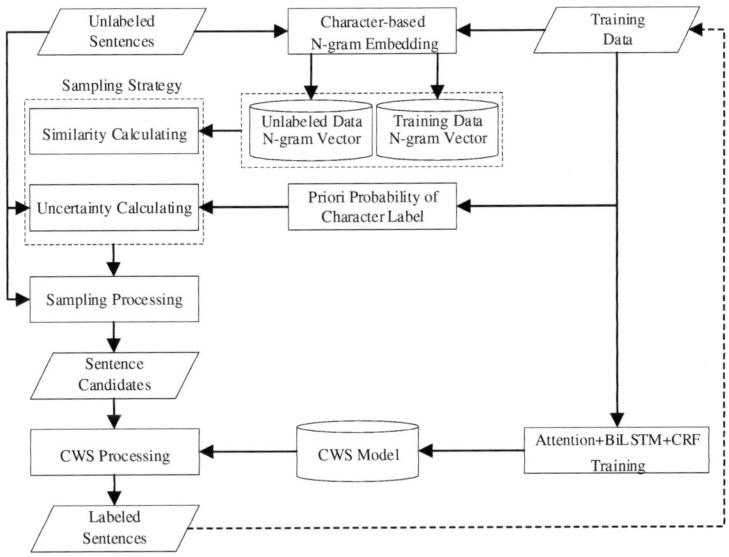

Figure 1. Framework of Semi-supervised Learning for CWS.

The contributions of this paper could be summarized as follows.

- The semantic similarity is first introduced to semi-supervised learning for enhancing CWS, which is effective to measure the diversity of samples.
- For improving the performance of CWS, we integrate an attention mechanism into our CWS model, which is used to dynamically focus on the contextual information and capture useful features (Section 3.2.).
- We propose a new semantic similarity calculation method with the help of character-based n-gram embedding, which incorporates rich contextual information (Section 3.3.).

2. Related Work

Semi-supervised learning is widely used in word segmentation tasks. Zheng et al. [4] and Sun et al. [5] experimented with deriving some statistics-based features such as mutual information, accessor variety and punctuation variety from unlabeled data, and add them to supervised discriminative training. Yang et al. [6] adopted partial-label learning with conditional random fields to make use of the valuable knowledge contained by unlabeled data. Melamud et al. [7] presented a neural model for learning a generic context embedding function from large corpora and applied it to several NLP tasks. Furthermore, Peters et al. [8] directly used large-scale unlabeled corpus to pre-train language models. By using language models, character-level knowledge can be integrated into context vectors and the potential meaning of character knowledge can be well encoded.

One of semi-supervised learning methods in NLP is bootstrapping, which generate more training instances by automatically labeling large-scale data. This is also the main idea of our method. Yang et al. [9], Zeng et al. [10] and Zhang et al. [11] experimented with co-training for semi-supervised Chinese word segmentation. Huang et al. [12] adopted self-training to increasing the quality of the automatically parsed data. Liu et al. [13] combined self-training and character clustering to solve the problem of domain adaptability. Inspired by the previous work, we propose a new and novel sampling strategy for self-training.

3. Methodology

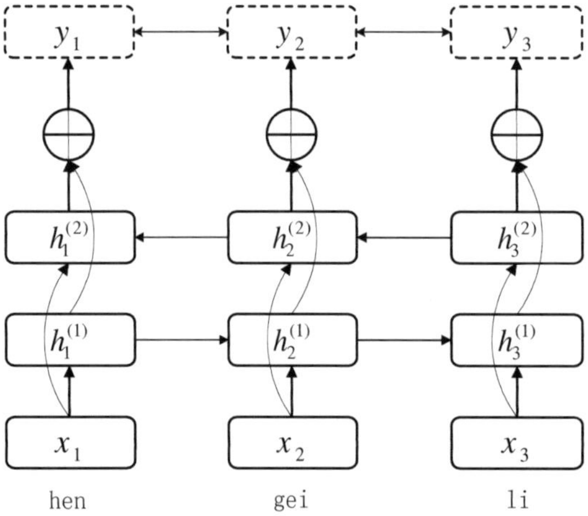

Figure 2. BiLSTM+CRF architecture for CWS. Character embeddings are given as input; a bidirectional LSTM produces context-dependent representations; the information is passed through a hidden layer and an output layer. The outputs are confidence scores for CRFs.

3.1. Semi-supervised Learning for CWS

Semi-supervised self-training learning can be regarded as a learning method which continuously increases training data. Our method consists of two main stages. In the first stage, we train an initial segmentation model (Attention+BiLSTM+CRF) with a small amount of labeled data, which is used to annotate unlabeled samples (See section 3.2.). In the second stage, we obtain samples from large-scale unlabeled sentences according to sampling strategy (See section 3.3). These samples after annotated are added to training data for the next iteration. We can obtain the optimal word segmentation model until the iterations converge. Figure 1 illustrates the framework of our semi-supervised learning for CWS.

3.2. The Initial CWS model

3.2.1. BiLSTM+CRF Architecture for CWS. CWS task is usually solved by character-level sequence labeling algorithm. Specifically, each character in a sentence is labeled as one of $L = \{B, M, E, S\}$, indicating the begin, middle, end of a word, or a word with a single character. For a given sentence $X = (x_1, x_2, \dots, x_n)$ containing n characters, the aim of CWS task is to predict label sequence $y^* = (y_1, y_2, \dots, y_n)$. The BiLSTM+CRF architecture (our baseline) for CWS is illustrated in Figure 2.

Similar to other methods using neural networks, the first step of the BiLSTM+CRF based CWS method is to represent characters in distributed vectors. Formally, we lookup embedding vector from embedding matrix for each character x_i as $e_i \in R^{d_e}$, where d_e is a hyper-parameter indicating the size of character embedding.

In order to incorporate information from both sides of the sequence, we use bidirectional Long Short-Term Memory (BiLSTM) as feature layers. The update of each BiLSTM unit can be described as follows:

$$h_i = \overrightarrow{h_i} \oplus \overleftarrow{h_i} \quad (1)$$

where $\overrightarrow{h_i}$ and $\overleftarrow{h_i}$ are the hidden states at position i of the forward and backward LSTMs respectively; $\oplus$ is concatenation operation.

Following Lample et al. [14], we employ conditional random fields (CRF) layer to inference labels, which is beneficial to consider the dependencies of adjacent labels. For example, a B (begin) label should be only followed by a M (middle) label or E (end) label. Given that y is a label sequence $y_1, y_2, \dots, y_n$, then the CRF score for this sequence can be calculated as:

$$s(X,y) = \sum_{i=0}^{n} A_{y_i,y_{i+1}} + \sum_{i=1}^{n} P_{i,y_i} \tag{2}$$

where A is a matrix of transition scores such that $A_{i,j}$ represents the score of a transition from the tag i to tag j. y_0 and y_n are the start and end tags of a sentence, that we add to the set of possible tags. P is the fractional matrix of the BiLSTM network's output. $P_{i,j}$ corresponds to the j-th label of the i-th word in a sentence.

The output from the model is the tagging sequence with the largest score s(y). A softmax over all possible tag sequences yields a probability for the sequence y:

$$p(y|X) = \frac{e^{s(X,y)}}{\sum_{\tilde{y} \in Y_X} e^{s(X,y)}} \tag{3}$$

Finally, we directly maximize the log-probability of the correct tag sequence:

$$\log(p(y|X)) = s(X,y) - \log\left(\sum_{\tilde{y} \in Y_X} e^{s(X,\tilde{y})}\right) = s(X,y) - \log add_{\tilde{y} \in Y_X}(X,\tilde{y}) \tag{4}$$

While decoding, the output sequence we predict will have the highest score given by the following:

$$y^* = argmax_{\tilde{y} \in Y_X} s(X,\tilde{y}) \tag{5}$$

We use the Viterbi algorithm to find the optimal labeled sequence during training.

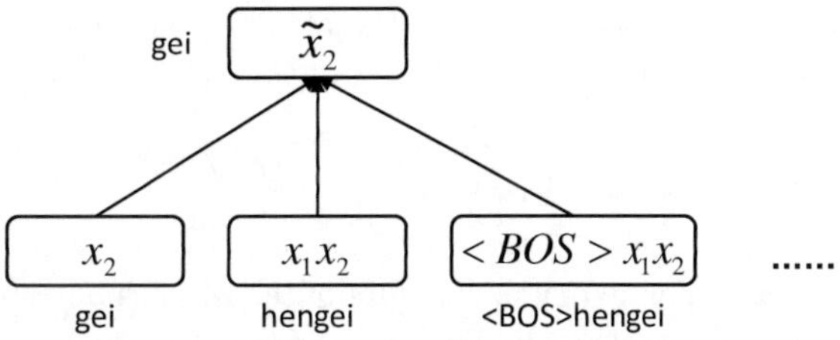

Figure 3. The attention mechanism. If the character at the current moment is "gei", the attention mechanism gives attention to the n-gram contextual embedding of this character.

3.2.2. Attension+BiLSTM+CRF Architecture for CWS. It is necessary to use an empirically good segmentation model in the process of semi-supervised word segmentation. Therefore, we improved the performance of baseline system by integrating an attention mechanism. The attention mechanism is illustrated in Figure 3. During training, the attention mechanism selectively pays attention to the character-based n-gram contextual embeddings. The input of the BiLSTM+CRF architecture computed as a weighted sum of the character-based n-gram contextual embeddings:

$$\tilde{x} = \sum_{j=1}^{N} \alpha_{ij} x_j \tag{6}$$

The weight of each annotation α_{ij} is computed by:

$$\alpha_{ij} = \frac{\exp(e_{ij})}{\sum_{k=1}^{T} \exp(e_{ik})} \tag{7}$$

where

$$e_{ij} = a(x_j) \tag{8}$$

here, x_j represents the character-based j-gram contextual embeddings, function $a(\cdot)$ is tanh function. α_{ij} is trainable parameter which denotes weight, $\tilde{x}$ denotes the output of the attention mechanism.

Because of the sparsity of the n-gram contextual embeddings of characters, here we set N to 3. Specifically, we only focus on the unigram, bi-gram and tri-gram contextual embeddings of characters. In order to get these contextual embeddings, we first deal with the sentence. After processing, this sentence "woshizhongguoren." (I am Chinese.) will be "wo shi zhong guo ren .", "<BOS> wo woshi shizhong zhongguo guoren ren. .<EOS>" and "<BOS0><BOS>wo <BOS>woshi woshizhong shizhongguo zhongguoren guoren. ren.<EOS> .<EOS><EOS0>" respectively. Then we train these vectors readily using word2vec [15] toolkit on the resulting n-gram sequences.

3.3. Sampling Strategy

3.3.1. Similarity between unlabeled sentences and training data. To extract useful samples from the large-scale unlabeled corpus, we propose to calculate the character-based semantic similarity between unlabeled sentences and training set sentences. In general, the more a sample sentence differs from the training set sentences, the more semantic and structural features it contains. Thus we seek to select sample sentences that are very different from training set sentences and they can then be easily deduced from the similarity between unlabeled sentences and training set sentences. Formally, the character-based semantic similarity is defined as follows:

$$\mathrm{Sim}(s, \tilde{s}) = \sum_{T=1}^{3} w_T |\vec{s}, \vec{\tilde{s}}| \tag{9}$$

where $\vec{s}$ denotes unlabeled sentence vector and $\vec{\tilde{s}}$ indicates training set sentence vector. $|\vec{s}, \vec{\tilde{s}}|$ denotes the distance between two vectors, which can be calculated via cosine distance and Euclidean distance. w_T is the weight of the distance calculated by three sentence vectors. We define sentence vector $\vec{s}$ as:

$$\vec{s} = \frac{1}{n} \sum_{i=1}^{n} e_i^T \tag{10}$$

where e_i^T represents the n-gram contextual embedding of the i-th character. Here are three kinds of sentence vectors due to the different values of T, which is set to 1, 2, and 3 respectively. We obtain these vector representation according to section 3.2.2. n denotes the length of a sentence. $\vec{\tilde{s}}$ is similar to $\vec{s}$.

3.3.2. The Uncertainty of Annotation. Higher the uncertainty of a sample sentence's annotation, more useful features the sample sentence contains. In our work, therefore, we choose the samples with higher uncertainty in annotation task from unlabeled data.

Additionally, the word segmentation can be represented as a two-class problem according to one factor, that is, whether the character is the right boundary of a word in a sentence. Specifically, the labels B, M, E, S can be divided into two categories: B and M can be grouped together, denoted as N, indicating that the character isn't the right boundary of a word; Also, E and S can be grouped and denoted as Y, indicating that the character is the right border of a word and need to be divided.

We use information entropy to measure the uncertainty of sequence annotation and it is formally denoted as:

$$\mathrm{E(c)} = -\sum_{i=N,Y} p_i log_2 p_i \tag{11}$$

where, c is a character from unlabeled sentences, $p_N = p_B + p_M$, $p_Y = p_E + p_S$, p_B represents the posterior probability that c is marked as B. p_M, p_E and p_S are similar to p_B.

3.3.3. Sampling Strategy. Due to the lack of labeled data, CWS models cannot learn enough features. In general, if a sample sentence is more different with sentences of the training set, the sample sentence contains more semantic features. Therefore, selecting useful sample sentences relies on two aspects: The uncertainty of annotation and the semantic similarity between unlabeled sentences and training set sentences. In general, the scoring model for selecting sample sentences is finally defined as:

$$D(s) = \lambda \sum_{i=1}^{n} E(c_i) + (1 - \lambda) minSim(s, \tilde{s}) \quad (12)$$

where, s represents a sentence in unlabeled data, n indicates the character number of the sentence, c_i represents the character in the unlabeled sentence, $\tilde{s}$ is a training set sentence, λ is a weight parameter. The higher the score, the more valuable the sample is.

4. Experiments

4.1. Datasets

We evaluate our proposed methods on three prevalent CWS datasets, including NLPCC 2016 Weibo dataset [16] and PKU, MSR from SIGHAN2005 [17]. Table 1 gives the details of three datasets. We use 10% data of shuffled training set as development set for PKU and MSR datasets. All datasets are preprocessed by replacing numbers and continuous English characters with special flags. The performance of our methods is evaluated by F-score.

In our experiments, we use about 2.2 million Weibo unlabeled data built from the Internet. Moreover, we utilize about the same 2.3 million Sogou Lab's news corpus for PKU and MSR datasets.

Table 1. Statistical information of three labeled datasets.

Datasets		Sentence	Words	Chars	Word Types	Char Types
Weibo	train	20,135	421,166	688,743	43,331	4,502
	test	8,592	187,877	315,865	27,804	3,911
PKU	train	19,056	1,109,947	1,826,448	55,302	4,685
	test	1,944	104,372	172,733	1,915	2,932
MSR	train	86,924	2,368,391	4,050,469	88,114	5,150
	test	3,985	106,873	184,355	3,953	2,836

4.2. Parameter Settings

The Hyper-parameters used in the experiments are shown in Table 2. We use different training batch sizes for three datasets. Specifically, we set batch sizes of PKU and MSR datasets to 512 and 256 respectively, and 128 for Weibo dataset.

Table 2. Hyper-parameters settings.

Hyper-parameter	value
Unigram dimension	100
Bi-gram dimension	100
Tri-gram dimension	100
LSTM hidden size	100
LSTM input size	100
Learning rate	0.01
Dropout	0.5

4.3. Experimental Results

4.3.1. The Results of Initial Word segmentation model. The initial word segmentation model for the semi-supervised method in this paper is an attention-based natural segmentation model (Attention+BiLSTM+CRF). Due to the sparsity of data, our experiments only focus on the unigram, bi-gram and tri-gram vector representation of characters. The results of our experiments are shown in Table 3. We can see that the experimental result in first group is lowest, which demonstrates the effectiveness of character contexts. To prove the effectiveness of the attention mechanism, we add two groups of comparative experiments by concatenating contextual embeddings. As shown in Table 3, the average performance is boosted by 0.30% on F-score from baseline to attention-based model, and 0.10% on F-score from concatenation-based model to attention-based model when set N as 3, which indicates that the attention mechanism is effective to integrate contextual information. In addition, we observe that our attention-based segmentation model can achieve an improvement on small dataset (e.g. Weibo and PKU) and have less effect on dataset with enough training data (MSR). In order to ensure the precision of word segmentation model, we finally choose Attention+BiLSTM+CRF (N=3) as our initial word segmentation model.

Table 3. Segmentation results of the initial word segmentation model on three datasets, evaluated by F-score.

Models		Weibo	PKU	MSR
	P	93.45	96.10	97.06
Baseline(BiLSTM+CRF)[a]	R	93.74	95.90	96.98
	F	93.59	96.00	97.02
	P	93.75	96.29	97.26
Concatenation+BiLSTM+CRF(N=2)[b2]	R	93.85	96.15	97.14
	F	93.80	96.22	97.20
	P	93.85	96.42	97.36
Concatenation+BiLSTM+CRF(N=3)[b3]	R	93.97	95.22	97.30
	F	93.91	96.32	**97.33**
	P	93.85	96.40	97.28
Attention+BiLSTM+CRF(N=2)[c2]	R	94.05	96.26	97.16
	F	93.95	96.33	97.22
	P	94.02	96.65	97.35
Attention+BiLSTM+CRF(N=3)[c3]	R	94.14	96.25	97.31
	F	**94.08**	**96.45**	**97.33**

[a]: Our baseline neural word segmentation, which uses the character embeddings as input.
[b2]: The concatenation of the unigram and bi-gram embeddings are used as the input.
[b3]: The concatenation of the unigram, bi-gram and tri-gram embeddings is used as the input.
[c2]: The Attention-based segmentation model that focus on the unigram and bi-gram embeddings.
[c3]: The Attention-based segmentation model that focus on the unigram, bi-gram and tri-gram embeddings.

4.3.2. The Results of Semi-supervised Learning Segmentation. We first investigated the impact of λ in formula (12) over segmentation performance. The parameter λ is searched ranging from 0.1 to 1 with a step size of 0.1. We put λ into the formula (12) and select the most valuable samples according the sampling strategy. In each iteration, 1000 samples were selected. The number of iterations was 5. According to Figure 4, it can be seen that the segmentation effect is best when λ is set to 0.5 in two cases. Two kinds of methods for calculating distance have little difference in Figure 4. The cosine distance gives a little improvement compared to Euclidean distance.

To verify the effectiveness of our sampling strategy, we conduct a comparative experiment which choose different number of samples randomly. We apply the sampling strategy proposed in Section 3.3 to the unlabeled data. The annotation work of unlabeled data is based on the attention-based segmentation model. Experiment results are shown in Figure 5. It can be seen that with the increase of the number of datasets, the performance of our methods becomes better, stable and convergent. It shows that our proposed method can learn continuously knowledge from selected instances. Our method achieves better performance compared to the method with selecting randomly samples, demonstrating that sampling strategy is helpful to improve the segmentation accuracy.

Furthermore, we add a comparative experiment, which uses the word-based semantic similarity. This method calculates similarity based on word-level. For unlabeled sentences, we use initial word segmentation model to annotate them. The experimental results are shown in Table 4. Compared the word-based semantic similarity representations, the character-based method outperforms the former on three datasets. The results show that character-based models have the potential of capturing morpheme patterns, thereby improving generalization ability of word segmentation models.

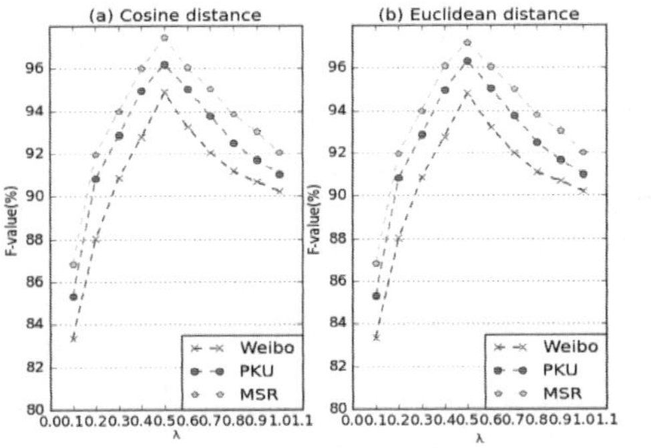

Figure 4. Performances with the different λ on Weibo, PKU and MSR datasets.

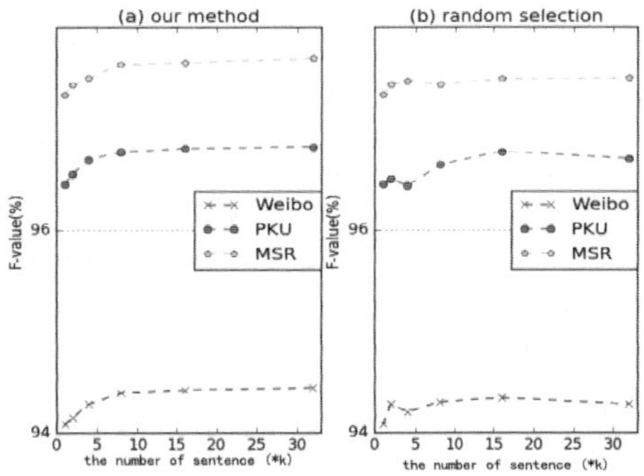

Figure 5. F1-value on three data sets when adding different numbers of samples.

Table 4. Comparison with word-based and character-based semantic similarity.

Methods		Weibo	PKU	MSR
Character-based	P	94.38	97.90	97.75
	R	94.50	97.74	97.63
	F	**94.44**	**96.82**	**97.69**
Word-based	P	94.27	96.82	97.64
	R	94.43	96.72	97.56
	F	94.35	96.77	97.60

Table 5. Comparisons with other methods on three datasets.

Models	Weibo	PKU	MSR
Chen et al. [1]	94.24	94.32	96.04
Yang et al. [2]	**95.50**	96.30	97.50
Zheng et al. [5]	-	96.67	97.50
Our work (words-based)	94.35	96.77	97.60
Our work(character-based)	94.44	**96.82**	**97.69**

Our final results compare favourably to the best neural models, including that using adversarial learning [1], and that leveraging abundant pretraining [2]. As shown in Table 5, our method gives the best accuracies on all corpora except for Weibo, where it underperforms the method of Yang et al. [2]. Compared with the state-of-the-art semi-supervised method [5] in the previous, our semi-supervised method is more stable, and the performance becomes better with the increase of unlabeled data. Experimental results verify that our method is simple but robust, which is of practical importance.

5. Conclusion and Future Work

In this paper, we propose a novel semi-supervised CWS method based on semantic similarity. Specifically, we propose a sampling strategy, which help us select more valuable sample sentences from massive unlabeled sentences to extend the training data. At the same time, we introduce character-based semantic similarity to the sampling strategy. Furthermore we introduce an attention-based neural segmentation, which dynamically focuses on the useful contextual information. The experimental results on the Weibo dataset fully show that our method achieve better word segmentation by frequent iteration. More importantly, it also shows that our sampling strategy effectively choose samples with richer features. In addition, experimental results on PKU and MSR datasets show that our method has cross domain robustness.

In the future research, we hope to use the method proposed in this paper to automatically acquire training corpora. Moreover, we wish to optimize the semi-supervised learning sampling strategy so that it can have better domain adaptability.

References

[1] Chen X, Shi Z, Qiu X, et al. Adversarial multi-criteria learning for chinese word segmentation[C]// Meeting of the Association for Computational Linguistics. 2017:1193-1203.
[2] Yang J, Zhang Y, Dong F. Neural word segmentation with rich pretraining[J]. 2017:839-849.
[3] Rei M, Crichton G K O, Pyysalo S. Attending to characters in neural sequence labeling models[J]. 2016.
[4] Zheng B, Che W, Guo J, et al. Enhancing LSTM-based word segmentation using unlabeled Data[J]. 2017.
[5] Sun W, Xu J. Enhancing chinese word segmentation using unlabeled data[C]// Conference on Empirical Methods in Natural Language Processing. Association for Computational Linguistics, 2011:970-979.
[6] Yang F, Vozila P. Semi-supervised chinese word segmentation using partial-label learning with conditional random fields[C]// Conference on Empirical Methods in Natural Language Processing. 2014:90-98.
[7] Melamud O, Goldberger J, Dagan I. Context2vec: Learning generic context embedding with bidirectional LSTM[C]// Signll Conference on Computational Natural Language Learning. 2016:51-61.
[8] Peters M E, Ammar W, Bhagavatula C, et al. Semi-supervised sequence tagging with bidirectional language models[J]. 2017:1756-1765.
[9] Yang F, Vozila P. An empirical study of semi-supervised chinese word segmentation using co-training[C]//Proceedings of the 2013 Conference on Empirical Methods in Natural Language Processing. 2013: 1191-1200.
[10] Zeng X, Wong D F, Chao L S, et al. Co-regularizing character-based and word-based models for semi-supervised Chinese word segmentation[C]// Meeting of the Association for Computational Linguistics. 2013:171-176.
[11] Zhang L, Wang H, Sun X, et al. Exploring representations from unlabeled data with co-training for Chinese word segmentation[C]//Proceedings of the 2013 Conference on Empirical Methods in Natural Language Processing. 2013: 311-321.
[12] Huang Z, Harper M, Petrov S. Self-training with products of latent variable grammars[C]// Conference on Empirical Methods in Natural Language Processing. Association for Computational Linguistics, 2010:12-22.
[13] Liu Y, Zhang Y. Unsupervised domain adaptation for joint segmentation and POS-tagging[C]// COLING. 2012.
[14] Lample G, Ballesteros M, Subramanian S, et al. Neural architectures for named entity recognition[C]// Conference of the North American Chapter of the Association for Computational Linguistics: Human Language Technologies. 2016:260-270.
[15] Mikolov T, Corrado G, Chen K, et al. Efficient estimation of word representations in vector space[C]// International Conference on Learning Representations. 2013:1-12.
[16] Qiu X, Qian P, Shi Z. Overview of the NLPCC-ICCPOL 2016 shared task: chinese word segmentation for micro-blog texts[J]. 2016.
[17] Sproat R, Emerson T. The first international chinese word segmentation bakeoff[C]// Sighan Workshop on Chinese Language Processing. Association for Computational Linguistics, 2003:133-143.

Convolutional Neural Network Based Structured Data Field Granularity Classification Method

Pin Wu, Quan Zhou, Qiu Wei, Zhidan Lei and Xiaoqiang Li

School of Computer Engineering and Science, Shanghai University, No. 99 Shangda Road, Baoshan District, Shanghai

jzhou8763@shu.edu.cn

Abstract. The data warehouse provides data support for enterprise decision-making and online analysis. In the process of building a data warehouse, many heterogeneous source system data needs to be integrated. In the integration process, these heterogeneous data needs to be classified and put into different topics. The diversity of the source systems of large and medium-sized enterprises poses difficulties for granularity at the field-level automated classification. However, the accuracy of previous methods cannot satisfy users. Therefore, this paper proposes a neural network-based classification technology to classify the data in the granularity field. This method adopts a sampling method to construct the characteristics of the field and innovates a novel classification framework based on the database field on the basis of the CNN network. Accurately achieve 89% by testing the data in the TPC-DS's dimension tables, and achieve 93% accuracy in real-world data testing. This method was validated in the actual environment of the three banks in China and achieved satisfactory results.

1. Introduction

Decision-making plays a crucial role in the success or failure of a business. The data warehouse is a structured data environment for decision support systems (DSS) and online analytical application data sources [1]. The quality of data warehouses is particularly important. Mouncey P [2] believed that the premise of improving the quality of data warehouse and business information is that the quality of information stored in the database is as measurable as the quality of the assembly line. So our efforts are to improve the quality of data warehouses.

Improving data quality requires the correct classification of data. Data warehouses are characterized by integration, topic-oriented, stability, and time-varying [3]. On the one hand, the data warehouse is integrated, and the data in the data warehouse comes from scattered operational data. The required data is extracted from the original data, processed and integrated, and unified and integrated before loading into the data warehouse. The data in the data warehouse is processed, summarized, and organized on the basis of the extraction and cleaning of the original scattered database data. The inconsistency in the source data must be eliminated to ensure that the information in the data warehouse is about the entire enterprise. On the other hand, data warehouses are topic-oriented. The data organization of the operational database is oriented to the transaction processing task, and the data in the data warehouse is organized according to a certain subject domain. Topics are the key areas of concern for users when making decisions using data warehouses. A topic is often associated with multiple operational information systems.

Building a data warehouse generally includes three key stages. They are data acquisition, data storage, and data access. There are many problems to be solved in the data acquisition phase for the characteristics of the data warehouse. Facing the task of high quality data, we have to classify heterogeneous data properly.

id	name	age	address
1	Jack	25	Dong Xin Road, Shanghai, China
2	Wendy	22	Oxford Street, London, United Kingdom
3	Iris	12	Fifth Avenue, New York, United States
4	Bella	13	Champs-Elysees, Paris, France
5	Cherry	18	Shinjuku Street, Tokyo, Japan

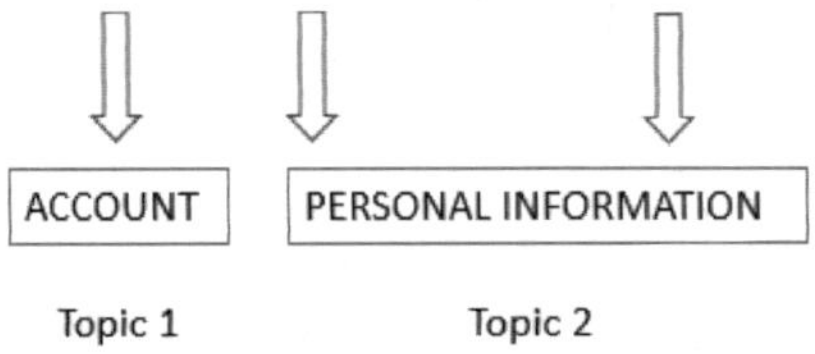

Figure 1. Topic classification at field granularity

As shown in Figure 1, our task is to map each field in the table to the one of the topics or subclasses in the data warehouse. The name field is mapped to the ACCOUNT topic, and the age and address fields are mapped to the PERSONAL INFORMATION topic. The data types in the fields are diverse, including character, numeric, and so on. This brings with it the difficulty of classification tasks.

The previous technology could not meet people's requirements. Now it takes a lot of labor to classify it. Therefore, this article reviews the previous technologies and finds their flaws. Then put forward our approach. Then describe in detail the architecture and implementation of the method. Finally, verify through experiments.

2. Related Works

People have been working hard to build data warehouses efficiently and continuously improving the quality of data warehouses. One of the important goals is to achieve automated classification.

2.1. Data Warehouse

The father of data warehouse Bill Inmon proposed the concept of data warehouse in 1990 [4]. Data warehousing is a group of decision-based technologies designed to enable knowledge workers to make better and faster decisions. In recent years, there has been explosive growth both in the number of products and services provided and in the industry's acceptance of these technologies. The first stage of building a data warehouse requires a variety of data extraction and cleaning tools. One of the most important tasks to use these tools before loading is to sort the data and load it into the corresponding theme, as shown in the blue box in Figure 2. Currently available tools and methods cannot achieve high-precision classification results.

Golfarelli M et al. [5] proposed a general method framework for data warehouse design based on dimension fact model (DFM). After analyzing the existing information system and collecting user requirements, the concept design is semi-automatically started from the operation of the database solution. . Fang Y [6] described the data quality problems in the data warehouse environment and the importance of ensuring data quality for data quality, and finally provided a data quality maturity model (DQMM) and method for controlling data quality of data warehouses. . Chen Z Q [7] proposed a new data warehouse model based on topic graph technology, which automatically generates and merges thematic maps. Batra V S et al. [8] proposed the use of DAGs to generate ETL scripts that can be used to populate newly proposed data warehouses based on existing data in existing schemas.

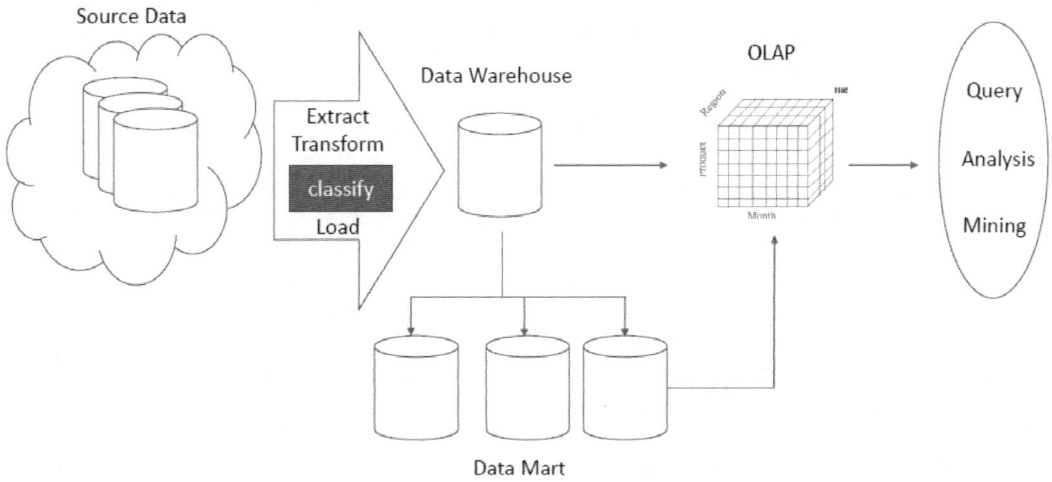

Figure 2. Data Warehousing Architecture

2.2. *Sampling Methods*

Due to the large amount of source data for large and medium-sized companies, analyzing the source data in full amount will lead to inefficiency. Therefore, we try to find a good sampling method to better characterize the source data. People have done a lot of work in the sampling field. Random sampling requires strict adherence to the principle of probability. Each sampling unit has the same probability of being extracted and can be reproduced. Random sampling is often used when the overall number is small, and its main feature is to extract one by one from the population. Random sampling can be divided into simple random sampling, systematic sampling, stratified sampling and cluster sampling.

Olken F et al. [9] analyzed the basic operations of large-scale database auditing and statistical analysis, and discussed sampling on this basis. A method of obtaining a sample from the result of a relational query without first executing the query is given. Specifically, check for simple random sampling of selections, projections, connections, joints and intersections. Kadilar C et al. [10] proposed ratio estimators by adjusting Ray and Singer's estimator types to find conditions that make each proposed estimator more effective than other estimators. The advantage of simple random sampling is that it is simple to operate, but it is not suitable for oversized data.

Earlier, Neyman J. et al. [11] proposed stratified sampling. The core idea is to divide the population into non-intersecting layers, and then to extract a certain number of individuals from each level according to a certain proportion, and to combine the individuals taken from each layer together as a method of sample. The smaller the variation within the layer, the better, and the greater the variation between the layers, the better. Tong C et al. [12] described some refinement strategies for layered experimental design, such as Latin hypercube, orthogonal array, and factorial design to solve the adequacy of a given sample to give acceptable statistical estimates. The problem identified. The advantage of this method is that it reduces the sampling error, the sampling method is flexible and the different layers can be analyzed independently. The disadvantage is that if the stratification variables are not properly selected, the intra-layer variability is large and the inter-level mean numbers are similar, stratified sampling will lose its significance.

2.3. *Machine Learning Classification Methods*

Once we have good characterization data, we will consider what method to classify these characterization data. With the rapid development of computing power, storage space, and networks, the amount of data that humans accumulate is rapidly growing. Classification is a very important task in data mining and is currently used most commercially. Classification algorithms are used to solve classification problems. The method is an important research area in machine learning. Classification is very common in our real life. Classification is one of the most useful technologies. Its application covers all areas of society. Common applications include: user portraits, credit assessment, recommendation systems, trend prediction, image classification, and text classification.

Breiman L. [13] proposed that random forests are a combination of tree predictors, so each tree relies on the value of an independent sampled random vector, and the distribution of all trees in the forest is the same. Zhu J [14] developed a new algorithm that directly extends the AdaBoost algorithm to multiple classes without reducing it to multiple types of problems. We have proved that the proposed multi-class AdaBoost algorithm is equivalent to a forward-stage additive modeling algorithm, which can minimize the novel exponential loss of multi-class classification. Landgrebe D [15] introduced the current decision tree classifier (DTC) design methodology and investigation of various existing problems. Decision tree is an algorithm model that is more widely used in the industry. The advantage is that the computational complexity is not high, and the reuse is high; the output result is easy to understand; the intermediate value is insensitive; the irrelevant feature data can be processed; and the continuous features can be processed well. The disadvantage is that it is prone to overfitting problems; it is more sensitive to noise data. Friedman J H [16] proposed that the gradient promotion of the regression tree provides a competitive, highly robust and interpretable procedure for regression and classification, and is particularly suitable for mining less than clean data.

2.4. Neural Network Classification Methods

The classification model of machine learning still has the defect of insufficient accuracy for large-scale data. In 2006, Hinton used the pre-training method to alleviate the problem of local optimal solution, pushed the hidden layer to the 7th layer [17], and the neural network had a “depth” in real sense, thus opening up a boom of deep learning. In the structure of the fully connected DNN, the underlying neurons and all upper neurons can form connections. The potential problem is the expansion of the number of parameters. For CNN, not all upper and lower neurons can be directly connected, but through the "convolution kernel" as an intermediary. It is precisely because the CNN model limits the number of parameters and taps this feature of the local structure. In the recurrent neural network RNN, the output of the neuron can be directly applied to itself at the next timestamp, ie, the input of the i-th layer neuron at time m, in addition to the output of the (i−1) layer neuron at this moment, Its own output at (m−1). The RNN can be seen as a neural network that passes in time. For t, the gradient it produces disappears after it has spread several layers to history on the time axis. In order to solve the disappearance of the gradient in time, the field of machine learning develops. The long-term memory unit LSTM realizes the memory function in time through the switch of the door and prevents the gradient from disappearing. The CNN and RNN are often connected to the full-connection layer before they are output by the upper layer, so they can complete the classification task.

Yoon Kim proposed in 2014 that TextCNN is an algorithm that uses convolutional neural networks to classify text. The algorithm transforms sentences into similar images through word vectors, and then obtains local information through convolutions to obtain word-based feature maps. It has achieved good results in sentiment analysis and question-answering classification tasks. FastText is a fast text classifier developed by Facebook that provides simple and efficient text classification and representation of learning [18]. The input to the FastText model is a sequence of words (a piece of text or a sentence), and the output is the probability that the sequence of words belongs to a different category. The words and phrases in the sequence form a feature vector. The feature vector is mapped to the middle layer through a linear transformation and then mapped to the label by the middle layer. FastText uses a nonlinear activation function when predicting labels, but does not use nonlinear activation functions in the middle layer. Pengfei Liu and Xipeng Qiu et al. [19] use a multitasking learning framework to learn multiple related tasks together. Based on recursive neural network, three different information sharing mechanisms are proposed to simulate texts with task-specific and shared layers, thus completing a very good text classification task.

3. Our method

3.1. Data Flow of Classification

The model we implemented was before the source data was loaded into the data warehouse, that is, in the blue box in Figure 3. It is possible to achieve highly automated loading into data warehouses through our highly accurate and efficient models.

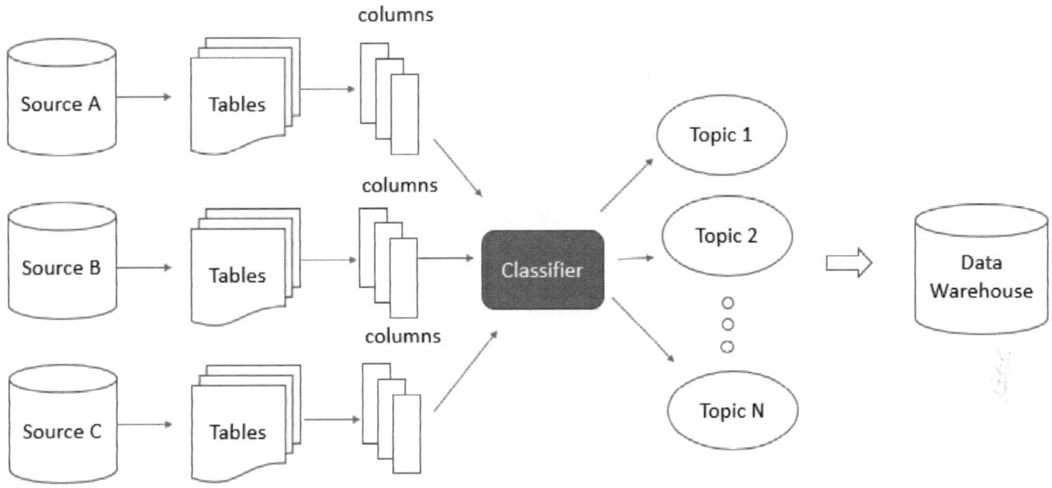

Figure 3. Data Flow of Classification

In the real world, large and medium-sized enterprises have a large number of heterogeneous source systems. For example, in China, a medium-sized bank has more than 300 source systems. Here, we use A, B, and C as the representative. The core of this paper is the processing in the blue box, which will be described in detail in the next section.

3.2. Architecture of Our Method

The method we propose can be used to classify data on the one hand, as shown in Figure 4. On the other hand, since there are a large number of fact tables in the real world, the data of its fields is foreign key ID data that has no practical significance. Therefore, for such data, we design the method as shown in Figure 5.

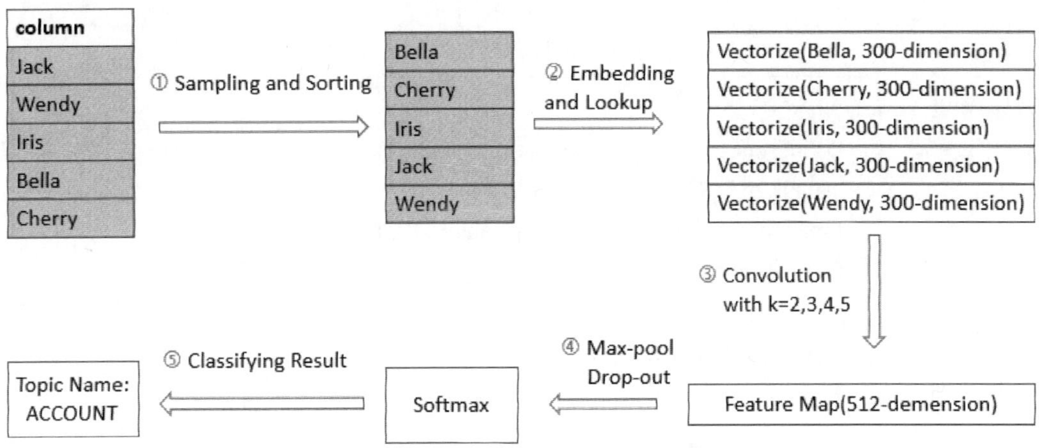

Figure 4. Architecture of Our Method for Data-Based

Since our task is to classify the data of the granularity at field-level to topic classification, we first select the data in one field of a table in the source system. Secondly, these data are sampled in a hierarchical manner and 100 samples are extracted. The 100 samples are then sorted lexicographically. The sorting process allows the data to have certain contextual information. Then, the words in the sample are mapped into a 300-dimensional vector space. Next, it is convolved with a convolution with coefficient k of 2,3,4,5. The 128-dimensional feature map generated by this process is concated into a 512-dimensional feature map. Maxfielding and dropout (0.5 in Training, 1 in Testing) are performed on the feature map and input to the softmax function. Finally, the classification result is obtained. That is, we need to obtain the topic classification.

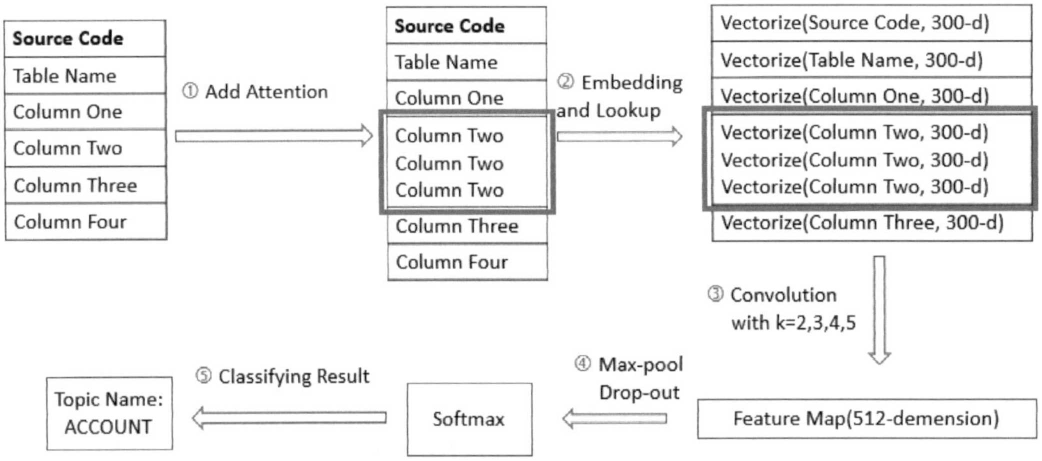

Figure 5. Architecture of Our Method for Name-Based

The name-based approach differs from the data-based approach in that the input to the neural network is not the data from the field, but the table's metadata or structure, the main active system name, the table name, and the field name. We enter the entire field names in order. We introduce a focus mechanism to reinforce each field, and when this is categorized, the red box in Figure 5 is used.

3.3. Algorithm of Our Method

To better illustrate our approach, we list our algorithm in Table 1.

Table 1. Algorithm of Sampling and Sorting.

Input	Column Data, Topic Dictionary
Output	Topic Name
1	DataFrame = GetColumnData(Column Data)
2	DataFrame = Sort(Sampling(DataFrame))
3	EmbeddingData = Embedding(Lookup(DataFrame))
4	For i in enum(kernels):
5	FeatureMap.append(Conv2d (EmbeddingData, i))
6	FeatureMap.concat(2)
7	TopicName = LookupDicts(Dropout(Softmax(Maxpooling())))
END	

4. Experiments and Discusses

4.1 Datasets and Environments

The configuration of our main equipment is a machine with a 4-core, 64G memory and a GTX-1080 GPU 8G memory. Our experimental data selects two types, one is the test data generated by the TPC-DS tool. TPC-DS is a set of decision support system test benchmarks, mainly for the retail industry. Test data is highly similar to real-world data and has various business models. See http://www.tpc.org. The data size of the test can be customized. The data has a fact table and a dimension table, so it is very consistent with our experimental tasks. We generated 1TB of test data in this experiment. We select 90% as the training set and 10% as the test set.

The other one selects data from three Chinese banks. Since the table names and field names given by TPC-DS are very regulated, if we put them into our method, the model would get 100% accuracy very quickly. However, in reality, the heterogeneity of the source system results in very diverse table names and field names. Therefore, we choose real-world data to complete the classification task by data-based method.

4.2 Experimental Result

We evaluate our method with accuracy, recall rate and F1 value. The results of the test are shown in Table 2.

Table 2. Experimental Result

Algorithm Name	Accuracy	Recall	F1Value
RandomForestClassifier	0.68	0.53	0.57
AdaBoostClassifier	0.73	0.73	0.58
DecisionTreeClassifier	0.85	0.85	0.85
GradientBoostingClassifer	0.89	0.85	0.86
Our Methods	0.93	0.91	0.92

4.3 Discusses

It is not difficult to find from the experimental results that our method obtains relatively high performance indicators. To get good results, you should choose the right way to process data and choose the right algorithm to learn the data according to different data structures and different tasks. As we discussed in this article is a classification task. Different feature extraction methods are selected according to the different characteristics of the fact table and the dimension table. For the fact table, we need to extract the characteristics of the source system data, such as the source system name, table name, field name, field type and field length to complete the fact table classification task. For the dimension table, we select the data in the source system data to complete the classification task. For data that has semantic information and context, such as table names and field names, using CNN can achieve better classification results. For a large number of practically meaningful data, the choice of ensemble learning in machine learning can achieve very good classification results and rapid acquisition of model features.

5. Conclusions

The data loading stage of the data warehouse requires high-accuracy classification tools to accomplish the task of topic classification and secondary classification. The method proposed by us has high-performance classification of data with a granularity field. Accurately achieve 89% by testing the data in the TPC-DS's dimension table, and achieve 93% accuracy in real-world data testing. The preliminary results can be drawn from the experimental results. Selecting the appropriate feature extraction method for the field data in the database and using the computational framework of the convolutional neural network can obtain the ideal classification effect. At the same time, the great potential of using neural networks to solve traditional problems can be seen. This research result can also be applied to the application of automatic ER map construction and other databases. The method we propose can save a lot of manpower for large and medium-sized enterprises that use data warehouses, and can contribute to obtaining higher quality data. The future work is to adopt a better field characterization method to further improve the neural network classification accuracy, achieve an accuracy rate that exceeds the manual classification, so as to achieve full automation of data warehouse loading.

Acknowledgments
Thanks to the professors and students of Shanghai University for their valuable suggestions. Thanks for the support of the hardware and software environment provided by Beyondsoft (Shanghai) Co., Ltd. Thanks to three Chinese banks for providing demand analysis and real-world data for verification.

References
[1] Widom J. Review - An Overview of Data Warehousing and OLAP Technology[J]. 1999.
[2] Mouncey P. Improving Data Warehouse and Business Information Quality[J]. Interactive Marketing, 2001, 2(3):295-297.
[3] Mathur N, Mathur N, Mathur N. An Overview of Data Warehouse and OLAP Technology Used in Banking System[J]. International Journal of Emerging Trends & Technology in Computer Science, 2016, 33(2):76-79.
[4] Inmon W H. Using Oracle to Build Decision Support Systems[M]. John Wiley & Sons, Inc. 1990.
[5] Golfarelli M, Rizzi S. A methodological framework for data warehouse design[J]. 1998, 2:3-9.
[6] Fang Y. Data Quality Managements in Data Warehouse[J]. Computer Engineering & Applications, 2003.
[7] Chen Z Q, Wang Z Y, Xia L X. Data Warehouse Model Based on Topic Map[J]. Information Science, 2008, 26(11):1689-1693.
[8] Batra V S, Bhide M A, Mohania M K, et al. Data warehouse data model adapters[J]. 2017.
[9] Olken F, Rotem D. Simple Random Sampling from Relational Databases.[C]// Vldb'86 Twelfth International Conference on Very Large Data Bases, August 25-28, 1986, Kyoto, Japan, Proceedings. DBLP, 1986:160-169.
[10] Kadilar C, Cingi H. Ratio estimators in simple random sampling[J]. Applied Mathematics & Computation, 2004, 151(3):893-902.
[11] Neyman J. On the Two Different Aspects of the Representative Method: the Method of Stratified Sampling and the Method of Purposive Selection[J]. Journal of the Royal Statistical Society, 1934, 97(4):558-625.
[12] Tong C. Refinement strategies for stratified sampling methods ☆[J]. Reliability Engineering & System Safety, 2017, 91(10):1257-1265.
[13] Breiman L. RANDOM FORESTS--RANDOM FEATURES[J]. Machine Learning, 1999, 45(1):5--32..
[14] Zhu J, Zou H, Rosset S, et al. Multi-class AdaBoost[J]. Statistics & Its Interface, 2009, 2(3):349-360.
[15] Landgrebe D. A survey of decision tree classifier methodology[J]. IEEE Transactions on Systems Man & Cybernetics, 2002, 21(3):660-674.
[16] Friedman J H. Greedy Function Approximation: A Gradient Boosting Machine[J]. Annals of Statistics, 2001, 29(5):1189-1232.
[17] Hinton GE, Salakhutdinov RR. Reducing the dimensionality of data with neural networks[J]. Science, 2006, 313(5786):504-507.
[18] Joulin A, Grave E, Bojanowski P, et al. Bag of Tricks for Efficient Text Classification[J]. 2016:427-431.
[19] Liu P, Qiu X, Huang X. Recurrent neural network for text classification with multi-task learning[C]// International Joint Conference on Artificial Intelligence. AAAI Press, 2016:2873-2879.

IOP Conf. Series: Materials Science and Engineering **435** (2018) 012034 doi:10.1088/1757-899X/435/1/012034

New Improved Training for Deep Neural Networks Based on Intrusion Detection System

Ilyas Benmessahel[1, a,*], Kun Xie[1,b] and Mouna Chellal[2, c]

[1] College of Computer Science and Electronics Engineering, Hunan University, Changsha, China.
[2] School of Information Science and Engineering, Central South University, Changsha, China.

[a] * ilyasbenms@hnu.edu.cn; [b] kunxie@hun.edu.cn; [c] mounachellal@gmail.com

Abstract. All Network intrusion detection is designed for detecting, preventing, and repelling network security breaches and it has become an urgent issue. Maintaining a safe and secure network requires an efficient and flexible solution called an intrusion detection system. This paper reports an advanced intrusion detection method created with a deep learning approach. Evolutionary operators can reduce the probability of stagnation in local solutions due to high local optima avoidance and have thus superseded conventional training algorithms, such as back propagation (BP). Combining a deep neural network (DNN) and an evolutionary algorithm (EA) may solve problems or outperform DNN in solving existing problems. We develop a hybrid training method that combines simulated annealing (SA) and BP to improve the performance of DNN (SABP-DNN). The NSL-KDD dataset is used to verify the accuracy and efficiency of the proposed method. The proposed method is also compared with the original DNN based on PB (PB-DNN) and DNN based on SA (SA-DNN). We confirm that the proposed method presents a strong potential to become an alternative solution to IDS through experiments and comparisons with existing methods.

1. Introduction

The increase in the number of local networks has led to continuous evolution of Internet data and the availability of massive amounts of network data has promoted the development of information technology, which requires careful attention. Consequently, this evolution, in turn, has increased the vulnerability of systems to various threats [1]. Any intrusion can have disastrous consequences. For example, personal data can be destroyed, corrupted, or illegally accessed as a result of confidentiality breaches. Furthermore, integrity breaches can lead to the alteration of personal data. Computer network security has become a promising tool to provide secure channels. One of the promising tools that can detect attacks is the intrusion detection system (IDS). Cybersecurity infrastructures employ IDS as an essential component and use it to protect systems and infrastructures from various threats.

IDS is a permanent monitoring scheme applied in computer or network systems to monitor targeted systems, collect audit data, analyze the gathered information to determine unusual activities and establish response plans [2]. IDS should either model any type of attacks or anomalous events that can affect the network under consideration (signature based) or build a general model that describes normal traffic (anomaly based).

Network IDS comprises a set of single-purpose sensors or host computers that are positioned at different points in a network or interconnected set of networks to monitor transmitted traffic. These units examine network traffic packets in real time, perform local analysis for a captured packet, and finally take action, such administrator notification [3]. NIDS involves two major detection techniques, namely, misuse-based detection and anomaly-based detection [4]. Misuse detection systems, such as works in [5, 6], define the behavior of an intruder and scan predetermined attack data to match signatures. Anomaly detection systems, such as works in [5, 7], set only normal (typical) or expected behavior and flag any deviation from this behavior. However, these two detection systems possess weaknesses. Misuse detection systems cannot detect new forms of attack because their signatures are not yet available for pattern matching. Anomaly detection systems produce a high false alarm rate when adopted in most existing approaches because building practical normal behavior to protect targeted systems is difficult [8].

Many techniques have been utilized to enhance the performance of IDS. However, security experts still strive to achieve IDS with improved performance, the highest detection rate and the lowest false alarm rate. Machine learning has been impeccably accomplished in many areas of computer science, such as speech, visual and face detection, but not sufficient in intrusion detection [3].

Recently, developed techniques have elicited considerable attention in the search for a solution to accommodate concerns regarding intrusion detection. These techniques include artificial neural networks [5], evolutionary algorithms (e.g., genetic algorithm (GA) [9], particle swarm optimization (PSO) [10], and ant colony optimization (ACO) [11]) and deep learning approaches (e.g., deep neural network (DNN), restricted Boltzman machine, deep Boltzman machine, and convolutional neural network (CNN)) [12]. The present study is focused on two of these methods: deep neural network (DNN) and simulated annealing (SA).

DNN and SA have not been sufficiently explored in literature. Furthermore, we selected these methods due to the fact that DNN employs consecutive layers of information processing stages in a hierarchical manner for pattern classification and feature or representation learning [10]. Moreover, weight learning tasks are considered essential in the application of DNN. In other hand, SA has shown good performance in solving continuous optimization problems due to the search mechanisms used in its mutation operators [13]. Thus, a natural question arises: can we benefit from the advantages of SA techniques in overcoming the training phase of DNN by injecting search strategies of SA into the training operator of DNN algorithms?

Although DNN can solve various learning tasks, it is difficult to train [14]. Deep-learning algorithms, which inspired by the neurons that make up the human brain, have gained numerous attention in scientific community due to their successful application in area. Recently proposed optimization techniques for training DL use layer-wise pre-training. DNN training requires powerful optimization techniques. Therefore, several conventional algorithms, such as BP, are used to solve this issue. However, the BP algorithm is sometimes susceptible to convergence into local optima because it is used as a local search method in which the final solution depends strongly on the initial weights [8]. The SA algorithm has been used to solve the global optimization problem. The primary advantage of SA over other methods is its capability to avoid local minima [15]. The algorithm functions by selecting the neighbourhood solution within a search space, accepting the worse solution, and moving to inferior solutions to escape entrapment (i.e., local optimal solution). Thus, the efficiency and robustness of the SA algorithm have been verified through the solution of numerical optimization samples.

We developed a new hybrid algorithm called SABP-DNN that combines SA and BP to build an advanced training algorithm for DNN. The proposed algorithm was applied to detect intrusion in the NSL-KDD dataset. This work exploited the local search capability of BP and the global search capability of SA. We injected SA into DNN to facilitate training and improve the detection rate. Consequently, combining SA with BP can reinforce the strengths of the two methods because of their computational advantages and can thus reduce the computational time of the training phase. The well-known intrusion detection dataset (NSL-KDD) was used to test the performance of the proposed method. The main contributions of this study can be summarized as follows:

- An intrusion detection model based on DNN-trained SA combined with BP was presented and proven feasible its usefulness, as evidenced by the increased detection rate for NSL-KDD data.
- The principles of SA were examined to optimize the DNN weight and determine its effectiveness in the IDS field.
- The efficiency of DNN was evaluated by using the NSL-KDD dataset. The SABP-DNN model outperformed SA-DNN and BP-DNN in the detection task.
- The proposed SABP-DNN was compared with other techniques by using the NSL-KDD dataset to confirm the applicability of the proposed approach.

The rest of this paper is structured as follows: Sections 2 provide a brief introduction to the concepts of ANN and outline the mathematical overview of the DNN and SA, respectively. Section 3 and Section 4 discusses the use of SA combined with BP in DNN training. Section 5 discusses an overview of the architecture of the proposed IDS. Section 6 presents the experimental setup and results. Section 7 summarizes the conclusions and directions for future work.

2. Background

2.1. Deep neural network

The term DNN can have several meanings, deep is usually related to the hidden layers contained in a neural network and formally defined as a learning model with many layers of nonlinear transformations. Deep learning is one of the newest techniques of machine learning, it allows computers to acquire human brain abilities in natural functionality such as learning using examples. A DNN is comprised of a collection of advanced deep connected neurons listed in a distinctly layered topology, namely, an input layer, one or multiple hidden layers, and a single output layer [16]. In DNN, which is one of the mainstream deep learning methods, data moves forward these artificial deep neurons from the input units, through the hidden units which processes an aspect of the data, and eventually reaching the output units. The sample DNN includes of three hidden layers, one input layer, and one output layer for a global of five layers. DNNs are typically trained by updating and adjusting neurons weights and biases through the use of the supervised learning BP algorithm [14]. Each node in layer performs a specific calculation, which is defined as:

$$v^{\{l\}} = f\left(z^{\{l\}}\right) = f\left(\sum_{\{j=1\}}^{K} \left\{w_{\{ij\}}^{\{l\}} * I_i^{\{l\}} + \beta_i^{\{l\}}\right\}\right) \qquad (1)$$

Where $z^l \in R^{N_l}$ is the stimulation vector, v^l is is the activation vector, $w_{ij} \in R^{N_l * N_l}$ the connection weight matrix, and $\beta_i \in R^N$ is the threshold (or bias) vector. $N \in R^{N_l * N_l}$ is the number neurons at layer l. $f()$: $R^{N_l} \rightarrow R^{N_l}$ is the activation function applied to the stimulation vector element wise. In most applications, the sigmoid function and the hyperbolic tangent function is the most used as activation function for ANN. Rectified Linear Unit function is popular activation functions for DNN.

2.2. Simulated Annealing Algorithm

SA is a generic probabilistic meta-heuristic for complex system optimization. SA was first proposed by Kirkpatrick et al. [17] in 1983. The SA algorithm simulates the annealing process in metallurgy. Annealing is a technique that involves heating a solid to the highest level and controlling cooling until the energy state of internal atoms is reduced to a minimum and crystal growth has initiated.

Here, we present a general outline of the algorithm. First, a random solution is generated and then the cost is calculated using one of the defined functions. A random neighbor solution is generated and the cost of the neighbor solution is calculated. The following comparisons are established: If the cost of the neighbor solution is 'smaller' than the cost of the current solution, then the new solution is accepted. If the cost of the neighbor solution is 'larger' than the cost of the current solution, then 'maybe' the new solution is accepted. Nevertheless, it may be accepted after checking the Boltzmann probability factor:

$$P(x) = e^{-\delta f(x)/_{k*T}} \tag{2}$$

where f(x) is the energy function, k is Boltzmann's constant, and T is temperature [13]. The steps are repeated until the algorithm reaches an acceptable solution or the maximum allowable iteration.

3. Design of proposed training method

The training process aims to generate a convenient network approach for the corresponding incoming input, and this procedure is achieved through an iterative learning process. The learning process can be considered as a modification of the randomly generated neural network weights based on the neural network response to a set of training input patterns. Therefore, DNN training can be interpreted as constructing a predictive model (function) and optimizing it in an n-dimensional space. This optimization problem can be solved by SA, which is a stochastic global optimization method suitable for nonlinear function optimization.

The design of the proposed method uses DNN: -input layer one hidden layer with 10 neurons -one hidden layer with 20 neurons -one hidden layer with 10 neurons -output layer. The most important step in DNN training is the use of SA, which is a problem caused by the optimal combination of connection weight values between DNN and SA. The problem of training DNN should be approached suitably for the SA algorithm.

The main goals of using SA to train DNN are to optimize weight complexity, minimize approximation error, and achieve the required accuracy. These purposes can be accomplished through a cost-based process by computing the cost value of each vector solution. The optimal solutions with minimal cost value are selected. The solution vector is represented using values ranging from -1 to 1, and the number of features in each solution vector is computed as follows:

$$Indv_{nbr} = (n_i * n_h) + (2 * n_h) + 1 \tag{3}$$

where nh, ni are the number of hidden neurons and input features, respectively.

In this study, the cost function refers to the representation and measures of the represented solution that used in the optimization algorithm. Cost function evaluates the efficiency of a single solution in a population or the chance of finding a solution and reaching high coverage. The optimization process terminates when the termination criterion is satisfied, which is the allowable number of iterations or a satisfactory cost level, has been achieved. One of the cost functions is adopted in this study is as follows:

$$MSE = \frac{1}{R} * \sum_{i=1}^{R} o - y \tag{4}$$

where o is the desired output, y is the actual output, and R is the number of training samples. Optimization can be terminated on the basis of two situations: First, the maximum allowable iteration is reached. Second, the cost function threshold is attained. Consequently, the optimal state of network has been achieved.

4. Improved training DNN with SA and BP

The improved training algorithm combines SA with BP; therefore the fundamental idea behind this hybrid algorithm is to exploit the global search capability of SA and the local search capability of BP. SA has a strong exploration capability for finding a global optimal solution, which means that SA has stable capabilities in global convergence but arrives at the global optimum. Thus, its search process slows down. By contrast, the BP algorithm can determine whether the global optimal solution is weak or strong in finding a locally optimal solution. Thus, the BP algorithm is capable of attaining rapid convergence around the global optimum for finding the local optimum.

The main principle behind combining SA with BP is as follows: SA is used to explore the global optimum during the initial stage of optimum searching, and BP is used to search around the global optimum to find the best local optimum.

A collection of random solutions is generated to initiate searching process and training. SA compares each network fitness value during each pass through a dataset. The solution with the optimal cost value is considered as the global best optimum. The BP algorithm then exploits the global best optimum to search for the local best optimum. The general training pseudo-code of SABP-DNN is presented in Algorithm 1.

Algorithm 1 Combining SA with BP training DNN pseudo code

```
1: Initialize training parameters: lb, ub, K, n.
2: Creating a set of random solutions X = x1,x2,,,xn in range of [ub,lb].
3: for each solution do do
4:      Calculate the MSE f(X) for solutions via equation 4.
5:      if f(x) < fbest then
6:      fbest = f(x)
7:      end if
8: end for
9: Until (minimum T or K=0)
10: for each solution do
11:     Use the BP algorithm to calculate a new solution x0.
12:     Calculate the MSE of f(x),f(x0).
13:     if f(x0) < f(x) then
14:     fbest = f(x0),x = x0
15:     else
16:     Calculate the difference Δf = f(x0) − f(x)
17:     Random value r(0 , 1)
18:     if r > e−Δf/kT then
19:     fbest = f(x0),x = x0
20:     else
21:      fbest = f(x)
22:     end if
23:     f = fbest
24:     Reduce the temperature gradually T = α ∗ T
25:     end if
26: end for
27: Save the best solution with fbest
28: Go to step 9 if the end criterion is not satisfied.
```

5. OVERVIEW OF PROPOSED MODEL

As shown in Figure 1, the architecture of our proposed IDS mainly consists of three components: The data input module and two basic modules, which comprise the DNN network and SA module.

Figure 1 shows that the data are not directly fed to the DNN module. The data must be pre-processed by normalizing, filtering, and extracting features from the NSL-KDD dataset [18]. The training and testing dataset will be provided as the input to the next phase for various processes by the DNN module.

In this study, the framework of the proposed system is based on the two stages of network training and detection. The DNN training stage uses the SA algorithm such that network weights are treated as solutions that have evolved in the SA algorithm. Once the evolutionary process of finding a suitable solution for DNN is completed, a selected solution representing a set of weights, which are regarded as the optimal weights, is achieved. The evolution process stops when termination is satisfied.

Finally, in the detection stage, the testing data are loaded into predicted models to predict whether the output exhibits normal or abnormal behavior. The detection stage uses the model established during the training stage to map the predicted output with the matched classes.

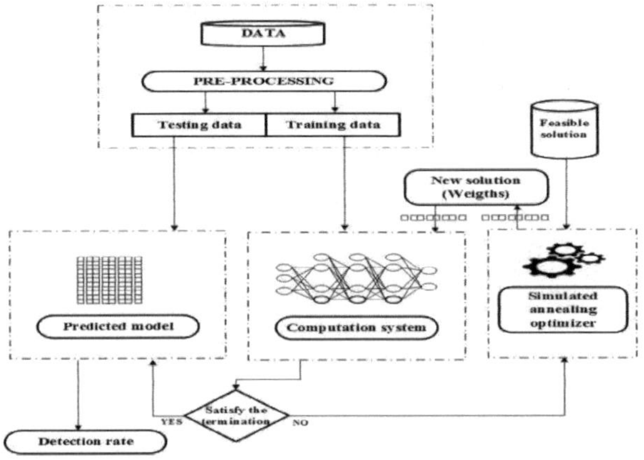

Figure 1. Framework of the proposed IDS.

6. EXPERIMENTAL SETUP AND RESULTS

6.1. NSL-KDD DATASET

NSL-KDD was developed in a suitable form by decreasing redundant instances that influence the evaluation of IDS results [19]. NSL-KDD, which is constructed from records selected from the whole KDDCup99 dataset, addresses many issues related to the KDDCup99 dataset by eliminating all duplicate and redundant instances in the whole dataset.

The NSL-KDD dataset includes the same features as the KDDCup99. Each NSL-KDD instance has 41 features that are labelled as either normal or an attack, with one specific attack type [1]. The whole set is grouped into training and test sets whose attack instances are classified into four major classes, i.e., DoS, U2R, R2L, and Probe, in accordance with purpose. Table 1 shows the distribution of the records in the NSL-KDD dataset for the training and testing sets.

Table 1. Distribution of records in the NSL-KDD dataset.

	Normal	Attack	Total
Training set	13449	11743	25192
Testing set	9714	12830	22544

6.2. Evaluation Metrics

This proposed approach was implemented and evaluated in Visual Basic 2010 on a personal PC with Core I5 2.4 GHz CPU and 4 GB RAM. This study does not focus on setting parameter values. Nevertheless, the selection of these parameters remains an open question. The performance of the proposed models is highly dependent on the proper setting of DNN and SA parameters. All feature records in the experiment were represented at the same predefined boundary [0, 1]. Min-max normalization is the most popular method to fit data in the same boundaries. The performance of the proposed model is assessed via four main parameters: recall (R), precision (P), accuracy (ACC), and F-measure (F). The result of the proposed model requires a high R, high ACC, and low false-alarm rate [2].

6.3. Experimental Results

The performance of the proposed DNN-based IDS was evaluated through a series of experiments. Two experiments were designed to study the performance of the DNN-IDS model for binary classification (normal, anomalous) and compare it with other existing approaches using deep learning.

Minimum and maximum MSE and accuracy were evaluated from all the simulations. Initially, the performance of the model depended on the value of the hyper-parameter initiated.

The first phase of our experiment was performed with the NSL-KDD dataset. Here, 20% of the whole NSL-KDD training dataset was used because these datasets have a large number of records that may require a long computation time with ordinary CPUs and machines or may exceed memory requirements. The training stage was completed with the selected training dataset. On the other hand, the size of the randomly selected testing set is only 80% of the whole NSL-KDD testing dataset.

Firstly, the Learners (BP) were tuned by selecting the hyper-parameters necessary to improve performance. Moreover, we attempted to tune BP-DNN by finding the optimal learning rate within the range of 0.6 to 0.02. We minimized MSE and maximized accuracy when training a model. Table 2 shows MSE of BP-DNN methods during training with NSL-KDD training dataset and the result of the testing with NSL-KDD testing dataset where MSE and training time are taken as performance evaluation factor. Table 3 shows performance of BP-DNN in terms of ACC, RECALL, PRECISION, FPR and F-measure.

Table 2. Results of the experiment based on MSE for BP-DNN.

Learning rate	Training set			Testing set	
	MSE		Time	MSE	Accuracy
	MIN	MAX			
0.6	0,80628	0.96472	25960	0.51880	52.30
0.5	0.16171	0.96472	27278	0.23939	87.51
0.4	0.82659	0.96472	29686	0.50701	47.69
0.3	0.20567	0.96472	27478	0.21357	88.16
0.2	0.11717	0.96472	27329	0.22201	88.94
0.1	0.04025	0.96472	27683	0.12703	92.68
0.08	0.05152	0.96472	26586	0.12372	92.78
0.06	0.03015	0.96472	29153	0.15449	92.00
0.04	0.02786	0.96472	27877	0.16073	91.51
0.02	0.02224	0.96472	28570	0.15430	90.80

Table 3. Experimental results of BP-DNN for NSL-KDD.

Metric	ACC	DR=RECALL	PRECISION	FPR	F-measure
results	92.78	85.78	79.41	0.020	91.32

Secondly, the Learners (SA) were tuned by selecting the hyper-parameters necessary to improve performance. Moreover, we optimized SA-DNN by finding the proper value of Alpha (α) within the range of 0.9 to 0.1. We minimized MSE and maximized accuracy when training a model. Table 4 shows the MSE of SA-DNN methods during training with the NSL-KDD training dataset and the result of the testing with the NSL-KDD testing dataset, where MSE and training time are regarded as performance evaluation factors. Table 5 shows performance of BP-DNN in terms of ACC, RECALL, PRECISION, FPR and F-measure.

Table 2 shows that in the training phase, a low learning rate was followed by a decrease in MSE and an increase in ACC. However, in the testing phase, when we reduced the learning rate to 0.009, the results of each change were matched by a change in MSE and accuracy. Therefore, all of the resulting changes were not as good as the learning rate of 0.08. Table 5 shows that the changing of (α) value leads to the changing of MSE and ACC values. Alpha decreases from 0.7 to 0.3 where MSE reaches its minimal value. Consequently, ACC achieves good result when MSE is minimal.

Table 4. Results of the experiment based on MSE for SA-DNN.

Alpha	Training set			Testing set	
	MSE		Time	MSE	Accuracy
	MIN	MAX			
0,9	0.27312	0.85425	24667	0.27958	84.57
0.8	0.22926	0.85425	27278	0.22541	86.88
0.7	0.23744	0.85425	25229	0.32475	82.05
0.6	0.22063	0.85425	24798	0.28204	70.46
0.5	0.21523	0.85425	25109	0.32514	82.08
0.4	0.19672	0.85425	23413	0.19908	87.96
0.3	0.16918	0.85425	23388	0.18335	93.10
0.2	0.17396	0.85425	24736	0.22996	89.76
0.1	0.27549	0.85425	24429	0.27973	85.05

Table 5. Experimental results of SA-DNN for NSL-KDD.

Metric	ACC	DR=RECALL	PRECISION	FPR	F-measure
results	93.10	88.00	97.25	0.022	92.17

Comparison of the results of the training and testing phases related to the proposed approach showed that SA-DNN achieved acceptable performance in terms of ACC (93.10), RECALL (88.00) and time of training compared with BP-DNN. BP-DNN performed better than SA-DNN in terms of PRECISION (97.41) and FPR (0.020). However, determining which method performs better by using ACC, RECALL, PRECISION, and FPR is impossible because all the results are close. The metric of F-measure provides a better measure of the accuracy of a model compared with all of the metrics because both of precision (P) and recall (R) are considered for calculating F-measure. The goal is to achieve a high F value to provide accurate results, which means all instances are classified correctly. SA-DNN obtained (92.17) better results than BP-DNN (91.32). Therefore, SA provides accurate results. For the overall evaluation, the learning rate of 0.08 and alpha of 0.3 of BP-DNN and SA-DNN, respectively, reveal that these two methods showed the best performance. Thus, they were selected for further evaluation.

Although the results of BP-DNN and SA-DNN were not good enough, we evaluated the result of the SABP-DNN model based on the two to show the performance of the proposed model. The performance of the SABP-DNN algorithm was evaluated using the best parameter found in the two previous experiments. Table 6 shows the MSE of the model with 0.08 learning rate and 0.3 alpha. Table 7 shows performance of SABP-DNN based on ACC, RECALL, PRECISION, FPR and F-measure.

Table 6 shows that the result of SABP-DNN in terms of MSE was 0.03356 in 21842 ms. this result shows an improvement compared with SA-DNN and BPDNN, which obtained 0.05152 in 26586 ms and 0.16918 in 23388 ms, respectively.

Table 6. Results of the experiment based on MSE for SABP-DNN.

Learning rate / alpha	Training set			Testing set	
	MSE		Time	MSE	Accuracy
	MIN	MAX			
0.08 / 0.3	0,03356	0.08539	21840	0.08097	95.56

Table 7. Experimental results of SABP-DNN for NSL-KDD.

Metric	ACC	DR=RECALL	PRECISION	FPR	F-measure
results	95.56	91.88	98.73	0.010	94.17

Table 7 also shows an improvement in the results in terms of ACC, RECALL, PRECISION, FPR, and F-measure, which gained values of 95.56, 91.88, 98.73, 0.010, and 94.17, respectively.

Table 8 shows that our SABP-DNN approach performed better than the existing approaches.

Table 8. Performance of our proposed method SABP-DNN compared to other methods for KDD99 dataset.

Author/reference	Technique	Accuracy (%)	DR (%)	FAR (%)
[7] (2016)	ANN	95.04	-	1.48
	LR	92.75	-	18.84
	NB	95.00	-	5
[11] (2014)	SVM	-	66.70	5.53
	CSOACN	-	80.10	2.84
	CSOAC	-	78.18	2.77
[18] (2012)	SSOWLS	93.30	-	-
	SSO	89.60	-	-
	PSO	88.50	-	-
[4] (2015)	DNN	75.75	76.00	0.86
[17] (2016)	ADBCC	92.71	91.79	3.50
[8] (2017)	GSPSO-ANN	95.26	-	-
	PSO-ANN	92.06	-	-
	GS-ANN	92.81	-	-
In this work(2018)	SABP-DNN	95.65	91.88	0.01

Overall, the SA algorithm is highly recommended for use in intelligent hybrid optimization schemes, such as hybridization with BP, for finding optimal DNN weights. This recommendation is based on the high exploratory behavior of the algorithm, which prevents finding the local optima through training DNNs. The SA algorithm employs a random neighbor search, which allows changes in acceptance probability to minimize loss function and inferior solutions to escape obtaining the local optima. The highly exploitative behavior of the algorithm is another reason for the rapid convergence of an SA-based trainer toward the global optimum for different datasets. Meanwhile, the BP algorithm can determine if the global optimal solution is weak and finds a locally optimal solution. Moreover, the BP algorithm can attain rapid convergence around the global optimum for finding the local optimum. Combining SA with BP can reinforce the strengths of each method because of their computational advantages and can reduce the computational time of DNN training.

7. Conclusion

This study demonstrated that the deep learning method of DNN can be successfully applied in intrusion detection. This deep learning model learns high-dimensional representations and efficiently performs attacks detection. SA can be combined with BP to train and adjust a DNN and to learn similarity representation over nonlinear and high-dimensional input data. Doing so can perfectly facilitate attack detection. In the IDS environment, the deep learning approach can be used to successfully develop an advanced detection model for detecting potential attacks. The main advantage of combining SA with BP is the use of distinct patterns to explicitly avoid being trapped in the local minima. Hence, many of the issues of premature convergence can be prevented, and good data generalization can be achieved.

Experiments on the NSL-KDD dataset showed that our proposed approach, SABP-DNN, effectively trains DNN compared with BP-DNN and SA-DNN training methods. The statistical results showed that SABPDNN is robust because the variance values are small. The experimental results for the NSL-KDD dataset showed that DNN can learn a good generative model and performs well in intrusion detection. Deep learning approaches provide new design ideas and methods for future IDS research. The results of this study prove the potential applicability of DNN as an alternative solution for developing practical IDSs.

References

[1] I Ahmad, A. B Abdullah, and A S Alghamdi. Application of artificial neural network in detection of probing attacks. *IEEE Symposium on Industrial Electronics Applications*, volume **2**, pages 557–562, Oct 2009.

[2] T Ma, Y Yu, F Wang, Q Zhang, and X Chen. A Hybrid Methodologies for Intrusion Detection Based Deep Neural Network with Support Vector Machine and Clustering Technique, pages 123–134, 2018.

[3] E Hodo, X Bellekens, A Hamilton, C Tachtatzis, and R Atkinson. Shallow and deep networks intrusion detection system: A taxonomy and survey. *arXiv preprint arXiv:1701.02145*, 2017.

[4] T. A. Tang, L. Mhamdi, D. McLernon, S. A. R. Zaidi, and M. Ghogho. Deep learning approach for network intrusion detection in software defined networking. In *2016 International Conference on Wireless Networks and Mobile Communications (WINCOM)*, pages 258–263, Oct 2016.

[5] H Berlin, L Djionang, and G Tindo. A new networks intrusion detection architecture based on neural networks. *Global Journal of Computer Science: Network, Web and Security*, **17**(1):19–27, 2017.

[6] Q Xu, L Yang, Q Zhao, and Z He. A novel intrusion detection mode based on understandable neural network trees. Journal of Electronics (China), 23(4):574–579, Jul 2006.

[7] N Moustafa and J Slay. The evaluation of network anomaly detection systems: Statistical analysis of the unsw-nb15 data set and the comparison with the kdd99 data set. *Information Security Journal: A Global Perspective*, **25**(1-3):18–31, 2016.

[8] T Dash. A study on intrusion detection using neural networks trained with evolutionary algorithms. *Soft Computing*, **21**(10):2687–2700, 2017.

[9] G Ke and Y H Hong. The research of network intrusion detection technology based on genetic algorithm and bp neural network. In *Frontiers of Manufacturing Science and Measuring Technology IV*, volume **599** of *Applied Mechanics and Materials*, pages 726–730. Trans Tech Publications, 10 2014.

[10] C Qiu and J Shan. Research on intrusion detection algorithm based on bp neural network. *International Journal of Security and Its Applications*, **9**(6):247–259, 2015.

[11] W Feng, Q l Zhang, G zh Hu, and J Huang. Mining network data for intrusion detection through combining svms with ant colony networks. *Future Generation Computer Systems*, **37**(Supplement C):127 – 140, 2014.

[12] M E Aminantoa and K Kimb. Deep learning in intrusion detection system: An overview. 2016.

[13] L Rere, M I Fanany, and A M Arymurthy. Simulated annealing algorithm for deep learning. *Procedia Computer Science*, **72**:137–144, 2015.

[14] T Nikoskinen. From neural networks to deep neural networks. 2015.

[15] H Shi and W Li. Evolving artificial neural networks using simulated annealing based hybrid genetic algorithms. *JSW*, **5**(4):353–360, 2010.

[16] A Koutsoukas, J. Monaghan, X l Li, and J Huan. Deep-learning: investigating deep neural networks hyper-parameters and comparison of performance to shallow methods for modeling bioactivity data. *Journal of Cheminformatics*, **9**(1):42, Jun 2017.

[17] S Kirkpatrick, C Gelatt, and M Vecchi. Optimization by simulated annealing. *Science*, **220**(4598):671–680, 1983.

[18] Y Y Chung and N Wahid. A hybrid network intrusion detection system using simplified swarm optimization (sso). *Applied Soft Computing*, **12**(9):3014 – 3022, 2012.

[19] M Tavallaee, E Bagheri, W Lu, and A Ghorbani. A detailed analysis of the kdd cup 99 data set, CISDA'09, pages 53 –58, Piscataway, NJ, USA, 2009. IEEE Press.

A Novel Customized FARS Recommendation Algorithm based on Search

Yao L [a], Yin Q Q [a], Zhong Y [b]

[a] College of Mathematics and Econometrics, Hunan University, Changsha 410082, China.
[b] College of Information Science and Engineering, Hunan University, Changsha 410082, China.

*E-mail: yao@hnu.edu.cn

Abstract: Search engines and recommendation systems work in different ways, but they are designed to handle the same problem: information overload. This paper combines the advantages of the two and proposes a new search-based recommendation algorithm: focused analysis recommendation system (FARS). Inspired by LDA , the FARS model was designed as a three-layer (user-interest-item) Bayesian structure. Based on user-specified keywords, FARS can only perform modeling analysis on relevant users, and return the distribution of interest of these users, which are estimated by the Gibbs sampling method. We treat the top-ranking items in each interest vector as related, and then sort them according to the probability that an item is selected by this user group. Finally, a specified number of recommended items is returned based on the above sorting result. The experimental results shows that the proposed model can outperform on Precision@n in comparison with the other baseline models. To some extent, it provides some inspiration for combining search and personalized recommendations.

1. Introduction

With the emergence of Internet, the problem of information overload[1] has become increasingly fierce. In order to enable users to get the information they need faster and better, the corresponding solution has been upgraded from the search engine to the recommendation system. Search engine is an information search tool that allows individuals with clear goals to search for information they need; Recommendation system (usually known as personalized recommendation system) is a kind of information filtering tool, by analyzing the user's historical behavior data to model the user's interest, actively provides users with information that interests them (note that the user does not need to have a clear goal in this process).

Both search engine and recommendation system are tools to help users get useful information, but in some cases, the results returned by these two methods may not be satisfactory to the user. In practice we found that the user are interested in a type of movie that he has never watched before and wanted to learn more about it. The user needs mentioned above, however, are not likely to excavate from his behavior records. Moreover, the evaluation criteria of the recommendation system include coverage and diversity, so the items they cover are often too extensive and may not even touch the keywords provided by users. Search engine will only return relevant introductions of this movie, video resources, movie review information, and the like.

To solve the above problem, we have carefully analyzed the performance in automatically mining user interests of recommendation system and the focused analysis function of search engine. Combining the advantages of two methods, this paper proposes a new search-based recommendation algorithm: focused analysis recommendation system(FARS). The proposed algorithm can return a collection of recommended items by accurate analysis of user data that contains or relate with keywords.

The rest of the paper is organized as follows: Section 2 describes related research work; Section 3 is about the proposed model and theoretical basis; Section 4 is the experimental results; and the end of the article is related references.

2. Background and related work

2.1. Recommendation system

Recommendation system is one of the hot areas of research and is also popular in industry in recent years. As a practical tool, it has been used in many different e-commerce and social networks[2]. There are many types of data that can be analyzed, but we mainly study the collaborative filtering algorithm[3] which makes recommendations by modeling user behavior data. This recommendation algorithm can be divided into: neighborhood-based[4,5] and model-based[6,7].

Neighborhood-based algorithm is recommend to the user based on the similarity measure results. It can be grouped into two general classes: user-based collaborative filtering algorithms(userCF) and item-based collaborative filtering algorithms (itemCF).

- userCF groups users based on similarity, and members in the same user group are, to a large extent, interested in similar items. So this method will recommend to users similar items that users prefer[4].
- itemCF is grouping items based on the assumption that two items that many people like at the same time are likely to belong to the same class. When the user is interested in an item, the method will recommend the most similar/related items to the user[5].

Model-based algorithm uses user behavior data to learn a model, and then use it to predict user behavior. Like LDA-based recommendation algorithm[7,8] attempts to construct a recommendation model by treated user interest as an implicit feature, which is regarded as the key to associating users with the items they are interested in. This algorithm not only has a good theoretical basis, but also can obtain more flexible recommendation results simply by changing the number of latent features in the algorithm.

2.2. LDA-based Algorithm

LDA(Latent Dirichlet Allocation) is a probabilistic generation model[9]. This model can mine the internal correlation of vocabulary in the corpus, find the hidden semantics of the document, and promote the topic discovery of the text. The advantages above make LDA widely used in document analysis, classification, or clustering[10].

In 2005, LDA was originally introduced into the recommendation system. It was used to model the implicit interrelationships between items, and generate dynamic recommendations for users[7]. Since then, LDA has become a popular optimization tool in the recommendation system. It is used to explore commonalities in data, like user-generated content[11-13](eg, tags, reviews) and the item information[14,15](eg, text Themes, movie categories).

Until 2011, LDA was used to mine user interest, and construct a social media user recommendation system that recommends to a user new friends having similar interests[8]. Later, Chang T M et al. used LDA to model user data and group them, to make better recommendations for a specific user group[16]. In addition, LDA is used to explore the user's potential interests for generating a personalized recommendation of specific user[17].

Rather than simply recommendation based on a single algorithm, some recommendation systems use a hybrid approach by combining LDA-based recommendation algorithm and other. For example:

LDA is used to model the user and the item simultaneously, find out their neighborhood sets respectively, and then recommend according to their relevance[18]; In order to obtain a better recommendation result, some scholars to combine LDA-based recommendation algorithm with matrix decomposition[19].

However, an over-specialization recommendation system makes it difficult for users to customize recommendations according to his or her needs[2]. Similar to other recommendation algorithms, LDA-based recommendation algorithm and its variants cannot return a personalized recommendation list for a single user or user group with the same interests. For the aforementioned example of that moviegoer, this user will never receive a recommendation that he or she have not seen, even if it is the most anticipated movie for user. In order to solve this problem, the search-based related research is considered in the proposed algorithm. The following is a detailed introduction to related work in search field.

2.3. Personalized Search

Search results that include only keywords are not what we need. So this section will make a introduction to some of the personalized search related to the work of this article. For movie websites, there may be multiple reasons why users be interested in a movie, such as a certain director, actor, or the theme of the movie. In order to return better search results, personalized search needs to figure out the user’ s true intentions.

Concept-Based retrieval method[20] can better respond to personalized needs in real-time, by embeds feature selection methods into the search process. Lee N et al. proposed a robot-based recommendation service[21] that can generate flexible responses to different language expressions to achieve real-time human-computer interaction recommendations. According to the user's search history, determine the user's true potential intention by LDA, Yan R et al. sorts related document sets according to KL distances, and returns the personalized search document list to the user[22]. Wu J T et al. expands the keyword-related training text sets by incrementally expanding the seed words, and determines keywords in these text sets corresponding to the query words. This method can improve the accuracy of topic clustering and topic diversity[23]. Although these methods achieve a certain degree of personalization, they are not enough to solve the problems mentioned above.

Therefore, we proposes a new algorithm that makes use of the complementarity of search engines and recommendation systems. This algorithm can perform focused analysis on certain relevant aspects according to the user-specified keywords, and return an appropriate recommendation list that is what the user needs. For the case mentioned in the example, we need only to set the appropriate number of interest categories, the proposed algorithm can return a suitable set of recommended items related to this movie.

3. FARS Model

As discussed in the introduction section, our problem statement is that given a collection C of user data, the proposed model can return a recommended list of the relevant aspect specified by the user using a set of keywords S.

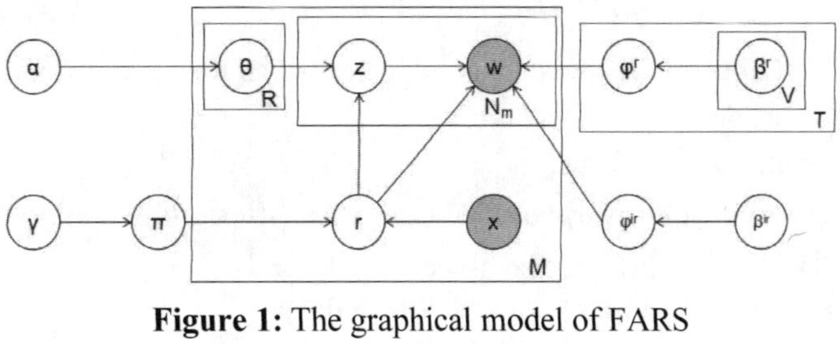

Figure 1: The graphical model of FARS

Inspired by the idea[24] that the items that a user is exposed to is a latent variable that may change for various user/item combinations, we treat the state that whether a user is related to a keyword as a potential random variable. Following previous work, we assume that each user's behavior is a probabilistic event. In other words, the process of a user select an item is actually randomly selecting a point of interest, and then draw an item randomly from the corresponding interest. The element definitions of the FARS model given in Table 1, and the graphical model of FARS is given in Figure 1. For a given keyword, it is easy to determine whether a user data contains the keyword. This state is represented by the variable x: where $x = 1$ means that a user's data contains the keyword, otherwise $x = 0$. The potential random variable that is used to represent the user data and keyword related state is represented by r: where $r = 1$ means the user data is relevant to the target and $r = 0$ is irrelevant.

Table 1: Element Definitions of the FARS Model

T	the number of interest
M	the number of user
N_m	the number of items in the m-th user's data
V	the number of different items
R	the number of the relevance status
m, z, v	user data, interest, item
w_i, z_i	item in position i (item i), interest of item i
$\boldsymbol{w}, \boldsymbol{z}, \boldsymbol{r}$	all items, assigned interest and relevance status
x, r	keyword indicator, relevance status
x_m, r_m	keyword indicator, relevance status of the m-th user's data
$\alpha, \beta^{ir}, \beta^{r}$	Dirichlet prior for θ, φ^{ir}, φ^{r}
γ	Beta prior for π
θ	multinomial distribution over interest
π	bernoulli distribution over relevance status
φ^{r}	multinomial distribution over interest-item
φ^{ir}	multinomial distribution over irrelevance interest-item
$\boldsymbol{z}^{\neg i}$	all assigned interest except the item i
$\boldsymbol{\beta}^{r(\neg v)}$	all selected items except v under relevance status r
$\boldsymbol{\beta}_t^{r(\neg v)}$	all selected items except v under relevance status r and interest t
$N_{r,m,v}$	the frequency of item v in the m-th user's data under relevance status r
$N_r^{R(\neg m)}$	the number of user's data under relevance status r except the m-th user
$N_{r,v}^{RW}$	the number of item v under relevance status r
$N_{r,m,t}^{RMT(\neg i)}$	the number of items under relevance status r and interest t in the m-th user's data except item i
$N_{r,t,v}^{RTW}$	the number of items v under relevance status r and interest t

3.1. Generative Process

Then when the keyword is given, each data can be identified as relevant or irrelevant to target. Then for each user $\mathrm{m} \in \{1,2,...,\mathrm{M}\}$:

(1) Draw a prior distribution $\pi \sim \mathrm{Beta}(\gamma)$;

(2) Draw relevance status r based on keyword indicator x and $\mathrm{Bernoulli}(\pi_{\mathrm{m}})$;

(3) If the document is relevant to the targeted aspect, that is $r = 1$:

(a) Draw $\theta \sim \mathrm{Dirichlet}(\alpha)$ as a user - interest distribution;

(b) Draw a interest $\mathrm{z} \sim \mathrm{Multinational}(\theta^{\mathrm{r}})$;

(c) Draw a item distribution $\varphi^{\mathrm{r}} \sim \mathrm{Dirichlet}(\beta^{\mathrm{r}})$;

(d) Emit a item $w_i \sim \mathrm{Multinational}(\varphi_z^{\mathrm{r}})$;

(4) In contrast the document is irrelevant to the targeted aspect, if $r = 0$:

(a) Draw $\varphi^{\mathrm{ir}} \sim \mathrm{Dirichlet}(\beta^{\mathrm{ir}})$ as a item distribution of a irrelevant interest to the keyword;

(b) Emit a item $w_{\mathrm{i}} \sim \mathrm{Multinational}(\varphi_z^{ir})$.

As shown above, there are two related states for each user. When $r = 1$, it means that the user is related, so we need to add this user's behavior data to the candidate data set for analysis. The analysis process is similar to the topic modeling of LDA: a topic z is chosen from φ^{r}, after that a item w_{i} is emitted from the selected topic by φ^{r}. When $r = 0$, that is to say this user data is not related to the specified aspect. In this case, we only draw $\varphi^{\mathrm{ir}} \sim \mathrm{Dirichlet}(\beta^{\mathrm{ir}})$ as a item distribution of a irrelevant interest to the keyword, and a item w_{i} is emitted from φ^{ir}. In other words, the items in irrelevant user are drawn from only one (irrelevant) interest.

The generative process of FARS above can be used to explain the relationship among user data, interests, and items. However, in order to calculate the probability that a user is interested in an item, we also need to compute the user-interest distribution θ and the interest-item distribution φ based on the given Dirichlet parameters α and β.

3.2. Parameter Estimation

Since random variables θ and φ are not easily found directly, they are generally estimated using an approximate methods, like variational reasoning EM (expectation maximization)[9] and Gibbs sampling[25]. The former is computationally efficient, but the derivation is complicated; On the contrary, the latter has low computational efficiency, but it is easy to implement, and more tolerant of local optima. Therefore, we choose Gibbs sampling algorithm to estimate θ and φ that are polynomial distribution parameters.

First, we sample the relevance status r for user m, by the following conditional distribution:

$$\mathrm{P}(\mathrm{r_m} = \mathrm{c} | \mathrm{x_m} = d, r_m^{\neg i}, \pi, \gamma, \beta^{\mathrm{r}}, \beta^{\mathrm{ir}}) \propto \begin{cases} \dfrac{N_c^{R\neg m} + \gamma}{N_*^{R\neg m} + |R|\gamma} \times \dfrac{\Pi_v^V \Gamma(\beta_v^r N_{c,v}^{RW\neg m} + \beta_v^r + N_{c,m,v})}{\Gamma(\Sigma_v^V(\beta_v^r N_{c,v}^{RW\neg m} + N_{c,m,v}) + \beta_*^r)}, & c = 1 \\ \dfrac{N_c^{R\neg m} + \gamma}{N_*^{R\neg m} + |R|\gamma} \times \dfrac{\Pi_v^V \Gamma(N_{c,v}^{RW\neg m} + \beta_v^{ir} + N_{c,m,v})}{\Gamma(\Sigma_v^V(N_{c,v}^{RW\neg m} + N_{c,m,v}) + \beta_*^{ir})}, & d = c = 0 \\ 0, & otherwise \end{cases} \quad (1)$$

Next, we sample a interest of the item in position i is:

$$\mathrm{P}(\mathrm{z_i} = \mathrm{t} | \mathrm{z}^{\neg i}, \mathbf{r}, \mathbf{w}, \alpha, \beta^{\mathrm{r}}, \beta^{\mathrm{ir}}) \propto \begin{cases} \dfrac{N_{r_i,m,t}^{RMT\neg i} + \alpha}{N_{r_i,m,*}^{RMT\neg i} + |T|\alpha} \times \dfrac{\beta_{t,v}^r N_{r_i,t,w_i}^{RTW\neg i} + \beta_{t,v}^r}{\beta_{t,v}^r N_{r_i,t,*}^{RTW\neg i} + |\beta_{t,*}^r|}, & r_i = 1 \\ \dfrac{N_{r_i,m,t}^{RMT\neg i} + \alpha}{N_{r_i,m,*}^{RMT\neg i} + |T|\alpha} \times \dfrac{N_{r_i,t,w_i}^{RTW\neg i} + \beta_{t,v}^{ir}}{N_{r_i,t,*}^{RTW\neg i} + |\beta_{t,*}^{ir}|}, & r_i = 0 \end{cases} \quad (2)$$

The user-interest distribution θ and the interest-item distribution φ of related user groups as following, and the procedure of learning FARS by Gibbs sampling show as Figure 2.

$$\theta = \frac{N^{RMT\neg i}_{r_i,m,t} + \alpha}{N^{RMT\neg i}_{r_i,m,*} + |T|\alpha}$$
$$\varphi = \frac{N^{RTW\neg i}_{r_i,t,w_i} + \beta^{r}_{t,v}}{N^{RTW\neg i}_{r_i,t,*} + |V|\beta^{r}_{t,*}} \tag{3}$$

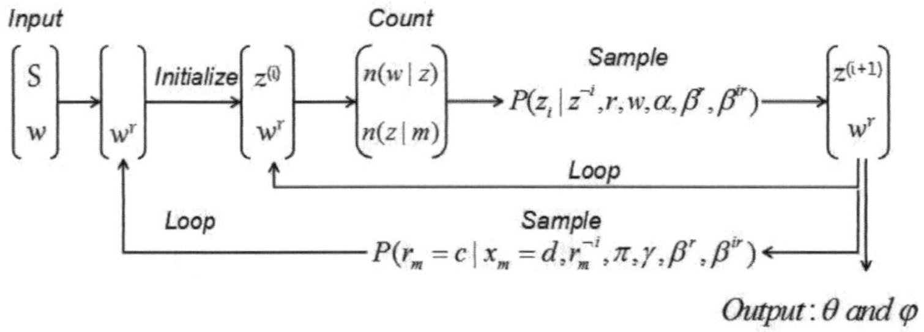

Figure 2: The procedure of learning FARS by Gibbs sampling

The process diagram of FARS is shown in Figure 3, and the textual description of FARS is described as follow:

1. For a given keyword, the proposed algorithm will determine the user's relevance;
2. Expand the keyword set to better express the interest classification related to the keywords, by use the commonality of relevant users;
3. Sample the relevant user using the Gibbs sampling formula above, to find more relevant user sets;
4. Reduce words that are not representative enough in the keyword set, add some better words to it, and iterate until convergence
5. According to the sampling results, estimate the user-interest distribution θ and the interest-item distribution φ;
6. Sort related items by the product of θ and φ and return a list of recommendations.

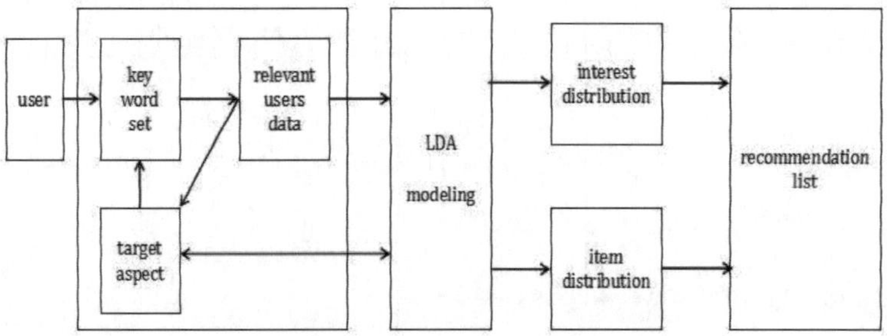

Figure 3: The process diagram of FARS

4. Results and Evaluation

Data and Keyword: MovieLens dataset is used in our experiments, these data were created by 671 users between January 09, 1995 and October 16, 2016. Users were selected at random for inclusion. All selected users had rated at least 20 movies. Instead of modeling with tags or contexts, our method takes only the user data and itemID to construct a recommendation model. Therefore we removed other information in the data set, such as: rating, time stamp, and so on. We will choose from several films to be analyzed as keywords, some of them are more frequent, like **Titanic**(1997) (movieID is 1721); some are infrequent, like **Safe**(1994) (movieID is 190).

4.1. Experimental Setting

Parameter Setting: For the hyper-parameter setting, we place: γ=1, α=1, β^{ir}=0.001 ; and β^{r}= {0,1} (for a keyword $s \in S$, the β^{r}_{s}=1). The models for comparison are also with the same setting.

Baseline Models for Comparison: To evaluate our proposed FARS, we compare it with the following baseline models.

Model1: We use LDA for user interests generation via analyze all user data[17], and inspection to all resulting interest class from LDA, to find the subset of relevant items. Return a set of recommended items by their probability.

Model2: Group users according to conditions that the user data includes keywords, and using LDA to model interest in relevant user groups. Return the most popular movies in this user group.

Model3: We runs TF-IDF only on the user data that contains one or more keywords from S, and uses this result to expand the keyword to obtain the keyword set S'. Return the most popular ten movies in the relevant user group[21] just like Model2.

Since LDA is a full-analysis model, it analyzes all user data and does not directly generate keyword-related interest categories and item recommendations. Therefore, in the experiment, we set the number of interest categories T of Model1 to 15 or 20, find the appropriate item from its related subset, and return a recommendation list according to the sort of probability; In other methods, the value of interest classification T is set to 5-10. Since we do not have a priori information about the number of user interests, the size of T is determined based on the popularity of the movie being searched (eg, popular movies such as **Titanic** may be seen in most user data). In general, T is set to 5 for infrequent targets, and T is set to 10 for other more frequent targets. Note that we will use the same T value to compare the results given by these methods.

4.2. Experimental Results and Evaluation

According to the characteristics of top-k recommendation, Precision@n(or P@n for short) were used to assess the recommendation's effect in the experiment. P@n can give the precision results at different rank position n. Specifically, we use a normalized form of precision that can estimate the proportion of related items in the recommendation list.

$$P_{(i)}@n = \frac{N_{(i)}(relevanted@n)}{N(items@n)} \tag{4}$$

In Equation 4, $P_{(i)}@n$ indicates the Precision@n for model(i). $N_{(i)}(relevanted@n)$ is the number of relevant items in the recommendation list returned by model(i). $N(items@n)$ is the length of the recommended list. The average performances of different models about P@n at the rank position of 5 and 10 are displayed in Figure 4 and Figure 5.

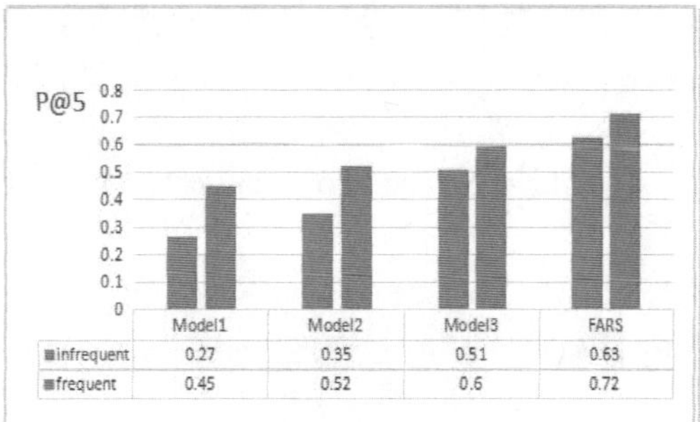

	Model1	Model2	Model3	FARS
infrequent	0.27	0.35	0.51	0.63
frequent	0.45	0.52	0.6	0.72

Figure 4: P@5

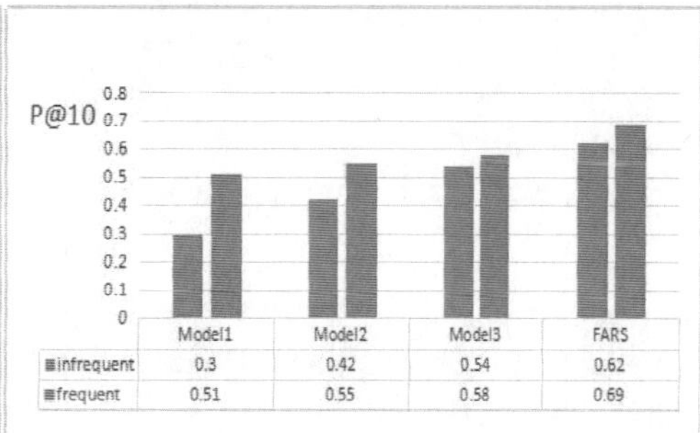

	Model1	Model2	Model3	FARS
infrequent	0.3	0.42	0.54	0.62
frequent	0.51	0.55	0.58	0.69

Figure 5: P@10

As can be seen from the picture, both P@n at the rank position of 5 or 10, and in terms of infrequent or frequent, our proposed FARS model is significantly outperforms other models. Compared to other models, FARS can identify user data related to the keyword through a random variable. This variable can continuously iterate while modeling user interest, and optimize the model results, which is also the main reasons for the higher scores.

Among them, Model3 obtains the second best scores. Because Model3 can use TF-IDF for keyword expansion, and analyze the subset of user data associated with the expanded set of keywords. Model3 can find more or better interest for the relevant aspect than Model2, because Model2 discards many relevant user data, resulting in information loss. Model2 can rules out the irrelevant user data that can interfere with good topic identification. But then it also loses relevant information as we discussed above. Thus, which results in it has better performance compared to model1, and inferior than Model3.

Compared whit FARS and Model3, the experimental results of the previous two models in infrequent aspect are worse. Because there are few user data containing the keywords about infrequent aspect, the amount of information contained in these data is insufficient to illustrate the true interest of such user groups; That is to say, FARS and Model3, to a certain extent, extend the set of user-specified keywords, which makes more data for analysis. The worst performance is model1. Because, instead of directly analyze the user's interest for the target aspect, Model1 is required to conduct modeling analysis on all raw data. When it comes to some infrequent movie, this model returned user interest are often too coarse and may not even be on target.

By comparing P@5 and P@10, it is easy to find that we proposed FARS recommends related items ranked relatively high. That is, the algorithm can better identify related items and recommend them to query users preferentially. Although Model2 and Model3 scored similarly in the frequent aspect, the difference was that Model3 had a slightly higher P@5 score than P@10, indicating that the expanded keyword set could have a certain impact on the ranking of related items.

5. Conclusion

In this paper, we studied the problem of search-based customize recommendation, and propose FARS model. Instead of analyzing user interest based on all data, FARS combines user requirements (search based on keywords) and recommendations (analysis of preferences of related user groups). Compared with the baseline model, our proposed model can not only return more relevant recommendation items, but also return related item rankings more forward.

Although the proposed model performs well, there are still many other optimizations that can be done: visualization, interactivity, interpretability, and so on. More specifically. Visualization: Shows users all possible results in an intuitive way, and the proportion of various results. Interactivity: If a user is interested in a certain type of item in the recommendation list, he can perform deeper or shallower analysis on such items without having to re-enter other keywords. Interpretative: Make some explanation for the recommendation result so that the user can judge whether the item is worth choosing according to his own needs.

Acknowledgment

This research was financially supported by the National Natural Science Foundation of China under Grant No. 61502159.

References

[1] Yang C C, Chen H and Hong K 2003 Visualization of large category map for internet browsing. *Decision Support Systems*, **35**(1), 89-102.

[2] Adomavicius G, Tuzhilin A 2005 Toward the next generation of recommender systems: a survey of the state-of-the-art and possible extensions. *IEEE Transactions on Knowledge & Data Engineering*, 176, pp.734-749.

[3] Schafer J B, Frankowski D, Herlocker J and Sen S 2007 Collaborative filtering recommender systems. *The adaptive web* Vol.**22**, pp.291-324. Springer-Verlag.

[4] Wang J, Vries Arjen P D, Reinders Marcel J T 2006. Unifying user-based and item-based collaborative filtering approaches by similarity fusion. *Proc. of, ACM SIGIR Conference on Information Retrieval* pp.501-508.

[5] Sarwar B, Karypis G, Konstan J and Riedl J 2001 Item-based collaborative filtering recommendation algorithms. *International Conference on World Wide Web* Vol.**4**, pp.285-295. ACM.

[6] Jenatton R, Roux N L, Bordes A and Obozinski G 2012 A latent factor model for highly multi-relational data. *International Conference on Neural Information Processing Systems* Vol.**4**, pp.3167-3175. Curran Associates Inc.

[7] Jin X, Zhou Y, Mobasher B 2005 A maximum entropy web recommendation system: combining collaborative and content features. *Eleventh ACM SIGKDD International Conference on Knowledge Discovery and Data Mining*, Chicago, Illinois, Usa, August pp.612-617. DBLP.

[8] Pennacchiotti M, Gurumurthy S 2011 Investigating topic models for social media user recommendation. *International Conference on World Wide Web*, WWW 2011, Hyderabad, India, March 28 - Aprilpp.101-102. DBLP.

[9] Blei D M, Ng A Y, Jordan M I 2003 Latent dirichlet allocation. *J Machine Learning Research Archive* Vol.**3**, pp.993-1022.

[10] Blei D 2011 Probabilistic topic models. *ACM SIGKDD International Conference Tutorials* Vol.**27**, pp.1-1. ACM.

[11] Xu Y, Chen Z, Yin J, Wu Z and Yao T 2015 Learning to Recommend with User Generated Content. *WAIM* Vol.**9098**, pp.108-113.

[12] Krestel R, Fankhauser P, Nejdl W 2009 Latent dirichlet allocation for tag recommendation. *ACM Conference on Recommender Systems*, Recsys 2009, New York, Ny, Usa, October pp.61-68. DBLP.

[13] Zhai C, Peng J 2017 Mining Latent Features from Reviews and Ratings for Item Recommendation. *International Conference on Computational Science and Computational Intelligence* pp.1119-1125. IEEE.

[14] Chang T M, Hsiao W F 2013 Lda-based personalized document recommendation.

[15] Wang S, Wen W D 2017 A hybrid recommendation method based on features. *Computer & Digital Engineering*.

[16] Chang T M, Hsiao W F, Lin Y L 2014. LDA-based group document recommendation.

[17] Xie W B, Dong Q, Gao H 2014 A probabilistic recommendation method inspired by latent dirichlet allocation model. *Mathematical Problems in Engineering*, 201424.

[18] Gao M, Jin C Q, Qian W N, Wang X L and Zhou A Y 2014 Real-time and personalized recommendation on microblogging systems. *Chinese Journal of Computers*.

[19] Cao Y, Li W, Zheng D 2018 A hybrid recommendation approach using lda and probabilistic matrix factorization. *Cluster Computing* **1**, 1-11.

[20] Egozi O, Markovitch S, Gabrilovich E 2011 Concept-based information retrieval using explicit semantic analysis. *Acm Transactions on Information Systems*, **292**, 1-34.

[21] Lee N, Kim E, Kwon O 2017 Combining tf-idf and lda to generate flexible communication for recommendation services by a humanoid robot. *Multimedia Tools & Applications*, **774**, 1-16.

[22] Yan R, Li S j 2018 Document Retrieval Algorithm Based on Query Intent Identification and Topic Modeling. *Computer Engineering*, **443**: 189-194.

[23] Wu J T, Li T N 2017 Keyword recommendation method and system based on latent dirichlet allocation model. WO 2017114019 A1.

[24] Liang D, Charlin L, Mcinerney J and Blei D M 2015 Modeling user exposure in recommendation. 951-961.

[25] Griffiths T L, Steyvers M 2004 Finding scientific topics. *Proc Natl Acad Sci USA*, 101 Suppl 11, 5228-5235.

IOP Conf. Series: Materials Science and Engineering **435** (2018) 012036 doi:10.1088/1757-899X/435/1/012036

Single Image Depth Level Estimation Using Dark Channel Prior

Yichao Cao[1,2] and Xiaobo Lu[1,2,*]

[1] School of Automation, Southeast University, Nanjing 210096, China
[2] Key Laboratory of Measurement and Control of Complex Systems of Engineering, Ministry of Education, Southeast University, Nanjing 210096, China

*E-mail: xblu2013@126.com

Abstract. In this paper, we present a simple image depth level estimation algorithm. From the dark channel prior theory, an estimate of the air transmittance in the image is calculated. In wild surveillance, the disparity in the image poses a huge challenge for smoke detection and other video analysis tasks. Appropriate depth level estimation provide significant prior knowledge for subsequent identification and detection. For landscape images, we can approximate the air transmittance to depth information for histogram analysis. The depth value is segmented by a multi-threshold segmentation algorithm, and the resulting image can be used for forest fireworks detection and the like. This method does not rely on samples and classifiers, and the algorithm does not require training. The final experimental results show that the depth level estimation of a single landscape image based on the dark channel prior can achieve good results.

1. Introduction

In optics, the optical phenomenon known as depth of field, is the distance about the Plane of Focus where objects appear acceptably sharp in an image. In some cases, it may be desirable to have the entire image sharp, and a large depth of field is appropriate. Depth of field is an important geometric relations within images taken by camera, and is of great significance in image processing and video analysis. In real life, even in sunny days, there are some particles in the air, so people can have a deep concept when observing the scene.

In applications such as forest fire prevention, depth estimation has an essential impact on algorithm performance. If the algorithm parameters of different depth of field are the same, it is bound to cause two situations: the shaking of trees near the field of view is easy to trigger false detection because the pixel area is large, and the distant smoke is easy to miss due to the small pixels occupying the image. The depth of field estimate for this scene is different from the general depth of field estimate. Instead of calculating a particularly accurate depth of field image, it estimates the region of interest through a relative depth map.

While there is much prior work on estimating depth based on stereo images or motion [1], there has been relatively little on estimating depth from a single image. Most work on estimating depth has focused on using methods such as stereovision or structure from motion [2], which require two (or more) images.

Obtaining accurate depth of field estimates from a single image is undoubtedly very difficult. In video content analysis, appropriate depth level estimation provide significant prior knowledge for subsequent identification and detection. The subsequent analysis algorithm can only focus on the required depth level area.

The method proposed in this paper is shown in Figure 1. First, the dark channel image of the source image is calculated. Secondly, the air transmission is calculated according to the atmospheric light and image forming model. Finally, the air transmission is threshold-divided to obtain the hierarchical depth estimation.

2. Related Work

Directly related to our work are several methods of estimating depth from a single image. Saxena et al. [3] use a Markov Random Field (MRF) to infer a set of "plane parameters" that capture both the 3-d location and 3-d orientation of the patch. However, the method for inferring detailed 3-d structure from a single still image rely on small homogeneous regions in the image and suffers in less controlled settings. Hoiem et al. [4] labels regions of the input image into coarse categories: "ground", "sky", and "vertical" which is similar to our work. These labels are then used to "cut and fold" the image into a pop-up model using a set of simple assumptions. But this method is not expected to work on our wide scenes, their boundaries are not clear.

Ladicky et al. [5] proposed a new pixel-wise classifier that can jointly predict a semantic class and a depth label from a single image. The main weaknesses of the method are the inability to deal with low resolution images, very large requirements in terms of hardware and apparent inability to locate the objects more precisely for semantic classes with high variance.

Konda et al. [6] presented a deep learning approach for joint estimation of disparity and motion based on learning about interrelations between images from multiple cameras, multiple frames in a video, or the combination of both.

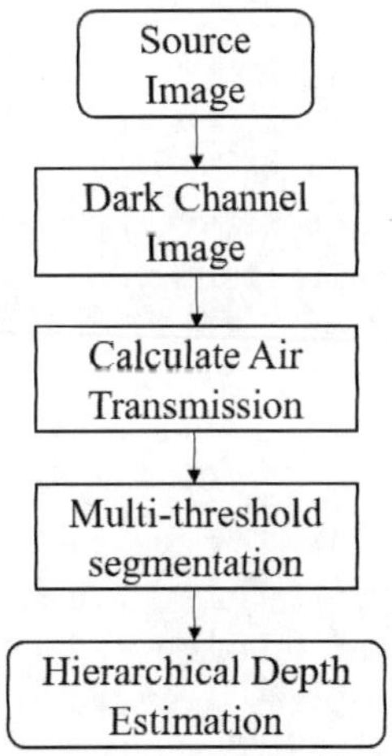

Figure 1. The flowchart of hierarchical depth estimation

3. Dark Channel Prior

The dark channel prior theory is based on statistics on a large number of outdoor pictures. The study found that [7], in most non-sky local areas, some pixels always have at least one color channel with a very low value. In other words, the minimum value of the light intensity in the area in the picture is a small number.

3.1. The dark channel prior theory

For any input image J, its dark channel can be expressed as:

$$J^{dark}(x) = \min(\min(J^c(y)))_{c\epsilon\{r,g,b\},y\epsilon\Omega(x)} \quad (1)$$

Where J^c is a color channel of input image J, and Ω(x) is a local patch centered at x. In the actual calculation, we first find the minimum value of the three channels of each pixel, store it in a grayscale image with the same size as the original image, and then filter the grayscale image with minimum value. The radius is determined by the size of the window, generally with $\text{WindowSize} = 2 * \text{Radius} + 1$.

The dark channel prior theory states that $J^{dark}(x)$ should be a small number close to zero. The pixel value of the dark channel of the near image in the actual scene is very low. We compared some outdoor forest surveillance images. As shown in Figure 2, the dark channel values of the image pixels at different distances in the picture differ greatly. The main idea of this paper is to make a deep estimate by analyzing this difference.

3.2. Estimating the transmission

In computer vision and computer graphics, the graph formation model described by the following equations is widely used:

$$\mathrm{I(x)} = J(x)t(x) + A(1 - t(x)) \tag{2}$$

Where I(x) is the image captured by the camera, J(x) is the ideal image, A is the global atmospheric light component, and t(x) is the transmission. The known condition is I(X), the purpose is to calculate J(x). Obviously, this is an equation with infinite solutions, so some priori constraints are needed.

Figure 2. The dark channel image uses a window size of 15×15 and a minimum filter radius of 7 pixels during the calculation.

In our approach, we select the top 0.1% brightest pixels in the dark channel which are the most opaque. Among these pixels, the highest intensity pixel mean in the input image is selected as atmospheric light A. Assume that the transmission t(x) is constant in each window. According to the dark channel prior theory mentioned above, the estimated value of transmission can be derived:

$$\tilde{t}(x) = 1 - \min(min_c(\frac{I(y)}{A_c}))_{y\in\Omega(\mathrm{x})} \tag{3}$$

In practice, even on sunny days, the atmosphere is not absolutely free of any particles. Therefore, when we use camera to capture images, the haze in the image still exists. In addition, the existence of haze is the basis of human perception of the depth of the scene. This phenomenon is called aerial perspective [8, 9], which also is the basis for our deep estimation method.

As shown in figure 3, the calculated air transmission map is approximately similar to the true deep value of the image. But note that what we get is relative image depth, not an absolute depth value. In order to adapt to different scenarios and obtain robust depth estimation, we need to do multi-threshold segmentation on the obtained relative image depth.

Figure 3. The calculated air transmission map

4. Multi-threshold Segmentation

In the calculation of the air transmission, the image we get is a grayscale image. In order to further segment the image, we need to use multi-threshold segmentation technology.

4.1. Single threshold segmentation

Suppose the grayscale space of an image is represented by L. Use i for each grayscale value in the grayscale space. The number of occurrences of each grayscale value in the grayscale histogram in the image is denoted by n_i and the total number of pixels of one image is denoted by N, then $N = n_1 + n_2 + \ldots + N_L$. For each grayscale value, the probability of appearing on the entire image is f_i.

If a single threshold segmentation is employed, the grayscale set is divided into G_0 and G_1. And $G_0 = \{0,1,2,\ldots,t\}, G_1 = \{t+1, t+2, \ldots, L-1\}$. The total mean value of the image is μ_T, and the mean values of the regions corresponding to G_0 and G_1 in the image are μ_0 and μ_1 respectively:

$$\mu_0 = \sum_{i=0}^{t} \frac{i f_i}{\omega_0} \tag{4}$$

$$\mu_1 = \sum_{i=t+1}^{L-1} \frac{i f_i}{\omega_1} \tag{5}$$

$$\mu_T = \sum_{i=0}^{L-1} i f_i = \omega_0 \mu_0 + \omega_1 \mu_1 \tag{6}$$

Where ω_0 and ω_1 are the probability ratios of the two types of pixels G1 and G2. The inter-class variance σ_b^2 and the intra-class variance σ_w^2 of the two regions are respectively:

$$\sigma_b^2 = \omega_0(\mu_0 - \mu_T)^2 + \omega_1(\mu_1 - \mu_T)^2 \tag{7}$$

$$\sigma_w^2 = \omega_0\sigma_0^2 + \omega_1\sigma_1^2 \tag{8}$$

When using the OTSU method [10] to segment an image, the optimal threshold t maximizes the inter-class variance or minimizes the intra-class variance. This threshold t is the segmentation threshold t_{opt} we are looking for. Because in the OTSU method it can be inferred that the relationship between the inter-class variance and the intra-class variance is inversely proportional. Another way to use the maximum inter-class variance to determine the optimal threshold t_{opt} is to use the smallest intra-class variance for optimal threshold discrimination:

$$t_{opt} = \underset{0 \le t \le L-1}{Max}\{\sigma_b^2(t)\} = \underset{0 \le t \le L-1}{Min}\{\sigma_w^2(t)\} \tag{9}$$

The principle of maximum inter-class variance is consistent with the principle of minimum intra-class variance, but because the maximum inter-class variance is relatively simple in calculation, the principle of maximum inter-class variance is usually used to calculate the optimal threshold.

4.2. Multi-threshold segmentation

It is natural to extend the OTSU method to multi-threshold, that is, the multi-threshold OTSU algorithm. Use i for each grayscale value in the grayscale space. The number of occurrences of each grayscale value in the grayscale histogram in the image is denoted by n_i and the total number of pixels of one image is denoted by N, then $N = n_1 + n_2 + \ldots + N_L$. For each grayscale value, the probability of appearing on the entire image is f_i.

For the multi-threshold segmentation algorithm, the image needs to be segmented into n classes, and it is obvious that n-1 thresholds are needed to segment the image. Assume that n-1 thresholds are t_1, t_2, ..., t_{n-1}. The grayscale set can be divided into $G_0 = \{0,1,2,\ldots,t_1\}$, $G_1 = \{t_1+1, t_1+2,\ldots,t_2\}$,..., $G_{n-1} = \{t_{n-1}+1, t_{n-1}+2,\ldots,L-1\}$. The total mean value of the image is μ_T, the mean values of the regions corresponding to n classes in the image are μ_0, μ_1,..., μ_{n-1}, and the probability of occurrence corresponding to n classes is F_0, F_2, ..., F_{n-1}.

Similar to the single threshold segmentation method, the optimal threshold for multi-threshold segmentation algorithm needs to follow the principle of maximum inter-class variance or minimum intra-class variance. The optimal thresholds can maximize the inter-class variance:

$$arg_t \max(\sigma_b^2 = \sum_{i=0}^{n-1} F_i(\mu_i - \mu)^2) \tag{10}$$

5. Experimental Results

In our experiments, the patch size is set to 15 × 15 for a 1400 × 800 image. It takes about 0.4 seconds to process a 1400 × 800 pixel image on a PC with a 2.2 GHz Intel Core i5-5200U Processor.

The depth level of field estimate for this scene is different from the general depth of field estimate. One advantage of this estimation algorithm is that the execution speed is fast, and in a practical application, it is necessary to calculate only once for a fixed scene. Another advantage of this method is that it does not rely on samples and classifiers, and the algorithm does not require training.

As shown in Figure 4, (a) is the input image and (b) is the calculated atmospheric transmission map. The range of variation is different for different lighting and scenes. And (c) is a double-threshold segmentation picture. After the double threshold segmentation, the picture can be basically divided into three parts: sky, distant view and close view. This basically satisfies the general visual recognition needs, such as forest fireworks detection and other tasks.

Figure 4. (a) Input images. (b) Air transmission map. (c) Multi-threshold segmentation image.

6. Discussions and Conclusions

In this paper, we present a simple image depth level estimation algorithm from single image. This method does not rely on samples and classifiers, and the algorithm does not require training. Multi-threshold segmentation image can provide significant prior knowledge for subsequent identification and detection.

Acknowledgments

This work was supported by the National Natural Science Foundation of China (No.61374194), the National Key Science & Technology Pillar Program of China (No.2014BAG0IB03), Key Research and Development Program in Jiangsu Province (No.BE2016739) and a Project Funded by the Priority Academic Program Development of Jiangsu Higher Education Institutions.

References

[1] Scharstein D, Szeliski R, Zabih R. A Taxonomy and Evaluation of Dense Two-Frame Stereo Correspondence Algorithms[C]// Stereo and Multi-Baseline Vision. IEEE, 2002:7-42.

[2] D. A. Forsyth and J. Ponce. Computer Vision: A Modern Approach. Prentice Hall, 2003.

[3] Saxena A, Sun M, Ng A Y. Learning 3-D Scene Structure from a Single Still Image[C]// IEEE, International Conference on Computer Vision. IEEE, 2007:1-8.

[4] Hoiem D, Efros A A, Hebert M. Automatic photo pop-up [J]. Acm Transactions on Graphics, 2005, 24(3):577-584.

[5] Shi J, Pollefeys M. Pulling Things out of Perspective[C]// IEEE Conference on Computer Vision and Pattern Recognition. IEEE Computer Society, 2014:89-96.

[6] Konda K, Memisevic R. Unsupervised learning of depth and motion [J]. Computer Science, 2013.

[7] He K, Sun J, Tang X. Single Image Haze Removal Using Dark Channel Prior [J]. IEEE Trans Pattern Anal Mach Intell, 2011, 33(12):2341-2353.

[8] James M K, James K, Goldstein E B. Sensation and Perception [J]. Leonardo, 1983, 15(2):160.

[9] Preetham A J, Shirley P, Smits B. A practical analytic model for daylight[C]// Conference on Computer Graphics and Interactive Techniques. ACM Press/Addison-Wesley Publishing Co. 1999:91-100.

[10] Otsu N. A Threshold Selection Method from Gray-Level Histograms [J]. IEEE Trans.syst.man. & Cybern, 2007, 9(1):62-66.

IOP Conf. Series: Materials Science and Engineering **435** (2018) 012037 doi:10.1088/1757-899X/435/1/012037

Multi-objective Satellite Scheduling Approach for Very Large Areal Observation

Yingjie Xu, Xiaolu Liu, Renjie He, Yingguo Chen and Yuning Chen

College of System Engineering, National University of Defense Technology, China

xvyingjie1@163.com

Abstract. This paper presents a multi-objective satellite scheduling problem for very large areal observation in response to various requests. The objectives are to maximize the total profits of generated observation schedule and the high efficiency of the satellite utilization simultaneously. To address the satellite scheduling problem, we first demonstrate a detailed problem description and then transform the problem into set covering problem within several criteria and constraints. Based on that, a mathematical model is established. To solve the problem, a new three-phase solving framework is proposed. In the discretizing phase, an area discretization method is adopted to establish the evaluation system. In the target decomposing phase, area target is decomposed into strips and corresponding visible time windows are calculated. In the scheduling phase, a multi-objective genetic algorithm is introduced to generate an optimal observation schedule taking account of the distinct aspect of objectives. Through extensive computational experiments on realistically generated problems in various scenarios including real-world data from China's satellite platform, the effectiveness and reliability of the proposed solving framework are verified.

1. Introduction

More and more attention is attached to environmental activities in recent years, such as territorial mapping, environmental monitoring, land desertification monitoring, tropical rain forest protection, Antarctic ice cover monitoring [1] etc. Earth observation satellites (EOSs) play an increasingly significant role in these activities. EOSs are space platforms which are able to acquire images of specified areas on the Earth's surface in response to various observation requests from various customers such as government agencies and research institutes [2]. They can provide images of the observation targets in different time points and in different imaging angles. These images are essential and reliable foundation of analysis of the environmental condition and taking opportune measures in case of emergency [3]. Handling the scheduling problem for the observing process and acquiring adequate images will give impetus to the promotion and development of environmental activities. Nevertheless, traditional techniques and methods have failed to cope with the observation scheduling in these environmental activities which involve very large area targets and to generate a satisfactory observation schedule effectively and efficiently. Therefore, solving the satellite scheduling for very large area observation is in urgent need according to the realistic and practical circumstances and the multi-objective satellite scheduling problem for very large areal observation is researched in this paper.

2. Literature review

Generally, the observation targets for satellites can be classified into spot targets and polygon targets (i.e. area targets). Spot targets can be completely captured by one pass according to the observation scope of satellite sensor while polygon targets need to be captured by multiple passes of multiple satellites in a cooperative manner. To the best of our knowledge, most literature focused on spot targets or comparatively small polygon targets given the NP-hard nature. Several researches have been devoted to the multi-satellite scheduling problem for area targets. Bai et al. [4] regarded spot target as a special polygon and proposed an integrated model to deal with two kinds of targets and a multi-diversifications simulate annealing algorithm was applied for the satellite observation scheduling problem. Zhu et al. [5] found an approach to reasonably utilize the existing satellites to rapidly image the affected area during a short time period, taking two factors into consideration simultaneously in orbit design, i.e., the maximum observation coverage time and the minimum orbital transfer fuel cost. Wang et al. [6] further developed satellite scheduling problem of an earth observing satellite constellation and proposed a priority-based heuristic with conflict-avoided, limited backtracking and download-as-needed features to select and timetable the observation activities in a very short time. Additionally, Gao et al. [7] analysed advantages and disadvantages of traditional methods for area target scheduling of remote satellite and developed a fast dynamic algorithm for area target scheduling. For improving the reliability of multiple earth-observing satellites, especially in emergent scenarios such as obtaining photographs on battlefields or earthquake areas, Zhu et al. [8] proposed a novel dynamic fault-tolerant scheduling model for real-time tasks running on multiple observation satellites and developed a novel fault-tolerant satellite scheduling algorithm named FTSS. Taking into account the dynamic imaging requirement, N. Niu et al. [9] proposed a satellite scheduling model to address dynamic imaging tasks triggered by emergent disasters and designed a dynamic heuristic algorithm embedding in a greedy criterion to solve the scheduling problem for area targets. Considering dynamically occurring events, Skobelev et al. [10] developed a multi-agent planning system for a group of Earth remote sensing satellites with dynamic balancing of interests of satellites, data receiving points and observation area agents.

These problems researched above are all involved with multiple satellites and area targets but are devoted to comparative small area targets and few to very large area targets. The original intention for our research is to solve the difficulties encountered in the satellite observation of environmental activities involving very large area. The proposed problem is much more sophisticated and hard to handle with more considerable enlargement of the search space and candidate solutions. Therefore, with the increasing number of requests from various customers, generating a feasible observation schedule for very large area with multiple satellites could be time-consuming and difficult. On this condition, current researches are not applicable for our problem whether in terms of mathematical models or proposed algorithms. Therefore, a new adequate solving method is essential for multiple satellite scheduling problem for large area observation.

3. Problem definition& formulation

3.1. Problem definition

Before addressing the proposed problem, several terms are defined:

- Visible time window(*VTW*): a period of time when the required area target is visible to the satellite.
- Strip: the projection on the surface of the earth of a single observation scope of a specific satellite when it flies over the target.
- Schedule: observation plan towards a specific area target. In an observation schedule, several elements are included: involved satellites, selected *VTW*s, attitudes of the involved satellites during selected *VTW*s, etc.

In this work, the problem is referred to the multi-satellite scheduling for very large area observation in response to various requests subject to constraints from both satellites and requirements. Earth observation scheduling employing multiple satellites involves optimization and coordination. The scheduling process is aimed to optimally select which available targets the satellites will observe and determine when and how these will take place with multiple satellites in a cooperative manner. Consequently, the problem of satellite scheduling for very large area observation is a combinatorial optimization problem.

Moreover, the image of the required area target can only be acquired during a specific visible time window(*VTW*). On account of the maneuvering ability of a specific satellite, there could be more than one *VTW* when a satellite flies over the area target for one time, as depicted in Figure 1. Each *VTW* is corresponding to a specific strip with a certain roll angle within the field of view of a specific satellite. Field of view (*FOV*) refers to the spatial range that the remote sensor can observe.

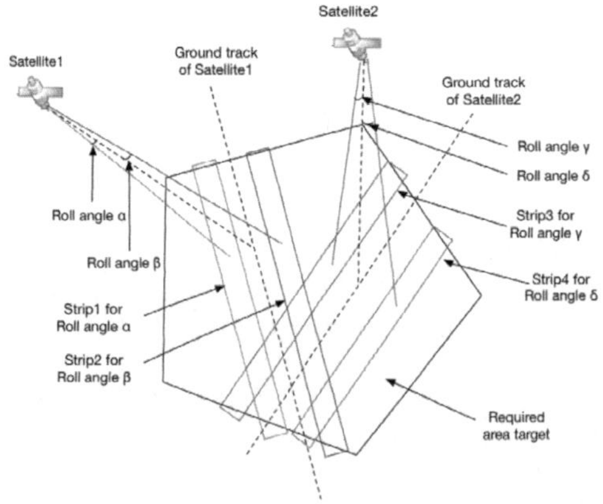

Figure 1. Multiple observation opportunities for targets.

Each acquisition of each strip generates a different profit. The profit of a specific schedule is proportional to the percentage of the target area covered by the selected strips and the main objective of optimization is total profit maximization. Meanwhile, it is realistically necessary to maximize the utilization efficiency of the satellites, such as minimize the number of selected strips and the overlaps. Thus, the multiple satellites scheduling problem of very large area observation is consequently a *VTW* selecting problem. The scheduling process is aimed to select an optimal set of *VTW*s from candidate ones within the fields of view of multiple satellites maximizing the observing profit and efficiency in response to the requests of various customers.

3.2. Problem formulation

3.2.1. Sets and Parameters

- $T = \{t_1, \ldots, t_{|T|}\}$, the ordered set of requests from various customers. For t_q, the following attributes are defined:

q is the required observing large area target of request t_q.
H_q is the required resolution of the images of request t_q.
B_q is the required start time of the scheduling for request t_q.
E_q is the required end time of the scheduling for request t_q.

- $S = \{s_1, \ldots, s_{|S|}\}$, the set of satellites. For s_i, the following attributes are defined:

O_i is the set of orbits of satellite s_i. o_{ij} is the jth orbit of satellite s_i.
M_i is maximal data storage capacity for satellite s_i.
V_i is maximal reserved electrical power for satellite s_i.
G_i is maximal manevering capacity for satellite s_i.

- $W_i = \{w_{i1},\ldots,w_{i|Wi|}\}$ is the set of *VTW*s of satellite s_i and w_{ijk} is the kth *VTW* on orbit o_{ij} of satellite s_i. For w_{ijk}, the following attributes are defined:

b_{ijk}, e_{ijk} are the start and end times of w_{ijk}.
m_{ijk} is the occupied data storage for the image corresponding with w_{ijk}.

v_{ijk} is the consumed electrical power for acquiring the image corresponding with w_{ijk}.
g_{ijk} is the attitude of satellite s_i during the w_{ijk}.
GSD_{ijk} is the resolution of the acquired image during the w_{ijk}.
q_{ijk} is the observing target during the w_{ijk}.

- Function applied in the modeling process is defined as follows:

f is a piecewise linear profit function defined by points (0, 0), (0.4, 0.1), (0.7, 0.4), (1, 1). And this function is associated with the summed contribution area of all scheduled strips for the target area (deducting the overlapped area among all the strips). The more contribution the strips make to the coverage for the target area, the more profit is generated.

p is a profit function intended for evaluating the contribution area for a specific strip that is corresponding to a *VTW*.

l is a function for calculating the proportion of observation overlaps in an observation schedule.

- Decision variable:

$$x_{ijk}=\begin{cases}1 & \text{if } w_{ijk} \text{ is selected in the observation schedule}\\ 0 & \text{otherwise}\end{cases} \quad (1)$$

3.2.2. Assumptions. Based on the practical applications in reality, some assumptions are adopted.

- The satellites mentioned in this paper are referred to the imaging satellites.
- In this paper, the download process of the acquired images falls outside the scope of our study.
- Once a visible time window is scheduled, the acquisition process cannot be interrupted or cancelled until it is finished. At any time, there is at most one task being processed on a satellite.
- The satellite scheduling problem for large area is over-constrained, and not all the requests can be fully satisfied. Thus, the circumstance that some tasks are not assigned or partially observed is acceptable.

3.2.3. Mathematical formulations

$$max\sum_{i=1}^{|S|}\sum_{j=1}^{|O_i|}\sum_{k=1}^{|W_{ij}|}f(x_{ijk}\cdot\Delta p) \quad min\sum_{i=1}^{|S|}\sum_{j=1}^{|O_i|}\sum_{k=1}^{|W_{ij}|}x_{ijk} \quad min\sum_{i=1}^{|S|}\sum_{j=1}^{|O_i|}\sum_{k=1}^{|W_{ij}|}l(x_{ijk}) \quad (2)$$

$$s.t.$$

$$\sum_{k=1}^{W_{ij}}x_{ijk}\leq 1, \forall i\in[1,|S|], \forall j\in[1,|O_i|] \quad (3)$$

$$\sum_{j=1}^{|O_i|}\sum_{k=1}^{|W_{ij}|}(m_{ijk}\cdot x_{ijk})\leq M_i, \ \forall i\in[1,|S|] \quad (4)$$

$$\sum_{j=1}^{|O_i|}\sum_{k=1}^{|W_{ij}|}(v_{ijk}\cdot x_{ijk})\leq V_i, \ \forall i\in[1,|S|] \quad (5)$$

$$\sum_{i=1}^{|S|}\sum_{j=1}^{|O_i|}\sum_{k=1}^{|W_{ij}|}(e_{ijk}-b_{ijk})\leq\delta \quad (6)$$

$$b_{ijk}\geq B_{qijk}, \forall i\in[1,|S|], \forall j\in[1,|O_i|], \forall k\in[1,|W_{ij}|] \quad (7)$$

$$e_{ijk}\geq E_{qijk}, \forall i\in[1,|S|], \forall j\in[1,|O_i|], \forall k\in[1,|W_{ij}|] \quad (8)$$

$$g_{ijk}\geq G_i, \forall i\in[1,|S|], \forall j\in[1,|O_i|], \forall k\in[1,|W_{ij}|] \quad (9)$$

$$GSD_{ijk}\geq H_{qijk}, \forall i\in[1,|S|], \forall j\in[1,|O_i|], \forall k\in[1,|W_{ij}|] \quad (10)$$

The objective function (2) is to maximize the summed reward, to minimize the number of selected VTWs and to minimize the overlaps of the observation schedule for a specific target area respectively. Constraint (3) regulates that for each target area, at most one visible time window can be selected on its every satellites orbit owing to the long-strip imaging mode. Constraint (4) restricts that the satellite memory storage is always within its safe range. Constraint (5) insures that the satellite is charged with adequate power to support the process of image acquisition. Constraints (6) reveals the duration for each image acquisition must be restricted to the maximal working duration for each image acquisition for the sake of normal operation according the long-strip imaging mode. Constraints (7) and (8) ensure that the selected visible time window must be included in a given schedule horizon. Constraint (9) points out the restriction of maneuvering capability for a specific satellite. Constraint (10) reveals that the resolution of the images is supposed to satisfy the requirements from customers.

4. Multi-objective satellite scheduling method for very large areal observation

The main solving process for multi-satellite scheduling problem for very large area observation can be divided into three phases: the discretizing phase, target decomposing phase and the scheduling phase, as illustrated in Figure 2.

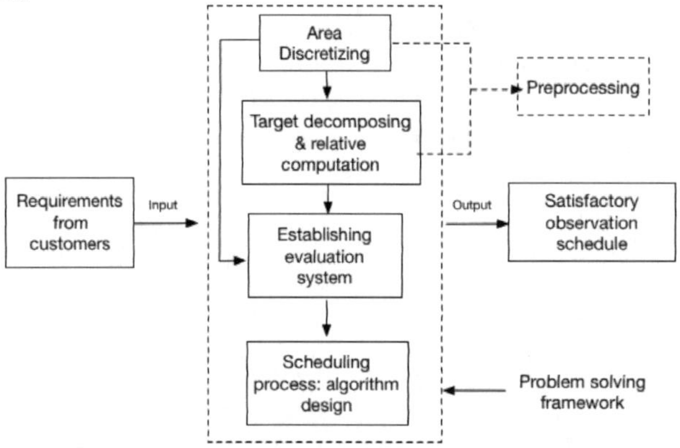

Figure 2. Main framework of the proposed problem.

4.1. The discretizing phase

As illustrated above, the objective of the multiple satellite scheduling problem for large area observation is to maximize the summed coverage of the strips in a generated observation schedule. Thus, the area calculation of the strips in a specific observation schedule is essential. Since our problem is involved with multiple satellites and very large areas, applying the geometric methods will lead to considerable computational time and the loss of accuracy.

Hence, we develop a discretization-based area calculation method to reduce the time and space complexity in the calculation process. The area target is discretized equidistantly with spots, and the set of interpolated spots represents the whole area target, as depicted in Figure 3. Meanwhile, the visible time window is represented as a strip. As one strip can cover a portion of the area target, each strip can cover some interpolated spots. Consequently, the evaluation system can be established. The profit of a specific observation schedule can be measured by the number of the interpolated points of the strips located in the whole area target. Despite the circumstance that different strips may cover several points repeatedly, the discretization-based calculation method largely simplifies profit measurement of observation schedules.

4.2. Target decomposing phase

Considering the area target is far much larger than single observation scope for the satellites, it is a common measure to decompose the area into strips. Accordingly, a visible capacity based area target decomposition method is proposed. The concrete decomposition method is depicted below.

STEP 1. For each satellite i, compute FOV for each area target r and the bounding rectangle of the overlapping region of the FOV and target.

The bounding rectangle is associated with the intersection points e between the FOV and the area target. The satellite observes the two intersection points on the length in the earliest time t^e and the latest time t^l. The satellite observes the two intersection points on the width with the smallest roll angle r^- and the largest roll angle r^+.

STEP 2. Set the excursion parameter $\triangle\lambda$.

Decompose the bounding rectangle of each area target into strips with $\triangle\lambda$. The excursion parameter can adjust the extent to which the strips overlap considering the requirement and preference of the customers. In terms of the satellites equipped with optical cameras, $\triangle\lambda$ is represented by a specific roll angle which can be changed.

STEP 3. Cut the strips.

After **STEP 3**, strips of equal length are obtained. They are supposed to be cut according to target boundary in order to improve the effective coverage.

STEP 4. Compute the *VTW*s of the corresponding strips.

VTWs are calculated by judging the visibility between the satellites and the area target in a specific time span [T_{begin}, T_{end}] and search the visible time spans as the visible time window [TW_{begin}, TW_{end}].

After the 5 steps, strips of a specific target area are obtained, as depicted in Figure 4.

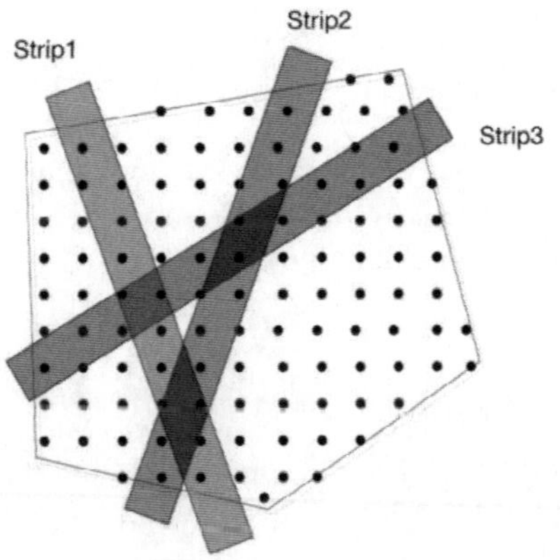

Figure 3. Area target discretization.

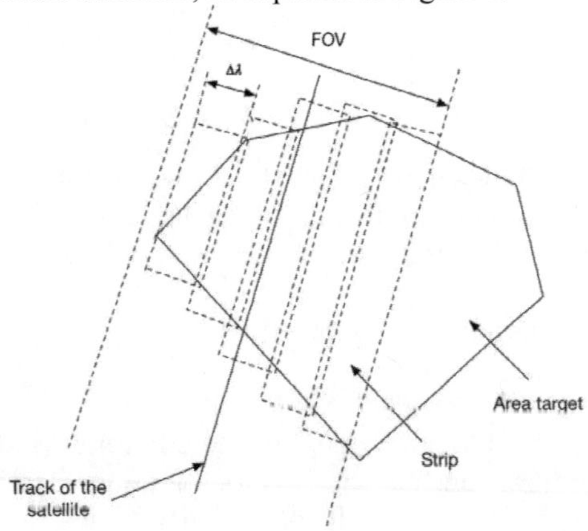

Figure 4. The set of generated strips.

4.3. The scheduling phase

Based on the two preconditioning phases, the multi-satellite scheduling problem for very large area observation can be transformed into a multi-objective set covering problem subject to several constraints. The definition of the complex set covering problem is presented as follows.

Given a set of points $I_q = \{i_{q1},...,i_{qn}\}$ and a set of m subsets of I_q, A=$\{I_{q1}, I_{q2},...,I_{qm}\}$, find an optimal collection C of the sets from A such that C covers the elements in I_q to the maximal extent and contains the minimal number of elements and overlapping portions. I_q is the set of points representing the area target q, and I_{qj}, j is the set of points that a specific strip can cover in the area target q. I_{qj} is relative exclusively to each *VTW*. I_q and I_{qj} are generated in the front phases.

Much interest and effort has been devoted to the set covering problem; the proposed algorithms up to now can be categorized into two classes: heuristic and exact approaches. Considering the considerable strips yielded by the large area target and the high combinatorial nature of the scheduling problem, heuristic solving approach is more appropriate for our problem. In order to generate an observation plan taking count of the efficiency and accuracy simultaneously, in view of the characteristics of the multiple satellite scheduling problem that the problem is virtually a *VTW* selecting problem, a multi-objective genetic algorithm is developed based on NSGA-II. Applying fast non-dominated sorting operator, crowded-comparison operator and elite strategy selection operator in our approach, we further develop the encoding and fitness evaluation formulations.

Note that due to the maneuver constraints, satellites are in a long-strip imaging mode while observing a very large area target. It means that one visible time window can be selected into the schedule when the satellite passes over the target once. Accordingly, the encoding of individuals is constructed by the number of the satellite orbits Q_{num} in a given schedule horizon. The structure of each chromosome in the population is illustrated as Figure 5. Each chromosome reaches a length of the Q_{num} and each gene on the chromosome represents a *VTW* on a specific orbit. Sequence of the genes on a chromosome represents the order of the observation.

t_1	t_2	...	t_{Qnum-1}	t_{Qnum}

Figure 5. The encoding of chromosome in the proposed algorithm.

As depicted above, each target area is represented by a set of interpolated points and the visible time window is represented as a strip. The coverage rate of the obtained observation schedule and overlaps can be measured by the quantity of points both inside the target area and the visible time windows in the observation plan. The fitness evaluation formulations of our algorithm are presented as follows.

$$F_1(schedule)=f\left(\frac{N\left(\left(I_{q1} \cup I_{q2} \cup \ldots \cup I_{qQnum}\right)\cap I_q\right)}{N(I_q)}\right) \tag{11}$$

$$F_2\ (schedule)=N(schedule) \tag{12}$$

$$F_3(schedule)=\frac{\bigcup_{i=1}^{Qnum}\bigcup_{j=1,j\neq i}^{Qnum}(I_{qi}\cap I_{qj})}{N(I_q)} \tag{13}$$

In the formula above, F_1 defines the optimization objective to maximize the total profit. F_2 defines the objective to minimize the number of selected *VTW*s. F_3 defines the objective to minimize the overlaps in an observation schedule. And N is a number calculating function.

5. Computational experiments

5.1. Design of the test scenarios

To the best of our knowledge, little concentration and effort has been devoted to the multiple satellite scheduling problem for large area observation. Besides, different satellites in different countries are designed for different missions, which leads to their capabilities, adopted technologies and management differ significantly. It is accordingly unreasonable to test a satellite scheduling algorithm by a common benchmark designed for a specific satellite. There is no competing algorithm and instances exiting for our problem. Hence, it is extremely hard to compute the optimal solutions due to the high combinatorial nature. Additionally, the greedy-based heuristic algorithm is widely accepted in the practical circumstances for our problem because it can guarantee the efficiency of the scheduling process and a generated feasible solution. We therefore conducted various scenarios to evaluate the performance of the greedy-based heuristic algorithm and our algorithm. And this computational experiment was coded in C# and executed on a PC with Intel (R) Core (TM) i5-3317U2.99 GHz CPU speed and 4 GB RAM. The configurations of the scenarios are depicted as follows.

Table 1. Location & size of three area targets

Target	Latitude& Longitude	Size(km^2)
African Savannah	(5.33,30.26),(5.33,9.32),(5.17,9.32),(-5.17,30.26)	123210
Amazon Forest	(0.53,50.50),(0.53,75.26),(9.06,75.26),(-9.06,-50.50)	106154
Southern China	(32.58,114.91),(32.58,105.09),(22.09,105.09),(22.09,114.91)	109780

Given that nowadays EOSs play an increasingly important role in the environment protection and the forests in the world are precious presents from the nature that are supposed to be supervised on end in case of emergency, high vegetation covered areas are selected to be the observation targets. Three high vegetation covered areas are selected from African Savannah, Amazon Forest and Southern China respectively. And the geographical location and basic information of the three target are illustrated as Table 1. For each target area, the scheduling is conducted in 5 independent one-day horizons, from 2018-5-22 0:00:00 to 2018-5-27 0:00:00. And 14 optical satellites from Chinas satellite platform are involved.

Based on several tests on the impact of different parameters of the GA, the values for some parameters are fixed as follows: population size equals to 100, maximal iteration equals to 500, mutation probability equals to 0.1, crossover probability equals to 0.9, maximal working duration for each image acquisition equals to 10 min and excursion parameter equals to 1.8. Figure 6 illustrates the distinct results of the proposed algorithm performance in different mutation probability(p_m) and crossover probability(p_c) settings.

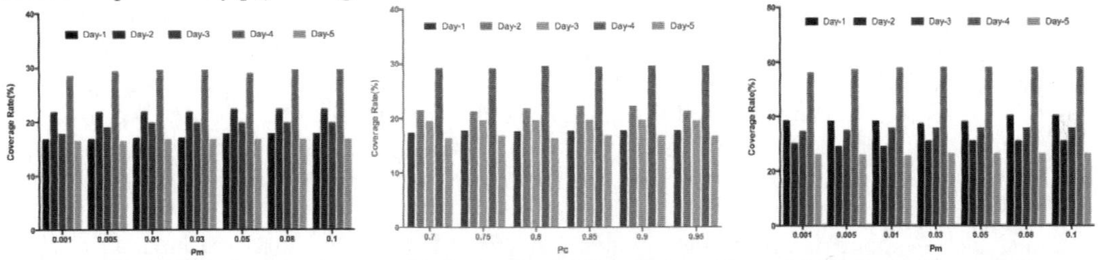

(a)Scenario African Savannah (b) Scenario African Savannah (c) Scenario Amazon Forest

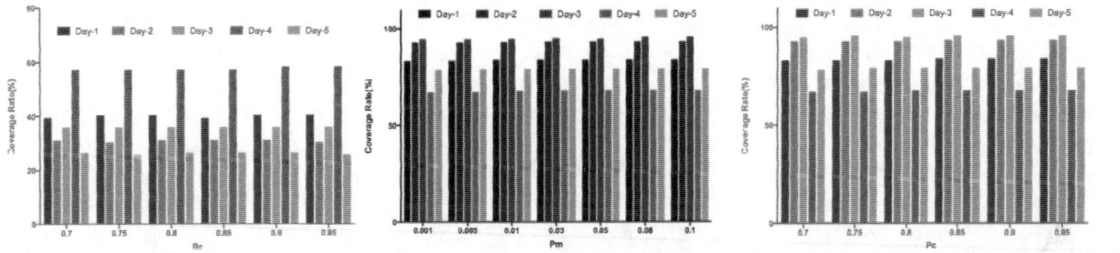

(d) Scenario Amazon Forest (e) Scenario Southern China (f) Scenario Southern China

Figure 6. The results of the proposed algorithm in different parameter settings.

5.2. Experimental results

Table 2. Comparison between the proposed algorithm and greedy based heuristic algorithm

Performance		Scenario_African Savannah					Scenario_Amazon Forest					Scenario_Southern China				
		D1	*D2*	*D3*	*D4*	*D5*	*D1*	*D2*	*D3*	*D4*	*D5*	*D1*	*D2*	*D3*	*D4*	*D5*
Proposed Genetic Algorithm	$DT_{1(s)}$	190	388	397	410	408	420	414	492	769	569	777	734	591	375	555
	$CR_{1(\%)}$	17.7	22.2	19.6	29.5	16.7	40.2	30.9	35.5	58.1	26.3	83.8	93.2	95.4	67.7	79.1
	Sec(s)	88.8	109.3	76.2	71.4	85.8	176.2	126	157.5	159.8	136.3	156.7	151.5	223.2	133.7	150.4
Greedy-based heuristic algorithm	$DT_{2(s)}$	189	378	392	410	402	410	404	490	749	570	781	720	588	368	547
	$CR_{2(\%)}$	4.2	17.1	13	26.9	10.2	31.2	22.7	28.3	53	22.1	80.9	92.5	93.5	58.7	78.3
	Sec(s)	31	80	82	70	66	141.4	136.4	107.4	192.9	204.7	91.1	85.1	56.8	32.8	56.8

Table 2 presents the results of the proposed multi-objective genetic algorithm and the greedy search algorithm for 3 different scenarios in aspect of coverage(the main objective). The result of each scenario is the average of twenty runs. The column *Performance* indicates that these two algorithms are evaluated in three main aspects, *DT*, *CR* and *Sec*. DT refers to duration time of the observation process of the generated optimal schedule, indicating the efficiency of the image acquisition for a requested large area target. *CR* points out the maximal coverage rate of the target areas in a specific schedule horizon, indicating the completion of the requests from various customers. Here we use 0 as the minimal possible reward of each observation schedule and 100% as the maximal. The fractions occur in the rewards are because of the partially imaged area targets. Due to the constrained capacity of the involved satellites from Chinas satellite platform, low completion of the tasks is acceptable. *Sec* reports the computational time in seconds. DT_1, CR_1, Sec_1 evaluate the performance of the proposed genetic algorithm while *DT2*, *CR2*, Sec_2 evaluate the performance of the heuristic greedy search. Different scenarios are denoted in accordance with the location of the requested areas, which are Africa, Amazon and China. The column *D1*, *D2*, *D3*, *D4* and *D5* describe the 5 independent one-day schedule horizons, from 2018-5-22 0:00:00 to 2018-5-27 0:00:00. Each scenario is evaluated in these 5 schedule horizons.

As depicted in Table 2, the proposed genetic algorithm improves the performance of the image acquisition to a large degree compared with the greedy search which is commonly used in practical circumstances. In order to reveal the improvement for the scenarios precisely, three histograms are applied as illustrated in Figure 7. For each instance, the total profit (i.e. coverage rate) is presented on y-axis and the schedule horizon is presented on x-axis. Surprisingly, the increment of the performance does not cost considerable computational time. The maximal computational second reached to 223.2, which is quite within the scope of our acceptance. Meanwhile, as illustrated in Figure 8, the multi-objective genetic algorithm presents better performance in reducing the overlaps and improving utilization efficiency.

Note that the rewards obtained by the 14 involved satellites for Scenario China have an obvious advantage over the rewards of other two scenarios under the condition that all the scenarios are scheduled in a same given time horizon. This may be because 14 involved satellites come from Chinas satellite platform and are designed to observe the whole land of China and its surroundings. On the other hand, this indicates that the observation performance by multiple satellites still needs to be improved, possibly by means of coordination and cooperation of satellites from satellite platforms in other countries.

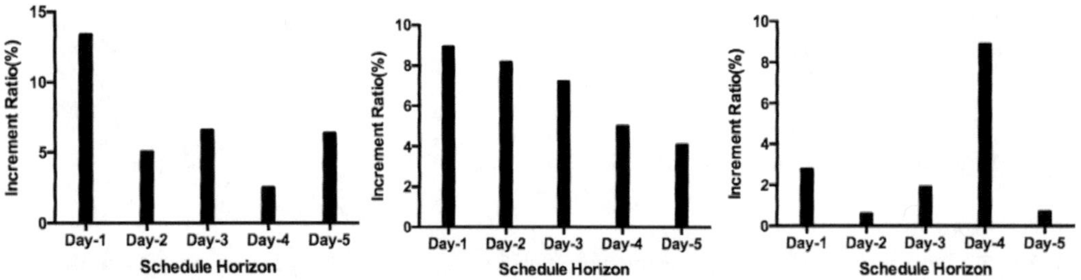

Figure 7. The performance increment of multi-objective genetic algorithm versus greedy-based heuristic algorithm in the aspect of coverage.

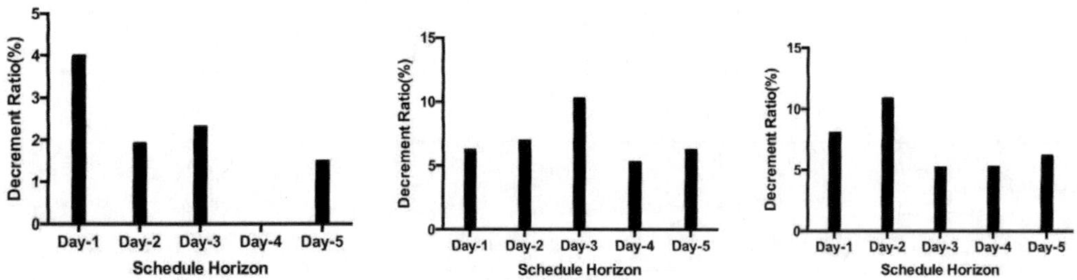

Figure 8. The decrement of multi-objective genetic algorithm versus greedy-based heuristic algorithm in the aspect of observation overlaps.

Remarkably, the computational time consumed by our algorithm is never larger than 5 min for the tested scenarios, which reveals the outstanding performance in terms of speed and reliability in practice. This allows useful information and images acquired by ground scheduling personnel in order to support large area supervision such as worldwide forests.

6. Conclusions

The main contributions of our research are summarized as follows: (1) The multi-satellite scheduling problem for large area observation is introduced for the first time and a multi-objective mathematical model is established. (2) A new three-phase solving framework is proposed. In the discretizing phase, an area discretization method is adopted. In the target decomposing phase, area target is decomposed into strips and corresponding visible time windows are calculated. In the scheduling phase, a multi-objective genetic algorithm is introduced to generate an optimal observation schedule. (3) Extensive computational experiments have been conducted on realistically generated problem instances to evaluate the performance the proposed multi-objective genetic algorithm compared with greedy-based heuristic algorithm. The results indicate the superiority of the proposed algorithm in terms of reliability and speed over the other algorithm.

Acknowledgments

This work was supported by the National Natural Science Foundation of China (Nos. 61473301, 71501180).

References

[1] R. Xu, H. Chen, X. Liang, H. Wang, Priority-based constructive algorithms for scheduling agile earth observation satellites with total priority maximization, Expert Systems with Applications 51 (2016) 195 – 206.

[2] G. Denis, A. Claverie, X. Pasco, J.-P. Darnis, B. de Maupeou, M. Lafaye, E. Morel, Towards disruptions in earth observation new earth observation systems and markets evolution: Possible scenarios and impacts, Acta Astronautica 137 (2017) 415 – 433.

[3] G. Denis, H. de Boissezon, S. Hosford, X. Pasco, B. Montfort, F. Ranera, The evolution of earth observation satellites in europe and its impact on the performance of emergency response services, Acta Astronautica 127 (2016) 619 – 633.

[4] B. C. Bai, Y. W. Chen, H. E. Ren-Jie, R. Qi-Ming, Multi-satellite scheduling toward spot and polygon observing requests, Journal of Astronautics 30(2) (2009) 755–759.

[5] K.-J. Zhu, J.-F. Li, H.-X. Baoyin, Satellite scheduling considering maximum observation coverage time and mini- mum orbital transfer fuel cost, Acta Astronautica 66 (2010) 220 – 229.

[6] P. Wang, G. Reinelt, P. Gao, Y. Tan, A model, a heuristic and a decision support system to solve the scheduling problem of an earth observing satellite constellation, Computers & Industrial Engineering 61 (2011) 322–335.

[7] P. Gao, Y. Tan, F. Yao, A fast dynamic algorithm for area target scheduling of remote satellite, in: Electrical & Electronics Engineering, 2012, pp. 71–73.

[8] X. Zhu, J. Wang, X. Qin, J. Wang, Z. Liu, E. Demeulemeester, Fault-tolerant scheduling for real-time tasks on multiple earth-observation satellites, IEEE Transactions on Parallel & Distributed Systems 26 (2015) 3012–3026.

[9] X.N.Niu,X.Zhai,H.Tang,L.Wu,Multi-satellite scheduling approach for dynamic areal task striggered by emergent disasters, in: ISPRS - International Archives of the Photogrammetry, Remote Sensing and Spatial Information Sciences, volume XLI-B1, 2016, pp. 475–481.

[10] Skobelev,E.Simonova,A.Zhilyaev,V.Travin,Application of multi-agent technology in the scheduling system of swarm of earth remote sensing satellites, Procedia Computer Science 103 (2017) 396 – 402. XII International Symposium Intelligent Systems 2016, INTELS 2016, 5-7 October 2016, Moscow, Russia.

IOP Conf. Series: Materials Science and Engineering **435** (2018) 012038 doi:10.1088/1757-899X/435/1/012038

Particle Filter via Chopthin Resampling for Target Tracking with Glint Noise

C Liu[1,*], S C Yang[1] and L D Wang[2]

[1]Army Engineering University, No.97 Heping West Road, Shijiazhuang, Hebei, 050003, China
[2]National Key Laboratory of the Complex Electromagnetic Environment Effect of Electronic Information System, Luoyang, Henan, 471003, China

[*]Corresponding e-mail:1561174321@qq.com

Abstract. Traditional resampling methods in particle filter suffer from the disadvantage of the low filtering accuracy, the severe fluctuation of effective sample size (ESS) and the great influence of parameter setting. A particle filter algorithm based on chopthin resampling is proposed in this paper. Traditional resampling methods produce a set of equally weighted particles and only occur when ESS below a threshold. In contrast to this, the proposed algorithm enforces an upper bound on the ratio between the weights and can be implemented in every step. Simulation shows that the new method outperforms traditional resampling methods in filtering accuracy, ESS stability and computational efficiency, especially when sample size or observation noise is small.

1. Introduction

In target tracking, changes in the target aspect with respect to the radar can cause random wandering of the angle tracking, which is also called angle fluctuations or target glint. The glint is heavy-tailed and non-Gussian and may severely affect the tracking accuracy, especially that of large targets at short ranges. Kalman filters, such as extended Kalman filter (EKF), unscented Kalman filter (UKF) and cubature Kalman filter (CKF), have been widely applied in target tracking. However, their performance degrades when the observation noises are not Gaussian.

One possible alternative approach to this problem is particle filter (PF). Since the publication of [1], PF has become popular for nonlinear state-space models where the noises of the model are non-Gaussian. During the past two decades, PF has been successfully applied in target tracking. Using the PF method, various distributions of interest are approximated by the generated particles as well as associated weights. Almost all PF methods are based on three operations: particle propagation, weight computation and resampling. Particle generation and weight computation amount to the generation of particles and assignment of weights, whereas resampling replaces one set of particles and their weights with another set.

Resampling, which is one of the key steps of PF, is highly important to PF. Without resampling, PF will suffer from weight degeneracy, i.e., only a few particles are significantly weighted, whereas the weight of most of the particles is almost zero. This means that the obtained estimates will be inaccurate and will have unacceptably large variances. Consequently, various resampling schemes (traditional resampling as they may be called), include multinomial resampling [1], stratified resampling [2], systematic resampling [3] and residual resampling [4], have been proposed.

Generally speaking, resampling is a process in which one samples from the original random measure $\chi_t = \{x_t^i, w_t^i\}_{i=1}^n$ to create a new random measure $\tilde{\chi}_t = \{\tilde{x}_t^i, \tilde{w}_t^i\}_{i=1}^N$. In traditional resampling methods, the weights of the particles after resampling are all equal. However, the benefits of resampling are accompanied with potential side effects such as sample impoverishment and the speedup of implementation of PF. Thus, traditional PF often only perform the resampling step if the effective sample size (ESS) drops below a fixed threshold. This avoids resampling if the weights are relatively even and thus reduces the noise being introduced through the resampling. On the other hand, this causes the ESS to fluctuate over time.

Recently, a new resampling algorithm called chopthin was proposed [5]. In contrast to standard resampling methods, the algorithm does not produce a set of equally weighted particles; instead it merely enforces an upper bound η on the ratio between the weights. Chopthin will not alter the weights much if they are relatively even, thus it can be executed at every step of a particle filter. Simulation studies show that chopthin consistently outperforms the other resampling methods and implicitly controls the ESS.

This paper is organized as follows. After Section 1 which gives the introduction, a brief review of PF and resampling is presented in Section 2 and the chopthin resampling is discussed in Section 3. Then, Section 4 provides the numerical simulation result via comparison of chopthin resampling with numerous typical resampling methods and finally Section 5 summarizes the results and proposes some recommendations for further research.

2. State estimation by the particle filter

Consider a discrete time nonlinear stochastic system given by the state equation (1) and the measurement equation (2):

$$x_t = f(x_{t-1}, u_t), t = 1,2,\ldots \tag{1}$$

$$y_t = h(x_t, v_t), t = 1,2,\ldots \tag{2}$$

Where the vectors $x_t \in \mathrm{R}^n$ and $y_t \in \mathrm{R}^m$ represent the state of the system and the measurement at time t, respectively; $u_t \in \mathrm{R}^n$ and $v_t \in \mathrm{R}^m$ are independent state and measurement noises, with known probability density function (pdf) $p(u_t)$ and $p(v_t)$; $f: \mathrm{R}^n \times \mathrm{R}^n \to \mathrm{R}^n$ and $h: \mathrm{R}^n \times \mathrm{R}^m \to \mathrm{R}^m$ are known functions; and the pdf $p(x_0)$ of the initial state x_0 is known. The system can alternatively be described by the transition pdf $p(x_t|x_{t-1})$ and the measurement pdf $p(y_t|x_t)$, which can be obtained from equation (1) and equation (2), respectively.

The objective of PF here is the sequential estimation of filtering distribution $p(x_t|y_{1:t})$, which can be expressed in terms of the filtering distribution $p(x_{t-1}|y_{1:t-1})$, i.e., in a recursive form by

$$p(x_t|y_{1:t}) \propto \int p(y_t|x_t) p(x_t|x_{t-1}) p(x_{t-1}|y_{1:t-1}) \mathrm{d}x_{t-1} \tag{3}$$

Where the symbol $\propto$ signifies "proportional to". This update can be implemented analytically in a very few cases, therefor, one resorts to approximations. The basic idea of PF is to represent the belief $p(x_{t-1}|y_{1:t-1})$ by a set of N weighted samples $\chi_{t-1} = \{x_{t-1}^i, w_{t-1}^i\}_{i=1,\ldots,N}$, i.e.,

$$p(x_{t-1}|y_{1:t-1}) \approx \sum_{i=1}^{N} w_{t-1}^i \delta(x_{t-1} - x_{t-1}^i) \tag{4}$$

Where $\delta(\cdot)$ is the Dirac delta impulse, and w_{t-1}^i is the weight associated with the system state w_{t-1}^i. As $N \to \infty$, the discrete weighted approximation in equation (4) approaches to the true posterior density $p(x_{t-1}|y_{1:t-1})$.

First basic step of PF is to draw particles x_t^i from $p(x_t|y_{1:t})$, which is referred to as particle propagation or importance sampling because a particle x_{t-1}^i is moved forward in time and is a parent of x_t^i [6]. However, $p(x_t|y_{1:t})$ is unkown in most cases, therefor, one has to use an importance density function $\pi(x_t)$. A simplest and widely used choice is

$$\pi(x_t) = p(x_t|x_{t-1}) \tag{5}$$

The second basic step is computing weights of the particles. As particles generated from $\pi(x_t)$ is different from $p(x_t|x_{t-1})$, the particle weights have to be modified according to

$$w_t^i \propto w_{t-1}^i \, p(y_t|x_t^i)\,p\left(x_t^i|x_{t-1}^i\right)\Big/\pi(x_t^i) \tag{6}$$

Subsequently, the particle weights are normalized so that they sum up to one, i.e.

$$w_t^i = w_t^i \Big/ \sum_{i=1}^{N} w_t^i \tag{7}$$

PF using only the aforementioned two steps has sample degradation problem, i.e. a few particles have most of the weight and the rest of the particle weights are negligible. For this reason, PF needs a third step, referred to as resampling. Formally, resampling is a process in which one modify the random measure $\chi_t = \{x_t^i, w_t^i\}_{i=1}^n$ to a new random measure $\tilde{\chi}_t = \{\tilde{x}_t^i, \tilde{w}_t^i\}_{i=1}^N$. Then, the random measure χ_t is replaced with $\tilde{\chi}_t$. In the process, particles with low importance weights are most likely removed and those of large weights are replicated. Although the number of resampled particles N is not necessarily equal to the number of propagated numbers n, traditional resampling methods keep $n = N$ and the weights after resampling are equal. However, resampling can also delete good samples from the sample set, causing sample impoverishment. Thus, resampling should only be applied when necessary and therefore it is important to find a criterion when to perform a resampling step. Effective sample size (ESS) as a mean to measure the variation of the weights is defined as [7]

$$N_{eff} = 1\Big/\sum_{i=1}^{M}(w_t^i)^2 \tag{8}$$

Where N_{eff} varies in the $[1, M]$ range. Resampling is required when N_{eff} falls below a selected threshold N_{thr}. This avoids resampling if the weights are relatively even and thus reduces the noise being introduced through the resampling. This results in ESS to fluctuate over time.

The general algorithm of the PF can be summarized in table 1 as follows:

Table 1.SIR particle filter.

Input: target number of particles N; ESS threshold N_{th}; resampling scheme r; observations $y_{1:T}$
Output: weighted particles $(w_i, x_t^i)_{1:n_t}$, for $t = 1, \cdots, T$
Sample $x_0^i \sim p(x_0)$, $i = 1, \cdots, N$
$w_i = 1/N$, $i = 1, \cdots, N$
Let $n_0 = N$
for $t = 1, \ldots, T$ do
Sample $x_t^i \sim p(x_t|x_{t-1}^i)$, $i = 1, \cdots, n_{t-1}$ $w_i = w_i p\left(y_t|x_t^i\right)$, $i = 1, \cdots, n_{t-1}$
Normalize $w_i = w_i / \sum_{i=1}^{n_{t-1}} w_i$
if ESS$(w) \le N_{th}$, then
Run r to get a set of particles $(w_i, x_t^i)_{1:n_t}$
end if
end for

3. Chopthin resampling

Recently, a new resampling algorithm called chopthin was proposed in [5]. In contrast to standard resampling methods, the algorithm does not produce a set of equally weighted particles; instead it merely enforces an upper bound η on the ratio between the weights. As mentioned in Section II, traditional resampling method only occurs when ESS drops below a fixed threshold. In contrast to this, chopchin can be executed at every step of a particle filter. This is because chopthin will not alter the weights much if they are relatively even.

Denote the particle set before resampling as $\chi_t = \{x_i, w_i\}_{i=1}^{n}$ and after resampling as $\tilde{\chi}_t = \{\tilde{x}_j, \tilde{w}_j\}_{j=1}^{N}$. Let G be the σ-field generated by $w_1,\ldots,w_n$ and let C^i be the number of times that the i th particle is resampled. The constraints that chopthin should satisfy are as follows:

(1) $E(C^i \tilde{w}_i | G) = w_i \;\; \forall i$ (Unbiasedness)

(2) $\sum_{i=1}^{n} C^i = N$ (Target count)

(3) $\sum_{i=1}^{n} w_i = \sum_{j=1}^{N} \tilde{w}_j$ (Conserve weight)

(4) $\tilde{w}_i / \tilde{w}_j \le \eta \;\; \forall i, j$ (Bounded ratio)

Property (1) ensures that the total weight of a particle is equal before and after resampling. Property (2) ensures that N particles are returned after resampling. Property (3) ensures that the sum of all the weights before and after resampling are equal. Property (4) bounds the ratio of weights returned from chopthin.

The general algorithm of the chopthin can be summarized in table 2 as follows:

Table 2. Generic chopthin.

Input: particle weights $(w_i)_{1:n}$;maximal weight ratio η ;target number of particles N ;function
  $h_a^\eta : [0,\infty) \to [0,\infty)$
Output: ancestors $I \in \{1,\ldots,n\}^N$,weights $\tilde{w} \in [0,\infty)^N$

Let a be a solution to $\sum_{i=1}^{n} h_a^\eta(w_i) = \mathrm{N}$ //Find a
Let $L = \{\mathrm{j}: w_j < a\}$ and $U = \{\mathrm{j}: w_j \ge a\}$
Let $I = ()$ and $\tilde{w} = ()$
Draw $u \sim U(0,1)$
for $i \in L$ do //Thin
  $u = u + h_a^\eta(w_i)$
  if $u \ge 1$ then
    Append (i) to I and (a) to $\tilde{w}$
    $u = u - 1$
  end if
end for
Let $N_L = length(I)$, $N_U = N - N_L - \sum_{i \in U} \lfloor h_a^\eta(w_i) \rfloor$ and $\varsigma := \sum_{i \in L} w_i - aN_L / \sum_{i \in U} \{h_a^\eta(w_i)\}$
Systematic $[\{\{h_a^\eta(w_j)\}; j \in U\}; N_U]$;returns m^j for $j \in U$
for $i \in U$ do //Chop
  $c = \lfloor h_a^\eta(w_i) \rfloor + m^i$
  append $(i,\ldots,i) \in \mathrm{N}^c$ to I
  $\hat{w}_i = w_i + \varsigma\{h_a^\eta(w_i)\}$

append $(\hat{w}_i/c,\ldots,\hat{w}_i/c)\in \mathrm{R}^c$ to $\tilde{w}$
return $I,\tilde{w}$
end for

The key steps of generic chopthin include find a, thin and chop. The parameter a is a threshod of weights, i.e., particles with weights below a will be thinned and particles with weights above a will be chopped. For a particle with weight w, the expected number of offspring from chopthin will be $h_a^\eta(w)$, which is a given function that depend on a and η. Hence, first key step of chopthin is choose $h_a^\eta(w)$ and find a such that

$$\sum_{i=1}^{n} h_a^\eta(w_i) = N \tag{9}$$

One such choice given by [5] is

$$h_a^\eta(w) = \begin{cases} w/a & \text{if } w<a \\ 1 & \text{if } a\le w<\eta a/2 \\ 2w/\eta a & \text{if } w\ge \eta a/2 \end{cases} \tag{10}$$

We use table 3 to solve equation (9) using equation (10) for a.

Table 3. Fast determination of a.

Input: particle weights w_i ;maximal weight ratio η ;target number of particles N
Output: $a>0$ such that $\sum_{i=1}^{n} h_a^\eta(w_i)=N$

$w^u=w^l=w$, $s^l=0$, $c^m=0$, $s^u=0$, $c^u=0$
while $w^u\ne\varnothing$ or $w^l\ne\varnothing$ do
  if $|w^l|\ge|w^u|$, then sample a uniformly from w^l and let $b=\eta a/2$
  else, sample b uniformly from w^u and let $a=2b/\eta$
  end if
  $h=s^l/a+\sum_{v\in w^l}\min(v/a,1)+c^m+\sum_{v\in w^u}\max(v/b-1,0)+s^u/b-c^u$
  if $h=N$,then return a
  if $h>N$,then $w^l=\{v\in w^l;v>a\}$, $s^l=s^l+\sum_{v\in w^l} v\mathrm{I}(v\le a)$, $w^u=\{v\in w^u;v>b\}$
  else $c^m=c^m+\sum_{v\in w^l}\mathrm{I}(v\ge a)$, $s^u=s^u+\sum_{v\in w^l} v\mathrm{I}(v\ge b)$, $c^u=c^u+\sum_{v\in w^l}\mathrm{I}(v\ge b)$, $w^u=\{v\in w^u;v<b\}$
  end if
return $a=(s^l+2s^u/\eta)/(N-c^m+c^u)$

It is proved by [5] that the expected effort of algorithm in table 3 is linear in n, and overall the expected computational effort is $O(\max(n,N))$. Algorithm 3 works by determining which weights are above or below a and which weights are above or below ηa ,without fully knowing a yet.

Second key step of chopthin is Thinning, which determines the new weight and the number of offspring for particles with weights below a. With this step, particles with small weights will have 0 or 1 offspring, which is determined by systematic resampling on $\{h_a^\eta(w_i): w_i<\mathrm{a}\}$. This step returns N_L particles with weight a such that $E(N_L|G)=\sum_{i:w_i<a} w_i/a$. The total weight of the surviving particles is aN_L. Thus, the total sum of the weights may have changed. We compensate for this using ζ in the chopping step, thus ensuring property (3).

Last key step, the chopping step, determines how the particles with weights above a are subdivided. Before chopping, the original weight w_i is adjusted to $\hat{w}_i$ using ζ , where

$\hat{w}_i = w_i + \zeta\{h_a^{\eta}(w_i)\}$ and $\{x\}:=x-\lfloor x \rfloor$. This adjustment ensures that the totals sum of the weights is conserved, property (3) and that the chopped weights are unbiased. Each large weight will receive $c^i = \lfloor h_a^{\eta}(w_i) \rfloor + m^i$ offspring. The m^i are determined by a second systematic resampling step on the fractional parts $\{h_a^h(w_i)\}$. Further, the entire algorithm will return exactly N particles. Thus property (2) is satisfied.

4. Simulation

In this section, the performance of the proposed approach is examined by simulating a bearings-only target tracking problem. The kinematics of the target can be modeled by the following nonlinear equation:

$$x_t = \begin{bmatrix} 1 & T & 0 & 0 \\ 0 & 1 & 0 & 0 \\ 0 & 0 & 1 & T \\ 0 & 0 & 0 & 1 \end{bmatrix} x_{t-1} + \begin{bmatrix} T^2/2 & 0 \\ T & 0 \\ 0 & T^2/2 \\ 0 & T \end{bmatrix} \begin{bmatrix} v_{1,t} \\ v_{2,t} \end{bmatrix} \tag{11}$$

Where the state of the target is $x_t = [x_{1,t}, x_{2,t}, x_{3,t}, x_{4,t}]$, $x_{1,t}$ and $x_{3,t}$ denote positions, $x_{3,t}$ and $x_{4,t}$ denote velocities in the x and y directions at the Cartesian coordinates, respectively; T is the sampling period; the process noise $v_{1,t} \sim N(0,\sigma_{v1}^2)$ and $v_{2,t} \sim N(0,\sigma_{v2}^2)$ are independent Gaussian white noise.

A radar located at the origin is used to measure the angle between the target and the X-axis. Hence, the measure equation is described as

$$\theta_t = \arctan(x_{3,t}/x_{1,t}) + w_t \tag{12}$$

Where w_t denote the glint noise, which is often modelled as a mixture of a Gaussian distribution with high probability and a heavy-tailed Laplacian distributed noise with low probability, i.e.,

$$p(w) = (1-\varepsilon)p_{\mathrm{G}}(w) + \varepsilon p_{\mathrm{L}}(w) \tag{13}$$

Where ε is the glint probability and the subscripts G and L stand for Gaussian and Laplacian, respectively, i.e.,

$$p_{\mathrm{L}}(w) \triangleq L(w;0,\sigma_L) \triangleq \frac{1}{2\eta} e^{-|w|/\sigma_L} \tag{14}$$

$$p_{\mathrm{G}}(w) \triangleq N(w;0,\sigma^2) \triangleq \frac{1}{\sqrt{2\pi}\sigma} e^{-w^2/2\sigma^2} \tag{15}$$

The numerical values of the parameters of the dynamic system are as follows:

Sampling period T =1 s. Process noise standard deviation $\sigma_{v1} = \sigma_{v2}$ =0.001 m/s^2. Measurement noise parameters ε =0.05, σ^2 =0.4 mrad^2, σ_L =3 mrad. Initial state of the target x_0 = [-0.05 m, 0.001 m/s, 0.7 m, -0.055 m/s]^T, associated covariance $P_{0|0}$ =diag[0.25 m2, 2.5 × 10-5 m^2/s^2, 0.09 m^2, 10^{-4} m^2/s^2]^T. The initial state estimate $\hat{x}_{0|0}$ =[0, 0, 0.4 m, -0.05 m/s]^T.

Numerous typical resampling methods are compared with chopthin resampling via simulations in terms of ESS and estimation accuracy. The typical resampling methods used for simulation include systematic, multinimial, stratified, residual. The detailed pseudocodes for selected algorithms are presented in [8].

The simulation adopts different resampling methods but the same observation at all time-steps and the same starting number of particles 2000 are adopted for each filter. All resampling algorithms maintain a constant number of particles that is equal to 2000.

The tracking scene in one trial is given in figure 1. The mean tracking position error of a total 100 trials are given in figure 2. The tracking position error is the Euler distance between the estimate and the true position of the target which is defined as

$$error = \sqrt{(x_{1,t} - \hat{x}_{1,t})^2 + (x_{3,t} - \hat{x}_{3,t})^2} \tag{16}$$

Where $[\hat{x}_{1,t}, \hat{x}_{3,t}]^{\mathrm{T}}$ is the x-y position estimate of the target at time t. Figure 3 shows the effect of different resampling methods on ESS in one trial during 100s. N_{eff} used in the traditonal resampling methods is set to 1000, η used in the chopthin is $3+\sqrt{8}$.

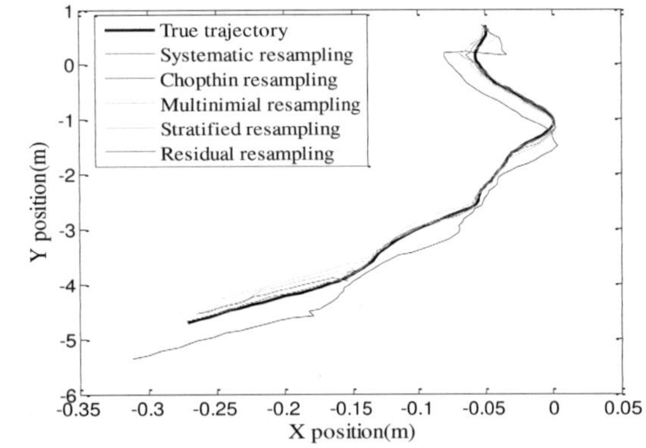

Figure 1. Target tracking scene in one trial

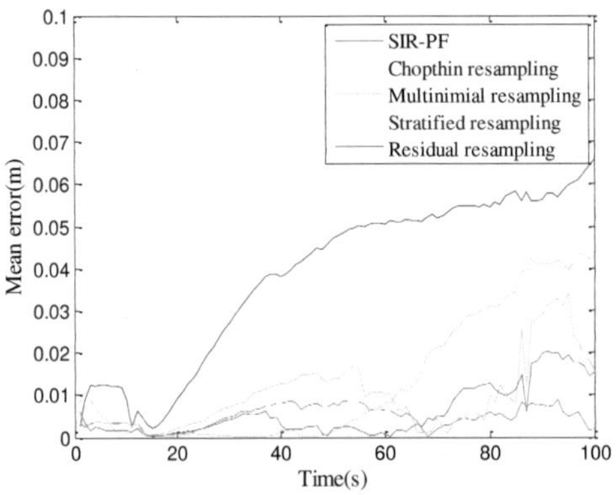

Figure. 2. Mean tracking error of 1000 trials.

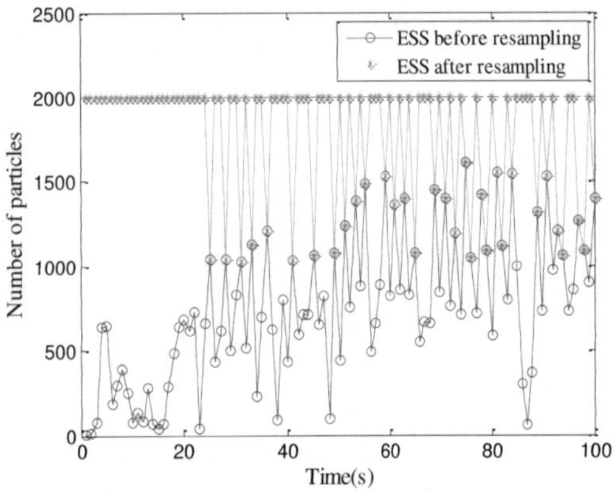

(a)

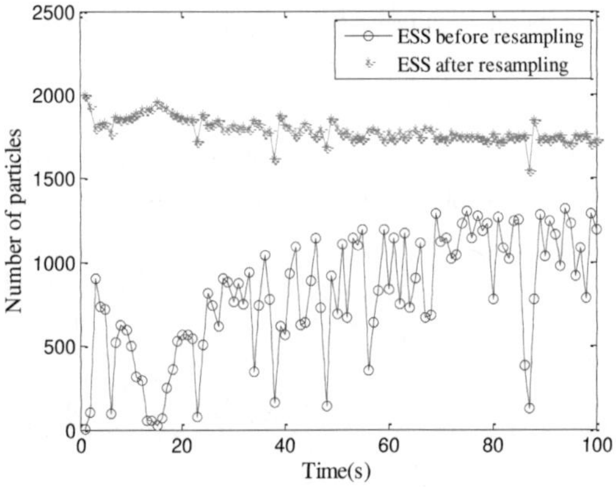

(b)

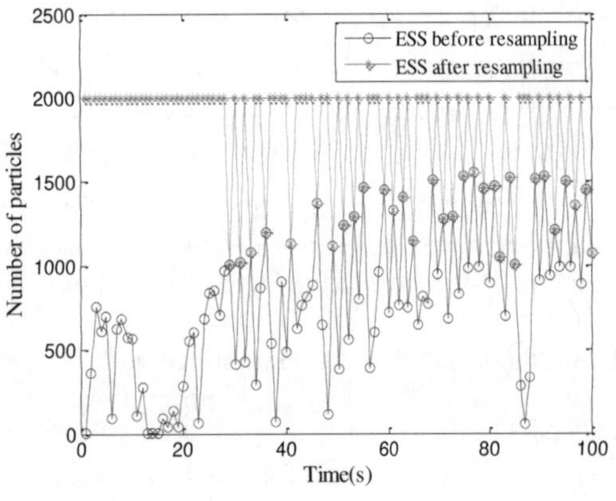

(c)

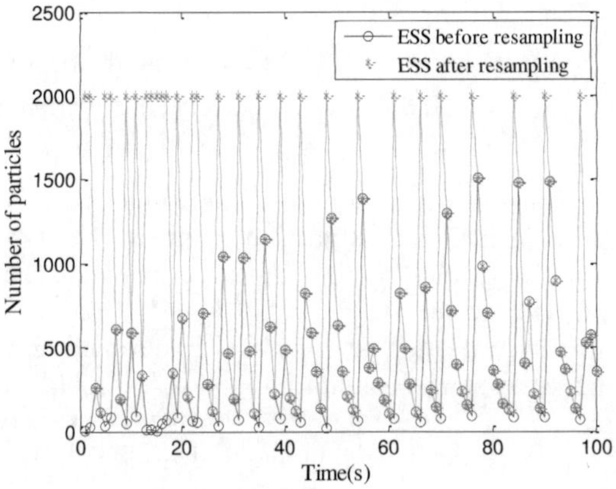

(d)

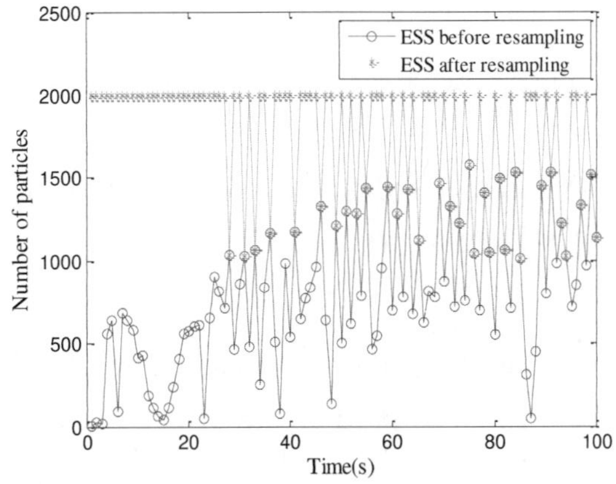

(e)

Figure 3. ESS before and after resampling for different resampling methods.

Overall, the simulation results demonstrate that, first of all, the traditional resampling methods can rarely produce very different results except systematic resampling, which is worse than other methods, second, the chopthin methods consistently outperforms traditional resampling methods, and third, chopthin resampling implicitly controls the ESS than traditional resampling methods, as resampling for traditional resampling methods only occurs if the ESS has dropped below N_{eff} .

5. Conclusion

Resampling is an essential procedure for particle filtering, which is of both theoretical and practical importance. In this paper, a new target tracking algorithm via particle filter with chopthin resampling is proposed. Simulations demonstrated that the chopthin methods consistently outperforms traditional resampling methods used in particle filters. The simulations also demonstrated that chopthin can be used at every iteration with no detrimental effects. We have also shown that imposing a bound on the ratio between weights implicitly controls the ESS.

References

[1] N. Gordon, D. Salmond, and A. Smith, "Novel approach to nonlinear/non-Gaussian Bayesian state estimation," IEE Proc., F Radar Signal Process., vol.140, no.2, pp. 107-193, 1993

[2] G. Kitagawa, "Monte Carlo filter and smoother and non-Gaussian nonlinear state space models," J.Comput.Graph.Stat., vol. 5,no. 1,pp. 1-25,1996

[3] J. Carpenter, P. Clifford, and P. Fearnhead, "An improved particle filter for nonlinear problems," IEE Proc., Radar Sonar Navigat., vol. 146, no. 1, pp. 2–7, 1999.

[4] E. R. Beadle and P. M. Djuric′, "A fast-weighted Bayesian bootstrap filter for nonlinear model state estimation," IEEE Trans. Aerosp. Electron. Syst., vol. 33, no. 1, pp. 338–343, 1997.

[5] Gandy, D.H. Lau,"The chopthin algorithm for resampling", IEEE Transactions on Signal Processing, 2016,64(16):4273-4281

[6] D. B. Rubin, "A noniterative sampling/importance resampling alternative to the data augmentation algorithm for creating a few imputations when fractions of missing information are modest: The SIR algorithm," J. Amer. Statist. Assoc., vol. 82, no. 398, pp. 543–546, 1987.

[7] Kong, J. S. Liu, and W. H. Wong, "Sequential imputations and Bayesian missing data problems," J. Amer. Statist. Assoc., vol. 9, no. 425, pp. 278–288, 1994.

[8] Li, T., Bolic,M., and Djuric, P.M., 2015. Resampling methods for particle filtering: classification, implementation, and strategies. IEEE Signal Processing, 2015, 32(3):70-86

IOP Conf. Series: Materials Science and Engineering **435** (2018) 012039 doi:10.1088/1757-899X/435/1/012039

Weighted-Diffusion Distance Measurement Ranking for Pedestrian Re-Identification

Hua Han[1,*], Yujin Zhang[1] and Chunyuan Wang[1]

[1]School of Electronic and Electrical Engineering, Shanghai University of Engineering Science, Shanghai 201620, P. R. China

[*]2070967@mail.dhu.edu.cn

Abstract. Pedestrian re-identification (Re-ID) is a very difficult problem due to complex and changeable environment. Although we can extract discriminative features, simple metric algorithms tend to yield their unreliable matching results. In order to achieve higher matching rate, we propose a more reliable histogram comparison method called weighted-diffusion distance. When calculating the cumulative difference between spatial vector histograms on the basis of high-dimensional diffusion distance, lacking the representation of differences between internal elements results in less desired accuracy. Thus, greater weight should be given to those of the corresponding element values that are same or similar. After these operations, the proposed weighted-diffusion distance is computed. Then we use it to complete multi-feature fusion, and establish a similarity measurement function, which is used to measure the similarity between two pedestrians. At last, pedestrian Re-ID experiments with different cameras are performed and the results show that the proposed method can achieve much better results than some known methods.

1. Introduction

Pedestrian re-identification (Re-ID) is an important issue in multi-camera tracking applications [1-2]. Because of highly complex environment characterized by such as different illuminations, pedestrian gaits, and scales occlusion, it is usually very difficult to achieve excellent Re-ID matching rate. The fundamental question is whether a pedestrian detected in different views or at different time instants can be linked to a same individual [3]. The past decade has witnessed many outstanding solutions in this field. They can be divided into two classes: feature and metric ones.

Feature-class methods, from the appearance of targets (pedestrians in this paper), select or design a discriminating feature descriptor, which has a certain intra-class invariance, and strong inter-class distinction. The features used in pedestrian Re-ID are mainly color, structure, shape, gradient, texture, key point, and area descriptor. Before feature extraction, a pre-processing process, such as foreground segmentation and human body parts detection, can be carried out. The different features extracted can usually be cascaded or combined to form a more discriminative feature. Symmetry-Driven Accumulation of Local Features (SDALF) [4] divides a human body into head, torso and legs after extracting the foreground, and uses the weighted HSV (Hue, Saturation and Value) histogram, maximally stable color region and recurrent highly-structured patches to re-identify a person. Custom Pictorial Structures (CPS) [5] complete Re-ID after body parts detection by using a human body structure. Histogram Plus Epitome (HPE) [6] uses the appearance model that is similar to that in [4]. It is proposed to condense a set of frames of the same individual into a highly informative signature.

Hybrid Spatiogram and Covariance Descriptor (HSCD) [7] combines Multi-Channel based Spatio-Histogram (MCSH) and Multi-Statistics on Pyramid of Covariance (MSPC) to complete pedestrian Re-ID. Encoded Local Descriptors Fisher Vectors (eLDFV) [8] uses a Fisher vector to encode luminance and gradient information of an image and combines it with SDALF. Structural Constraints Enhanced Feature Accumulation (SCEFA) [9] uses a hierarchical weighted HSV histogram, bidirectional matching color region feature, and Gabor texture patterns to describe the key points. Effectiveness Salience and Dense Correspondence (eSDC) [10] searches distinct individual features by using unsupervised learning methods and combines it with SDALF. Mean Riemannian Covariance Grid (MRCG) [11] uses an appearance model that relies on no body parts. It is an extension of Spatial Covariance Regions (SCR) [12] that is the covariance of a vector of eleven cues derived from equalized RGB colors. An MRCG descriptor is computed as a mean of gallery examples but only applicable to multi-shot Re-ID modalities.

Metric-class methods focus on distance measurement. The essence of metric learning is to estimate the scale or weight of each component when calculating the distance between feature vectors, and to select or highlight the strongly distinguishing features. Ensemble of Localized Features (ELF) [13] selects color and texture features via AdaBoost classifiers. Ensemble RankSVM (ERSVM) [14] integrates multiple eigenvector weights by using SVM classifiers based on relative distance ordering. In [15] a metric learning framework is used to obtain a robust Mahalanobis metric for Large Margin Nearest Neighbor with Rejection (LMNN-R). Probabilistic Relative Distance Comparison (PRDC) [16] introduces a novel comparison model. Its goal is to make the probability of a correct match pair with a smaller distance greater than that of an incorrect match pair. Local Fisher (LF) [17] carries out principal- component-analysis to reduce dimensionality of a high dimension feature, then performs dimensionality reduction mapping based on local Fisher discriminant analysis, and finally obtains very high Re-ID rate.

Compared with feature-class methods, metric-class methods have lower requirements for feature selection, and they can usually achieve better Re-ID effects. But their training time complexity and space size can be tremendous. There is an over-fitting phenomenon when training samples are insufficient. When a scene and library change, they need to be re-trained. The feature-class methods require manual search and design of relatively better features, and their performance depend on the used features. This work designs excellent features and descriptions under the idea of feature-class methods. Achieving the desired results requires us to compute weighted-diffusion distance quickly and accurately.

2. Measurement Ranking by Weighted-Diffusion Distance

Most of traditional metric algorithms for similarity measurement are optimized based on Euclidean or cosine distance. However, either distance has its shortcoming. The former cannot be restored from a high-dimensional eigenvector to a low-dimensional geometric mechanism. It is sensitive to the deformation of an image, and lacks any spatial information between pixels. The cosine distance has some disadvantages due to the large difference in position determination and insensitivity to numerical values, which can result in many errors. Diffusion distance [18], to be defined next, does not bear any mentioned shortcomings. It is different from the case where the similarity is determined by Euclidean distance. The similarity based on diffusion distance depends on the speed of diffusion and number of diffusion paths. It appears to be superior in matching pedestrians. Its principle can refer to [19]. Assume that there are two m-dimensional distributions $h_i(\mathbf{x})$ and $h_j(\mathbf{x})$, and $\mathbf{x}$ is an m-dimensional real matrix. High-dimensional diffusion distance can be defined as follows:

$$K(h_i,h_j)=\sum_{l=0}^{L}k(|d_l(\mathbf{x})|) \tag{1}$$

Where

$$d_0(\mathbf{x})=h_i(\mathbf{x})-h_j(\mathbf{x}) \ , \ l=0 \tag{2}$$

$$d_l(\mathbf{x}) = [d_{l-1}(\mathbf{x}) * \phi(\mathbf{x},\sigma)] \downarrow_{2/3}, \ (l = 1, \cdots, L) \tag{3}$$

$$\phi(\mathbf{x},t) = \frac{1}{(2\pi)^{m/2} t} e^{-\frac{\mathbf{x}^T\mathbf{x}}{2t^2}} \tag{4}$$

In (3), $\downarrow$ denotes downsampling. If the sampling value is too small, some useful information may be missing; If the sampling value is too large, it reduces computational efficiency. Based on the above considerations, the size of this sample is set to the empirical value $\downarrow_{2/3}$, that is, two thirds of the sample value. L is the number of Gauss Pyramid layers and σ is the standard deviation in the Gauss function $\phi(\mathbf{x},\sigma)$. Using L_1 norm, (1) can be simplified to:

$$K(h_i,h_j) = \sum_{l=0}^{L} |d_l(\mathbf{x})| \tag{5}$$

2.1. *Weighted-Diffusion Distance*

From (3), the amount of computation of downsampling in d_l exponentially reduces; Secondly, a Gaussian filter convolution operation is a linear one, which has obvious advantages in terms of computational efficiency. Despite this, from (1)-(2), when representing a cumulative difference between spatial vector histograms $h_i(\mathbf{x})$ and $h_j(\mathbf{x})$, the optimized diffusion distance misses the differences between internal individual elements. Although (5) can be applied directly to the similarity measurement, it ignores the influence between the corresponding elements of a spatial vector on similarity, thus resulting in lower accuracy. Since each element in a different histogram represents a different characteristic attribute, for example, the z-th element in a wHSV histogram represents the total number of pixels in the z-th region or represents the color value in the image. Therefore, if similarity measurement is carried out, it is necessary to consider not only the effects of cumulative differences, but also the effects of similarity between two corresponding elements. Assume that $h_1(\mathbf{x}) = [3,7,10]$, $h_2(\mathbf{x}) = [1,1,5]$, and $h_3(\mathbf{x}) = [3,6,2]$. To simplify the calculation, let $\phi(\mathbf{x},\sigma) = 1$. Without using downsampling, we have $K(h_1,h_2) = \sqrt{65} = K(h_1,h_3)$. It is not possible to distinguish which two histograms are more similar. Through observation we can find that $h_1(\mathbf{x})$ and $h_3(\mathbf{x})$ have more of the same or similar elements than $h_1(\mathbf{x})$ and $h_2(\mathbf{x})$. Intuitively, $h_1(\mathbf{x})$ and $h_3(\mathbf{x})$ should be more similar than that $h_1(\mathbf{x})$ and $h_2(\mathbf{x})$ are.

Through the above analysis, in the calculation of the distance between two space vector histograms, we should not give the same weight to all internal elements, but greater weight to those same or similar corresponding elements. Therefore, in this work, we re-optimize the diffusion distance based on the former one optimized by Gaussian pyramid, and propose a new distance concept called weighted-diffusion distance. Hence, (2) is rewritten as

$$d_0(\mathbf{x}) = \sqrt{\sum_{p=0}^{n} (h_{ip}(\mathbf{x}) - h_{jp}(\mathbf{x}))^2} \tag{6}$$

Where *ip* and *jp* represent the *p*th element in histograms *i* and *j*, respectively. $p \in \{0,1,\cdots,n\}$. Equations (2) and (6) represent similar meanings that the distance between two histogram differences is equivalent to the accumulation of the corresponding element differences in each pair of histograms. Then we rewrite (6) as:

$$d_0(\mathbf{x}) = \sqrt{\sum_{p=0}^{n} \frac{r_p}{R} (h_{ip}(\mathbf{x}) - h_{jp}(\mathbf{x}))^2} \tag{7}$$

$$r_p = e^{-|h_{ip} - h_{jp}|/\sigma}$$

Where R is a normalization factor, $R = \Sigma r_p$, r_p is the weight, σ is the adjustment factor. According to (10) given the prior $h_1(\mathbf{x})$ - $h_3(\mathbf{x})$, we have $K(h_1,h_2) = 2.337$ and $K(h_1,h_3)$ =1.025 via (5). Intuitively, $h_1(\mathbf{x})$

and $h_3(\mathbf{x})$ should be more similar than $h_1(\mathbf{x})$ and $h_2(\mathbf{x})$. It can be seen that the weighted-diffusion distance based on the quadratic optimization of the proposed diffusion distance enables the similarity measurement to exhibit better discriminability. Next we use the weighted-diffusion distance to measure and rank the distance among features. The greater distance indicates greater difference among individuals. On the contrary, the smaller distance, the smaller difference. Therefore, we can choose the foremost ranked candidate to re-identify him/her as the same individual.

2.2. Multi-Feature Fusion Based on Weighted-Diffusion Distance

In our work, the features we used include local RGB, local wHSV, Histogram of Oriented Gradient (HOG) [20], and global Maximally Stable Color Regions (MSCR) [21]. The similarity measurement function constructed by any of the above mentioned features alone fail to achieve desired matching effects. Therefore, we propose a multi-feature fusion method based on weighted-diffusion distance to complete pedestrian Re-ID. Extract the histograms of the above features, and then calculate the corresponding weighted-diffusion distance. In order to prevent big difference in the weighted-diffusion distance calculated from different features. The calculated weighted-diffusion distance of each feature will be normalized. In pedestrian Re-ID, there are usually two sets of images: target I_A to be re-identified and candidate target I_B.

We use I_A and I_B to denote two image sets, calculate the weighted-diffusion distance based on the above mentioned features and normalize them. Use (5) and (7) to calculate the matching distance $d(I_A, I_B)$ between the pedestrian targets of I_A and I_B:

$$d(I_A, I_B) = \alpha \cdot d_{MSCR}(I_A, I_B) + \beta \cdot d_{RGB}(I_A, I_B) + \gamma \cdot d_{wHSV}(I_A, I_B) + \lambda \cdot d_{HOG}(I_A, I_B) \quad (8)$$

Where α, β, γ, and λ represent the weights of the different feature histograms, respectively, satisfying $\alpha+\beta+\gamma+\lambda=1$. Rank all matching distances, and choose the candidate target with the smallest matching distance as the final re-identified one.

3. Experiments and Results

VIPeR is a challenging pedestrian Re-ID database that has been widely used for benchmark evaluation [2]. It contains 632 pairs of person images, captured by a pair of cameras in an outdoor environment. Some pairs contain large variations in background, illumination, and viewpoint. All are scaled to 128×48 pixels. A widely adopted experimental protocol is to randomly divide all pairs into one half for training and another half for test, and repeat the procedure several times to obtain average performance. Although our method has achieved a desired result, we have to do some specific numerical comparison experiments to test its performance. Gray et al. [22] propose a standard protocol to compare results by using cumulative match curve (CMC) by introducing the VIPeR dataset for Re-ID. Next, we use CMC to evaluate the performance of our method.

3.1. Statistical Evaluation of Experimental Results

In order to verify the performance of the proposed distance in pedestrian Re-ID, we have done many experiments based on VIPeR. In the experiment, 50 pairs of pedestrian targets are randomly selected. Five trials are conducted. Finally, an average CMC curve is given. The results are shown in Figure 1.

Figures 1(a)-(e) show the CMC curves generated after the first to fifth experiments, where the horizontal axis in Figure 1 represents a rank score, and the vertical axis is matching rate. The matching rate of the first point on the average curve can be up to 63.6%, the second one 72.4%, and the one 77.6%. The results are in line with our expectations. Yet we need to know how our methods performs in comparison with the existing algorithms. The comparison with the state-of-the-art is given next.

3.2. Comparison with the State-of-the-Art

We have done many comparison tests with the state-of-the-art methods. These experimental comparisons are done for the algorithms in both feature-class and metric-class.

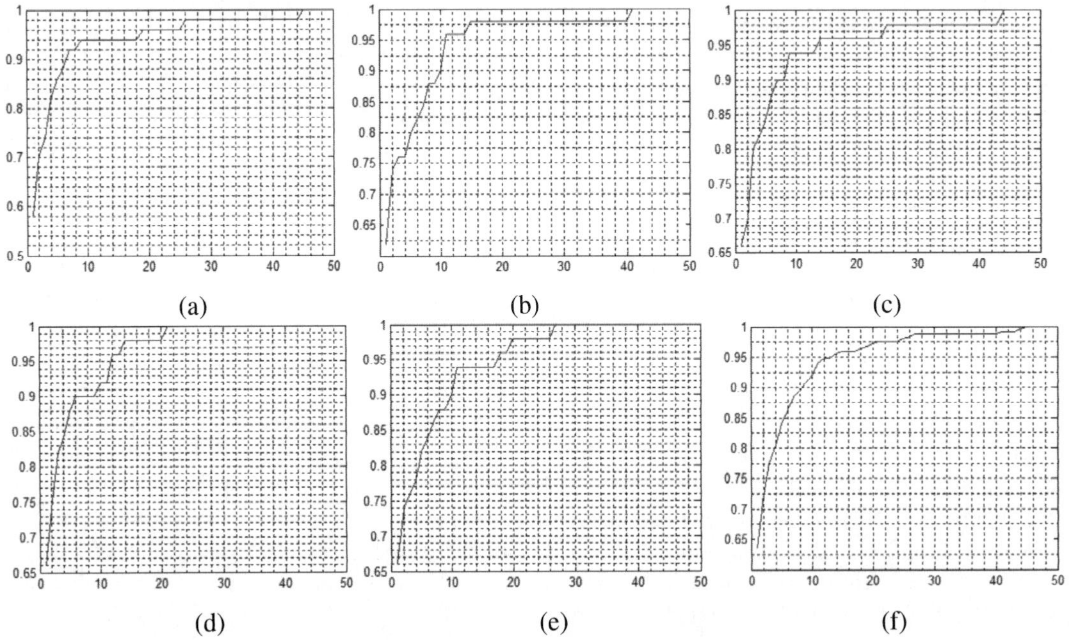

Figure 1. CMC curves of experiments based on VIPeR. (a) CMC curve generated after the first experiment, (b) CMC curve generated after the second experiment, (c) CMC curve generated after the third experiment, (d) CMC curve generated after the fourth experiment, (e) CMC curve generated after the fifth experiment, and (f) The average CMC curve after 5 trials.

1) Comparison with feature-class algorithms

In this paper, we compare the proposed with HSCD [7], eSDC [10], SCEFA [9], eLDFV [8], CPS [5], and SDALF [4]. The experimental results can be found in Figure 2.

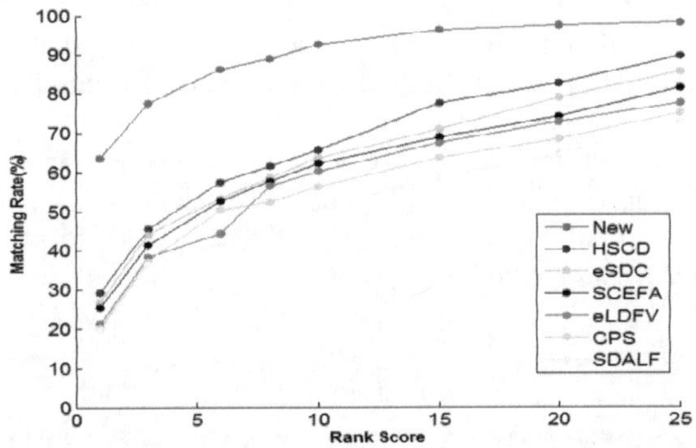

Figure 2. CMC curves of comparison with feature-class algorithms

From Figure 2, the matching rate of SDALF is relatively low. After dividing the pedestrian body into three parts: head, torso, and leg, SDALF combines wHSV, MSCR and repetitive texture block to describe the pedestrian target. Finally, the Euclidean distance is used to measure the similarity. The advantage of this widely used algorithm is the better feature selection, but it uses the traditional Euclidean distance for the measurement part. This is the reason of its lower matching rate. CPS has the advantage of being able to detect various parts of a body according to the pedestrian body structure. It can extract the foreground features better. eLDFV, combined with SDALF, uses a Fisher vector to encode the brightness and gradient information of an image. Its matching rate at the first point in its CMC curve can reach 21.42%. But its metric algorithm needs to be improved. SCEFA uses key point detection based on Gabor texture and two-way color feature matching. The use of multiple features is its biggest advantage. eSDC is also a combination of SDALF and unsupervised learning methods to find better features of an individual. But it lacks the consideration of how to improve the metric algorithm. HSCD combines two excellent statistical descriptions. They are spatial histogram and regional covariance. A variety of mutually complementary statistical vectors are extracted from several statistical areas. This method is the best among the other feature-class algorithms. Its matching rate at the first point can reach 29.32%. Nevertheless, the matching rate of the above algorithms is still lower than ours. Based on multi-feature fusion and diffusion distance similarity measurement, the first point matching rate of our method can reach 63.6%, which is much higher than the best one (29.32%) from all the others, the second one is 72.4%, and the third one is 77.6%, which all beat the state-of-the-art methods. This represents a very significant improvement in this field. Next, we compare the proposed method with algorithms of the metric-class.

2) Comparison with metric-class algorithms

Our approach well outperforms the feature-class algorithms. Now we compare it with the state-of-art metric-class algorithms. They are ELF [13], ERSVM [14], PRDC [16], LMNN-R [15], COS [23], and Histogram Intersection (HI) [24]. The experimental results can be found in Figure 3.

From Figure 3, the matching rate of ELF is relatively low. It uses AdaBoost classifiers to select color and texture features. This method makes full use of the selection of classifiers, but it does not improve greatly in terms of matching rate. ERSVM has a certain similarity to PRDC. They use SVM classifiers, which are sorted by relative distance, to integrate multiple weights of feature vectors. This method has obvious advantages in the measurement algorithm, but it needs improvement in terms of feature selection. LMNN-R also uses PRDC to do relative distance comparison. It uses an LMNN classifier. But, in the feature design, it needs to be improved. In the metric-class, the best performance algorithm is HI. It is a kind of kernel functions based on an implicit correspondence relation. It solves the classification problem of unordered, variable-length vectors. It is positive definite. Because of its advantages, the matching rate at the first point is as high as 47.38% in pedestrian Re-ID. Therefore, under the same conditions, the metric-class methods have some certain advantages in matching rate compared with the mentioned feature-class methods. From Figure 3, our algorithm is able to beat them in terms of matching rate.

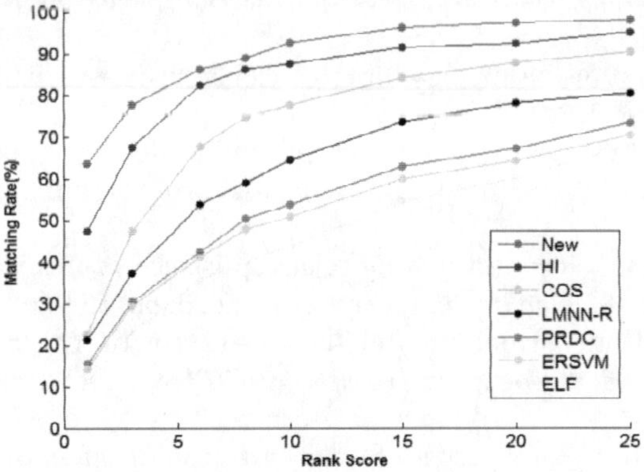

Figure 3. CMC curves of comparison with metric-class algorithms.

In summary, our algorithm is clearly outstanding in comparison with both feature-class and metric-class methods. This is because of proper features' fusion and the advantage of the weighted-diffusion distance in similarity measurement.

4. Conclusion

This paper has presented a new method to measure the similarity among histograms of pedestrians. This resultant measurement is called weighted-diffusion distance. It can be used to perform histogram comparison well. In order to obtain higher matching rate, we assign different internal elements of the histograms with different weights in the defined high-dimensional diffusion distance. In order to obtain more discriminative features, we extract multiple features of pedestrians. Then we use the proposed distance to do multi-feature fusion, and establish the similarity measurement function, which is in turn used to measure the similarity of two pedestrians. Experimental results show that the proposed method can achieve drastically better results than its six peers, thereby representing one of major breakthrough results in the field.

Acknowledgments
This work was supported in part by the National Nature Science Foundation of China (No. 61305014, 61473077, 61473078), China Scholarship Council (No. 201508310033), "Chen Guang" project supported by Shanghai Municipal Education Commission and Shanghai Education Development Foundation (No. 13CG60).

References
[1] D. Figueira, M. Taiana, J. C. Nascimento, and A. Bernardino 2016 A Window-Based Classifier for Automatic Video-Based Reidentification *IEEE Transactions on Systems, Man & Cybernetics: Systems*, vol. 46, no. 12, pp. 1736-1747.
[2] D. Tao, L. Jin. Y. Wang., and X. Li, "Person reidentification by minimum classification error-based KISS metric learning," *IEEE Transactions on Cybernetics*, vol. 45, no. 2, pp. 242-252, Feb, 2015.
[3] T. Zhao and R. Nevatia 2004 Tracking multiple humans in complex situations, *IEEE Transactions on Pattern Analysis and Machine Intelligence*, vol. 26, no. 9. pp. 1208 -1221.
[4] N. Martinel and C. Micheloni 2012Re-identify people in wide area camera network, *2012 IEEE Computer Society Conference on Computer Vision and Pattern Recognition Workshops*, Providence, USA, pp. 31-36.
[5] I. Kviatkovsky, A. Adam, and E. Rivlin 2013 Color invariants for person reidentification, *IEEE Transactions on Pattern Analysis and Machine Intelligence*, vol. 35, no.7, pp. 1622-1634.
[6] L. Kai, Z. Zhao, and A. Cai 2015 Datum-adaptive local metric learning for person re-identification, *IEEE Signal Processing Letters*, vol. 22, no. 9, pp. 1457-1461.
[7] Y. Cai, W. Chen, K. Huang, and T. Tan 2007 Continuously tracking objects across multiple widely separated cameras, *8th Asian Conference on Computer Vision*, Tokyo, Japan, pp. 843-852.
[8] C. Motamed and O. Wallart 2007 A temporal fusion strategy for cross-camera data association, *Pattern Recognition Letters*, vol. 28, no.2, pp. 233-245.
[9] S. F. Tahir and A. Cavallaro 2014 Cost-effective features for reidentification in camera networks, *IEEE Transactions on Circuits and System for Video Technology*, vol. 24, no. 8, pp. 1362-1374.
[10] W.-S. Zheng, S. Gong, and T. Xiang 2013 Reidentification by relative distance comparison, *IEEE Transactions on Pattern Analysis and Machine Intelligence*, vol. 35, no. 3, pp. 653-668.
[11] N. Martinel and C. Micheloni 2012 Re-identify people in wide area camera network, *IEEE Conference on Computer Vision and Pattern Recognition Workshops(CVPRW)*, Providence, USA, pp. 31-36.
[12] I. Kviatkovsky, A. Adam, and E. Rivlin 2013 Color invariants for person reidentification, *IEEE Transaction on Pattern Analysis and Machine Intelligence*, vol. 35, no. 7, pp. 1622-1634.
[13] B. Wang, G. Wang, K. L. Chan, and L. Wang 2014 Tracklet association with online target-specific metric learning, *Proceedings of the IEEE Computer Society Conference on Computer Vision and Pattern Recognition (CVPR)*, Columbus, OH, USA, pp. 1234-1241.
[14] A. Li, L. Liu, S. Liu, S. Yan, and K. Wang 2015 Clothing attributes assisted person reidentification, *IEEE Transactions on Circuits and Systems for Video Technology*, vol. 25, no. 5, pp. 869-878.
[15] X. Wang, W.-S. Zheng, X. Li, and J. Zhang 2016 Cross-Scenario Transfer Person Reidentification, *IEEE Transactions on Circuits and Systems for Video Technology*, vol. 26, no. 8, pp. 1447-1460.

[16] X. Liu, H. Wang, J. Wang, and X. Ma 2017 Person re-identification by multiple instance metric learning with impostor rejection, *Pattern Recognition*, vol. 67, pp. 287-298.

[17] J. Lin 1991 Divergence measures based on the Shannon entropy, *IEEE Trans. on Information Theory*, vol. 37, no. 1, pp. 145-151.

[18] H. Ling and K. Okada 2006 Diffusion distance for histogram comparison, *Proceedings of the IEEE Computer Society Conference on Computer Vision and Pattern Recognition (CVPR)*, New York, pp. 246-253.

[19] Q. J. Wang and R. B. Zhang 2008 LPP-HOG: A new local image descriptor for fast human detection, *IEEE International Symposium on Knowledge Acquisition and Modeling Workshop*, Wuhan, China, pp. 640-643.

[20] M.-Y. Zeng, Z.-M. Wu, C. Tian, Y. Fu, and F.-R. Jie 2014 Fusing appearance statistical features for person re-identification, *Journal of Electronics & Information Tech.*, vol. 36, no. 8, pp. 1844-1851.

[21] A. Mignon and F. Jurie 2012 PCCA: A new approach for distance learning from sparse pairwise constraints," *IEEE Conf. on Computer Vision and Pattern Recognition (CVPR)*, Providence, USA, pp. 2666-2672.

[22] M. Kostinger, M. Hirzer, P. Wohlhart, P.M. Roth, and H. Bischof 2012 Large scale metric learning from equivalence constraints, *IEEE Conference on Computer Vision and Pattern Recognition (CVPR)*, Providence, USA, pp. 2288-2295.

[23] H. Zhao, et al. 2015 Single Image Fog Removal Based on Local Extrema, *IEEE/CAA Journal of Automatica Sinica*, 2(2), 158-165.

[24] X. Feng, et al. 2016 Coupled Cross-correlation Neural Network Algorithm for Principal Singular Triplet Extraction of a Cross-covariance Matrix, IEEE/CAA Journal of Automatica Sinica, 3(2), 147-156.

IOP Conf. Series: Materials Science and Engineering **435** (2018) 012040 doi:10.1088/1757-899X/435/1/012040

A Camera Calibration Method with Large Oblique Angle Machine Vision

Meng Chen[1], YueLei He[1], PeiYi Gong[2], ZaiWei Li[1] and HongYao Lu[1]

[1] School of Urban Transportation, Shanghai University of Engineering Science, Shanghai 201620, China
[2] Shanghai Xintie Mechanical & Electrical Technology co., LTD, Shanghai 201100, China

Meng Chen: cm109544@163.com,YueLei He: hyldoc@163.com

Abstract. In order to ensure the safety of the subway when it stops at the platform, it is necessary to detect the limits of the subway platform. However, the currently used station limit detection tools have large detection errors and low efficiency, and cannot meet the detection requirements. In this paper, a non-contact platform limit detection system based on machine vision is studied, and a new set of comprehensive calibration methods is proposed for the calibration of the camera in the system at large angles. After collecting the images of large multi array points set by camera, the effective calibration area is selected and the reference circle center is extracted from the connected domain and centroid center algorithm. After collecting the images of large multi array points set by camera, the effective calibration area is selected and the reference circle center is extracted from the connected domain and centroid center algorithm. The base circle center is used as the reference point to carry out perspective transformation, and the change matrix is obtained and tested. Finally, the maximum error of pixel calibration is 1.639pixel. It is proved by laboratory and field experiments that the measurement system of this method can obtain the contour geometry of the platform section accurately, and the gauge accuracy is within ±1mm after repeated measurements, and the accuracy of platform height and platform distance can be within ±3mm.

1. Introduction

Subway platform clearance is the space required to ensure the safety of vehicles running, stopping and passenger riding in the platform area[1]. Therefore, the penetration limit detection of platform clearance is an important part of metro vehicle outage operation. In recent years, with the rise of machine vision technology, non-contact measurement based on machine vision has gradually replaced the traditional tool measurement in the field of subway limit detection field. In high-precision vision measurement, the accuracy of the system will directly affect the results of the system measurement. Therefore, the calibration of system accuracy is a crucial step in the construction of the system.

At present, the commonly used plane calibration boards are chessboard format [2], black and white format[3], black and white dot array[4].These calibration plates are regularly distributed with some characteristic points. Among them, there is a classical Zhang Zhengyou calibration method for checkerboard[5].The use of Harrris corner detection to find the relationship between the world coordinates and the image coordinates reduces the complexity of the algorithm, and the calibration plate is easy to produce and has been widely used. However, this method is based on multi-vision

IOP Conf. Series: Materials Science and Engineering **435** (2018) 012040 doi:10.1088/1757-899X/435/1/012040

vision and multiple checkerboard images for calibration. It is not suitable for camera calibration based on monocular vision and single image. Because of its advantages of convenient recognition and high positioning accuracy, circular marking points are widely used in the field of stereo vision. At present, there are many methods for extracting the center point of the dot array calibration plate. The literature[6-8] uses the ellipse fitting method to extract and mark the center of the ellipse image; In literature[9], the 3 ellipse boundary information on the calibration plate is searched through the algorithm and the public tangent line of the ellipse is searched to extract the center point. In the above calibration algorithm, the angle between the camera and the target is small, the acquired target image is often not greatly distorted, and the feature points are easy to extract, which is not suitable for camera calibration of the system.

Because of the particularity of the rail platform, the camera of the system is to collect the profile profile from a close range and wide perspective, therefore, when the indoor scene is simulated, the target image of the system camera is distorted greatly. This paper presents a high-angle limit detection camera calibration method. Based on a large-scale multi-array calibration plate, an effective calibration area is selected and preprocessed, and then the center of circle algorithm is used to identify the reference circle in the positioning calibration plate and the center coordinates are extracted. The reference circle feature point transforms the image as a reference point for image transformation to complete camera calibration. Finally, the indoor and outdoor test verification of system accuracy.

2. Calibration model establishment area selection

2.1. Establish a calibration model

According to the triangulation principle, the camera calibration model with large inclination angle can be established, as shown in figure 1. When the camera is calibrated, the structured light band emitted by the laser is in the same cross-section as the large multi-array calibration plate. At the same time, adjusting the light and the various components on the instrument to ensure that the parameters in the acquisition process remain unchanged, so as to ensure the consistency of the calibration board images collected multiple times, to avoid the effects of instrument operation, lighting and other external factors on the system calibration accuracy.

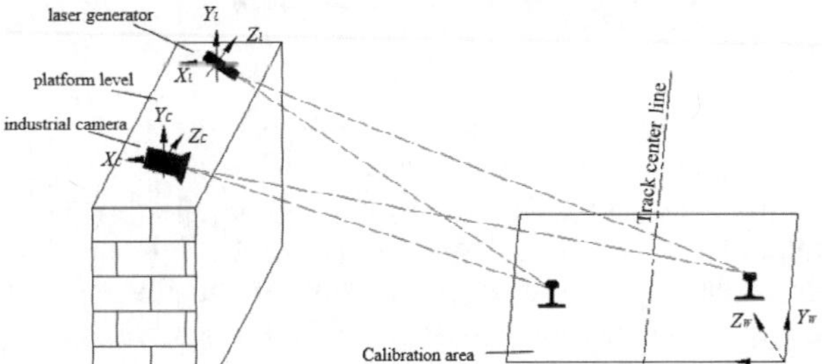

Figure 1. Camera calibration model

2.2 Determine the calibration area

The calibration plate, as shown in figure 2, is shown in Figure 2. The overall size is 2400*2400mm, and the calibration board is square array with 3600 reference circles with a diameter of 20mm. The adjacent base circle center distance is 40mm, and the overall machining precision of the calibration board is more than 0.02mm. In addition, the reference circle is set to black, and the metal background is set to silver-white, so that there is a clear gray difference between the two, and the quality of the divided reference circle is improved.

Figure 2. Calibration plate

Figure 3. Baseline in the effective calibration area

Determine the effective calibration area based on the station profile actually shot by the system camera as shown in the white reference circle in figure 3. Placing some white interference on the selected calibration plate area to simulate on-site lighting conditions and then image processing. The selection of this area avoids the influence of changes in the edge position on the calibrated pixel accuracy during the calibration of the circle center distance of the reference circle.

3. Reference circle center extraction and image transformation

3.1 demarcated area preprocessing

For the selected calibration area, image preprocessing is first performed using Gaussian filtering [10] for smoothing, then using local thresholding [11] for binarization and finally using Sobel operator [12] for edge extraction. The processing result is shown in figure 4.

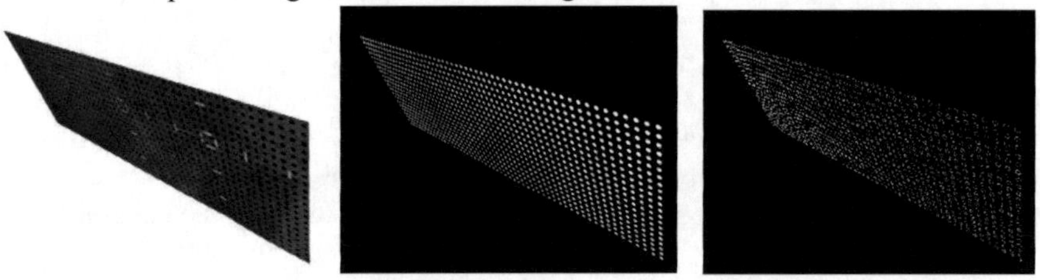

(a) Gaussian filter (b) Binarization (c)Edge extraction

Figure 4. Image Preprocessing

3.2 Center extraction

The connected area generally refers to an image area consisting of foreground pixel points with the same pixel value and adjacent positions in the image.

Commonly used adjacencies include two kinds: 4 neighborhoods and 8 neighborhoods. 4 neighborhoods for a total of 4 points. If there are pixels, then the four-neighborhood is the top, bottom, left, and right adjacent points of the point. In addition to the four-neighborhood point, the eight-neighborhood includes four diagonal points. Based on the 8-neighborhood labeling algorithm and the centroid localization algorithm, this paper proposes a centroid algorithm and combines the large multi-array calibration plate used to achieve the extraction of the reference circle center. The specific algorithm steps are as follows:

Step1: Read the calibration image after extracting the edges, and traverse the entire image from top to bottom and from left to right.

Step2: Give each pixel a label and assign the same label to the set of pixels in the same connected region.

Step3: The resulting connected area is marked with a circumscribed rectangle until all connected fields are marked.

Step4: Judge whether there is a non-reference circle in the connected area of the marker. If yes, set the connected area size and delete the non-reference circle connected area; if not, enter Step5;

Step5: Based on the principle of graph centroid, the obtained circle-connected circle is marked and encoded by the circle center, and the coordinate data file is output.

Figure 5 is the result of the center of the circle.

(a) Bounding rectangle positioning (b) Center extraction and encoding

Figure 5. The center of circle results

3.3 Image transformation

The large multi-plane plane calibration plate used in this paper establishes a two-dimensional world coordinate system. Set the coordinate point of any point T on the calibration plate as (X,Y) the image coordinate corresponding to the projection is $T'(x,y)$,The coordinate system relationship between the object point and the image point is shown in figure 6.

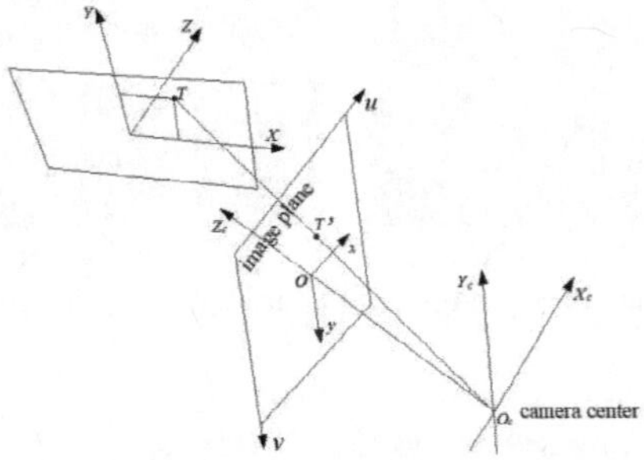

Figure 6. Coordinates of object points and image points

Perspective transformation [13] essentially projects an image onto a new viewing plane. The general formula is:

$$[x \quad y \quad z]=[X \quad Y \quad Z]\begin{bmatrix} a_{11} & a_{12} & a_{13} \\ a_{21} & a_{22} & a_{23} \\ a_{31} & a_{32} & a_{33} \end{bmatrix} \tag{1}$$

$z=a_{13}X+a_{23}Y+a_{33}Z$, $[x, \quad y, \quad z]$ is a homogeneous coordinate,the transformed image pixel coordinates are: $\begin{bmatrix} \frac{x}{z}, & \frac{y}{z}, & 1 \end{bmatrix}$. Transformation matrix $\begin{bmatrix} a_{11} & a_{12} & a_{13} \\ a_{21} & a_{22} & a_{23} \\ a_{31} & a_{32} & a_{33} \end{bmatrix}$ can be split into four parts: $\begin{bmatrix} a_{11} & a_{12} \\ a_{13} & a_{14} \end{bmatrix}$ is used for image linear transformation, $[a_{31} \quad a_{32}]$ is used for image translation, $[a_{13} \quad a_{23}]^T$ is used to generate an image perspective transformation. The transformed coordinate value (u,v) can be obtained by the following formula:

$$X=\frac{x}{z}=\frac{a_{11}X+a_{21}Y+a_{31}}{a_{13}X+a_{23}Y+a_{33}} \tag{2}$$

$$Y=\frac{y}{z}=\frac{a_{12}X+a_{22}Y+a_{32}}{a_{13}X+a_{23}Y+a_{33}} \tag{3}$$

Selecting the four corner points of the outer circle reference circle in the calibration area: $P_1(x_1,y_1)$, $P_{192}(x_{192},y_{192})$, $P_{1158}(x_{1158},y_{1158})$, $P_{1176}(x_{1176},y_{1176})$, the coordinates in the corresponding world coordinate system are: $P_1'(x_1,y_1)$, $P_{192}'(x_{192},y_{192})$, $P_{1158}'(x_{1158},y_{1158})$, $P_{1176}'(x_{1176},y_{1176})$. According to equations (2), (3) and written in a matrix multiplied form:

$$\begin{bmatrix} x_1 & y_1 & 1 & 0 & 0 & 0 & -X_1x_1 & -X_1y_1 \\ 0 & 0 & 0 & x_1 & y_1 & 1 & -Y_1x_1 & -Y_1y_1 \\ x_{192} & y_{192} & 1 & 0 & 0 & 0 & -X_{192}x_{192} & -X_{192}y_{192} \\ 0 & 0 & 0 & x_{192} & y_{192} & 1 & -Y_{192}x_{192} & -Y_{192}y_{192} \\ x_{1158} & y_{1158} & 1 & 0 & 0 & 0 & -X_{1158}x_{1158} & -X_{1158}y_{1158} \\ 0 & 0 & 0 & x_{1158} & y_{1158} & 1 & -Y_{1158}x_{1158} & -Y_{1158}y_{1158} \\ x_{1176} & y_{1176} & 1 & 0 & 0 & 0 & -X_{1176}x_{1176} & -X_{1176}y_{1176} \\ 0 & 0 & 0 & x_{1176} & y_{1176} & 1 & -Y_{1176}x_{1176} & -Y_{1176}y_{1176} \end{bmatrix} \begin{bmatrix} a_{11} \\ a_{21} \\ a_{31} \\ a_{12} \\ a_{22} \\ a_{32} \\ a_{13} \\ a_{23} \end{bmatrix} = \begin{bmatrix} X_1 \\ Y_1 \\ X_{192} \\ Y_{192} \\ X_{1158} \\ Y_{1158} \\ X_{1176} \\ Y_{1176} \end{bmatrix} \quad (4)$$

(4) is marked as: $M * A = Q$, so the transformation matrix is $A = M^{-1} * Q$, According to the above operation steps, an image perspective transformation matrix M is obtained:

$$M = \begin{bmatrix} 1.024 & -0.223 & -35.519 \\ -1.772 & 2.842 & 125.307 \\ -0.001 & 0.001 & 1 \end{bmatrix} \quad (5)$$

The image and center extraction of the transformed calibration board are shown in figure 7.

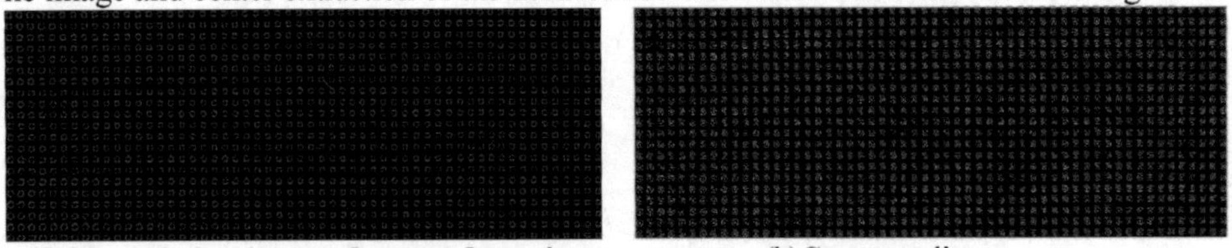

(a) Calibrated plate image after transformation (b)Center coding

Figure 7. Image results after transformation

4. Experimental verification

The accuracy of the calibration was verified from two aspects. On the one hand, the difference between the theoretical coordinate and the actual coordinate of the center of the reference circle in the transformed calibration board image was used. On the other hand, the system was used to detect the limit indicator of the platform in the field.

4.1 Calibration accuracy verification

The theoretical circle of the reference circle center in the corrected calibration area is recorded as C_i, the actual coordinates are recorded as $C_i^{'}$,The absolute error value of the two is calculated by Δ_i:

$$\Delta_i = \left| C_i - C'_i \right| , \text{ where } i = 1, 2, ..., 1176 \quad (6)$$

According to the result of equation (6), the absolute difference between the corrected center of the reference circle theory and the actual coordinates can be obtained as shown in figure 8.

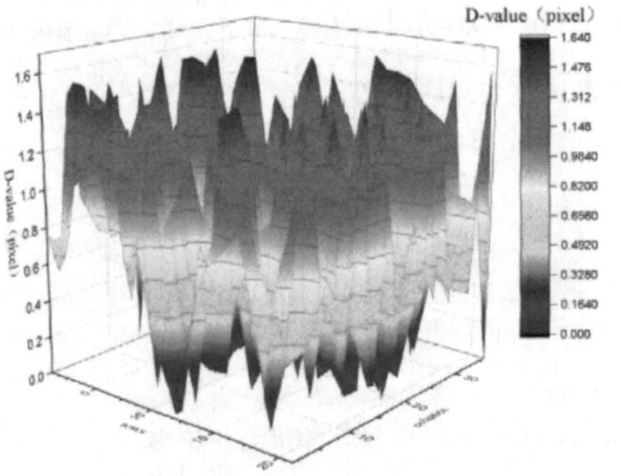

Figure 8. The absolute error distribution of the center coordinates

IOP Conf. Series: Materials Science and Engineering **435** (2018) 012040 doi:10.1088/1757-899X/435/1/012040

The maximum absolute coordinate difference is $\Delta_{\max}=1.639\,pixel$, the standard deviation is $\sigma=\sqrt{\frac{1}{1176}\sum_{i=1}^{1176}(\Delta_i-\overline{\Delta}_i)}=0.419$ ($\overline{\Delta}_i$ is the difference between the theoretical coordinates of the center of the circle and the actual coordinates) .Thus, the calibration accuracy of the system calibration method used in this paper is better.

4.2 Measured data analysis

In order to further verify the robustness and system accuracy of the calibration method in this paper, a field test was conducted at a subway station in Shanghai. figure 9 shows the image of the field test.

Figure 9. System field test

Three geometric dimensions of gauge, platform height (rail top to platform distance) and platform distance (distance between track center line to the edge of rubber strip) are detected. Through the subway station platform detection tool gauge gauge (accuracy ± 0.5mm), platform ruler (accuracy ± 1mm) and this system to test 50 platform sections, and compare the detection values and trends, verify the effectiveness of the system data detection.

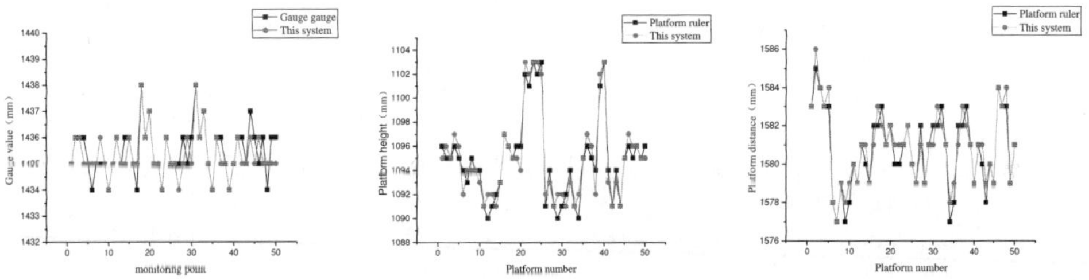

(a) gauge (b) Platform height (c) Platform distance

Figure 10. Data analysis (Unit: mm)

From figure 10, it can be seen that the measurement data of the platform meter and the track ruler are basically the same, and the maximum difference is 1mm, which satisfies the requirements of the Urban Rail Transit Engineering Measurement Specification (GB50308-2008). From the test data of platform distance and platform height, it can be seen that the trend of data of platform meter and platform ruler is relatively consistent, the maximum error is below 2mm, and the maximum detection error satisfies the requirements of Metro Design Code (GB50157 - 2013). The data analysis of the above detection results verifies the reliability of calibration accuracy of the calibration method described in this paper, and fully satisfies the requirements for the detection of the limit of the MTR in each life cycle limit.

5. Conclusions

In this paper, a comprehensive calibration method based on large multi array calibration plate is proposed, the key calibration area is selected, and the image preprocessing is carried out by Gauss fuzzy. The domain core method is proposed to segment the reference circle and extract the center of the circle. The feature value is found out according to the center of the circle and the perspective transformation matrix is obtained for image transformation. Through repeated experimental tests and field test analysis, it is concluded that the calibration method of this paper has high robustness, the maximum pixel error is 1.639pixel and in the field test, the track gauge detection accuracy can reach within ±1mm, the platform distance and the platform The height is within ±3mm, meeting the specification measurement requirements. In addition, unlike the traditional testing tools, the instrument is portable and detachable. At the same time, the operator does not have to enter the platform rail line operation, and the measuring speed is fast, the accuracy and safety are high, and it can meet the time and precision requirements of the subway outage operation.

References

[1] Feng Wang, HuiLin Yu and XiaoHua Zhao, et al. Discussion on subway station platform clearance [J]. *railway standard design*, 2009, (2): 81-82.
[2] HaiTao Niu, XunJie Zhao. A new camera calibration method using checkerboard template [J]. *infrared and laser engineering*, 2011, (01): 133-137.
[3] HaiGang Meng. Improvement of CCD camera calibration method based on plane constraint [D]. *Jilin University*, 2009.
[4] FanLu Wu, JianJun Liu , Xin Ren and ChunLai Li. Calibration method of deep space exploration panoramic camera based on circular landmarks[J]. *Acta Optica Sinica*, 2013,(11): 147-153.
[5] Ying Lu, HuiQin Wang, Wei Ju and JunJie Li.Camera calibration algorithm based on Harris-Zhang Zhengyou plane calibration method[J]. *Journal of Xi'an University Of Architecture And Technology (NATURAL SCIENCE EDITION)*,2014,46(06):860-864+870.
[6] Andrew Fitzgibbon, Maurizio Pilu and Robert B Fisher. Direct least square fitting of ellipses[J]. *IEEE Transaction on Pattern Analysis and Machine Intelligence*, 1999, 21 (5): 476-480.
[7] Qiao Yu, S. H. Ong. Arc-based evaluation and detection of ellipses[J]. *Pattern Recognition*, 2007, 40 (7): 1990-2003.
[8] RuiXue Xia, RongSheng Lu, Ning Liu and JingTao Dong. Automatic extraction of feature point coordinates based on dot array targets[J]. *China Mechanical Engineering*, 2010,(16): 1906-1910.
[9] Peng Xu, JianYe Wang and YanRu Wang. Accurate calculation of coordinates of target center point in camera calibration[J]. *Infrared and Laser Engineering*, 2011,(07): 1342-1346.
[10] YaoGui Wang. Image Gauss smoothing filter analysis [J]. *Computer and information technology*, 2008 (08): 79-81+90.
[11] LiangLiang Wang, Li Wang, XiaoRong Gao and ZeYong Wang.Two improved local threshold segmentation algorithms[J].*Modern Electronic Technique*,2009,32(14):78-80.
[12] ChunJian Hua, XueMei Xiong and Ying Chen.Feature extraction of circular arc profile based on sobel operator[J].*Laser & Optoelectronics Progress*,2018,55(02):239-246.
[13] LiGang Li, Bo Liu, HongJian You, HaiLiang Peng and YiJun Wu.Analysis and comparison of geometric precision correction algorithms for spaceborne remote sensing images[J].*Acta Photonica Sinica*,2006(07):1028-1034.

IOP Conf. Series: Materials Science and Engineering **435** (2018) 012041 doi:10.1088/1757-899X/435/1/012041

Study on Vibration Acceleration Prediction Model of Track Inspection Vehicle Based on BP Neural Network

JinKai Wang[1], YueLei He[1], HongYao Lu[1] and ZaiWei Li[1]

[1]School of Urban Railway Transportation, Shanghai University of Engineering Science, ShangHai 201620, China

JinKai Wang: 309905411@qq.com, Hongyao Lu: luhongyao1@163.com

Abstract. In order to study the mapping relationship between the track irregularity parameters, vehicle running speed and the vehicle vibration acceleration detected by the track inspection vehicle, and to provide better data support for diseases detection of track inspection vehicle and maintenance of rail line, this paper develops a BP (Back Propagation) neural network model that combines the big data analysis technology to predict the vibration acceleration of the track inspection vehicle. The sample data for this model comes from the inspection data of a passenger transport special line in East China. Based on grey relational analysis method, this research preprocesses the correction of mileage deviation for the sample data and analyzes the correlation degree between seven track irregularity parameters and two kinds of vehicle vibration accelerations, the results show that all input layer indicators of the neural network model have different degrees of influence on the output layer indicators. And the predicting results with test data shows that the average accuracy of the whole sample for vertical vibration acceleration and lateral vibration acceleration is 85.75% and 90.25%.

1. Introduction

With the increase of the speed and transporting volume of high-speed trains in China, rail diseases are increasing and the causes are complex. It has become an important basic work to dynamically detect rail status by track inspection vehicles for learning track quality in time and guiding the maintenance and repair of rail lines. Among the detecting items of track inspection vehicles, measurement of vehicle vibration acceleration is one of the important means for monitoring and evaluating rail regularity and passenger comfort. The historic track dynamic detection indicates that the generation of vertical vibration acceleration and lateral vibration acceleration of the vehicle is closely related to the quality of the track geometry and the running speed of the train. Generally, significant track irregularities may cause great vehicle acceleration response and different wavelengths of track irregularities could lead to various vibration accelerations of vehicle under different running speeds [1]. Therefore, analyzing and mining the detection data of track inspection vehicle and studying the mapping relationship between rail geometry status, train running speed and the vibration acceleration of track inspection vehicle is of great significance to detecting rail diseases, guiding line maintenance and improving the train running safety and degree of comfort.

At present, some achievements have been made in predicting the vehicle vibration acceleration referring to track irregularity stimulation and vehicle running speed. Literature [2] analyzes the influence of track irregularity type, amplitude and wavelength on vehicle acceleration based on measured track data. Literature [3] presents a prediction model of RBF (Radial Basis Function) neural network that can predict the changing trend of vibration acceleration based on the vehicle vibration acceleration data calculated by dynamic simulation software; Literature [4] develops a correlation model between track irregularity and vehicle vibration based on principal component analysis and BP neural network. Literature [5] classifies vehicle inspection data by the magnitude of amplitude and develops a classification prediction model of vehicle vibration acceleration with an improved decision tree algorithm. At present, track inspection vehicle is the main tool for track disease detection, but there are relatively few studies on its vibration acceleration prediction model. And there are still some defects in the detection of vibration acceleration: vibration acceleration is closely related to the running speed of the train. However, the track inspection vehicle usually works under the speed that is lower than the operating speed of high-speed trains, which means the detection values cannot fully represent the actual operating situation of the trains and it could lead to a certain effect on the analysis of vibration acceleration diseases. Meanwhile, the high-speed detection of the whole line is difficult to realize and often leads to errors. And the service life and operating cost of track inspection equipment will be affected to some extent.

This paper proposes a preprocessing method and a correlation analyzing method to detected data of track inspection vehicle for a passenger transport special line in East China. Based on BP neural network algorithm, the study develops a nonlinear mapping model between the rail geometric parameters, running speed and the vibration acceleration of track inspection vehicle, and the predicting accuracy of the model is verified with the actual sample data. Its accuracy meets the engineering standard and could provide more references for inspection and analysis of rail diseases and rail line maintenance.

2. Data preprocessing and Analysis of track inspection vehicle based on grey correlation analysis

2.1. Correction of mileage deviation of track inspection vehicle data

The location on the rail line of detection data is indexed by mileage. But the weak GPS signal, the slides between wheels and rails, malfunction grating encoder and other factors could inevitably lead to the mileage deviation that the detected mileage does not correspond to the standard mileage. The deviation is inconvenient for the on-site maintenance and has a certain impact on the analysis of detection data. So it is necessary to correct mileage deviation for inspected data [6]. In order to provide accurate data support for the development of BP neural network prediction model, this study corrects the mileage deviation of the detected sample data referring to the mileage correction model proposed by the literature [6].

The basic idea of grey correlation analysis is to measure the degree of correlation between factors by geometric relations or the similarity of curves. The more similar the curves change, the greater the greater the correlation degree between the corresponding time series is, conversely, the correlation degree is smaller. Each index of the test data of track inspection vehicle forms the waveform independently and when considering mileage deviation, they are mostly equivalent [7]. This paper selects the K22~K23 km section of a passenger transport special line in East China as the experimental object and takes the sequence of track gauge data detected on 14th February 2017 and 11th March 2017 as the experimental data to carry out the grey correlation analysis as shown in table 1.

Table 1. Partial track gauge sequence of inspection vehicle data

Mileage/m	K	X_0	X_1	X_2	X_3	X_4	X_5	X_6	X_7
22029.75	1	0	0.82	0.76	0.75	0.72	0.7	0.66	0.69
22030.00	2	0	0.9	0.82	0.76	0.75	0.72	0.7	0.66
22030.25	3	0.06	1.03	0.9	0.82	0.76	0.75	0.72	0.7
22030.50	4	0.17	1.08	1.03	0.9	0.82	0.76	0.75	0.72
22030.75	5	0.17	1.03	1.08	1.03	0.9	0.82	0.76	0.75
22031.00	6	0.14	0.88	1.03	1.08	1.03	0.9	0.82	0.76
22031.25	7	0.09	0.75	0.88	1.03	1.08	1.03	0.9	0.82

Mileage/m	K	X_0	X_1	X_2	X_3	X_4	X_5	X_6	X_7
22031.50	8	0.04	0.65	0.75	0.88	1.03	1.08	1.03	0.9
22031.75	9	0	0.65	0.65	0.75	0.88	1.03	1.08	1.03
22032.00	10	0	0.68	0.65	0.65	0.75	0.88	1.03	1.08
22032.25	11	0.09	0.78	0.68	0.65	0.65	0.75	0.88	1.03
22032.50	12	0.26	0.8	0.78	0.68	0.65	0.65	0.75	0.88
22032.75	13	0.34	0.8	0.8	0.78	0.68	0.65	0.65	0.75
22033.00	14	0.3	0.76	0.8	0.8	0.78	0.68	0.65	0.65
22033.25	15	0.23	0.77	0.76	0.8	0.8	0.78	0.68	0.65
22033.50	16	0.22	0.77	0.77	0.76	0.8	0.8	0.78	0.68
22033.75	17	0.3	0.69	0.77	0.77	0.76	0.8	0.8	0.78
22034.00	18	0.13	0.57	0.69	0.77	0.77	0.76	0.8	0.8

In table 1, X_0 is the data sequence of track gauge in the first inspection, it is set as the reference sequence $X_0 = \{ x_0(j) \}$, j = 1, 2, ···n; Xi is the data sequence of track gauge in the second inspection, it is set as the compare sequence $X_i = \{ x_i(j) \}$,i = 1, 2, ···m, j = 1, 2···n; Sequence K represents the order of detecting points, K = 1, 2, ···n.

The calculation steps are as follows:

The first step, calculate the gray correlation coefficient between $x_0(j)$ and $x_i(j)$ by formula 1-1:

$$\xi_i(j) = \frac{\min_i \min_j |x_0(j)-x_i(j)| + \rho \max_i \max_j |x_0(j)-x_i(j)|}{|x_0(j)-x_i(j)| + \rho \max_i \max_j |x_0(j)-x_i(j)|} \tag{2.1}$$

In the formula 2-1, $|x_0(j)-x_i(j)|$ is expressed as the absolute difference between sequence X_0 and sequence X_i at point j, denoted as $\Delta_i(j)$; $\min_i \min_j \Delta_i(j)$ and $\max_i \max_j \Delta_i(j)$ are respectively expressed as the minimum difference and maximum difference between sequence X_0 and sequence X_i. Identification coefficient $\rho \in (0,1]$, here takes $\rho = 0.1$。

The second step, evaluate the relational parameters between X_j and X_0, the calculating formula of the grey relational degree is:

$$\gamma_i = \frac{1}{n}\sum_{k=1}^{n} \xi_i(k),\ k = 1, 2, \ldots, n \tag{2.2}$$

Table 2 lists the partial calculation results as follows:

Table 2. Grey relational degree γ_{0j} （ρ =0.1）

γ_{01}	γ_{02}	γ_{03}	…	γ_{105}	γ_{106}	γ_{107}	…	γ_{118}	γ_{119}	γ_{120}
0.658	0.660	0.660	…	0.666	0.668	0.661	…	0.632	0.631	0.629

In the development of systems, if the trend of the two factors is consistent or synchronization is high, it can be said that there is a high degree of correlation between the two. Table 2 shows that the value of γ_{106} is the biggest, so the similarity between sequence X_{106} and X_0 is the best one. It means when moved the compare sequence backward by 106 detecting points, it has the highest correlation with the reference sequence X_0. Since the distance between each two inspection points of the track inspection vehicle is 0.25m, the mileage data corresponds to the second detection data should be moved backward by 26.5 m as a whole. That is the corrected mileage Δ=26.5 m. The corresponding curves before and after the correction are shown in figure 1 and 2.

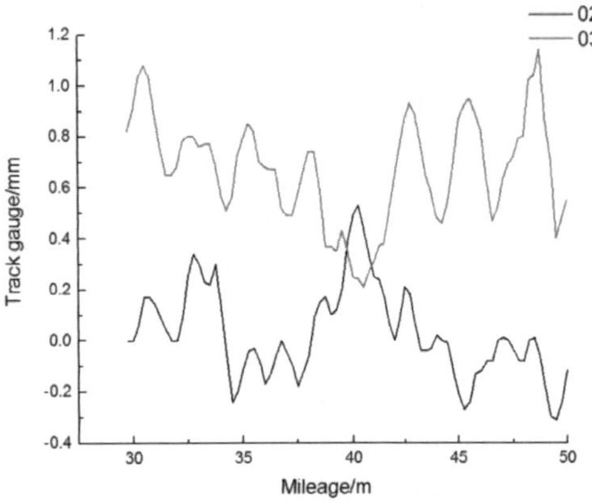

Figure 1. Comparison of the track gauge waveform between twice inspections in the 30-50m section of K22 km

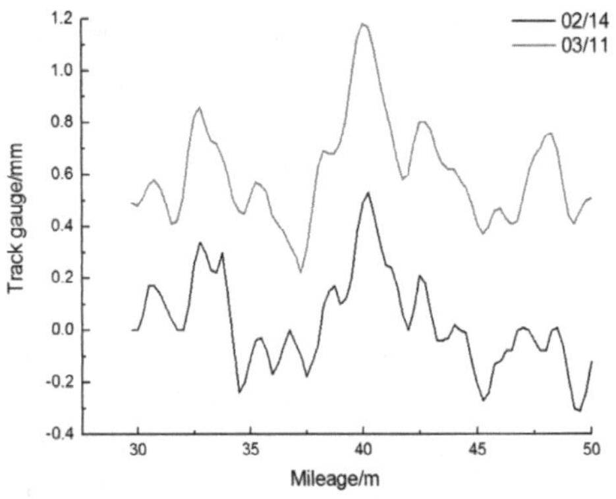

Figure 2. Comparison of the mileage corrected track gauge waveform between twice inspections in the 30-50m section of K22 km

The above results indicate that the correction of mileage deviation may improve the quality and availability of detection data. Based on this algorithm, the mileage deviation correction and collation of the detected data from February 2017 to July 2007 were carried out month by month to support the development of BP neural network predicting model.

2.2. Correlation analysis between track irregularity parameters and vibration acceleration

Grey correlation analysis can be used to analyze the degree of correlation between the various factors in the system, so as to judge the main and secondary factors that lead to the development and change of the system, it is a quantitative comparative analysis method for the dynamic development of the system. In order to develop the predicting model effectively and accurately, this paper selects track inspection vehicle data mentioned before as the experimental data to carry out the grey correlation analysis between lateral vibration acceleration and vertical vibration acceleration and track irregularity with the specific steps as follows:

The first step, set the vertical vibration acceleration and lateral vibration acceleration sequences as the reference sequences Y1 and Y2; set track irregularity sequences (track gauge, track alignment, before and after, the level of, superelevation, twist of track, curvature) as compare sequences Yi , the sequence matrix is:

$$Y_1 = \{y_1(j)\}, j = 1, 2, \cdots n; \tag{2.3}$$

$$Y_2 = \{y_2(j)\}, j = 1, 2, \cdots n; \tag{2.4}$$

$$Y_i = \{ y_i(j) \}, \text{i is the sequence number of data}, i = 3, \cdots 11; j = 1, 2, \cdots n \tag{2.5}$$

The second step, there are different dimension and dimension units between various types of track irregularity parameters. In order to eliminate the dimensional impact between the parameters and to enhance the comparability between the data indicators. This paper normalizes the original data by (0, 1) normalization methodology, the formula is as follows:

$$x^* = \frac{x-Min}{Max-Min} \tag{2.6}$$

Max and Min are the maximum and minimum values in each series.

Substitute the series Y1, Y2 and Yi into equations 1-6, the new normalized matrix is as follows:

$$Z_1 = \{ z_1(j) \}, j = 1, 2, \cdots n; \tag{2.7}$$

$$Z_2 = \{ z_2(j) \}, j = 1, 2, \cdots n; \tag{2.8}$$

$$Z_i = \{ z_i(j) \}, \text{i is the sequence number of data}, i = 3, \cdots 11; j = 1, 2, \cdots n \tag{2.9}$$

Take the previous example of the grey correlation analysis, substitute the sequence matrix Z_1 and Z_i , Z_2 and Z_i into equation 1-1 and 1-2, take identification coefficient $\rho = 0.5$, the grey correlation analysis results are shown in Table 3 and Table 4.

Table 3. Grey correlation Analysis of Vertical Vibration acceleration and track irregularity (ρ =0.5)

longitudinal irregularity		Track alignment		Track gauge	Horizontal irregularity	Super elevation	Curvature	Track twist
Left	Right	Left	Right					
0.87671	0.89739	0.87597	0.87266	0.78578	0.85880	0.69151	0.69037	0.82515

Table 4. Grey correlation analysis of horizontal vibration acceleration and rail irregularity (ρ =0.5)

longitudinal irregularity		Track alignment		Track gauge	Horizontal irregularity	Super elevation	Curvature	Track twist
Left	Right	Left	Right					
0.87675	0.89739	0.87597	0.87274	0.79599	0.86005	0.70544	0.70862	0.81789

In the development of BP neural network, the reason why the track irregularity parameters are selected as the input index of the input layer is that various track irregularities have different degrees of influence on the vibration acceleration of track inspection vehicles. Longitudinal (high and low) irregularity and track alignment diseases can increase the impact dynamics of passing trains, causing shaking and vibration of vehicles, it could result to the excitation frequency which has a significant effect on the vibration acceleration. From Table 3 and Table 4, it can be seen that the correlation coefficients of experimental rail line between either the longitudinal (high and low) irregularity or track alignment and the vibration acceleration are all above 0.87. Track twists and horizontal irregularity can cause tilting and rolling vibration of the vehicle, resulting in changes in the wheel-rail interaction force, which have a certain impact on the vibration acceleration. The correlation coefficients of these two parameters and vibration acceleration in the experimental rail line are all above 0.81. Significant deviation of track gauge, superelevation, and curvature can cause vehicle vibration and affect the operating safety of the trains. The correlation coefficients between each of the three parameters of the experimental rail line and the vibration acceleration are all above 0.69.

3. BP neural network prediction model based on inspection data of track inspection vehicle

BP neural network has excellent ability of multidimensional function mapping. This algorithm does not need to determine the mathematical equation of the mapping relationship between input and output in advance, only through the training of the sample data and learning some rules can obtain the result closest to the expected output value when the input value is given. This study collates 4000 groups of sample data and 500 groups of test data based on the existing inspected data. They are used for training the vibration acceleration prediction model and for analyzing its errors.

3.1. Development of BP neural network model

In this paper, the design of BP neural network structure is realized by toolbox function of MATLAB platform and by using single hidden layer neural network structure to develop the prediction network model of the vehicle vibration acceleration. The input layer is of 9 parameters of track irregularity and the running speed of track inspection vehicle, that is, the input layer contains 10 neurons; the number of the hidden layer neurons is determined to be 15 after debugging analysis and S-type transfer function (Sigmoid) is used. The output layer adopts linear transfer function (Purelin) and two output neural nodes are set to output vertical and lateral vibration acceleration respectively. The complete neural network is designed as shown in figure 3. Through trial calculations, the learning rate is set as 0.5; the error is set as 0.0001g and the maximum training step number is 1000.

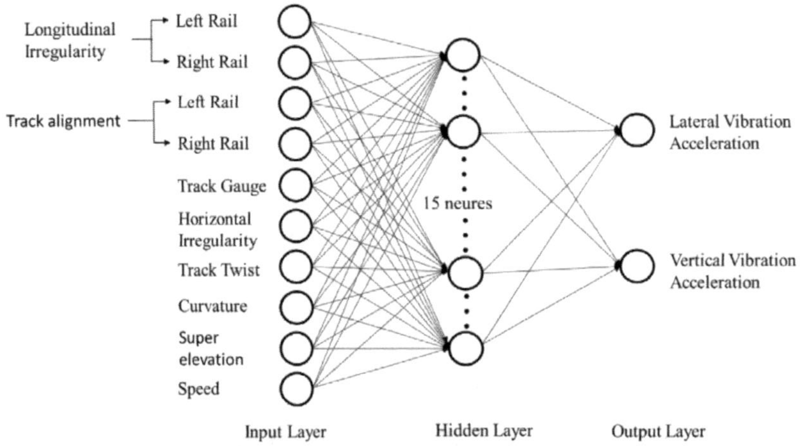

Figure 3. Debugging finished network structure diagram of BP neural network

3.2. The training and error analysis of prediction model

In this section, the selected sample data is edited to form a matrix data text, it is input into the designed BP neural network for training after the normalization. After the training of BP neural network, the average relative error of the sample can reach to 0.000085g, it is less than the expected error of 0.0001g. After calculated the test sample on the trained network, the prediction data of 0-25m range from K66 km point is randomly selected to compare with expected results, as shown in figure 4 and figure 5:

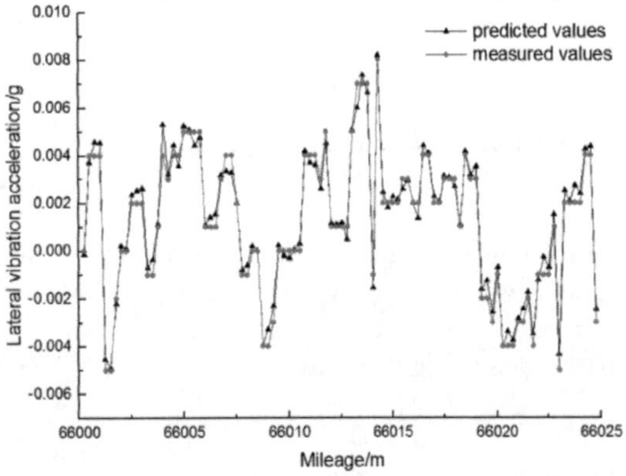

Figure 4. Effect diagram of lateral vibration acceleration prediction

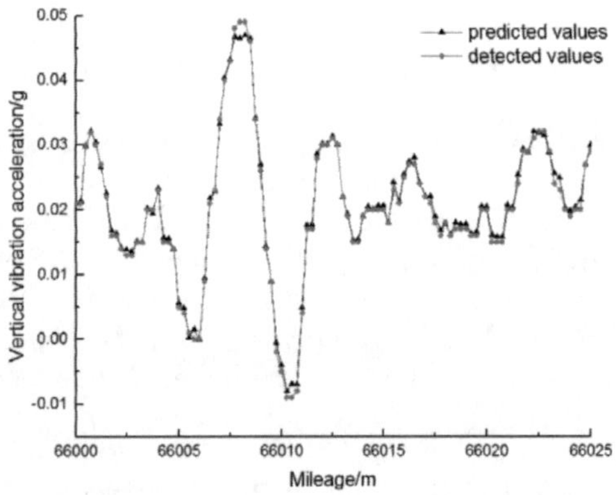

Figure 5. Effect diagram of vertical vibration acceleration prediction

Figures 4 and 5 indicate that the acceleration values of the vehicle vibration predicted by the neural network developed in this paper are close to or equal to the measured values. The BP neural network can well simulate the mapping relationship between track irregularity parameters, speed and vehicle vibration acceleration, so as to achieve a good prediction effect. Combined with the analysis of 500 groups of test samples, it is known that the maximum absolute error of vertical vibration acceleration prediction is 0.010 g, the average accuracy of the sample in whole is 85.75%; the maximum absolute error of the lateral vibration acceleration prediction is 0.006 g, and the average accuracy of the sample in whole is 90.25% (accuracy = 1- relative error).

4. Conclusions

In order to develop and apply a prediction model of vibration acceleration for track inspection vehicle based on track irregularity parameters and running speed, this paper selects and analyzes the track inspection vehicle data from a passenger transport special line in East China. The main works are summarized as follows:

(1) Using the grey relational algorithm correction model proposed in the existing research, the correction of the mileage deviation existing in the track inspection vehicle data was corrected and preprocessed.

(2) Based on the grey correlation analysis method, the correlation degree between the selected input layer indicators and output layer indicators of BP neural network was analyzed. Various track irregularity parameters (input layer) have a good correlation with vertical and lateral vibration accelerations (output layers) of the track inspection vehicle, and are suitable to be selected as the input layer variables of the BP neural network.

(3) Based on the detection data of track inspection vehicles, this study trained a BP neural network model that can predict vehicle vibration acceleration with track irregularity parameters and running speed. Through the computation with test data, the average accuracy of the predicted vertical vibration acceleration and the lateral vibration acceleration are respectively 85.75% and 90.25%. It proves that this model has good applicability and reliability for engineering application.

References

[1] SongLiang Lian, HuiMing Lu and WenZhong Yang, et al. Analysis program of track irregularity [J]. *China railway science*, 2006, 27(1):68-71.

[2] Jian Fang, XiaoYan Lei and SongLiang Lian, et al. Simulation research on influence of curvilinear track irregularity on vehicle dynamic response [J]. *Journal of Railway Engineering*, 2011, 28(5):45-50.

[3] LiMing Li, XiaoDong Chai and ShuBin Zheng. Prediction of vibration state of rail transit vehicles based on radial basis function neural network [J]. *Research of Urban Rail Transit*, 2017, 20(12):18-21.

[4] LiuBin Niu. A Study on the relationship between track irregularity and vehicle body vibration based on BP neural network [J]. *China Railway Science*, 2016, 37(2):26-32.

[5] JianLei Feng. Research and application of decision tree classification algorithm based on track inspection data [D]. Beijing Jiaotong University, 2009.

[6] XinYue Xu. Research on track state prediction model based on track inspection data [D]. Beijing Jiaotong University, 2007.

[7] Kui Wang. Research on the method of correcting vehicle inspection waveforms by using grey relational analysis [J]. *China Railway*, 2011(11):64-66.

IOP Conf. Series: Materials Science and Engineering **435** (2018) 012042 doi:10.1088/1757-899X/435/1/012042

Neural Network Based Temperature Field Mapping Model for CRTS II Type Ballastless Track

HaoMin Liu, HongYao Lu, YueLei He and ZaiWei Li

School of Urban Railway Transportation, Shanghai University of Engineering Science, Shanghai 201620, China

HaoMin Liu: lhm152104@163.com, HongYao Lu: luhongyao1@163.com
YueLei He: hyldoc@163.com, ZaiWei Li: lzw_5220964@163.com

Abstract. In order to investigate the mapping relationship between the meteorological parameters and the internal temperature field of the track plate, this paper takes the CRTS II type ballastless track of a passenger dedicated line in East China as the research object, and builds an on-line monitoring system for the temperature field of the intelligent track plate, and carries out real-time data on the ambient temperature, solar radiation, wind speed and the internal temperature of the track plate. Based on the BP neural network method, the variation rule of the temperature field inside the track plate is studied, and the mapping relationship between the environmental meteorological parameters and the temperature of the track plate is established. The prediction results show that the prediction accuracy of the nonlinear mapping model is 92.4% and the prediction accuracy of the temperature gradient is 81.5%, and the prediction accuracy is 96% and 93% respectively for the 13:00-16:00 time section of the overall temperature and the maximum positive temperature gradient of the track plate.

1. Introduction

Ballastless track structure is a multi-layer continuous band structure composed of reinforced concrete. The ballastless track is exposed to the atmosphere for a long time, and bears the environmental temperature, solar radiation and wind speed and other environmental factors. In the process of heat exchange with the outside world, due to the poor thermal conductivity and volume sensitive material, the temperature field of track plate structure is nonlinear. Different from the ballastless track plate, the longitudinal continuous ballastless track plate will produce large deformation and stress under the action of temperature load. The overall temperature of the track plate will be uniformly raised and down to make the track plate produce the whole expansion deformation, but because of the poor thermal conductivity of the concrete structure, the track plate has a temperature change along its thickness direction. The temperature gradient is the main cause of the defects such as the warping deformation of the track plate and the off seams of the bottom of the plate [1-3], which seriously affects the structural stability and service performance of the ballastless track, and directly threatens the safe operation of the train.

At present, the test and analysis of the temperature field of ballastless track under different conditions have made some achievements. SenRong Wang [4] measured the temperature of the rail plate by placing the plate overhead on the board and simulated the warpage stress of the track plate according to the measurement results. Yu Liu [5] made a long-term field observation on the early longitudinal connection stage of the CRTS II type ballastless track at Beijing-Shanghai high-speed railway. Based on the least square method, the prediction model of the surface temperature of the track plate was established. ZuMin Ou [6] used the meteorological data to describe the boundary conditions of environmental factors based on the heat transfer theory, established the temperature field equation of the ballastless track structure in the atmospheric environment, and analyzed the temperature distribution of the track structure caused by the changes of meteorological data. At present, although there are some research results in the temperature monitoring experiment and temperature field analysis of ballastless track, there are still some limitations. On the one hand, there are few studies on the combination of the measured ballastless track temperature measured by the actual operating high-speed rail line and the field weather data around the line. On the other hand, a large number of thermal parameters are involved in the theory of heat conduction analysis. There are many assumptions and simplification conditions, and the key parameters are difficult to apply to the environment complex. Study on the temperature field of ballastless track in mixed and changeable areas. Therefore, the monitoring and analysis of the internal temperature of the track structure and the meteorological data around the line is of great significance to the study of the mapping relationship between the meteorological parameters and the internal temperature field in the ballastless track, the accurate analysis of the distribution of the temperature field inside the track and the calculation of the temperature stress of the ballastless track.

In order to study the mapping relationship between environmental meteorological parameters and temperature field of CRTS II type ballastless track structure under service condition, The CRTS II type ballastless track of a passenger dedicated line in East China is taken as the research object in this paper. Based on BP neural network, the mapping model of temperature field of environmental meteorological parameters and ballastless track structure is set up. The validity and accuracy of the temperature field mapping model of ballastless track structure are verified by the field measured temperature distribution data. It can provide theoretical reference and data support for structural parameter design and maintenance plan of high-speed railway ballastless track.

2. Temperature field monitoring of CRTS II type slab ballastless track

The monitoring site is located on a dedicated passenger line in East China. The line operates at a speed of 300 km/h. This area belongs to the north subtropical monsoon climate zone. Due to the alternating cold and warm air, the weather is complex and changeable. The selected time for this monitoring is from June 2017 to August 2017, which is the summer high temperature weather in East China. The monitoring contents include ambient temperature (℃), solar radiation (W/m2), wind speed (m/s), and temperature inside the track plate (℃). According to the actual line operation requirements, by establishing a track plate temperature field test equipment, synchronous real-time on-line monitoring of the meteorological parameters around the operating line and the internal temperature of the CRTSII type track plate was realized. While satisfying the requirements of the test, the number of sensors should be as small as possible to reduce the impact of the installation of the sensor on the performance of the track structure. The monitoring system framework is shown in figure 1.

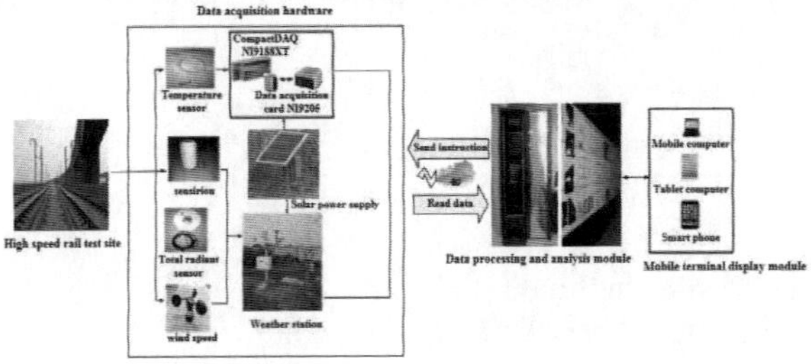

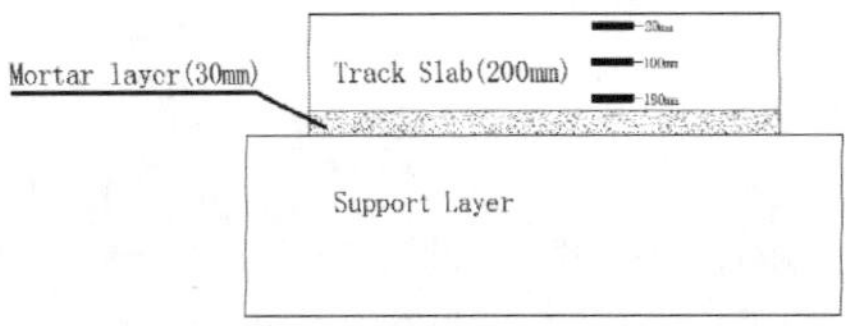

Figure 1. frame chart of monitoring system

Figure 2. Measurement Points layout

The temperature measurement points inside the track plate should be located in the areas with larger stress and displacement of the track plate, and areas sensitive to temperature changes. The temperature measurement points in the middle position of the track plate are selected, and the temperature sensor is used for sampling, the measuring range is -20-150℃, and the measuring accuracy is ±0.01℃. While satisfying the requirements of the test, the number of sensors should be as small as possible to reduce the impact of the installation of the sensor on the performance of the track structure. Therefore, three temperature sensors are arranged vertically along the track structure at a certain depth interval, as shown in figure 2.The temperature is monitored at the depth of 20mm, 100mm and 180mm on the surface of the track plate. The temperature of 20mm and 180mm on the upper surface of the track plate is used to analyze the vertical temperature gradient of the track plate.

$$gradT = \frac{T(y_1)-T(y_2)}{|y_1-y_2|} = \frac{T_{20}-T_{180}}{|y_{20}-y_{180}|} \tag{1}$$

In the formula, and are temperature from 20mm and 180mm on the surface of track plate respectively and are the vertical distance from the upper surface of the track plate respectively. The temperature of 100mm on the upper surface of the track plate is used to analyze the overall temperature variation of the track plate.

3. Monitoring data analysis

The analysis of monitoring data is shown in figure 3, which shows the amplitude of meteorological data from June 2017 to August.

The temperature and solar radiation of the test site are all high, the solar radiation is up to 1136W/m2, the maximum temperature of the environment is up to 43.7℃, and there is a large temperature difference between day and night, and the overall wind speed amplitude is not changed during the test. The overall wind speed is stable between 2~6m/s and the maximum wind speed is up to 10.3m/s. During the monitoring period, the amplitudes of temperature and solar radiation fluctuated more in June, and the amplitudes in July and August were relatively stable. At the end of August, there was a trend of decreasing temperature in late summer, and the amplitude of solar radiation and temperature showed a downward trend..

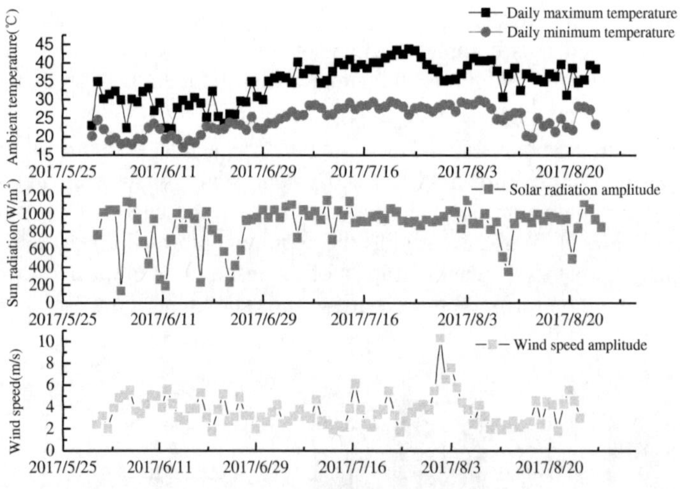

Figure 3. Monitoring site amplitude statistics

Based on the monitoring data of the weather station, the internal temperature of the track plate structure on the day when the maximum ambient temperature of 43.7℃ during the monitoring period was selected was used for data analysis. As shown in figure 4.

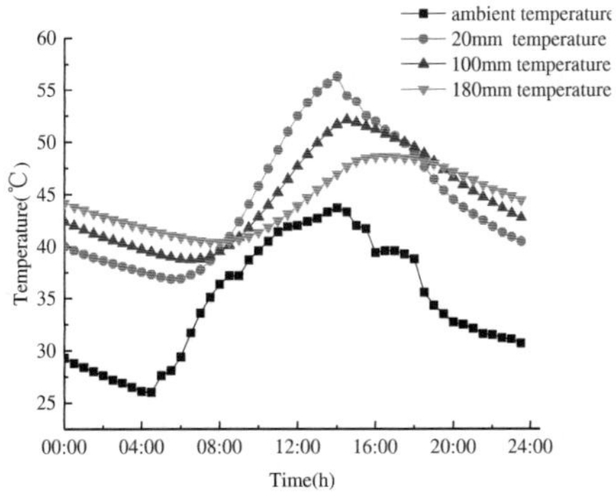

Figure 4. The variation of the temperature of the track plate and the ambient temperature

The change trend of the internal temperature and the ambient temperature of the track plate structure is basically the same, and the overall temperature of the track plate is high throughout the day. The maximum ambient temperature occurred at 14:00 on July 24, 2017, reaching 43.7℃, and the maximum temperature at the 20mm from the surface of the track plate exceeded 56℃ on the day and was 12.61℃ higher than the ambient temperature. During the day, the track plate absorbs heat from the outside world. The temperature at 20mm is higher than the temperature at 180mm. The night cooling track plate releases heat from inside to outside. The temperature at 180mm is higher than 20mm, and the minimum temperature at 180mm is 40.43℃.

Because of the low thermal conductivity of the concrete structure, the temperature has a slow transmission within the coagulant structure, and the time with the depth of the temperature extremum of different depth of the track plate increases with the depth. The minimum 20mm temperature on the track surface appears around 5:30, and the maximum value appears at 14:00. The minimum 180mm temperature on the surface of the track plate relative to 20mm appears at 7:30, which is about 2 hours behind. The maximum value appears at 16:30 or about 2.5 hours.

4. Mapping model of BP neural network

4.1 The principle and algorithm of BP neural network

Neural network is a widely parallel interconnection network composed of adaptive simple units, which can be used to deal with classification and fitting problems. At present, BP neural network algorithm is the most widely used network model. The BP network is a multi-layer feed-forward neural network. The full connection method is adopted between layers, and there is no mutual connection between the same layers. The hidden layer may have one or more layers. Ability to learn and store multidimensional input-output pattern mappings. The topological structure of BP neural network model includes input layer, hidden layer and output layer. The typical model is shown in figure 5.

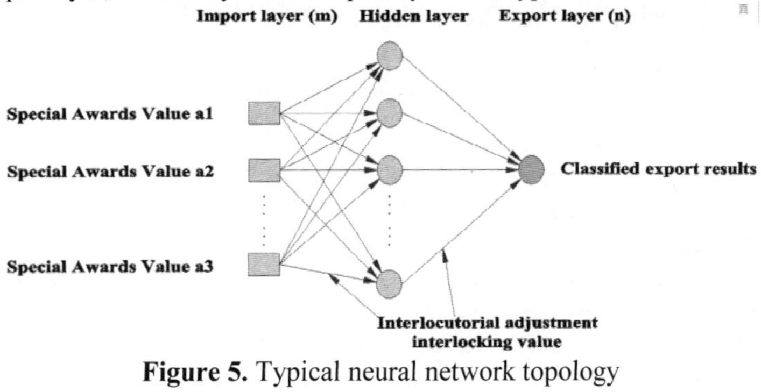

Figure 5. Typical neural network topology

The neural network learning adopts improved BP algorithm, and the learning process consists of forward calculation process and error back propagation process. In the forward calculation process, the input information is calculated layer by layer from the input layer through the hidden layer and passed to the output layer. The state of each layer of neurons only affects the state of the next layer of neurons. If the output layer can not get the desired output, then it goes to the process of error back propagation. The error signal returns along the original connecting path. By modifying the weights of the neurons in each layer, the network system error is minimized. In the end, the input and output mapping relationships that are desired to be established are approximated.

4.2 Mapping model construction

In heat transfer theory, there are mainly three kinds of heat transfer methods: conduction, convection, and radiation. The heat transfer modes on the surface of the track plate are mainly heat convection and heat radiation. Solar radiation is the main source of thermal radiation. The wind speed and the temperature affect the convective heat transfer on the surface of the track plate .The difference in the daily temperature is related to the highest and lowest temperature, which characterizes the radiation heat transfer intensity. Therefore, weather parameters such as ambient temperature, solar radiation, wind speed, and daily temperature difference are selected as input parameters, and the overall temperature and temperature gradient of the orbital plate are used as output parameters.

The 4225 track plate temperature data and corresponding meteorological parameters actually measured during the test period were integrated, of which 80% were used as training samples and the remaining 20% were used for test data. The raw data is normalized as follows before model training.

$$x'_{\mathrm{n}} = \frac{x_r - x_{r,\min}}{x_{r,\max} - x_{r,\min}} \times \left(x_h - x_l\right) \tag{2}$$

x'_{n} :normalized input data, x_r : original data, $x_{r,\max}$, $x_{r,\min}$: maximum and minimum values for the input, $x_h = 1$, $x_l = 0$.

The BP learning method is used to train the neural network model. The training functions of trainlm, trainscg, trainrp and traincgb4 are used in turn, and the mean square error (MSE) and Pearson correlation coefficient (PCC) are used to evaluate the performance of the network. The number of neurons in the hidden layer is 4 based on the formula (3), and a neural network mapping model is set up. The number of iterations is set to 300 times.

$$m = \sqrt{n + l} + \alpha \tag{3}$$

m: the number of hidden layer nodes, n: the number of input layer nodes, l: the number of nodes in the output layer, :the integer between 1 and 10. Run the network randomly 5 times for training function, and get intermediate results, as shown in Table 1. The trainlm training function shows better network structure training ability, and the prediction results show a strong correlation with the measured values (0.8 < PCC<1.0).

Table 1. The training results of different training functions

Serial number	Training function	MSE	PCC
1	trainlm	0.018441	0.9484
2	trainscg	0.020609	0.7427
3	trainrp	0.021499	0.6987
4	traincgb	0.020467	0.8044

The determination of the number of neurons in the neural network structure model is very important to the establishment of the model. Using the mean square error (MSE) to evaluate the network performance of the model to determine the number of neurons in the hidden layer. The number of hidden layer neurons varies in the range of 1~30.

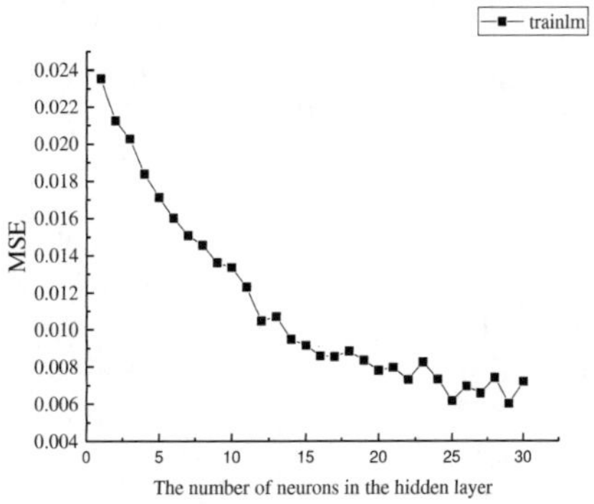

Figure 6. The variation of MSE with the number of neurons in the hidden layer

As shown in figure 6, when the neural network model is trained, the mean square error of the trainlm training function varies with the number of neurons in the hidden layer. When the number of hidden neurons increases from 1 to 15, the mean-square error decreases rapidly. When the number of hidden neurons exceeds 22, the mean square error of the trainlm training function fluctuates significantly. When the number of neurons in the hidden layer increases from 1 to 30, the mean-square error is smallest when the number of hidden neurons is 25. When the number of neurons in the hidden layer exceeds 25, small fluctuations in the mean square error occur. Therefore, a 4-25-2 mapping model neural network structure as shown in figure 7 is established.

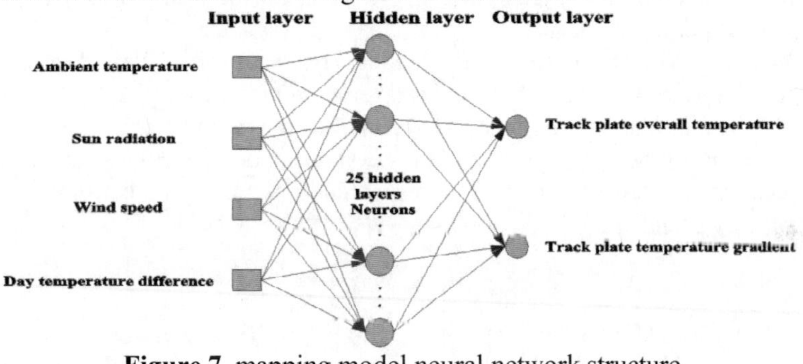

Figure 7. mapping model neural network structure

4.3 Mapping model validation

In order to verify the accuracy of the established mapping model, the test data was input into the trained model for calculation. The overall temperature prediction accuracy reached 92.4%, and the temperature gradient prediction accuracy reached 81.5%. Because the maximum overall temperature and the positive temperature gradient of ballastless track in our country generally appear in the 13:00-16:00 period of 5-8 months, the environmental meteorological parameters corresponding to the monitoring data of 13:00-16:00 are randomly selected into the established neural network mapping model. The prediction results are compared with the measured results, as shown in table 2.

Table 2. Comparison of prediction results of neural network mapping models (excerpt)

Time	Track slab overall temperature			Track slab temperature gradient		
	Forecast result /℃	Measured result/℃	Error /%	Forecast result /℃	Measured result/℃	Error /%
13:00	42.28128	44.21	4	52.15699	53.5	2
13:30	43.42063	45.09	3	53.26323	56.375	5
14:00	45.32934	45.61	0.6	51.71079	55	5
14:30	42.16435	45.895	8	38.47776	41.1875	6

Time	Track slab overall temperature			Track slab temperature gradient		
	Forecast result /℃	Measured result/℃	Error /%	Forecast result /℃	Measured result/℃	Error /%
15:00	44.21851	46.25	4	45.33099	49.125	7
15:30	45.66035	46.4	1.5	42.84183	45.375	5
16:00	45.02496	46.345	2	36.92999	40.0625	7

The neural network structure of the mapping model can accurately discover the nonlinear mapping relationship between meteorological parameters and the overall temperature and temperature gradient of the track plate.

5. Conclusions

1) This paper studies and analyzes the mapping relationship between environmental meteorological parameters and the temperature field of CRTS II type ballastless track structure in eastern China under service status, and draws the following conclusions:
2) An on-line temperature monitoring system for intelligent orbital plates was established to realize real-time simultaneous online monitoring of the temperature around the line and on the CRTS II type ballastless track;
3) In the high temperature weather conditions in East China, the interior temperature and the ambient temperature of the track plate structure are basically the same. Due to the low thermal conductivity of the concrete structure, there is a significant hysteresis at different moments when the extreme temperature of the track plate occurs.
4) Nonlinear Mapping Model of Temperature Field Inside Orbital Plate Based on BP Neural Network The prediction accuracy of the overall temperature reached 92.4%, and the accuracy of the temperature gradient prediction reached 81.5%. For the period from 13:00 to 16:00 when the extreme temperature of the orbital plate and the maximum positive temperature gradient occur, the prediction accuracy reaches 96% and 93%, respectively, which can meet the actual needs of the high-speed railway project.

References:

[1] Bin Wu, Shen Liu and ZhiPing Ceng, et al. Study on the temperature field characteristics of CRTS II type ballastless track [J]. Chinese Railway Engineering Journal, 2016, (3): 29-33.
[2] Yao Zhong, YueLei He, Hong Yao LU and ZaiWei Li. The analysis of interlayer seams of the ballastless track under continuous high temperature [J]. Journal of Railway Science and engineering, 2018,15 (5): 1128-1133.
[3] PingRui Zhao, XueYi Liu and Guan Liu.Experimental study of temperature gradient in track plateunder outdoor conditions in Chengdu area[J] Journal of Modern Transportation, 2014, 22(3):148-155.
[4] SenRong Wang, Li Sun and QiuYi Li, et al. Temperature measurement and temperature stress analysis of ballastless track [J]. Journal of railway engineering, 2009, 26 (2): 52-55
[5] Yu Liu, Pan Chen and Guotang Zhao. CRTS II type ballastless track structure early temperature field characteristics [J]. Chinese Railway Science, 2014, 35 (1): 1-6.
[6] ZuMin Ou, Lu Sun and Fun Cheng. Calculation and analysis of temperature field of ballastless track based on meteorological data [J]. Acta Railway Journal, 2014, (11): 106-112.

IOP Conf. Series: Materials Science and Engineering **435** (2018) 012043 doi:10.1088/1757-899X/435/1/012043

Moving a Chair through a Door: A Tutorial on Local Spatial Reasoning in Algorithmic Robotics

Godfried Toussaint*

New York University Abu Dhabi, Saadiyat Island Campus, Abu Dhabi, United Arab Emirates

*gt42@nyu.edu

Abstract. Spatial reasoning problems in robotics may be categorised into two types: *global* and *local*. A typical problem of the global kind involves determining if a robot can navigate along a specified terrain from one specified (starting) point to another (target) point by means of a suitable path, subject to a variety of constraints. A quintessential example of local spatial reasoning is the determination of whether an object can pass through a specified opening. This paper presents a tutorial on the local spatial reasoning problem of deciding whether a given rigid convex object can pass through a specified convex aperture, and if possible, planning a sequence of motions to accomplish the task, in a computationally efficient manner. The exposition of this material is at the freshman and sophomore undergraduate levels, and is well suited to motivate computer science students in discrete mathematics courses.

1. Introduction

A typical problem in robotics is concerned with whether a robot can navigate in a space filled with obstacles, from some starting location S to a target location T, while avoiding collisions with the obstacles, and perhaps satisfying additional constraints such as minimizing the total distance travelled, minimizing the cost incurred, or avoiding slopes that are too steep [1]. At different locations during such a global planned navigation path the robot may encounter specific local challenges such as grasping an object with a fingered-hand [2], [3], or jaw-grippers [4], [5], navigating through a narrow sharp turn at a corner of a corridor [6], or passing through a door [7] or window [8]. Such problems are referred to as *local spatial reasoning* (or *planning*), and comprise the topic of this tutorial.

2. Two-Dimensional Space

Let us first look at the problem in Edwin A. Abbott's *Flatland*—the two-dimensional space we call the plane. Furthermore, let us begin our investigation with a rather strange-looking chair—a convex polygon. Geometrically our problem can now be described as follows. Let the plane be divided into two half-planes by a vertical infinite line L, and refer to figure 1. This line might represent the wall between a room and a balcony. Let A denote the left half-plane (room) and B the right half-plane (balcony). Let C be a convex polygon (the chair) situated in A. Let us delete a line segment D from line L. This gap represents the door. What can we say about the necessary and sufficient conditions under which C can be moved from A to B? Before we proceed we should be clear about what we mean by *move*. To start with, no restrictions are placed on the type of motions allowed. In other words, in our attempt to move C from A to B we may enlist any family of continuous rigid motions consisting of combinations of translations and rotations. Of course we are not allowed to 'lift' C into the third dimension. Neither is it permitted to squeeze, mutilate, spindle or otherwise deform C, as one might do

IOP Conf. Series: Materials Science and Engineering **435** (2018) 012043 doi:10.1088/1757-899X/435/1/012043

with a real chair in the heat of frustration, if one cannot manage to transfer the chair to the balcony. In other words, here we consider C to be a *rigid* body. Before proving the first result we introduce some definitions.

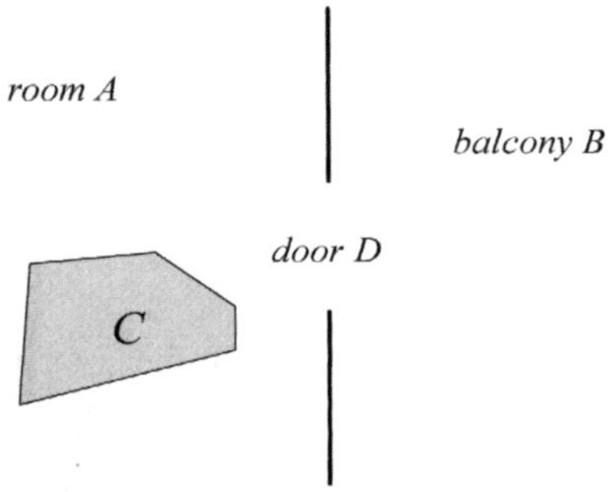

Figure 1. Moving a convex polygon (chair) through a door in the plane.

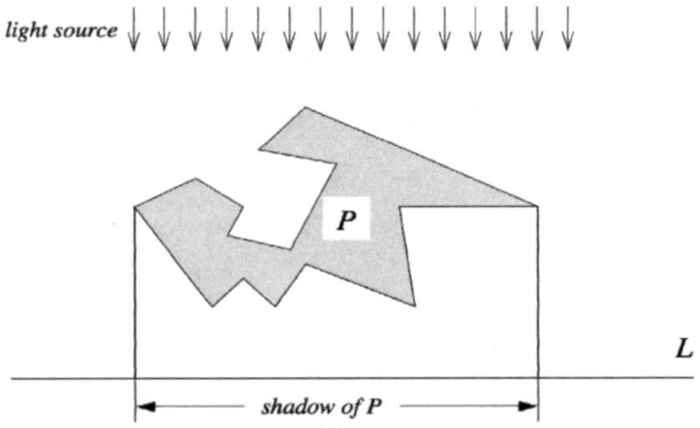

Figure 2. The shadow (projection) of a simple polygon determined by parallel light rays.

Let P be a polygon in the plane lying on one side of a line L and refer to figure 2. Consider a light source infinitely far away from P on the same side of L as P. Furthermore assume that the light rays are parallel to each other and orthogonal (at right angles) to L. Then the polygon casts a shadow on L. The length of this shadow depends on the orientation of P with respect to L. We define the *width* of a polygon P, denoted by $W(P)$, as the length of the shortest shadow made by P while P is rotated through all possible orientations.

Consider two infinite half-rays r_1 and r_2 emanating from some common point O, such that they are not collinear, and refer to figure 3 (left). These two rays divide the plane into two unbounded regions. The region with an acute angle at the base of the two half-rays is called a *cone*. Let H_i, $i = 1, 2$ denote the half-plane determined by a line through O that is orthogonal to r_i and contains r_i. The intersection of H_1 and H_2 is called the *dual cone*. If we let $\underline{H_i}$, $i = 1, 2$ denote the half-plane determined by a line through O that is orthogonal to r_i, but does not contain r_i, then the intersection of $\underline{H_1}$ and $\underline{H_2}$ is the *external dual cone*. Note that if the angle of a cone is small then the angle of the corresponding dual cone is large and vice versa.

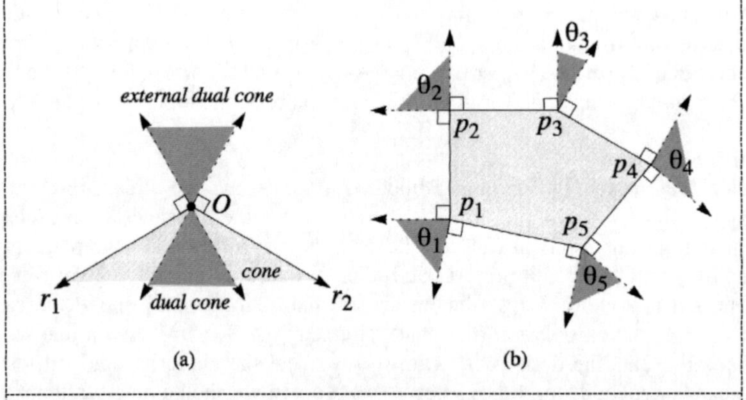

Figure 3. (a) A cone, its dual, and external dual (shaded dark grey), and (b) the external dual cones of the vertices of a convex polygon.

Now consider a convex polygon P consisting of n vertices $p_1, p_2, \ldots, p_n$. Each vertex p_i and its incident edges give rise to a *dual cone* C_i and a corresponding external dual cone. Let the angle of the external dual cone of vertex p_i be denoted by θ_i, and refer to figure 3 (b). Then it follows that the sum $\theta_1 + \theta_2 + \ldots + \theta_n = 2\pi$. One may think of continuously "wrapping" a tangent line around the polygon always keeping the line in contact with the polygon and observing the outer normal vector to a point of tangency. During this process the tangent line rotates only about the vertices of the polygon and at each vertex p_i the normal to the tangent rotates by and angle θ_i. Finally, when the tangent line has completed the wrap-around, the normal vector will have rotated by a complete revolution of 2π.

With these preliminaries we are ready to return to the problem of moving a chair through a door. The door D on the wall L, as illustrated in figure 1, is a gap in this line. Let us refer to the length of this gap as the *width* of the door and denote it by δ. A proof is now presented of Theorem 1. A proof of this theorem, based on the principle of continuity, for the case when C is a *smooth* convex body (with unique tangents at every boundary point of C) was given by Gilbert Strang [9]. The proof presented here is a modification of his proof to handle convex polygons with a finite number n of vertices.

Theorem 1: *A convex chair can be moved through a door if, and only if, the width of the chair does not exceed the width of the door.*

Proof: The proof of the *if* statement of the theorem is straightforward. It suffices to show that if the width of the chair C does not exceed the width of the door δ, then C can be moved from room A to balcony B. We will present a constructive proof of this claim by describing a sequence of translations and rotations that will accomplish this task. Imagine a light source of parallel rays which are orthogonal to L illuminating the polygon and creating a shadow on L. Assume the wall is a vertical line in the plane and refer to figure 4. First move C by a rotation R that minimizes the length of its shadow on the wall, i.e., the width $W(C)$ of the polygon. Then move C by a vertical translation T_1, such that the shadow is aligned with the door D. Since, by assumption, $W(C)$ does not exceed the width of D, we can complete the trajectory with one final translation T_2 in the horizontal direction. Thus, if the width of the chair does not exceed the width of the door, then C can always be moved from A to B by a sequence of motions consisting of at most one rotation and at most two translations. Furthermore, the computational complexity of the planned trajectory is dominated by computing the width of the convex polygon, and may be computed using $O(n)$ operations [5].

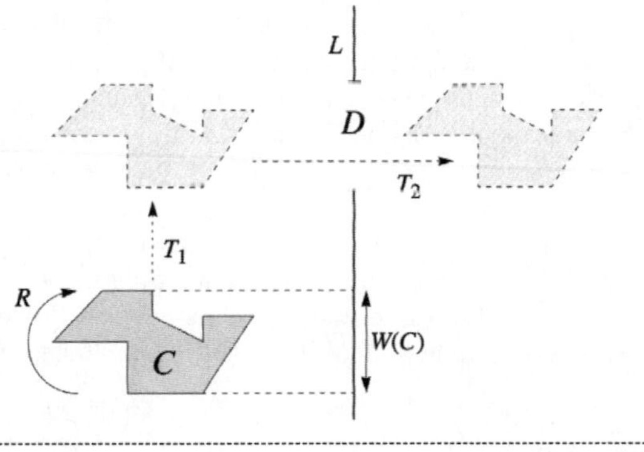

Figure 4. Moving a chair through a door with one rotation and two translations.

The proof of the *only if* statement of the theorem is more subtle. It must be shown that if the width of the chair exceeds the width of the door then the chair *cannot* be moved from A to B by *any sequence of combined rotations and translations*. An alternate equivalent statement of this claim is that if the chair *can* be moved from A to B (by some sequence of translations and rotations *unknown* to us), then the width of the chair cannot exceed the width of the door. Let us prove this version of the statement. Therefore assume there exists such a family of rigid motions that takes C from A to B. Before and after this sequence of motions, C does not intersect D. As C moves through D we look at the intersection I of C with D. Consider two lines tangent to C through the end points of I as well as unit outer-normal vectors at the end points of I orthogonal to the tangent lines. In figure 5, the diagrams (a) and (b) illustrate the position of C, along with the tangents and corresponding tangent-normal vectors, just after C begins to penetrate D, and just before C has gone completely through D.

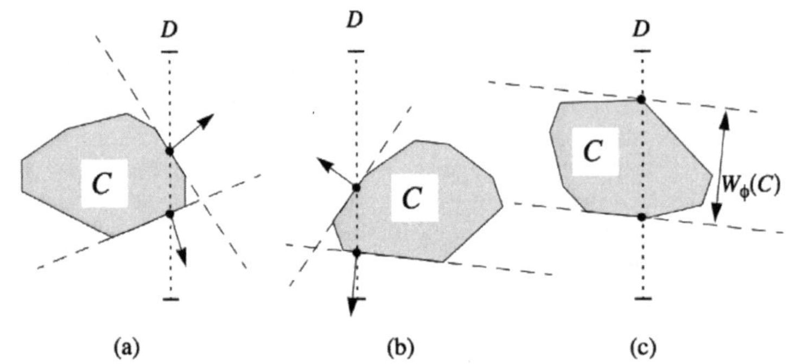

Figure 5. The tangents at the door intersections as the chair (a) enters the door, (b) leaves the door and (c) a point in time when the two tangents are parallel.

Two types of events occur, in a sufficiently small time interval Δt, while C passes through D: either (1) D intersects only the interior of two edges of C, or (2) one or two vertices of C pass across D. In the type-1 event the tangent lines and their normal vectors are distinct, and do not change throughout Δt. In the type-2 event the tangent lines and their normal vectors are not well defined when a vertex of C lies in D, and change their orientation after the vertex has gone through D. During the motion, as C passes through D we maintain the sum of the angles of the external dual cones of the vertices of C that are on or to the right of D. This sum is zero when C is left of D and equals 2π when it lies to the right of D. Therefore there exists a vertex of C such that when it crosses D it causes the sum to jump from a value less than π to a value greater than π (see figure 5 (c)). This implies that when this vertex intersects D there exists a pair of tangent lines in direction φ such that one line contains this vertex and is parallel to the other line. It follows that if C can be moved from A to B by some motion, however complex, it can do so with a single translation (after pre-positioning) in direction φ. But the width of the door δ, the length $l(\mathrm{I})$ of the intersection of C with D, the width of C in direction φ, denoted by $W_\varphi(C)$, and the width $W(C)$ of the chair C are all related by the inequalities: $\delta \geq l(\mathrm{I}) \geq W_\varphi(C), \geq W(C)$. Therefore, the width of the chair C cannot exceed the width of the door δ. Q.E.D.

Note that Theorem 1 applies only to convex objects. If an object is not convex, such as a boomerang for example, then it is possible that it can move from A to B even though the width of the boomerang exceeds the width of the door by an arbitrarily large amount. Consider the example in figure 6. The width of the boomerang is much greater than the width of the door but it can still pass through the door by means of three translations and two rotations. One translation and rotation are used to preposition the boomerang. Then one translation moves one of the arms across the door, a second rotation repositions the boomerang so that it is ready to continue, and a final third translation completes the trajectory.

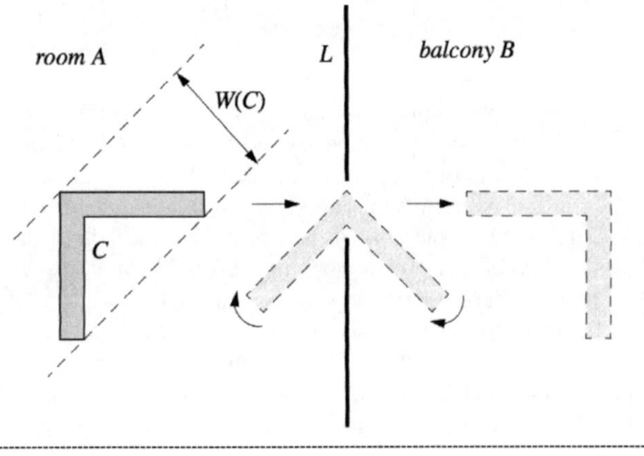

Figure 6. Moving an object C that has the shape of a boomerang, from A to B via a translation, followed by a rotation, followed by a second translation, even though the width of the boomerang is greater than the width of the door.

3. Three-Dimensional Space

Theorem 1 establishes in two dimensions that if a convex polygon can be moved through a door by some sequence of continuous rigid motions, then it must possess a shadow that fits in the door. It is natural to ask if the same property holds in 3-dimensional space: if a *convex polyhedron* can be moved through a convex 2-dimensional door in a wall, does it follow that the polyhedron must possess a shadow that fits in the door? Surprisingly, the answer to this question is negative, thus highlighting the difference that exists between two and three dimensional space.

We now describe a convex polyhedron P and a convex door D (that Gilbert Strang [9] attributes to Harold Stark) with the property that although no shadow of P fits into D, the polyhedron can nevertheless be moved through the door. The shape of the convex door is a circular disk such as a port-hole in a ship, a docking port in the international space station, or an access hole in a sewer. As a relevant aside, it is useful to remark here that the reason sewer access-holes tend to be round, and not square, is that a square cover possesses a shadow that fits in the hole, and therefore such a cover could easily fall into the sewer possibly injuring a worker. A round access-hole on the other hand contains a lid with a diameter slightly greater than the diameter of the hole. Therefore it does not possess any shadows that fit in the hole, and therefore cannot fall into the hole. As we shall see however, such conclusions cannot be drawn when the cover is not a circular disc in spite of it being convex. The example object that is to be moved from A to B is a polyhedron with five faces, well known since antiquity, namely a pyramid. Let d denote the diameter of the circular door. For the moment assume the pyramid has the standard shape, such as the ancient pyramids of Egypt, *i.e.,* the top or apex of the pyramid projects downwards to the center point of the square base. Such a pyramid will actually not accomplish our task, but is analyzed first since it helps to understand the correct solution. Later we will modify this pyramid. Let the length of the side of the square base of the pyramid also be d, the same as the diameter of the circular door, and let its height denoted by d^* be slightly greater than $d/2$, the radius of the door. In other words, $d^* = d/2 + \varepsilon$ where ε is some relatively small positive constant.

It will be established that this polyhedron contains no shadow that fits in the door. First, without loss of generality, place the pyramid with its base on the xy-plane such that the sides of the base are parallel to the x and y axes. The points on the surface of the unit direction sphere may be used to describe all directions in which light can produce shadows. Consider for the moment only the base of the pyramid as an infinitely thin 2-dimensional plate and ask for which directions of light rays this plate contains a shadow that fits in the door. Before proceeding further it is useful to introduce the notion of the *diameter* of a set. When the set is a circle, the diameter is a well-known concept, but what about the diameter of an arbitrary set? Let x and y be two points contained in the set in question and let $d(x, y)$ denote the distance between x and y. The *diameter* of the set is the greatest value that $d(x, y)$ can be take over all points x and y in the set. Now let us return to the base of the pyramid and examine its shadows in different directions. From the top (or $+z$ direction) the shadow is a square with diameter equal to $\sqrt{2}\, d \approx 1.414d$. From any other direction, excluding the equatorial directions (along the xy-plane, the shadow is either a rectangle or a rhombus with diameter greater than d. For the equatorial directions other than parallel to the x or y axes, the shadow is a line segment with length greater than d and less than $\sqrt{2}\, d$. Hence the only directions which yield a shadow with diameter not exceeding the diameter of the door are the x and y directions. In these directions the base of the pyramid has a shadow which is a line segment of length precisely d. Thus, the only way the pyramid has any chance of having its shadow fit in the door is for its base to be perfectly aligned to a center line of the door (diameter of the circle). Figure 7 illustrates front and side views of one such possible position. However, since the height of the pyramid is greater than the radius of the circular door, a portion of the top of the pyramid creates a shadow on a portion of the wall outside of the door. Therefore this pyramid cannot yield any shadow that fits in the door. Unfortunately, neither is it possible for the object be moved through the door because the only way it can do so, requires that first the entire base of the pyramid translate through the door, and this is not possible because the top of the pyramid blocks such a translation. To fix this problem the shape of the pyramid is modified as follows. Rather than use an Egyptian-style pyramid we use a pyramid in the architectural style of the leaning Olympic Stadium in Montreal, so that the top of the pyramid projects to a point outside of its base. We accomplish this by translating the top vertex of the pyramid in a direction parallel to the base and sufficiently far so that the entire base can translate through the door before being blocked by the top of the pyramid. Such a pyramid is illustrated in figure 8. The height of the new pyramid is the same as that of the old pyramid and for exactly the same reasons as before, this new pyramid has no shadows that fit in the door. Nevertheless, this new pyramid can be completely moved through the door.

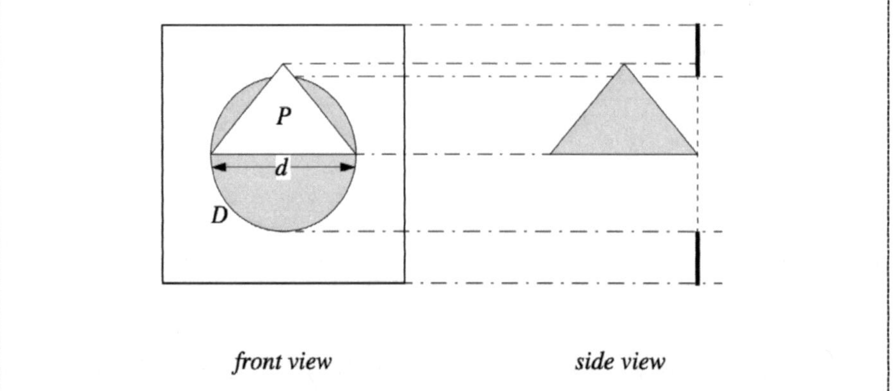

Figure 7. No shadow of the Egyptian-style pyramid fits in the circular door because the height of the pyramid is greater than the radius of the circular door.

First the continuous rigid motion proposed by Gilbert Strang is explained. Once the pyramid is prepositioned, this motion consists of two translations and one rotation. Subsequently, it will be shown that a different motion can be accomplished with just two translations, dispensing with the rotation altogether. The first solution is illustrated in figure 9. After prepositioning the pyramid so that its base is parallel to the *xy*-plane and it lies flush with the diameter of the door, translation T_1 moves the entire base of the pyramid in a direction parallel to the *xy*-plane through the door until the top of the pyramid comes into contact with the door at two points. A rotation R_1 is then carried out about the axis determined by these two points, through an angle that positions the upper triangular face of the pyramid parallel to the *xy*-plane. Finally, since the top of the rotated pyramid is no longer blocked by the door, translation T_2 in a direction parallel to the *xy*-plane moves the entire pyramid across the door. This completes the argument that Theorem 1 does not hold in 3-dimensional space.

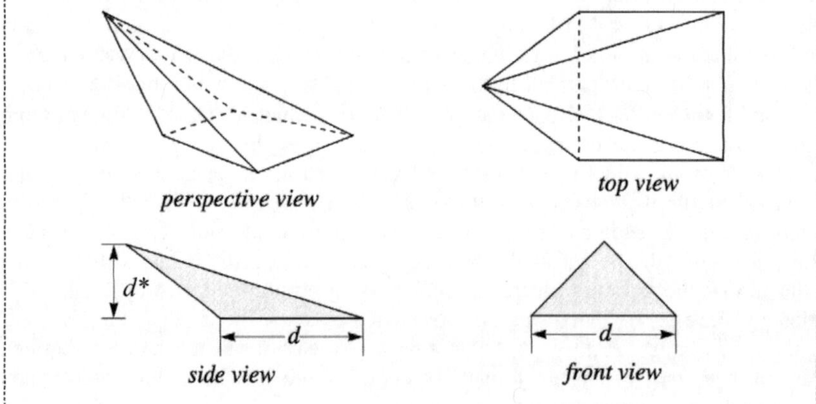

Figure 8. The shape of the leaning pyramid.

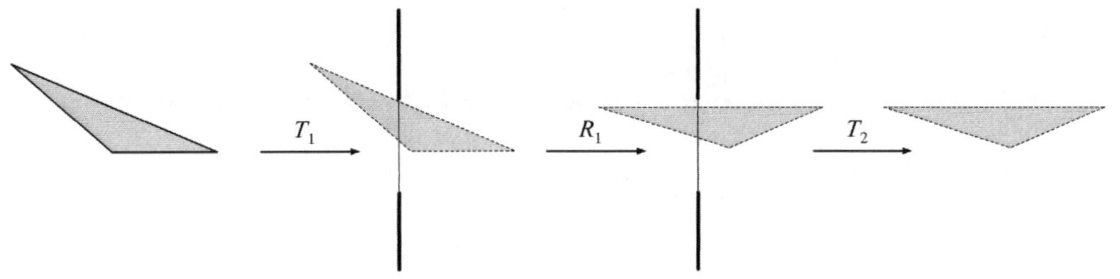

Figure 9. Side view of the motion of the pyramid through the door via one translation, followed by one rotation, followed by another translation (after prepositioning).

It will be shown now that, after prepositioning the leaning pyramid, two translations suffice to move the pyramid across the door, and hence the rotation portion of the motion described by Strang [9] is not necessary. The first translation T_1 is as before: the pyramid is translated with its base parallel to the *xy*-plane and flush with the center line of the door as illustrated in figure 10 until the top of the pyramid collides with the edge of the door. Once this position is reached, there exists a set of directions that will permit a second translation T_2 to move the entire pyramid across the door. To see this consider the polyhedral four-sided unbounded cone determined by the four rays emanating from the top vertex of the pyramid, in the directions of the four vertices of the base of the pyramid. The interior of this cone does not intersect any portion of the wall. Any translation of the pyramid in a direction defined by a ray from the apex of the pyramid that is obtained by a linear combination of the four rays that determined the cone will maintain the pyramid in the interior of this cone. Therefore such a translation will not cause any collisions with the boundary of the door.

4. Moving Non-convex Polygons

For more challenging problems students can explore similar questions for non-convex polygons such as *monotone* polygons, *star-shaped* polygons, and arbitrary *simple* polygons [10]-[17].

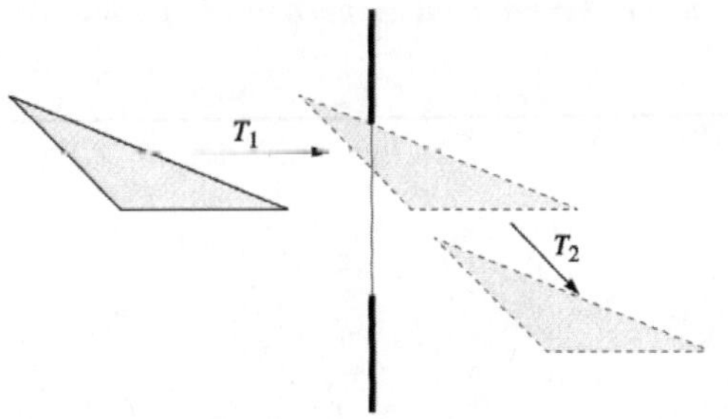

Figure 10. View of pyramid moving through the door via two translations (after prepositioning).

Acknowledgments
This research was supported by a grant from the Provost's Office of New York University Abu Dhabi, in Abu Dhabi, the United Arab Emirates.

References.

[1] Lozano-Perez T and Wesley M 1979 An algorithm for planning collision-free paths among polyhedral obstacles *Communications of the ACM* vol 22 pp 560-70
[2] Davidson C and Blake A 1998 Caging planar objects with a three-finger one-parameter gripper *Proc IEEE International Conf on Robotics and Automation*, Leuven, Belgium, pp 2722-27

IOP Conf. Series: Materials Science and Engineering **435** (2018) 012043 doi:10.1088/1757-899X/435/1/012043

[3] Markenscoff X and Papadimitriou C H 1989 Optimum grip of a polygon *International J Robotics Research* vol 8 no 2 pp 17-29

[4] Bose P, Bremner D and Toussaint G T 1996 All convex polyhedra can be clamped with parallel jaw grippers *Computational Geometry: Theory and Application* vol 6 pp 291-302

[5] Houle M E and Toussaint G T 1988 Computing the width of a set *IEEE Trans Pattern Analysis & Machine Intelligence* vol 10 no 5 pp 761-65

[6] Gerver J L 1992 On Moving a sofa around a corner *Geometriae Dedicata* vol 42 pp 267–83

[7] Yap C-K 1987 How to move a chair through a door *IEEE J Robotics and Automation* vol 3 no 3 pp 172-181

[8] Bhattacharya B K 1991 An optimal algorithm to translate a convex polyhedron through a two-dimensional convex window *CVGIP: Graphical Models and Image Processing* vol 53 no 3 pp 269-70

[9] Strang G 1982 The width of a chair *The American Mathematical Monthly* vol 89 no 8 pp 529-34

[10] Toussaint G T 2017 Mobility of objects in space: Some recent results and open problems *J Mathematics - ISJ* vol 3 pp 19-26

[11] Houle M and Toussaint G T 2017 On the separability of quadrilaterals in the plane by translations and rotations *Contributions to Algebra and Geometry* vol 58 no 2 pp 267-76

[12] Toussaint G T and ElGindy H 1984 Separation of two monotone polygons in linear time *Robotica* **2** pp 215–20

[13] Toussaint G T 1985 Movable separability of sets In *Computational Geometry* North-Holland Publishing pp 335-75

[14] Toussaint G T and Sack J-R 1983 Some new results on moving polygons in the plane *Proc Robotic Intelligence and Productivity Conf* Detroit Michigan pp 158-163

[15] R. Dawson R 1985 On the mobility of bodies in R^n *Math Proc Cambridge Philosophical Society* vol 98 pp 403-12 corrigendum 1986 vol 99 pp 377-79

[16] Toussaint G T 1989 On separating two simple polygons by a single translation *Discrete and Computational Geometry* vol 4 no 3 pp 265-78

[17] Pollack R, Sharir M and Sifrony S 1988 Separating two simple polygons by a sequence of translations *Discrete and Computational Geometry* vol 3 pp 123-36

IOP Conf. Series: Materials Science and Engineering **435** (2018) 012044 doi:10.1088/1757-899X/435/1/012044

Intelligent Assessment method of Air-Defense & Anti-Missile Command Model Based on Genetic Algorithm Optimization

LIN Chi , LI Song , WU Kai and LIU Tang

Graduate College, Air Force Engineering University, Xi'an 710051, China

E-mail :fkfd2016@163.com

Abstract. In order to solve the problems existing in the traditional assessment methods of Air - Defense and Anti-Missile command models, based on the requirements of the command model assessment, an intelligent Air-Defense and Anti-Missile command model intelligence assessment method based on genetic algorithm optimization was proposed. On the one hand, in order to meet the needs of the intelligent assessment, the evaluation knowledge bases is established based on the Air-Defense and Anti-Missile operational command process; on the other hand, basing on the historical data the objectivity of the assessment is improved by adjusting the index weights dynamically with the entropy weighting method and the intelligent optimization method. At the same time, the fuzzy relation matrix is obtained by real-time data processing of the command model and then using the multiple fuzzy comprehensive evaluation method to get the result. Finally, the feasibility of the method is verified by examples.

1. Introduction

Modeling and simulation is an effective method to study the Air-Defense and Anti-Missile command system. Whether or not the established model is true and reliable is the key to the success of the research, the evaluation of the model is an effective solution to the problem[1]. On the one hand, there is little research on the command model assessment in China, and the traditional assessment method has some disadvantages such as strong subjectivity, which cannot effectively solve the problem of the model assessment. To solve the problem of lack of objectivity due to excessive intervention of human factors, the literature[2] uses the intelligent means such as Expert System(ES) to realize the evaluation intelligence and improve the assessment results; on the other hand, the index is the key to assess the merits of the assessment method, and the importance of the indicator is differently, the relative weight of each indicator needs to be determined before the assessment[3]. In view of the shortcomings of AHP and other single-class methods in index weighting, combining with the characteristics of the model, a genetic algorithm is used to intelligently optimize the weights of the indexes.

2. Air-Defense and Anti-Missile operations command process and command model analysis

The Air-Defense and Anti-Missile warfare has changed the previous pattern of Air-Defense and Anti-Missile warfare relying solely on its own defense systems, and has turned into an integrated networked model for coordinated operations. The Air-Defense and Anti-Missile command system is the key node for carrying out integrated networked operations. The new-generation command system must have the ability to coordinate the control of two combat operations. Although the command process of Anti-Missile warfare calls for the accuracy of the command system is much higher than Air-Defense combat, both is based on the target data provided by the early-warning detection system to conduct a series of command, control and decision-making. The specific Air-Defense and Anti-Missile warfare command process is shown in Figure1.

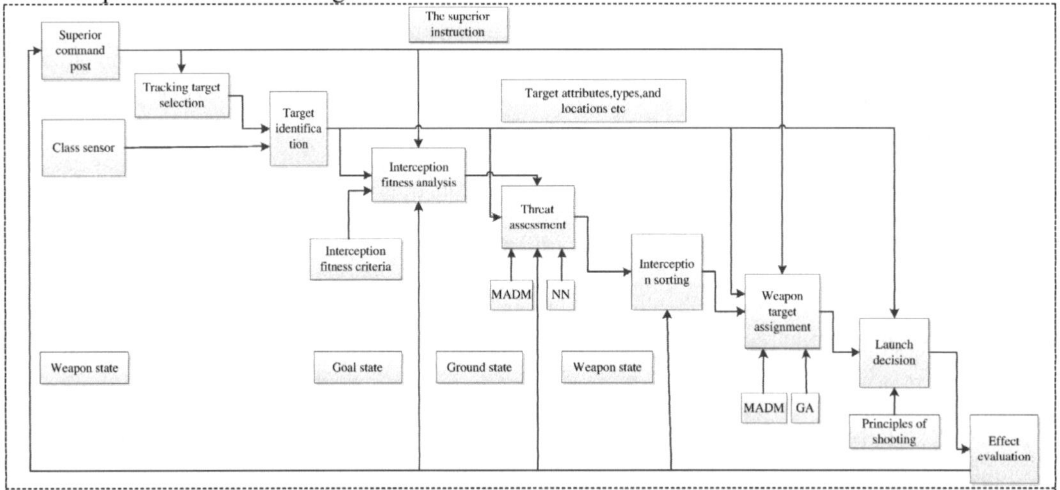

Figure 1. The operational command processes.

The process includes eight sub-processes: target tracking selection, target identification, interception suitability analysis, threat assessment, interception sorting, goal allocation, launch decision and effect assessment. At the same time, on the one hand the command process of Air-Defense and Anti-Missile warfare involves the types of equipment, the interaction of information, the complex coordination of operations, the rules of battle, the rules of engagement and the optimal design of operational logic are non-unique. On the other hand, Air-Defense and Anti-Missile operations include various types of combat aircraft, cruise missiles, unmanned aerial vehicles, ballistic missiles and other weapons. The combat targets are unknown and dynamic, so it is difficult to predict the time, space, mode of operations and the intension. And the use of combat resources has the characteristics of nonlinearity, dynamic programming and multi-objective and multi-attribute decision-making[4]. These characteristics determine that the air defense and missile defense allegation model assessment is a difficult problem that is multi-scale, multi-parameter, qualitative and quantitative. Now the evaluation in China is basically based on the evaluation of the model's own software and the realization of functions. There are few studies on the scientific nature of the Air-Defense and Anti-Missile command model that are close to operational requirements.

3. Intelligent assessment framework design

3.1. Intelligent assessment process

The intelligent assessment mainly aims at constructing corresponding evaluation knowledge bases for evaluation objects[5]. The evaluation situation knowledge base is established according to the corresponding air conditions, and the evaluation task knowledge base is established basing on the different tasks faced during the command process at each stage. Then, an knowledge bases of evaluation rules is generated by combining specific air conditions and specific tasks in each phase to obtain evaluation rules. Each rule mainly corresponds to the data and parameters generated by the corresponding command sub-process in the case of specific air conditions, that is, the corresponding command model processing generates the data and parameters. During the entire process, the data and parameters that associated with each command model are recorded and detected through the evaluation platform, and compared with the corresponding evaluation rules to evaluate and acquire the assessment data, and the evaluation results are calculated through multiple fuzzy comprehensive evaluations[6] , In the index weights, using the entropy weight method and historical evaluation data to dynamically adjust the weights and optimize the genetic algorithm. The framework is shown in Figure2.

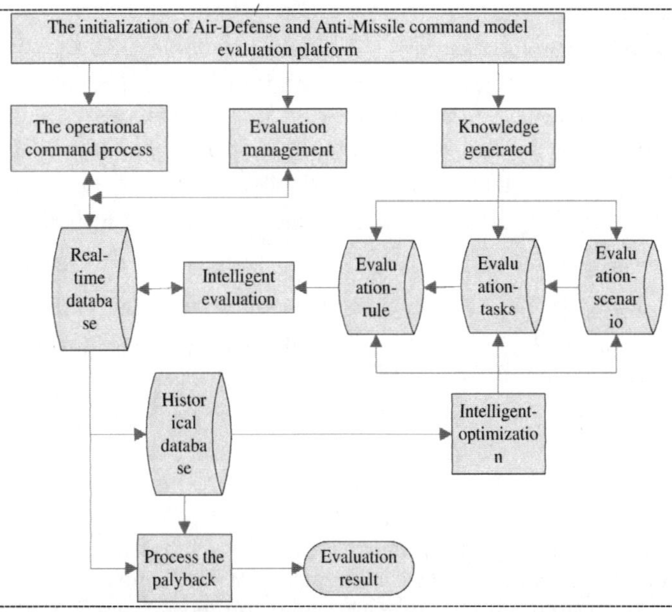

Figure 2. The intelligent assessment framework.

3.2. Evaluation knowledge bases construction

According to the analysis of the process, the model assessment knowledge bases is constructed, which is divided into the assessment situation knowledge bases, the assessment task knowledge bases and the evaluation rule knowledge bases. The content of the situation knowledge bases are classified on the basis of the air-raid weapons, which contain aerodynamic targets, ballistic missile targets and the situation combining aerodynamic targets with ballistic missile targets. According to the specific conditions, they can be divided into large-scale air strikes, medium-scale air strikes, small-scale air strikes and other specific situations. The evaluation task knowledge bases are mainly divided into sub-processes such as target identification, threat assessment, interception sorting and target allocation according to the various stages of the process. The evaluation rule knowledge bases is constructed according to the corresponding air conditions and tasks in each stage. The evaluation rules are the basis for assessment, including the data and parameter changes in the sub-processes. The model verification and evaluation platform analyzes the rationality of the model by performing real-time dynamic observation and data collection of the corresponding sub-processes and comparing them with the evaluation rules.

3.3. Multiple fuzzy comprehensive evaluation

The basic principle of multi-fuzzy comprehensive evaluation is to quantify indicators with fuzziness using fuzzy transformation in fuzzy mathematics. The literature [7] pointed out that the evaluation considers each index related to the object being evaluated, uses the principle of fuzzy linear transformation and the principle of maximum membership degree to quantify the index, and then carries out comprehensive evaluation to realize the quantitative and qualitative analysis of the evaluation object. The main steps of this method are as follows:

- build an evaluation index system: the evaluation index system U is determined by each model and evaluation rules, which is composed of k indicators, $U=(u_1, u_2,\cdots, u_k)$. For more complex tasks, it can be re-decomposed to form the simple and independent evaluation sub-processes.

Each sub-processes corresponds to a subset of evaluation factors $u_p(p=1,2,\cdots,k)$, consisting of q factors, $u_p=(u_{p1},u_{p2},\cdots,u_{pq})$.

- establish evaluation result set: setting up the evaluation result set $V=(v_1,v_2,\cdots,v_l)$ according to the evaluation rules, l is the number of comments, and the specific evaluation result is determined by the percentage system or the level according to the specific circumstances.
- determining the fuzzy relationship matrix: it is to determine the fuzzy relationship matrix from the lowest evaluation indicator set u_p to the evaluation result set $R_p=(r_{ij})_{q\times l}$. The elements in the matrix are all in the [0,1] interval, indicating the degree of membership u_{pi} to v_j. First, determine the single factor evaluation value r_{ij} of the factor u_{pi} for the comment v_j. When the membership degree is judged, c_{ij} shows the number of votes for the item i to obtain the comment j is multiplied by the scale value corresponding to the quantitative membership of the indicator[8], that is:

$$r_{ij}=\frac{c_{ij}}{\sum_{k=1}^{l}c_{ik}} \tag{1}$$

- determining the index weight: the index weight is an important factor of the evaluation . According to the relative importance of the elements in the evaluation index set U and the evaluation result data, the weight of each evaluation index is calculated layer by layer, and intelligent optimization is performed to obtain the weight set $W=(w_1,w_2,\cdots,w_k)$, and $w_p=(w_{p1},w_{p2},\cdots,w_{pq})$.
- single-level fuzzy comprehensive evaluation: the weighted average method, that is, the matrix multiplication, is used as a fuzzy synthesis operation model to calculate the underlying fuzzy evaluation results $B_p=w_p\times R_p$. If the result of the evaluation is represented by a quantitative percentile score, the underlying fuzzy evaluation result is $B_p=100w_p\times R_p$.
- multiple fuzzy comprehensive evaluation: the weighted average method is used to weight and summarize multiple evaluation results layer by layer to obtain the final evaluation result $S=W\times B^T$, where $B=(B_1,B_2,\cdots,B_k)$.

4. Intelligent optimization of evaluating indicator weights

4.1. Weight calculation based on entropy weight method

Entropy is an effective method for evaluating the attributes and importance of indicators and mainly used in information theory, which can be used to measure the size of information. The greater the entropy value is, the smaller the amount of information that the corresponding evaluation index can provide, and the smaller the influence of the index on the evaluation decision. Therefore, the calculation of the entropy weight in this paper is mainly based on the real-time and historical correlation evaluation data of the command model evaluation index, and the appropriate adjustment according to the data changes[9]. There are n evaluation objects, each evaluation object contains m evaluation indexes, and the matrix $Y=(y_{ij})_{n\times m}$ is obtained by dimensionless processing of the evaluation data. The steps are as follows:

- normalization process:

$$z_{ij}=\frac{y_{ij}}{\sum_{i=1}^{n}y_{ij}},\ j=1,2,\cdots,m \tag{2}$$

- the calculation of the evaluation index entropy is as follows (in the formula: k is the adjustment coefficient, $k = 1/\ln n$, so k=0.2569):

$$H(x_j) = -k\sum_{i=1}^{n} z_{ij} \ln z_{ij}, j = 1,2,\cdots,m \tag{3}$$

- converting entropy into weight($0 \le d \le 1, \sum_{j=1}^{m} d_j = 1$):

$$d_j = \frac{1 - H(x_j)}{m - \sum_{j=1}^{m} H(x_j)}, j = 1,2,\cdots,m \tag{4}$$

4.2. Weight optimization based on genetic algorithm

The genetic algorithm is derived from the survival of the fittest, and mainly through the simulation of this evolutionary principle to achieve the best solution in the search domain[10], according to the characteristics of the real-time command model assessment, in its evaluation the weights are dynamically adjusted and intelligently optimized at the same time. The specific steps are as follows:

- determining the subjective weights $W_s = (w_{s1}, w_{s2} \cdots w_{sm})$
- determining the objective weights $W_o = (w_{o1}, w_{o2} \cdots w_{om})$
- using genetic algorithm to optimize indicator weights: Assume that the subjective weight ratio is $X = (x_1, x_2, \cdots, x_m)$, and the objective weight ratio is $X' = (x'_1, x'_2, \cdots, x'_m)$, where $x_i + x'_i = 1, (i = 1,2,\cdots,m)$. Then, the vector W is obtained by synthesizing the subjective and objective weights with the optimized parameter vector X.

The Air-Defense and Anti-Missile warfare command process involves many types of equipment and complex air conditions, such as diversification of air strike weapons and complex operational coordination relationships. Therefore, when assessing command models, it is necessary to thoroughly consider the impact of objective factors on the assessment results. On the other hand, it is the complexity of the combat command process that makes it difficult to predict the target of operations, time and the strength. In many cases, the commander needs to make decisions based on experience, which has a great influence on the subjective evaluation of the command model assessment, and how to coordinate the subjective and objective aspects of the command model assessment is particularly important. Therefore, in order to ensure the minimum deviation of the evaluation results, a non-linear optimization is performed using a genetic algorithm, the degree of dispersion is represented by the discrimination degree of the evaluation data, and the weight for obtaining the least dispersion is calculated. By minimizing the deviation between the subjective and objective weights, the credibility of the assessment results can be improved to a greater extent. The specific function is as follows:

$$\begin{cases} F = \min\sum_{i=1}^{n}\sum_{j=1}^{m}[y_{ij} \times w_{sj} \times x_j - y_{ij} \times w_{oj} \times x'_j]^2 \\ s.t. \quad x_j + x'_j = 1 \\ 0 \le x_j \le 1 \\ 0 \le x'_j \le 1 \end{cases} \tag{5}$$

5. Example verification and analysis

Take the threat assessment model of Air-Defense and Anti-Missile operations under specific conditions as an example, which mainly includes analysis of threat factors, such as air strike target type and capability estimation, importance analysis of covered objects, analysis of target flight time, target altitude, and comparing analysis of target route shortcuts and so on. And other processing processes such as threat calculation, each processing process also includes related logic processing and operations. In this paper, the assessment objects are seven different types of threat assessment models and basing on the corresponding sub-processes and the acquisition and comparison of relevant data parameters producing by the seven types of threat assessment models, an evaluation index system U is constructed, which consists of eight evaluation index subsets, $U=(u_1, u_2,\cdots, u_8)$. Each assessment index includes a number of single-factor indicators that form a multi-level assessment index system, and uses the percentage system to represent the evaluation results. The evaluation results of the seven models are calculated by single-level fuzzy comprehensive evaluation are shown in table 1.

Table 1. Single-level fuzzy comprehensive evaluation results.

Models	u_1	u_2	u_3	u_4	u_5	u_6	u_7	u_8
1	86	89	81	75	78	80	92	79
2	70	88	79	86	69	74	70	78
3	74	45	69	67	44	57	53	70
4	82	97	70	83	89	83	85	86
5	60	77	51	56	65	42	67	45
6	79	91	81	77	78	79	92	75
7	68	73	76	86	91	82	91	67

Firstly, the subjective and objective weights of the evaluation indicators are determined by the Analytic Hierarchy Process(AHP) and the entropy method respectively, and then the subjective and objective biases are minimized and optimized by the genetic algorithm to obtain the combined weights. The population size is chosen 150, the crossover probability is 0.007, and the mutation probability is 0.003. After the evolution of 600 generations, the optimized weight ratio is obtained, and then the weights optimized by the genetic algorithm are obtained[11]. The weights of the indexes basing on the method are as follows:

$$W_s = (0.0609, 0.2500, 0.0652, 0.1800, 0.1851, 0.1108, 0.0710, 0.0770)$$
$$W_o = (0.1506, 0.1109, 0.1110, 0.1280, 0.1209, 0.1257, 0.1420, 0.1109)$$
$$W = (0.0619, 0.1750, 0.0970, 0.1731, 0.1605, 0.1229, 0.0995, 0.1100)$$

Based on the index weights obtained above, the evaluation results obtained from multiple fuzzy comprehensive evaluations are shown in Table 2. At the same time, the average of the evaluation results obtained by multiple fuzzy comprehensive evaluations under different index weights is taken as the ideal value, and the error of the evaluation results of various methods are obtained and is shown in Table 3.

Table 2. Multiple fuzzy comprehensive evaluation results.

Model	AHP	Entropy Weight Method	Objective and Subjective Bias Minimization
1	82.185	82.724	81.957
2	78.841	76.312	78.020
3	55.928	60.143	57.808
4	87.075	84.301	85.632
5	61.216	58.047	59.056
6	82.200	81.578	81.657
7	80.376	79.442	80.373

Table 3. Errors between assessment results.

Errors	AHP	Entropy Weight Method	Objective and Subjective Bias Minimization
MAE	1.019	1.092	0.238
MRE	1.508	1.586	0.331
MSE	1.538	1.600	0.070

6. Conclusion

Based on the current development and application of artificial intelligence in various fields, this paper proposes and designs a corresponding intelligent assessment method based on the characteristics of the command model assessment to improve the subjectivity of the experts. The treatment uses genetic algorithms for intelligent optimization. The feasibility of the method is verified by examples and the data obtained is analyzed. According to the results of data analysis, it can be intuitively understood that the intelligent assessment method is more reasonable and the error of the evaluation result is the smallest. However, the research is in the initial stage and there are still some problems existing. For example, the design of the current assessment knowledge bases or the lack of support for actual combat data; and there needs more in-depth study for the specific command model to ensure that the establishment of the evaluation index system is more reasonable and so on, so there are some study should be made in the follow-up.

References

[1] LI Song,LIU Jin-mang,ZHANG Jin-cheng,etc Research on Model Validation for the Distributed Antiaircraft C^3I[J] Systems Engineering and Electronics,2003,2511):1421-1423
[2] ZHANG Gui-chen Design and implementation of an automatic grading system for practical examination in marine powerstation[J] Navigation of China,2010,33(3):27－30
[3] ZHUANG Wang-yu, LING Dan, ZHAO Jin,etc Research on Weights of Agility Appraisement Indexes[J] Journal of University of Electronic Science and Technology of China,2006,35(6):985-988
[4] LOU Shou-chun Command and Control Models of Surface-to-Air Missile Firing[M] Peking:National Defense Industry Press,2009,6:197-234
[5] DUAN Zun-lei, REN Guang, CAO Hui,etc Intelligent assessment of virtual engine roomcollaboration based on genetic algorithm optimization[J]. Journal of Harbin Engineering University 2017,38(4):514-522
[6] ZHANG Pan, WANG Bo, QIN Xiao- xia Research and Development of Evaluation system Based on Neural Network and Expert System[J] Journal of Chongqing Jianzhu University,2004,26(1):130-131
[7] CHEN Shui-li,LI Jing-gong,WANG Xiang-gong. Fuzzy Set Theory and Application[M].Peking: Science Press,2016:78-80
[8] LIN Chi,LI Song Agility Evaluation of Air-Defense & Anti-Missile Command System Based On Subjective & Objective Comprehensive Index Weighting[J] Journal of Air Force Engineering University:Natural Science Edition,2018,19(4):21-26
[9] LI Shi-wu,YIN Yan-na,WANG Lin-hong etc Driving Fatigue Quantization Based on Entropy Weight Method[J] Journal of South China University of Technology(Natural Science Edition) 2017,45(8):50-55
[10] GAO Shi-yi,LUO Xiao-hua,LU Yu-feng,etc Functional coverage convergence technique based on genetic algorithm[J] Journal of Zhejiang University(Engineering Science)2015,49(8):1509-1514
[11] HE Yan-ping,HE Hui,ZHANG Xiao-lei. Evaluation index of population diversity in genetic algorithm[J] Journal of Har-bin Institute of Technology,2009,41(3):67－70

IOP Conf. Series: Materials Science and Engineering **435** (2018) 012045 doi:10.1088/1757-899X/435/1/012045

Armed helicopter Combat power Index Assessment based on BP Neural Network

DONG Ze-wei*, LIU Xiao-qin, GAO Fu and TANG Xiao-chuan

Army Aviation Academy, Beijing ,101123

*E-mail: hktk0000@163.com

Abstract. According to the use characteristics of armed helicopter in wartime, this paper analyzes and studies the influencing factors of combat power index of armed helicopter, and explains each factor. Based on BP neural network, the evaluation model of combat power index of armed helicopter is established. The BP neural network was trained with 30 groups of training samples determined by experts. The results show that the evaluation method of combat power index based on BP neural network is feasible and effective, reduces the subjective factors in the evaluation process, and makes the evaluation results more accurate and reliable.

1. Introduction

Helicopter is the main battle equipment for the army to carry out three-dimensional attack and defense, and its weapon consists of artillery / gun, air-to-air missile, air-to-surface missile and rocket. The so-called combat power index of armed helicopter is to measure the ability of armed helicopter to carry out combat task by exponential method.

The battlefield situation is changing rapidly and the commander is able to grasp the information of the attack damage situation and the overall combat capability of the armed helicopter in a timely and accurate manner is the key to the victory of the war. The combat power index of the armed helicopter is an important parameter index to measure the combat damage situation and combat capability of the armed helicopter. At present, analytic hierarchy process (AHP), fuzzy comprehensive evaluation (FIC) and power exponent method are generally used to evaluate [1]. However, due to their subjectivity and lack of self-learning ability, they are easy to be judged randomness in practical evaluation. The subjective uncertainty and fuzzy cognition of the participants are restricted by many factors.

BP neural network (back propagation neural network) is a multilayer forward network with one-way propagation. By adjusting the scale of BP neural network (number of input neurons, number of output neurons, number of hidden layers and number of hidden layer neurons) and connection weights in the network, nonlinear classification can be realized. Moreover, any nonlinear function [2] can be approximated with arbitrary accuracy.

2 .Analysis of influencing factors of combat power index

There are many factors affecting the combat power index of armed helicopter, which is a multi-index decision problem. The establishment of evaluation factor set is based on the target tree to determine all the factors that affect the evaluation target.

(1) Communication command capability: the ability to transmit operational information quickly, safely and reliably, and to receive or send instructions effectively to combat operations, recorded as.

Communication command capability mainly includes communication distance, communication quality, communication security, timeliness and so on.

(2) Maneuverability: the ability of helicopter gunships to perform tactical maneuvers flexibly and to achieve free transformation of multiple methods of warfare. The maneuverability mainly includes the maximum speed, the raising limit, the endurance performance, the turn slope and so on.

(3) Strike damage ability: timely target fire strike, effectively damage or destroy the enemyundefineds living force, so that they lose combat capacity, recorded as. The ability of attacking damage mainly includes attack range, attack precision, launching speed, warhead performance, which is mainly related to weapon range, precision, firing speed, fire control system level, projectile capacity, warhead penetration and damage probability and so on.

(4) Electronic self-defense capability: refers to the ability of helicopter gunships to detect enemy targets in advance and are not easy to be found and not to be hit in the course of combat, and to ensure that airborne equipment is not vulnerable to harm under complex jamming conditions. The ability of electronic self-defense mainly includes detection and recognition, anti-interference, alarm, stealth and so on.

(5) Security capability: the ability of an armed helicopter to continue to perform a mission in a complex environment or after being hit by an enemy. The safety protection capability mainly includes the environmental reliability and the protective device resistance to destroy.

3. Combat power index evaluation model

3.1 The Establishment of Model

The combat power index evaluation model of armed helicopter based on BP neural network can be constructed as follows:

(1) The number of neurons in the input layer is determined. According to the analysis of the factors affecting the combat power index of armed helicopters, the main influencing factors include communication command ability, maneuvering ability, ability of attacking damage, capability of electronic self-defense and capability of security protection. The number of neurons in the combat power index of armed helicopter is 5. 5%.

(2) Determining the number of neurons in the output layer[3]. The evaluation results are used as the output layer of the network, and the number of neurons is 1. 5%.

(3) The determination of the number of hidden layers. The more hidden layers, the slower the learning speed of neural network is. According to Kosmogorov theorem, the three-layer BP network can approximate any continuous function under the condition of reasonable structure and proper weight, so the three-layer BP network with relatively simple structure is selected.

(4) The determination of the number of neurons in the hidden layer. There is no uniform rule for determining the number of neurons in the hidden layer. In general, the number of neurons in the hidden layer is determined according to the convergence performance of the network. In this case, the number of neurons in the input layer is calculated according to the empirical formula mx2n1, which is the number of neurons in the input layer. The number of hidden layer neurons is initially set at 11 (the generalization ability of visual network is adjusted during training).

(5) The determination of activation function. The Sigmoid function is chosen as the activation function of the hidden layer and output layer of BP neural network.

$$f(x)=\frac{1}{\left(1+e^{-x}\right)} \tag{1}$$

(6) Determine the structure of the evaluation model. Based on the above analysis, the combat power index evaluation model of armed helicopter based on BP neural network is established, as shown in figure 1.

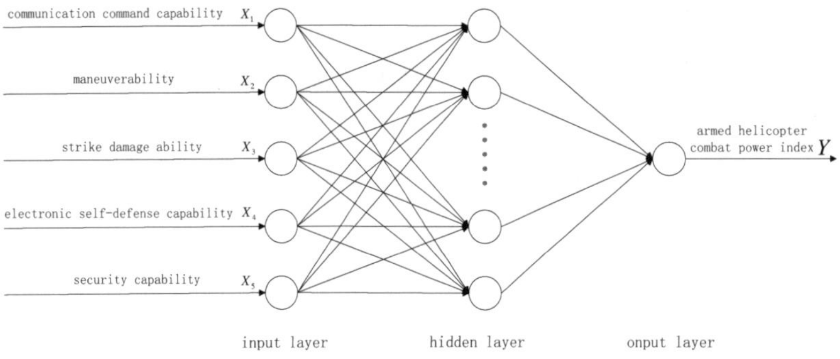

Figure. 1 Armed helicopte Combat power index evaluation model

3.2 Selection of training samples
The performance of the network is closely related to the selection of samples, and it is directly related to the reliability of the evaluation results. In order to make the evaluation result accord with the objective reality and truly reflect the combat power of the armed helicopter, three principles should be followed when selecting the sample [4]: (1) network training, the law extracted from the network training should be contained in the sample, so the sample must be representative; (2) the selection of samples should pay attention to the equilibrium of sample types; (3) the organizations that try to make the number of samples of each category about the same number of samples should pay attention to the cross input of samples of different classes, or select input samples randomly from the training set.

3.3 Normalization of samples
In this paper, the method of linear transformation is used to convert the original training sample into a feasible training sample (the value of each influencing factor is taken between [0 / 1]), which can be divided into two situations:

(1)when the influence factor value is bigger, the combat power index is bigger, according to formula (2) carries on the normalization.

$$Y = \frac{X - \min}{\max - \min} \tag{2}$$

(2) when the factor value is smaller and the combat power index is larger, normalization is carried out according to formula (3).

$$Y = \frac{\max - X}{\max - \min} \tag{3}$$

In the formula: the original training sample value max (or min) is the maximum (or minimum) value of the current gunship helicopter for the same influence factor. For the values which are not easy to be quantified, the values of Maxima and min are obtained by using the method of expert scoring.

3.4 Network training
The standard BP algorithm uses the steepest gradient descent method to revise the weights, and the training process gradually reaches the minimum point from a certain starting point along the slope of the error function to zero error [5]. Because the standard BP algorithm is related to the order of input samples, the convergence rate is slow, it is easy to fall into local minima, and it is easy to cause oscillation effect. In order to overcome the shortcomings of the algorithm, an improved algorithm with additional momentum is adopted in this paper. The method is to add a value proportional to the change of the last weight (threshold) to the change of each weight (or threshold) on the basis of the backpropagation method, and to produce a new weight (or threshold) change according to the backpropagation method. The weight adjustment formula with additional momentum factor is:

IOP Conf. Series: Materials Science and Engineering **435** (2018) 012045 doi:10.1088/1757-899X/435/1/012045

$$\Delta w(k+1) = (1-m_c)\eta \nabla f(w(k)) + m_c(w(k) - w(k-1)) \tag{4}$$

In the formula, it is a weight vector, a training number, $(0 \leq 1)$ a momentum factor (generally about 0.95), a learning rate and a gradient of the error function. It can be seen from formula (4) that its essence is to transfer the influence of the last weight (or threshold) by a momentum factor. With the improved BP algorithm program, the input matrix of the training sample can be input into the computer, and the network model can learn the training sample repeatedly until the precision of the network model meets the requirements. The threshold and connection weights of each neuron after learning are outputted.

4 .Case analysis

In order to verify the accuracy and validity of the combat power index evaluation model of armed helicopter, a certain type of armed helicopter is taken as an example. Identify 30 groups of training samples by organizing experts and normalize them, as shown in Table 1. In this paper, we use the MATLAB neural network toolbox to study the training network, select the first 25 groups of samples as the learning samples of the neural network, and use the additional momentum method to train the network. The learning rate is 0.95, the momentum factor (MC) is 0.98, and the training error is 0.001. After 4695 times of training, the network achieves the precision requirement. The latter five samples are used as test samples and input into the trained neural networks. The test results are shown in Table 2.

Table 1 Training samples standardized results

No.	x_1	x_2	x_3	x_4	x_5	x_6	Expert assessment result
1	0.85	0.77	0.75	0.33	0.83	1.00	0.780
2	1.00	0.85	0.85	1.00	0.67	1.00	0.891
3	0.43	0.92	0.70	0.68	1.00	1.00	0.722
4	0.88	0.85	0.50	1.00	1.00	0.77	0.714
5	0.92	0.75	0.45	0.42	0.50	1.00	0.696
6	1.00	0.60	1.00	0.84	1.00	0.84	0.790
7	1.00	0.85	0.79	0.50	0.83	1.00	0.840
8	0.75	0.58	0.22	1.00	0.58	0.83	0.570
9	0.92	0.75	1.00	0.60	1.00	1.00	0.860
10	0.85	0.50	0	0.85	0.83	1.00	0.540
11	0.82	1.00	1.00	0.90	0.82	0.89	0.920
12	1.00	0.680	1.00	0.60	0.50	0.83	0.776
13	0.92	0.55	1.00	0.76	0.92	1.00	0.820
14	0.50	0.56	0	0	0.83	0.67	0.250
15	1.00	0.91	0.63	1.00	0.50	1.00	0.842
16	1.00	0.75	0.87	0.55	0.84	0.83	0.770
17	1.00	0.85	0.64	1.00	0.92	0.95	0.840
18	0.84	0.59	0.41	0.75	0.92	0.75	0.590
19	0.87	0.69	1.00	1.00	0.60	0.93	0.810
20	0.92	0.78	0.78	0.45	1.00	0.92	0.780
21	0.85	0.34	0.35	0.58	0.83	0.67	0.420
22	1.00	0.50	0	0.83	0.75	0.92	0.580
23	0.84	1.00	0.60	0.85	1.00	0.81	0.780
24	0.68	0.75	0.75	0.85	0.33	1.00	0.710
25	1.00	0.93	1.00	0.84	0.50	0.92	0.880
26	1.00	0.62	0.72	0.67	0.83	0.67	0.580
27	0.52	0.92	0.87	1.00	0.77	0.83	0.752
28	0.55	0.58	0.50	0.75	0.84	0.92	0.570
29	0.68	0.68	0.69	0.83	0.92	1.00	0.710
30	0.68	0.27	0.25	0.83	1.00	0.83	0.424

Table 2 BP neural network test results

No.	Armed helicopter combat power index		Error%
	Expert assessment results	Network output results	
26	0.580	0.5622	3.07
27	0.752	0.7739	2.77
28	0.570	0.5492	3.65
29	0.711	0.7374	3.86
30	0.424	0.4387	3.47

It can be seen from Table 2 that the output data of the network basically meet the scheduled requirements, and from the test results, the network output data are basically consistent with the expert evaluation results of the combat power of a certain type of armed helicopter, and the accuracy is relatively high. It shows that it is feasible and effective to evaluate the combat power index of armed helicopter by using this model.

5 .Conclusion

According to the decision demand of wartime maintenance support, this paper synthetically considers all kinds of influence factors and interrelations, and makes use of the "black box" characteristic of neural network to find out more objectively the implicit relation between each factor and the influence on the combat power of armed helicopter. BP neural network is used to evaluate the combat power index of armed helicopter. The evaluation method reduces the subjectivity of the evaluation process and no longer depends on the expert to determine the weight of the influencing factors, which makes the evaluation results more accurate and credible, and has important practical significance to improve the command and decision-making ability of the commander.

References

[1] WANG Feng, HE Jun, WANG Qiang, Evaluation of Operational Effectiveness for ARM System Based on BP Neural Networks[J],journal of projectiles roktets missiles and guidance ,2006, 26(3): 315-317.

[2] FENG Hui, MAO Hong-bao, LONG Guang-zheng, Operational Efficiency Evaluation of Air Defense Weapon System Based on Vague Sets[J], Modern Defence Technology ,2016, 44(2): 26-31.

[3] LI Guo-hong, SUN Jian, XIA Wei-peng, Evaluation of Operational Capability for Surface-to-Air Missile Weapon System Based on Neural Network[J], FireControl & Command Control, 2011, 36(8): 110-113.

[4] YAO rui-pu,SHEN Hui-zhang, Multicriteria Decision Making Method Based on Vague Sets and Fuzzy Linear Ranking[J], Computer Engineering and Applications, 2009, 45(28): 39-40.

[5] YANG Zuo-bin,GONG Rui,LIU Guang-ming, Antijamming Effectiveness Evaluation of Fire Control Radar Based on BP Neural Network[J], Modern Defence Technology,2010,38(1) : 9-12.

A Comprehensive Cost Function Path Planning Algorithm for Sliding Prediction Based on Terrain Slope

Lanfeng ZHOU[1,*], Lina YANG[2] and Weijie QIAN[3]

School of Computer Science and Information Engineering, Shanghai Institute of Technology, No. 100 Haiquan Road, Fengxian District, Shanghai, China

*E-mail: lfzhou@sit.edu.cn

Abstract. In this paper, the terrain gradient algorithm and the slip prediction algorithm are studied in the three-dimensional path planning process. The slope, aspect and slip of the terrain affect the path planning to a certain extent. At the same time, the ant colony algorithm The selection of each of the relevant parameters was explained. By introducing the cost function of terrain passability, combined with the ant colony algorithm, the path planning algorithm was optimized to improve the efficiency of path planning.

1. Introduction

S.H.Liao et al. [1] mentioned the research methods and development trends of robot path planning technology. According to different understanding of environmental information, robot path planning can be divided into model-based global path planning and model-based local path respectively. planning. The global path planning refers to the planning of the path of the robot under the condition that all the operating environments have been mastered. It is also called the static path planning. Sensor-based local path planning refers to the path planning in the case where all or unknown operating environment information is unknown. The path planning is also called dynamic path planning.

Global path planning algorithms mainly include A* algorithm [2], genetic algorithm [3], neural network method [4], etc. Local path planning algorithms mainly include artificial potential field method [5], fuzzy logic control method, hybrid method, etc. These algorithms have been well used in the two-dimensional environment, but their own characteristics and deficiencies have made their application to 3D path planning problems limited.

Ant colony algorithm is a heuristic algorithm, which has positive feedback, self-organization, distributed computing and strong robustness. Traditional ant colony algorithm has slow convergence speed in solving path planning, and it is easy to fall into local optimum. The resulting path is not the best. Literature [6] proposes an improved ant colony algorithm based on gradient optimization Adadelta algorithm to increase the randomness of the initial pheromone due to the shortcomings of traditional ant colony algorithm in slow path planning and lack of initial pheromone in the 3-D environment path of intelligent machines. The literature [7] adopts different expectation value mechanisms to update the information hormone adaptively in the way of volatility coefficient, and add the inflection point parameter as one of the criteria for the evaluation path. Literature [8] introduced the concepts of optimal solution and worst solution, changed the way of updating global pheromone, and improved the efficiency of global search. Based on the above research foundation, this paper designs the heuristic search in ant colony algorithm. At the same time, it considers the updating strategy of three-dimensional topographic gradient pheromone, and introduces the integrated cost function f(p,n) of slip information into ant colony algorithm. The fitness value function. Matlab software was used to simulate the experiment. The experimental results were analyzed to verify the feasibility and applicability of the improved algorithm.

2. Ant Colony Algorithm

Let the number of ants in the ant colony be m, and the distance between the node i and the node j be denoted as di,j. In the movement process, the ant k selects the next move point according to the pheromone concentration of the path, and the initial time The location path pheromone is all the same constant. At a certain moment t, the probability of selection of ant k from node i to node j follows the following formula [9].

$$p_{i,j}^{k}(t)=\begin{cases}\dfrac{\tau_{i,j}^{\alpha}(t)\,\eta_{i,j}^{\beta}(t)}{\sum_{z\in allowed_k}\tau_{i,z}^{\alpha}(t)\,\eta_{i,z}^{\beta}(t)}, & j\in allowed_k\\ 0, & \text{otherwise}\end{cases} \tag{1}$$

Among them, $\tau_{i,j}(t)$ is the concentration of pheromone on the path of node i to node j at time t, $\eta_{i,j}(t)$ and is the heuristic information of node i to node j. It is generally $\eta_{i,j}(t)$ =1/di,jdefined as, α is the pheromone heuristic factor, and β is the expectation. Heuristic factors, which are used to adjust the degree of influence on the decision., allowedk denotes the next set of feasible nodes.allowedk={0,1,.....,n}-Tabuk,Tabuk(k=1,2,...,m)is a collection of taboos.

In order to continuously screen the path information, the pheromone volatile factor $\rho(0<\rho<1)$ was introduced into the ant colony algorithm. When the ant completes a cycle, the pheromone on the path is updated according to the following formula.

$$\begin{cases}\tau_{i,j}(t+1)=(1-\rho)\,\tau_{i,j}(t)+\rho\,\Delta\tau_{i,j}(t)\\ \Delta\tau_{i,j}(t)=\sum_{k=1}^{m}\Delta\tau_{i,j}^{k}(t)\end{cases} \tag{2}$$

$\Delta\tau_{i,j}^{k}(t)$ indicates the amount of pheromone remaining on the path ij after the current cycle of the ant k, and $\Delta\tau_{i,j}(t)$ represents the sum of the pheromone amount remaining on the path ij for all ants in the current cycle.

$$\Delta\tau_{i,j}^{k}=\begin{cases}\dfrac{Q}{L_k}, & (i,j)\in L_k\\ 0, & \text{otherwise}\end{cases} \tag{3}$$

Among them, Q represents a constant of the initial pheromone quantity, Lk represents the path length of the kth ant in this cycle.

3. Terrain Passivity Cost Function Based on Slip Degree

Slip is a measure of the ability of a lunar rover to move or advance in a given terrain. It is defined as the difference between the control speed and the actual speed in the direction of each degree of freedom of the car body with the control speed as a standard. Because of different terrain types, the terrain has different slip characteristics. Therefore, each terrain has a separate slip model. The slip model is a non-linear approximation function of the terrain slope mapping to slip. Therefore, based on the terrain slope information, the slip can be predicted, the slip cost of the terrain is obtained, and the slip cost of each grid in the map is generated. Sliding maps. The slip cost of each grid can be used as an input to the terrain passivity cost function.First, the terrain classification algorithm is used to classify the terrain based on the terrain texture and color information. Then, a neighborhood analysis method based on the analysis window is used to extract deeper terrain information, such as slope, terrain roughness and topographic relief, etc., which are used to describe the terrain. Characteristic terrain factor;According to the terrain type of known terrain type, the terrain slip model S=S(Xlongit,Xlateral), Among them, Xlongit, Xlateral, Respectively, the terrain is based on the horizontal and vertical slopes of the robot's movement direction. Finally, the slip prediction algorithm is used to predict slip.

The terrain pass-through cost function is a function for evaluating terrain passability. This function considers factors such as terrain slope, terrain roughness, and ladder obstacles. Here, the sliding cost considerations are taken into consideration to form a slip-based map. The terrain passes the cost function. Use this function to evaluate the path's passability between two points.

In path navigation, combining ant colony algorithm is the commonly used 3D path planning algorithm. Here, the above-mentioned terrain cost function is comprehensively considered to form a comprehensive cost function f(p,n), after modeling, sensing data, etc. are not deterministic factors, and the ant colony algorithm is adopted to adapt the value function, and an ant colony algorithm is used for path planning.

A generalized cost function that incorporates the extended slip practicable cost function ftrav trianle into the uncertain information such as the consideration from the node p to the node n edge of the undirected weighted graph created from the grid environment is:

$$f(p,n) = f_1 \bullet f_{trav}(p,n) + f_2 \bullet f_{risky}(p,n) + f_3 \bullet f_{guide}(p,n) + f_4 \bullet f_{smooth}(p,n) \tag{4}$$

Here, ftrav(p,n) is the passability cost function from node p to node n, frisky(p,n) Potential risk cost function, fguide(p,n) Path-directed cost function (whose cost value is negative), fsmooth(p,n) Path smoothness cost function, F1, f2, f3 and f4 are the weights of ftrav(p,n), frisky(p,n), fguide(p,n) and fsmooth(p,n) respectively. Then, the extended slip feasibility passability cost function ftrav trianle is incorporated into the comprehensive cost function f(p,n).

The integrated cost function f(p,n) incorporating the slip information is incorporated into the fitness function of the ant colony algorithm:

$$f'(p,n) = f(p,n) + f_0(p,n) \tag{5}$$

$f'(p,n)$ is an improved fitness function of the algorithm function, f(p,n) is based on the slip prediction integrated cost function, and $f_0(p,n)$ is the basic ant colony algorithm fitness function.

4. Four Experiments and Analysis

A. Initialize the program and create an environmental model of the 3D terrain.

B. Set the starting point and end point of the path search in the 3D model, determine the search direction of the ants, and place all the ants on the starting point. Initialize the iteration number NC_max of the ant ant colony algorithm and the number of ants M.

C. Apply the passability cost function to the fitness value function of the algorithm to calculate the fitness value of each path

D. The ants begin to search for the next node, and each time a search is completed, the pheromone on the path is updated iteratively accordingly.

E. Repeat the above steps to determine whether the final number of iterations has been reached. If yes, end the search and output the optimal 3D path. Otherwise, continue execution.

In order to verify the pass-through cost function of terrain gradient and slip prediction in this chapter, the ant colony algorithm is used to further optimize the path algorithm and simulate it with Matlab software. The simulation environment is $21 \times 21 \times 21$ a three-dimensional topographic map. As shown in Figure 1, the corresponding search starting point and target point are set. According to the ant colony algorithm parameter selection idea, the number of populations is set to 21, the maximum number of iterations is 200, and the α value is 1; The β value is 1.5 and the ρ value is 0.25.

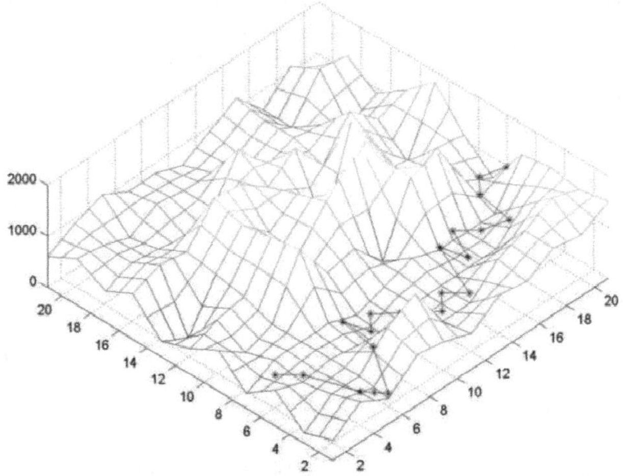

Figure 1. Path planning diagram of the basic algorithm.

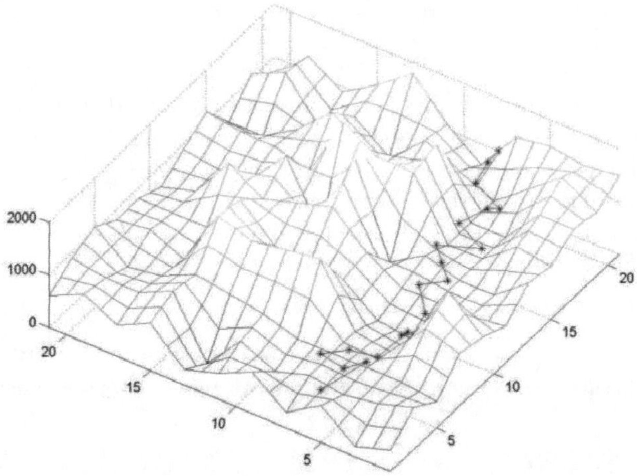

Figure 2. Path planning diagram of improved algorithm.

The results show that a comprehensive algorithm that incorporates the terrain pass-through cost function can effectively obtain the optimal path, can accurately predict the pathability of the path between two terrain points, reduce the running time and number of iterations of the algorithm, and improve the search. Ability and convergence speed.

Table 1 .Comparison of Path Length Results of Two Algorithms (/km).

Compare items	Experiment 1	Experiment 2	Experiment 3	Experiment 4	Average
Basic ant colony algorithm	48.23	47.38	46.59	48.65	47.71
Improved ant colony algorithm	37.32	36.42	34.31	37.56	36.40

Four groups of experimental data were randomly selected for comparison. The comparison of the optimal path length obtained by the two algorithms in Table 1 shows that the path length of the improved algorithm has been improved to some extent, and the passability has been introduced. The function makes the algorithm pay more attention to the tortuosity of the path and avoiding the robot search to generate the invalid movement distance in the process of searching for the path. It can be seen from Table 2 that the algorithm has obtained certain time efficiency in finding the optimal path. improve. Because on the basis of considering the terrain passability, the efficiency of the path planning of the improved algorithm is more stable, and subsequent search will not be performed on some unreasonable paths.

Table 2. Comparison of Finding Path Time Results for Two Algorithms (/s).

Compare items	Experiment 1	Experiment 2	Experiment 3	Experiment 4	Average
Basic ant colony algorithm	43.27	44.38	43.59	44.65	43.97
Improved ant colony algorithm	34.21	32.42	30.20	35.46	33.07

The number of iterations required by the obtained path length of the algorithm reflects the convergence performance of the algorithm to a certain extent. Through the number of iterations of the optimal path length obtained by the two algorithms of Table 3, it can be seen that the convergence speed of the algorithm is also obtained. A great improvement.

Table 3. Comparison of Finding Path Iterations for Two Algorithms.

Compare items	Experiment 1	Experiment 2	Experiment 3	Experiment 4	Average
Basic ant colony algorithm	211	209	205	206	207.75
Improved ant colony algorithm	172	168	163	166	167.25

The number of iterations required by the obtained path length of the algorithm reflects the convergence performance of the algorithm to a certain extent. Through the number of iterations of the optimal path length obtained by the two algorithms of Table 3, it can be seen that the convergence speed of the algorithm is also obtained. A great improvement.

The results show that a comprehensive algorithm that incorporates the terrain pass-through cost function can effectively obtain the optimal path. The improved algorithm can effectively consider the pathability of the path between the two terrains and carry out path planning, and the improved algorithm is effective. Reduces the running time and number of iterations, and improves the ability to search for optimal paths.

5. Conclusion

In this paper, some preliminary explorations of the three-dimensional path planning problem are carried out, and some researches are carried out based on the path planning of the basic ant colony algorithm. For the basic ant colony algorithm, there is a long search time in the path planning process. On the general deficiencies of the path quality effect, the terrain slope between nodes and the slippage problem generated during the movement of the robot are considered in the three-dimensional terrain, and a terrain passability cost function is introduced to the basic ant colony. The algorithm fitness has been improved. The experimental results also show that the improved algorithm has been effectively improved in the length and time of the path search. It can be seen that the improved algorithm in this paper has certain feasibility in the actual process.

Acknowledgments

This work was supported by the National Natural Science Foundation ofChina(41671402).

References

[1] Z.P.Xie , C.H.Cai i. Game map shortest path search design and implementation. Computer Engineering and Application.2015.24:115~117.

[2] D.J.Wang .Path planning of indoor mobile robots based on improved A* algorithm[J].Journal of Tsinghua University,2012, 52(8): 1085-1089.

[3] T.F.Shi .Application of Improved Genetic Algorithm in Path Planning of Mobile Robots[J].Computer Simulation,2011,2(28):193-195.

[4] W.Qian ,A.G. Song , H.T.Zhang, P.W.Xiong. Robot path planning method based on adaptive fuzzy neural network[J].Journal of Southeast University(Natural Science Edition)2012,42(4):637-642

[5] P.Biber,S. Fleck and T. Duckett. 3D Modeling of Indoor Environments for a Robotic Security Guard. IEEE Computer Society Conference on Computer Vision and Pattern Recognition.2005,3(6):124~1305.

[6] K.L.Wang ,S.W.Dang ,F.J.He ,H.J.Huang .Three-dimensional path planning of autonomous mobile robot based on improved ant colony algorithm[J].Industrial Control Computers,2017,30(11):85-87.

[7] X.F.Wan , W.Hu , W.Y.Fang . Robot path planning based on improved ant colony algorithm[J]. Computer Engineering and Applications,2014,50(18):63-66.

[8] C.Zhang ,Y.Z. Ling , M.Y.Chen .Improved ant colony algorithm for path planning of mobile robots[J].Journal of Electronic Measurement and Instrumentation,2016,30(11):1758-1764.

[9] Y.Q.Chen . Improved ant colony algorithm and its research in path planning of mobile robot[D]. Donghua University, 2010.

IOP Conf. Series: Materials Science and Engineering **435** (2018) 012047 doi:10.1088/1757-899X/435/1/012047

The Spacetime Perspective on AI-assisted Decision Making

Thang Nhut Nguyen*

California State University Long beach, USA

*Thang.Nguyen@csulb.edu

Abstract. We consider human decisions as shaped by five P's: People, Principle, Policy, Process and Practice. We examine and characterize three continua in terms of the five P's: (1) the *natural continuum* from particle to transcendental system where human species (people) is the major component, (2) the *hardware-software continuum* from bit to complex computer application created by humans (people), and (3) *human decision spacetime* as "E^4={rational, non-rational, irrational, time} → problem identification and investigation → solution" in which human (people) decisions govern and move their creations such as business, market and economy. We use analogy and analogical reasoning among components of the continua to formulate automation for AI-assisted decision making.

1. Five P's of business: People, Principle, Policy, Process and Practice

Decisions and actions are entities that move business and government from a point A to point B, e.g. one situation to another. For business and government environments, we consider decisions as shaped by the five P's as shown in Figure 1. Each P is complex in its own right. The P in the middle (People) creates and affects all other P's and itself.

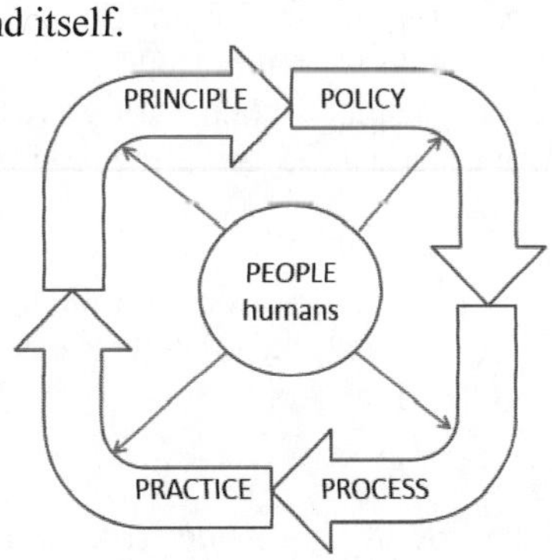

Figure 1. The five P's

People is the most intelligent and most important entity of the environment in which they live and work. People identify *principles* that they draw from nature and life, adopt them, observe them, and experience them. A simple principle is "common sense". Another simple but highly difficult principle to observe is "ethics" which passively guides human behavior. People create *policies* (e.g. rule, regulation, law, etc. in policy making) as norm or guidance to scope their decisions. One simple policy on financial derivatives is the cap on futures and options trading in Baring Bank Singapore [1]. People develop *processes* (including *planning*) and models to arrive at their decisions and actions. They build theories and devise methods such as those found in Management Science/Operations Research or

Business Analytics. They exercise best *practices* (execution) for business growth and profit. For the most part, the P's in business are grossly flat and linear in descriptive, prescriptive and predictive modeling.

1.1. Problematic P's

Note that four of the five P's, Principle, Policy, Process and Practice, commonly exist in corporate and government documents. They aim at expressing full rationality (plain reason/logic and data-driven), however, at times, rationality is not fully observed. It can be highly influenced by non-rationality and irrationality [2]. The non-rationality or irrationality occur due to a number of reasons. One of which is the disrespect to one or more of the other P's by the first (People) resulting in anomalies. Signs and symptoms of anomalies might be hidden, covered up, undetectable, and unpredictable [3) [4].

Examples of corporate fiascos' undetected anomalies caused by decision making in business institutions are: *People* can turn bad due to greed or other reasons (e.g. Nicholas Leeson, GM of Barings Bank Singapore played the Martingale game and exhausted the Bank's assets in 1995). *Principles* may not be observed or are intentionally ignored (e.g. leverage level of Lehman Brothers reached 40 to 1 resulted in huge debt, low rating and insufficient cash flow). *Policies* may be narrow, incomplete or intentionally violated and *Processes* may be ill-defined or intentionally skipped (e.g. Andrew Fastow, CFO of Enron violated the 3% independent investment rule in thousands of special purpose entities). *Practices* may be badly executed or intentionally abused (e.g. the subprime lending market went uncontrollably due to NINJA loans (No Income No Job or Asset). These have resulted in corporate fiascos leading to market turmoil and economic meltdown as we have experienced in 2008 [5].

1.2. Incomplete reforms

Many previous and major corporate fiasco such as Barings Bank, Enron and Lehman Brothers have been largely and carefully investigated, e.g. for why and how they happened, e.g. violations and issues on the P's and lessons learned. They were well discussed in the literature [6] [7] [8].

They have resulted in reforms which have been proposed for future corporate fiasco avoidance or prevention purpose. For example, after Enron and MCI WorldCom, at the policy level, we had Sarbanes-Oxley Act (SOX) in 2002 and after Lehman Brothers, Dodd-Frank Act in 2009. Ethics in institution became a hot issue in the literature and business schools. However, debacles caused by fiascos kept reoccurring. After Barings Bank's collapse in 1995, another rogue trader of Societe Generale, Jerome Kerviel, single-handedly lost some $7 billion in 2008 in practically the same fashion. After SOX, Freddie Mac committed accounting fraud in 2003. Reforms in terms of policy, process and practice appear insufficient. In addition, at times, the decisions on corporate fiascos appear arbitrary, e.g. both Bear Stearns and AIG, were bailed out by the Fed and US Treasury but Lehman Brothers was not. In fact, there is current effort to repeal Dodd-Frank Act.

2. Approaches to decision making

Traditional decision making under uncertainty and under risk is based on probability theory, utility theory, expected utility theory, and recently prospect theory [9], and other methods in disciplines such as MS/OR [2]. Two characteristics of prior theories are observed: the domain is the set D of relevant decisions to be mapped to [0, 1], and the goal is optimization.

Herbert Simon gave an account of rational decisions and decision making methods and a plethora of diverse treatments which were called theories of the firms, including his own. He differentiated three categories of decisions: rational, non-rational and irrational. Since then, in the 1980's and 1990's, there has been effort to account for the role of emotion and intuition (irrational and non-rational) in decisions under risk and uncertainty. In the 2000's, decision making became the subject of neuroscience research [10], cognitive science [11], cognitive neuroscience [12]. Neuroeconomics was introduced [13]. Kahneman formalized decision making in his Thinking Fast and Slow, and which argued the linkage between neuroscience to psychology [14].

Note that in theory and practice, we have been taught in textbooks and best practices to follow the following general decision model: "*Euclidian problem space E^3 over time E^1 as in E^3 X E^1 → linear decision alternatives → solution*" for deliberation on a choice to be selected. In this model, the thinking process is actually effective on a narrowly defined problem since it involves only a small collection of alternatives. However, a selected decision is good only in the problem space identified or scoped. Its impact to other problems spaces (lager, smaller or different from or relevant to the original problem space) is of limited understanding where it might be relevant or unpredictable.

To overcome at least the above issues and maybe others, we propose to reverse the problem space model sketched in the previous paragraph. We look at a decision continuum space in which everything is subject to consideration and investigation: a thought, a sign, a symptom, a problem, a decision, a solution, etc. as in "*decision spacetime (E^4={rational, non-rational, irrational, time} → problem identification and investigation → solution*", where decision spacetime is scoped and defined analogously to Minkowsky-Riemann-Einstein spacetime [15] sketched in Figure 2 below. Thus, in spacetime jargon, a decision is an event, thinking is light, and a thought is the occurrence of a flash light. *A thought* involves neurological as well as psychological involving the P's. We claim that *decisions* could be looked at in decision spacetime.

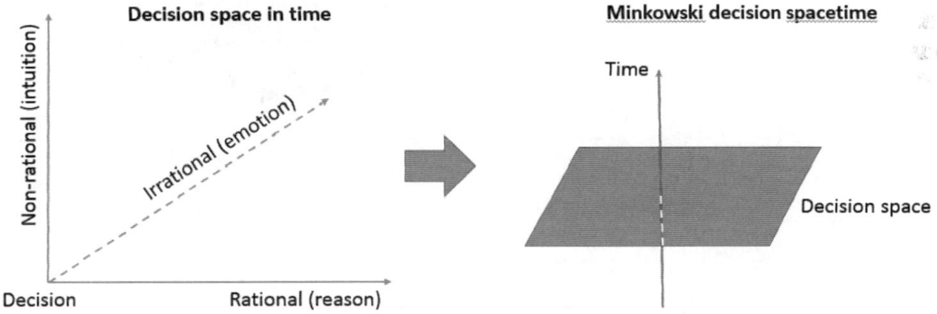

Figure 2. Decision spacetime

3. Characterizing decision spacetime

3.1 Decision cone

The first insight is that a decision made is similar to a mass dropped on the surface of water in a pond. It creates the circular effect which ripples through time (Figure 3.A). It is seen as a cone (Figure 3.B). In this way, the *decision cone* is similar to the *light cone* in Minkowski's environment, in which human thought is the flash light traveling through spacetime.

With the light cone concept, there are the *future cone* (the upper cone of Figure 3.C) and *past cone* (lower cone). The light travels within the light cone is called worldlines (at the speed less than speed of light), on the surface of the cone is spacelines (at the speed of light) and outside is timelines (at speed higher than the speed of light). The connection between the past cone and future cone illustrates a causal relationship. Nothing can pass from inside to outside area. Such a jump is an anomaly or singularity.

In the decision analogy, the *past cone* houses past decisions, and the *future cone* the consequence and impact, and future decisions. Different than the light cone concept, the human thought process as decision making can go from inside to outside. It is because a thought is a networked neurons where the synapses can intertwine in an unpredictable manner, specifically it can involve from rationality (pre-frontal cortex) to irrationality (emotion, limbic system) or non-rationality (intuition, gut feeling). Human intention, desire, greed, ambition, etc. can be well hidden in their decision's spacetime, unreachable by anyone. This would offer an explanation on why Nicholas Leeson of Barings Bank, Andrew Fastow of Enron and Richard Fuld of Lehman Brothers committed fraudulent acts.

IOP Conf. Series: Materials Science and Engineering **435** (2018) 012047 doi:10.1088/1757-899X/435/1/012047

3.2 Decision curvature

We argue that a critical decision is similar to a heavy mass creating a curvature (Figure 3.D) in decision spacetime similar to how the sun creates a gravitational curvature, around which the earth and other planets are orbiting. The curvature of a heavy or critical decision affects other decisions. Decisions can be the matter expressed as mass *m* (or force or energy). It implies that we can express, analogously, the *decision (e.g. matter) tells the institution, market or economy (entities in decision spacetime) how to curve*, and the *institution, market and economy tells decision how to move*.

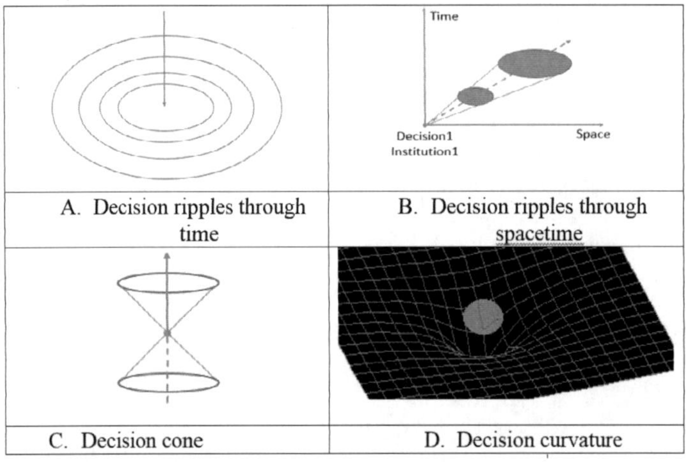

Figure 3. Decision in spacetime and curvature

3.3 Decision impact

The cone will introduce a quadratic *term similar to* $\Delta e = (dx^2 + dty^2 + dz^2 - cdt^2)$ called Minkowski interval (where c is the speed of light) as opposed to the linear distance (difference) of all decision probabilities used in decision science for decision making. When c=1, it is equivalent to a decision having no action (equivalent to a Newtonian free fall). The closer a particular institutionN (Figure 3.B) is to the point of reference of decision1 (institution1) the stronger and the faster the impact affecting it from decision1 in the future cone. Another insight is that we need to watch and measure the combined effect in the collection of different decisions affecting the current one (decision1) in the past cone. The above discussion suggest two measures: Minkowski interval and combined measure of decision impact.

4. Analogy and analogical reasoning for potential AI processes and algorithms

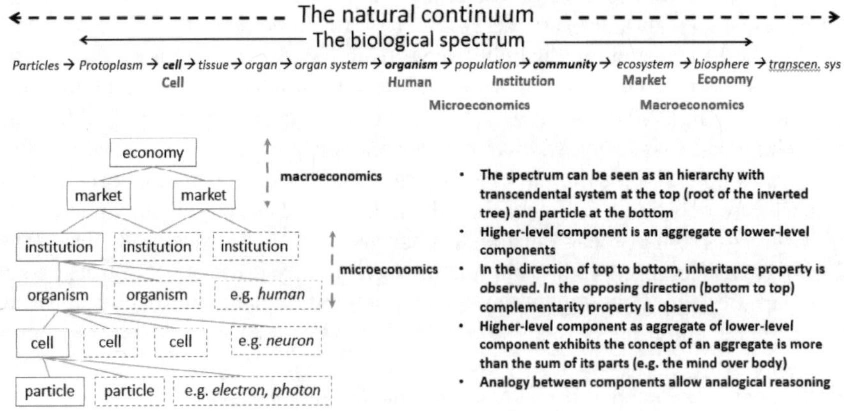

Figure 4. The biological spectrum/hierarchy

4.1. Human decisions from biological spectrum perspective

Placing decision spacetime in the natural continuum where any entity occurs suggests that there are analogies among different components of von Bertalanffy-Boulding's biological spectrum/hierarchy [16] [17] from protoplasm to biosystem (Figure 4). The analogies must be there since all components are part of the hierarchy. The human brain is an aggregate of lower components (particles, cells, brain tissues, etc.) whereas the mind is capable of creating abstract concept such as institution (a kind of community), market (a kind of ecosystem) and economy (a kind of biosystem). Decisions are made by the human brain/mind. Furthermore, the human component is to maintain balance (constancy in cell, homeostasis in human, stability in institution, equilibrium in market or balance in economy) and control in the other P's it creates. It implies that any deviations from balance and control would expose the signs and symptoms of anomalies in decisions made by humans (people).

4.2. The P's by analogy and analogical reasoning

First, we identify three core principles underlying the balance and control of each component: its environment, goal and control: (1) *Environment* in which each component lives and works, e.g. cells (Cells box of Figure 5) are in the milieu interieur (internal environment) [18], humans of institutions are in the data environment (Humans box), institutions are in the information environment (Institutions box), markets are in the product environment (Market box), and economy in the financial environment (economy), (2) *Goals* to be achieved by each environment (in each of the above boxes). e.g. constancy in cells, homeostasis in humans [19], stability in institutions, equilibrium in markets and balance in economy, and (3) *Control* to be exercised by each component, e.g. Ashby's molecular cybernetics in cell component [20], Wiener's cybernetics in humans [21], Beer's managerial cybernetics in institutions [22], Grubstrom's market cybernetics [23], and Hoffman's economic cybernetics [24].

Second, we recognize the analogous organizations among components: either hierarchical, networked or both. These organizations exist in the human body, human brain, in institutions, market and economy. And, *third*, the analogy between operations in each of the component involving their supporting entities, e.g. DNA in cells is analogous to internal biochemical and electrical rule in human body, corporate policy in institution, regulation and law in market and economics. Below is an analogy example between cancer in human and fiasco in institution in terms of fiascos.

When abnormal cells in the body grow uncontrollably, they exist as tumor. The tumor can turn malignant and can proliferate to other organs and affect the homeostasis of human body. By the time cancer is discovered it is in later phase, and possibly causes death. The autonomic nervous system (ANS) is aware of most of what is going on with cancer cells. However ANS could not detect cancer cells for a couple of reasons and it does not report to human conscience. One of the reasons, for example, is that the cancer cells have on their membrane a special protein which makes them look like friends rather than foes to the white cells. Analogously, in an institution, we might have one employee and/or a group of employees who turn greedy. They can grow uncontrollably. They cover up their decisions to commit fraud. By the time top management is aware of what is going on, it is too late.

The first insight from the above analogy is that, based on the fact that ANS does not have complete knowledge of cancer and does not report to higher level, we want our institution to be more than the ANS-equivalent functionality. We want our model to have a Management by Exceptions (MBE), which can fully detect and report anomalies and singularities. It means we have to look at decision sin spacetime, and build an MBE or rebuild an existing one, which can do more than data mining from our data warehouse. This is to assist higher management in spotting anomalies hidden, covered up or undetectable by the current IT applications.

4.3. AI applications

We can look at human decisions in its *human decision continuum* (darkened triangular area) as it interacts with the two continua as shown in Figure 5. The *natural continuum* (top thick bar, from particle to transcendental system) which the biological spectrum is a part of (detailed in Figure 4) offers an understanding on how human decisions are made from neurological perspective (low level) in the brain

IOP Conf. Series: Materials Science and Engineering **435** (2018) 012047 doi:10.1088/1757-899X/435/1/012047

to the psychological perspective (high level) in the mind. Human decisions create everything in the *hardware-software continuum* (bottom thick bar, from bit to complex applications, AI included). We postulate that humans must have been patterning their decisions after the autonomous processes in the biological spectrum components, and implementing them on the hard-ware-software continuum.

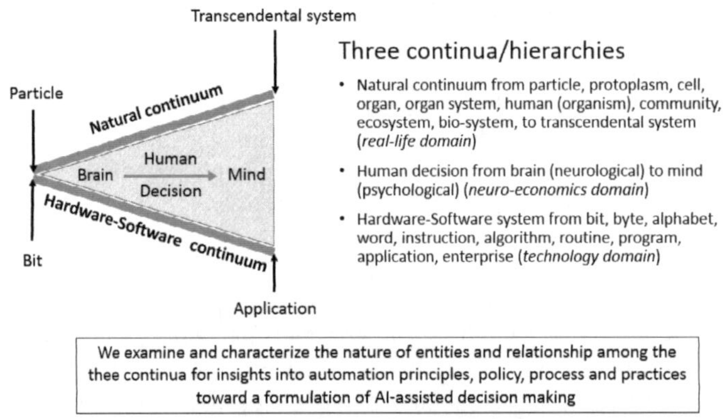

Figure 5. The three continua

The above argument and example attempt to illustrate, from a curved spacetime perspective, the *why* and *how* a corporate fiasco may occur as the results of disrespect to the P's for the purpose of prevention. Our general scheme to address the AI-assisted solution is as follows.

- *Ability to recognize anomalies in situation perception of critical signs and symptoms.* In Barings case, executives and top management missed all the signs of Nicholas Leeson being in a path to destruction because the only thing they saw was the bonus. There is a need to increase monitoring and reporting capability of the signs and symptoms of anomalies by technological advances to enhance human shortcomings [25] [26].
- *Ability to measure decision impact using Minkowski interval as metric, beyond the narrowly defined problem space with a set of alternative D.* We extend the set D of singleton decisions with corresponding measure to the σ-field $\{D, 2^D, m\}$ on D with Dempster-Shafer's belief measure *m*, where 2^D is the powerset of D, where the rule of combination can be exercised. The above gives a suitable scope for developing formulas, equations, algorithms for potential AI applications.
- *Consideration on long-term decision impact to subsequent consequences in curved spacetime.* People tend to formulate immediate or short-term reactions to perceived or reasoned consequences (therefore flat). Long-term consequences are in curved spacetime, more difficult to detect or predict them. In Enron case, it was the complexity of SPE structural and functional organization, in addition to the double role of Arthur Anderson, both auditing and consulting (a violation of policy), which blinded the process and practice.

5. Concluding remarks

We have presented in this paper a collection of systemic concepts and a sketch towards addressing critical decisions from a preventive perspective, differently from traditional approaches: (1) descriptive, (2) prescriptive, and (3) predictive. Two disciplines, healthcare and criminology, and recently cyber security, are probably more advanced and more developed than business and management disciplines, in terms of prevention involving the five P's: people, principles, policy, process and practice. In fact, healthcare has seen an abundant of medical research and development. So are criminology and cybersecurity. We thought lessons could be learned from reviewing the above disciplines for potential analogy, where applicable, among their five P's and ours. After all these disciplines all belong to the two continua (natural continuum and hard-ware-software continuum). From the spacetime perspective, AI-assisted decisional algorithms for prevention will be further detailed. This is a work-in-progress, far from reaching its end. We ask for the indulgence of readers and we seek feedback on the validity of the proposed concepts and descriptive sketch of prevention for further development.

IOP Conf. Series: Materials Science and Engineering **435** (2018) 012047 doi:10.1088/1757-899X/435/1/012047

References

[1] Leeson, Nicholas. 2012. *Rogue Trader*, Little Brown Book Group.

[2] Simon, H. 1979. Rational Decision Making in Business Organizations, *The American Economic Review*, Vol. **69**, No. 4 (Sep. 1979

[3] Ketz, J. E. 2003. *Hidden Financial Risk: Understanding Off-Balance Sheet Accounting*, Wiley, 2003.

[4] Le Maux, J. and Morin, D. 2011. "Black and White and Red All Over: Lehman Brothers' Inevitable Bankruptcy splashed across its Financial Statements". *International Journal of Business and Social Sciences*, Vol **2**, No. 20.

[5] Shell, A. 2015. "Lehman Bros. collapse triggered economic turmoil", http://abcnews.go.com/Business/lehman-bros-collapse-triggered-economic-turmoil/story?id=8543352

[6] Powers, W. Jr. 2002. Report of Investigation, http://i.cnn.net/cnn/2002/LAW/02/02/enron.report/powers.report.pdf

[7] Dharan, B. and Rapoport N. (Eds). 2009. *Enron and Other Corporate Fiascos*, Foundation Press.

[8] Kirkpatrick, G. 2009. "The Corporate Governance Lessons from the Financial Crisis". Financial Market Trends, OECD, Vol. 1

[9] Kahneman D. and Tversky, A. 1979. A. Prospect Theory: An Analysis of Decision under Risk. *Econometrica*. **47**(2).

[10] Damasio, A. 2005. *Descartes' Error: Emotion, Reason and the Human Brain*. Penguin Book.

[11] Frydman. 2016. The Psychology and Neuroscience of Financial Decision Making, *Trends in Cognitive Science*. DOI: https://doi.org/10.1016/j.tics.2016.07.003, 2016

[12] Naqvi, N., Shiv, B. and Bechara, A. 2006. The Role of Emotion in Decision Making: A Cognitive Neuroscience Perspective, *Current Directions in Psychological Science*, Vol. **15**, No. 5. 2006.

[13] Glimcher P.W. Camerer C.F., Fehr, E. and Poldrack R. A. 2013. *Neuroeconomics: Decision Making and the Brain*. Elsevier.

[14] Kahneman D. 2011. *Thinking Fast and Slow*. Farrar, Strauss and Giroux.

[15] Chappell, J. M., Hartnett, J. M., Chappell, J. G., Hartnett, N., Iqbal , A. and Abbott, D. 2016. *Exploring the origin of Minkowski spacetime*, https://arxiv.org/pdf/1501.04857.pdf

[16] von Bertalanffy, L.V. (1950). An Outline of General System Theory, http://www.isnature.org/Events/2009/Summer/r/Bertalanffy1950-GST Outline_SELECT.pdf

[17] Boulding, K. E. 1956. "General System Theory – The Skeleton of Science" in *Management Science*, Vol **2**, No. 2.

[18] Gross, C. G. 1996. "Claude Bernard and the constancy of the internal environment", *Neuroscientist* **4** (1)

[19] Cannon, W. 1963. *The wisdom of the body*, The Norton Library, Norton & Company

[20] Ashby, W. R. 1957. *An Introduction to Cybernetics*, Chapman and Hall LTD.

[21] Wiener, N. 1948. *Cybernetics or Control and Communication in the animal and the machine*, The Technology Press.

[22] Beer, A. 1995. Brain of the Firm, 2e. Classic Beer Series. ISBN: 978-0-471-94839-1.

[23] Grubbstrom, R. 1969. *Market Cybernetics Processes*, Stockholm, Almqvist & Wiksell.

[24] Hoffman, R. 2010. "A Cybernetic Approach to Economics", in *Cybernetics and Human Knowing*. Vol. **17**, no. 4

[25] Nguyen, T. 2014. A Different Approach to Information Management by Exceptions: Towards Bankruptcy Prevention, *Information & Management*, **51** (Issue 1)

[26] Nguyen, T. 2016. *Preventing Corporate Fiascos: A Systemic Approach*, Palgrave MacMillan.

IOP Conf. Series: Materials Science and Engineering **435** (2018) 012048 doi:10.1088/1757-899X/435/1/012048

The Method of Machining Process Parameter Generation Based on MBD

Pengpeng Ding[1], Xiaohu Zheng[1, *], Jie Zhang[1], Shimin Liu[1] and Changwei Ren[1]

[1]College of Mechanical Engineering, Donghua University, Shanghai 201620, China

[*]E-mail: xhzheng@dhu.edu.cn

Abstract. The steering engine room is one of the key structural components of the carrier rocket. It has many types of parts features and a large number of process parameters. Moreover, it needs a long processing cycle. The processing parameters are generally set manually by the technicians based on experience, and the design information obtained from the parts cannot directly provide a reference for the generation of subsequent process parameters. In order to correlate the non-geometric information extracted from the MBD (Model Based Definition) model with the process parameters generation process and make the MBD model run through the entire process parameter generation process, a method of process parameter generation based on MBD is proposed in this paper. The feature information of MBD model can be identified with feature recognition algorithm. The non-geometric information of MBD model can be extracted by information extraction algorithm. In addition, processing parameters can be generated automatically by using BP neural network according to the acquired information. This method can shorten the generation time of the entire process parameters, reduce the production cycle and increase the production efficiency.

1. Introduction

Carrier rockets are of great significance for the security. In order to meet the demands of increasing launch requirements, higher requirements are placed on the production cycle and production efficiency of launch vehicles. The steering engine room of carrier rocket is one of the key structures, which has irregular surface shapes, staggered features, and complex feature structures. Nowadays, identification and extraction of parts information by means of manual methods takes a long time. According to part machining information, manual selection of processing parameters will increase production time and reduce processing efficiency. These processing methods relying on experience and knowledge not only seriously affect the efficiency of the design and the quality of machining, but also cannot meet the requirements for integrated manufacturing of different types of products. Therefore, it is necessary to combine the actual manufacturing process with digital technology and propose a method that can automatically extract part information and automatically generate process parameters from the design to the manufacturing process to change the current status of low digitization and intelligence in the development and production process.

Domestic and foreign experts have conducted in-depth research on the extraction and processing parameters of non-geometric information till this moment. Qiao et al. proposed an approach to region-based feature recognition for structural parts [1]. Zubair et al. proposed the volume decomposition method to recognize feature automatically of regular features for symmetrical and non-symmetrical cylinder part [2]. Sui proposed the method to parts information extraction and storage of 3D-CAPP system [3]. Fang et al. proposed the method to coordinate measuring information extraction of parts based on MBD dataset [4].These research mainly from the MBD model to identify the part features or to extract the part information. The digitized information obtained is not directly used as a basis for generating the process parameters. Vafadar et al. proposed a simulation model using genetic algorithms to find the optimal process parameters [5]. Torres-Treviño et al. proposed a hybrid system that sets processing parameters from experimental data [6]. Roy et al. proposed a method to optimization of process parameters in CNC lathe by using taguchi coupled genetic algorithm [7]. Kant uses artificial neural network and genetic algorithm to predict and optimize the surface roughness [8]. The above studies mainly use genetic algorithms or neural networks to realize the generation and optimization of process parameters. These methods input the existing processing information into the trained model to generate the process parameters, but do not fully consider the source of processing and manufacturing information. It makes the extracted non-geometric information and process parameter generation process not directly related. In summary, the entire manufacturing process from part information extraction to process parameter generation is not closely related and there is no suitable method to solve this problem.

In order to associate the non-geometric information extracted from the part with the process of the process parameters generation and make the MBD model of the part run through the entire process parameter generation process, this paper proposes a method for generating process parameters based on MBD. This method can identify feature and extract information of parts based on the MBD model to obtain the design information of parts. Also, it can use BP neural network to generate process parameters directly with the acquired non-geometric information. The workload of the process personnel can be reduced, the time for generating the process parameters can be reduced, and the processing efficiency can be improved through this method.

2. The framework of process parameter generation

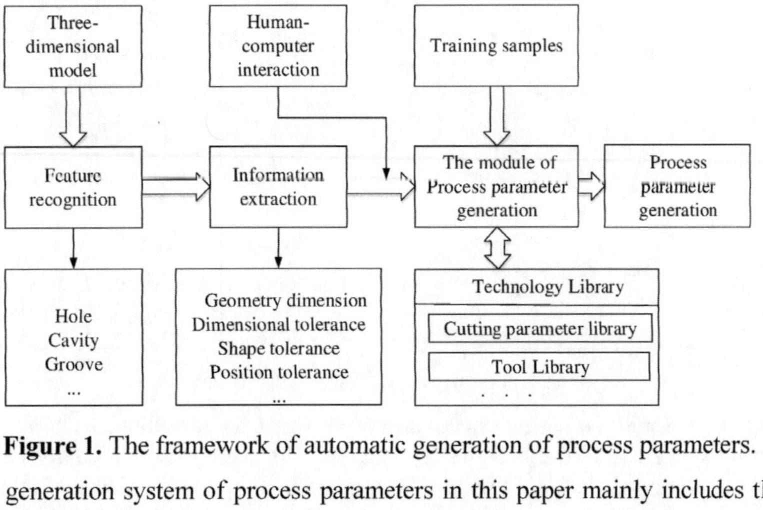

Figure 1. The framework of automatic generation of process parameters.

The automatic generation system of process parameters in this paper mainly includes three parts, i.e. feature identification, information extraction and automatic process parameter generation. The designer designs the model in the three-dimensional software, and then uses the feature recognition algorithm to identify the typical features in the model, such as hole features, groove features, and cavity features. After identifying the feature of the part, the non-geometric information of the part is extracted through the information extraction algorithm, including the geometric dimensions, dimensional tolerances, shape and position tolerances, and surface roughness of the part. After the feature recognition and information extraction are completed, the two types of different information data need to be unified and stored. Then, the craftsman formulates the process and determines the working step according to the obtained geometric information and non-geometric information. The processing features, such as drilling or milling planes, are determined by human-computer interaction and the non-geometric information, which is extracted before, are used as inputs for the module of process parameter generation.

The module of process parameter generation uses the BP neural network to train the existing process parameter samples, and then obtains the corresponding network structure model. Finally, the obtained data are input into the BP neural network to generate processing parameters directly for machining. The framework of the system is shown in Figure 1.

3. Specific generation process of process parameter

3.1. The method of feature recognition based on MBD

During the automatic generation of the entire process parameters, feature recognition and information extraction are the most basic parts. The design information of the parts mainly comes from manual identification in the current production process. However, the work of converting design information into manufacturing information is heavy. The method based on artificial identification has the disadvantages of low efficiency and great manpower. It also increases preparation time and extends manufacturing cycle. Feature recognition and information extraction are the key technologies of integrated CAD, CAM and CAPP [9], which can replace manual information extraction and automatically transform design information into manufacturing information, so as to reach the goal of reducing preparation time and improve process efficiency.

This paper divides these features into concave and convex features to identify feature information in the MBD model. In order to make the software automatically identify the concave and convex manufacturing features in the part model, the following feature recognition method is proposed, as shown in Figure 2.

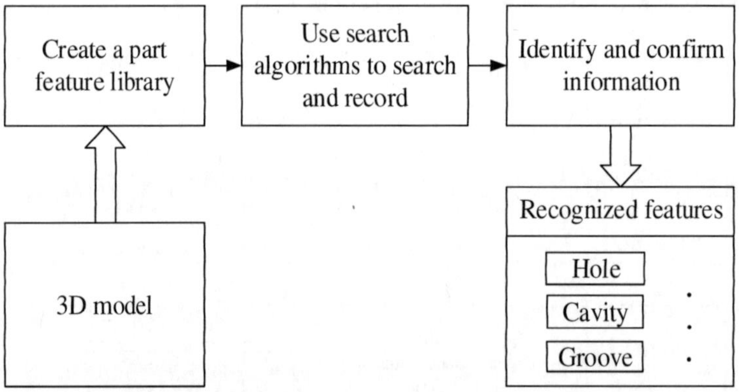

Figure 2. The process of the feature recognition.

First, the feature recognition module needs to create a part feature library. The feature of the part can be divided into hole features, closed cavity features, slot features, open cavity features and boss features. After completing the typical feature classification, we can use adjacency relationship model of the three-level attribute including top layer, geometric layer and newly added outer adjacency layer [10] to record the required information for manufacturing feature recognition in matrix form and store this information in the part feature library. Different from the recognition mode of sub-graph segmentation-Feature database matching method, we use the descriptive information of the feature model in the Part feature library to establish the necessary conditions for searching. The search algorithm is used to select and record the feature surface satisfying the conditions based on the search condition. Finally, all the features can be identified and confirmed. The result of identifying a groove feature in the steering engineer room is shown in Figure 3.

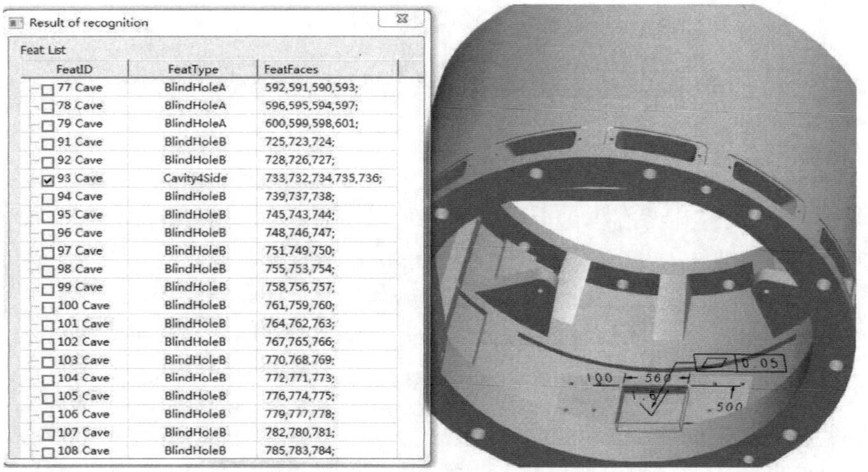

Figure 3. The result of identifying a groove feature.

3.2. The theory of information extraction based on MBD

After using the feature recognition algorithm to obtain the feature information in the part MBD model, the non-geometric information can be acquired by extraction algorithm [11]. MBD model contains a lot of non-geometric information. The information play important role in the entire process, which affect the processing parameters in the manufacturing process, such as tool parameters and cutting parameters. Therefore, this algorithm can extract non-geometric information during the manufacturing process and show the information in a 3D solid model in order to guide the production process .The running process of entire algorithm is shown in Figure 4.

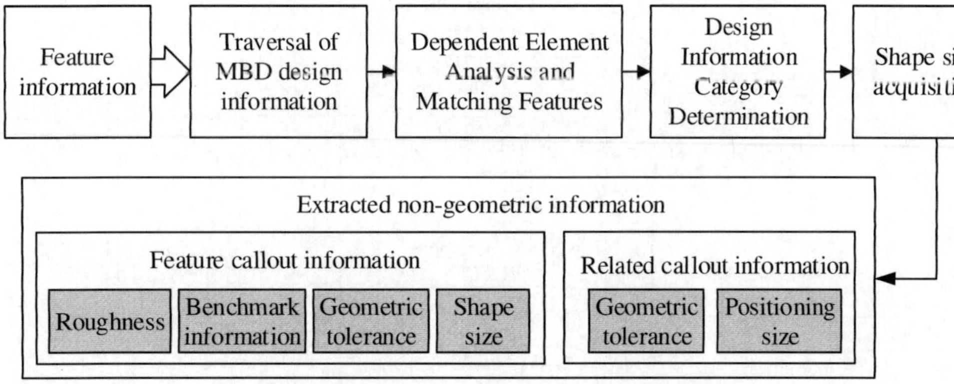

Figure 4. The algorithm process of information extraction.

First, all the annotation information is obtained from the part MBD model, and then the attached elements of all the acquired annotation information are converted into unified geometric element information. Next, the geometric element information is matched with the geometric information in the feature to obtain the feature information associated with the annotation. Then, the category of three-dimensional annotation is divided according to the associated feature information. For example, the dimension is classified into the position size and the shape size and the tolerance annotation information is classified into the correlation tolerance and the independent tolerance. Finally, the rules for defining the shape and size of information are made and the annotation information is output in more detail based on the features. After the annotation information is extracted and classified, it is stored in the corresponding feature data structure. The non-geometric information extracted from a groove feature of the steering engine cabin is shown in Figure 5.

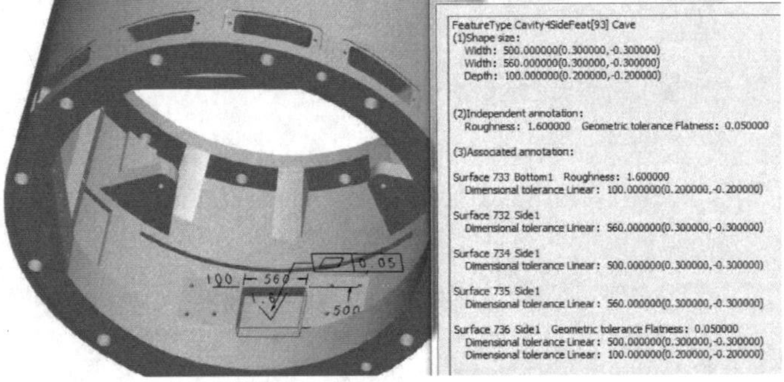

Figure 5. The extraction results of non-geometric information.

3.3. Automatic generation of process parameters

The characteristics of the parts can be obtained through the research of the front feature recognition part. Information extraction module can obtain non-geometric information such as feature size information, tolerance information and surface roughness. The acquisition of feature information and design information provides the basis for the generation of process parameters.

The machining processing parameters play an important role in the mechanical manufacturing process. It affects directly the machining quality and efficiency of the parts. If traditional manual methods are used to determine the process parameters, there will be low efficiency and errors maybe occur. Therefore, automatic generation of process parameters based by using BP neural network is proposed in the process parameter generation section. The process of automatic generation of process parameters is shown in Figure 6.

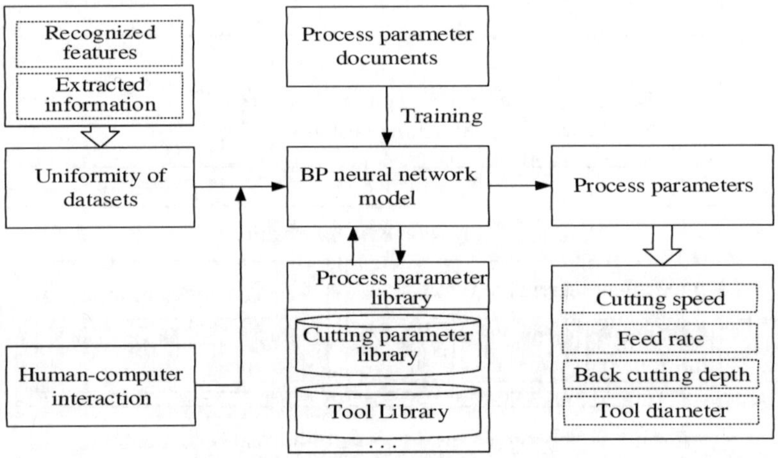

Figure 6. The process of automatic generation of process parameters.

First, the data of feature recognition and information extraction need be unified and these data information need be stored. Next, the craftsman formulates the process planning according to the acquired information and selects the working step in the form of human-computer interaction. The BP neural network model trains the existing process parameter files continuously and generates a network model finally. During the training process, the process parameter database including cutting parameter database and tool database is generated at the same time. The data of parameter database will become more and more accurate as the number of training becomes more and more. Finally, the operator input the step name and non-geometric information to the established network model and then the machining parameters will be output automatically.

3.3.1. Uniformity of datasets. The features identified in the feature recognition module, such as closed cavities, slots, open cavities and bosses, are geometric solid features, while the information of information extraction module including geometry dimension, dimensional tolerance, geometric tolerance, datum and surface roughness is non-geometric information. They belong to two different types of data information. If different types of data are to be input to a neural network, the two types of data must be uniformly expressed.

3.3.2. Process parameter generation with BP neural network. The unified expression of feature information and non-geometric features is to use the data they contain as the input of BP. It will facilitate the generation of the following process parameters. Before the data is input into the BP neural network, the system established a human-machine interface. The craftsman formulates the process route and selects the work step based on the feature information and the non-geometric information, which is input to the interactive interface. The information of current machining step and design information is expressed in a unified manner and input to a well-established neural network model, and then the required process parameters will be generated. Here is a set of step information and corresponding design information. This set of data is used as the input parameter of the BP neural network, as shown in Table 1.

Table 1. The input parameters of BP neural network.

No.	feature	Step name	Length (mm)	Width (mm)	Depth (mm)	Tolerance type	Tolerance size(mm)	Surface roughness(um)
1	groove	Milling plane	560	500	100	Flatness	0.05	1.6

BP neural network [12] is a kind of multi-layer feedback network which is trained according to error back-propagation algorithm. It is one of the most widely used neural network models. BP neural network can learn and store a large number of the mapping relationship of input and output without revealing the mathematical equations describing the mapping relationship beforehand. Its learning rule is to use the steepest descent method to adjust the weights and thresholds of the network continuously by back propagation, which can make the sum of the squared error of the network minimally. As shown in Figure 7, it mainly includes input layer, hidden layer and output layer. The input layer mainly includes the step name, geometric size, dimensional tolerance, shape tolerance, surface roughness and other information of the part. The obtained output layer has the processing parameters such as cutting speed, feed rate, tool diameter and other parameters.

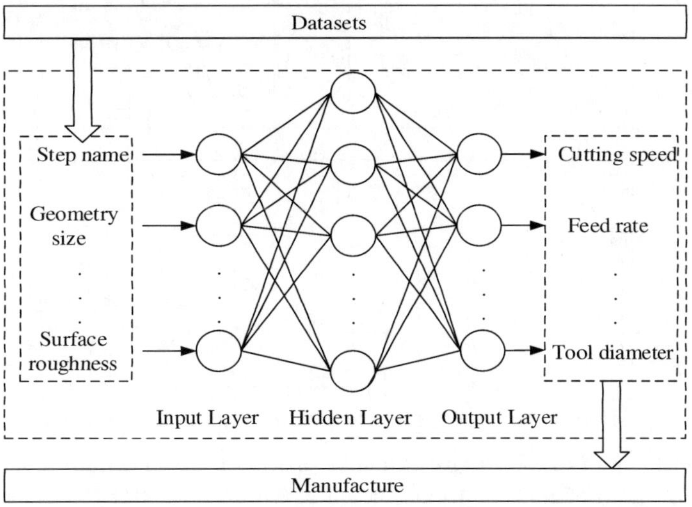

Figure 7. BP neural network model structure.

Before the prediction of the neural network data, there must be a network model established. Therefore, the data of the files in the existing library must be trained to establish a suitable model. The training process is as follows:

(1)Initialize the neural network model and set the threshold and weight of the process parameter training model.

(2)Set the corresponding output expectations and select 30000 input samples randomly in the existing technology library for input to the model for training.

(3)Calculate the input and output of each layer neuron, and then constantly modify the weights and thresholds of each neuron according to the input and output of each node.

(4)Calculate the error of the output parameters such as cutting parameters, feed amount and tool diameter.

(5)Determine whether the network error meets the requirements. The algorithm ends when the error reaches the requirement of the precision. Otherwise, it needs to select the next learning sample, return to the third step and enter the next round of learning.

Through the above steps, appropriate thresholds and weights can be obtained, and then a suitable neural network model can be established. Inputting the data of Table 1 into the established BP neural network model can generate cutting parameters and tool parameters, and the output process parameters are listed in Table 2. The output parameters can be provided for next manufacturing.

Table 2. The output parameters of BP neural network.

No.	v (m/min)	f_z (mm/z)	a_p (mm)	a_e (mm)	d (mm)	Z
1	250	0.03	2	2	10	4

4. Calibration and correction of process parameters

These generated machining parameters, such as cutting parameters and tool parameters, have a great influence on the machining time and machining quality of the parts of steering engine room. The machining time and machining quality of parts need to meet certain requirements, otherwise the process parameters should be corrected. The effect of process parameters on machining time and machining quality is shown in Figure 8.

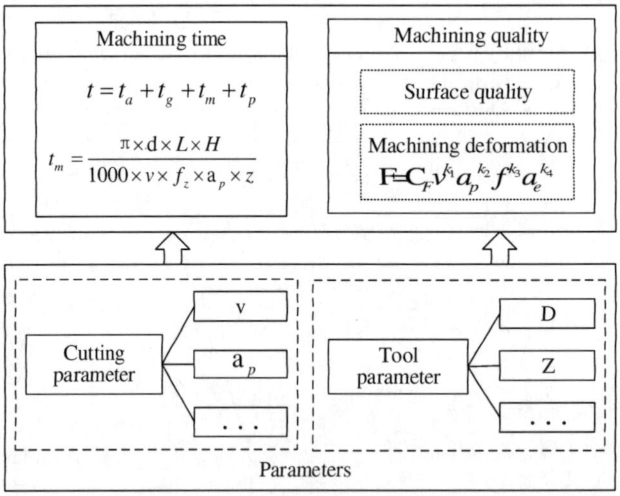

Figure 8. The effect of process parameters on machining time and machining quality.

The entire machining time of the part includes the information acquisition time of the part, the generation time of the process parameters, the cutting processing time and other auxiliary time, as shown in formula (1). In order to reduce the processing production cycle, it is necessary to reduce the consumption time of each phase. The design information of the parts is extracted using the corresponding algorithm, and the process parameters are generated through the BP neural network. These processes are realized automatically, which can shorten the preparation time of the early stage. The cutting time is related to the value of the process parameters, such as formula (2). We can generate appropriate machining process parameters through the neural network to reduce the cutting time. If the calculated machining time is more than the time determined by the manually selected process parameters, the process parameters must be regenerated. However, only pursuing the reduction of cutting time may increase the value of cutting force, which leads to the deformation of parts and affects the machining quality of parts. Therefore, we must make the cutting force of the part lower than the maximum allowable cutting force before reducing the cutting time, as in formula (3). If the conditions of cutting force are not satisfied, process parameters must be regenerated.

$$t = t_a + t_g + t_m + t_p \tag{1}$$

Among them, t_a is the time of the feature recognition and information extraction, t_g is the time of the process parameter generation, t_m is machining time of the working step, and t_p is the auxiliary time.

$$t_m = \frac{\pi \times \mathrm{d} \times L \times H}{1000 \times v \times f_z \times \mathrm{a}_p \times Z} \tag{2}$$

Among them, d is the tool diameter, L is the milling length of the workpiece, H is the milling depth, v is the milling speed, f_z is the feed per tooth, a_p is the axial depth of cut, and Z is the number of cutter teeth.

$$\mathrm{F}=\mathrm{C}_F v^{k_1} a_p{}^{k_2} f^{k_3} a_e{}^{k_4} \le \mathrm{F}_{\max} \tag{3}$$

Among them, C_F is the coefficient that depends on the processing material and the cutting condition. v is the cutting speed. a_p is the axial cutting depth. f is the feed. a_e is the radial cutting depth. k1, k2, k3, and k4 are undetermined parameters. $\mathrm{F}_{\max}$ is the maximum allowable cutting force.

We used the identified groove characteristics as an example to compare the machining time consumed in two different ways. The machine type is DMC 70V hi-dyn. The tool parameters and cutting parameters are shown in Table 2. The tool material is cemented carbide. The machining size of the groove is $560 \times 500 \times 100$. Under the constraint condition of cutting force, we calculated the processing time of two different process parameters according to formula (2). The results are shown in Table 3.

Table 3. A table with comparison of cutting time.

ways	Cutting parameters			Tool parameters			t_m (s)
	v (m/min)	f_z (mm/z)	a_p (mm)	a_e (mm)	d (mm)	Z	
Manual setting	250	0.03	2	2	10	4	29.32
BP—NN	254	0.0247	2.515	1.88	10	4	27.87

From Table 3, it can be seen that the time of cutting process is 29.32 second by manually selecting process parameters, while that is 27.87 second by BP neural network to generate process parameters. Therefore, the latter consumes less time. Meanwhile, the use of feature recognition and information extraction methods can reduce the processing time of machining preparation. Consequently, for mass production of the parts of steering engine room, it can reduce the time of the entire machining and improve the machining efficiency by using the method of process parameter generation proposed in this paper.

5. Conclusion

In this paper, a method of process parameter generation based on MBD is proposed. It is analysed from three parts of feature recognition, information extraction and process parameter generation, which makes the MBD model run through the process of process parameter generation. The characteristic information of the MBD model is quickly identified by the feature recognition algorithm and the MBD information of the model is obtained by using the algorithm of information extraction. Moreover, the process parameters are generated quickly and accurately by using BP neural network according to the characteristic of recognition and the extracted non-geometry information. The results show that this method can reduce the preparation time of processing and generate the process parameters quickly. It can reduce the whole processing time and improve the production efficiency. The next work is to apply this method to the different parts and improve the result of its parameter generation continuously.

Acknowledgments

The authors would like to thank the support from National Natural Science Foundation of China (No.51475301), the Fundamental Research Funds for the Central Universities (No.2232016D3-30) and Natural Science Foundation of Shanghai (No.17ZR1400600).

References

[1] Qiao L and Wu J 2012 An approach to region-based feature recognition for structural parts *International Conference on Manufacturing Science and Engineering (ICMSE)* **482-484** 2114-17

[2] Zubair A F, Abu M and Mohd S 2018 Automatic feature recognition of regular features for symmetrical and non-symmetrical cylinder part using volume decomposition method *Engineering with Computers* 1-21

[3] Sui X L, Zhang G L, Wang Q, Sun Q Y and Zhang X W 2013 Parts information extraction and storage of 3D-CAPP system oriented Virtual Manufacturing *2nd International Conference on Mechanical Engineering, Industrial Electronics and Informatization (MEIEI)* **427-429** 2645-49

[4] Fang Y X, Liu E F, Gao T and Huang F S 2013 Coordinate measuring information extraction of parts based on MBD dataset *Computer Integrated Manufacturing Systems* **19** 1532-40

[5] Vafadar A, Hayward K, and Tolouei-Rad M 2017 Drilling reconfigurable machine tool selection and process parameters optimization as a function of product demand *Journal of Manufacturing Systems* **45** 58-69

[6] Torres-Treviño L M, Escamilla-Salazar I G, González-Ortíz B and Praga-Alejo R 2013 An expert system for setting parameters in machining processes *Expert Systems with Applications* **40** 6877-84

[7] Roy S, Paulo Davim J and Kumar K 2016 Optimization of process parameters using taguchi coupled genetic algorithm: Machining in CNC lathe *Mathematical Concepts and Applications in Mechanical Engineering and Mechatronics* 67-93

[8] Kant G and Sangwan K S 2015 Predictive modelling and optimization of machining parameters to minimize surface roughness using artificial neural network coupled with genetic algorithm *15th CIRP Conference on Modelling of Machining Operations* **31** 453-58

[9] Wang Q and Yu X 2014 Ontology based automatic feature recognition framework *Computers in Industry* **65** 1041-52

[10] Zhou Y P, Zhang J and Qin W 2016 Feature recognition methodology for large thin-wall part of spacecraft *Computer Integrated Manufacturing Systems* **22** 2505-12

[11] Qu L G, Zhou H J and Dong Z L 2015 Technological Research on Information Extraction of Part Process Model Based on Machining Feature *Applied Mechanics and Materials* **696** 287-92

[12] Zhou Z H 2016 Machine Learning (Beijing: Tsinghua University Press) p 101

IOP Conf. Series: Materials Science and Engineering **435** (2018) 012049 doi:10.1088/1757-899X/435/1/012049

Supervised Learning of Single-Layer Spiking Neural Networks for Image Classification

Qiang Ma, Xianghong Lin and Xiangwen Wang

College of Computer Science and Engineering, Northwest Normal University, Lanzhou 730070, China

Email: linxh@nwnu.edu.cn

Abstract. The traditional artificial neural networks encode information through the spike firing rate. Spiking neural networks fall into the third-generation artificial neural network models, which use the precisely timed spike trains to encode neural information. The computational models can accurately simulate the neural network activities of human brain, and provide powerful capabilities of signal processing to solve the complex problem. In this paper, we propose a supervised learning algorithm for single-layer spiking neural networks based on the spike train kernel function, which can implement the complex spatio-temporal pattern learning of spike trains. Furthermore, a pattern classifier based on single-layer spiking neural networks is constructed for image recognition problem. We test the learning performance of the proposed algorithm by the image classification task on the LabelMe dataset. The experimental results show that the proposed algorithm has got good image classification accuracy for the test dataset, and the different sizes of receptive fields influence classification accuracies significantly.

1.Introduction

The traditional artificial neural networks usually use the spike firing rate to encode external information [1]. However, the precisely timed spike train encoding strategy is used in networks of spiking neurons which are more effective than the traditional neural network models [2]. Spiking neural networks can simulate different neuronal signals and arbitrary continuous functions, so they are very suitable for processing the nervous signals in the brain. Spiking neural networks have been applied to many fields now, such as image recognition, natural language processing, and medical systems. In recent years, some progress has been made in the supervised learning algorithms for spiking neural networks [3]. Researchers have presented many supervised learning algorithms of networks of spiking neurons, such as the gradient descent rule [4], ReSuMe (Remote Supervised Method) based on spike timing-dependent plasticity [5]. In addition, Mohemmed et al. [6] proposed SPAN (Spike Pattern Association Neuron) algorithm based on the kernel function convolution and use the Widrow-Hoff rule to adjust synaptic weights. Yu et al. [7] apply traditional Widrow-Hoff rule to spiking neural networks and proposed a PSD (Precise-Spike-Driven) supervised learning algorithm, in which only the input spike trains are converted.

Image recognition and classification problems are very important for the future development of spiking neural networks. Spiking neurons conform to the characteristics of biological neural networks and can be used to explain the activities of complex biological networks [8]. There are spatio-temporal pattern recognition problems are solved by the learning methods for spike times encoding in spiking neural networks [9] [10]. Wu et al. [11] constructed a multi-layer neural network with integrate-and-fire neurons, which is applied to image processing such as image edge detection and image segmentation. The results show that their work has better image processing performance. Combining temporal encoding with Hebbian reinforcement learning method, Meftah et al. [12] put forward a spiking neural network model for image segmentation. Bawane et al. [13] proposed a network of leaky integrate-and-fire neurons for character recognition, extracted the features from the image using the extended histogram of gradients and other methods, and added the features to the classifier to classify the characters. Combining the K-means classifier, Mukhopadhyay et al. [14] proposed a spiking neural network classifier based on a memristive crossbar network. The results show that it has good fault tolerance due to the robustness of the spiking neural networks. In this paper, we propose a supervised learning method for single-layer spiking neural networks based on spike train kernel and apply it for image recognition and classification problems.

2. Spike Train Kernel Supervised Learning Rule

The spike train $s=\{t^f \in \Gamma : f=1,...,N\}$ represents the ordered trains of spike times fired by spiking neurons in time interval $\Gamma=[0,T]$. A spike train can be expressed as:

$$s(t)=\sum_{f=1}^{N}\delta(t-t^f) \tag{1}$$

where N is the number of spikes, $\delta(x)$ is the Dirac delta function, and t^f is the fire time of the *f*th spike.

Due to the spike train is a set of discrete time, in order to facilitate the analysis and calculation, we choose a specific kernel function $\kappa(\cdot)$ to transform the discrete spike train into a unique continuous function:

$$\tilde{s}(t)=s(t)*\kappa(t)=\sum_{f=1}^{N}\kappa(t-t^f) \tag{2}$$

For convenience, we use the linear Poisson neuron model to describe the relationship between the input spike trains and the output spike train [15]. The transformed postsynaptic spike train $\tilde{s}_o(t)$ can be expressed as the linear summation of the transformed presynaptic spike trains $\tilde{s}_i(t)$ with corresponding synaptic weight w_{io}:

$$\tilde{s}_o(t)=\sum_{i=1}^{N_I}w_{io}\tilde{s}_i(t) \tag{3}$$

where $\tilde{s}_o(t)$ and $\tilde{s}_i(t)$ is the transformed postsynaptic spike train and presynaptic spike train, N_I is the number of presynaptic neurons.

The synaptic weights of spiking neural networks are adjusted by supervised learning algorithm, and finally the network outputs the target spike trains. Firstly, we define the spike train error function according to the transformed actual output spike train $\tilde{s}_o^a(t)$ and target output spike train $\tilde{s}_o^d(t)$. The network error at time t is the square difference between $\tilde{s}_o^a(t)$ and $\tilde{s}_o^d(t)$ for all output neurons, it can be defined as:

$$E(t)=\frac{1}{2}\sum_{o=1}^{N_O}\left[\tilde{s}_o^a(t)-\tilde{s}_o^d(t)\right]^2 \tag{4}$$

where N_O is the number of output neurons. Therefore, the total error of spiking neural networks in the time interval Γ is expanded as $E=\int_\Gamma E(t)dt$.

The delta update rule is applied to adjust all synaptic weights. The weight adjustment for presynaptic neurons i to postsynaptic neurons o is calculated as:

$$\Delta w_{io} = -\eta \nabla E_{io} \tag{5}$$

where η denotes the learning rate and ∇E_{io} represents the gradient value of the spike train error function E for the synaptic weight w_{io}. It can be expressed as the integral of the derivative of the error function $E(t)$ with respect to the weight w_{io} in the time interval Γ:

$$\nabla E_{io} = \int_{\Gamma} \frac{\partial E(t)}{\partial w_{io}} dt \tag{6}$$

Using the chain rule, the derivative of the error function $E(t)$ to the synaptic weight w_{io} at time t can be derived:

$$\frac{\partial E(t)}{\partial w_{io}} = \frac{\partial E(t)}{\partial \tilde{s}_o^a(t)} \frac{\partial \tilde{s}_o^a(t)}{\partial w_{io}} \tag{7}$$

According to Eq. (4), the first partial derivative term on the right-hand part of Eq. (7) is calculated as:

$$\frac{\partial E(t)}{\partial \tilde{s}_o^a(t)} = \frac{\partial \left[\frac{1}{2} \sum_{o=1}^{N_O} \left[\tilde{s}_o^a(t) - \tilde{s}_o^d(t) \right]^2 \right]}{\partial \tilde{s}_o^a(t)} = \tilde{s}_o^a(t) - \tilde{s}_o^d(t) \tag{8}$$

According to Eq. (3), the second partial derivative term of the right-hand part of Eq. (7) is computed as:

$$\frac{\partial \tilde{s}_o^a(t)}{\partial w_{io}} = \frac{\partial \left[\sum_{i=1}^{N_I} w_{io} \tilde{s}_i(t) \right]}{\partial w_{io}} = \tilde{s}_i(t) \tag{9}$$

Combining Eq. (8) and Eq. (9), the derivative of the error $E(t)$ on the weight w_{io} at time t is equal to:

$$\frac{\partial E(t)}{\partial w_{io}} = \frac{\partial E(t)}{\partial \tilde{s}_o^a(t)} \frac{\partial \tilde{s}_o^a(t)}{\partial w_{io}} = \left[\tilde{s}_o^a(t) - \tilde{s}_o^d(t) \right] \tilde{s}_i(t) \tag{10}$$

According to Eq. (6) and Eq. (10), the gradient ∇E_{io} is calculated as follow:

$$\nabla E_{io} = \int_{\Gamma} \left[\tilde{s}_o^a(t) - \tilde{s}_o^d(t) \right] \tilde{s}_i(t) dt = \int_{\Gamma} \tilde{s}_o^a(t) \tilde{s}_i(t) dt - \int_{\Gamma} \tilde{s}_o^d(t) \tilde{s}_i(t) dt \tag{11}$$

Based on the derivation process discussed above, we get a spike train kernel supervised learning rule for single-layer spiking neural networks, the synaptic weight adjustment is expressed as:

$$\Delta w_{io} = -\eta \left[\int_{\Gamma} \tilde{s}_o^a(t) \tilde{s}_i(t) dt - \int_{\Gamma} \tilde{s}_o^d(t) \tilde{s}_i(t) dt \right] = \eta \left[\sum_{g=1}^{N_d} \sum_{f=1}^{N_i} \kappa(t_d^g - t_i^f) - \sum_{h=1}^{N_o} \sum_{f=1}^{N_i} \kappa(t_a^h - t_i^f) \right] \tag{12}$$

where η is the learning rate, and $\kappa(\cdot)$ represents the kernel function.

3. Pattern Classifier for Image Recognition

3.1. Spiking Neuron Model

The spike response model is applied to analyse the network dynamic characteristics and learning process. Assuming that N_I is the total of synaptic inputs, N_i is the number of spikes in the *i*th spike train. w_{io} is synapse weight from *i*th input neuron to the *o*th output neuron. t_i^f is the *f*th spike firing time of the input neuron *i*. The potential of the neuron is defined as the sum of the postsynaptic potential induced by all input spikes from presynaptic neurons. When the internal state $u(t)$ exceeds the neuron threshold θ, the output neuron firing a spike, and the internal state decreases instantly to the resting potential, this phase is called repolarization. The potential of the postsynaptic neuron at time *t* is calculated as:

$$u(t) = \sum_{i=1}^{N_I} \sum_{f=1}^{N_i} w_{io} \varepsilon(t - t_i^f) + \rho(t - t^f) \tag{13}$$

In Eq. (13), $\varepsilon(s) = (s/\tau)\exp(1 - s/\tau)H(s)$ represents the spike response function, $\rho(s) = -\theta\exp(-s/\tau_R)H(s)$ represents the refractoriness function, where τ and τ_R are time constants, $H(s)$ is the Heaviside function.

3.2. Spiking Neural Network Architecture

Similar to the architecture of traditional artificial neural networks, spiking neural networks have feedforward, recurrent, and other network architectures. In this paper, we adopt a feedforward network architecture with input layer and output layer, which is divided into encoding stage and learning stage for image recognition and classification. Fig. 1 shows the network architecture for image recognition and classification.

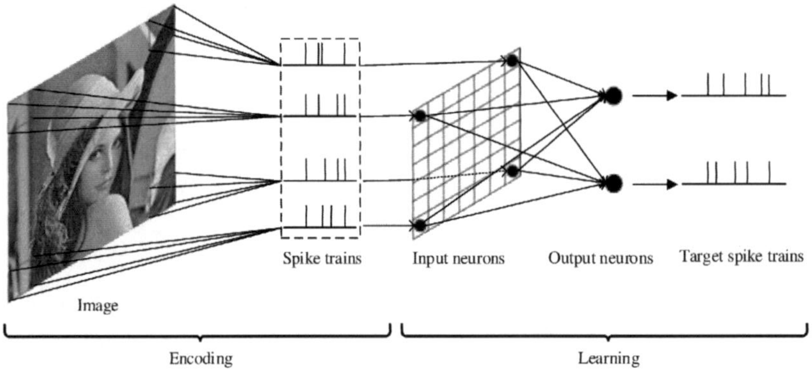

Figure 1. The network architecture for image recognition and classification

The spiking neural network model is applied to image recognition and classification problems. In the coding stage, the latency-phase encoding method is used to transform the pixels of image receptive field into precisely timed spike trains, where the spike trains are used to represent external image stimuli information [16]. In the learning stage, each spike train corresponds to an input neuron and is inputted into the spiking neural network. The synaptic weights are learned by the spike train kernel supervised multi-spike learning method. The spiking neural network outputs the target spike pattern for given images.

4. Simulations and Results

In order to demonstrate the learning performance of the proposed algorithm, we perform a set of experiments in this section. The first experiment verifies the learning performance of learning algorithm for spiking neural networks on the Lena and Baboon grayscale images recognition tasks. In the second experiment, we select the outdoor road images and the outdoor city street images from LabelMe dataset [17] and perform an image classification problem to test the classification ability of the proposed supervised learning algorithm with different sizes of receptive fields.

In order to determine the accuracy of image recognition and classification, we introduce the similarity measure function C to measure the similarity between the actual output spike train and the target output spike train after spiking neural network learning [18]. C is defined as:

$$C = \frac{\tilde{s}_o^d(t) \cdot \tilde{s}_o^a(t)}{\left|\tilde{s}_o^d(t)\right|\left|\tilde{s}_o^a(t)\right|} \tag{14}$$

where $\tilde{s}_o^d(t)\cdot\tilde{s}_o^a(t)$ represents the inner product of $\tilde{s}_o^d(t)$ and $\tilde{s}_o^a(t)$, $|\cdot|$ represents the Euclidean norm of spike train. When two spike trains are completely identical, the value of C equals to 1. When the correlation decreases gradually, the value of C gradually tends to zero.

4.1.Image Recognition Task

In order to test the learning performance of spike train kernel supervised learning algorithm, we choose the Lena image and the Baboon image for recognition task. The original 256 × 256 image is encoded into spike trains by the latency-phase encoding with 8 × 8 receptive field. Each spike train corresponding to an input neuron inputted into the network and sets the target output spike train of two images. The target output spike train of the Lena image is {20, 40, 60} ms, while the target output spike train of the Baboon image is {40, 60, 80} ms.

Fig. 2 shows the learning results of spike train kernel supervised learning algorithm on Lena image and the Baboon image recognition task. Fig. 2(a) and Fig. 2(b) show the original Lena image and Baboon image, respectively. Fig. 2(c) and Fig. 2(d) show the learning process of Lena image and Baboon image, respectively, where $*$ is the target output spike train, and $\cdot$ is the actual output spike trains of the neuron. It can be seen that with the increasing of iterations, the actual output spike trains are closer to the target output spike train for both Lena image and Baboon image. Finally, the output spike train is consistent with the target output spike train. Fig. 2(e) and Fig. 2(f) illustrate the change of recognition accuracy on Lena image Baboon image respectively in the learning process. It can be seen that with the increasing of iterations, the value of C increases gradually for both Lena image and Baboon image. After 32 and 38 learning epochs respectively, the value of C reached to 1. It indicates that the actual output spike train is equal to the target spike train. This experiment shows that the proposed method can well perform image recognition task.

(a) Original Lena image.

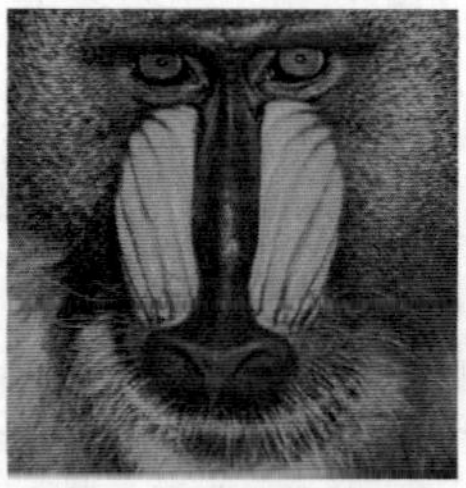

(b) Original Baboon image.

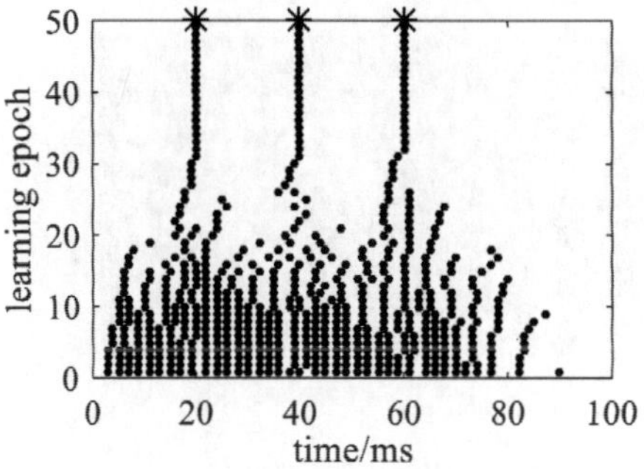

(c) Learning process of Lena image.

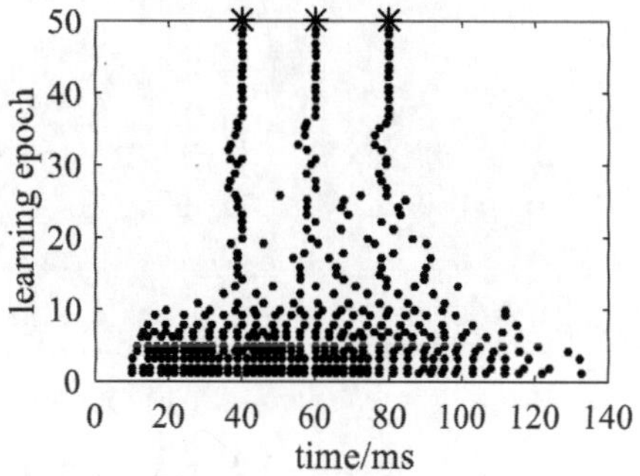

(d) Learning process of Baboon image.

IOP Conf. Series: Materials Science and Engineering **435** (2018) 012049 doi:10.1088/1757-899X/435/1/012049

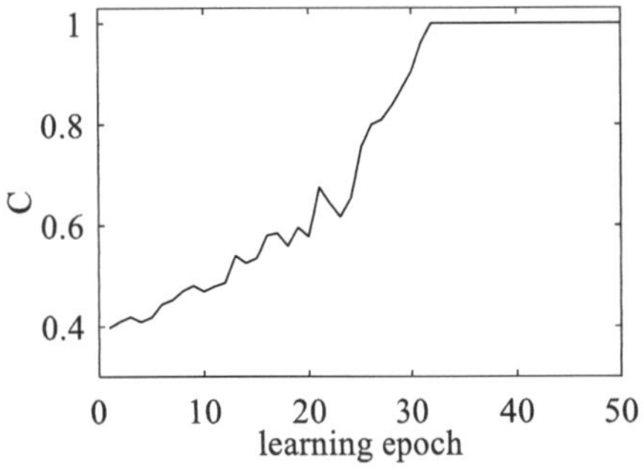

(e) Recognition accuracy on Lena image.

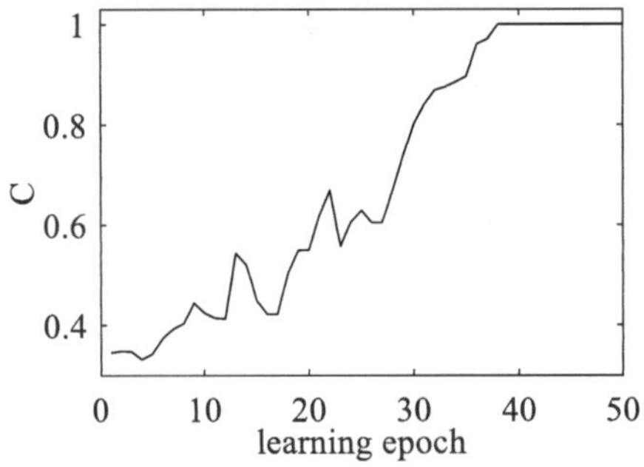

(f) Recognition accuracy on Baboon image.

Figure 2. The learning process and results for Lena and Baboon image recognition.

4.2.Image Classification Experiment

In this experiment, we test the learning performance of our proposed learning method by an image classification problem. We choose the outdoor road images and the outdoor city street images from the LabelMe dataset in the experiment. Each kind of images includes 20 samples, in total of 40 samples. Fig. 3 shows some typical outdoor road images (Fig. 3(a)) and outdoor city street images (Fig. 3(b)). In our experiment, we choose 10 samples randomly from the outdoor road images and the outdoor city street images respectively to constitute the training set, and the remaining samples are constituted the testing set. The original 256 × 256 images are encoded into spike trains by the latency-phase encoding. In addition, we need to set the target output spike trains of two kinds of images. The target output spike train of the outdoor road images is set as {20, 40, 60, 80} ms, while the target output spike train of the outdoor city street images is set as {40, 60, 80, 100} ms.

(a) Outdoor road images.

(b) Outdoor city street images.

Figure 3. Some typical images from the LabelMe dataset.

Table 1 shows the image classification results on the LabelMe dataset with different size of receptive field. The number of input neurons is equals to the size of image divided by the size of receptive field. The size of receptive field takes 2×2, 4×4, 8×8, 16×16, 32×32 and 64×64 in total of six values. As seen from the table, with the increasing of the size of receptive field, both the training accuracy and the test accuracy are firstly increased, and then decreased. When the size of receptive field is 8×8, the training accuracy is 100% and the test accuracy is 94.77%. It shows that the receptive field can't be too big or too small, the appropriate size of receptive field will obtain higher training accuracy and test accuracy. The experimental results show that the proposed spike train kernel supervised learning algorithm can be well applied to image classification problem and achieve high classification accuracy.

Table 1. The image classification results with different sizes of receptive fields.

Number of input neuron	Receptive filed size	Training accuracy	Testing accuracy
16384	2×2	$94.33\% \pm 0.08$	$74.82\% \pm 0.17$
4096	4×4	100%	$85.34\% \pm 0.12$
1024	8×8	100%	$94.77\% \pm 0.04$
256	16×16	100%	$93.33\% \pm 0.09$
64	32×32	100%	$91.71\% \pm 0.11$
16	64×64	$98.68\% \pm 0.06$	$83.18\% \pm 0.07$

5. Conclusion

This paper uses the spike train kernel to transform the discrete spike train into a unique continuous function, and then proposes a spike train kernel supervised learning algorithm for single-layer spiking neural networks. The proposed supervised learning method is firstly successfully applied on Lena image and the Baboon image recognition task. It is also applied on an image classification problem on the LabelMe dataset with different size of receptive field. The experiment results show that the proposed algorithm can effectively solve image recognition and image classification problems. In future research, we will study more efficient supervised learning algorithms for spiking neural networks and apply them to more complex spatio-temporal pattern recognition problems.

Acknowledgment

This research is supported by the National Natural Science Foundation of China under Grant no. 61762080, and the Medium and Small Scale Enterprises Technology Innovation Foundation of Gansu Province under Grant no. 17CX2JA038.

References

[1] Haykin S S 2009 Neural Networks and Learning Machines. Upper Saddle River: Pearson Education

[2] Ghosh-Dastidar S and Adeli H 2009 Spiking neural networks *International Journal of Neural Systems* 19(4) 295-308

[3] Lin X H, Wang X W, Zhang N, et al. 2015 Supervised learning algorithms for spiking neural networks: a review *Acta Electronica Sinica*, 43(3) 577-586

[4] Florian R V 2012 The chronotron: a neuron that learns to fire temporally precise spike patterns *PLoS One* 7(8) e40233

[5] Ponulak F, Kasiński A 2010 Supervised learning in spiking neural networks with ReSuMe: sequence learning, classification, and spike shifting *Neural Computation* 22(2) 467-510

[6] Mohemmed, A, Schliebs S, Matsuda S and Kasabov N 2012 SPAN: spike pattern association neuron for learning spatio-temporal spike patterns *International Journal of Neural Systems* 22(04) 1659-85

[7] Yu Q, Tang H, Tan K C, et al. 2013 Precise-spike-driven synaptic plasticity: learning hetero-association of spatiotemporal spike patterns *PLoS One* 8(11) e78318

[8] Paugam-Moisy H and Bohte S 2012 Computing with spiking neuron networks *Handbook of natural computing* pp 335-376

[9] Kasabov N K 2014 NeuCube: A spiking neural network architecture for mapping learning and understanding of spatio-temporal brain data *Neural Networks* 52 62–76

IOP Conf. Series: Materials Science and Engineering **435** (2018) 012049 doi:10.1088/1757-899X/435/1/012049

[10] Tapson J C, Cohen G K, Afshar S, et al. 2013 Synthesis of neural networks for spatio-temporal spike pattern recognition and processing *Frontiers in Neuroscience* 7(7) 153
[11] Wu Q X, McGinnity T M, Maguire L P, et al. 2008 Processing visual stimuli using hierarchical spiking neural networks *Neurocomputing* 71(10-12) 2055-2068
[12] Meftah B, Lezoray O and Benyettou A 2010 Segmentation and edge detection based on spiking neural network model *Neural Processing Letters* 32(2) 131-146
[13] Bawane P, Gadariye S, Chaturvedi S, et al. 2018 Object and character recognition using spiking neural network *Materials Today Proceedings* 5(1) 360-366
[14] Mukhopadhyay A K, Chakrabarti I, Basu A, et al. 2018 Power efficient spiking neural network classifier based on memristive crossbar network for spike sorting application *arXiv preprint* 1802, 09047
[15] Carnell A and Richardson D 2005 Linear algebra for times series of spikes. *European Symposium on Artificial Neural Networks* pp 363-368
[16] Hu J, Tang H, Tan K C, et al. 2013 A spike-timing-based integrated model for pattern recognition *Neural Computation* 25(2) 450-472
[17] Russell B C, Torralba A, Murphy K P and Freeman W T 2008 LabelMe: A database and web-based tool for image annotation *International Journal of Computer Vision* 77(1-3) 157-173
[18] Schreiber S, et al. 2003 A new correlation-based measure of spike timing reliability *Neurocomputing* 52 925-931

Speech Emotion Feature Analysis Based on Emotion Fingerprints

Yuantao Jiang[1], Kaifa Deng[2*] and Chunxue Wu[1]

[1] School of Optical-Electrical and Computer Engineering, University of Shanghai for Science and Technology, Shanghai, China
[2] Shanghai University of Engineering Science, Shanghai, China
Corresponding author: Kaifa Deng, Shanghai University of Engineering Science, Shanghai, China

Tel.:+86-176-2160-7346,E-mail:j17621607346@163.com, *dengkaifa@126.com, wcx@usst.edu.cn

Abstract: The speech recognition has caught the attention of more and more researchers as the important branch of artificial intelligence. However, speech recognition just can recognize the neutral meaning and ignore the important emotion information. So,this paper proposes an emotion fingerprint extraction algorithm. The speech signal is decomposed by lifting wavelet packets. And combining singular value entropy and sample entropy as the feature vector to characterize the speech time-frequency matrix coefficient. Then adopting the statistical values to extract emotion fingerprints. The experiments show that this algorithm can reflect the emotion information of the speech fully and distinguish several major emotions well.
Keyword: speech recognition, emotion recognition, emotion fingerpringting, lifting wavelet packets, singular value entropy , sample entropy

1.Introduction

Artificial intelligence is one of the hottest technologies recently. Speech recognition has also developed rapidly as an important branch of it. At present, speech recognition is the neutral semantic recognition, which extracts the acoustic feature parameters of human language and recognizes the meaning of the speech according to the recognition models. However, the emotion information which is one of the most important features of language is ignored. One of the founders of artificial intelligence, Prof. Minsky of the Massachusetts Institute of Technology in the United States, pointed out that the question is not whether smart machines can have emotions, but whether intelligent machines can realize intelligence in "*The Society of Mind*"[1].And Damasio's research shows that human intelligence not only shows normal rational thinking and logical reasoning ability, but also shows normal emotional ability[2]. The emotion is an important symbol of human intelligence. So, adding emotion features to speech recognition is an important research topic[3]. Emotion recognition is a classification problem of pattern recognition, which includes three mainaspects: speech emotion feature extraction, selection and speech emotion recognition[4]. The principle of speech emotion recognition is as figure 1.

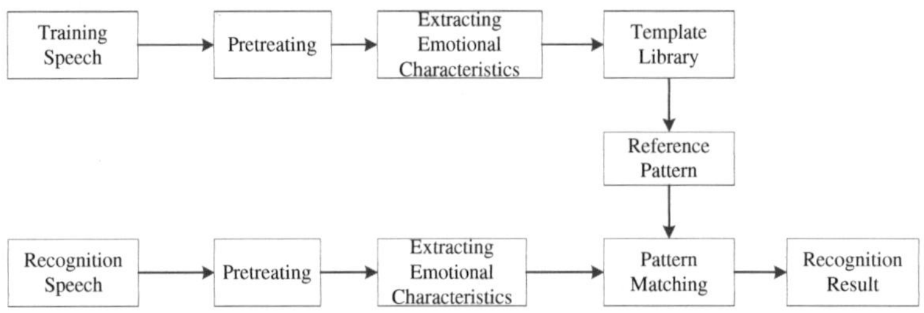

Figure 1. The principle of emotion recognition.

The audio fingerprint technology as the important audio retrieval technology has become more and more important. It is a compact digital signature which is extracted from audio data to characterize the acoustic characteristics of the audio fully. Great audio fingerprint should meet the following four conditions: robustness, compactness, distinguishability, simplicity[5].

2. Speech emotion feature analysis

The parameters of speech can describe emotion features and express human's emotions.And the parameters include acoustic characteristics, time-frequency characteristics and statistical characteristics of speech signals. And the time domain waveforms of different emotions are shown as figure 2.

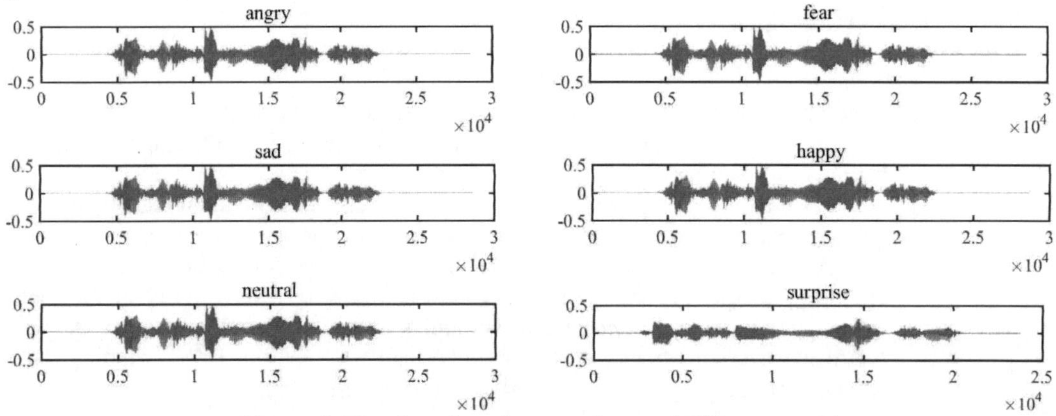

Figure 2. The time domain wave forms of different emotions.

Many literatures have done a lot of research on emotion feature parameters and they mainly focus on amplitude, time, fundamental frequency, formant,energy, etc[6]. At present, speech features can be simply divided into three categories: rhythm features, spectrum signature, and other features.

1). Prosodic features

The prosodic features is the physical property of the speech. Because the prosodic features express more speech emotion information in sentences. Prosodic features include (1).The pitch frequency characteristics;(2).Formant characteristics;(3) Energy characteristics;(4).Time characteristics[7,8].

2). Spectrum features

Compared to the prosodic features, speech does not change significantly in a short time. The classical spectrum features include short-term Fourier transform, linear predictor coefficients (LPC), Mel frequency cepstrum coefficients (MFCC), and perceptual linear predictive cepstral coefficients (perceptual linear predictive cepstral coefficients PLP), line spectrum pair (LSP),short time coherence (SMC).

3) Other features

In addition, there are other speech feature extraction methods in speech emotion recognition, such as teager energy operator (TEO), empirical mode decomposition (EMD),Fractal dimensions, deep learning, etc.

As for the differences in the language and research environment, the research methods and emphasis are also different. And the majority classification of emotions include angry, happy, sad, neutral, surprised, fear.

3. Emotion fingerprint extraction algorithm

The traditional fingerprint extraction algorithm has limitations which is base on time-frequency analysis. The two-dimensional time-frequency information matrix dimension from the time-frequency transformation of the original signal is too large. It is necessary to process the time-frequency matrix information multiple times to make it effective. Therefore, considering the nonlinear parameter methods to extract the signal features. In this paper, the singular value entropy and sample entropy are introduced in the emotion fingerprint extraction. The feature vector is based on the lifting wavelet packet transform and the combination of singular value entropy and sample entropy is adopted to characterize the time-frequency matrix coefficients of emotions.

3.1 Lifting Wavelet Packet Transform

The lifting wavelet construction method proposed by Sweldens[9] and it's different from the traditional Mallat wavelet construction method. It doesn't depend on the translation and stretching of the mother wavelet, nor does it require spectrum analysis, so it is particularly suitable for structure of wavelets in Finite Area, Surface and Non-uniform Sampling. And the lifting wavelet packet transform not only overcomes the shortcomings of traditionals, such as the heavy computation, the inability to accurately reconstruct the original signal, but also has the advantages of traditionals, such as fast operation speed,no extra space, and the integer wavelet transform, and easy parallel computing. It is suitable for speech recognition which requires higher real-time signal processing.

Constructing biorthogonal wavelet packets with a lifting scheme is simpler and more efficient. The high-frequency components of signal can be obtained by interpolation subdivision, and constructinga scale function to obtain the low-frequency components. The lifiting wavelet packet is divided into: splitting, predicting, updating[9]. And the inverse transform of the lifting algorithm can be implemented by changing the plus or minus sign in the forward calculation formula. The entire lifting wavelet transform process is shown as figure 3.

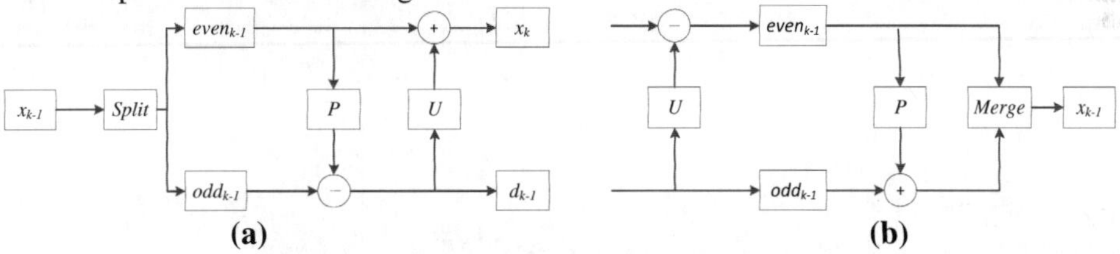

(a) **(b)**

Figure 3. Lifting wavelet packet transform process.

3.2 Singular Value Decomposition and singular value entropy Calculation

Singular values are inherent in the matrix and reflect the information contained in the matrix fully. The singular value entropy of wavelet packet is based on the singular value decomposition theory. The signal after wavelet packet transform is decomposed into a series of singular values and then utilizing statistical properties of information entropy to analyze the uncertainty of a set of singular values and then give a deterministic measure of the complexity of the speech signal. Therefore, using the singular value of the wavelet packet space feature matrix A as the characteristics of the speech signal to construct the feature vector[10].

Assuming A is a $m\times n$ $(m>n)$ matrix with a rank r $(r\leq n)$, there are $m\times m$ orthogonal matrices U and $n\times n$ orthogonal matrices V to make $U \wedge V^T = A (or\ U \wedge V^H = A)$. And $\wedge$ is $m\times n$ non-negative diagonal matrix.

$$A = \begin{bmatrix} R & 0 \\ 0 & 0 \end{bmatrix} \tag{1}$$

In (1), $R = diagram(\sigma_1, \sigma_2, \dots, \sigma_r)$ and the diagonal elements are the singular values of matrix *A*. And if these non-zero singular values are combined into a feature vector $X(X=(\sigma_1, \sigma_2, \dots, \sigma_r)$. From the properties of the matrix singular value, this feature vector can characterizes the characteristics of the wavelet packet coefficient matrix. The coefficient matrix can reflect the nature and characteristics of the signal. So the feature vector can be used to represent the emotion characteristics.

So, the singular value contains the characteristics of the emotion, and the difference in the singular value reflects the different characteristics of the different frequency bands. Each wavelet coefficient has the different frequency component, and the singular value is also different, normalizateing each component then get:

$$T_i = \frac{\sigma_i^2}{E} \tag{2}$$

In (2), $E = E_1 + E_2 + \dots + E_n$. Then the singular value entropy of each wavelet coefficient can be obtained.

$$H = -\sum_{i=1}^{n} p_i \, log \, p_i \tag{3}$$

In (3), $p_i = \frac{T_i}{T}, T = \sum_{i=1}^{n} T_i$.

3.3 Sample entropy of speech signal

Add 0.37 seconds sliding time window to calculate the sample entropy of the speech signal. Then calculating the sample entropy of the next 0.37 seconds time window until get the sample entropy of the speech signal of the last time window, thereby the time sequence of the sample entropy is obtained,that is, $X = \{x_{(1)}, x_{(2)}, \dots, x_{(N)}\}$. Sample entropy calculation steps as follows[11][12].

(1). Given a mode dimension m, the original sequence constitutes the m-dimensional vector.

$$X_{(i)} = \{x_{(i)}, x_{(i+1)}, \dots, x_{(i+m-1)}\} \tag{4}$$

In (4), *i=0,1,2,...,N-m+1*.

(2). Defining the distance between $X_{(i)}$ and $X_{(j)}$.

$$d(i,j) = max_{k=1-m-1} |x(i+k) - x(j+k)| \tag{5}$$

In (5), *k=0,1,2,...,m-1*.

(3). Given a threshold *r*, counting the number of $d(i,j) < r$ for each *i* and the ratio of the number to the distances total number *N-m+1* and denoted as $B_i^m(r)$.

$$B_i^m(r) = \frac{L[d_{(i,j)}<r]}{N-m+1} \tag{6}$$

In (6), $L[d_{(i,j)}<r]$ is the number of $d_{(i,j)}<r$. $1 \le j \le N-m$ and $j \neq i$.Then calculating the average of all of *i*.

$$B^m(r) = \frac{1}{N-m+1} \sum_{i=1}^{N-m+1} B_i^m(r) \tag{7}$$

(4). Repeat steps (1)~(3) for *m+1* to obtain $B^{m+1}(r)$.

(5). Theoretically, the sample entropy of this sequence is

$$SampEn(m,r) = lim_{n\to\infty}[-ln\frac{B^{m+1}(r)}{B^m(r)}] \tag{8}$$

When *N* takes a finite value, the estimated value of the sample entropy can be obtained.

$$SampEn(m,r,N) = -ln\frac{B^{m+1}(r)}{B^m(r)} \tag{9}$$

The parameters *m*, *r*, and *N* in the formula are referenced in [9]. And in this paper, *m=2*, *r=0.2SD*, *N=1024* and *SD* is the standard deviation of the original data.

From the above theoretical analysis, it can be found that the singular value entropy of each wavelet packet coefficient contains more comprehensive characteristics, and the difference of singular values reflects different characteristics of different frequency bands. The stable estimates can be obtainde from the sample entropy analysis and it can represent signal complexity and irregularity and has good noise immunity and anti-jamming capability. So, the emotion fingerprintings which is extracted by the combination of singular value entropy and sample entropy can fully reflect the emotion characteristics of speech.

4. Experiment and result analysis

4.1 Emotion fingerprinting extraction process

The emotion fingerprinting extraction process is shown as the figure 4:

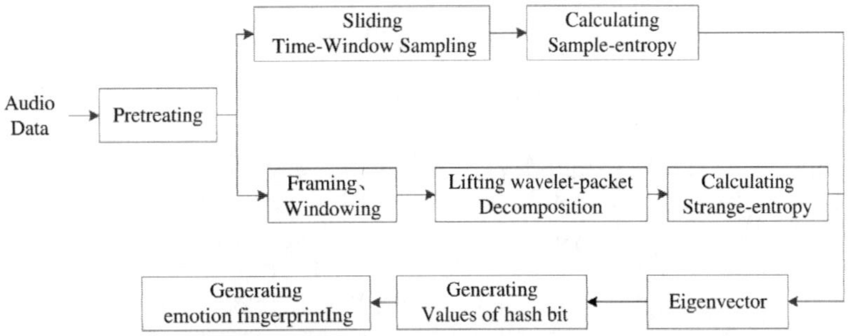

Figure 4. Emotion fingerprinting extraction process.

(1). Converting speech to 16 bit/sample and sample rate is 44. 1 kHz mono signal, then adopting endpoint detection to distinguish speech signal and non-voice signal[13]. The endpoint detection results shown in figure 5.

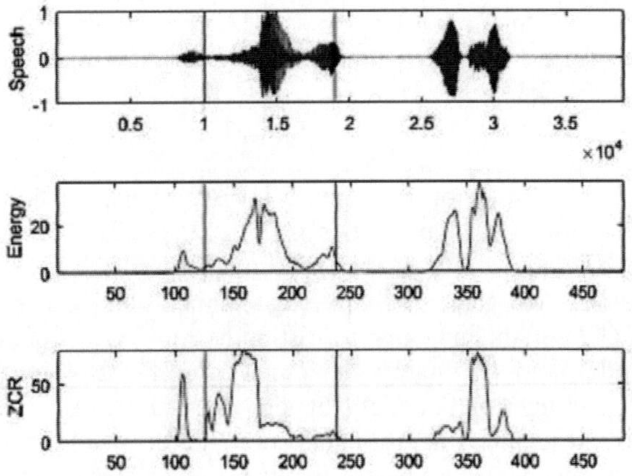

Figure 5. Endpoint detection processing

(2). Calculating sample entropy $SampE_{n0}$, $SampE_{n1}, ..., SampE_{nn}$ according to formula (8) for all sample points.

(3). In the experiment, the frame length is 2048 and adopting the Hanning window to smooth the frame edge. The overlap factor $P = 28/32$.The Hanning window formula is as follows:

$$W_{(n)} = \begin{cases} 0.5 \times [1 - cos(2\pi n/(N-1))], 0 \leq n \leq N-1 \\ 0 \quad , \quad else \end{cases} \tag{10}$$

(4). Adopting the lifting wavelet packet to perform 5-layers decomposition on each frame,then extracting the decomposition coefficients of the fifth layer from low-frequency to high-frequency bands. Then reconstructing the obtained wavelet packet coefficients for the feature matrix *A* of the lifting wavelet packet.

(5). Performing singular value decomposition on the feature matrix *A*, and the feature vector $\sigma = (\sigma_0, \sigma_1, ..., \sigma_n)$ formed by the singular value can be obtained. Then calculating singular value entropy according to formula (3).

(6). Combining the singular value entropy and sample entropy as the feature vector

$[SampE_n(\sigma_1), SampE_n(\sigma_2), \ldots, SampE_n(\sigma_n)]$ of the emotion fingerprint.

(7) Extracting the hash value to generate the emotion fingerprint. The statistics is an effective method for extracting relevant feature values. This algorithm calculates the following entropy statistics.

$$H_i = \sum_{n=1}^{j} |SampE_n(\sigma_n)|^2 \quad (11)$$

$$H_{sumk} = \sum_{i=1}^{m} H_i \quad (12)$$

$$H_{avg} = \frac{1}{t}\sum_{k=1}^{t} H_{sumk} \quad (13)$$

In the formula (11),(12),(13), $SampE_n(\sigma_n)$ is the value of singular value entropy and sample entropy of the *n*th wavelet packet coefficient of the *k*th frame's *i*th subspace. H_i represents the entropy value of the *k*th frame's *i*th subspace. H_{sumk} represents the entropy value of the *k*th frame , H_{avg} represents the average entropy value of all frames.

(5). Extracting audio fingerprints by hash bit values. Comparing the energy value E_{sumk} of each frame with the average of energy value E_{avg} of all frames, one bit of hash value of each frame is generated according to the formula (12).The splicing of all generated hash bit values constitutes the emotion fingerprint.

$$H_{(k)} = \begin{cases} 1, H_{sumk} > H_{avg} \\ 0, else \end{cases} \quad (14)$$

Then the emotion fingerprint of the audio signal which length is *N* can be expressed as:

$$S_{(t)} = \sum_{t=1}^{N} (k_1 h_1(t) + k_2 h_2(t) + \cdots + k_n h_n(t)) \quad (15)$$

In (15), h_i is the hash value of the emotion fingerprint extracted for each frame signal, and K_i (i=1, 2, 3, 4, 5) is the synthesis coefficient.

4.2. Experiment Analysis

This paper take the casia's Chinese emotion database for experiments[14]. There are four speechers include two men and two women. And the number of speeches are 50 with angry, fear, happy, neutral, sad, and surprise respectively and the time of each speech is 1 second.

The emotion fingerprint is extracted with the algorithm of this paper for these speeches. The fingerprint distance is an important indicator to distinguish different emotions. If the speech with different emotion has similar emotion fingerprint, it will cause higher bit error rate(*BER*) in emotion recognition. And the emotion fingerprint of angry, happy, sad, fear, neutral, surprise are shown as figure 6. And taking the *BER* to represent the fingerprint distance and the result is shown as table 1. Assuming that the length of the two fingerprints is *B* bits, the number of mismatched bits is *n*, and the *BER* presses as the formula (16).

$$BER = \frac{100}{B}\sum_{n=0}^{B-1} \begin{cases} 1, h'(n) \neq h(n) \\ 0, h'(n) = h(n) \end{cases} \quad (16)$$

In (16), $h'(n)$ and $h(n)$ represent different fingerprints respectively.

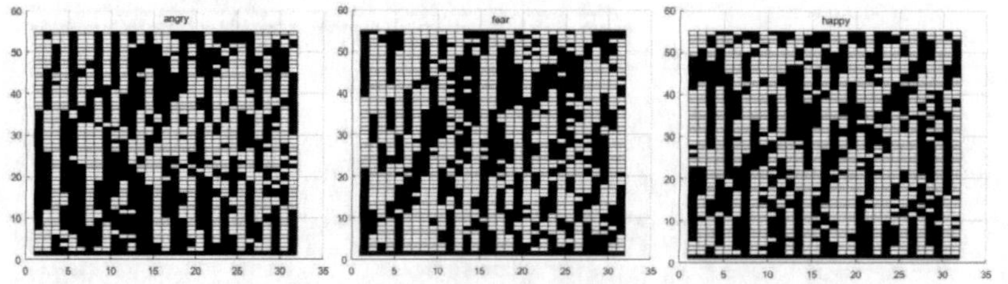

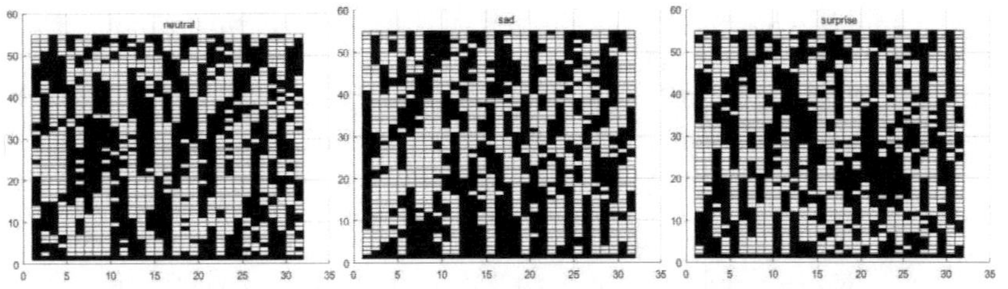

Figure 6. The emotion fingerprinting of the different emotions

Table 1. The distance of the different emotions

Emotion	Angry	Fear	Happy	Neutral	Sad	Surprise
Angry	0	0.5431	0.6553	0.6089	0.5645	0.6643
Fear	0.5431	0	0.6479	0.6728	0.6469	0.7213
Happy	0.6553	0.6479	0	0.7070	0.6145	0.6329
Neutral	0.6089	0.6728	0.7070	0	0.6935	0.7142
Sad	0.5645	0.6469	0.6145	0.6935	0	0.6534
Surprise	0.6643	0.7213	0.6329	0.7142	0.6534	0

Then the emotion fingerprint of different people with the same emotion is shown as figure 7, the *BER* is also adopted to represent the fingerprint distance and the result is shown in table 2.

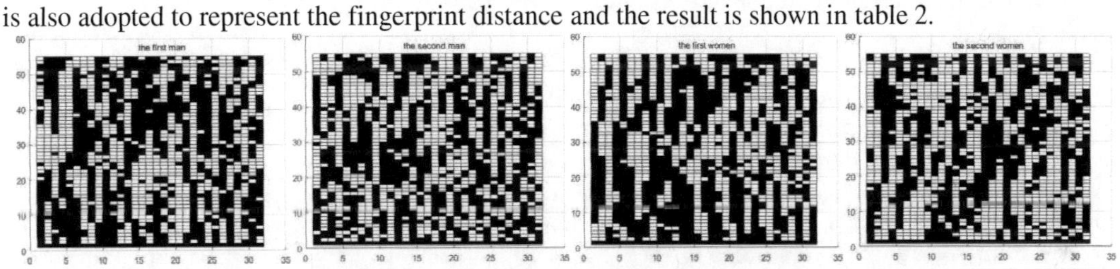

Figure 7. The emotion fingerprinting of different person.

Table 2. The *BRE* of emotion fingerprinting by different person.

Persons	Woman1	Woman2	Man1	Man2
Woman1	0	0.2416	0.3024	0.2977
Woman2	0.2416	0	0.2843	0.2861
Man1	0.3024	0.2843	0	0.3146
Man2	0.2977	0.2861	0.3146	0

The results show that the bigger the *BER*, the greater the difference between different emotions, the better it can distinguish between different emotion. From figure 6 and table 2, it is evident that the *BER* of the fingerprint comparison between the same speecher's different emotion is between 0.5431 and 0.7213, beyond the threshold 0.35 proposed in [15], which shows that this algorithm has a strong distinguishability in different emotions.

And from figure 7 and table 2, the *BER* of fingerprints comparison between the same emotion of different speechers is between 0.2416 and 0.3146, and the distance between the fingerprintings is small, so judging these are the same emotion basically. And the *BER* is that the emotion fingerprint includes the semantics characteristic parameter and the acoustic characteristics of men and women. Therefore, the emotion fingerprint also provides a good theoretical basis for speech recognition which combine the semantics characteristic and emotion characteristic.

5. Conclusion

In this paper, the singular value entropy and sample entropy are introduced into the emotion fingerprint, adopting a feature vector combine the singular value entropy and sample entropy to characterize the time-frequency matrix coefficients of the speech emotion based on the lifting wavelet packet transform. extracting the emotion fingerprints by statistical value calculations. Experiment analysis shows that there are significant differences in emotion fingerprints of different emotion, so emotion fingerprints can characterize the emotion information in the speech well and distinguish different emotion clearly. Emotion fingerprints with emotion information are smaller than original signals, occupy less memory, and are easier to process. It provides a good theoretical basis for speech recognition combined with semantic and emotion recognition. So considering the speech recognition algorithm recognise emotion and semantic information synchronously according to the audio fingerprint algorithm is the future work.

Acknowledgments

The authors would like to appreciate all anonymous reviewers for their insightful comments and constructive suggestions to polish this paper in high quality. This research was supported by the ShanghaiScience and Technology Innovation Action Plan Project (16111107502, 17511107203),the National Natural Science Foundation of China (61502220),the Zhejiang Provincial Natural Science Foundation of China(No.LY14F020044),the program for tackling key problems in Henan science and technology (No.172102310636) and the Nanjing Leading Science and Technology Entrepreneurial Talents Introduction Program Funded Project (2014A090002).

Reference

[1] Minsky M. The Society of Mind[J]. Personalist Forum, 1987, 3(1):19-32.

[2] Zimmerman, Corinne. Descartes' Error: Emotion, Reason and the Human Brain[J]. Psychosomatics, 1996, 310(36):151–153.

[3] Ververidis, D., Kotropoulos, C., & Pitas, I. (2004). Automatic emotional speech classification. *Proc.ieee Int.conf.acoustics Speech & Signal Processing Montreal Canada, 1*(1), I-593-6 vol.1.

[4] Nwe, T. L., Foo, S. W., & Silva, L. C. D. (2003). Speech emotion recognition using hidden markov models. *Speech Communication, 41*(4), 603-623.

[5] Cano, P., Batle, E., Kalker, T., & Haitsma, J. (2002). *A review of algorithms for audio fingerprinting*.

[6] Mencattini, A., Martinelli, E., Costantini, G., Todisco, M., Basile, B., & Bozzali, M., et al. (2014). Speech emotion recognition using amplitude modulation parameters and a combined feature selection procedure. *Knowledge-Based Systems, 63*(3), 68-81.

[7] Zhao, X., Zhang, S., & Lei, B. (2014). Robust emotion recognition in noisy speech via sparse representation. Neural Computing & Applications, 24(7-8), 1539-1553.

[8] Ooi, C. S., Seng, K. P., Ang, L. M., & Li, W. C. (2014). A new approach of audio emotion recognition. Expert Systems with Applications, 41(13), 5858-5869.

[9] Sweldens, W. (1998). The lifting scheme: a construction of second generation wavelets. *Siam J.math.anal, 29*(2), 511-546.

[10] Xie, H. B., Zheng, Y. P., & Guo, J. Y. (2009). Classification of the mechanomyogram signal using a wavelet packet transform and singular value decomposition for multifunction prosthesis control. *Physiological Measuremsent, 30*(5), 441-457.

[11] Zurek, S., Guzik, P., Pawlak, S., Kosmider, M., & Piskorski, J. (2012). On the relation between correlation dimension, approximate entropy and sample entropy parameters, and a fast algorithm for their calculation. *Physica A Statistical Mechanics & Its Applications, 391*(24), 6601-6610.

[12] Liang, J., & Yang, Z. (2015). A Novel Wavelet Transform – Empirical Mode Decomposition

Based Sample Entropy and SVD Approach for Acoustic Signal Fault Diagnosis. *International Conference in Swarm Intelligence* (Vol.9142, pp.232-241). Springer International Publishing.

[13] Li, Q., Zheng, J., Tsai, A., & Zhou, Q. (2002). Robust endpoint detection and energy normalization for real-time speech and speaker recognition. *Speech & Audio Processing IEEE Transactions on, 10*(3), 146-157.

[14] MINGYU YOU, GUO-ZHENG LI, JACK Y. YANG, & MARY QU YANG. (2010). An enhanced lipschitz embedding classifier for multi-emotion speech analysis. Internationoal Journal of Pattern Recgnition & Artificial Intelligence, 23(08), 1685-1700.

[15] Cotton, C. V., & Ellis, D. P. W. (2010). Audio fingerprinting to identify multiple videos of an event. *IEEE International Conference on Acoustics Speech and Signal Processing* (pp.2386-2389). IEEE.

IOP Conf. Series: Materials Science and Engineering **435** (2018) 012051 doi:10.1088/1757-899X/435/1/012051

Attention-based Hierarchical LSTM Model for Document Sentiment Classification

Bo Wang[1,*] and Binwen Fan[2]

[1,2]Harbin Institute of Technology, Shenzhen, Shenzhen, China

[*]wb_hitsz@foxmail.com

Abstract. Document sentiment classification is a fundamental task in data mining, contains extensive underlying commercial value. With the development of deep learning, we can extract features in an automatic way, instead of design it by oneself. Which can help us use semantic information to classify the document in a better way. Base that, in this paper, we present a hierarchical network structure according to the structure in real document. Based on LSTM to encode semantic information; then combine with attention mechanism to improve the accuracy of classification. And last, conduct experiment on two dataset, analyse the accuracy result of different model, and study some tricks in parameter selection.

1. Introduction

Document sentiment classification [1] is used to analyse public opinion towards commodity, current event and so on, which is an important task in field of Nature Languages Processing. For this task, dominant approaches exploit machine learning algorithm to build document classifier[2], which need researchers to design hand-crafted features[3, 4]. With the development of deep-learning approaches, many methods appears such as recurrent neural network [5] and convolutional neural network [6] to do this work. Recently years, attention mechanism [7] appears, and achieves state-of-art in many different fields.

In this paper, we will introduce an attention-based hierarchical bidirectional long short-term memory (LSTM) to achieve the task of document sentiment classification. Make use of information of document structure [8, 9] in review document, combine attention mechanism. And we will demonstrate the effectiveness of hierarchical network structure and attention mechanism in task of document sentiment classification.

2. Model

2.1. Long-short term memory network

Long short-term memory (LSTM) [10] has been widely replace the recurrent neural network (RNN). The LSTM uses gate mechanism to update information of memory cell. There are three types of gates, the forget gate f_t, the input gate i_t and output gate o_t. They control how to update the cell state together. The architecture of LSTM is shown in Figure 1.

At time t, we need to compute how much past information would be forget with previous hidden state h_{t-1} and current input information x_t. Forget gate f_t is computed by:

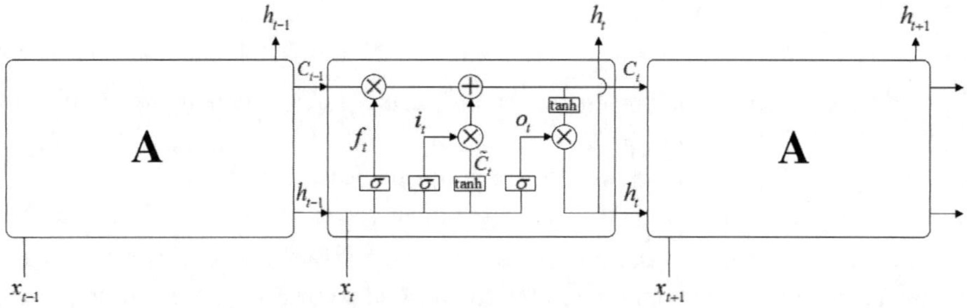

Figure 1. The architecture of LSTM

$$f_t = \sigma(W_f \cdot [h_{t-1}, x_t] + b_f) \quad (1)$$

Where σ denote the sigmoid activation function. Then use input gate and candidate cell state to get the new cell state: x_{t-1} x_{t+1}

$$i_t = \sigma(W_i \cdot [h_{t-1}, x_t] + b_i) \quad (2)$$

$$\tilde{C}_t = \tanh(W_c \cdot [h_{t-1}, x_t] + b_c) \quad (3)$$

$$C_t = f_t * C_{t-1} + i_t * \tilde{C}_t \quad (4)$$

At last, output the cell state and hidden state of current unit. Cell state has been computed before, and now use output gate which control how much information will be output to hidden state to decide the hidden state, the output gate and hidden state are updated as:

$$o_t = \sigma(W_o \cdot [h_{t-1}, x_t] + b_o) \quad (5)$$

$$h_t = o_t * \tanh(c_t) \quad (6)$$

2.2. Attention based hierarchical LSTM

In this part, we will focus on document structure. We divide a document into word-level and sentence-level two part, use LSTM compute the sentence representation and document representation separately. The structure viewed in Figure 2 below.

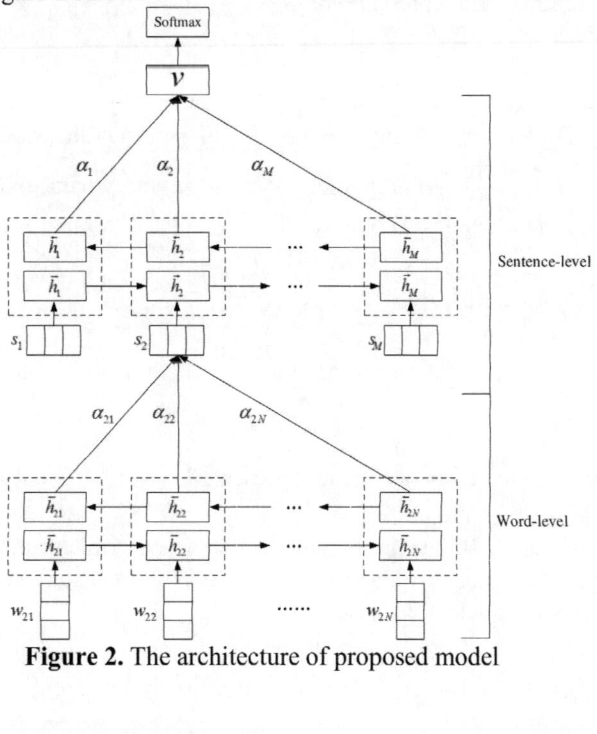

Figure 2. The architecture of proposed model

2.3. Word-level

Assume that there are M sentences in a document, and N words for each sentence. We use s_m $m \in [1, M]$ to denote sentences of document. And w_{mn}, $n \in [1, N]$ to denote words in sentence m. Now we will present how to build the model to represent document vector from the raw corpus.

Firstly, we need to represent the words which is the least unit in a document in a vector representation by the embedding matrix W_e, so we get input vector of word-level encoder. We use the bidirectional LSTM to encoder words information in both forward and backward direction, which could cover the contextual information more effectively. And we use h_{mn} (the brief representation of hidden state) to represent both forward and backward direction encoder information [$\vec{h}_{mn}$, $\overleftarrow{h}_{mn}$], which represent the whole sentence information of s_m centres on w_{mn}.

By now, we have get the sentence vector representation, every h_{mn} can be used to represent the information of s_m. Generally speaking, the hidden state h_{mn} of the last word or mean vector of all words will be used to represent the sentence information. In our model, we will introduce the attention mechanism to extract more important information in all hidden state.

$$u_{mn} = \tanh(W_n h_{mn} + b_n) \tag{7}$$

$$\alpha_{mn} = \frac{\exp(u_{mn}^T u_w)}{\sum_n \exp(u_{mn}^T u_w)} \tag{8}$$

$$s_m = \sum_n \alpha_{mn} h_{mn} \tag{9}$$

In word attention, we introduce u_{mn} and α_{mn} to represent the hidden representation of h_{mn} and importance of each hidden representation. Firstly, we feed h_{mn} into a one-layer MLP to get hidden representation u_{mn}, then use a softmax layer to compute the importance of u_{mn}. And last we combine u_{mn} and α_{mn} to get the sentence representation s_m.

2.4. Sentence-level

We have get sentence representation according to above process. Then we will get document representation in a similar way. Firstly we use h_m to represent forward and backward direction encoder information [$\vec{h}_m$, $\overleftarrow{h}_m$]. After to get more important sentence information, we introduce attention mechanism in sentence-level continue.

$$u_m = \tanh(W_m h_m + b_m) \tag{10}$$

$$\alpha_m = \frac{\exp(u_m^T u_s)}{\sum_m \exp(u_m^T u_s)} \tag{11}$$

$$d = \sum_m \alpha_m h_m \tag{12}$$

This time we use d to represent document representation. And introduce u_m and α_m represent the hidden representation and importance.

From the above, we have got the representation of document, which can be regard as feature vector of document. And now we will add a linear layer to transform the feature vector into another vector v whose length is the number of class, followed:

$$v = W_d d + b_d \tag{13}$$

Then add a softmax layer to convert v to a probability value:

$$p = \frac{\exp(v)}{\sum \exp(v)} \tag{14}$$

In training process, we use cross-entropy loss function compute training loss:

$$loss = -\sum_{i} \log(p_{ij}) \tag{15}$$

And j is the real label of document i.

3. Experimental and analysis

3.1. Dataset

For test the effectiveness of our model, we choose two datasets as experiment data, Yelp and IMDB. The statistics of datasets shown in Table 1. The Yelp is an unbalance dataset, and have five levels from 1 to 5, higher is better. The IMDB is a balance dataset have positive and negative two classes.

Table 1. Experimental data

dataset	documents	class	length/mean	vocabulary
Yelp	230907	5	151.2	204982
IMDB	50000	2	349.2	149393

All datasets divided into train data and test data with 9:1 randomly.

3.2. Hyper parameter setting

For raw document, we first use NLTK tools to split sentences and replace low-frequency word appear less than five times, after use Word2vec tool to train our word embedding as the initial value of word vector.

We set hyper parameters after some tests. In experience, we set dimension of word embedding is 200, the hidden state of LSTM unit is 64 and the activation function of LSTM is tanh. We set the max length of document is 100 and max number of sentence is 10. We use Adam gradient descent algorithm as optimizer. We use mini-batch to train our model, and set mini-batch size is 64. We use early-stopping strategy to avoid over-fitting.

3.3. Results and analysis

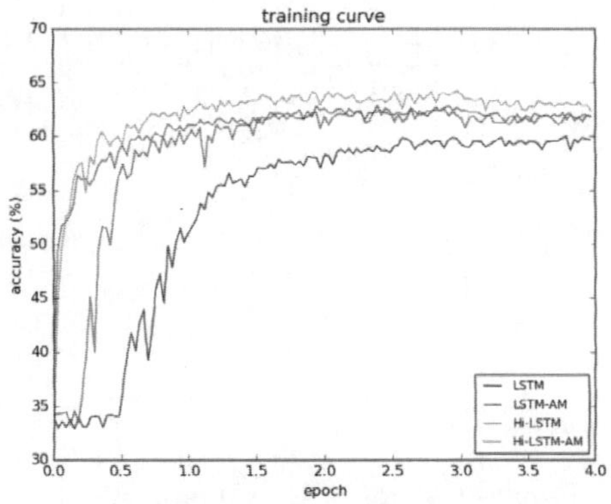

Figure 3. Model training curve

We present the results of accuracy between different models in Table 2 and the training curve is depict in Figure 3. From the Table 2, our hierarchical attention model achieves better result than hierarchical model and LSTM attention model. Compare with use LSTM only, add both an attention model and a hierarchical model will contribute to make a little improvement for classification accuracy. And from the training curve we can see, model with attention mechanism will convergence faster. Besides, when training process get to third epoch, the accuracy begin to drop, that is because the over-fitting of model, so we introduce early-stopping strategy, when monitor training get into over-fitting phase, stop the training process ahead of time to avoid over-fitting.

Table 2. Accuracy for experiment in different model

model/dataset	Yelp	IMDB
LSTM	59.91	89.30
LSTM-AM	62.80	90.35
Hi-LSTM	62.56	90.70
Hi-LSTM-AM	**64.44**	**91.51**

3.4. Word embedding dimension

In this experiment, we fixed number of hidden units as 64, choose the tanh as activation function to test the effect of word embedding. We present results in Table 3.

Table 3. Experiment of word embedding dimension

Database/embed dim	50	100	200	300
Yelp	64.13	64.38	**64.44**	64.37
IMDB	91.16	91.35	**91.51**	91.43

The dimension of word embedding represent the semantic features of the word, generally speaking, the more features, the better efficiency of experiment have. However, from the table, when embedding dimension increase from 50 to 200, the accuracy of classification have some growth. But when embedding dimension increase to 300, the accuracy go down. That is because a large number of features will contribute to distinguish different words in some degree, but it will reduce the character of features.

3.5. Number of hidden units

This time we will study the effect of number of hidden units, word embedding dimension fixed as 200 and activation function choose tanh. As we know, the hidden units is the storage of long term information. So as we can see results from the Table 4, the accuracy will increase follows the increase of number of hidden units. But the speed of increase will be slow.

Table 4. Experiment of number of hidden units

Database/number hidden	32	64	128	256
Yelp	63.28	64.44	64.63	**64.66**
IMDB	90.90	91.51	91.65	**91.75**

3.6. Activation function

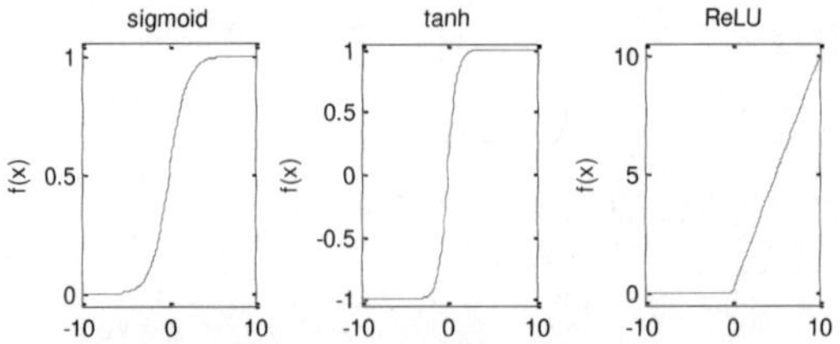

Figure 4. Three of activation function

At last, let us study the effect of different activation function. We used to use sigmoid and tanh as activation function. Recent years, an activation function named Relu appears, which have the advantages of simplify computation and prevent vanish of gradient in positive interval. This time we fixed word embedding dimension as 200 and number of hidden is 128. The Figure 4 shows different activation function curves.

Table 5. Experiment of activation function

database/activation function	sigmoid	tanh	Relu
Yelp	61.94	64.63	**65.25**
IMDB	90.15	91.65	**91.90**

From the Table 5, Relu function achieve best result during three function. Compare with sigmoid function, tanh function solve the zero-mean problem, prevent 'shift' during training process. For Relu, the function always keep a gradient in are interval, as for sigmoid and tanh, there are vanishing gradient problem happened when input value too large or too small. But the Relu still need to be improved in negative interval.

4. Summary

In this paper, we present an attention based hierarchical LSTM model. And conduct experiments on two database with classification accuracy to demonstrate the effectiveness of our model. Analysis how word embedding dimension, number of hidden units and activation function effect the classification results, demonstrate the effectiveness of hierarchical network structure and attention mechanism in task of document sentiment classification.

Acknowledgments

This work was supported by Special funds for the development of strategic emerging industries in Shenzhen (no. JCYJ20150625142543448).

References

[1] Pang, Bo, and Lee, Lillian. "Opinion Mining and Sentiment Analysis." Foundations & Trends in Information Retrieval 2.1–2(2008):1-135.

[2] Pang, Bo, L. Lee, and S. Vaithyanathan. "Thumbs up?: sentiment classification using machine learning techniques." Proceedings of the Conference on Empirical Methods in Natural Language Processing (EMNLP), Philadelphia, July 2002, pp. 79-86.

[3] Ifrim, Georgiana, and G. Weikum. "The Bag-of-Opinions Method for Review Rating Prediction from Sparse Text Patterns." Proceedings of the 23rd International Conference on Computational Linguistics (Coling 2010), Beijing, August 2010, pages 913–921.

[4] Paltoglou, Georgios, and M. Thelwall. "A Study of Information Retrieval Weighting Schemes for Sentiment Analysis." Proceedings of the 48th Annual Meeting of the Association for Computational Linguistics, pages 1386–1395, Uppsala, Sweden, 11-16 July 2010.

[5] Vo D T, Zhang Y. Target-dependent twitter sentiment classification with rich automatic features. International Conference on Artificial Intelligence. AAAI Press, 2015:1347-1353.

[6] Kalchbrenner, Nal, Grefenstette, Edward, Blunsom, Phil. A Convolutional Neural Network for Modelling Sentences. Proceedings of the 52nd Annual Meeting of the Association for Computational Linguistics, pages 655–665, Baltimore, Maryland, USA, June 23-25 2014.

[7] Xu K, Ba J, Kiros R, et al. Show, Attend and Tell: Neural Image Caption Generation with Visual Attention. Proceedings of the 32nd International Conference on Machine Learning, 2015, p. 2048-2057.

[8] Yang, Zichao, et al. "Hierarchical Attention Networks for Document Classification." Proceedings of NAACL-HLT 2016, pages 1480–1489, San Diego, California, June 12-17, 2016.

[9] Tang, Duyu, B. Qin, and T. Liu. "Document Modeling with Gated Recurrent Neural Network for Sentiment Classification." Proceedings of the 2015 Conference on Empirical Methods in Natural Language Processing, pages 1422–1432, Lisbon, Portugal, 17-21 September 2015.

[10] Hochreiter S, and Schmidhuber J. "Long short-term memory." Neural Computation. Vol.9.8(1997): 1735-1780.

Research on LK Optical Flow Algorithm with Gaussian Pyramid Model Based on OpenCV for Single Target Tracking

Jing Bai[1,a] **and Lili Huang**[2,b]

[1]School of Information Science and Technology, East China Normal University, 200062.
[2]School of Information and Technology, Shanghai Jian Qiao University, 201319, China

[a] Phone: 18016297367, E-mail: 15008@gench.edu.cn
[b] Phone: 15001817383, E-mail: 1577339166@qq.com

Abstract. There is a problem that the tracker of TLD algorithm in computer vision tracking will lose tracking target or follow the wrong one when the target is fast moving or blocked. To solve this problem, this paper proposes an improved LK optical flow algorithm with Gaussian pyramid model for single target tracking, through theoretical analysis and formula deduction. This paper applied this algorithm and the traditional LK optical flow algorithm with the unified video to compare the experimental results with the OpenCV visual library in Visual Studio 2012. The result shows that this algorithm has the effect of improving tracking accuracy and adaptability.

1. Introduction

Visual tracking is a hot topic in the field of computer vision. It is usually applied to lane departure detection, vehicle speeding detection, license plate recognition, behavior detection, falling objects detection and other scenes. However, there may be problems to visual tracking in some complex scenes during a long time, such as the target is moving fast or blocked and so on[1-2]. TLD (Tracking-Learning-Detection) visual tracking algorithm combines detection and tracking while adding a learning mechanism, which making the overall target tracking is more stable and effective[3].

However, the tracker of the TLD algorithm is implemented with the LK optical flow algorithm. It will lose tracking target or follow the wrong one when the target is fast moving or blocked. This paper proposes an improved optical flow algorithm with Gaussian Pyramid model as the algorithm of TLD tracker, which improves the accuracy and adaptability of tracking.

2. Implementation of tracking algorithm

The TLD algorithm is a visual tracking algorithm which was proposed by Zdenek Kalal from 2008 to 2010 and constantly improved. The TLD algorithm is composed of tracker, detector and the online learning module. The tracker tracks the motion of the target between successive frames. The tracker can achieve its function when the target is always visible. The tracker can predict the region of the target of the current frame from the result of the previous frame, thus obtaining a positive sample trajectory. The detector corrects the tracker. Detector traverses the image of the current frame to find all possible positions similar to the target object, and obtains the positive and negative samples from the detection result. It takes the most trusted position of all detection windows as the TLD output. The learning module generates the corresponding classifier by iterating and training according to the positive and negative samples obtained from tracker and detector so as to improve performance and correct the corresponding parameters[4]. The process of TLD algorithm is shown in Figure 1.

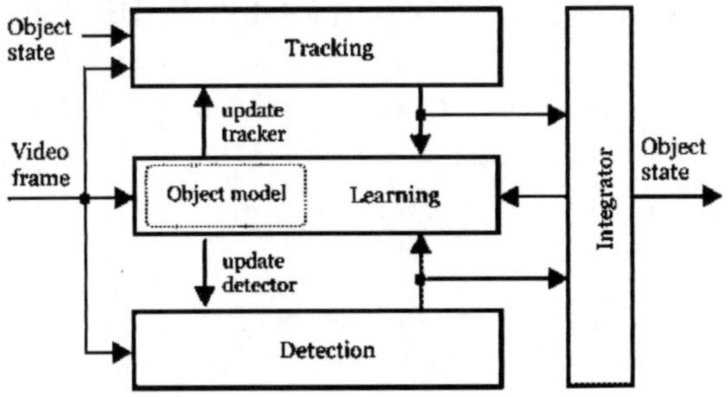

Figure 1. Process of TLD algorithm

2.1. LK (Lucas-Kanade) optical flow algorithm

The tracker of TLD is based on the LK optical flow algorithm which requires the following three assumptions:

- The brightness is constant. It is assumed that the surface color of the target is constant during the movement, that is, the brightness of the pixels in the two frames in the image is constant.
- Time is continuous or the movement is a movement of small amplitude. Changes in the motion of target in the image are slow, that is, the displacement of the target is small between the two continuous frames.
- The optical flows are consistent. In the image, the motion of adjacent pixels on the surface of the same target is consistent, and these pixels must be aggregated in the same region[5].

Assuming that the optical flows in the grayscale region centered on the M point are the same, the calculation formula of the optical flow can be converted to the formula (1) by giving different weights to different pixels.

$$\sum_{x\in\Omega} W(x,y)^2[\nabla I \cdot V_m + I_t]^2 = 0 \tag{1}$$

Ω represents a grayscale region centered on M points, and W(x,y) represents the weight function of the region. From the formula to draw a conclusion that the closer to the M, the higher the weight, the farther away from the M, the lower the weight.

The solution of the formula (1) can be calculated by the formula (2).

$$A^T W^2 A v = A^T W^2 b \tag{2}$$

Assuming that there are N*N pixels in the region, the matrix variable s of the formula (2) can be expressed as formula (3), formula (4), and formula (5).

$$A = \begin{bmatrix} \nabla I(x_1,y_1) & \nabla I(x_1,y_2) & \dots & \nabla I(x_1,y_N) \\ \nabla I(x_2,y_1) & \nabla I(x_2,y_2) & \cdots & \nabla I(x_2,y_N) \\ \vdots & & \dots & \vdots \\ \nabla I(x_N,y_1) & \nabla I(x_N,y_1) & \cdots & \nabla I(x_N,y_N) \end{bmatrix} \tag{3}$$

$$W = \begin{bmatrix} w(x_1,y_1) & w(x_1,y_2) & \dots & w(x_1,y_N) \\ w(x_2,y_1) & w(x_2,y_2) & \cdots & w(x_2,y_N) \\ \vdots & & \dots & \vdots \\ w(x_N,y_1) & w(x_N,y_2) & \cdots & w(x_N,y_N) \end{bmatrix} \tag{4}$$

$$b = \begin{bmatrix} I_t(x_1,y_1) & I_t(x_1,y_2) & \dots & I_t(x_1,y_N) \\ I_t(x_2,y_1) & I_t(x_2,y_2) & \cdots & I_t(x_2,y_N) \\ \vdots & & \dots & \vdots \\ I_t(x_N,y_1) & I_t(x_N,y_1) & \cdots & I_t(x_N,y_N) \end{bmatrix} \tag{5}$$

Solve equation to get the formula (6).

$$v = (A^T W^2 A)^{-1} A^T W^2 b \tag{6}$$

Among, $A^T W^2 A$ is a matrix of 2*2. Convert the form to get the formula (7).

IOP Conf. Series: Materials Science and Engineering **435** (2018) 012052 doi:10.1088/1757-899X/435/1/012052

$$A^TW^2=\begin{bmatrix}\sum W^2I_x^2 & \sum W^2I_xI_y\\ \sum W^2I_xI_y & \sum W^2I_y^2\end{bmatrix} \quad (7)$$

If the A^TW^2A of formula 7 is non-singular, it must have two characteristic roots λ_1 and λ_2 which are not equal to 0. When we set the $\lambda_1>\lambda_2$, we can get the following conclusion:

- If $\lambda_1>\tau$, $\lambda_2>\tau$, we can calculate the optical flow using the formula (6).
- If $\lambda_1>\tau$, $\lambda_2<\tau$, we can't get the complete optical flow information.
- If $\lambda_1<\tau$, $\lambda_2<\tau$, we can't calculate the optical flow.

2.2. Gauss Pyramid LK optical flow algorithm

2.2.1 Principle of algorithm When calculate the optical flow, we first need to set up a window of adjacent areas. Calculation of the optical flow is more robust when the window is large, while calculation of the optical flow is more correct when the window is small. The reason is that when the motion of each part of the image is not consistent, if the window is too large, it is easy to go against the assumption which the optical flows must be consistent. This may not be realistic. So, smaller the window is and fewer pixels thw window contains, more accurately we calculate. There is a problem that the above mentioned assumptions can not be achieved, when the movement range is large. To solve this problem, this paper introduces the Gaussian Pyramid model. Set a fixed window, and then generate a pyramid image. In each layer of the pyramid, calculate optical flow with the same size (resolution) of window. When the size of image is small, the window appears large. Then the optical flow can track the fast target. While, when the size of image is original size, the optical flow window is relatively small and the optical flow is more accurate.

Image pyramid is a set of images, and all multi-resolution images come from the same original image. This paper uses a down-sampled Gaussian Pyramid model. First, calculate the optical flow of the top level image of the pyramid. Then calculate the initial optical flow of the second-level according to the top level optical flow result, and further estimate the exact value of optical flow. Finally, calculate the initial optical flow of the next level, using the calculated second-level optical flow result. and calculate its exact value. Put the calculated exact value into the next level for calculation until the bottom of the pyramid[6]. The principle is shown in Figure 2.

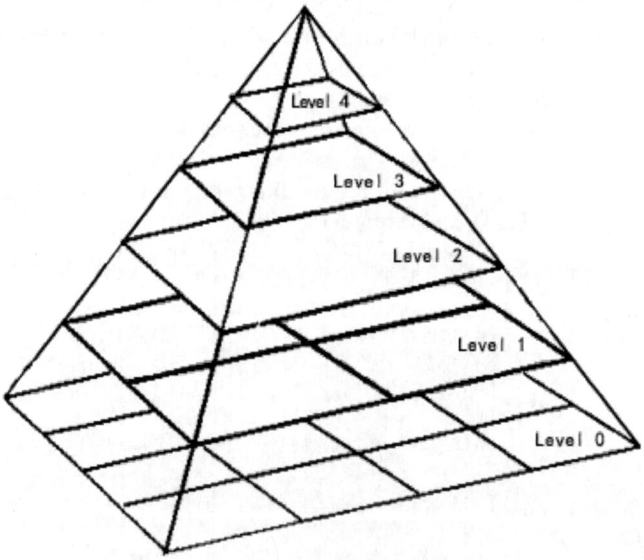

Figure 2. Principle of Gaussian Pyramid

Assume the original image is I (x, y) whose size is ×N , it is to say that the image of Level 0 is I^0 which is at the bottom of pyramid. The size of I^0 is the same as the size of the original image, whose size is M×N too. It is mean that I^0 =I.

The way, the Gaussian pyramid model decreases downward, is as shown in formula (8).

$$I^i(x,y) = \sum_{m=-2}^{2}\sum_{n=-2}^{2} w(m,n) I^{i-1}(2x+m, 2y+n) \quad (8)$$

Wherein, for the ith level, $I^i = \text{Reduce}[I^{i-1}]$.

w(m,n) is the generated kernel. The window function is a low-pass filter. The limitation of the generated kernel is to ensure the low-pass quality and to ensure that the brightness of the image, after expansion and contraction, is smooth without boundary cracks.

Pyramid is established in the way of downwards 1/2 sampling. Let L (L=1, 2,..) represent the number of levels in the pyramid model, and the images are expressed as I^1, I^2, …… The length and width of the L level image are represented by N^L and M^L. L level image of the pyramid (I^L) can be represented as a formula (9):

$$I^L = \frac{1}{4} I^{L-1}(2x,2y) + \frac{1}{8}[I^{L-1}(2x-1,2y) + I^{L-1}(2x+1,2y) + I^{L-1}(2x,2y-1) + I^{L-1}(2x,2y+1)] + \frac{1}{16}[I^{L-1}(2x-1,2y-1) + I^{L-1}(2x+1,2y-1) + I^{L-1}(2x+1,2y-1) + I^{L-1}(2x-1,2y+1)] \quad (9)$$

The formula (9) performs Gaussian low-pass filtering on the image while downwards sampling to prevent image distortion. At the same time, some pixel value around the image, formulas (10), must be defined to satisfy the formula (9).

$$I^{L-1}(-1,y) = I^{L-1}(0,y) \quad (10a)$$
$$I^{L-1}(x,-1) = I^{L-1}(x,0) \quad (10b)$$
$$I^{L-1}(-1,-1) = I^{L-1}(0,0) \quad (10c)$$
$$I^{L-1}(N^{L-1},y) = I^{L-1}(N^{L-1}-1,y) \quad (10d)$$
$$I^{L-1}(x,M^{L-1}) = I^{L-1}(x,M^{L-1}-1) \quad (10c)$$
$$I^{L-1}(N^{L-1},M^{L-1}) = I^{L-1}(N^{L-1}-1,M^{L-1}-1) \quad (10d)$$

2.2.2 Process of algorithm. Taking into account the amount of computation, the pyramid usually only calculated to 3 to 4 levels. The algorithm can deal with the problem of feature point tracking when optical flow is greater than the window, because of the moving image[7].To increase the calculation speed of the optical flow field, algorithm with Pyramid LK optical flow is usually used to estimate the optical flow of feature point[8].

Given a pixel u in image I, calculate the pixel v in image J that matches it, making v=u+d.

Create a pyramid model of I and J:

$$\{I^L\}_{L=0,1,\ldots\ldots,L_m} \text{ and } \{J^L\}_{L=0,1,\ldots\ldots,J_m} \quad (11)$$

Initialize the estimate of the optical flow:

$$g^{L_m} = [0\ 0]^T \quad (12)$$

Determine the coordinates of the pixel point u on the image:

$$I^L{:}\ u^L = [p_x\ p_y]^T = \frac{u}{2^L} \quad (13)$$

Calculate the partial derivative of image I^L in the x direction with u as the center :

$$I_y^L(x,y) = \frac{A(x,y+1) - A(x,y-1)}{2} \quad (14)$$

gradient matrix:

$$G = \begin{bmatrix} \sum I_x^2 & \sum I_x I_y \\ \sum I_x I_y & \sum I_y^2 \end{bmatrix} \tag{15}$$

Initialize LK optical flow vector : v= 〖[0 0]〗 ^T. Calculate the pixel difference of the image :

$$\delta I_k(x, y) = A(x, y) - J^L(x + g_x^L + v_x^{k-1}, y + g_y^L + v_y^{k-1}) \tag{16}$$

Calculate image mismatch vector :

$$b_k = \begin{bmatrix} \delta & \sum I_k I_x \\ \delta & \sum I_k I_y \end{bmatrix} \tag{17}$$

Calculate LK optical flow vector :

$$\eta_k = G^{-1} b_k \tag{18}$$

Estimate the next iteration :

$$v^k = v^{k-1} + \eta^k \tag{19}$$

The optical flow in the Lth level can be finally determined as :

$$d^L = v^k \tag{20}$$

Calculate the light flow at L-1 level :

$$g^{L-1} = [g_x^{L-1} \; g_y^{L-1}]^T = 2(g^L + d^L) \tag{21}$$

Calculate optical flow vector : $d = g^0 + d^0$. Then the corresponding point of the pixel u in the next frame image J is :

$$v = u + d \tag{22}$$

3. Test

This LK optical flow algorithm with Gaussian pyramid model for single target tracking, proposed in this paper, was implemented with the efficient and flexible library function of OpenCV 3.0.0 in Visual Studio 2012. The source video for testing used a self-recorded video with a handheld yoghurt bottle moving. There are some scenes such as target blocked, disappeared, and fast moving in this video.

The tracking effect of the traditional LK optical flow algorithm in the TLD algorithm is shown in Figure 3. The result shows that the tracker will lose target when blocked or partially disappeared, as shown in Figure 3b. When the integral image is restored, the tracker may have a tracking error when it is tracked again, as shown in Figure 3c. When the tracked target is moving too fast, the tracker may also have erroneous tracking, as shown in Figure 3d.

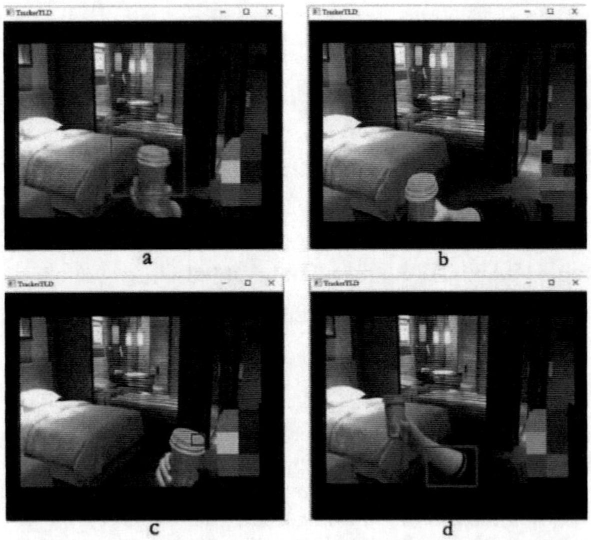

Figure 3. tracking effect of the traditional LK optical flow algorithm in the TLD algorithm

4. Conclusion

There is a problem that the traditional LK optical flow algorithm in TLD can not capture fast moving or blocked objects. To solve this problem, this paper proposes an improved LK optical flow algorithm with Gaussian pyramid model. The test results show that the algorithm enhances the accuracy and adaptability of TLD algorithm in tracking the target under fast motion in the long time tracking. After the blocked target is recovered, the target can be accurately repositioned. So the effect of this algorithm is obviously improved for single moving target.

References

[1] Xiaojun Jia, Qingcang Yu. The Image Processing Based on Open Source Computer Vision Library[J]. Computer Applications and Software, 2008,25(4):276～278

[2] Mei Fang, Qingcang Yu, Huaqiang Li. Digital Image Processing Based on OpenCV in C++ Builder[J]. Computer Engineering and Design, 2008,29(4):882～884

[3] Kalal Z,Mikolajczyk K, Matas J. Tracking-Learning-Detection[J]. IEEETransactions on Pattern Analysis & Machine Intelligence, 2012, 34(7):1409

[4] Wei Zhu, Chunguang Zhao, Leqing Ou, Shoufeng Wang. Examples of OpenCV Image Processing Program[M]. Beijing: Publishing House of Electronics Industry, 2016:233~240

[5] Xiaoyang Wu. Moving Target Detection and Tracking Based on OpenCV[D]. Hangzhou: Zhejiang University, 2008:48～55

[6] Boneng Lai, Xiyuan Chen, Qinghua Li. Speed Calculation with Pyramid Lucas-Kanade Optical Flow[J]. Measurement and Control Technology, 2016,35(6):145～148

[7] Hongyan Huang, Siwen Gao, Yanmei Yu, Xiaohai He. Visual Obstacle Avoidance System Based on Optical Flow Method[J]. Journal of Image and Signal Processing, 2016:5(2), 66-72

[8] Anurag Ranjan, Michael J Black. Optical Flow Estimation using a Spatial Pyramid Network[C]. 2017 IEEE Conference on Computer Vision and Pattern Recognition,2017

A Belief Rule Extraction Method Based on Rough Sets

YU Meng, HUANG Jian and KONG Jiangtao

National University of Defense Technology, Changsha, China

1546737193@qq.com

Abstract. Most of the existing research on belief rule-base focused on parameter optimization problem. However, there are few studies on the problem of establishing the initial belief rule-base due to the difficulty in obtaining expert experience and the unconformity of standards. Based on this, a method of extracting belief rule based on rough sets is proposed. First, the initial attribute index is used to divide the data set into equivalent classes. Then, the belief rule is extracted with the knowledge of rough sets. Finally, the extracted belief rule is used for reasoning and the index is optimized according to the result of reasoning. An example of pipeline leak detection is introduced in the experiment section. Experimental results show that the method is effective and feasible.

1. Introduction

In order to effectively use quantitative information and incomplete or uncertain information provided by experts to model complex problems, Yang et al.[1] proposed a belief rule-base inference methodology using the evidential reasoning approach (RIMER). RIMER combines D-S evidence theory, decision theory, fuzzy theory, and traditional IF-THEN rule base knowledge to model data with ambiguous or fuzzy uncertainty, incomplete or probabilistic uncertainties, and nonlinear features. At present, RIMER has been widely used in pipeline leak detection[2-3], fault diagnosis[4-7], intent identification[8-10], and capability evaluation[11-13].

RIMER mainly includes the expression of knowledge and the reasoning of knowledge. Among them, the expression of knowledge is realized by the belief rule-base (BRB) system, and the reasoning of knowledge is carried out by using the evidence reasoning (ER) algorithm. Due to the uncertainty of expert experience, the initial BRB systems is often inaccurate. To this end, the obtained data set can be used to optimize the inference parameters of the BRB system in combination with the parameter training method. At present, the research on RIMER is mainly concentrated in this field. Including Yang et al. proposed to use the FMINCON function in Matlab[14] for parameter training; B Qian et al. proposed a particle swarm optimization algorithm with improved velocity update way and repair methods (PSO_VR) to deal with the NNOP of SPO-BRB[15]; J Liu et al. proposed a new learning method based on the given sample data for optimally generating a consistent BRB[16]; R Chang proposed a new simple optimization algorithm to realize the self-learning capability of the rule base parameters[17]. However, in the process of establishing the initial BRB system, due to the difficulties in obtaining expert experience and the inconsistency of standards, it is often the case that the initial BRB system has a long establishment period and the initial BRB system does not match the actual system. How to quickly and effectively establish the initial BRB system has become an urgent problem to be solved, and at present there is relatively little research in this area.

Based on the above reasons, this paper proposes a method based on rough set[18-19] for the extraction of belief rule. Firstly, according to the attribute index, the data is divided into equivalence classes; then, the belief degree is calculated for the class of the same precondition attribute, and the belief rule is extracted; finally, the extracted belief rule is used for reasoning, and the initial attribute index is optimized according to the inference result, which improves the inference precision of the belief rule-base. An example of pipeline leak detection was introduced in the experimental part to verify the effectiveness of the method.

2. Related basic concepts

2.1. belief rule representation

The confidence rule is developed according to the IF-THEN rule, introducing a distributed confidence framework and weight parameters. The k th rule in the BRB system is as follows:

R_k :if x_1 is $A_1^k \wedge x_2$ is $A_2^k \wedge \ldots \wedge x_M$ is A_M^k

Then $\{(D_1, \beta_{1,k}),(D_2, \beta_{2,k}),\ldots,(D_N, \beta_{N,k})$

With a rule weight θ_k and attribute weight $\bar{\delta}_1, \bar{\delta}_2, \ldots, \bar{\delta}_M$ (1)

Where x_i (i=1, 2, ..., M) indicates the i th premise attribute, M indicates the number of premise attributes; A_i^k indicates the reference value the k th input of the i th rule, D_i (i=1, 2, ..., N) indicates the i th evaluation level of the result attribute , N indicates the number of evaluation levels; $\beta_{i,k}$ indicates the belief level of the i th evaluation level in the k th rule, k = 1, 2, L ..., L indicates the number of belief rules in the BRB; θ_k indicates the rule weight, $\bar{\delta}_i$ indicates the premise attribute weight.

2.2. Belief rule-base inference methodology

2.2.1 Calculation of the activation weight of the belief rule. The activation weight of the input information x for the k th rule can be calculated by the following formula, ie

$$\omega_k = \frac{\theta_k \prod_{i=1}^{M} (\alpha_i^k)^{\bar{\delta}_i}}{\sum_{l=1}^{L} \theta_l \prod_{i=1}^{M} (\alpha_i^l)^{\bar{\delta}_i}} \tag{2}$$

Where $\omega_k \in [0,1], k$ =1,2,..., L ; α_i^k (i=1,2,..., M) indicates the confidence of the i th input x_i in the k th rule with respect to the reference value A_i^k .

2.2.2 ER algorithm fusion. After calculating the activation degree of the rule, the ER algorithm can be used to fuse the rules in the BRB. First, convert the belief degree in the result attribute to the basic probability mass.

$$m_{j,k} = \omega_k \beta_{j,k} \tag{3}$$

$$m_{D,k} = 1 - \omega_k \sum_{j=1}^{N} \beta_{j,k} \tag{4}$$

$$\bar{m}_{D,k} = 1 - \omega_k \tag{5}$$

$$\tilde{m}_{D,k} = \omega_k (1 - \sum_{j=1}^{N} \beta_{j,k}) \tag{6}$$

Then, using the ER parsing algorithm to fuse the L rules, the formula is as follows:

$$\hat{\beta}_j = \frac{\mu \times \left[\prod_{k=1}^{L} (\omega_k \beta_{j,k} + 1 - \omega_k \sum_{i=1}^{N} \beta_{i,k} \right]}{1 - \mu \times \left[\prod_{k=1}^{L} (1 - \omega_k) \right]} - \frac{\mu \times \left[\prod_{k=1}^{L} (1 - \omega_k \sum_{i=1}^{N} \beta_{i,k}) \right]}{1 - \mu \times \left[\prod_{k=1}^{L} (1 - \omega_k) \right]} \tag{7}$$

$$\mu = \begin{bmatrix} \sum_{j=1}^{N} \prod_{k=1}^{L} (\omega_k \beta_{j,k} + 1 - \omega_k \sum_{i=1}^{N} \beta_{i,k}) \\ - (N-1) \prod_{k=1}^{L} (1 - \omega_k \sum_{i=1}^{N} \beta_{i,k}) \end{bmatrix}^{-1} \tag{8}$$

Among them, $\hat{\beta}_j$ indicates the belief level with respect to the evaluation result D_j.

2.3. Rough set theory

Rough set theory is a data analysis theory proposed by Polish mathematician Z.Pawlak in 1982. Its basic idea is to divide the instances in the database according to the attribute values of different attributes, and then establish rules for the relationship between the divided subsets[20-22].

The knowledge expression system is an ordered pair $S = \langle U, C, D, V, f \rangle$, in which U is a non-empty finite set , called a universe, C is a set of conditional attributes, D is a set of decision attributes, $A = C \cup D$ called a set of attributes. V is a collection of attribute values, the elements of the universe U are called objects or instances, f specifies the mapping between objects and attribute values.

In the knowledge expression system, for any attribute a in the attribute set A, if the instance x_i and x_j have the same value for the attribute a, we call x_i and x_j equal based on the attribute set and are indistinguishable. A collection of all equivalent records based on a set of attributes A is defined as an equivalence class. Records belonging to the same equivalence class are grouped into one class, called R a division based on an attribute set A, expressed as $U / ind(R) = \{R_1, R_2, \cdots, R_n\}$, $[x]_R$ is an equivalence class contained x in the $U / ind(R)$. When $X \subseteq U$,and X is the sum of $U / ind(R) = \{R_1, R_2, \cdots, R_n\}$, X is said to be R -definable, called the exact set of R , otherwise X is R undefined, also known as the rough set of R .

Rough sets can be characterized by upper and lower approximations. The lower approximation of X with respect to R is $R\_(X) = \bigcup \{Y \in U / R; Y \subseteq X\}$, and the upper approximation of X with respect to R is $R^{-}(X) = \bigcup \{Y \in U / R; Y \cap X \neq \Phi\}$. The positive domain for the decision attribute D with respect to R is $pos_R(X) = R\_(X)$, $X \in U / ind(D)$. When $R\_(X) \neq R^{-}(X)$, X is rough for R , the roughness is defined as $\alpha_R(X) = card(R\_(X)) / card(R^{-}(X))$.

3. Belief rule extraction method

3.1 Equivalence class division

First, establish the knowledge expression system $S=\langle U,C,D,V,f\rangle$, where $U=\{x_1,x_2,x_3,\cdots,x_n\}$ is a set containing all instances, called the universe. $C=\{c_1,c_2,c_3,\cdots,c_m\}$ is the set of conditional attributes, $D=\{d_1,d_2,d_3,\cdots,d_k\}$ is the set of decision attributes, $A=C\cup D$ is called the set of attributes, V is the set of attribute values, f specifies the instance The mapping relationship between the attribute value and the attribute value.

In order to divide the instance into equivalence classes, you need to first determine the attribute indicators. For attributes a_i, if you want to set the z_i level index, you can determine the index width

$$\mathrm{wide}_{\mathrm{i}}=\frac{\max(a_i)-\min(a_i)}{z\text{-}1} \tag{9}$$

It is thus determined that the indexs at each level are $\{\min(a_i),\min(a_i)+wide_i,\cdots,\min(a_i)+(z-1)wide_i\}$. For the attribute value v_i, it can be determined its index level is

$$j_i=round(\frac{v_i-\min(a_i)}{wide_i})+1 \tag{10}$$

For each attribute, the index is determined, and the index can be divided into $\sum_{i=1}^{m+k} z_i$ x equivalent classes. For the instance $x_i=\{v_1,v_2,\cdots,v_{m+k}\}$, the index level corresponding to each attribute is calculated separately, and the instance can be divided into its corresponding equivalence class.

3.2 Extracting belief rules

For a decision rule "if X then Y" (or denoted as $X\rightarrow Y$), where X is the set of objects with the same conditional attributes; Y is the set of objects with the same decision attributes. The confidence factor of the rule, or belief level, can be expressed as:

$$cf=\frac{card(X\cap Y)}{card(X)}(X\neq\phi) \tag{11}$$

$card(X)$ represents the number of elements of the set X.

The belief level of the decision rules with the same conditional attributes and different decision attributes is obtained separately, and then these decision rules are combined to form a belief rule in the following form:

$$\text{If C then }\{(d_1,cf_1),\ (d_2,cf_2),\cdots,(d_k,cf_k)\} \tag{12}$$

Where $C=\{c_1,c_2,c_3,\cdots,c_m\}$ is the set of conditional attributes.

3.3 Index optimization

Index optimization can be achieved by optimizing the index levels. The selection of the attribute index series can be determined by using the control variables and the principle of stepwise determination. Specific steps are as follows:

① Use the index level of the first attribute as the variable, and select an empirical value as the fixed value for the other attribute index series to determine the indexs at all levels;

② then use the data set for rule extraction;

③ The cumulative error is obtained by inference of the test set;

④ determining the optimal index level of the first attribute according to the obtained cumulative

IOP Conf. Series: Materials Science and Engineering **435** (2018) 012053 doi:10.1088/1757-899X/435/1/012053

error of reasoning;

⑤ The index level of the second attribute is taken as the variable. The optimal index level of the first attribute and the empirical value of the other attribute are used as fixed values to determine the indexs at all levels, and repeat ② to ④ to obtain optimal index level of the second attribute;

⑥ Similar to ⑤, determine the optimal index level of the third attribute;

......

And so on, determine the index levels for all attributes.

4. Experimental verification

4.1. Experimental conditions

In order to verify the effectiveness of the method, an example of oil pipeline leak detection is introduced. The pipeline leak detection problem is to be installed in a 100-kilometer-long oil pipeline in the UK. When a leak occurs in the pipeline, the (FD) and (PD) of the pipeline will change according to a certain pattern, which in turn affects the (LS). A belief rule base system is constructed by taking the FlowDiff and PressureDiff as inputs and the LeakSize as an output.

In the literature, the initial belief rule base system FD has 8 indicators, respectively {-10,-5,-3,-1,0,1,2,3}; PD has 7 indicators, respectively {-0.042 , -0.025, -0.01, 0, 0.01, 0.025, 0.042}; LS has 5 indicator levels, respectively {0, 2, 4, 6, 8}. The experiment collected real-time data of the 2008 group from normal to 25% leakage every 10 seconds as a cycle. The initial belief rule-base system output is shown in Figure 1.

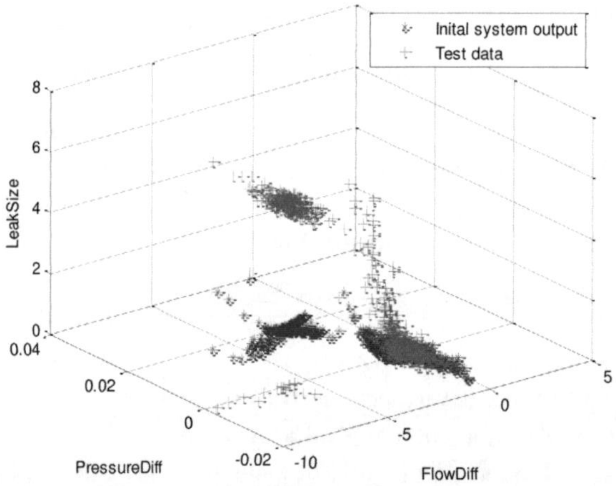

Figure 1. Initial confidence rule base output and test set data.

4.2. Experimental process

The framework of the experiment in this paper is shown in Figure 2. First determine the index level, calculate the indexs at each level, and then select 500 sets of data from 7:00 am to 7:33 am, 9:46 am to 10:20 am and 10:50 am to 11:08 am as the data set for rule extraction, and then the 2008 data is used as the test set. The extracted rules are used to perform the belief rule-base reasoning, and the inference cumulative error is obtained. Finally, the index is optimized until the optimal index is obtained, which makes the inference cumulative error minimum.

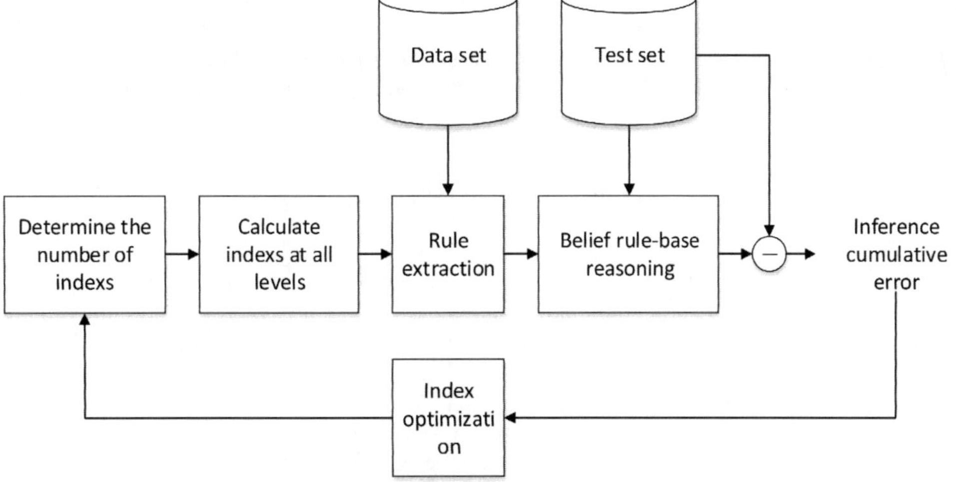

Figure 2. Experimental framework.

In the experiment, through the control variables, the index levels of the three attributes are determined step by step. The index of each index is extracted from 2 to 50, and then the test set is used to perform the belief rule-base reasoning. By comparing the final inference cumulative error, the optimal index level of the three attributes is determined to be 10, 3,10. The calculated index of FD is {-7.0500, -6.1888, -5.3277, -4.4666, -3.6055, -2.7444, -1.8833, -1.0222, -0.1611, 0.7000}, and the index of PD is {-0.0140, 0.0012, 0.0165. }, the indicators of LS are {0, 0.7131, 1.4262, 2.1393, 2.8525, 3.5656, 4.2787, 4.9918, 5.7050, 6.4181}. Finally, the extracted rule is used to perform the belief rule-base reasoning. The result is shown in Figure 3.

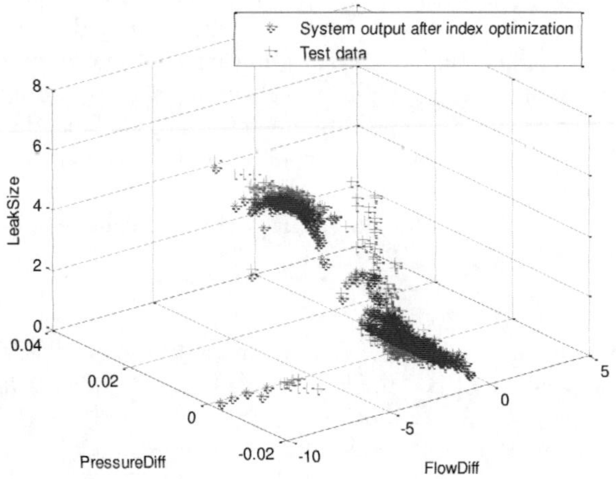

Figure 3. System output after optimization of indexs and test set data.

In order to reflect the superiority of the method, other methods were used in the experiment to compare with the method. In the previous research, most of the optimization algorithms are used to optimize the rule weight and precondition attribute weight of the belief rule-base based on the initial belief rule-base. In this paper, we use the fmincon in MATLAB to optimize the rule weight and precondition attribute weight of the initial belief rule-base. The optimization result is shown in Figure 4.

Comparing Fig. 1 and Fig. 3, it can be found that the reasoning error of the belief rule-base extracted by the index optimization is significantly smaller than the initial belief rule-base in Fig. 1, which shows that the method is effective and greatly reduces the inference error. Comparing Fig. 4, the inference error of the proposed method is also smaller than the belief rule-base after rule weight and the preconditioner attribute weight optimization, which indicates that the method has certain advantages over other methods.

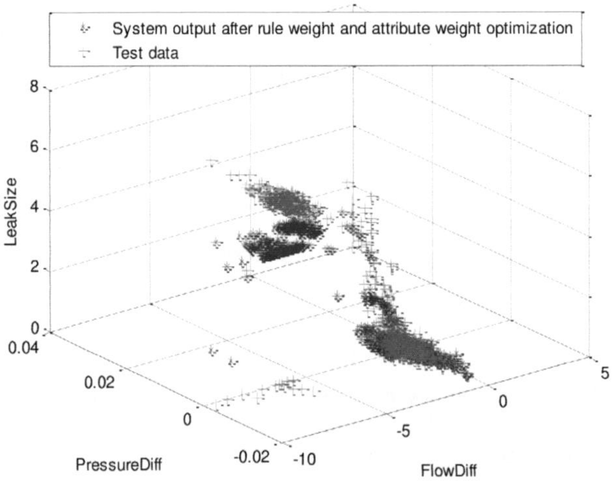

Figure 4. System output after rule weight and attribute weight optimization and test set data.

5. Conclusion

In this paper, the difficulty of obtaining expert experience and the inconsistency of standards in the process of establishing the belief rule-base are proposed. A method based on rough set is proposed. Firstly, the data is divided into equivalence classes by the initially determined attribute indexs; then the rules are extracted according to the relevant knowledge of the rough sets; finally, the extracted rules are used to perform the belief rule-base reasoning, and the attribute indexs are optimized according to the inference results. Through experiments, it is found that the method can extract better rules based on the data set, so that the belief rule no longer depends solely on expert experience. However, in the actual process, similar to other optimization methods, the quality of the rule extraction depends on the quality of the selected data set. How to avoid the selection of the "poor" data set requires further research.

References

[1] Yang J B, Liu J, Wang J, et al. Belief rule-base inference methodology using the evidential reasoning Approach-RIMER[J]. *IEEE Transactions on Systems Man & Cybernetics Part A Systems & Humans*, 2006, **36(2)**:266-285.

[2] Xu D L, Liu J, Yang J B, et al. Inference and learning methodology of belief-rule-based expert system for pipeline leak detection[J]. *Expert Systems with Applications*, 2007, **32(1)**:103-113.

[3] Zhou Z J, Hu C H, Yang J B, et al. Online updating belief rule based system for pipeline leak detection under expert intervention[J]. *Expert Systems with Applications*, 2009, **36(4)**:7700-7709.

[4] Zhao, Wang, Shicheng, et al. Real-time fault detection method based on belief rule base for aircraft navigation system[J]. *Chinese Journal of Aeronautics (English)*, 2013, **26(3)**:717-729.

[5] Feng Z, Zhou Z, Hu C, et al. Fault diagnosis based on belief rule base with considering attribute correlation[J]. *IEEE Access*, 2017, **PP(99)**:1-1.

[6] Zhou Z G, Liu F, Li L L, et al. A cooperative belief rule based decision support system for lymph node metastasis diagnosis in gastric cancer[J]. *Knowledge-Based Systems*, 2015, **85(C)**:62-70.
[7] Solic K, Ocevcic H, Golub M. The information systems' security level assessment model based on an ontology and evidential reasoning approach[M]. *Elsevier Advanced Technology Publications*, 2015.
[8] Yan X, Xu X, Sheng C, et al. Intelligent wear mode identification system for marine diesel engines based on multi-level belief rule base methodology[J]. *Measurement Science & Technology*, 2018, **29(1)**.
[9] Yu H E, Chang L L, Jiang J, et al. Intension Identification in Air Defense Based on Belief Rule Base Expert System under Expert Guidance[J]. *Fire Control & Command Control*, 2017.
[10] Zhao F J, Zhou Z J, Chang-Hua H U, et al. Aerial Target Intention Recognition Approach Based on Belief-Rule-Base and Evidential Reasoning[J]. *Electronics Optics & Control*, 2017.
[11] Zhang J, Jiang J, Chen Y, et al. A Belief Rule-Based Safety Evaluation Approach for Complex Systems[J]. *Mathematical Problems in Engineering*, 2015, 2015:1-10.
[12] Liu J J, Chang-Hua H U, Zhou Z J, et al. Life assessment approach of equipment based on belief-rule-base and evidential reasoning[J]. *Control Theory & Applications*, 2015.
[13] Long Q, Li Y, Sun Y, et al. An Approach of Converter Transformer Condition Evaluation Based on The Belief Rule Base Inference Methodology and Evidence Reasoning[C]// *International Conference on Civil, Transportation and Environment*. 2016.
[14] Yang J B, Liu J, Xu D L, et al. Optimization Models for Training Belief-Rule-Based Systems[J]. *IEEE Transactions on Systems, Man, and Cybernetics - Part A: Systems and Humans*, 2007, **37(4)**:569-585.
[15] Qian B, Wang Q Q, Hu R, et al. An effective soft computing technology based on belief-rule-base and particle swarm optimization for tipping paper permeability measurement[J]. *Journal of Ambient Intelligence & Humanized Computing*, 2017**(15)**:1-10.
[16] Liu J, Martinez L, Da R, et al. Optimization algorithm for learning consistent belief rule-base from examples[J]. *Journal of Global Optimization*, 2011, **51(2)**:255-270.
[17] Chang R, Zhang S. An Algorithm for Training Parameters in Belief Rule-bases Based on Gradient Methods with Optimization Step Size[J]. *Journal of North China Institute of Water Conservancy & Hydroelectric Power*, 2011.
[18] Pawlak Z. Rough set[J]. *International Journal of Computer & Information Sciences*, 1982, **11(5)**.
[19] Swiniarski R W, Skowron A. Rough set methods in feature selection and recognition[J]. *Pattern Recognition Letters*, 2003, **24(6)**:833-849.
[20] Saedudin R R, Mahdin H, Kasim S, et al. A Relative Tolerance Relation of Rough Set for Incomplete Information Systems[J]. 2018.
[21] Qian Y, Liang X, Wang Q, et al. Local rough set: A solution to rough data analysis in big data[J]. *International Journal of Approximate Reasoning*, 2018, **97**:38-63.
[22] Yang G J, Yang Y U. Selection of Rough Set and Attribute Reduction Based on BIC[J]. *Statistics & Information Forum*, 2018.

IOP Conf. Series: Materials Science and Engineering **435** (2018) 012054 doi:10.1088/1757-899X/435/1/012054

Intelligent Improvement of Pipeline Centralized Dispatch and Control

Lan Hao[1,*,a], Shao Tiemin[2], Li Bo[1,*,b], Diao Yu[1], Liu Miao[3] and Ai Sen[4]

[1]PetroChina Pipeline R&D Center, Langfang, China;
[2]PetroChina Oil & Gas Pipeline Control Center, Beijing, China;
[3]China Petroleum Longhui Automation Engineering Co., Ltd, Langfang, China.
[4]No.4 Branch Co. of China Petroleum Pipeline Engineering Co.,Ltd, Langfang, China.

[*]E-mail: [a]kjlanhao@petrochina.com.cn; [b]libocolby@yeah.net

Abstract. Under the requirement of CNPC in the construction of intelligent oil and gas pipeline, PetroChina Oil & Gas Pipeline Control Center has focused its work on the intelligent improvement of pipeline centralized dispatch and control system. The concept of Pipeline Intelligent Dispatch and Control is put forward, of which the framework, service objects, platforms and targets are established as well. Besides, the development strategy with 2 technical directions are proposed for the intelligent improvement of pipeline centralized dispatch and control, on the basis of learning from practices of power grid intelligent dispatch and petrochemical intelligent factory. In the first direction of informatization construction, there are 5 routes including central database, office automation, state/trend demo, professional simulation and industrial automation. In the second direction of intelligence promotion, there are 3 routes including optimization problem, prediction problem and recognition/judgment problem. In addition, the intelligent development is not a pronoun of staff reduction, but a personnel structural optimization. The cultivation and reserve of high-skilled talents should be emphasized as well as persistent technical innovations.

1. Introduction

With the rapid development of big data and artificial intelligence techniques, a new round of worldwide industrial transformation competition has been launched. By the grand strategic plans of Germany "Industry 4.0", American "National Strategic Plan for Advanced Manufacturing" and the domestic "Made in China 2025", a deep integration of traditional industrial manufacturing technology and electronic information is promoted, and the goal of advance towards automation, informatization and artificial intelligence is proposed.

Through decades of hard work, the transformation of domestic pipeline industry from traditional pipeline to digital pipeline has been accomplished, with the achievements such as digital design, mechanical construction, informational management, and electronic supplies purchase. Under the new requirement of China National Petroleum Corporation(CNPC) in the construction of intelligent oil and gas pipeline, PetroChina Oil & Gas Pipeline Control Center has focused its work on the intelligent improvement of pipeline centralized dispatch and control system, to upgrade the pipeline remote monitoring, enhance the decision support ability, and level up the operation and management.

2. Pipeline Intelligent Dispatch and Control

The intelligent development, as a new engine of industrial innovation, is carried out with various routes and targets in research and application for different industries. Through an investigation on the demand and technical characteristics of pipeline industry, the concept of Pipeline Intelligent Dispatch and Control is put forward as follows, based on the successful practices and experiences from the intelligent manufacturing, smart power grid, and intelligent petrochemical plant. While, the framework of pipeline intelligent dispatch and control system is obtained, and the service objects, working procedure, platforms and upgrading goals are shown in figure 1.

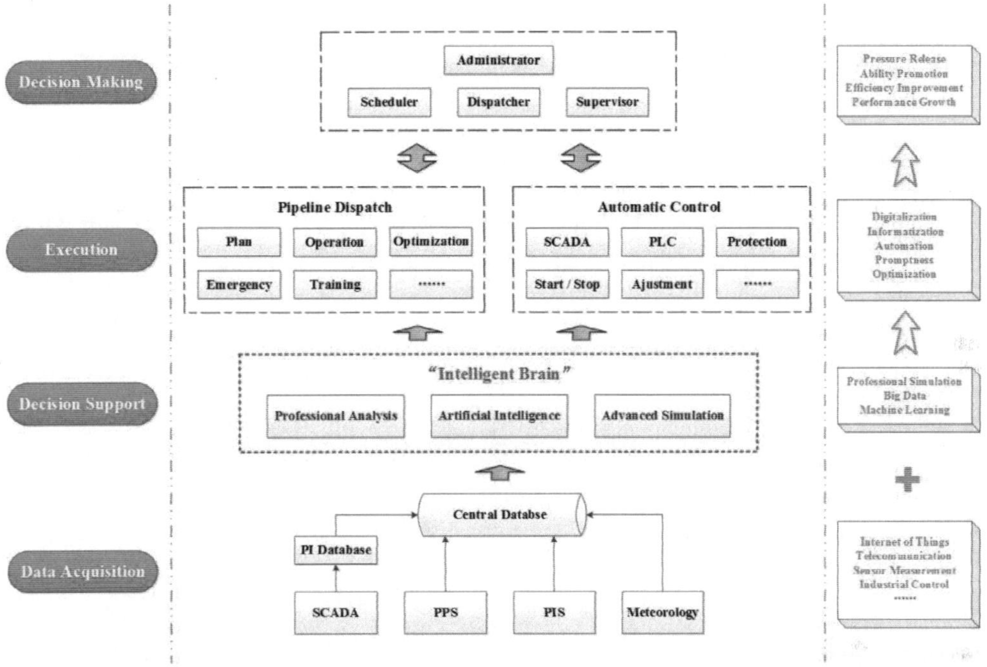

Figure 1. Framework of Pipeline Intelligent Dispatch and Control

Pipeline Intelligent Dispatch and Control, which is based on the pipeline automation and informatization by a comprehensive use of advanced techniques such as IoT, modern communication, sensor measurement, industrial control, simulation, optimization, big data and machine learning, is aimed to promote the visualization of working state, optimization of operation, automation of control and early warning of emergency, enhance the ability of self-learning, self-adaption and decision support of the whole system, improve the performance and efficiency of scheduler, dispatcher and supervisor, and further ensure the safe, economical and environment friendly operation of pipeline network.

In the stage of data acquisition, an enterprise-level central database is expected by the combination of meteorological data with the existing database from SCADA System, Plant Information System, Pipeline Production System and Pipeline Integrity Management System, to supply more comprehensive information for decision support[1].

In the stage of decision support, a virtual brain for intelligent dispatch and control is necessary to assist the working staffs with timely and optimal guide. The development of the virtual brain could be carried out in three aspects: firstly, enhance and improve the application of pipeline simulation to promote the transformation of empirical decision to scientific decision; secondly, explore the application of cutting-edge machine learning algorithms to discover knowledge and open mind by deeply mined information; thirdly, upgrade the pipeline simulation system to serve as a reliable and authoritative platform for various dispatch and control tests, by joint development of hydraulic thermal simulation and automatic control simulation, following the example of flight simulator in the aerospace industry.

In the stage of decision making, the work ability and efficiency of scheduler, dispatcher and supervisor will be promoted by the supports from the professional analysis, discovered knowledge and accurate simulation. It is beneficial to accelerate the progress in digitalization, promptness, automation, informatization and optimization of the pipeline centralized dispatch and control, through the interaction of the three stages mentioned above [2].

3. Development Strategy

Under the overall guideline of the strategy "Made in China 2025", the main work of PetroChina Oil & Gas Pipeline Control Center has been focused on energetically improving the intelligent level of transportation and operation control to fulfill and promote the construction of intelligent oil and gas pipeline. On the basis of learning from successful practices of power grid intelligent dispatch and petrochemical intelligent factory, combined with the features of pipeline operation, the following advices(shown in Figure 2) are put forward for the intelligent improvement of pipeline centralized dispatch and control.

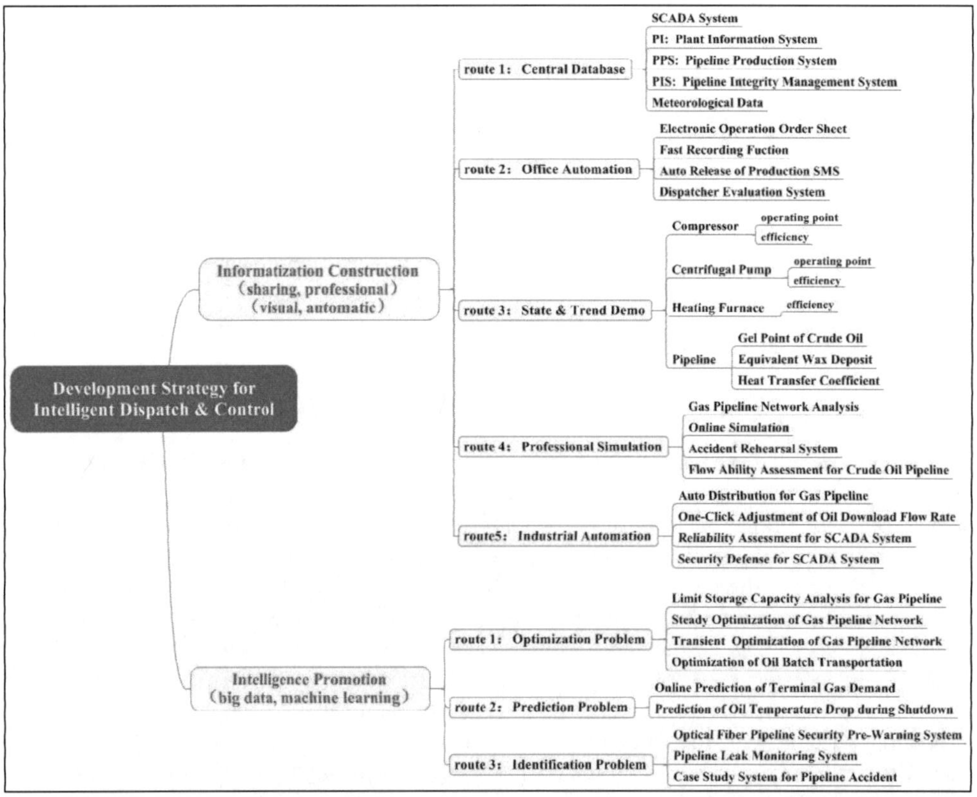

Figure 2. Development Strategy and Key Points of Pipeline Intelligent Dispatch and Control

3.1. Promoting the construction of pipeline informatization

Informatization construction is an important way of intelligent development. The key point is to open up information barriers to share resources, strengthen professional analysis to improve cognitive level, improve the visual effects to promote information presentation, and advance the automatic level to reduce labour intensity. The specific improvements could be carried out according to the following five aspects.

1) Promote the enterprise central database management.

On the one hand, based on the existing databases from SCADA, PI, PPS and PIS, some necessary meteorological data should be complemented to improve the comprehensiveness of basic data information; on the other hand, the interworking and resource sharing of basic database, simulation test database and accident database should be promoted, in order to provide comprehensive information guarantee for making overall plan and optimized decision.

2) Improve the level of office automation.

First, develop the electronic operation ticket system, to replace the current paper operation ticket and realize informatization management of operation examination and approval. Second, develop the quick recording function, to make a flexible log and classified memory for important events such as shutdown or abnormal production process. Third, develop the auto message function for production information, to simplify the tedious daily office work and better play the role of mobile terminal to operation management.

3) Strengthen the power equipment state monitoring and key parameters trend tracking.

First, improve the precision of the characteristic curve of power equipment, to improve the display of operating point for gas compressor and oil pump. Second, the trend of operating efficiency for compressor, pump and heating furnace should be presented in real time, to timely check their operating conditions and health. Third, strengthen the dynamic trend tracking of oil pipeline flow ability indicators such as oil gel point, pipeline friction resistance, wax deposition and heat transfer coefficient, to alert the abnormal condition in time and improve the pipeline safety pre-warning ability.

4) Strengthen the professional analysis and pipeline simulation.

First, improve the simulation model and pay more attention to the application of pipeline simulation, to give full play to the support role of auto-prediction function of online simulation software, and promote the change of experiential decision-making to scientific decision-making. Second, develop a virtual platform of accident rehearsal by using online/offline simulation software to replace and upgrade the current "orally ask and answer" rehearsal training. Third, develop an authoritative and comprehensive virtual platform for various pipeline simulations and intelligent tests including commissioning/transport/shutdown test, staff training, accident rehearsal and decision support, by upgrading the pipeline simulation system and promoting the cooperative development of hydraulic/thermal simulation and automatic control simulation like virtual PLC. Fourth, on the basis of existing offline evaluation software, develop an online system for flow ability evaluation of waxy crude oil pipeline and periodically release the evaluation results, to replace the current work mode of "manual input + post assessment" with the advanced mode of "auto input + real-time assessment".

5) Further improve the level of automatic control.

First, improve the PID control logic and algorithm to realize the automatic distribution of natural gas pipeline [3-4]. Second, optimize the control process and logic based on the experience of manual regulation, to realize the function of one-click adjustment of download flow rate for product oil pipeline. Third, promote the reliability evaluation and establish the white list protection strategy of pipeline SCADA, to scientifically assess the control safety and network security of PetroChina Oil & Gas Pipeline Control Center.

3.2. Exploring the application of artificial intelligence

Intelligent development is the inevitable trend and senior stage of the informatization construction. The artificial intelligent techniques, represented by big data, deep learning and intensive learning, is developing rapidly with tremendous and promising futures. However, a blind application and following-up should be prevented due to the high technical threshold of artificial intelligence. It is suggested to explore the application of big data and machine learning to gradually advance the system abilities of "self-learning, self-adaption and self-decision" and boost the pipeline informatization construction to a higher level, on the basis of necessary foundation reinforcement and clear goal recognition.

The intelligent development of pipeline dispatch and control system can be carried out from the following three aspects of optimization, prediction and recognition/judgment problems.

1) Optimization problem.

Convert the complicated engineer problem to a mathematical optimization problem, through rational selection of optimization objectives and constraint conditions. Attempt to use intelligent optimization techniques such as genetic algorithm, simulated annealing algorithm and ant colony algorithm, to provide an optimal plan for pipeline operation [5]. For instance: First, use the dynamic programming algorithm to automatically figure out the maximum, minimum and optimal storage capacities of natural gas pipeline, to avoid the tedious manual trial calculations by using TGNET software. Second, strengthen the application of the existing steady state optimization software, to promote the optimization of large-scale natural gas pipeline network and improve the quality and efficiency of the monthly plan. Third, commence investigations on the transient optimization of natural gas pipeline network and develop the unsteady optimization software, to guide the condition transition and emergency disposal. Fourth, upgrade the planning software of product oil pipeline network by using optimization algorithms, to realize automatic planning instead of the current manual scheduling with fussy and tedious trial processes.

2) Prediction problem.

Attempt to build mathematical prediction models by using machine learning algorithms, to find the rules from the known or historical data, and improve the prediction accuracy. For instance: First, based on a bulk of historical sales data, set up a gas demand forecast model for terminal customer, by a comprehensive use of recursive neural network and time series algorithms, to provide timely and reliable decision support information for the peak load regulation of natural gas pipeline. Second, based on the historical data of oil pipeline shutdown temperature drop, set up a mathematical forecast model by using deep neural network or support vector regression algorithms, as a supplement to the traditional prediction means of Stoner Pipeline Simulator, to improve the prediction accuracy and promote the online prediction.

3) Recognition and judgment problem.

Explore the application of data mining technologies, to find links, clues and answers from known information, and even get access to some unconventional and insightful understandings. For instance: First, enrich the training samples by enlarging the applications of Optical Fiber Pipeline Security Pre-Warning System and Oil Pipeline Leak Monitoring System, and try to use deep learning and enhance learning algorithms, to improve their accuracy and pre-warning ability[6]. Second, develop the case study system for pipeline accidents with functions of classified retrieval, statistical analysis and similar-case reminding, and use logistic regression and correlation analysis to make up for the shortcomings of traditional paper case report and simple electronic document management, in order to give full play to the early warning education of the historical accidents in pipeline dispatch and control.

3.3. Expanding the technical service team

The intelligent development is not a pronoun of staff reduction, but a personnel structural optimization. The cultivation and reserve of high-skilled talents should be emphasized as well as persistent technical innovations, during the promotion of the pipeline intelligent dispatch and control. For example: Facing the heavy tasks of model reconciliation and maintenance for the scale-expanding natural gas pipeline network, more skilled technicians are needed to improve the performance and application of the existing online simulation software such as Gregg, Atmos and SPS, to better support and guide the dispatchers in their daily operation and control. Besides, more professional talents of industrial automation are needed to upgrade the pipeline SCADA system to a higher smart level, to further relieve the working pressure and labour intensity.

4. Conclusions

Through investigations on the demand and technical features of pipeline industry, the concept of Pipeline Intelligent Dispatch and Control is first put forward, of which the framework, service objects, platforms and targets are established as well. Besides, the development strategy with 2 directions, 8 routes and 32 key points are proposed for the intelligent improvement of pipeline centralized dispatch and control, based on the experiences from power grid intelligent dispatch and petrochemical intelligent factory. In addition, the intelligent development is not a pronoun of staff reduction, but a personnel structural optimization. The cultivation and reserve of high-skilled talents should be emphasized as well as persistent technical innovations.

Acknowledgments
I would like to acknowledge the technical support from team members. I do appreciate the assistance from PetroChina Oil & Gas Pipeline Control Center. This paper was financially supported by the research program of Investigation on the Intelligent Improvement of Pipeline Centralized Dispatch and Control System.

References

[1] Aouadj M, Naceri F, Touileb M, et al. 2015, SCADA System for the Modeling and Optimization of Oil Collecting Pipeline Network: A Case Study of HassiMessaoud Oilfield, *Research Journal of Applied Sciences Engineering & Technology*.
[2] AI Muyang, LIU Jianjun, LI Bo, et al. 2015, Current status and prospect of steady operation optimization techniques for gas pipeline network, *Oil & Gas Storage and Transportation*.
[3] Luo Z, Lv H and Liu Z.2016,Study of Fuzzy PID Control System for the Product Oil Pipe-Convey, *Sixth International Conference on Instrumentation & Measurement, Computer, Communication and Control*.
[4] Demissie A, Zhu W and Belachew C T.2017, Amulti-objective optimization model for gas pipeline operations, *Computers & Chemical Engineering*.
[5] Ozdemir G, Aydemir Eand Olgun M O.2016,Forecasting of Turkey natural gas demand using a hybrid algorithm, *Energy Sources Part B Economics Planning & Policy*.
[6] Karkulali P, Mishra H, Ukil A, et al. 2016, Leak detection in gas distribution pipelines using acoustic impact monitoring, *Industrial Electronics Society*, *Conference of the IEEE*.

IOP Conf. Series: Materials Science and Engineering **435** (2018) 012055 doi:10.1088/1757-899X/435/1/012055

Long Tail Constraint on Non-negative Matrix Factorization

Quanye Jia[1], Rui Liu*[,1], He Zhang[1] and Lu You[1]

[1]School of Computer Science and Engineering, Beihang University, Beijing, China

*corresponding author: lr@buaa.edu.cn

Abstract. The topic distribution in text generally has long tail effect, but few people do research on how to dig out long tail topics from matrix factorization. So we propose a method, this is Non-negative Matrix Factorization with Long-tail Constraint (LTNMF). LTNMF adds the soft orthogonal constraints to the feature matrix to ensure the independence of the topics on the basis of the non-negative matrix factorization. The sparse constraints and long tail constraints are added to the topic document matrix to enhance the robustness of the model and the characterization of the long tail features of the topic distribution. The combination of soft orthogonal constraints, sparse constraints and long tail constraints enables the model to extract the long tail topic information in the data and ensure the quality of the topic. We use Sougou and 20newsgroup datasets to experiment, and the results show that LTMNF can dig more topic words and improve the accuracy and the standard mutual information of clustering in text classification.

1. Introduction

In text data mining, the document is generally represented by vector, then the document set can be represented as a matrix. With the gradual increase of the amount of information, the size of the matrix has become very large and takes a long time to compute. The factorization of the matrix can not only reduce the dimension, but also compress and cluster data to help people understand the data.

The long tail distribution can be expressed as a Pareto distribution in statistics. From the distribution curve we observe that the head part of the fabric is very wide, the tail distribution is relatively flat, like a long tail, so it is called the long tail. It reflects an unbalance relationship between things, that is, a few things occupy a big head, but a large number of small things can also cause some big effects.

The topic distribution obtained by the matrix factorization algorithm is also long tail distribution. Most of the documents have the correlation with a small amount of hot topics, and other topics (long tail topics) with only a few related documents, but the long tail topics of the document has more richness than those hot topics. So it is necessary to dig out more evenly distributed topic information.

The rest of this paper is organized as follows: In Section 2, we introduce the related work on topic models and long tail theory. In Section 3, we introduce the details on LTMNF. Experiment results are presented in Section 4. At last, we give the conclusion and future work.

2. Relate work

2.1. Matrix factorization

Meth Hofmann pointed out in 1999 that topic models can be interpreted as matrix factorization [1] [2]. The commonly used matrix decomposition methods include Principal Component Analysis (PCA) [3],

Independent Component Analysis (ICA) [4], Single Value Decomposition (SVD) [5], and so on. For the principal component analysis and singular value decomposition, the matrix obtained after decomposition may contain negative numbers, which has no practical significance.

In 1999, the Non-negative Matrix Factorization (NMF) algorithm [1] was first proposed by Lee D et al. in the famous scientific journal Nature. That is to say, the matrix obtained by the original data decomposition is nonnegative. NMF has attracted many researchers' attention, many scholars have proposed the algorithm of non-negative matrix factorization model under different constraint conditions. In 2004, Patrik proposed a non-negative matrix factorization algorithm [6] with sparse constraints. By adding L1 regularization constraints to feature topic matrix and topic document matrix, the sparse representation of the distribution of feature topics and subject documents was realized, but the classification effect could not be optimized. In 2008, Choi S [7] proposed the orthogonal non-negative matrix factorization, the full orthogonal constraint is added to feature topic matrix to keep independence between topics. He proved through the experiments that the method can achieve better results in the task of clustering and classification, and reduces the repetition ratio of the topic. Non-negative results has good explanation and the reduction effect also saves more storage space. Non-negative matrix factorization has been widely used in various fields, including text classification, image processing and clustering, and has achieved good results.

2.2. Long tail theory

In October 2004, the "connection" magazine editor Chris Anderson published "the long tail" [8] for the first time in this paper and formally put forward the theory of long tail in United States. He pointed out in the paper that under the condition of rich economy, people's attention to goods is gradually transferred from hot commodity to small public goods, which brings the market of small public goods. The aggregation of a large number of small markets has formed a large market and occupies a more important position in the whole industry. After a long time of research and practice, Anderson's view has been widely recognized, has made great success in the business, and become the focus of attention.

3. Method

3.1. Non-negative Matrix Factorization with Long-tail Constraint (LTNMF)

The matrix obtained by NMF is non-negative, NMF is used as the matrix factorization algorithm. The basic method of NMF [1] is: for a given non-negative matrix $D \in R_{M\times N}$ and a positive integer $k < \min\{M, N\}$, find two non-negative matrices $U \in R_{M\times K}$ and $V \in R_{K\times N}$ to make matrix $D \approx UV$.

$$\min_{U,V} f(U,V) = \frac{1}{2}\|D - UV\|_F^2$$
$$s.t. U \geq 0, V \geq 0 \quad (1)$$

Among them, D represents the feature (words) document matrix, U represents the feature topic matrix, V represents the topic document matrix, M represents the number of words in the dictionary, N represents the data set number of documents, K represents the number of topic. F represents Frobenius norm. Through statistical analysis, the weight of a small part of the topic (hot topic) in the document set is large, and most the weight of the topic (long tail topic) is small, such as the green curve shown in figure 1. In order to dig the long tail topic, we can reduce the gap between the topic weight. In the topic document matrix V, v_{ij} representing the weight of the i-th topic in the j-th document, then the r line in the matrix $V(v_r)$ can represent the weight vector of the r-th topic in the document set, and v_r^2 (or $v_r v_r^T$) can be used in order to quantify this value. Due to the weight difference in the document of each subject is large, we subtract the different value of w_r, to shorten the gap between the topic weight to improve the probability of mining the long tail topics. Assuming that the original topic weight is $v_r v_r^T$, the weight of optimization is $v_r v_r^T - w_r$ (the red curve shown in Figure 1), w_r (yellow in Figure 1) can be used by the function of characteristics of the long tail.

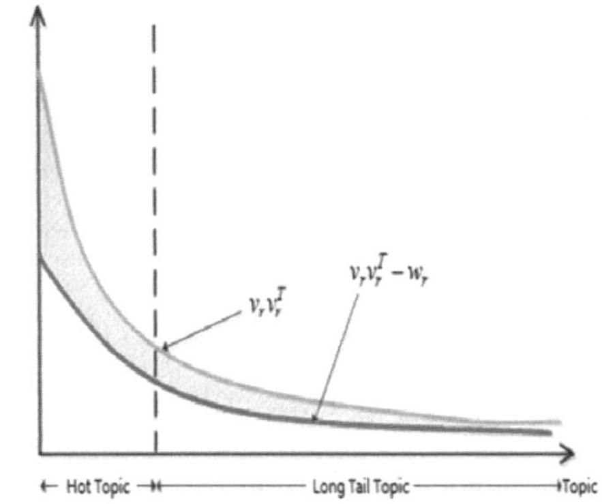

Figure 1. The Long tail phenomenon in the topic.

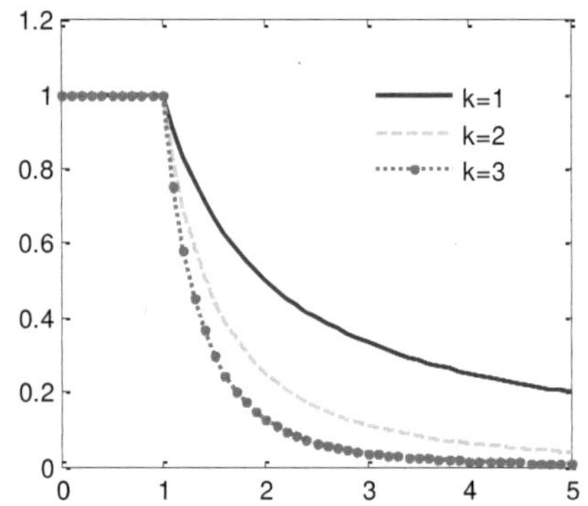

Figure 2. The curve of Pareto distribution.

For a given data set, the standard non-negative matrix factorization algorithm is difficult to obtain very clear local features, often accompanied by noise, and can't well dig the long tail topic features in data sets. According to the long tail theory, the distribution of topics in the document presents the probability distribution of Pareto distribution. Therefore, in order to better dig out the long tail information in the corpus, a reasonable long tail constraint is applied to the feature topic matrix, and the formal representation is $\sum_{r=1}^{K}\|v_r v_r^T - w_r\|_F^2$ (w is the weight vector of K dimension by Pareto distribution) and the soft orthogonal constraint [7] is introduced, that is $\|U^T U - I_K\|_F^2 \le t, (t \ge 0)$ (I_K represents identity matrix of order K), the feature topic vector is approximately orthogonal. In addition, due to the large scale of training data, the feature dimension is very high, but each data sample is only related to a small amount of features. In order to represent data better, the L2 regularization constraint ($\|V\|_F^2$) is added to the representation vector of the data to realize sparse representation of the topic document matrix.

Thus, for the feature document matrix $D = [d_1, d_2, \ldots, d_N]$, model LTNMF can be represented in the following mathematical equation (2):

$$\min_{U,V} \|D - UV\|_F^2 + \alpha\|U^T U - I_K\|_F^2 + \frac{\lambda}{2}\|V\|_F^2 + \frac{\beta}{4}\sum_{r=1}^{K}\|v_r v_r^T - w_r\|_F^2$$

$$s.t.\begin{cases} u_{mk} \ge 0, m = 1,2,\ldots,M; k = 1,2,\ldots,K \\ v_{kn} \ge 0, n = 1,2,\ldots,N; k = 1,2,\ldots,K \\ \alpha > 0, \lambda > 0, \beta > 0 \end{cases} \tag{2}$$

To determine the weight value of each topic in a document set, the Pareto distribution function with long tail features can be used, such as equation (3) (Figure 2):

$$\mathrm{F}(x) = \Pr(X > x) = \begin{cases} \left(\frac{x}{x_{min}}\right)^{-k}, x \ge x_{min} \\ 1, \quad x \le x_{min} \end{cases} \tag{3}$$

Among them, x is the random variable of Pareto distribution, x_{min} is the boundary of variable x of the normal distribution and the power law distribution, also means the starting point of the power law curve. k is the shape parameter (positive number), the greater the k value, the steeper the curve descent trend.

Pareto distribution is a continuous probability distribution, and what we need is a number of discrete points in the curve to generate the weight vector of the topic, and retain the distribution characteristics of the curve. Therefore, we select an area in the power law curve (such as equation (3)), and then get the value from equal distance as the W vector, and the generating rules can be expressed as the following mathematical equation (4):

$$w_r = \begin{cases} P\_BASE & ,r < P\_LONG \\ P\_BASE(\frac{(P\_MAX-P\_MIN)(r-P\_LONG)}{K\cdot P\_MIN} + 1)^{P\_K}, & r \geq P\_LONG \end{cases} \tag{4}$$

Among them, P_BASE is the unifying weight value of the hot topic, (P_MAX – P_MIN) / (K×P_MIN)×(r - P_LONG)+1 is the tail distribution equation of the input, which is x in the equation (3), P_MAX and P_MIN are the upper and lower bounds of x for power law curve selected in the equation (3), P_K is a shape parameter, K is the number of topic in data, P_LONG is the number of hot topic.

3.2. The algorithm of LTNMF

In equation (2), there are two unknown matrices of U and V, which can not directly get the global optimal solution through matrix operation directly. Therefore, this paper adopts the methods commonly used by most scholars at present, to fix U for V, to fix V for U, so that the iteration is alternately, and then the local optimal solution of the model is obtained, and the basic procedure of LTNMF is shown in Table 1.

Table 1. The basic procedure of LTNMF.

Algorithm 1: Solving algorithm of LTNMF
Input: $D \in \boldsymbol{R}^{M\times N}, \boldsymbol{\alpha}, \boldsymbol{\beta}, \boldsymbol{\lambda}$, *P_K,P_MIN,P_MAX,P_LONG,P_BASE*
Output: $U \in \boldsymbol{R}^{M\times K}, V \in \boldsymbol{R}^{K\times N}$
$U^{(0)} \in \boldsymbol{R}^{M\times K} \leftarrow random(0,1)$
$V^{(0)} \in \boldsymbol{R}^{K\times N} \leftarrow random(0,1)$
$W \in \boldsymbol{R}^{K\times N} \leftarrow pareto(P\_K, P\_MIN, P\_MAX, P\_LONG, P\_BASE)$
for t=1:T do
U=UpdateU(D,U,V)
V=UpdateV(D,U,V,W)
end for
return U,V

We can use gradient descent method to solve matrix U and V and get the update equation as follows, θ_u,θ_v are gradient descent steps.

$$U^* = U - \theta_u \frac{\partial L}{\partial U} \tag{5}$$

$$V^* = V - \theta_v \frac{\partial L}{\partial V} \tag{6}$$

According to the matrix operation rule $tr(AA^T) = \|A\|_F^2$, $tr(AB) = tr(BA)$, $tr(A) = tr(A^T)$, the upper equation (2) can be rewritten as:

$$L = tr((D-UV)(D-UV)^T) + \alpha \cdot tr\big((U^TU - I_K)(U^TU - I_K)^T\big) + \frac{\lambda}{2} tr(VV^T) + \beta \sum_{r=1}^{K} \|v_r v_r^T - w_r\|_F^2$$

$$= tr(DD^T) - 2tr(DV^TU^T) + tr(UVV^TU^T) + tr(U^TUU^TU) - 2\alpha \cdot tr(U^TU)$$

$$+\alpha \cdot tr(I_K I_K) + \frac{\lambda}{2} tr(V^TV) + \frac{\beta}{4} \textstyle\sum_{r=1}^{K} \|v_r v_r^T - w_r\|_F^2 \tag{7}$$

The objective function L seeks the partial derivative of the variable matrix U and V:

$$\frac{\partial L}{\partial U} = 2(UVV^T - DV^T) + \alpha(4UU^TU - 4U) \tag{8}$$

$$\frac{\partial L}{\partial V} = 2(U^TUV - U^TD) + \lambda V + \beta \begin{bmatrix} v_1 v_1^T - w_1 & \cdots & \mathbf{0} \\ \vdots & \ddots & \vdots \\ \mathbf{0} & \cdots & v_K v_K^T - w_K \end{bmatrix} V \tag{9}$$

We can assign $\theta_v = \frac{V}{2U^TUV+\lambda V+\beta\cdot diag(v_1v_1^T,\cdots,v_Kv_K^T)\cdot V}$, $\theta_u = \frac{U}{2UVV^T+4\alpha UU^TU}$ and get the following update rules:

$$u_{ij} \leftarrow u_{ij}\frac{(DV^T+2\alpha U)_{ij}}{(UVV^T+2\alpha UU^TU)_{ij}} \tag{10}$$

$$v_{ij} \leftarrow v_{ij}\frac{(U^TD)_{ij}+\beta w_i v_{ij}}{(U^TUV)_{ij}+\lambda v_{ij}+\beta v_i v_i^T v_{ij}} \tag{11}$$

4. Experiments

4.1. Experimental setup

4.1.1. Dataset. This paper selects Sogou and 20newsgroup two different news corpus for experiments.

4.1.1.1. Sogou news corpus: Sogou news corpus is a large amount of news corpus edited and annotated by Sohu. It is commonly used in Chinese classification and clustering tasks. As the size of the data set is too large, this paper selects 2500 documents randomly in 5 categories of finance, IT, health, sports and tourism, and each category has 500 documents. Then this paper pretreat the selection of data pretreatment, filter the weight of low frequency words and calculate all the TF-IDF value of the document. Then we get an original data matrix of 6413×2500 dimensional.

4.1.1.2. 20newsgroup: 20newsgroup is a commonly used data set in Natural Language Processing domain such as text classification. This article uses the version after Jason Rennie processing, which contains 18828 single annotation documents, almost evenly distributed in 20 news classes under 6 major classes. This paper uses the six major classes (alt.atheism, comp.os.ms-windows.misc, misc.for-sale, rec.motorcycles, sci.electronics, talk.politics.misc) to experiment. For each class selected in the experiment, this paper selects all the documents, preprocessing and computing TF-IDF. Finally, 19183×5514 matrix is obtained.

4.1.2. Baseline methods: In order to verify the effectiveness of the proposed model LTNMF, this paper selects latent Dirichlet distribution (LDA) [9], Hierarchical Latent Dirichlet distribution (hLDA) [10], non-negative matrix factorization (NMF), orthogonal sparse non-negative matrix factorization (OSNMF) as baseline methods to verify the effectiveness of the proposed model LTNMF.

*4.1.3. Parameter setup.*This paper used cross validation to determine the parameters of LTNMF under better results of accuracy rate. For LTNMF model, this paper sets K=100, P_K=-3.1, P_MAX=5, P_MIN=1, P_LONG=15, P_BASE=3 in Sougou dataset. This paper sets α=0.5, β=0.01, λ=0.001 in Sougou dataset and α=0.05, β=0.01, λ=0.005 in 20newsgroup dataset. The OSNMF model is regarded as the case of LTNMF under K=100, α=0.5, β=0.01, λ=0 condition in Sougou dataset and K=100, α=0.05, β=0.01, λ=0 condition in 20newsgroup dataset.

For the hierarchical topic model hLDA, this paper performs about 300 tests in each experiment by cross validation, and then selects the results according to the number of topics and the accuracy of clustering, and finally determines the values of each parameter in the hLDA. In Sougou dataset, this paper sets α=10, γ=0.1, η=10, level=5, 109 topics are obtained through 100 iterations. In 20newsgroup, this paper sets α=5, γ=0.01, η=1, level=5, 131 topics are obtained through 100 iterations.

4.1.4. Evaluation measures:

4.1.4.1. Average number of topic words(Average): For topic models, topics can be understood as a series of related words, which can be repeated when the keywords of the two topics are the same or the similarity is very high. Therefore, the more the number of topic words that do not repeat, the better the effect of the long tail topic mining.

4.1.4.2. Accuracy rate. According to the topic document matrix mined by the topic model algorithm, this paper uses K-means algorithm to cluster documents, and then compares the clustering categories with the actual categories, calculates the accuracy rate, and the equation is as follows:

$$Accurancy = \sum_{c=1}^{N_C} \frac{TP_c}{TN_c} \tag{12}$$

Among them, N_c is the total number of category, TN_c is the sample number of the real category c, TP_c is the sample number of the prediction category and the real category is c, and the conditional $TP_c \leq TN_c$, $\sum_{c=1}^{N_C} TN_c = TN$ is established, so $0 \leq Accurancy \leq 1$, and the higher the accuracy rate is better.

4.1.4.3. Standardized mutual information. Assuming that C represents a set of real clusters, C' is a set of predicted clusters, then the mutual information measure between them can be calculated by the following equation (13):

$$MI\left(C,C'\right)=\sum_{c_i\in C,c_j'\in C'} p\left(c_i,c_j'\right)\log\frac{p\left(c_i,c_j'\right)}{p\left(c_i\right)p\left(c_j'\right)} \tag{13}$$

Among them, $p(c_i)$ and $p(c_j')$ are the probability of selecting clusters c_i and c_j' from dataset C and C', respectively. $p(c_i, c_j')$ is the joint probability of two. Because the value of mutual information has no upper limit, it can not accurately evaluate the effect of the model. Therefore, the standardized mutual information is used to evaluate it. The equation (14) is as follows:

$$NMI(C, C') = \frac{MI(C,C')}{\max(H(C),H(C'))} \tag{14}$$

In the upper form, $H(C)$ and $H(C')$ represent the entropy of C and C', and it is easy to find $0 \leq NMI(C, C') \leq 1$. $NMI(C, C') = 0$ indicates that the two cluster sets are exactly the same. $NMI(C, C') = 1$ indicates that the two cluster sets are independent of each other, so the greater the value of $NMI(C, C')$, the better of the model.

4.2. Experimental results:

In this paper, we extract the feature-topic matrix U from each model in Sogou news corpus, take the first 10 and the first 20 words of every column u_r from large to small, and remove duplicates, and then calculate the average value, get the average number of keywords of each model, the results are shown in table 2.

Table 2. Comparison of the number of average topic words.

Model	LDA	NMF	hLDA	LTNMF
Sougou(top 10)	5.99	6.92	7.29	**8.23**
Sougou(top 20)	10.75	12.48	13.11	**14.06**
20newsgroup(top 10)	5.44	8.36	8.56	**9.19**
20newsgroup(top 20)	10.47	16.22	16.67	**17.16**

The number of average topic words on the above table shows that LTNMF is the best in the 4 models and is far more than LDA. Experimental results show that LTNMF can extract more topic words compared with other methods, which shows that long tail topic mining can really improve the effect of topic mining.

After modeling by the topic model, two matrixes, the feature topic matrix U and the topic document matrix V, can be obtained, which can be regarded as the projection of the original data in the low rank space after dimension reduction. This paper will apply matrix V to clustering algorithm, get the prediction labels of each document, then compare it with the real labels, get the accuracy rate and standard mutual information results respectively as the classification effect, as shown in Figure 3 and Figure 4. It can be seen that the LTNMF with long tail constraints can improve the accuracy and the standardized mutual information of clustering. At the same time, the importance of the long tail topic information is verified.

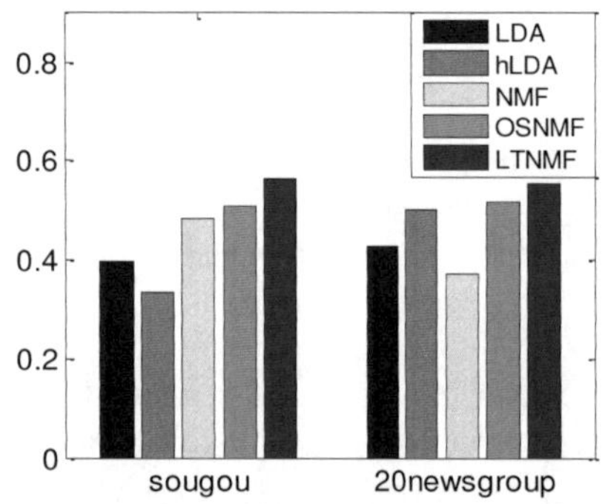

Figure 3. Comparsion of accurancy results of clustering for different models.

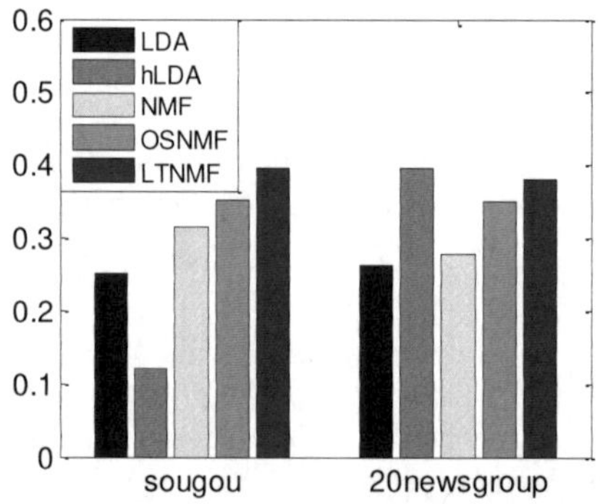

Figure 4. Comparsion of standardized mutual information results of clustering for different models.

5. Conclusion and future work

In this paper, we propose the LTNMF model, that is, added the long tail constraint to the non-negative matrix factorization. We use Sougou and 20newsgroup datasets to experiment, and the results show that LTNMF can dig more topic words and improve the accuracy and the standard mutual information of clustering in text classification.

Our proposed method is fully unsupervised. We would like to integrate supervisory information into existing long tail topic mining model LTNMF. Also, the Pareto distribution is used to fit the long tail distribution trend of the topic in the document. We would like to find another distribution function or the superposition of several distributions in place of the Pareto distribution.

References

[1] Lee D and Seung H S 2000 Algorithms for Non-negative Matrix Factorization *Neural information processing systems* **13** 535-41
[2] Lee D and Seung H S 1999 Learning the parts of objects by non-negative matrix factorization *Nature* **401** 788-91
[3] Abdi H and Williams L J 2010 Principal component analysis: Wiley interdisciplinary reviews *computational statistics* **2** 433-59
[4] Comon P 1994 Independent component analysis, a new concept? *Signal processing* **36** 287-314
[5] Golub G H and Reinsch C 1970 Singular value decomposition and least squares solutions *Numerische mathematic* **14** 403-20
[6] Hoyer P O 2004 Non-negative matrix factorization with sparseness constraints *The Journal of Machine Learning Research* **5** 1457-69
[7] Choi S 2008 Algorithms for orthogonal non-negative matrix factorization *Proc. of the IEEE Int. Joint Conf. on Neural Networks* (Piscataway: IEEE) pp 1828-1832
[8] Chris Anderson 2006 *The long tail: Why the Future of Business is Selling Less of More* (New York: Hyperion)
[9] Blei D M, Ng A Y and Jordan M I 2003 Latent dirichlet allocation *Machine Learning Research Archive* **3** 993-1022
[10] Blei D M, Jordan M I, Griffiths T L and Tenenbaum J B 2003 Hierarchical topic models and the nested chinese restaurant process *Int. Conf. on Neural Information Proc. Systems* vol 57 (MIT Press: Cambridge MA) pp 17-24

IOP Conf. Series: Materials Science and Engineering **435** (2018) 012056 doi:10.1088/1757-899X/435/1/012056

Robust Template Matching via Pruning Deep Feature

F Wang[1], J P Xiong[1], J Yin[1] and X M Zhang[1]

[1]Zhejiang Dahua Technology CO., LTD., Zhejiang Province, Hangzhou, China

Email: wzfeng1991@163.com

Abstract. The application environment with slightly diversity between template image and target image has been the mainstream in template matching over the past decade or so. This paper, however, will discuss template matching method in such scenarios with erratic weather, deformations, scaling etc. For the feature engineering, by choosing the appropriate layers in a CNN and pruning inefficient convolutional kernels in the layers we want, we construct a feature space that can be used to represent more complex features compared to the traditional computer vision feature engineering, such as corners, colours, edges etc. Meanwhile, the feature space and computational capacity can be greatly reduced by the pruning operation on convolutional kernels. For similarity measure, a distance penalty term on the feature of image patches will be added in the final score function to make our method robust to deformation and scaling. Furthermore, the key coefficients of penalty term have been opened so that our method can be adjusted based on the actual scenarios. Numerous experiments on the benchmark dataset are conducted accompanied by comparisons with a few recent proposed methods, e.g., BBS, DDIS, etc. The results have demonstrated well the robustness and accuracy of our method relative to the other methods. Note that, our method can reach the industrial standard with GPU acceleration.

1. Introduction

The study of template matching, as a fundamental issue in the fields of computational photography and computer vision, is always active over the years, and we note that some future developing trends in this study is motivated by the application environment for complex scenarios in industry. The major mission of this work is asked to solve a problem: how to increase the robustness of the template matching algorithm under illumination changes fleetly and background interface mightily, as illustrated in figure 1.

Based on the analysis of existing literature, the emphasis of template matching can be split into two parts[1]: feature engineering and similarity criterion. Feature engineering is the first step of computer vision tasks, and is also the most critical step. There could be the low-level features[2], such as edges, corner[3], which can be comprehended by humans, and the high-level features, such as hand-craft features or more abstract features learned by a CNN(convolutional neural network)[4] with a big training set. Similarity criterion is another research emphasis in template matching, which is used to find the minute location of the template in a big image. As pointed in [5], the most significant performance impact in a computer vision tasks is the quality of feature engineering relative to other modules, like pre-processing, regularization term, numerical optimization. In this work, we redefine the deep feature of a CNN to improve the positioning precision in complex environment, and redesign the calculation method which is more efficient for these high dimensionality feature space. Though template matching and object detection are two themes in the computer vision, they have some internal connections[6]. Both of them need to find the specific location of an object in a big image. The number of objects is settled in object detection, and a template in the work procedure of template matching could be just one of these objects. However, the feature engineering in template matching is consistent with object detection in essence. Inspired by the great success of deep learning achieved in object detection, such as Faster RCNN[7], YOLO[8], etc., we redefine and prune the basic convolutional network that make it applicable to template matching. Specifically, the descriptive ability of inter-layers or kernels in one layer will be analyzed in detail here. We find that the features described by the previous layers in VGG-Net[9] have a low-level characteristic, such as large edge, colour information, which can be comprehended by mankind. Whereas the last layers in VGG-Net[9] could describe more abstract features which contain distinct semantic information. These features with apparent semantic information may really suit object detection, but are not suitable for building the feature engineering in template matching and will be counterproductive in many cases. How intricate the feature is applied in a computer vision technique hinge on the specific case. Meanwhile, the kernels in one layer of a network structure appear in an endless variety of peculiarity. We can see through our studying and analysing, that twenty per cent of the kernels concentrate the vast majority energy in one layer. Those low-strength convolutional kernels are inadaptable to the problem of template matching. And, it means that we just need ten to twenty per cent of the convolutional kernels in one layer to construct the feature engineering. By designing the feature in this way, we can construct an more efficient feature engineering with lower storage space requirement and higher computation efficiency.

(a) (b)

GroundTruth · SSD · BBS · DDIS · Ours

(c) (d)

Figure 1. Template Matching under illumination changes and background interface: Template matching expression of different methods. (a). The dial plate in this image is marked as template (in green). (b)(c)(d), the performance of different template matching are marked in these complex luminance images.

As in the earlier study, comparing the template and the target image pixel by pixel is the most common approach used in similarity criterion. Nevertheless, such a mode can't handle robustly with the deformation[10]. To solve this problem, the features extracted by some algorithms will be mapped into a higher dimensional space in [11], meanwhile, the number of nearest neighbours will be obtained to establish a heat map[12]. The specific location of the template will be confirmed by this heat map. The feature engineering used in BBS[11] is low-level, and its' computational power is highly inefficient, in particular for a 1080p HD target image. The computing time will be spent 20 minutes or more with a hardware configuration of Intel Core i5, 8G memory. Although the problem about calculation efficiency has been solved by DDIS[13], it still can't work well under illumination changes fleetly and background interface mightily.

The application environments of template matching in the industry are indoors mostly with a good light condition. But we will mainly discuss template matching used in the outdoors with complicated scenarios, such as unpredictable weather, complex background and various industrial components. Meanwhile, the deformation and scaling partly are considered as well. Our contributions can be summarized as follows:

Firstly, we construct the feature engineering used in template matching with a pruning operation on CNN. The pruning operation mainly contains two parts. For the inter-layers in CNN, the ahead and middle layers that can extract texture or colour information of the image will be applied to final feature space. And the last layers with obvious semantic information have been discarded here. For convolutional kernels in inter-layers, more than eighty per cent of kernels will be pruned to optimize the feature space and accelerate the calculations.

Secondly, in order to overcome the unforeseeable elements in industry, such as deformation, scaling etc., a new similarity measure with adjustable coefficients based on the actual situations has been proposed here. For the high-dimensional feature space constructed by deep features, we use the product quantization based nearest neighbour method to solve this problem.

2. Proposed Framework

In this section, we will discuss how to prune the inter-layers and intra-layers in a CNN and how to design the similarity measure to strengthen the robust aimed at deformation and scaling.

2.1. Pruning Deep Features in Template Matching

2.1.1. Feature Description of Different Inter-Layers

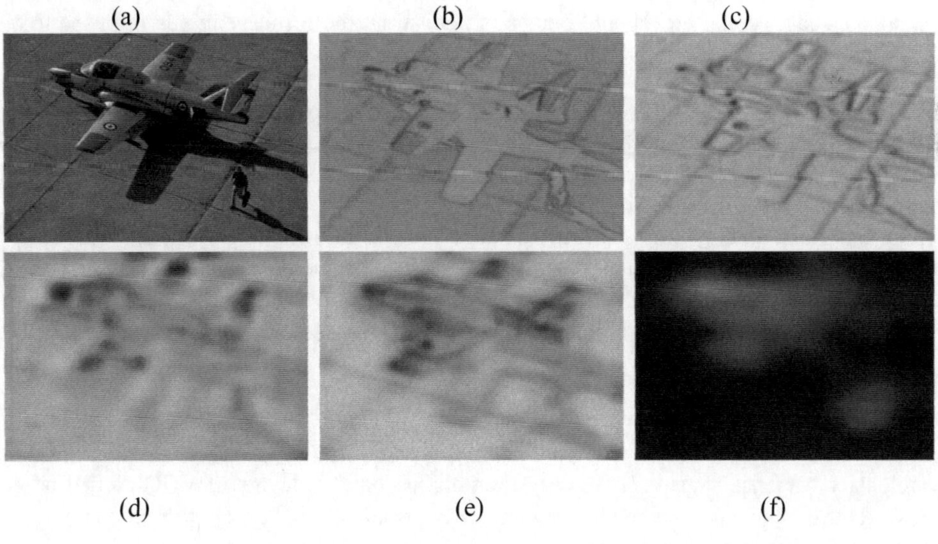

Figure 2. Visual comparison of different layers' intermediate results in VGG-19 for the 'trainer aircraft' image from 'BSD500' dataset. (a) is the original image. (b)-(f) reveal the feature map extracted from convolutional layer *conv1_2*, *relu2_2*, *conv3_4*, *relu4_4* and *conv5_1*.

Many pre-training convolutional neural network that involved in some object detection algorithms, such as VGG-Net[9] pre-trained by the large-scale ImageNet dataset[14], possess the ability to generate different levels of features in an input image. Taking VGG-Net[9] as an example, the distinction between features extracted from different layers will be discussed in detail here. The other networks, e.g. AlexNet[15], ResNet[16], can also be used in template matching.

With the increase of convolutional layers and pooling layers, the spatial resolution of the feature space produced by different layers in VGG-19 will be decrease in proportion. For example, the spatial size of feature maps on the *conv5_1* will be decrease to 7×7 in the VGG-Net[9], which is $1/32$ of the input image size 224×224, due to the pooling operation. In order to make these layers' output have the identical contribution to the final feature engineering, all these output features will be resized to the input image size by bicubic interpolation.

Figure 2 demonstrates visible feature maps of different layers in VGG-Net[9] pre-trained by the large-scale ImageNet dataset[14]. The shallow layers' feature description corresponding to figure 2 (b) and figure 2 (c) reserve abundant edge structure details and color information, and have good recognition ability to area with high contrast. These low-level feature maps are very close to the feature engineering produced by theoretical analysis or handcraft design, which is efficient for locating the template. Nevertheless, the feature maps will be shrunken and gather together to form more high-level abstract feature for the final layers of VGG-Net[9]. The feature map shown in figure 2 (f) reveals the specific locations of the training plane and the pilot, which exhibits the semantic information of templates obviously. These high-level feature maps with semantic information will be indispensable in the task of objection detection, image classification etc. However, the result, shown in the next section, reveals that these high-level features are unnecessary for template matching, and even might have negative impact on the feature engineering. The negative impact can be attributed to two aspects. On one hand, fine-grained features will be more useful for template matching, relative to the high-level feature with semantic information. On the other hand, because the basic network VGG-Net[9] used here is pre-trained by the large-scale ImageNet dataset[14] with category-level 1000 labels, the final feature maps will be invalid once the object in the template does not belong to the 1000 labels, which is obviously shown in figure 2 (f).

According to the above comprehensive analysis, the layers that could extract the fine-grained features, such as edge, color, etc. should be used to construct the feature engineering used in template matching, instead of the layers with high-level semantic characteristics.

2.1.2. Feature Description of Kernels in Intra-Layer. There are great redundancy kernels in the pre-trained VGG-Net[9] by observing, which are inadaptable to construct the feature space in template matching. In general, the smaller the magnitude of the convolutional kernel, the lower the intensity of the response to feature – the majority of these kernels' response will be decay to become a 'zombie' [17], which cannot be used to express the image edge of color information.

The effectiveness of convolutional kernels in CNN can be measured by the intensity of kernels' response, which can be expressed numerically as sum of absolute value. The convolutional kernel is noted as $\boldsymbol{K}\in R^{i\times j}$ (e.g. 3×3), and the magnitude of $\boldsymbol{K}$ is noted as $k_{i,j}$. And we measure the intensity of convolutional kernel as

$$S=\sum\left|k_{i,j}\right| \tag{1}$$

The intensity response statistics of 64×64 kernels in *conv2_1* layers of VGG-Net[9] is shown in figure 3, which have been normalized to 1. For a better visualization, kernels with intensity response 0, 0.2, 0.4, 0.6, 0.8, 1 are interpolated as shown on the second line of figure 2.

It can be observed that convolutional kernels in VGG-Net[9] blur image (for reducing details, similar to a normal distribution as shown in figure 3(f)) or sharpen image (for emphasizing the details or directions of object as shown in figure 3(e)). Meanwhile, there are plenty of 'zombie' kernels, as shown in figure 3 (b), (c), which have a low response to pattern information, such as edge, color. The results reveal that the proportions of kernels with strength less than 0.2 and 0.4 account for 79.91% and 97.36% respectively. These 'zombie' kernels will weaken predominantly valid features' influence in the engineering feature used in template matching, and augment the time complexity and the space complexity.

(a)

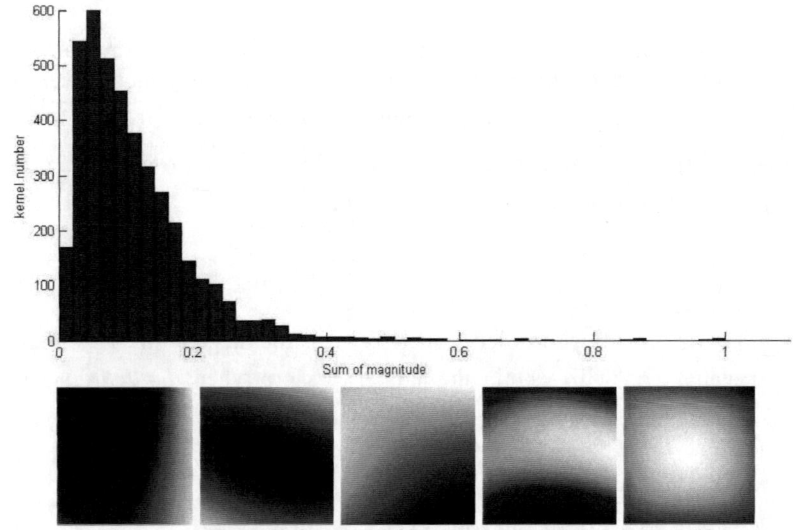

(b) (c) (d) (e) (f)

Figure 3. Statistical and visualization of kernels corresponding to *conv1_2* layer in the VGG-Net. (a) shows the statictics of intensity response, and (b)-(f) is interpolated visually for kernels with intensity response 0, 0.2, 0.4 ,0.6, 0.8, 1.

The experimental results showed that the majority of kernels in intra-layer are unnecessary for feature space used in template matching. By removing these 'zombie' kernels, we can improve the localization effect and reduce calculation amount which is especially effective in terms of similarity measure.

2.2. Similarity measure in NN-based Space

Plenty of location methods have been applied here, such as affine mapping[18][19], tone mapping[20]. An approximate solution, which decides whether the template and the target image window possess the same distribution, is elaborated here as a similarity measure[11]. Meanwhile, in order to cope with complex situation, a deformation penalty term is added, and a new searching strategy is adopted for the complex deep feature space constructed above.

The patches of the template image and the target image window is written as t_i, $s_j \in R^d$. Hence, the object function is measure the similarity between these two sets $T = \{t_i\}_{i=1}^N$ and $S = \{s_j\}_{j=1}^N$ in nearest neighbor space, where T denotes template image and S denotes target window image. Both experimental results and theoretic analysis indicate that the unique nearest neighbor of $t_i \in R^d$ can be found in $s_j \in R^d$ when the template image and target window image have the same or similar distributions. We take the feature patches of template and target window image as $t_i^f \in R^d$ and $s_j^f \in R^d$, so we define the nearest neighbor of t_i in S^f as:

$$s_j^f = \arg\min_{s \in S^f} dis(t_i^f, s) \qquad (2)$$

where $dis(t_i^f, s)$ is distance calculation formula of feature sets, such as l_1 or l_2 norm, and the total number of nearest neighbor of t_i^f in s_j^f is noted as $num(t_i, S^f)$. As t_i^f with the least nearest neighbor in S^f should have a greater weight, we adopt the reciprocal form of $num(t_i, S^f)$ in the final score function.

The practical engineering, template matching should be robust to deformation since the target objects often have shifted or rotated in reality, which is reflected as some jitter in feature space. Here, we note the location information of patches in template and target image as t_i^l, s_j^l respectively, and the distance between them is measured as $d = dis(t_i^l, s_j^l)$ which can be the Euclidean distance. So the penalty term of distance in the final similarity score is shown as follows:

$$Penalty(d) = a^{b^d} \tag{3}$$

where $a > 0$ and $b > 1$, which is decided concretely by the degree of deformation in reality. Specifically, the penalty term will restrain the similarity score when $d = dis(t_i^l, s_j^l)$ is large than a certain threshold. Hence, the penalty strategy is applied to (2), leading to a general score function as follows:

$$Score(T, S) = \sum_{t_i \in R^d} \frac{1}{num(t_i, S)} a^{b^d} \tag{4}$$

A similarity score map can be obtained by calculating score between template and image sliding window in the target image with formula (4), and the final template's location in target image can be determined by the highest scores in the score map.

The feature space constructed by numerous convolutional kernels of a CNNs can't be solved effectively by the conventional brute-force, KD-tree search strategy[21] for approximate nearest neighbor. Hence, product quantization based nearest neighbor method[22] is adopted here. In this method, the feature space will be divided into several subspaces by building codebooks and there is new searching strategy for it. After quickly positioning the certain subspace, the nearest neighbor feature we want can be obtained by traversal operation[23].

3. Experimental Results

In this section, we will discuss the performance of our method and make comparisons to the classical methods (SSD[1]), state-of-the-art methods (BBS[11], DDIS[13]). Meanwhile, an outdoor testing dataset is constructed here to evaluate the performance of these algorithms.

3.1. Data Preparation and Parameter Settings

A testing dataset to the template matching for outdoors is constructed here, and the ground truth location of template object in the test images is marked by professional workers. The entire data collections contain 1861 scenarios, and a template corresponds to three different testing images in different scenarios, including the change of weather, deformation, scaling etc. as illustrated in figure 1. All these high-resolution images with 1920×1080 format are captured by industrial camera. In the testing process, the ground truth in the first image, such as figure 1(a) will be treated as template, the other three images in figure 1 are the target images.

To note that, the experiment results of BBS[11] and DDIS[13] with default arguments are produced with codes from the homepage of their authors, and SSD[1] is implemented by ourselves. And the parameters of our methods mainly comprise three components:

1. Inter-layers: the shallow layers of CNN will be used here, such as *conv1_2, relu2_2, conv3_4* in VGG-Net.

2. Intra-layers: 10% of the convolutional kernels in a layer will be kept according to the strength of these kernels' response discussed in section two.
3. Similarity measure: the coefficients 'a' and 'b' in formula (4) are set to 1.09, 1.22 respectively for the scenarios with slight deformation or scaling.

IoU (Intersection-over-Union)[24] is utilized in our approach felicitously as an evaluation criteria. For the complicated scenarios, the location with IoU greater than 80% is to be regarded correct, and we will calculate accuracy of these template matching methods according to this stipulation.

3.2. Results and Comparison

Templates in the testing dataset can be various, such as instruments, character etc. figure 2 demonstrates the template matching results in the case of varying illumination. We can see that the complicated background including trees, telegraph pole can also cause large perturbations to the matching results.

The matching object in figure 4 is a Chinese character as shown in figure 4 (a), and the target images' background is complicated. For one thing, the matching process will become especially complicated with the different weather conditions, such as cloudy sky, clear sky, which is hardly modeled mathematically. For another, some unexpected watermarks might be encountered due to the industrial applications or emergency situations. These factors have negative effects on the robustness of the proposed algorithm. As shown in figure 4, the different weather conditions between (a) and (c) directly lead to the failure using SSD[1], BBS[11], and some watermarks lead DDIS[13] out of work. Many experimental results here explain that the huge gap between these template matching methods is mostly caused by the feature engineering used. The features used in conventional computer vision including corners, edge, colors don't possess the ability to describe the complicated scenarios, such as cloud in the sky, the complex Chinese character.

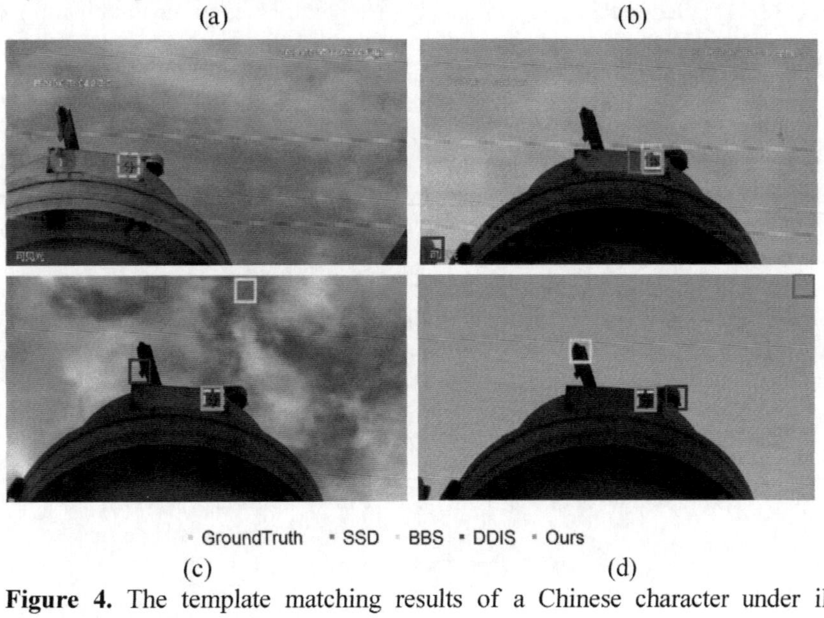

Figure 4. The template matching results of a Chinese character under illumination changes and background interface.

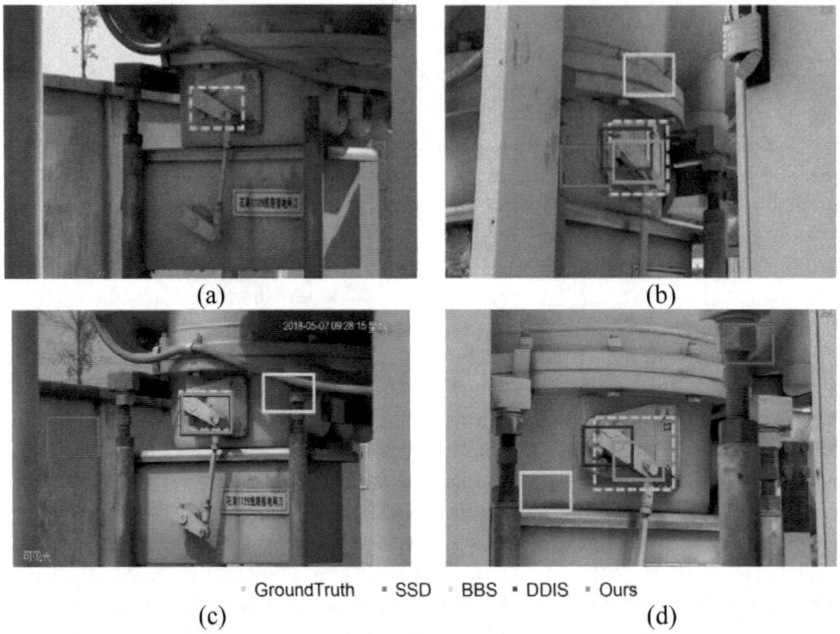

Figure 5. The template matching results of object with deformation and scaling under illumination changes and background interface.

The object template in target images with deformations and scaling is demonstrated figure 5. SSD[1], and BBS[11] failed in this case thoroughly mostly attributes to the illumination, deformation, scaling. Although both DDIS[13] and our method can locate the template in the target images, the localization effect of our method is more effective than DDIS [13] under the standard of IoU.

Table 1 provides the average accuracy and time-consuming of SSD[1], BBS[11], DDIS[13] and our method under the assessment criteria we make above. And our testing environment is a computer configured to I5 processor, 8G RAM, Windows 7 Ultimate, Matlab 2014a. SSD[1] and BBS[11] may not suit the complicated outdoors with an accuracy of 48.9% and 61.7% respectively. And DDIS[13] shows a notable improvements in accuracy compared to BBS[11]. Some of the scenarios that cannot be solved by our method mainly attributes to the illuminations change sharply relative to template image to the extent that also cannot be identified by human. In terms of time-consuming, BBS[11] is the slowest, and the run time of our method would take 23.7s averagely. However, we can use the graphics processor (GPU) to accelerate our method so that it can satisfy the requirements of applications in reality.

Table 1. The accuracy and time-consuming of different method.

Method	SSD	BBS	DDIS	Ours
Accuracy	48.9%	61.7%	82.1%	98.7%
Time (s)	1.2	1308.1	9.5	23.7

4. Conclusions

This paper proposes a high-efficiency template matching method to solve the complicated scenarios with illumination variation, deformation, scaling etc. with a pruning operation on CNN and new similarity criterion. The pruning operation contains two parts. On the one hand, within the template matching region we select the ahead and middle partially layers of a CNN to construct the feature engineering according to their character. And only 10% around convolutional kernels in a layer of CNN is used in the final feature space because more than 80% kernels' strength of response to the images' details could be canceled out. For the similarity measure, an adjustable parameter penalty term is proposed here to make our method robust to the deformation and scaling. The final experimental results show that under the scenarios with illumination variation, deformation, scaling, a good matching result can be gained with our method compared to the other template matching method, such as BBS[11], DDIS[13]. The follow-up will explore more complicated scenarios with noise[25], or improve efficiency of the algorithm[26] to make it more accommodative in industry.

References

[1] Ouyang W, Tombari F, Mattoccia S, Di Stefano L and Cham W K, Performance evaluation of full search equivalent pattern matching algorithms, *IEEE Trans. Pattern Anal. Mach. Intell. (PAMI)*, vol **4**, no. 1, pp. 127-143, 2012.

[2] Balntas V, Lenc K, Vedaldi A and Mikolajczyk K, HPatches: A Benchmark and Evaluation of Handcrafted and Learned Local Descriptors, *IEEE Int. Conf. Comput. Vis. Pattern Recognit. (CVPR)*, pp.3852-3861, 2017.

[3] Rublee E, Rabaud V, Konolige K and Bradski G, ORB: An efficient alternative to SIFT or SURF, *IEEE Int. Conf. Comput. Vis. Pattern Recognit. (CVPR)*, pp. 2564-2571, 2011.

[4] Schmidhuber J, Deep learning in neural networks: An overview, *Neural networks*, vol. **61**, pp. 85-117, 2015.

[5] Wang N, Shi J, Yeung D Y and Jia J, Understanding and diagnosing visual tracking systems, *IEEE Int. Conf. Comput. Vis. Pattern Recognit. (CVPR)*, pp. 3101-3109, 2015.

[6] Felzenszwalb P F, Girshick R B, McAllester D and Ramanan D, Object detection with discriminatively trained part-based models, *IEEE Trans. Pattern Anal. Mach. Intell. (PAMI)*, vol. **32**, no. 9, pp. 1627-1645, 2010.

[7] Ren S, He K, Girshick R and Sun, J., Faster r-cnn: Towards real-time object detection with region proposal networks. *IEEE Trans. Pattern Anal. Mach. Intell. (PAMI)*, vol. **39**, no. 6, pp. 1137-1149, 2015.

[8] Redmon J, Divvala S, Girshick R and Farhadi A, You only look once: Unified, real-time object detection, *IEEE Int. Conf. Comput. Vis. Pattern Recognit. (CVPR)*, pp. 779-788, 2016.

[9] Simonyan K and Zisserman A, Very deep convolutional networks for large-scale image recognition, Computer Science, *IEEE Int. Conf. Learning Representation (ICLR)*, 2015.

[10] Kim H Y and De Araújo S A, Grayscale template-matching invariant to rotation, scale, translation, brightness and contrast, *In Pacific-Rim Symposium on Image and Video Technology*, pp. 100-113, 2007.

[11] Dekel T, Oron S, Rubinstein M, Avidan S and Freeman W T, Best-buddies similarity for robust template matching, *IEEE Int. Conf. Comput. Vis. Pattern Recognit. (CVPR)*, pp. 2021-2029, 2015.

[12] Simakov D, Caspi Y, Shechtman E and Irani M, Summarizing visual data using bidirectional similarity, *IEEE Int. Conf. Comput. Vis. Pattern Recognit. (CVPR)*, pp. 1-8, 2008.

[13] Talmi I, Mechrez R and Zelnik-Manor L, Template matching with deformable diversity similarity, *IEEE Int. Conf. Comput. Vis. Pattern Recognit. (CVPR)*, pp. 1311-1319, 2017.

[14] Deng J, Dong W, Socher R, Li L J, Li K and Fei-Fei L, Imagenet: A large-scale hierarchical image database, *IEEE Int. Conf. Comput. Vis. Pattern Recognit. (CVPR)*, pp. 248-255, 2009.

[15] Krizhevsky A, Sutskever I and Hinton G E, Imagenet classification with deep convolutional neural networks, *In Advances in Neural Information Processing Systems*, pp. 1097-1105, 2012.

[16] He K, Zhang X, Ren S and Sun J, Deep residual learning for image recognition, *IEEE Int. Conf. Comput. Vis. Pattern Recognit. (CVPR)*, pp. 770-778, 2016.

[17] Osherov E and Lindenbaum M, Increasing CNN Robustness to Occlusions by Reducing Filter Support, *IEEE Int. Conf. Comput. Vis. Pattern Recognit. (CVPR)*, pp. 550-561, 2017.

[18] Korman S, Reichman D, Tsur G and Avidan S, Fast-match: Fast affine template matching, *IEEE Int. Conf. Comput. Vis. Pattern Recognit. (CVPR)*, pp. 2331-2338, 2013.

[19] Zhang C and Akashi I, Fast Affine Template Matching over Galois Field, *British Mach. Vis. Conf.*, pp. 121-1, 2015.

[20] Hel-Or Y, Hel-Or H and David E, Matching by tone mapping: Photometric invariant template matching, *IEEE Trans. Pattern Anal. Mach. Intell. (PAMI)*, vol. 36, no. 2, pp. 317-330, 2014.
[21] Silpa-Anan C and HartleyR, Optimised KD-trees for fast image descriptor matching. *IEEE Int. Conf. Comput. Vis. Pattern Recognit. (CVPR)*, pp. 1-8, 2008.
[22] Jegou H, Douze M and Schmid C, Product quantization for nearest neighbor search, *IEEE Trans. Pattern Anal. Mach*, vol. **33**, no. 1, pp. 117-128, 2011.
[23] Ge T, He K, Ke Q and Sun J, Optimized product quantization, *IEEE Trans. Pattern Anal. Mach. Intell. (PAMI)*, vol. 36, no. 4, pp. 744-755, 2014.
[24] Nowozin S, Optimal decisions from probabilistic models: the intersection-over-union case. *IEEE Int. Conf. Comput. Vis. Pattern Recognit. (CVPR)*, pp. 548-555, 2014.
[25] Elboher E and Werman M, Asymmetric correlation: a noise robust similarity measure for template matching, *IEEE Trans. Image Pro.*, vol. 22, no. 8, pp. 3062-3073, 2013.
[26] Pele O and Werman M, Robust real-time pattern matching using bayesian sequential hypothesis testing, *IEEE Trans. Pattern Anal. Mach. Intell. (PAMI)*, vol. **30**, no. 8, pp. 1427-1443, 2008.

Reliability Modeling of NC Machine tools Based on Artificial Intelligence

Guofa Li, Hongxiang Zhu, Jialong He*, Yongchao Huo and Jiancheng Zhang

College of Mechanical Science and Engineering, Jilin University, Changchun, Jilin, 130022, China

*Corresponding author: hjl.star@163.com

Abstract. The level of reliability for NC machine tools represents the development level of the country's manufacturing industry, and its reliability modeling is very important. In order to improve the reliability of NC machine tools, this paper proposes the application of artificial intelligence model for NC machine toolreliability. Firstly, the time history of the NC machine tool reliability is analyzed. Secondly, a reliability evaluation framework based on neural network and bayesian network, intelligent fault diagnosis based on deep learning, and intelligent fault prediction framework based on least squares support vector machine (LS-SVM) are established to achieve remote maintenance of NC machine tools. Finally, with the background of big data, the development vision about the application of artificial intelligence model for NC machine toolreliability is summarized and forecasted.
Keywords: Artificial intelligence; Reliability evaluation; Fault diagnosis; Fault prediction.

1. Introduction

Manufacturing and manufacturing technologies are moving towards the era of Industry 4.0. Intelligent manufacturing has become the main direction for industrial advanced countries to ensure their future competitiveness and status. The manufacturing industry of China faces the opportunities and challenges of innovation-driven development and intelligent transformation and upgrading. The NC machine tool is the "working machine" of equipment manufacturing industry, its reliability level represents the level of development of the manufacturing industry [1]. Its reliability analysis has always been the focus of researchers and users [2-4]. The era of artificial intelligence brings challenges and innovation opportunities to the traditional reliability analysis methods of NC machine tools.

The reliability evaluation technology of NC machine tools is to quantify the reliability level by establishing the reliability model of the machine tool. According to the actual situation, Peng fused the success or failure data, life data, and degenerate data at different system levels, and built a multi-level heterogeneous data index set. The system reliability evaluation and prediction method based on the bayesian network is proposed [5-7]. Graves uses the Bayesian approach to integrate the system's performance degradation data. Genetic algorithms are used to guide the allocation of scarce test resources to reduce the uncertainty, and the bayesian reliability evaluation method based on genetic algorithm is proposed [8]. Reese presented a bayesian model for assessing the reliability of multicomponent systems which the component, subsystem, or system level are integrated with prior information at any level. The model allows pooling of information between similar components, the incorporation of expert opinion, and straightforward handling of censored data [9]. However, the above method didn't consider the influence of the change of operating conditions on the reliability in the actual cutting process of NC machine tools.

With the development of computer technology and signal processing technology, fault diagnosis is developing in the direction of automation and intelligence [10-11]. On the basis of combining the characteristic information collected by various types of sensors for reciprocating compressors, Zhang and Jiang proposed a fault diagnosis method based on multi-source Information fusion to use a combination of multi-source information fusion technology, radial basis neural network, and weighted evidence fusion theory. Reciprocating compressor fault diagnosis method [12]. Che proposed an aero engine fault diagnosis model based on deep learning for a combination of deep neural network and decision fusion algorithms by analyzing a large number of engine performance parameters and working conditions of each component [13]. Lei introduced the characteristics of mechanical intelligent fault diagnosis based on big data. From the aspects of signal acquisition, feature extraction and fault identification, the research progress and development trend of mechanical intelligent fault diagnosis at home and abroad are summarized. Finally, the challenges of mechanical intelligent fault diagnosis theory and method in the context of big data are pointed out [14].

The application of fault prediction technology can prevent maintenance of NC machine tools before the fault occur and improve the reliability of NC machine tools.
Zhang proposed a new method for equipment degradation status recognition and remaining service life prediction based on the hybrid Gaussian output Bayesian belief network model [15]. Tian proposed a method for predicting the remaining service life based on artificial neural network equipment to use the current and previous checkpoint usage times and multiple online state measurement values as input, and the percentage of life as output, which greatly improved the reliability of the equipment and reduced the overall Maintenance costs [16]. Loutas proposed a method for estimating the remaining service life of rolling bearings based on the data-driving method of support vector machines [17]. Sikorska discussed the advantages and disadvantages of fuzzy empirical, neural networks, statistical model and physical models in predicting industrial residual life [18].

This paper discusses the time history of reliability of NC machine tools. A reliability evaluation based on neural network and bayesian network framework, an intelligent fault diagnosis based on deep learning, and an intelligent fault prediction framework based on LS-SVM are proposed.

2. The time history of the NC machine tool reliability

Reliability modeling is a prerequisite to obtain the reliability level of NC machine tools. As shown in Figure 1, the main processes of reliability modeling are reliability evaluation, fault diagnosis, and fault prediction in the life cycle of the NC machine tool. From the assembly completion of the NC machine tool to the end-of-life processing, real-time reliability evaluation of the NC machine tool can accurately obtain the reliability level and understand the dynamic characteristics of the NC machine tool. When the NC machine tool is fault, fault diagnosis is performed to determine the cause of the fault, the location of the fault, and the failure mode to achieve fast and accurate maintenance of the NC machine tool. The fault prediction can determine the time of the next failure for the NC machine tool in advance to perform preventive maintenance. It can reduce the loss caused by the failure of the NC machine tool and improve the reliability of the NC machine tool.

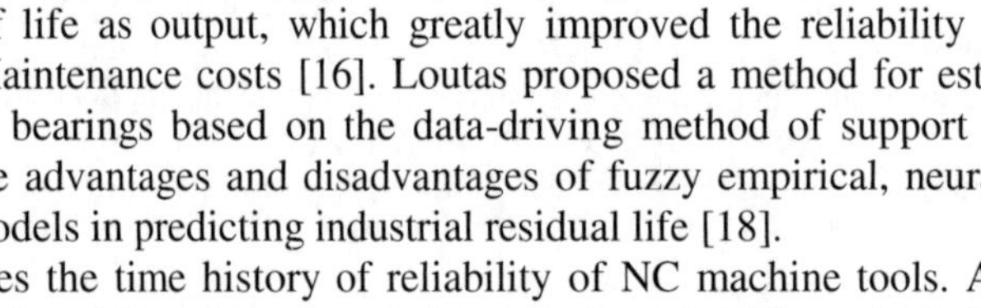

Figure 1. The life cycle of NC machine tools

3. Application of artificial intelligence model for NC machine tool reliability

The data fusion algorithms as bayesian network, deep learning, and support vector machine are the basic methods to realize the reliability analysis of NC machine tools based on artificial intelligence. According to the time history of NC machine tools, this paper proposes an intelligent reliability evaluation framework based on neural network and bayesian network, an intelligent fault diagnosis framework based on deep learning and an intelligent fault prediction framework based on LS-SVM.

3.1. The intelligent reliability evaluation framework based on neural network and bayesian network

The NC machine tool are complex electromechanical-hydraulic integrated systems. The reliability level is one of the most important indicators of its performance. It directly affects the quality and efficiency of processing. The intelligent evaluation of NC machine tools can reflect the reliability level of NC machine tools in real-time, and it has important guiding significance for the reliability design of NC machine tools. Therefore, it is of great significance for real-time reliability evaluation of NC machine tools.

3.1.1. Application of Neural Networks. Due to the large scale and variety of data are collected by sensors, the collected data is the high reproducible and the low data density. The neural network are used to fuse information gathered by multiple sensors, which can improve the credibility of the data.

The data collected by each sensor is very different. The mutual learning is performed through the neural network. The load information with different physical meanings from different sources such as cutting force, current and temperature is converted into values with the same meaning. As shown in Figure 2, the information collected by each sensor X is used as an input of the neural network after preprocessing, and the value Y with the same meaning is obtained after the mutual learning process.

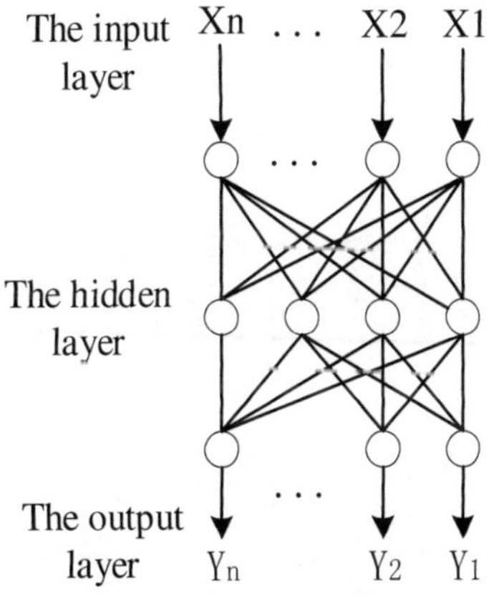

Figure 2. The flowchart of data fusion

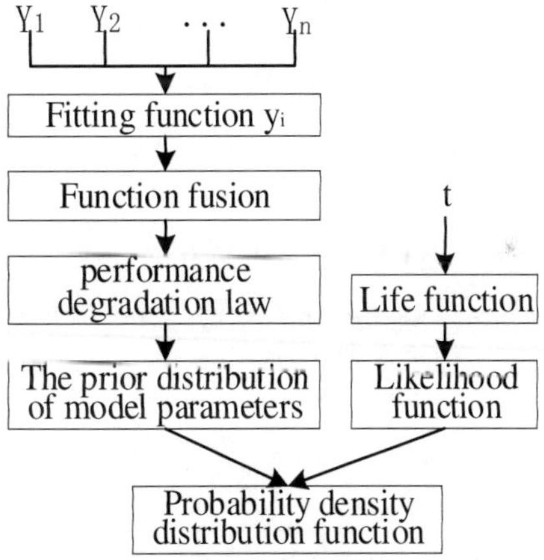

Figure 3. The flowchart of reliability evaluation

3.1.2 Application of bayesian network. As shown in Figure 3, the specific steps of the Bayesian network are as follows.

Step1: the value Y of the output layer for the neural network is fitted with the function, and the performance degradation law of the NC machine tool is obtained, which the prior distribution of reliability model parameters for the NC machine tool is obtained.

Step2: The likelihood function is obtained by using fault data.

Step3: The probability density function is obtained.

The bayesian network is used to fuse operating conditions and fault data, which achieve an accurate evaluation of the reliability for the NC machine tool.

3.2 The intelligent fault diagnosis framework based on deep learning

In the era of big data, failure modes are characterized by coupling, indefiniteness, and concurrency, and the traditional methods have weak self-learning capabilities to result in low fault recognition accuracy and weak generalization capability. Therefore, in the context of big data, intelligent fault diagnosis of NC machine tools requires new theories and methods.

The deep learning (DL) is a big data processing tool. It builds deep models and simulates the brain learning process to realize automatic feature extraction and fittness of complex mapping relationships. The intelligent fault diagnosis of NC machine tools is realized to excavate the intrinsic information rich in big data and improve the fault recognition accuracy.

The real-time acquisition of operating data for the NC machine tool represents the running state of the NC machine tool, which is the premise of fault diagnosis. The fault information of the NC machine tool is the result of the comprehensive action of vibration, sound source, acoustic emission, and thermodynamics. Therefore, the real-time acquisition of various signal dynamic data provides basic data for the intelligent fault diagnosis of NC machine tools.

The frequency spectrum of the training sample fault feature signal is obtained by using the collected data through a sparse autoencoder (SAE). The hidden layer number of the DL is determined, and the hidden layer output of each SAE is taken as the input of the next layer SAE until completing the training of SAEs, which the output layer is determined. The DL parameters are fine tuned by using the back propagation (BP) algorithm to complete the DL training, the detailed process of intelligent fault diagnosis for NC machine tools is shown in Figure 4.

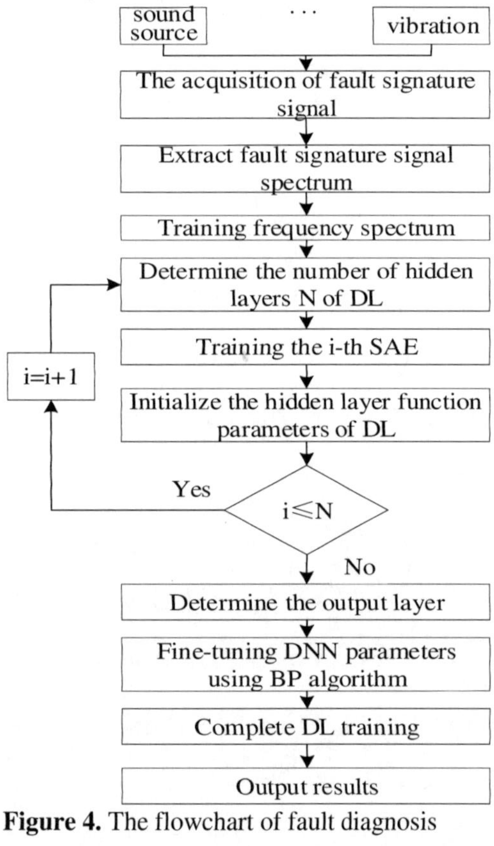

Figure 4. The flowchart of fault diagnosis

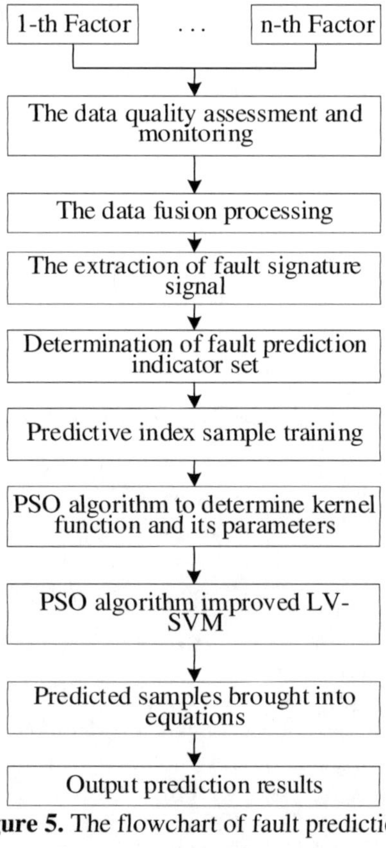

Figure 5. The flowchart of fault prediction

3.3 The intelligent fault prediction framework for NC machine tools based on LS-SVM

In the actual cutting process of NC machine tools, the operating environment is dynamically changing. Therefore, a multi-factor fusion forecasting model should be established to realize intelligent fault prediction under variable operating conditions.The hybrid prediction model considered the relationship of different components and the combination of cutting force, temperature, and other multi-condition factors fusion models can achieve more accurate and intelligent fault prediction of NC machine tools.

The collected real-time data on the multi-condition factors of the subsystem or key component for NC machine tools are evaluated and selected, which fusion processing is performed. The feature signal representing the fault information for the NC machine tool is extracted, the fault prediction index set is constructed, and the hidden fault state information in the data is fully mined. The sample size of the forecasting indicator was trained to determine the SVM kernel function and its parameters using the particle swarm algorithm, and the intelligent fault prediction model for the NC machine tool based on the LS-SVM was obtained as shown in Figure 5.

4. Conclusions

This paper introduces the time history of the reliability for the NC machine tool and proposes the reliability evaluation framework, fault diagnosis framework and failure prediction framework for the NC machine tool based on artificial intelligence.

(1) The neural networks are used to fuse information from a variety of different sources. An intelligent real-time reliability evaluation framework for NC machine tools based on the combination of neural networks and bayesian networks is proposed.

(2) In the era of big data, signals such as vibration, sound source, acoustic emission, and thermodynamics were fused after taking full account of the unique features of failure modes, and an intelligent fault diagnosis framework of NC machine tools based on deep learning was proposed.

(3) The information from a variety of different sources were fused to determine the fault prediction index set. An intelligent fault prediction framework for NC machine tools based on LS-SVM was proposed.

Acknowledgments

This work was supported by the High-level NC machine tools and basic manufacturing equipment (Grant No. 2017ZX04011005), Key Scientific and Technological Project of Jilin Province (Grant No.20170204007G X) and the Doctoral Fund of Ministry of Education of China (Grant No. 2017M621204).

References

[1] Yang Z, Chen C, Chen F,et al 2013 Progress in the Research of Reliability Technology of Machine Tools. Journal of Mechanical Engineering. **49** 130-139

[2] Merrick J R W, Soyer R, Mazzuchi T A. 2003 A Bayesian Semiparametric Analysis of the Reliability and Maintenance of Machine Tools. Technometrics. **45** 58-69.

[3] Zhang Z, Liu Z, Cai L, et al. 2016 An accuracy design approach for a multi-axis NC machine tool based on reliability theory. International Journal of Advanced Manufacturing Technology. **91** 1-20.

[4] W Peng, L Shen, Y Shen, Q Sun. 2018 Reliability Analysis of Repairable Systems with Recurrent Misuse-Induced Failures and Normal-Operation Failures. Reliability Engineering & System Safety. **171** 87-98

[5] Peng W, Huang H Z, Xie M, et al. 2013 A Bayesian Approach for System Reliability Analysis With Multilevel Pass-Fail, Lifetime and Degradation Data Sets. IEEE Transactions on Reliability. **62** 689-699.

[6] Peng W, Huang H Z, Li Y, et al 2013 Life cycle reliability evaluation of new products—A Bayesian model updating approach. Reliability Engineering & System Safety. **112** 109-119.

[7] Peng W, Liu Y, Li Y F, et al. 2014 A Bayesian optimal design for degradation tests based on the inverse Gaussian process. Journal of Mechanical Science & Technology. **28** 3937-3946.

[8] Graves T L, Hamada M S. A 2016 Demonstration of Modern Bayesian Methods for Assessing System Reliability with Multilevel Data and for Allocating Resources. Journal of Quality & Reliability Engineering. **2009**.

[9] Reese C S, Wilson A G, Guo J, et al. 2011 A Bayesian Model for Integrating Multiple Sources of

Lifetime Information in System-Reliability evaluations. Journal of Quality Technology. **9** 489-495.
[10] Niu G, Yang B S. 2010 Intelligent condition monitoring and prognostics system based on data-fusion strategy. Expert Systems with Applications. **37** 8831-8840.
[11] Niu G, Han T, Yang B S, et al. 2007 Multi-agent decision fusion for motor fault diagnosis. Mechanical Systems & Signal Processing. **21** 1285-1299.
[12] Zhang M, Jiang Z. 2017 Reciprocating compressor fault diagnosis technology based on multi-source Information Fusion. Journal of Mechanical Engineering. **53** 46-52.
[13]Che C, Wang H, Ni X, Hong J. 2018 Aeroengine fault fusion diagnosis based on deep learning. Journal of Beijing University of Aeronautic and Astronautic. **44** 621-628.
[14] Lei Y, Jia F, Kong D,Lin J, Xing S. 2018 Opportunities and challenges of machinery intelligent fault diagnosis in big data era. Journal of Mechanical Engineering. **54** 94-104.
[15]Zhang X, Kang J, Zhao J,et.al. 2014 Equipment degradation state identification and residual life prediction based on MoG-BN. Journal of vibration and shock. **33** 171-179.
[16]Zhigang Tian. 2012 An artificial neural network method for remaining useful life prediction of equipment subject to condition monitoring. Journal of Intelligent Manufacturing. **23** 227-237.
[17]Loutas T H, Roulias D, Georgoulas G. 2013 Remaining Useful Life Estimation in Rolling Bearings Utilizing Data-Driven Probabilistic E-Support Vectors Regression. IEEE Transactions on Reliability. **62** 821-832.
[18] Sikorska J Z, Hodkiewicz M, Ma L. 2011 Prognostic modelling options for remaining useful life estimation by industry. Mechanical Systems & Signal Processing, **25** 1803-1836.

IOP Conf. Series: Materials Science and Engineering **435** (2018) 012058 doi:10.1088/1757-899X/435/1/012058

A Passive Snake-like Arm Based on Wire-driven Mechanism for Laparoscopic Surgery

Tianyu Hu[1], Yan Wang[1], Baoping Ma[1], Fan Zhang[2,*], Jing Sun[3], Minhua Zheng[3] and Ruijun Pan[4]

[1]College of Mechanical Engineering, Shanghai University of Engineering Science, Longteng Road No. 333, Songjiang District, Shanghai, 201620, P. R. China
[2]Intelligent Robotics Research Center of Shanghai University of Engineering Science, Shanghai University of Engineering Science, Longteng Road No. 333, Songjiang District, Shanghai, 201620, P. R. China
[3]Department of General Surgery, Shanghai Minimally Invasive Surgery Center, Ruijin Hospital, Shanghai Jiao Tong University School of Medicine, Shanghai, 200025, P. R. China
[4]Department of Digestive Surgery, Shanghai Minimally Invasive Surgery Center, Ruijin Hospital, Shanghai Jiao Tong University School of Medicine, Shanghai, 200025, P. R. China

[*]Corresponding Author: pdszhangfan@aliyun.com

Abstract. In this paper, a passive snake-like arm based on wire-driven mechanism is proposed, which can be connected with remote center of motion (RCM) mechanism to form a laparoscope-holder robot and be used in laparoscopic surgery. The purpose of the snake-like arm is to replace traditional multi-joint manipulator, making it possible for the robot to assist the surgeon in the best position in a surgery. The snake-like arm consists of a main body composed of four arm units and a wire locking device with 12 wire locking units. The arm units are connected by hook hinges. Four classes force amplificatory structure is adopted in the wire locking unit. The detail structure and working principle of the wire locking device are proposed in this paper. The wire tensioning and guiding mechanism as well as the layout of wires are also introduced in this paper. Mechanical calculation of the force amplicatory structure is proposed in this paper to ensure that sufficient locking force can be provided. The workspace of the snake-like arm is drawn by simulation, and the large workspace allows the snake-like arm to be used in most laparoscopic surgeries.

1. Introduction

Laparoscopic surgery is a kind of surgery whereby the surgical operation is performed through small incisions placed on the patient's body by using some specialized instruments such as laparoscope and endoscope [1][2]. Compared to traditional open surgery, laparoscopic surgery has the advantages of low infection rate, less pain, less bleeding and quick recovery [3][4][5]. Therefore, laparoscopic surgery is called the "second revolution" of surgery. And with the continuous development and progress of robot technology, robots are gradually applied to laparoscopic surgery [6]. Compared to the hands-on surgery, robot is helpful to improve the following problems such as poor tracking performance and shaking. And the robot will not have fatigue phenomenon, thus avoiding the potential harm to patients caused by surgeons' fatigue due to long time surgery [7].

At present, one kind of laparoscopic surgery robots is laparoscope-holder robot, such as freehand and Soloassist. The robots always combine RCM mechanism [8] and multi-joint manipulator and hold a laparoscope to assist surgeons in surgeries. There are roughly eight kinds of institutions to implement RCM at present: Isocenters, Circular Tracking Arcs, Parallelograms, Synchronous Belt Transmmision, Spherical Linkages, Parallel Manipulators, Complaint Mechanisms and Passive RCM [7]. However, the mechanical arm that supports RCM mechanism is less researched, most of them are multi-joint manipulators connected by active or passive joints.

In a laparoscopic surgery, the best position for a doctor holding the laparoscope is in the rear of the surgeon, and the laparoscope should be held through the armpit of the surgeon. However, at present, most robot manipulators are difficult to achieve such operation because of their traditional driving and transmission systems. In order to solve this problem, this paper proposes a passive snake-like arm, which is used to replace traditional multi-joint manipulator and combines with RCM mechanism to form an laparoscope-holder robot. The snake-like arm is driven by wire and has the performance of rigid-flexible conversion. The high degrees-of-freedom (DOF) enables the end of the snake-like arm to reach the same point in different positions. This feature can effectively prevent interference between robot and surgeon, and even hold the laparoscope in the best position mentioned above. In Section 2, the structural design and wire layout of snake-like arm are proposed. Section 3 presents the mechanical calculation of the force amplicatory structure. Section 4 shows the working space of the snake-like arm. The conclusion of this paper is given in Section 5.

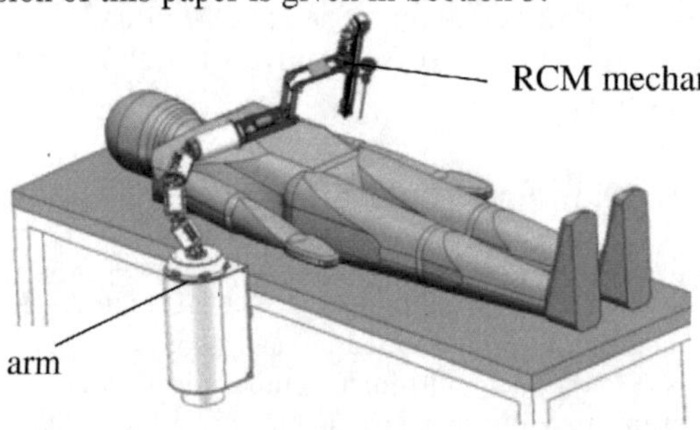

Figure 1. The overall structure of the laparoscope-holder robot

2. The structural design and wire layout of snake-like arm

The overall structure of the laparoscope-holder robot is shown in Figure 1, which contains a passive snake-like arm based on wire-driven mechanism (including the main body of the snake-like arm and a wire locking device) and RCM mechanism. Because the posture of the snake-arm is manually adjusted, the key part of the arm is the wire locking device.

2.1. The structural design of snake-like arm

2.1.1. The main body of the snake-like arm. The current design of snake-like arm generally consists of backbone (rigid or flexible), rigid universal or spherical joints, plate and wire [9]. The main body of the snake-like arm we designed [see Figure 2 (a)] use rigid backbone connected by hook hinges [see Figure 2 (c)], and contains four arm units [see Figure 2 (b)]. Each joint provides two DOFs, making the entire arm have eight DOFs. To reduce the weight of the snake-like arm, making the operation more flexible, the arm unit uses three supports to connect two guide plates instead of the whole cylinder and uses plastic as the shell. Twelve wires stretch out from the guide plate 4 and pass through the arm units through the wire guide holes. Every time a unit is passed, three wires are fixed on it. When the wires can move freely, hook hinges can rotate freely, and the surgeon can adjust the angle of the two adjacent units at will; When all wires are locked, all joints cannot move, thus fixing the posture of the snake-like arm. The main parameters of the arm are shown in Table 1.

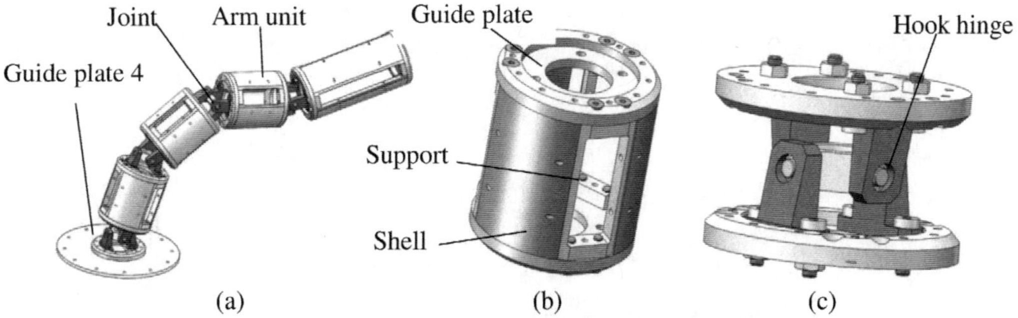

Figure 2. The main body of the snake-like arm

Table 1. Main parameters of snake-like arm

Main dimension parameter	
Length of the first three snake-like arm units (mm)	62
Length of the forth snake-like arm unit length (mm)	150
Diameter of Snake arm (mm)	55
Joint height (mm)	23
Maximum deflection Angle of Universal Joint (degree)	45°
Total length of snake-like arm (mm)	428

2.1.2. The wire-locking device. Compared with the motor driven snake-like arm, the passive snake-like arm proposed in this paper changes the motor driving device into a wire locking device, whose overall structure is shown in Figure 3. There are 12 wire locking units, each two locking units are arranged in a row, and 6 rows are evenly distributed inside the locking device with a circumference. The stepper motor and wire guide mechanism are placed on both sides of the frame respectively. Each wire will be connected to a locking unit and then penetrated into the main body of the snake-like arm through the wire guide mechanism.

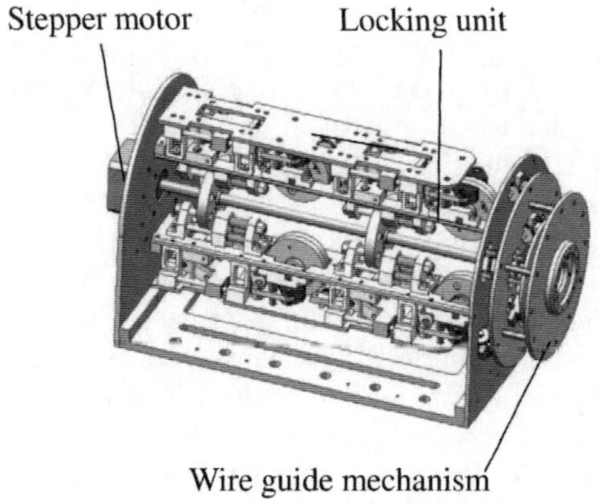

Figure 3. Wire locking device

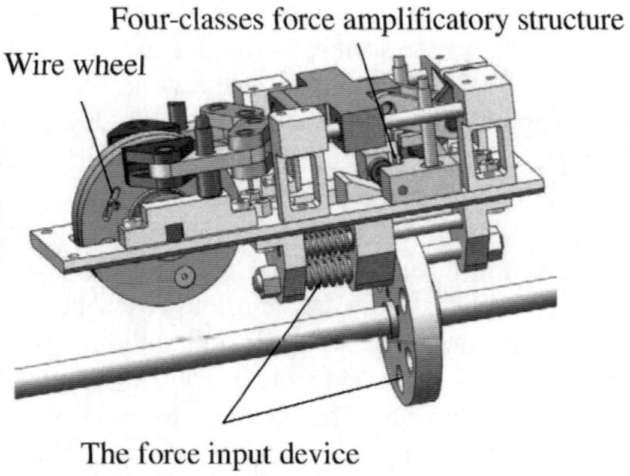

Figure 4. The structure of the locking unit

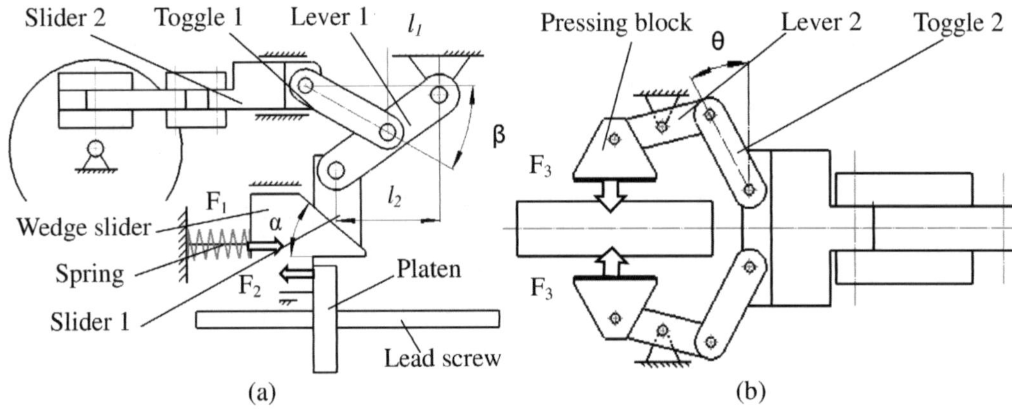

Figure 5. The working principle of the locking unit

Figure 4 shows the structure of the wire locking unit, mainly including the force input device, the four classes force amplificatory structure and the wire wheel. The force input device provides the force of locking wires and loosening wires by the compression spring and the motor respectively. The force amplificatory structure amplifies the force [10] four times to ensure that there is enough pressure to lock the wheel. Using springs instead of motor to provide input force is to maintain clamping force at all times without using force sensors, thus saving the manufacturing cost. Figure 5 shows the working principle of the locking unit. In the wire locking stage, the spring pushes the wedge slider and provides a pressure F_1, and F_1 is amplified to F_3 by a wedge-lever-toggle-toggle-lever four classes force amplicicatory structure. The wedge slider pushes the slider 1 upward to achieve the first amplification of the force; the slider 1 transfers the thrust to the toggle 1 through the permanent force increasing lever 1 to achieve the second amplification; the toggle 1 push the slider 2, making the pressure angle β reduce, and the third amplification of the force is achieved by the principle of angle amplification; the slider 2 drives the toggle 2 to reduce the pressure angle θ, achieving the fourth amplification again by the principle of angle magnification; lever 2 is an equal arm lever, which only changes the direction of the force. Finally, the pressing block is forced to tighten the wire wheel, making it impossible to rotate. Thus, wires are locked. In the loosing stage, the motor drives the lead screw to make the platen connected with the feed screw nut move to the left, pushing the wedge slider and recompressing the spring, so that the motion of the force amplificatory structure is opposite to that of the locking stage. The force on the wire wheel is removed, and the wire wheel can rotate freely, thereby enabling the wire fixed on it to move freely.

It is necessary to tighten the wire when using wire transmission. The locking unit implement the tension of the wire at the wire wheel. As shown in Figure 6(a), the wheel shaft is fixed and will not rotate, the front and rear half of the wire wheel are bolted to form a whole wire wheel rotating around the shaft. In the inner of the wire wheel, there is a power spring, and its inner and outer hooks are respectively stuck in the shaft and the wire wheel. The wire is fixed to the front half of the wheel through a wire fixing hole and twine around the wheel. In Figure 6(b), the wire wheel is rotated to make the spring tighten and store energy, so that the spring always gives the wire wheel a clockwise torque, and the wire wheel has a clockwise rotation trend under the action of this torque. Then the wire is wound counterclockwise around the wire wheel in the wire slot. When pulling the wire, the force on the wire gives a counterclockwise torque to the wire wheel. Through these two opposing torques, the steel wire gets tensioned.

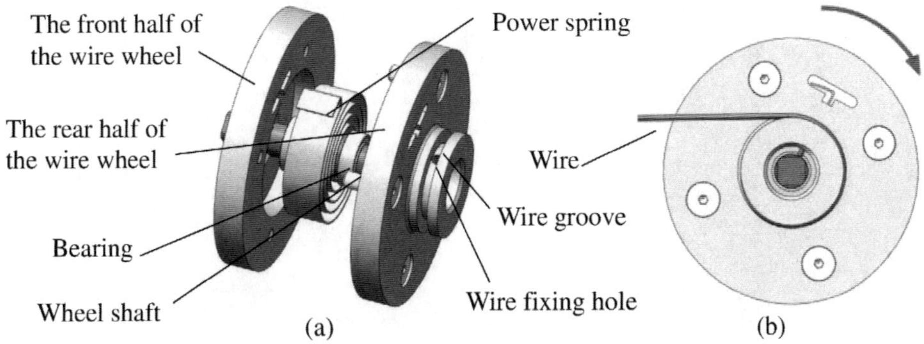

Figure 6. Wire tensioning mechanism

The diameter of the wire locking device is larger than that of the snake-like arm. To make wires penetrate into the snake-like arm without interference, a wire guide mechanism (see Figure 7) between the wire locking device and the snake-like arm is required. To avoid the permanent deformation of the wire in the transmission due to the excessive bending angle, there are 12 pulleys arranged on the guide plate 1 and 2. Also, there are 12 wire guide holes on the guide plate 3 and 4. A wire is led out by the locking unit, and penetrated into the snake-like arm by pulley 1, pulley 2, guide hole 1 and guide hole 2 respectively.

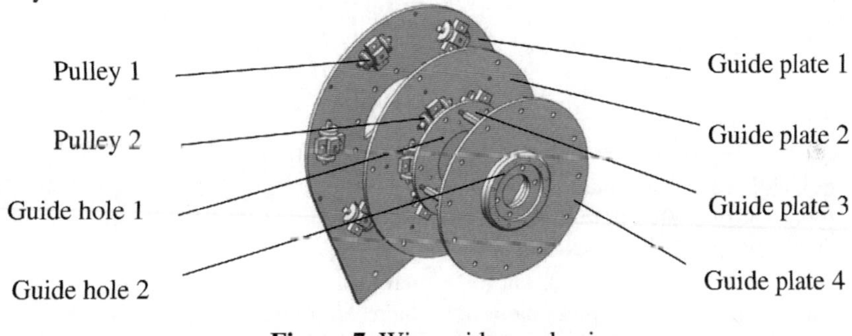

Figure 7. Wire guide mechanism

2.2. Layout design of wire

Figure 8 shows the wire layout in the wire locking device and the snake-like arm respectively. In Figure 8 (a), wires are pulled out from the left side of the front wire wheels and the right side of the rear wire wheels, the wires of the front wire wheel directly enters the guide wire mechanism, and the wires of the rear wire wheel passes through a pulley and then transmits to the wire guide mechanism. Figure 8 (b) shows the way of wire transmission in the main body of the snake-like arm. Each time a guide plate is passed, three wires distributed around the circle are fixed to it and the rest of the wires continues to go through the arm, making 12 wires pass through four arm units to achieve the wire transmission.

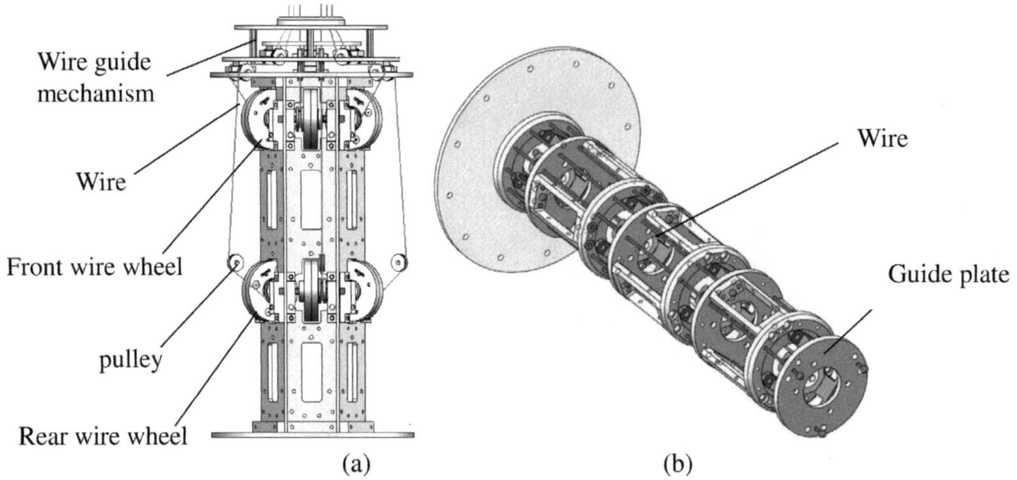

Figure 8. Layout design of wire

3. Mechanical calculation of the force amplicatory structure

Amplificatory coefficient represents the ratio of the output force to the input force. It is expressed with letter i. When the mechanical transfer efficiency and friction loss are not considered, the theoretical force augmentation coefficient and the theoretical output force are respectively expressed in symbol i_t and F_t. Through the mechanical analysis of the force amplificatory structure by Figure 5, easily obtain the result as follow.

Theoretical force increasing coefficient: $$i_t = \frac{l_2}{l_1 \tan\alpha \tan\beta \tan\theta} \quad (1)$$

Theoretical output force: $$F_t = F_1 \frac{l_2}{l_1 \tan\alpha \tan\beta \tan\theta} \quad (2)$$

Here α — dip angle of wedge slide
β, θ — theory of pressure angle of toggle mechanism
F_1 — the pressure of compression spring
l_1 — active arm length of permanent force increasing lever
l_2 — passive arm length of permanent force increasing lever

While in practical use, the actual amplificatory coefficient and the output force will be slightly different due to the existence of mechanical transfer efficiency and friction. They are expressed in symbol i_p and F_3. They can be calculated according to the literature as follows:

Actual force increasing coefficient: $$i_p = \frac{l_2}{l_1 \tan(\alpha+\varphi_1)\tan(\beta+\varphi_2)\tan(\theta+\varphi_3)}\eta \quad (3)$$

Actual output force: $$F_3 = F_1 \frac{l_2}{n l_1 \tan(\alpha+\varphi_1)\tan(\beta+\varphi_2)\tan(\theta+\varphi_3)}\eta \quad (4)$$

Here φ_1 — friction angle between oblique planes on wedge slider
φ_2, φ_3 — equivalent friction angle of the toggle, $\varphi = \arcsin\frac{2rf}{l}$

Where r — the radius of the hinge axis
l — the center distance of two hinge
f — friction factor of hinge, usually, $f = 0.1 \sim 0.15$
η — mechanical efficiency of a lever, usually, $\eta = 0.97$
n — number of pairs of terminal hinge pairs, $n = 2$

For this force amplificatory structure, after the wheel is pressed by the compression block, α=25°, β=15°, θ= 15°, l_1/l_2=2, f=0.1, r_1=r_2=2.5mm, φ_1=5°, φ_2=1.1°, φ_3=2.3°, η=0.97. The actual force amplificatory coefficient i_p can be obtained from the formula (3) and it is about 37. It can be seen that the force increasing effect of the force amplificatory structure is very obvious. This enables the block to be pressed more tightly on the wire wheel, which is beneficial to the fixing of the snake-like arm.

4. Workspace Analysis of snake-like arm

By importing the 3D model into MATLAB and simulating the snake arm with SimMechanics in MATLAB, the workspace of the snake-like arm is generated, as shown in Figure 9. The large workspace enables the snake-like arm to reach the position of trocar in most laparoscopic surgeries to meet the needs of surgery.

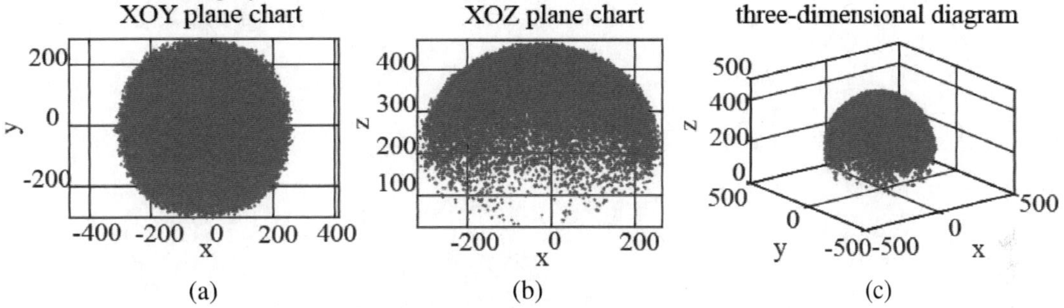

Figure 9. Workspace of snake-like arm

5. Conclusions

Based on the analysis of laparoscopic surgery, a kind of passive snake-like arm based on wire transmission is proposed to support RCM. The main body of the snake-like arm and the locking device of the wire are designed, and the input and output forces of locking wire are analyzed in order to ensure that the locking device can effectively lock the wire and make the snake-like arm become a rigid body to support the RCM. Finally, simulation analysis ensures that the snake arm can meet most of the surgical requirements.

Acknowledgments

The study is financially supported by Science and Technology Commission of Shanghai Municipality under grant No. 17441901200.

References

[1] Hong M B, Jo Y H 2014 Design of a novel 4-DOF wrist-type surgical instrument with enhanced rigidity and dexterity, *IEEE/ASME Transaction on Mechatronics*, 19(2): 500-511.

[2] Ki-YoungKim, Ho-SeokSong, Suk-HoonPark, Jung-JuLee and Yong-SanYoon 2010 Design and evaluation of a teleoperated surgical manipulator with an additional degree of freedom for laparoscopic surgery, *Advanced Robotics*, 24(12), 1695-1718.

[3] Munoz V F, Vara-Thorbeck C, Degabriel J G and Lozano J F 2000 A medical robotic assistant for minimally invasive surgery, *IEEE International Conference on Robotics and Automation*, Proceedings. ICRA. IEEE, Vol.3, pp.2901-2906 vol.3

[4] Zheng Li, Hongliang Ren, Philip Wai Yan Chiu, Ruxu Du and Haoyong Yu 2016 A novel constrained wire-driven flexible mechanism and its kinematic analysis, *Mechanism & Machine Theory*, 95, 59-75.

[5] Kaouk J H, Haber G P, Goel R K, Desai M M, Aron M and Rackley R R, et al 2008 Single-port laparoscopic surgery in urology: initial experience, *Urology*, 71(1), 3-6.

[6] Hockstein N G, Nolan J P, O'Malley B W and Woo Y J 2010 Robotic microlaryngeal surgery: a technical feasibility study using the davinci surgical robot and an airway mannequin,

Laryngoscope, 115(5), 780-785.

[7] Kuo C H and Dai J S 2009 Robotics for Minimally Invasive Surgery: A Historical Review from the Perspective of Kinematics. International Symposium on History of Machines and Mechanisms, *Springer Netherlands,* 337-354.

[8] Taylor R H and Stoianovici D 2003 Medical robotics in computer-integrated surgery, *IEEE Transactions on Robotics & Automation*, 19(5), 765-781.

[9] Dong X, Raffles M, Guzman S C, Axinte D and Kell J 2014 Design and analysis of a family of snake arm robots connected by compliant joints, *Mechanism & Machine Theory*, 77(7), 73-91.

[10] Qin P L, Wang Y, Zhong K M and Wang M D 2016 Symmetric beauty: three-steps force amplified clamping device based on lever-toggle mechanism driven by pneumatic muscle, *2016 International Conference on Computer, Mechatronics and Electronic Engineering*.

IOP Conf. Series: Materials Science and Engineering **435** (2018) 012059 doi:10.1088/1757-899X/435/1/012059

Power Saving LED Luminance Fuzzy Control with IOT Network

Yuan Xiaolei[1,*], Yang Chunlai[1], Yin Zhe[1], Yuan Hongbo[2] and Peng Gang[1]

[1]Hebei Electrical Power Research Institute, No. 238, South Tiyu Street, Shijiazhuang, Hebei Province, China
[2]Shijiazhuang Foreign Language Education Group, No. 318, Yucai Street, Shijiazhuang, Hebei Province, China

[*]hbdyyyxl@163.com

Abstract. To lower down illumination power consumption and enhance lighting effects, smart LED controller and luminance sensors were connected to a WIFI based IOT network, fuzzy luminance control was employed on the IOT net. The schematic topology of the IOT net was outlined. The smart LED controller used the dedicated IOT chip ESP 8266 as the main processor, which was connected to the WIFI network and drove LEDs with PWM signals. The luminance fuzzy control strategy was also implemented on ESP8266 in C language. The fuzzy control process was a typical feedback loop, which took the error (difference between the luminance measured value and the luminance set value) and error change as inputs and calculated the output (the PWM increment) according to fuzzy rules. To test system efficiency, step response experiment and power saving experiment were conducted, proving the system to have good control quality and capable of saving more power by means of intelligent control.

1. Introduction

LED has better luminous efficiency and is more power-saving than other light sources. In developed countries and southeast area of China, LEDs are widely used for illumination, but in other parts of China, luminous sources like fluorescent, incandescent lamps are still used in large quantity, pending to be replaced by LEDs.

To test the effect of LED illumination, we developed a WIFI based IOT [1][2][3] (internet of things) network, smart LED controller and illumination sensor were connected to the IOT network, fuzzy luminance control [4][5] was employed on the IOT net to save power and enhance lighting effect. Experiments were conducted which showed the system had very good control quality and power saving efficiency.

Detailed description and analysis of the system is as follows.

2. IOT topology for fuzzy luminance control

The IOT net used for LED control is composed of the main controller, the luminance sensor, and the smart LED controller. The main controller is a Linux router which connects all other IOT devices. The luminance sensor measures the environmental luminance and send it to the LED controller via IOT. The LED controller compares the measured luminance with the user set value, and regulates PWM outputs to drive LEDs using fuzzy logic feedback control. The IOT net can be connected to the

internet and the cloud. Using a smart android phone, users can control the LED luminance locally by Wifi or remotely through internet cloud. All communications are conducted using UDP protocol.
The topology of the IOT network is shown in Figure.1.

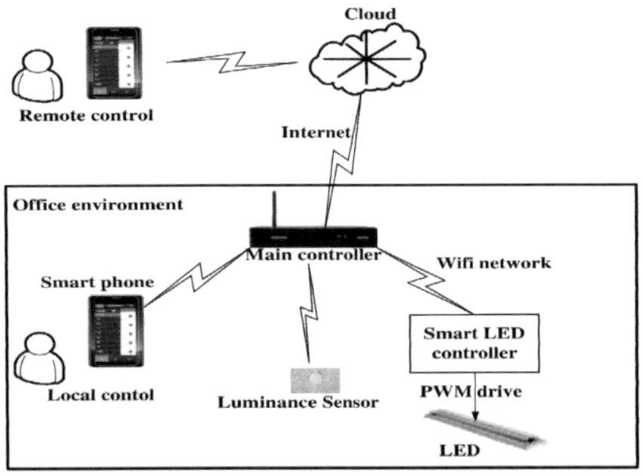

Figure 1. IOT network topology for LED luminance fuzzy control

3. Smart LED controller hardware architecture

The smart LED controller is the centre of the fuzzy luminance control. It uses ESP8266 IOT MCU to communicate via WIFI with other devices (smart phone, luminance sensor) using UDP protocol, implementing the feedback fuzzy control in C language.
The outlining of the smart LED controller hardware structure is shown in Figure.2.

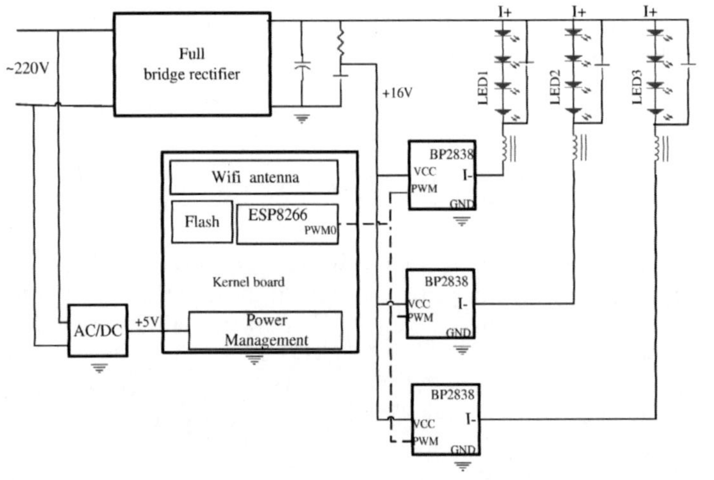

Figure 2. Smart LED controller hardware structure

3.1. Power supply

The full bridge rectifier provides high voltage DC power supply for LEDs, and through a RC-buck, it also provides a 16V DC power supply to the constant current chip BP2838. The AC/DC generates a 5V DC voltage which powers up the kernel board.

3.2. Kernel board

The kernel board consists of the ESP8266 MCU, 4M flash, power management unit and the Wifi antenna. Through Wifi based UDP communication, ESP8266 gets user commands from android phone, and set desired luminance level; it also gets measured environmental luminance value from the luminance sensor, it then implements a feedback fuzzy control strategy to regulate the PWM output, which is fed to the constant current source (BP2838) input, each BP2838 drives a LED, in this way, the luminance is regulated.

3.3. Constant current source

The constant current source (BP2838) gets its input from ESP8266's PWM output and regulates the LED current according to the PWM duty cycle. The luminance can be adjusted from 0% to 100% continuously. The source regulates current in a feedback way, the current precision can be as high as 2%.

4. Luminance fuzzy control algorithm

4.1. Control flowchart

The luminance fuzzy control algorithm flowchart is shown in Figure 3.

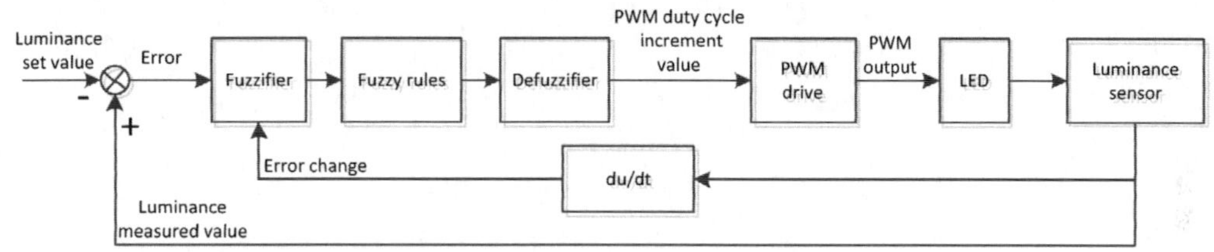

Figure 3. Fuzzy control algorithm flowchart.

The control process is a typical feedback loop [6][7]. The error and error change between luminance set value and measured value are two inputs of the fuzzy system, then fuzzy rules are applied to generate the output-the PWM duty cycle increment value. The current PWM duty cycle value is added to the increment value, the new duty cycle value is applied to drive the LED and regulates luminance. The system is a network-based control system, since the inputs (set value and measured value) are from the IOT net.

4.2. Fuzzy system

The fuzzy system has two inputs: error and error change, one output: PWM duty cycle increment. The membership functions of these 3 variables are shown in Figure 4 to Figure 6.

In these figures, terms NL, NB, NM, NS, ZO, PS, PM, PB, PL mean negative large, negative big, negative middle, negative small, zero, positive small, positive middle, positive big, positive large respectively.

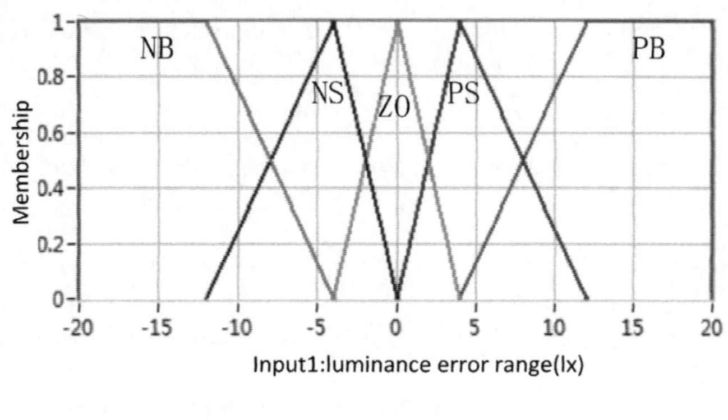

Figure 4. Luminance error membership function

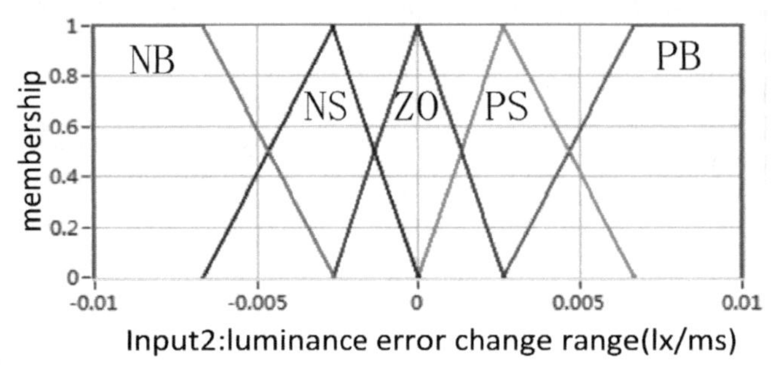

Figure 5. Luminance error change membership function

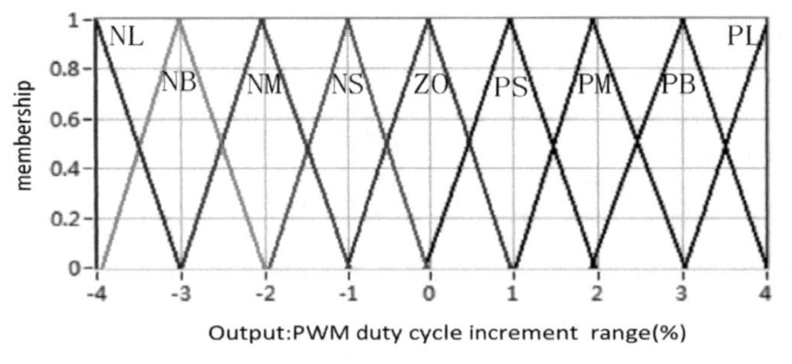

Figure 6. PWM duty cycle increment membership function

Based on these 3 membership functions, fuzzy rules are defined in Table.1

Table 1. Fuzzy rules table.

	EC:NB	**EC:NS**	**EC:ZO**	**EC:PS**	**EC:PB**
E:NB	DCI:PL	DCI:PB	DCI:PM	DCI:PS	DCI:ZO
E:NS	DCI:PB	DCI:PB	DCI:PM	DCI:PS	DCI:ZO
E:ZO	DCI:PM	DCI:PS	DCI:ZO	DCI:NS	DCI:NM
E:PS	DCI:PB	DCI:PB	DCI:PM	DCI:NS	DCI:ZO
E:PB	DCI:ZO	DCI:NS	DCI:NM	DCI:NB	DCI:NL

In Table 1, EC stands for error change, E stands for error, DCI stands for PWM duty cycle increment.

4.3. Algorithm realization
The algorithm is realized in C language in ESP8266 IOT MCU, it is a discrete control process and executes every 50 ms.

5. Experiments and analysis
After the establishment of the system, 2 experiments were conducted: step response experiment and power saving efficiency experiment. The former is to test the control quality, the latter is to test the power saving efficiency.

5.1. Step response experiment
Initial conditions: The PWM ran with a duty cycle of 85%, the luminance set value was 300 lx, the luminance measured value was 300lx.
Step response: Set value was changed to 320 lx. Within 500 ms, the measured luminance followed to 320 lx and stayed stable, the PWM duty cycle increased to 89% and also stayed stable. The transition

process was recorded and drawn in Figure 7. It can be seen there is almost no overshoot during the transition process, the transition was smooth, and the control quality was good.

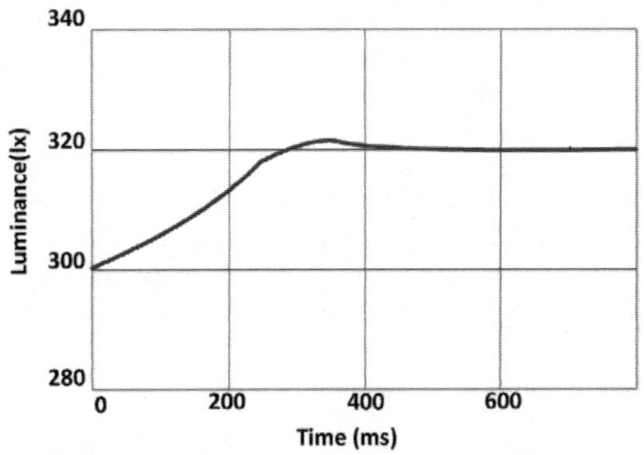

Figure 7. Luminance step response in fuzzy control

5.2. Power saving efficiency experiment

To test the power saving efficiency of the luminance fuzzy control, three rooms in our office building with similar natural illumination situations were prepared: in Room-1 four common 36W T8 fluorescent tubes without luminance control were installed; in Room-2 four common 18W LED T8 tubes without luminance control were installed; in Room3 four 18W LED T8 tubes controlled by above mentioned IOT network were installed. All light sources were switched on from 6:00 in the morning to 20:00 in the evening. The experiments lasted for 5 consecutive days. Total consumption of electricity and average luminance in each room were measured and recorded, the results were summarized in Table 2.

Table 2. Power saving efficiency experiment results.

	Room1	**Room2**	**Room3**
light sources	Fluorescent 4×36W	LED 4×18W	LED 4×18W
Experiment condition	Lights on from 6:00 to 20:00 each day. lasted for 5 consecutive days	Lights on from 6:00 to 20:00 each day. lasted for 5 consecutive days	Lights on from 6:00 to 20:00 each day. lasted for 5 consecutive days
Measured total Electricity consumption (kWh)	13.5	6.8	6.2
Luminance control	-	-	Fuzzy IOT control (set value: 320lx)
Measure maximum luminance(lx)	313	378	317
Measure minimum luminance(lx)	272	308	322
Measured average luminance (lx)	291	345	319

According to Chinese national standard [8] for office illumination, the luminance level should be above 300 lx in reading and writing circumstances. The following is our analysis of data in Table 2:

1) In Room1, fluorescent light sources could not provide required luminance, the average luminance was 291lx, below the national standard, and due to no control of luminance, the luminance level varied significantly from 272lx to 313lx, the power consumption was also the highest among the three rooms.
2) In Room2, LED sources were used but without luminance control. The power consumption was lowered by almost 50% compared with Room1, and the average luminance level raised greatly, exceeding the national standard by 15%, and it was too bright during noon hours. So if there was

no luminance control, it seemed to be a waste of energy. And the luminance level varied greatly in a day, it was not comfortable for users.

3) In Room3, with fuzzy luminance control, we got desired results. The luminance set value was set just a little above the national standard, and the measured luminance value followed the set value perfectly. The average luminance level was 319lx, and the variation of luminance was within 6lx. The illumination of the room was stable and users felt comfortable.

6. Conclusion

Intelligent IOT based LED fuzzy control can effectively maintain stable luminance level in office environment, saving power and giving users comfortable experiences. Users can control light sources locally and remotely, different scenario modes can be applied. The system build a basis for further cloud-based big data mining to optimize power saving strategy. If applied in large scale, energy saving effects can be enormous.

References

[1] He C L 2018 Design and realization of a greenhouse temperature intelligent control system based on NB-IoT *Journal of South China Agricultural University* **2** 117

[2] Sun Erjie, Sun Bingying, Shen Ruibao, Li Wei and Shi Zhen 2015 The underground garage lighting navigation system based on the internet of things *Microcontroller & embedded systems* **8** 15

[3] Sean Dieter Tebje Kelly, Nagender Kumar Suryadevara, Subhas Chandra Mukhopadhyay 2013 Towards the implementation of IoT for environmental condition monitoring in homes *IEEE Sensors Journal* **13** 3846

[4] Jia Yunhui 2017 Fire alarm system based on fuzzy control *Foreign Electronic Measurement Technology* **11** 94

[5] Cai Jiaming 2017 Temperature of Automobile air-conditioning Fuzzy Control Simulation *Electronic Test* **23** 42

[6] Lu X 2010 Fuzzy Control Technology Application and Research on Multi-Position Carrier-Based on Fuzzy-Control software *Journal of Agricultural Mechanization Research* **10** 164

[7] Wang Ruilan 2009 The Design of Industrial Temperature Controller Based on Fuzzy-control *Control & Automation* **13** 45

[8] Zhao J P 2013 *Standard for Lighting Design of Buildings GB-50034-2013* (Beijing: China Architecture & Building Press)

IOP Conf. Series: Materials Science and Engineering **435** (2018) 012060 doi:10.1088/1757-899X/435/1/012060

Modeling and Maneuverability Simulation for Vertical Plane of Autonomous Underwater Vehicle in Current

Yang Kun[1,2], Sui Haichen[1,2], An Li[3*], Jiang Yanqing[3], Wang Youkang[3], Sun Yeyi[3] and Li Ye[3]

[1] Tianjin Research Institute for Water Transport Engineering, M.O.T., Tianjin 300456, China
[2] Tianjin Key Laboratory of Surveying and Mapping for Water Transport Engineering, Tianjin Survey and Design Institude for Water Transport Engineering, Tianjin 300456, China
[3] Science and Technology on Underwater Vehicles Laboratory, Harbin Engineering University, 150001, China.

[*]anli0315@hrbeu.edu.cn

Abstract. As the large diving depth Autonomous underwater vehicle (AUV) developed, modeling and maneuverability simulation for AUV in vertical plane under current disturbance are increasingly important. The REMUS model is adopted to build the mathematical model of AUV. The observation data of the current in South China Sea are analysed to build the current model varied with depth, then the relative motion method is adopted to build the model of AUV in current. The simulation system of maneuverability is established with MFC and some simulation experiments including without current and in current are carried out. The results show that the motion of AUV can be well simulated and some suggestions are given for vertical plane control of AUV.

1. Introduction

The large diving depth Autonomous underwater vehicle (AUV), especially the full sea depth AUV, is currently a hot research field. The maneuverability of AUV in vertical plane is the main factor affecting the efficiency of the diving.

The accurate mathematical model is the basis of the maneuverability analysis. The precise dynamic model is convenient for the designer to grasp the motion characteristics of AUV. It also can help understand the law for movement and maneuverability of AUV and improve the stability and robustness of the controller. The models currently used include DTNSRDC (1979) standard submarine equations of motion [1] and REMUS model [2].

The research for maneuverability of AUV in vertical plane concentrates on the mobility without current. Hao discussed the effect of the longitudinal distance between barycenter and buoyant centre and the buoyancy to the maneuverability of the vertical plane based on the analysis of the motion equation of the vertical plane and the response equation of the pitch angle [3]. Liu et al carried out the unpowered diving prediction for "CR-02" AUV, and the accuracy is verified by sea trial [4].

In this paper, the maneuverability of AUV in the current varied with depth is studied.

2. Modelling of the AUV in vertical plane

The research object of this paper is a underactuated AUV with rudders developed by Harbin Engineering University, China. The AUV belongs to a propeller-rudder-drived AUV. The dimensions and specifications of AUV are shown in table 1.

Table 1. Dimensions and specifications of AUV

Length L	5600mm	Diameter of the hull D	1000mm
Weight m	2150kg	Maximum speed	5kn
Cruising speed		3kn	
Gravity center		$\begin{bmatrix} x_g & y_g & z_g \end{bmatrix}' = \begin{bmatrix} 0 & 0 & 0 \end{bmatrix}'$	
Buoyancy center		$\begin{bmatrix} x_b & y_b & z_b \end{bmatrix}' = \begin{bmatrix} -0.094 & 0 & -0.0223 \end{bmatrix}'$	
Inertia moment	$I_{xx} = 5.42 \times 10^2 kg \cdot m^2$	$I_{yy} = 7.58 \times 10^3 kg \cdot m^2$	$I_{zz} = 7.62 \times 10^3 kg \cdot m^2$

2.1. Coordinate system

The inertial coordinate system $\{I\}$ is established with the Earth defined as the origin, and the body-fixed coordinate system $\{B\}$ with origin chosen to coincide with the centre of mass of the AUV, as shown in figure 1.

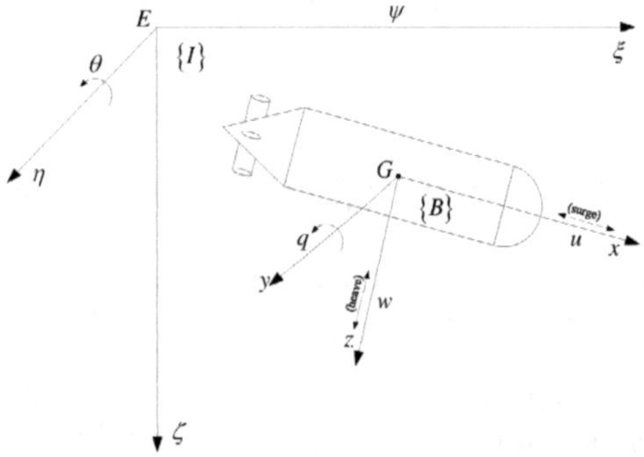

Figure 1. Coordinate system of the AUV in vertical plane

The AUV is equipped with a thruster and a pair of horizontal rudder. They help in producing the vertical plane control input for the three degrees of freedom including surge, heave and pitch.

2.2. Kinematics model

The coordinate transformation between inertial coordinate system and body-fixed coordinate system can be expressed as:

$$\begin{bmatrix} \dot{\xi}_G \\ \dot{\zeta}_G \\ \dot{\theta} \end{bmatrix} = \begin{bmatrix} \cos\theta & \sin\theta & 0 \\ -\sin\theta & \cos\theta & 0 \\ 0 & 0 & 1 \end{bmatrix} \begin{bmatrix} u \\ w \\ q \end{bmatrix} \tag{1}$$

Where, ξ_G, ζ_G and θ are the barycentre coordinates and pitch angle in inertial coordinate system, u , w and q are the linear velocity and angular velocity in body-fixed coordinate system.

2.3. Dynamics model

Based on the rigid momentum theorem, the nonlinear equations of motion for the vertical plane of AUV in matrix form can be expressed as follows [5]:

$$\begin{bmatrix} \dot{u} \\ \dot{w} \\ \dot{q} \end{bmatrix} = \begin{bmatrix} m - X_{\dot{u}} & 0 & mz_g \\ 0 & m - Z_{\dot{w}} & -mx_g - Z_{\dot{q}} \\ mz_g & -mx_g - M_{\dot{w}} & I_{yy} - M_{\dot{q}} \end{bmatrix}^{-1} \begin{bmatrix} \sum X \\ \sum Z \\ \sum M \end{bmatrix} \tag{2}$$

The force X, Z and M for the degrees of freedom surge, heave and pitch can be decomposed according to the source of force, including the forces and moments of hydrostatics, hydrodynamics, thruster and rudders in each direction. So they can be expressed as:

$$\begin{cases} \sum X = X_{u|u|}u|u| + (X_{wq} - m)wq + (X_{qq} + mx_g)q^2 + +X_{Prop} \text{-} (W - B)\sin\theta + X_{rudder} \\ \sum Z = Z_{w|w|}w|w| + Z_{q|q|}q|q| + (Z_{uq} + m)uq + Z_{uw}uw + mz_g q^2 + (W - B)\cos\theta + Z_{rudder} \\ \sum M = M_{w|w|}w|w| + M_{q|q|}q|q| + (M_{uq} - mx_g)uq + mz_g wq) + M_{uw}uw \\ \quad \text{-}(z_g W - z_b B)\sin\theta \text{-} (x_g W - x_b B)\cos\theta + M_{rudder} \end{cases} \tag{3}$$

Where, $X_{u|u|}, X_{wq}, X_{qq}, Z_{w|w|}, Z_{q|q|}, Z_{uq}, Z_{uw}, M_{w|w|}, M_{q|q|}, M_{uq}$ and M_{uw} denote the Hydrodynamic Coefficients of AUV, W denotes the weight of vehicle, B is the buoyancy.

The most hydrodynamic coefficients of AUV can be determined by the Planar Motion Mechanism (PMM) tests [6]. The dimensionless hydrodynamic coefficients obtained from PMM tests are shown in table 2. Then remaining hydrodynamic coefficients can be calculated from the empirical formula.

Table 2. The dimensionless hydrodynamic coefficients of AUV

1	$X'_{\dot{u}} = -1.58\text{e-}3$	5	$Z'_{w\|w\|}$=-1.06e-1	9	$M'_{w\|w\|}$=-2.63e-4
2	$X'_{u\|u\|}$=-6.86e-3	6	$Z'_{q\|q\|}$=-8.02e-3	10	$M'_{q\|q\|}$=-1.00e-5
3	$Z'_{\dot{w}} = -7.63\text{e-}2$	7	$M'_{\dot{q}} = -2.05\text{e-}3$		
4	$Z'_{\dot{q}} = -1.19\text{e-}2$	8	$M'_{\dot{w}}$ = -3.80e-3		

The output force of the thruster $\bullet_{Prop}$ can be obtained by interpolating the thrust curve.

The force and moment produced by rudder $\bullet_{rudder}$ can be obtained by calculating the force of lift and resistance [7].

3. Modeling in the current for the vertical plane

3.1 The current model in the vertical plane

According to [8], observation results of the current in the South China Sea indicates that the current velocity is stable for a period of time and varies along depth direction. In this paper, it is assumed that the velocity of the current increases evenly between 100m and 400m, the average velocity of the current at 100m is 1kn, the current velocity increases 0.025cm/s with the depth of 1m and the current direction is positive along the $E\xi$ axis.

Based on the previous analysis, the ocean current model varied with depth can be expressed as:

$$V_c = 0.514 + 0.00025 \times (d - 100) \tag{4}$$

Where, d denotes the depth from the surface of water.

3.2 Modelling in the current

It is assumed that the current velocity in the inertial coordinate system $\{I\}$ and body-fixed coordinate system $\{B\}$, shown in figure 2, can be expressed as $\mathbf{U_C} = (U_\xi \quad U_\zeta) = (U_\xi \quad 0)^T$ and $\mathbf{U_{CM}} = (u_c \quad w_c)$.

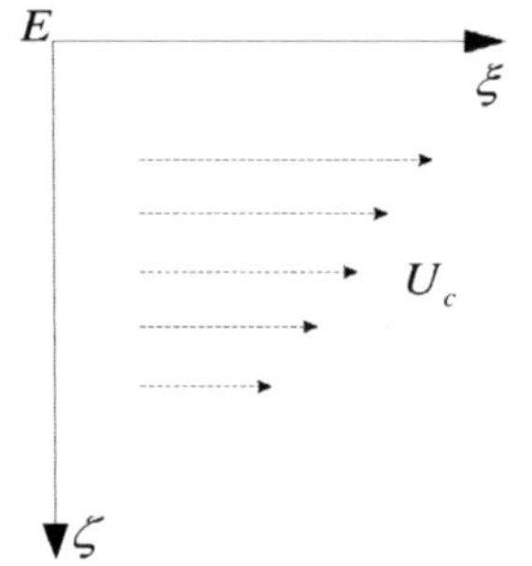

Figure 2. current in the vertical plane

So, the coordinate transformation can be expressed as:

$$\mathbf{U_{CM}} = \mathbf{R}^{-1}\mathbf{U_C} = \begin{bmatrix} \cos\theta & \sin\theta \\ -\sin\theta & \cos\theta \end{bmatrix}^{-1} \mathbf{U_C} \tag{5}$$

The current acceleration in the body-fixed coordinate system $\{B\}$ can be expressed as:

$$\dot{\mathbf{U}}_{\mathbf{CM}} = (\dot{\mathbf{R}})^{-1}\mathbf{U_C} + (\mathbf{R})^{-1}\dot{\mathbf{U}}_{\mathbf{C}} \tag{6}$$

Then, the relative velocity and acceleration can be expressed as:

$$\begin{bmatrix} u_r \\ w_r \end{bmatrix} = \begin{bmatrix} u - u_c \\ w - w_c \end{bmatrix}, \begin{bmatrix} \dot{u}_r \\ \dot{w}_r \end{bmatrix} = \begin{bmatrix} \dot{u} - \dot{u}_c \\ \dot{w} - \dot{w}_c \end{bmatrix} \tag{7}$$

Assuming that the hydrodynamic coefficients do not change, the relative motion method is adopted. Replacing $u, w, \dot{u}$ and $\dot{w}$ in formula (2) with $u_r, w_r, \dot{u}_r$ and $\dot{w}_r$, we can get the mathematical model of the AUV in vertical plane under the current disturbance .

4. The simulation system design and experiment

4.1 The simulation system design

The simulation system of maneuverability is established with MFC based on the mathematical model of the AUV. It contains five modules, which are the initialization module, the hydrodynamic coefficient module, the force module, the acceleration module and the state module. The calculation process is shown in figure 3.

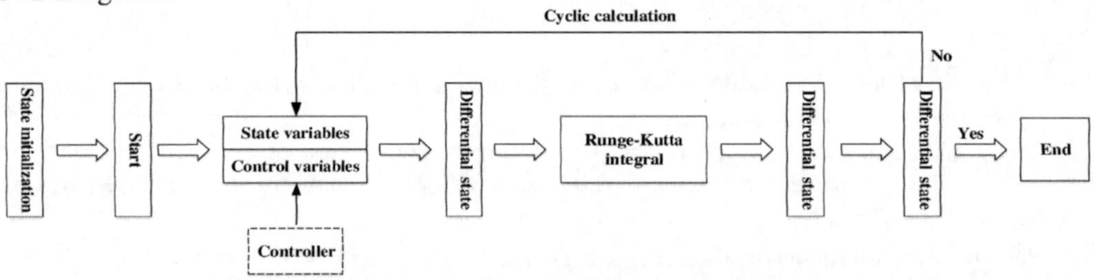

Figure 3. Flow chart of the simulation system

4.2. Simulation experiment

4.2.1. No current. After several simulation experiments, the thruster input voltage needed is obtained when the speed is stabilized at 1-5kn. The thruster voltage is set to a specific value, when the speed is stable, AUV will dive after the rudder is set as a certain angle. δ=10° and δ=15° are selected as the characteristic rudder angle to carry out the simulation experiment. The relationship curve of the maximum pitch angle and rudder angle at each speed is obtained, as shown in figure 4. From this figure, nder the condition of the simulation experiment, the law of floating and diving for AUV of underwater vehicles is basically the same; When the speed is the same, a bigger rudder angle meant bigger the stable pitch angle of the AUV; The maximum pitch angle of the AUV is 31.2° when the speed is 5kn and the rudder angle is 15° . The motion performance of the vertical plane is good.

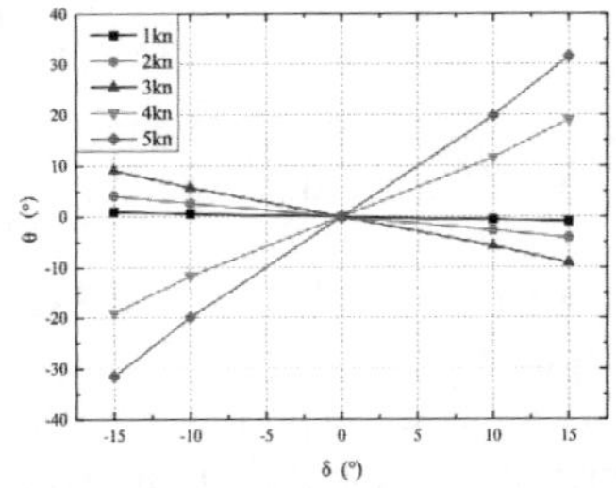

Figure 4. Relationship between the maximum pitch angle and rudder angle

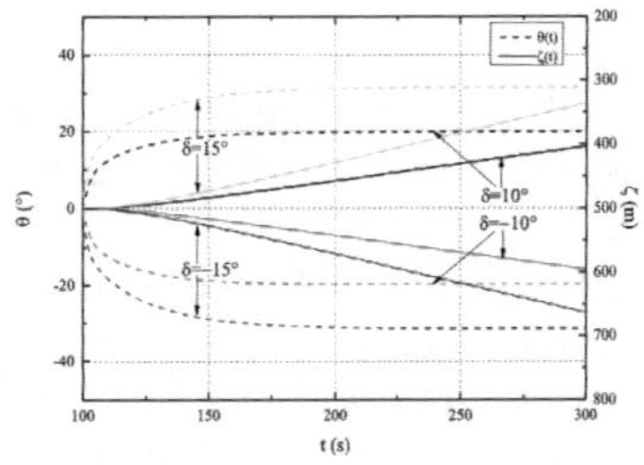

Figure 5. Response for step steering (5kn)

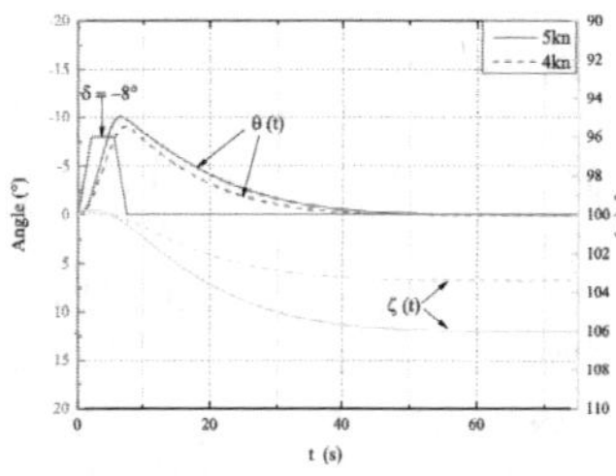

Figure 6. trapezoidal steering motion simulation

The simulation experiment for step steering on the typical stable speed 5kn is carried out. The response curve of pitch angle and depth about the time are obtained when the stable speed is 5kn and the rudder angle is ±10° and ±15°, as shown in figure 5. From the figures, under the typical speed, the AUV moves smoothly and its pitch angle is convergent, and the depth is divergent with time, which is consistent with the actual motion characteristics of AUV. When the rudder angle is same, a higher stable speed meant slower increase of pitch angle.

The trapezoidal steering motion simulation is carried out. The typical stable speed is 4kn and 5kn, rudder angle instruction is $\delta_0 = -8°$, pitch angle instruction is $\theta_e = -7°$ and rudder steering rate is $4° / s$. The response curve is shown in figure 6. From the figure, the execution time is about 4.1s and the transcendental pitch angle is about 3° when the stable speed is 5kn; the execution time is about 5s and the transcendental pitch angle is about 2.1° when the stable speed is 4kn. In this paper, the transcendental pitch angle of the AUV is larger. Enough attention should be paid to the steering in vertical plane.

4.2.2 With current. The simulation for step steering in the current is carried out. The thruster input voltage is 5V. The rudder angle is -10°. The current model uses the model described in 3.1 sections. From the figure 7, the speed of AUV is firstly rapidly increased then slowly increased. It will experience an unstable period at the beginning stage. Attention should be paid when controlling. From the figure 8, the curves of depth and pitch angle about the time in the current coincide with the curves without current, the final pitch angle converges to about 21°. This indicates that the current model built in this paper has little effect on the AUV pitch angle.

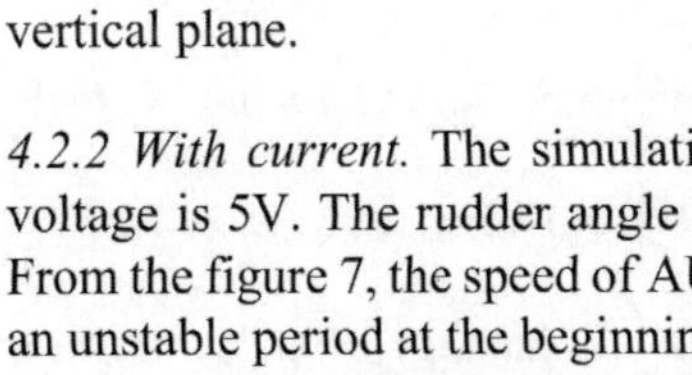

Figure 7. Response of speed in the current

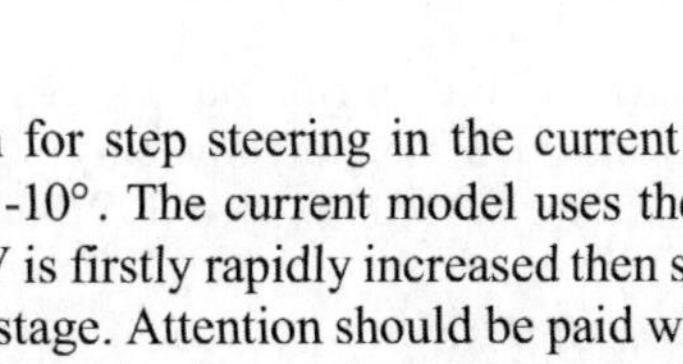

Figure8. Response for step operation of the rudder in the current

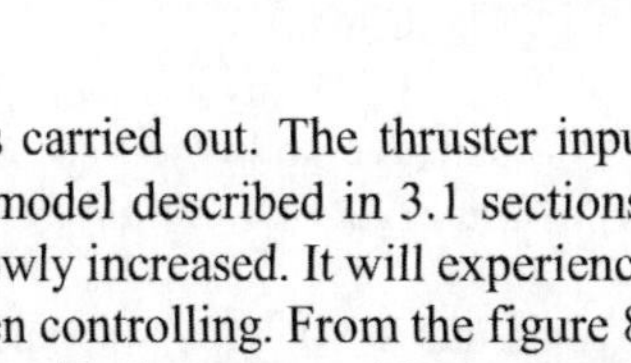

Figure 9. trapezoidal steering motion simulation in the current

IOP Conf. Series: Materials Science and Engineering **435** (2018) 012060 doi:10.1088/1757-899X/435/1/012060

The trapezoidal steering motion simulation in current is carried out. The thruster input voltage is 5V, The rudder angle instruction is $\delta_0 = -8°$. The pitch angle instruction is $\theta_e = -7°$. The response curve is shown in figure 9. From the figure, the response curves of the pitch angle and depth in the current are basically consistent with the curves without current, and the transcendental pitch angle has a slight decrease, and the depth also decreases slightly, which indicates that the current model has little influence on the trapezoidal steering of the AUV.

5. conclusion

In this paper, the model for AUV of the vertical plane in the current is built. And the simulation experiments are carried out including the no current and with current. The following conclusion can be drawn.

1) The simulation system built in this paper can be used to simulate the propeller-rudder-drived AUV motion in vertical plane well. The mathematical model is accurate and reliable.

2) The response of speed in the current varied with depth will experience an unstable period, attention should be paid when controlling. And the current model built in this paper has little effect on the AUV pitch angle.

Acknowledgments

This work was supported by the special fund project of the basic scientific research service of the Central Public Welfare Research Institute (No. TKS170224), "Research on underwater emergency search and detection based on AUV " (No. TKS180406), the National Key R&D Program of China (Grant No. 2017YFC0305700), Qingdao National Laboratory for Marine Science and Technology (No. QNLM2016ORP0406) and supported by the Fundamental Research Funds for the Central Universities (Grant No. HEUCFG201810).

References

[1] Vollmayr A N, Sosnowski S, Urban S, et al. Snookie: An Autonomous Underwater Vehicle with Artificial Lateral-Line System // Flow Sensing in Air and Water 2014 *Springer Berlin Heidelberg* 521-562

[2] Timothy J. Prestero, Development of a Six-Degree of Freedom Simulation Model for the REMUS Autonomous Underwater Vehicle 2001 *MIT/WHOI Joint Program in Oceanographic Engineering*

[3] Hao L, Analysis and research of influencing factors on Autonomous Underwater Vehicle maneuverability in vertical plane 2013 Tianjin University

[4] Liu Z, PREDICTION OF AUV "CR-02" SUBMERGE WITHOUT POWER 2002 *Robot* **24**:385-388

[5] Duan F, Pang S, Comparison of Two Six-Degree of Freedom Simulation Models for Mini Autonomous Underwater Vehicle 2012 *ASME 2012 International Conference on Ocean, Offshore and Arctic Engineering* **2012**:281-287

[6] Gao T, Wang Y, Pang Y, et al, A time-efficient CFD approach for hydrodynamic coefficient determination and model simplification of submarine 2018 *Ocean Engineering* **154**:16-26

[7] Liang X, Li Y, Peng Z, et al, Nonlinear dynamics modeling and performance prediction for underactuated AUV with fins 2016 *Nonlinear Dynamics* **84(1)**:237-249

[8] Yang Q X, Liang X F, Tian J W, et al, OBSERVATION ON THE CURRENT IN THE NORTHERN SOUTH CHINA SEA AND SPECTRUM ANALYSIS 2008 *Oceanologia Et Limnologia Sinica*

IOP Conf. Series: Materials Science and Engineering **435** (2018) 012061 doi:10.1088/1757-899X/435/1/012061

Design and Control of a Two Degree of Freedom Teleoperated Manipulator

Magdi Mohsen[1, *, a], Khalil Ibrahim[2, b] and Abdelfatah MMohamed[3,c]

[1]Egyptiandrillingcompany .
[2]Department of Mechanical Engineering, Faculty of Engineering.
[3]Electrical Engineering, Faculty of Engineering, Assiut University, Assiut, Egypt.

[*]E-mail: [a]eng.magdimohsen@gmail.com; [b]Khalil.ibrahim@aun.edu.eg;
[c]Abdelfatah.mohamed@aun.edu.eg

Abstract. This paper deals with the problem of position-tracking of bilateral teleoperated manipulators ,especially that performs difficult and critical tasks such as under water works usually done using ROV manipulators. A workspace-control system was developed to aid the operator in controlling and driving the slave manipulator .The proposed control technique referred to replacing tradition, master robot that was used for human input commands, with workspace station, consisting of a computer , mouse or joystick as an input device. A wireless UDP network was held between the master and slave manipulator . A PD controller with two different software architecture were applied. We proved experimentally that with parallel executing of two loops in the software architecture, we could achieve total delay time in the system around 90 milliseconds and get a good position tracking of the slave manipulator with system stability even with presence of network delays .

1. Introduction

Teleoperated manipulatoris a robot ora manipulator that is controlled remotely by a human operator. A teleoperation system usually composed of a master station , slave robot, and a communication channel to passing data between the two systems. Slave robot usually located in the task area, which could be unreachable or hazardous area for human operator. Master station or system is placed at the operator site .Teleoperation is utilized in wide variety of applications such as under water tasks, telesurgery, space exploration handling of radioactive sources in reactors [1-3].

As data is transmitted over the delayed network channel between the master and slave subsystems, a time delaysalways causes a serious problem that can destabilize stable system oraffecting the tracking performance of the whole teleoperated system [4-6]. Many researchers proposed different control methods to solve the problem of time delays ,and to enhance the tracking performance. These approaches have been reviewed in [4 , 7]. One of the common control schemes are the PD (proportional derivative) [8]. Also, more advanced approaches are proposed todeal with the uncertainty in teleoperation systems, such as robust control, adaptive control, and other control schemes [9-14].

Most teleoperation works depend on operator's skill. The operator's effort is higher in special works when the slave environment area is hard and had many hazards or with poor knowledge like under water tasks with ROV manipulators. Under water tasks, or tasks that require two degree of motion at same time i.e. linear motion control with constant orientation. The operator usually controls the slave joints, as well as orientation of the end-effector. So, the manipulator's control is not easy, even if the operator is an expert.

This paper presents a computer-based master control system to solve the problem of performing a difficult teleported task, and achieve good position tracking of slave manipulator under delayed networks .The new master system consisting of a computer ,mouse ,or joystick as an input device . The new system enables the operator to control the slave manipulator in workspace not in joint space.In addition, the workspace-control system provides the operator with a virtual simulated graph of slave manipulator. A PD controller with enhanced scheme for a UDP wireless communication channel has been implemented to develop the teleoperation system and the system performance areverifiedby experiments.

2. System descripetion

2.1. The slave manipulator

The slave manipulator which used in implementing the system is a SCARA robot (Selective Compliance Assembly Robot Arm). Many researches proposed to achieve the precise control of SCARA robot[14 ,15].In the experiments we fixed the vertical arm and performed the experiments on only the waist and elbow links (Axis1,Axis2).

2.2. The master device

Not like most of teleoperated systems that have a master manipulator used for generating angles commands and simulating feedback forces on slave environment ,we introduces alternative master system called (workspace control system).This system replaces the master manipulator with a joystick for input commands, and a computer for solving the forward and inverse kinematics of the slave arm . Then the controller calculates motors torque for each joint and send to slave local controller over the wireless network communication channel. The slave controller then controls the slave arm joints motors and returns the posture of the slave arm back to the master computer, which displays the slave arm status graphically.

2.3. Network media and protocol

In our setup we implemented a UDP (User Datagram Protocol) wireless network .Teleoperation system using UDP protocol is a better solution , as it does not require long acknowledgments packets , shorter timeouts , and shorter data frames . This makes UDP protocol faster than TCP/IP. With the choice of preferring fast communication over guaranteed data delivery, the problem of data packets loss appears, whichis considered a common problem in any digital network. Among the three common policies always applied when data packet is lost,we count on "using previous data packet" .

3. Control algorithms

It's known that, the trajectory tracking of the robotic systems is the best method for testing any control algorithm, and for a teleoperated delayed system, it was proven that its stability is highly dependent on system delay time [1].So, a model free PD control algorithm has been applied. Butwith different software and networking architecture, which reducestotal system delay time. These algorithms areshown in figure (1,2). In the first algorithm the PD controller, and the communication processes were in one loop, so the communication performance was affected by the time consumed in the controller calculations. This affects the controller performance and system overhaul delay time. In the second algorithm we made two different loops in the program work simultaneously. One for communication process and the other for whole controller calculations, inverse and forward kinematics, input device interface, and visual displaying .This reflects positively on the controller performance, stability ,and system delay time .All software programs are developed using Labview programming language.

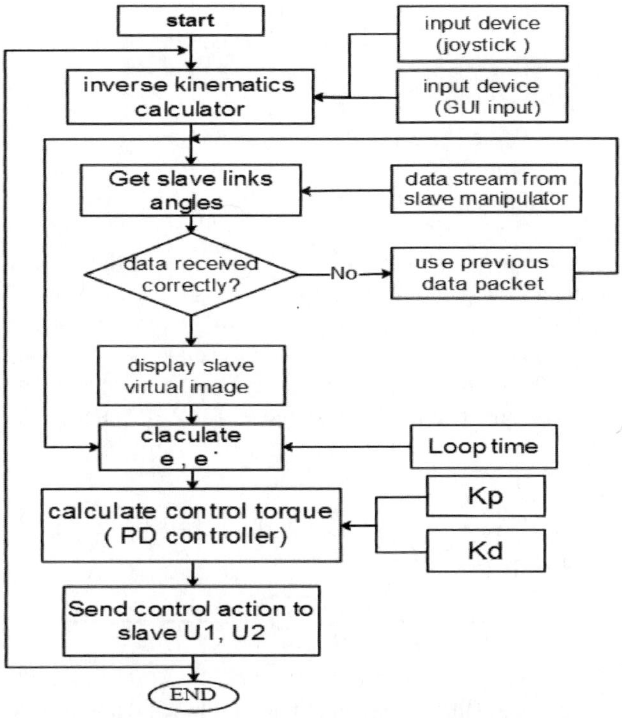

Figure 1. One loop software architecture

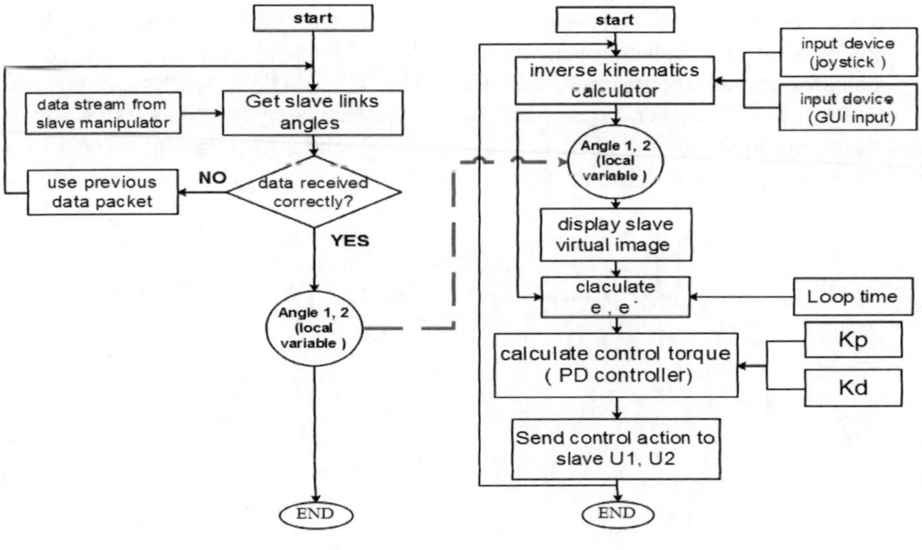

Figure 2. Two loops software architecture

4. The PD controller

The goal of the PD controller is to get a good tracking of slave manipulator with the operator input commands. For the two link used from SCARA robot the following parameters of joints are considered $\theta = [\theta_1\ \theta_2]$, where$\theta_1$ is the angle of joint1 and θ_2 is the angle of joint2. And considering the state variables asθ (t) and$\dot{\theta}$ (t). and the tracking error vectors are e(t) and$\dot{e}$ (t) as:

$$e(t) = \theta(t) - \theta_d(t) \quad (1)$$
$$\dot{e}(t) = \dot{\theta}(t) - \dot{\theta}_d(t) \quad (2)$$

Whereθ_d and$\dot{\theta}_d$ are vectors of the desired joint position and velocity, respectively.
The motors control torque for the PD controller is defined by:

$$U(t) = K_P e(t) + K_d \dot{e}(t) \quad (3)$$

WhereK_P and K_d are the proportional and the derivative gains of the controller, respectively. The block diagram of the bilateral teleoperated PD controller is shown in Figure 3. it is different from traditional PD control block diagram , as we can see intermediate block representing the delay times in the way to and from the slave manipulator (T_{in} ,T_{out}). These delays decrease the values of the constant control gains of the system. These gains should bekept within theiraccepted amount, as if they passes the critical limit the system becomes unstable. These values of gains are usually experimentallydetermined.

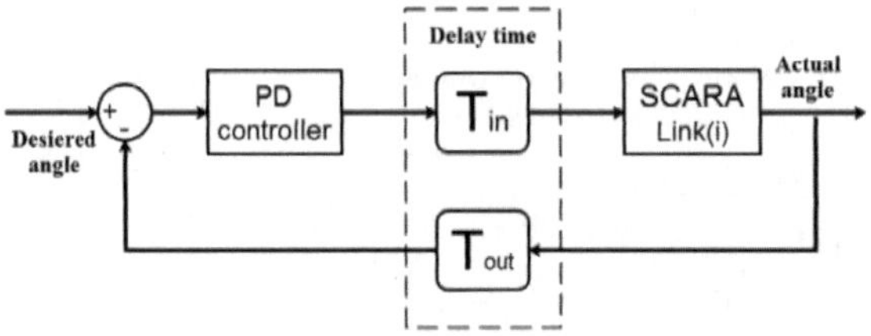

Figure 3. Bilateral telemanipulation block diagram.

5. Experimental results

The experimental results of the PD bilateral control algorithm are shown in this section , the different software and communication schemes performance will be compared, and the total average system delaysare computed. In the experiments, the controllers are tested with a sign wave input, and a step input signals. As mentioned earlier, experiments are performed on only two links, axis1, and axis2. A sinusoidal input trajectory with amplitude of 57.3° or1.0 rad, and a frequency of 0.16cycles/Sec was applied to all control algorithms. The desired trajectory becomes : $[\theta_d = \sin(0.16 * 2\pi t)]$for each joint.

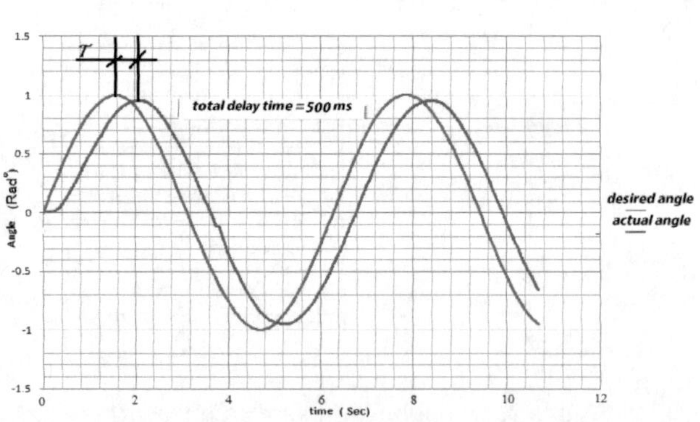

Figure 4. The trajectory tracking of PD controller with first scheme

The trajectory tracking of PD controller with first scheme is shown in Figure 4. In this Figure, The blue line describes the desired trajectory and the red line describes the actual trajectory. The proportional and derivative gains used in the experimentare 2 ,8 respectively. In Figure 4.we can see the lag between input and output signals, or we can called it sent wave and reflected wave . The total delay time between them was nearly 500 milliseconds. As the desired and actual angle measurement logged at master computer side, also assuming equal communication delay time for sending and receiving, so the actual delay time between master and slave time was (total delay time /2 = 250 ms). It is shown that the system is stable and tracking has been achieved with reasonable error and actual delay time round 250 milliseconds. A step input response of 1 rad for the first scheme is shown in Figure5 , shows a total delay time near 400 milliseconds..Figure 6 shows the trajectory of PD controller with second algorithm The blue line describes the desired trajectory and the red line describes the actual trajectory. The proportional and derivative gains used in the experiment are 2 and 8 respectively.

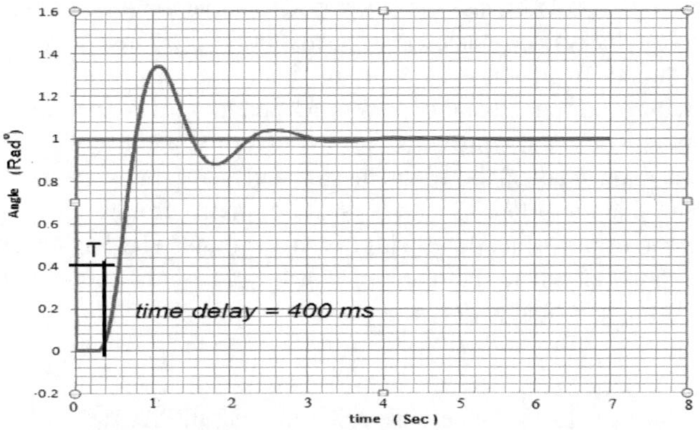

Figure 5. Step response for the first scheme

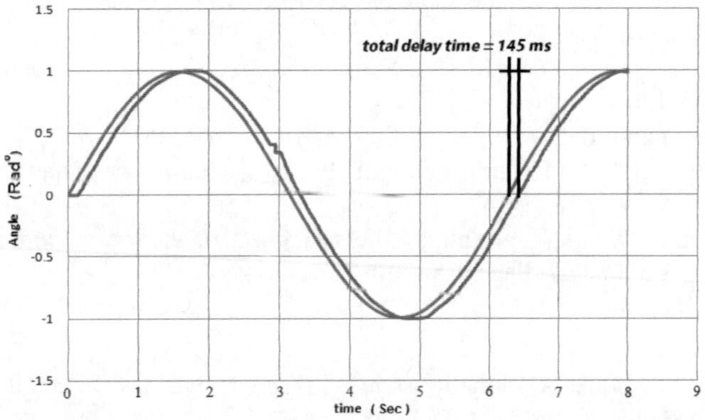

Figure 6. The trajectory of PD controller with second algorithm

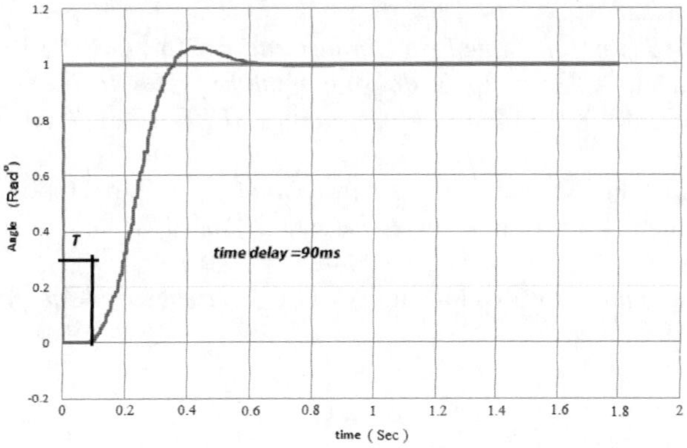

Figure 7. A step response for the second scheme.

As seen in figure 6 the total delay time was nearly 145 milliseconds. So the actual delay time between master and slave time was (total delay time /2 = 73ms) . It is shown that the system is stable and tracking has been achieved with reasonable error and actual delay time round 73 milliseconds. A step input response test for these scheme is also shown in Figure 7.It shows total delay time near 90ms .

6. Conclusion

A workspace control technique and PD controller with two parallel loops software architecture for simultaneous execution of the program tasks is developed to solve the problem of remote manipulation of critical and difficult teleoperated tasks that needs high precision and tracking of the slave manipulator under predefined path with fixed orientation. Thiswas very difficult before with joint space technique even with good trained and experiencedoperators. The developed system has achieved a good tracking characteristics and provided robust stability against network delays . Also we enhanced the communication channel between master and slave performance ,and achieved total delay time in the system round to 90 milliseconds .We used mechatronics laboratory SCARA robot as a testing platform, and the experiments done with only two joints. The ROV manipulation could be a good real industrial application ,where the operator will benefit by the new workspace and computer aided technique in performing critical underwater tasks..

References

[1] Ya-kun Zhang, Hai-yang Li, Rui-xue Huang, Jiang-hui Liu , August *ActaAstronautica*, Volume 137, Pages 312-319
[2] Satja Sivčev, Matija Rossi, Joseph Coleman, Gerard Dooly, Daniel Toal, Control Engineering Practice, Volume 74, May 2018, Pages 153-167
[3] Seung-Ju Lee, Sang-Chul Lee and Hyo-Sung Ahn, *Mechatronics*, Volume 24, Issue 5, August 2014, Pages 395-406
[4] Peter F. Hokayem and MarkW. Spong (2006, *Automatica* 42 2035 – 2057
[5] Da Sun, FazelNaghdy and Haiping Du , February , *Control Engineering Practice*, Volume 47, Pages 15-27
[6] Hongbing Li and Kenji Kawashima , February 2016, *Robotics and Computer-Integrated Manufacturing*, Volume 37, Pages 188-196
[7] Da Sun, FazelNaghdy and Haiping Du, *Annual Reviews in Control*, Volume 38, Issue 1, Pages 12-31
[8] Nuno, E., et al., 2008 A globally stable PD controller for bilateral teleoperators. *IEEE Transactions onRobotics,*. 24(3): p. 753-758.
[9] Yang, Y., C. Hua and X. Guan 2014,*Robotics and Computer-Integrated Manufacturing*, 2014. 30(2): p.180-188.
[10] Yang, Y., et al.2015, *Journal of the Franklin Institute*. 352(5): p. 1850-1866.
[11] Yang, Y., C. Hua, and X. Guan, 2016 , *Journal of the Franklin Institute*,
[12] Hua, C., Y. Yang and P.X. Liu,2015 , *Mechatronics, IEEE/ASME Transactions*. 20(5): p. 2009-2020.
[13] Hua, C.-C., Y. Yang and X. Guan,2013 ,*Neurocomputing*. 113: p. 204-212
[14] Claudio Urrea, Juan Cortés and José Pascal, 2016, *Journal of Applied Research and Technology*, Volume 14, Issue 6, Pages 396-404
[15] Niphun Surapong and Chowarit Mitsantisuk, 2016, *Procedia Computer Science*, Volume 86, Pages 116-119

IOP Conf. Series: Materials Science and Engineering **435** (2018) 012062 doi:10.1088/1757-899X/435/1/012062

Research on the Air Combat Countermeasure Generation of Fighter Mid-range Turn

Huiming JI, Minjian Yu and Jiayi YANG
Air Force Engineering University, Xi'an, China

*Email:jhm320826@163.com

Abstract. When a fighter implements air combat, the pilot maneuvers according to the acquired air situation and the guidance password of the ground commander, and combined with training experience. In order to make the pilot's mid-range turning maneuver more accurate and effective, this paper makes an in-depth study on the generation of countermeasures in air combat of fighter mid-range turning. Firstly, the influencing factors of the mid-range turning maneuvering countermeasures are analyzed, and the boundary function of the maximum attacking zone of the air-to-air missile and the fighter's comprehensive superiority function are obtained; Then, based on the main maneuvering action of the fighter, a medium-range turning maneuvering vector model is established, and the medium-range turning maneuver countermeasures of our aircraft are calculated; Finally, through simulation calculations, the validity of this model is analyzed and verified, which is of great significance for our fighter to constitute combat advantages and improve combat effectiveness.

1. Introduction

Generalized free air combat refers to that in air confrontation, both sides in air combat give full play to the limit performance of aircraft in three-dimensional space, reasonably use all kinds of airborne weapons, and flexibly and freely use tactical countermeasures according to the mastered air situation and combined with the guidance countermeasures of ground command post, so as to achieve attack conditions and attack. At present, great achievements have been made in the research of air combat maneuvering countermeasures at home and abroad. Main solving methods are: influence graph method [1], matrix game method [2], and artificial intelligence algorithm based on "IF-THEN" rule, risk decision [3], genetic algorithm [4], tactical immune maneuver system [5], etc. These methods mainly provide auxiliary decision-making for real-time confrontation in medium-short range free air combat, but seldom study the generation problem of turning maneuver countermeasures in medium-range air combat. In this paper, based on the boundary function of air-to-air missile attack zone and the superiority function of air combat situation, the medium-range turning maneuver vector model in free air combat is established, and the auxiliary guidance countermeasures are solved and generated, which can effectively solve the generation problem of medium-range turning maneuver countermeasures.

2. Analysis on influencing factors of middle-range turning maneuver countermeasures

In free air combat, the pilot will maneuver the fighter plane according to preset methods of warfare, combined with air combat situation and ground command and guidance strategy. The main factors that influence the generation of maneuver countermeasures of middle-range turning in our aircraft are the maximum attack area of missiles of IFF and the current situation of our aircraft.

2.1 Missile attack zone

The missile attack zone, also known as the kill envelope, refers to the space area that may hit the target determined by the missile performance under certain attack conditions, and it is possible to hit the target only when the missile is launched in the attack zone [6]. Document [7]describes the boundary function of the maximum attack zone of air-to-air missiles as follows:

$$D_{M\max} = f\left(h, h_m, v, v_m, n_y, q\right) \tag{1}$$

Where: $D_{M\max}$ is the boundary of the missile's maximum attack zone, h is the altitude of the aircraft and h_m is the target altitude; v is the carrier speed and v_m target speed; n_y is the target maneuver overload; q is the entry angle.

In order to simplify the middle-range turning maneuver model, the boundary of the missile's maximum attack zone can be regarded as an ellipse when it is expressed schematically, and it can be solved by polar coordinate equation: $L = L_{(\theta)}$ Wherein, θ represents the angle with the aircraft heading, and defines the right hemisphere as positive and the left hemisphere as negative; L is the boundary distance of the missile attack zone when the angle between the aircraft heading is θ, as shown in figure 1.

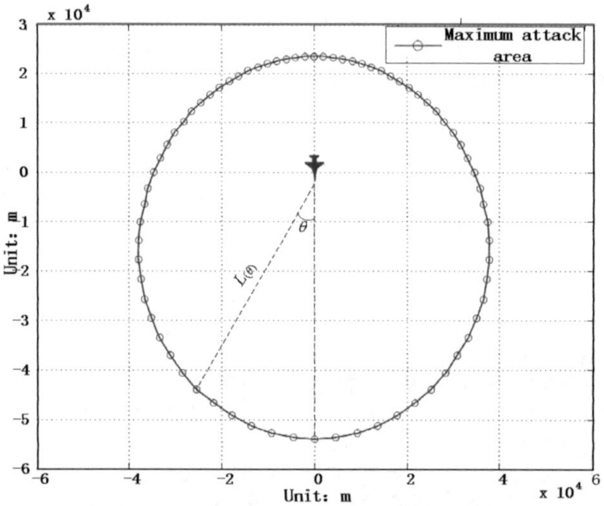

Figure 1. Boundary of missile maximum attack zone

2.2 Aircraft dominance function

The purpose of air combat countermeasures is to command fighter planes to fly according to predetermined altitude, course and speed, and the enemy constitutes favorable air situation and occupies favorable tactical position. Mathematically, according to the current air combat situation, it is our greatest air combat advantage to construct the corresponding aircraft superiority function and choose the corresponding aircraft maneuver. Literature [8]gives the angle, distance, speed and altitude superiority functions and weights them to get the air combat superiority comprehensive function:

$$S = \omega_a S_a + \omega_r S_r + \omega_v S_v \tag{2}$$

Where: ω_a, ω_r, ω_v is the weight coefficient corresponding to the aircraft angle superiority function S_a, distance superiority function S_r and speed superiority function S_v respectively, and $\sum \omega_i = 1$ $(i = a, r, v)$. In different air combat situations, the importance of each parameter of the aircraft superiority function is different, and the calculation formula of its weight coefficient [9] is as follows:

IOP Conf. Series: Materials Science and Engineering **435** (2018) 012062 doi:10.1088/1757-899X/435/1/012062

$$\omega_i = \omega_{i0} + k_i e^{-(\frac{D}{1.5D_{Mmax}})^2} \tag{3}$$

Where: ω_{i0} is the initial weight coefficient value of each parameter, and k_i is the adjustment coefficient of the weight coefficient, and $\sum k_i = 0, (i = a, r, v)$.

3. Medium range turning maneuver vector model

When modeling and studying the mid-range turning maneuver of fighter in free air combat, this paper, aiming at the difficulty of establishing aircraft coordinates in air combat, establishes a vector model, and expresses the relative relationship between enemy and aircraft through vector equations, which can optimize the computational complexity and improve the effectiveness of aircraft maneuver.

3.1 Introduction to vector model

In the agreed plane rectangular coordinate system, the situation between enemy and enemy aircraft can be measured by the relative distance and azimuth of the two aircraft. According to the rectangular coordinates and current heading of the two aircraft, through Euler equation transformation, the vector description of the IFF can be expressed by complex expressions, and then substituted into the vector equation, and the corresponding vector model can be established, thus solving unknown variables.

In this article, the units of physical quantities used are unified as international standard units. The basic units of distance, time and angle are meters, seconds and radians respectively.

Suppose the distance from A to B is S, and direction is K, then there are: $\overrightarrow{AB} = S(\cos K + i \sin K)$

Combined Euler formula: $e^{iK} = \cos K + i \sin K$, Then: $\overrightarrow{AB} = Se^{iK}$.

3.2 Construction of medium range turning maneuver vector model

At the initial moment, through the comprehensive superiority function of fighter aircraft, we calculate the course, speed and altitude that our aircraft should fly, and then adjust our aircraft to the enemy receiving state. According to the missile attack area range and angle superiority function of the enemy and mine, we calculate the turning time and turning elements in our aircraft, thus generating the guidance countermeasures of middle-range turning maneuver. The specific flow is shown in figure 2:

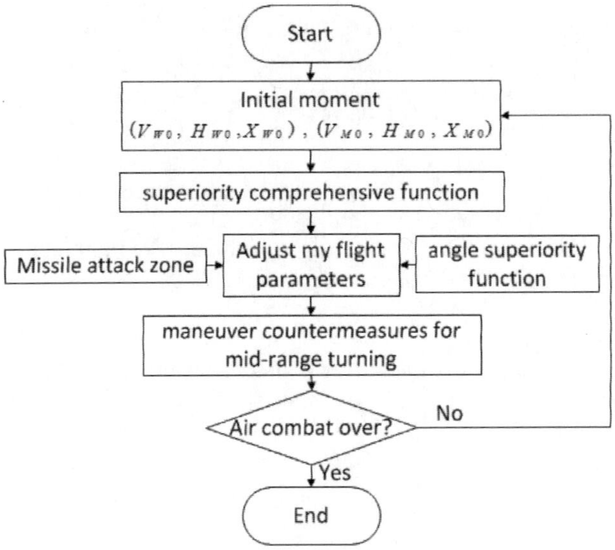

Figure 2. Middle-range turning maneuver countermeasure generation flow

At times t_0, our aircraft and the target aircraft are respectively located at $W_0\,(x_{W0}, y_{W0})$, $M_0\,(x_{M0}, y_{M0})$, V_W、V_M、H_W、H_M、K_W、K_M are respectively the speed, height and course of the enemy aircraft and me. After time Δt, our aircraft has passed W_1 and implemented medium-range turning maneuver to reach W_2, and the target aircraft has reached M_1. At this time, our aircraft is outside the maximum attack zone of the target aircraft missile, while the target aircraft is in mine, so our aircraft occupies an advantage. Launching missiles here or turning into close combat can effectively improve the operational efficiency of our aircraft. As is shown in figure 3.

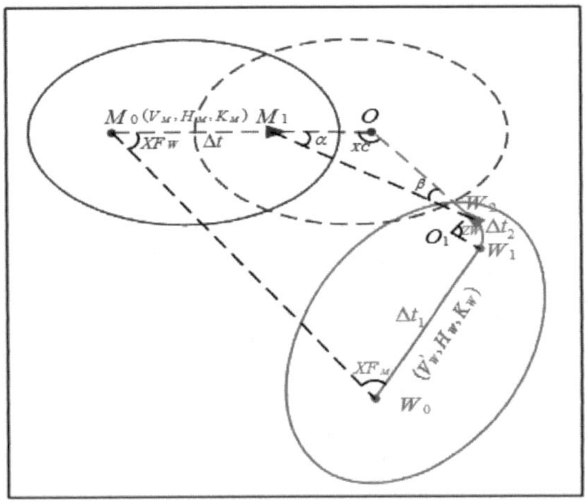

Figure 3. Schematic diagram of pitch turning maneuver

$$\overrightarrow{M_1W_2} = \overrightarrow{M_1M_0} + \overrightarrow{M_0W_0} + \overrightarrow{W_0W_1} + \overrightarrow{W_1O_1} + \overrightarrow{O_1W_2} \tag{4}$$

$$S_1 e^{i(K_M + n_1 \alpha)} = -V_M \cdot \Delta t e^{iK_M} + S_0 e^{iK_{MW}} + V_W' \cdot \Delta t_1 e^{iK_W} + R_1 e^{i(K_W - \frac{\pi}{2} n_1)} + R_1 e^{i(K_{Wjd} + \frac{\pi}{2} n_1)}$$

Where: S_0 refers to the distance between my and the enemy aircraft at the initial moment, n_1 refers to the turning direction, the turning radius is R_1, K_{Wjd} refers to the course of our aircraft connecting with the enemy, and V_W' refers to the speed of our aircraft in the period of turning and laying potential. Suppose the middle distance of our aircraft to turn, and the incoming angle of the enemy is XC.
Set up the time equation:

$$\Delta t = \Delta t_1 + t_{ZW} = \Delta t_1 + \frac{n_1 \cdot (K_{Wjd} - K_W) \cdot R_1}{V_W'} \tag{5}$$

Where: t_{ZW} represents the turning time

In conjunction with (4)(5), we can calculate the timing Δt_1 and the speed V_W' of the aircraft in the middle of our machine. According to the variation of the maneuvering movement of our aircraft, we can get the mid-range turning maneuver countermeasures of our aircraft.

3.3 Generation of maneuver countermeasures for mid-range turning

In air combat, the ground command post can generate the maneuver countermeasures of mid-range turning in real time by using the advantages of strong computing power and the vector model of mid-range turning established above according to the information of IFF received, thus improving the operational efficiency of our aircraft. The maneuver countermeasures of mid-range turning mainly include our aircraft response course K_W, response speed V_W, maneuver, mid-range turning maneuver time Δt_1 and turning to change course.

4. Simulation analysis

In order to verify the validity of this model, simulation is made for an air combat method. At the initial moment, Enemy aircraft and mine is located at $(10Km, 10Km)$ and $(82.75Km, 4Km)$ respectively, and the flight parameters of enemy aircraft and mine are $(8000, 800, 90)$ and $(8000, 1100, 270)$ respectively. The gravity acceleration $g = 10m/s^2$ is specified, and the simulation step size is $3s$. The simulation results are shown in figure 4.

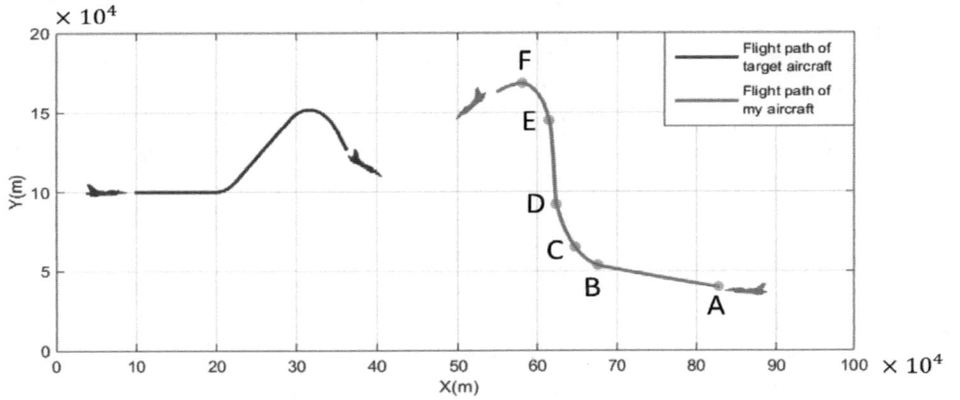

Figure 4. Flight path of our aircraft and target aircraft.

And the table 1 lists the maneuver strategy instructions of our aircraft at six special nodes, as follows:

Table 1. Instructions for maneuver countermeasures of our aircraft at special nodes

Position	Time(s)	Response speed	Response heading	Maneuver action	Maneuver timing(s)	Course of Maneuver to change
A	0	1100	274	accelerate, right turn	83	287
B	45	1000	275	uniform, right turn	42	287
C	57	1000	356	uniform, left turn	38	300
D	69	1000	354	uniform, left turn	27	300
E	87	800	-	slow down, left turn	-	-
F	117	800	262	uniform, left turn	5	255

So it can be preliminarily concluded that the course, speed and mid-range turning maneuver timing of our aircraft obtained by model calculation and simulation are more accurate, which can make our aircraft maneuver more effective, air combat advantage more obvious and pose a threat to the enemy. And it is that which proves the effectiveness of the model.

5. Conclusion

In free air combat, the automatic generation of middle-range turning maneuver countermeasures still plays a very important role. This paper combines the missile maximum attack area with the superiority function of fighter planes, analyzes the factors that affect the middle-range turning maneuver, and establishes the vector model of middle-range turning maneuver. through simulation analysis, it verifies the validity of the model, which is significant to improve the accuracy of middle-range turning maneuver countermeasures in free air combat.

IOP Conf. Series: Materials Science and Engineering **435** (2018) 012062 doi:10.1088/1757-899X/435/1/012062

References

[1] Kirtanen K,Karelahti J and Raivio T 2006 *Journal of Guidance,Control and Dynamics* vol 29 pp 1080-91
[2] Hui Y N Zhu H Y and Shen L C 2009 *Ordnance Industry Automation* vol 28 pp 4-7
[3] Zhou S Y Wu W H Zhang N and Zhang J 2012 *Aeronautical Computing Technique* vol 42 pp 27-31
[4] Guo H Zhou D Y and Zhang K 2010 *Electronics Optics and Control* vol 17 pp 28-32
[5] Krishna Kumar K and Kaneshige J 2003 *Proc. of the 41st Aerospace Sciences Meeting &Exhibit* pp 1-10
[6] Gu J J Liu W H and Jiang W Z 2015 *Systems Engineering and Electronics* vol 37 pp 1306-12
[7] Wang Y H Gao X G and Zhen J 2010 *Fire Control and Command Control* vol 35 pp 32-36
[8] Han Q S and Yu M J 2015 *Journal of Air Force Engineering University* vol 16 pp 12-15
[9] He X Jing X N and Feng C 2017 *Journal of Air Force Engineering University* vol 18 pp 36-41

IOP Conf. Series: Materials Science and Engineering **435** (2018) 012063 doi:10.1088/1757-899X/435/1/012063

Collaborative Consistency Control Algorithm Research of Unmanned Aerial Vehicle Formation

Liu Yong[1,2,*] and Liu Wei[2]

[1]Changchun University of Science and Technology
[2]Unit 92941 of PLA

[*]E-mail: m18829236694@163.com

Abstract. In this paper, the cooperative consistency control algorithm of unmanned aerial vehicle is studied. The formation of unmanned aerial vehicles has the characteristics of controlling uncertainty. This makes the controller have certain uncertainty which may cause the multi-agent system to be difficult to agree. In view of this feature, the intelligent algorithm for collaborative consistency control of unmanned aerial vehicle formation is designed with the properties of the Laplace matrix of the multi-agent network which is composed of a series of intelligent bodies with autonomous capability and the interaction between them. The simulation results verify the feasibility of the intelligent algorithm.

1. Raise of problem

Unmanned aerial vehicle formation flight has a broad application prospect[1-3]. Its consistency and collaborative control technology has always been the focus of research, and it can provide a strong support control algorithm for formation and track planning. In this paper, the collaborative control algorithm is based on multi-agent system. This system is composed of a series of intelligent bodies with autonomous capability and the interaction between them, and it can also be described as a multi-agent network. In this system, a single agent is a node in a network, and the interaction between the intelligent bodies (information interaction) is the connection between network nodes. Intelligent body can be unmanned aerial vehicle, sensor node and so on[4-8].

At present, in the research on the collaborative control of unmanned aerial vehicles, all the cooperative controllers designed by unmanned aerial vehicles can be executed. In practice, the controller has some uncertainty due to hardware (such as A/D, D/A conversion), software (such as the calculation phase error) and other reasons [9]. This makes the controller have certain uncertainty which may cause the multi-agent system to be difficult to agree. Therefore, this paper studies the control problem of unmanned vehicle with control uncertainty and gives the control model of the controller. In order to facilitate the discussion and consider the universality of the model, the intelligent body is replaced by the unmanned aerial vehicle for analysis modeling and simulation calculation.

2. Model building

Considering the following N linear time-invariant smart bodies composed of a multi-agent system, the dynamic model of the intelligent body is described as follows:

IOP Conf. Series: Materials Science and Engineering **435** (2018) 012063 doi:10.1088/1757-899X/435/1/012063

$$\begin{aligned} \dot{x}_i &= Ax_i + Bu_i \\ y_i &= Cx_i \quad i = 1, 2, \ldots, N \end{aligned} \tag{1}$$

Among them, $x_i \in \mathbb{R}^n$, $u_i \in \mathbb{R}^p$, $y_i \in \mathbb{R}^q$ are the state of the ith intelligence, controlling the input and output. A, B and C are a constant matrix of dimensions.

For a given multi-agent system, a system agreement means that for any finite initial value $x_i(0)$, if there is a controller which makes:

$$\lim_{t\to\infty} \left\| x_i(t) - x_j(t) \right\| = 0, \quad \forall i, j = 1, 2, \ldots, N \tag{2}$$

In this case, the communication topology between the intelligent body is described by a directed graph containing a directed spanning tree [10].

Based on the neighbor relative status information, a distributed collaborative controller with uncertainty is constructed:

$$u_i = \sum_{j=1}^{N} a_{ij} \left(K + \Delta K_i \right) \left(x_j - x_i \right), \quad i = 1, 2, \ldots, N \tag{3}$$

Where $K \in \mathbb{R}^{p\times n}$ is the feedback control matrix to be designed, c is the coupling strength parameter to be designed, and a_{ij} is the element of the adjacency matrix $\mathcal{A}$ of communication topological $\mathcal{G}$. ΔK_i represents the uncertainty in the controller of the intelligent body i ,which has the following form:

$$\Delta A_i = DF_i H \tag{4}$$

In this case, the matrix D and H are the constant matrices with the appropriate dimension, which represent the uncertain structure.

$$F_i^T F_i \le \rho_i^2 I \tag{5}$$

Where, $i = 1, 2, \ldots, N$ is the normal number giver.

It can be seen from the above that different intelligent bodies have the same nominal feedback matrix in the multi-agent system, but the controllers with different norm are uncertain. The main research of this paper is to design the appropriate feedback matrix (2) so that the multi-agent system can reach agreement under the controller.

According to the above analysis, the closed-loop dynamic equation of the intelligent body system composed of N linear time-invariant intelligent bodies is as follows:

$$\dot{x} = \left(I_N \otimes A - \left(I_N \otimes B \right) \left(I_N \otimes K + \tilde{K} \right) \left(\mathcal{L} \otimes I_n \right) \right) x \tag{6}$$

Where, $\Delta\tilde{K} = diag\left\{ \Delta K_1, \Delta K_2, \ldots, \Delta K_N \right\}$, $\mathcal{L}$ is the Laplacian matrix of figure $\mathcal{G}$ [11-13]. $x = [x_1^T, x_2^T, \ldots, x_N^T]^T$

According to the literature, there is a matrix $M \in \mathbb{R}^{N\times N-1}$ for the Laplacian matrix $\mathcal{L} \in \mathbb{R}^{N\times N}$ of the graph, which makes $\mathcal{L} = ME$, among which, $E \in \mathbb{R}^{N-1\times N}$ is as follows:

$$E = \begin{bmatrix} 1 & -1 & 0 & \cdots & 0 \\ 0 & 1 & -1 & \cdots & 0 \\ \vdots & \vdots & \ddots & \vdots & \vdots \\ 0 & 0 & 0 & 1 & -1 \end{bmatrix} \tag{7}$$

Further, if the graph contain a directed spanning tree, the matrix M is column full rank, and the eigenvalue of matrix EM is the non-zero eigenvalue of $\mathcal{L}$, $\mathrm{Re}(\lambda(EM))>0$. For matrix EM, if figure G contains a directed spanning tree, there is a symmetric positive definite matrix Q and a positive scalar parameter α, which makes:

$$(EM)^T Q + QEM > \alpha Q \tag{8}$$

Where, $0<\alpha<2\min\{\mathrm{Re}(\lambda(EM))\}$.

Make $\xi=(E\otimes I_n)x$, $\xi_i = x_i - x_{i+1}$, $i=1,2,\ldots,N-1$. The system closed-loop dynamic equation can be reconstituted in the following form.

$$\dot{\xi}=\left(I_{N-1}\otimes A - EM\otimes BK-(E\otimes B)\Delta\tilde{K}(M\otimes I_n)\right)\xi \tag{9}$$

Where, b is the matrix that satisfies c. When $\xi\to 0$, $x_1=x_2=,\ldots,=x_N$, the system agrees.

Therefore, the system consistency and cooperative control problem is transformed into system stability problem. Based on this, the following algorithm is introduced to design the appropriate feedback matrix K to make the system agree. The topology G contains a directed spanning tree. The collaborative controller (2) can be designed as follows:

(1) Given the parameters $M=\mathcal{L}E^T\left(EE^T\right)^{-1}$ and $0<\alpha<2\min\{\mathrm{Re}(\lambda(EM))\}$, the solution can get a feasible solution $Q>0$;

(2) For a given parameter $c>0$, the following linear matrix inequalities are solved to obtain a feasible solution $P>0$.

$$\begin{bmatrix} A^TP+PA-c\alpha PBB^TP & PBD & H^T \\ D^TB^TP & \frac{1}{\beta\rho_0^2\varphi_1}I & 0 \\ H^T & 0 & \frac{\varphi_2}{\phi} \end{bmatrix}<0 \tag{10}$$

Where, $\beta=\max\{\lambda(EE^T)\}$, $\phi=\max\{\lambda(M^TM)\}$, $\varphi_1=\max\{\lambda(Q)\}$, $\varphi_2=\min\{\lambda(Q)\}$, $\rho_0=\max\{\rho_1,\rho_2,\ldots,\rho_N\}$;

(3) The design feedback matrix is $K=cB^TP$.

Assume that the topology diagram contains a directed a spanning tree, for a given parameter $c>0$, linear matrix inequality (5) is true, the design of coordinated controller (2) to agree on the multi-agent system.

3. The simulation result

Simulation and verification of the above models and methods are carried out. Assuming a multi – intelligence system with four intelligent bodies, the matrix of the intelligent body system with controller uncertainties is described as follows.

$$x_i=\begin{bmatrix} x_{i1} \\ x_{i2}\end{bmatrix},\ A=\begin{bmatrix} -3 & -9 \\ 9 & 3\end{bmatrix},\ B=\begin{bmatrix} -1 & 0 \\ 5 & 1\end{bmatrix},\ D=H=\begin{bmatrix} 1 & 0 \\ 0 & 1\end{bmatrix} \tag{11}$$

$F_i, i=1,2,3,4$ is randomly selected in $(0,1]$.

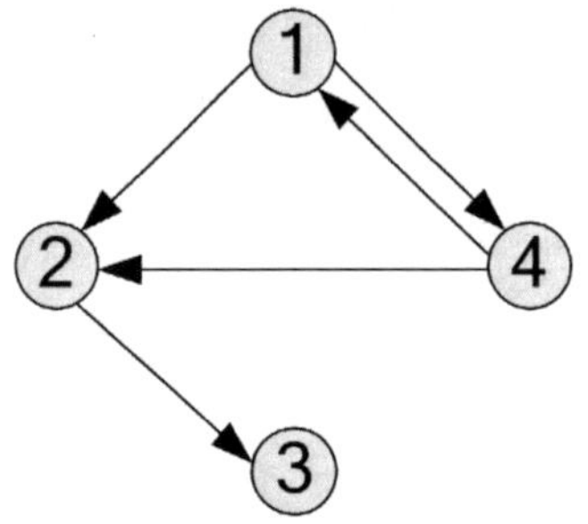

Figure 1. Communication topology.

The communication topology of the intelligent body is shown in figure 1. As can be seen from the figure, the communication topology contains a directed spanning tree. The Laplacian matrix of the topology diagram is described as follows.

$$\mathcal{L} = \begin{bmatrix} 1 & 0 & 0 & -1 \\ -1 & 2 & 0 & -1 \\ 0 & -1 & 1 & 0 \\ -1 & 0 & 0 & 1 \end{bmatrix} \tag{12}$$

You have a column full rank matrix M that makes $\mathcal{L} = ME$ .Where

$$E = \begin{bmatrix} 1 & -1 & 0 & 0 \\ 0 & 1 & -1 & 0 \\ 0 & 0 & 1 & -1 \end{bmatrix}, \; M = \begin{bmatrix} 1 & 1 & 1 \\ -1 & 1 & 1 \\ 0 & -1 & 0 \\ -1 & -1 & -1 \end{bmatrix} \tag{13}$$

So you get $\beta = 3.4142$, $\phi = 7.0340$.

Make α=3.5 , a feasible solution of the linear matrix inequality (3) is

$$Q = \begin{bmatrix} 0.9406 & -0.0006 & -0.4304 \\ -0.0006 & 0.3064 & 0.2037 \\ -0.4304 & 0.2037 & 0.7363 \end{bmatrix} \tag{14}$$

So you get $\varphi_1 = 1.2975$, $\varphi_2 = 0.1762$.

Make c=10， $\rho_0 = 9063$. To solve matrix inequality (5), we can find a feasible solution for P.

$$P = \begin{bmatrix} 7.7553 & 0.1602 \\ -0.1602 & 1.1270 \end{bmatrix} \tag{15}$$

According to the algorithm, the feedback matrix can be designed as follows:

$$K = \begin{bmatrix} -29.9477 & 20.2827 \\ -0.5608 & 3.9444 \end{bmatrix} \tag{16}$$

The state trajectories of each agent are shown in figure 2 and figure 3. The trajectories of consensus error of each agent are shown in figure 4 and figure 5. The simulation results show that the multi-agent system is in aggrement.

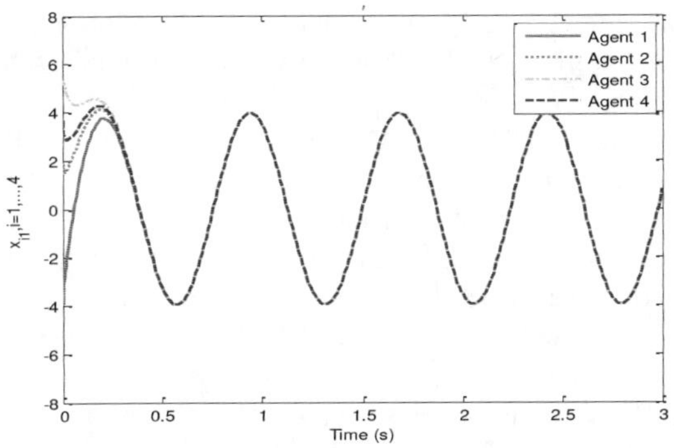

Figure 2. Trajectories of $x_{i1}, i = 1,...,4$

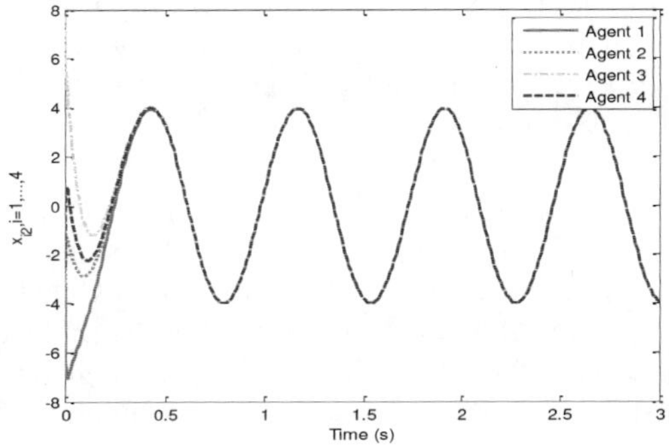

Figure 3. Trajectories of $x_{i2}, i = 1,...,4$

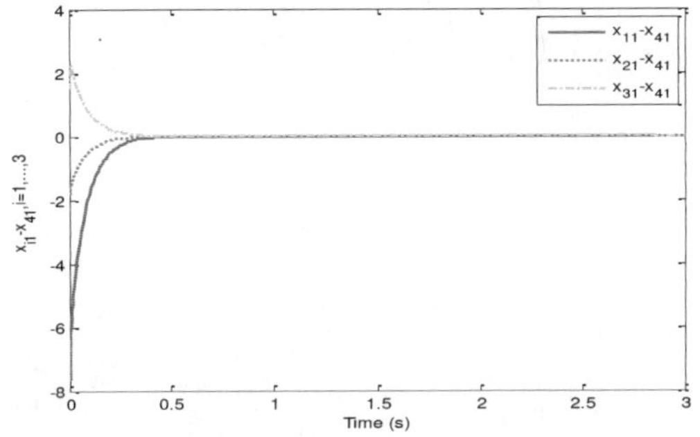

Figure 4. Trajectories of consensus error

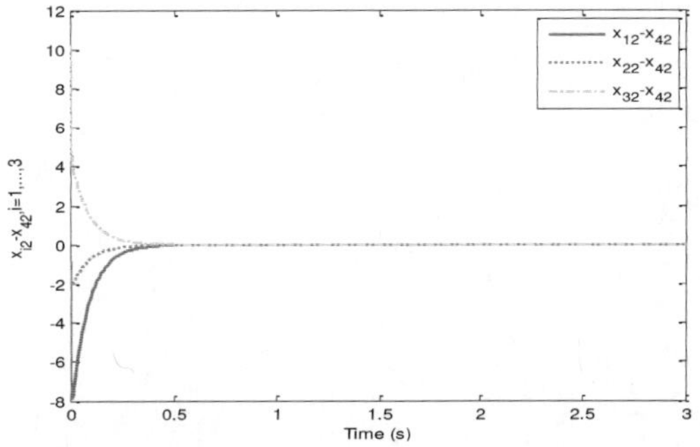

Figure 5. Trajectories of consensus error

4. The conclusion

In this paper, a class of unmanned aerial vehicle formation coordination problem with controller uncertainty is solved by using the special property of topology diagram and linear matrix inequality method. The design algorithm of formation cooperative controller is given, and the simulation results prove the correctness of the design method.

Reference

[1] Wang Xiao-fang, ZHENG Yi-yu, LIN Hai 2014 *J.* Control Method for Missile Formation Flight. Transactions of Beijing Institute of Technology. **34 (12)** 1272-77.

[2] Peng Mingxing, Wu Yunjie, Zhu Futao 2017 *J.* Research on Control Method for Anti-ship Missiles Formation Flight. Journal of System Simulation.,29(1): 212-217.

[3] MEI Quan, YAO Pei-yang. An Overview of Consensus-Based Formation Control for Multi-Agent System. Electronics Optics & Control. **24(1)** 54-57.

[4] CHEN, Wu, ZHANG Xin, JIN Xin, LI Zehong, HONG Liang 2017 *J.* A cooperative information consensus algorithm for multi-agent system. Acta Aeronautica et Astronautica Sinica. **38(12)** 1-13.

[5] SONG Jun-hong, SONG Shen-min, XU Sheng-li 2016 *J.* A Cooperative Guidance Law for Multiple Missiles to Intercept Maneuvering Target. Journal of Astronautics. **37(12)** 1306-14.

[6] Zhang Bo , Luo Jianjun , Yuan Jianping 2010 *J.* A Satellite Formation Cooperative Control Strategy Based on Information Consensus. ACTA AERONAUTICA ET ASTRON AUTICA SINICA. **31(5)** 1004-13.

[7] PENG Chen, LIU Xing, WU Sen-tang, LI Meng 2010 *J.* Consensus problems in distributed cooperative terminal guidance time of multi-missiles. Control and Decision. **25(10)** 1557-66.

IOP Conf. Series: Materials Science and Engineering **435** (2018) 012063 doi:10.1088/1757-899X/435/1/012063

[8] Wang Zhong-Lin, Liu Zhong-Xin, Chen Zeng-qiang, Sun Qing-lin, Wu Yin, Ma Zhen-guang 2014 *J*. A kind of new type controller for multi-agent leader-follower formation. CAAI Transactions on Intelligent Systems. **9(3)** 298-306.
[9] WANG Wu, YANG Fu-Wen. CONTROL FOR LINEAR TIME-DELAY UNC ERTAIN SYSTEMS WITH CONTROLLER GAIN VARIATIONS.
[10] Shaolei Zhou , Wei Liu , Qingpo Wu . Gaoyang Yin 2015. Leaderless Consensus of Linear Multi-agent Systems: Matrix Decomposition Approach . 2015 7th International Conference on Intelligent Human-Machine Systems and Cybernetics (IHMSC).
[11] Horn R A, Johnson C R M 1985 *Matrix analysis* (New York: Cambridge University Press)
[12] Godsil C, Royle G M 2001 *Alrbraic graph theory* (New York: NY: Spriing-Verlag)
[13] Jialing Zhou, Jianying Yang. Zhongkui Li 2017 *J*. Simultaneous attack of a stationary target using multiple missiles: a consensus-based approach. SCIENCE CHINA Information Sciences. **60** 14

A Novel Network Proactive Defense Model: Anomaly Driven Dynamic Cooperative Defense Model

Li Lixun[1,2], Zhang Bin[1,2], Wu Haoming[1,2]and Li Yingbo[1,2]

[1] Zhengzhou Information Science & Technology Institute, ZhengZhou, China
[2] Henan Province Information Security Key Laboratory, ZhengZhou, China

Corresponding author's e-mail address: lilixun_1994@126.com

Abstract. To defend the more complex and stealthy network attacks, an anomaly driven dynamic cooperative defense model (ADCD) is proposed in this paper. ADCD takes anomalies as defense drive factors to identify network attack types and evaluate system security state. And according to the anomaly processing result, ADCD can combine network cooperative defense (NCD) with moving target defense (MTD) to defend known and unknown network attacks. Firstly, detail explanations of ADCD is given, including its elements, operations, relations and rules. Then, a formal analysis of ADCD is conducted based on finite state machine (FSM) to prove its security. Finally, A typical application scenario where ADCD is realized as a mutiagent system is displayed to verify the effectiveness of ADCD in defending complex and multi-step attacks.

Key words. Network security, Proactive defense, Formal model, Intelligent agent, Multiagent system

1. Introduction

With the development of information technology, network security is becoming increasingly important. Network cooperative defense (NCD) refers to conducting targeted network security policy by collaborating security devices to detect, analyze and deal with various network attacks according to the security mission demands predetermined. [1] NCD is very efficient when attacks are detected, but it does not work well if attacks are too stealthy to be detected. To cope with the more complicated and secretive attacks, researchers proposed moving target defense technology (MTD), which aims at increasing attackers' difficulty and cost by reducing information system's definiteness, stability and isomorphism using dynamic mutations. [2,3] Now, lots of researches proving the advantages of MTD in defending attacks undetectable have been put forward, but MTD still faces two challenges in practice:

- How to collaborate NCD and MTD to realize complementary advantages for the better defense effect? Some of new MTD technologies such as the port hopping are based on dynamic mutations to reduce the success probability of attackers but not to protect information systems by eliminating vulnerabilities[4]. While existing NCD technologies such as the security operation center (SOC) and the unified threat management (UTM) are based on the intrusion detection and vulnerability mining to defend attacks[1,5]. These two different defense methods depend on their different foundations, so it is a challenge to create a cooperative mechanism between them.
- How to determine the key parameters such as the mutation object and the mutation period in

MTD according to the security state of information system protected? Most of MTD technologies focus on conducting dynamic mutations on specific system elements such as softwares or operation systems[6], but not to take flexible driven mutations according to system's security state to choose the best mutation parameter. This non-driven mutation mechanism is lack of flexibility and target when faced with complicated and stealthy attacks, and may increase the risk of been attack for the unintentional exposure of system vulnerabilities, as well as increasing the system's running cost[6]. So, considering the balance of security and stability, it is necessary to figure out how to take driven mutations.

As for the first challenge, both MTD and NCD are to defend network attacks, although they have different foundations. And most of network attacks, whether it is known or unknown, would cause anomalies on network traffic or host behavior. Therefore, it is worth exploring to create a new cooperative mechanism between MTD and NCD based on anomaly processing.

To deal with the second challenge, when the cooperative mechanism of MTD and NCD is constructed, the mutation parameters in MTD such as the mutation object and period could be determined by combining the anomaly processing result in NCD with mutation parameter choosing method in MTD.

Therefore, we propose a novel network proactive defense model called anomaly driven dynamic cooperative defense model (ADCD). ADCD takes anomaly as defense drive factor to identify network attack types and evaluate system's security state, and combines NCD with MTD technologies based on the anomaly processing result in the previous step to realize complementary advantages in defending known and unknown network attacks.

2. ADCD

2.1. Concepts

The basic idea of anomaly driven dynamic cooperative defense model (ADCD) is the combination of anomaly driven and policy response. The anomaly driven refers to that the inputs of ADCD are anomalies such as abnormal file operations or abnormal network communications detected by monitors. Then the anomalies would be processed by the misuse detection, the anomaly detection and the attack effect evaluation, in which the misuse detection and anomaly detection are to identify the attack types while the attack evaluation is to determine the degree of negative impact caused by attack. And these three processes could provide references for the next defense policy together. It is worth addressing that The detection efficiency and accuracy of ADCD can be improved by using the misuse detection and the anomaly detection at the same time.

The policy response refers to that ADCD would output defense policies as long as anomalies input into it, no matter whether the attack types causing anomalies are known or unknown. That is to say, ADCD would output precise policies such as the access control and anti-virus responding to known attacks, while it may also output MTD policies, which take the anomaly module such as the IP address and the software as the mutation object in response to unknown attacks.

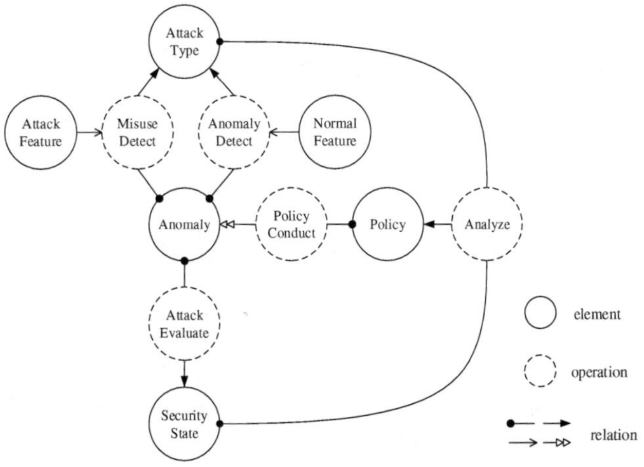

Figure 1 ADCD model

As shown in Figure 1, there are 6 kinds of elements, 5 kinds of operations and 4 kinds of relations in ADCD, and the followings are their explanations.

2.2. Elements

Anomaly (*A*). *A* refers to the system's anomalies detected in a period, $A=\{id,a_1,a_2,\ldots,a_n\}$, $n\in \mathrm{N}$ (N is natural number). *id* is the event number to mark anomalies detected in the same period; a_i $(0\le i\le n)$ is the single anomaly such as the abnormal file operation.

Attack feature (*AF*). *AF* refers to the attack feature which is already known, $AF=\{af_1,af_2,...,af_n\}$, $n\in \mathrm{N}$. af_i $(0\le i\le n)$ is the single attack feature such as a hash code of a known virus or an API calling sequence of a known malware.

Normal feature (*NF*). *NF* refers to the feature of modules of components in a security system, $NF=\{nf_1,nf_2,...,nf_n\}$, $n\in \mathrm{N}$. nf_i $(0\le i\le n)$ is the single normal feature such as an API calling sequence of a known security software or an operation log of a legitimate user.

Attack types (*AT*). $AT=\{id,at_1,at_2,...,at_n\}$, $n\in \mathrm{N}$. *id* is the event number same as in *A*. at_i $(0\le i\le n)$ is the single attack type such as the famous virus WannaCry or Stuxnet.

System security state (*S*). *S* reflects the degree of negative effect caused by attack, $S=\{id,value\}$. *id* is the event number same as in *A*. *value* is a quantitative value of security state calculated by attack effect evaluation.

Defense policy (*Pol*). $Pol=\{id,subject,target,action,constraint\}$. *id* is the event number same as in *A*. *subject* is the policy executor. *target* is the defense object. *action* is executor's action, *constraint* is the limitations of policy. Noting that the *action* in the *Pol* contains both NCD technologies and MTD technologies. In this paper, we choose XML to describe *Pol* for its universality. Here is an example:

```
<?xml version="1.0" standalone="yes"?>
<eventid>0002</eventid>
<policyid>0010</policyid>
<policyname>IP mutation</policyname>
<policybody>
  <subject>Router</subject>
  <target>Communication parameter</target>
  <action>IP Random Mutation</action>
  <constraint>
    <mutationtarget>IP address</mutationtarget>
    <mutationperiod>60s</mutationperiod>
```

<mutationspaee>65536</mutationspaee>
</constraint>
</policybody>
<decision>true</decision>

The meaning of the *Pol* above is: this IP mutation policy should be conducted by routers to take IP random mutation on communication parameters, and the mutation target is IP address, the mutation period is 60 seconds, the mutation space is a network address space of B class (the B class address space contains 65536 addresses).

2.3. Operations

Misuse detection (*MD*). *MD* is to fast detect known attack types by comparing the extracted features in anomalies detected with the *AF* stored in database.

Anomaly detection (*AD*). *AD* is to detect the stealthy attacks missed by the *MD*. In the *AD* process, the feature vector should be constructed according to the anomalies detected, and then running the anomaly detection algorithm based on the extracted feature vector and the *NF* stored in database. *AD* could detect known attacks and part of unknown attacks.

Attack effect evaluation (*AE*). *AE* is to evaluate the negative effect caused by attacks, and to get a quantitative result of attack effect, which is an important reference for the defense policy making.

Association analysis (*A&A*). *A&A* is to gather the results of *MD*, *AD* and *AE* so that we can conduct a further analysis to get the optimal dynamic defense policy. If the attack is detected as a known one, ADCD would output a precise defense policy to defend. While ADCD would output a MTD policy to defend if the attack is detected as an unknown one, and the mutation object should be the module or the component detected as abnormal (e.g. the file operation API should be the mutation object once some abnormal file operations are detected). As for the rest mutation parameters, we can combine the system security state with other MTD researches[7] to determine.

Policy conduct (*PC*). *PC* is to distribute the policies output from *A&A* to related defense subjects such as the firewall and anti-virus software for further implementation.

2.4. Relations

There are 4 basic relations between elements and operations.

Drive relation: An element can drive an operation to work, and the formal definition is: $Drive(\text{element} \rightarrow operation)$. There are 6 drive relations in ADCD: $Drive(A \rightarrow MD)$, $Drive(A \rightarrow AD)$, $Drive(A \rightarrow AE)$, $Drive(S \rightarrow A \,\&\, A)$, $Drive(AT \rightarrow A \,\&\, A)$, $Drive(Pol \rightarrow PC)$.

Support relation: An element is the support of an operation, in another word, an operation cannot work without an element. The formal definition is: $Support(element \rightarrow operation)$. There are 2 support relations in ADCD: $Support(\text{AF} \rightarrow \text{MD})$ and $Support(NF \rightarrow AD)$.

Output relation: An operation can output an element, and the formal definition is $Output(\text{operation} \rightarrow element)$. There are 4 output relations in ADCD: $Output(MD \rightarrow AT)$, $Output(AD \rightarrow AT)$, $Output(AE \rightarrow S)$, $Output(A \,\&\, A \rightarrow Pol)$.

Restrain and eliminate relation: The *PC* operation can restrain and eliminate the element *A*, in another word, ADCD can output and conduct a policy that can restrain and eliminate the *A* input into the model. There is only 1 *restrain and eliminate* relation in ADCD, whose formal definition is $RestrainandEliminate(PC \rightarrow A)$.

2.5. Rules

It is important to ensure the security of ADCD when the operations are running, so there are 6 rules in ADCD. The rules are described by the Datalog language[8], whose grammars are:

(1) RuleName: RH $\Leftarrow$ RB

(2) RB=RE_1 op RE_2 op … op RE_n

IOP Conf. Series: Materials Science and Engineering **435** (2018) 012064 doi:10.1088/1757-899X/435/1/012064

RH is the header of a rule. RB is the body of a rule, and RB can consist of several rule expressions (RE) and operators (op). There are 3 kinds of operators: and ($\wedge$), or ($\vee$), not ($\neg$).

Rule 1 Attack type output rule. The *MD* and *AD* can determine attack types only with the drive of *A* and the support of *AF* or *NF*. The formal definition is: $Output(MD \to AT) \Leftarrow Drive(A \to MD) \wedge Support(AF \to MD); Output(AD \to AT) \Leftarrow Drive(A \to AD) \wedge Support(NF \to AD)$.

Rule 2 System security state output rule. The *AE* can evaluate system's security state only with the drive of *A*. The formal definition is: $Output(AE \to S) \Leftarrow Drive(A \to AE)$.

Rule 3 Defense policy output rule. The *A&A* can output *Pol* only with the drive of *AT*, while the drive of *S* is optional. The formal definition is: $Output(A \& A \to Pol) \Leftarrow Drive(AT \to A \& A) \vee (Drive(AT \to A \& A) \wedge Drive(S \to A \& A))$.

Rule 4 Restrain and eliminate rule. The *PC* can restrain and eliminate *A* only when the *Pol* is distributed and conducted and the *id* of *Pol* should be the same as *A*'s. The formal definition is: $RestrainandEliminate(PC \to A) \Leftarrow Drive(Pol \to PC) \wedge Pol.id{=}A.id$.

Rule 5 Event stable rule. The *id* of an attack event causing the anomalies should be stable after the 5 operations of ADCD. The formal definition is: $AT.id{=}A.id \Leftarrow (Drive(A \to MD) \wedge Output(MD \to AT)) \vee (Drive(A \to AD) \wedge Output(AD \to AT))$; $S.id{=}A.id \Leftarrow Drive(A \to AE) \wedge Output(AE \to S)$; $Pol.id{=}AT.id{=}S.id \Leftarrow Drive(AT \to A \& A) \wedge Drive(S \to A \& A) \wedge Output(A \& A \to Pol)$; $A.id{=}Pol.id \Leftarrow Drive(Pol \to PC) \wedge RestrainandEliminate(PC \to A)$.

According to the rules above, we can further derive the following 4 theorems:

Theorem 1. If there is an *AT* and *AT*'s *id* is id', there must be an *A* and *A*'s *id* is also id'. The formal definition is: *if* $\exists AT$ and $AT.id = id'$, so $\exists A$, *s.t.* $A.id = id'$.

Proof. $\because \exists AT.id = id'$, and $\because AT.id{=}A.id \Leftarrow (Drive(A \to MD) \wedge Output(MD \to AT)) \vee (Drive(A \to AD) \wedge Output(AD \to AT))$, $\therefore \exists A$, *s.t.* $A.id = id'$.

Theorem 2. If there is a *S* and *S*'s *id* is id', there must be an *A* and *A*'s *id* is also id'. The formal definition is: *if* $\exists S$ and $S.id = id'$, so $\exists A$ *s.t.* $A.id = id'$.

Theorem 3. If there is a *Pol* and *Pol*'s *id* is id', there must be an *AT* and a *S*, and *AT*'s *id*, *S*'s *id* and *Pol*'s *id* is the same. The formal definition is: *if* $\exists Pol$ and $Pol.id = id'$, so $\exists AT$ and $\exists S$, *s.t.* $AT.id = S.id = id'$.

The proofs of theorem 2 and theorem 3 is similar, we haven't described them in order to avoid repetition in this paper.

Theorem 4. If there is an *A* and *A*'s *id* is id', there must be a *Pol* and *Pol*'s *id* is also id'. The formal definition is: *if* $\exists A$ and $A.id = id'$, so $\exists Pol$, *s.t.* $Pol.id = id'$。 Noting that theorem 4 conforms to the ADCD's policy response concept that ADCD would output defense policies as long as anomalies input into it, no matter whether the attack types causing anomalies are known or unknown.

Proof. $\because \exists A.id = id'$, and $AT.id{=}A.id \Leftarrow (Drive(A \to MD) \wedge Output(MD \to AT)) \vee (Drive(A \to AD) \wedge Output(AD \to AT))$, and $S.id{=}A.id \Leftarrow Drive(A \to AE) \wedge Output(AE \to S)$, $\therefore \exists AT, S$, and $AT.id = S.id = id'$. $\because Pol.id{=}AT.id{=}S.id \Leftarrow Drive(AT \to A \& A) \wedge Drive(S \to A \& A) \wedge Generate(A \& A \to Pol)$, $\therefore \exists Pol$, $Pol.id = id'$.

Rule 6 Defense feedback rule. If the policy can't restrain or eliminate the anomalies and the defense feedback factor is still low than the threshold, the anomalies existed would drive the *MD*, *AD* and *AE* and make the defense feedback factor add 1. The formal definition is: $Drive(A \to MD) \wedge Drive(A \to AD) \wedge Drive(A \to AE) \wedge Feedbackfactor + 1 \Leftarrow \neg RestrainorEliminate(Pol \to A) \wedge Feedbackfactor < threshold$. Rule 6 creates a beneficial system feedback and also introduces a security mechanism avoiding a deadlock of ADCD. And if in the situation that the *Feedbackfactor* is larger than the threshold, we could adjust some modules in the model such as the anomaly detection algorithm or eliminate the anomalies manually.

3. Security Analysis

The state of ADCD will change with the conduction of its operations, and because of the finite operations and elements of ADCD, there are also finite states of it. Therefore, we carry out the security analysis of ADCD based on finite-state machine (FSM) as follows.

3.1. FSM of ADCD

3.1.1. State variables. ADCD is defined as a FSM tuple: M={V, I, T, v_0, v_n}. V={v_0, v_1, …, v_n}, which is the set of states that will transfer from one to another under the drive of transfer functions, and ADCD can only be in a certain state at any moment. I is the input set. T is the set of transfer functions, which represent the mapping relation of I and V and can be described as $V \times I \rightarrow V$. v_0 is the original state and v_n is the final state.

For any v_i belonging to V, $v_i=\{A, AT, S, Pol\}$, and A, AT, S, Pol are anomaly, attack type, security state and policy. From the definition we know that v_i can describe the running situation of ADCD, i.e., whether there is anomaly input, whether the attack type is determined and attack effect is evaluated, and whether there is a dynamic defense policy output.

According to the definitions above, and combining the elements and operations of ADCD, we can get all of the states and inputs of ADCD as show in table 1 and table 2.

Table 1. States and inputs of ADCD

state	Explanation	Formalization
v_0	Anomaly input	$\{A,\varnothing,\varnothing,\varnothing\}$
v_1	Known attacks determined	$\{A, AT^k,\varnothing,\varnothing\}$
v_2	Unknown attacks determined	$\{A, AT^u,\varnothing,\varnothing\}$
v_3	Attacks and security state determined	$\{A, AT, S,\varnothing\}$
v_4	Defense policy output	$\{A, AT, S, Pol\}$
v_5	Anomaly eliminate	$\{\varnothing,\varnothing,\varnothing,\varnothing\}$

Note: AT^k refers to known attacks, AT^u refers to unknown attacks, $AT=AT^k \cup AT^u$.

Table 2. States and inputs of ADCD

input	explanation
i_0	Misuse detection
i_1	Anomaly detection
i_2	Attack effect evaluation
i_3	Association analysis
i_4	Policy conduction

3.1.2. Transfer function

T_0: $v_0+i_0 \rightarrow v_1/v_2$. Conduct the misuse detection after anomaly is input, which aims to determine attack types quickly based on the attack features.

T_1: $v_2+i_1 \rightarrow v_1/v_2$. Conduct a further anomaly detection if attack types can't be determined by misuse detection, and T_1 is based on normal features.

T_2: $v_1+i_2 \rightarrow v_3$. Take the attack effect evaluation after the attack being determined as known, to calculate the degree of negative influence caused by anomalies.

T_3: $v_2+i_2 \rightarrow v_3$. Take the attack effect evaluation after the attack being determined as unknown too, which is in order to provide references about mutation parameter determination based on the result of evaluation.

T_4: $v_3+i_3 \rightarrow v_4$. Conduct the association analysis according to the result of attack type determination and attack effect evaluation, to output precise defense policy.

T_5: $v_4+i_4 \rightarrow v_0/v_5$. Conduct the defense policy.

According to the 6 transfer functions above, we can get the state transfer graph as shown in Figure 2.

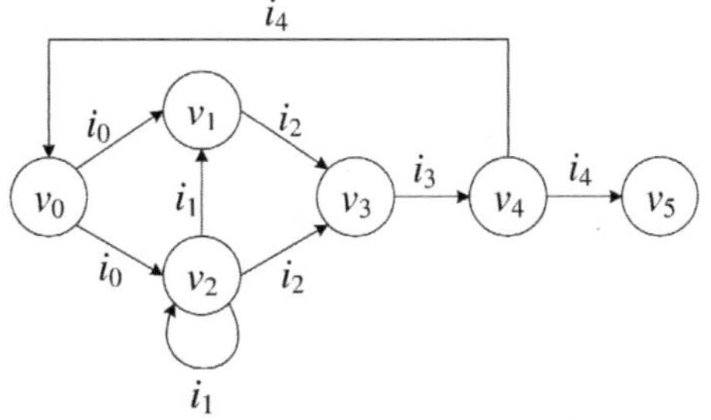

Figure 2. State transfer graph

3.2. Analysis based on FSM

3.2.1. Basic definition. **Secure state.** As for the state v_i={A, AT, S, Pol}, if the drive element (e.g. A is the drive element of AT) exists for any AT, S or Pol, and drive element's id is the same as its output element's id, so we call the state v_i is secure. According to the relevant theorems of ADCD, this secure state definition can be formalized as: $\forall v_i$={A, AT, S, Pol}, v_i is secure when $v_i \mapsto$ Theorem 1 $\wedge$ $v_i \mapsto$ Theorem 2 $\wedge$ $v_i \mapsto$ Theorem 3. The symbol $\mapsto$ means "satisfy".

It is obvious that the original state $v_0=\{A,\varnothing,\varnothing,\varnothing\}$ is secure according to the definition.

Secure transfer function. For any transfer function T_i, it is secure only when its state output is secure for any secure state input.

3.2.2. Analysis of ADCD. There are 6 transfer functions, T_0~T_5, in the FSM of ADCD. According to the operations and rules of ADCD, we know that T_0, $T_1 \mapsto$ Rule1, $T_2 \mapsto$ Rule2, $T_3 \mapsto$ Rule2, $T_4 \mapsto$ Rule3, $T_5 \mapsto$ Rule4 $\wedge$ Rule6. Based on the 6 rules, we can prove the security of these transfer functions. The security of T_0 and T_5 are proven in this paper for instance, the rest transfer functions' security proof is similar.

Proof of T_0's security. Assuming that the FSM of ADCD is in the original state $v_0=\{A,\varnothing,\varnothing,\varnothing\}$, and the *A* in v_0 is any kind of anomalies. Now we conduct the misuse operation, i.e., input i_0 into the FSM. According to Rule 1, *AT* will be output. So the FSM will go into state $v_1=\{A, AT^k,\varnothing,\varnothing\}$ or $v_2=\{A, AT^u,\varnothing,\varnothing\}$. And according to Rule 5, there is *AT.id=A.id*. Furthermore, we can draw the conclusion: as for v_1 and v_2, $v_1,v_2 \mapsto$ Theorem 1 $\wedge$ $v_1,v_2 \mapsto$ Theorem 2 $\wedge$ $v_1,v_2 \mapsto$ Theorem 3. Therefore, the state v_1 and v_2 are secure. According to the Secure transfer function definition, T_0 is secure.

Proof of T_5's security. Assuming that the FSM of ADCD is in the secure state $v_4=\{A, AT, S, Pol\}$. According to the secure state definition, there is an *A* existed, and *A.id =Pol.id*. Based on Rule 4 and Rule 6, the FSM will go from v_4 into $v_5=\{\varnothing,\varnothing,\varnothing,\varnothing\}$ or $v_0=\{A,\varnothing,\varnothing,\varnothing\}$. It is obvious that $v_0,v_5 \mapsto$ Theorem 1 $\wedge$ $v_0,v_5 \mapsto$ Theorem 2 $\wedge$ $v_0,v_5 \mapsto$ Theorem 3. Therefore, according to the Secure transfer function definition, T_5 is secure.

Based on the above analyses, the original state of ADCD is secure and every transfer function is secure too. According to the theory of FSM, it is proven that ADCD is secure.

4. Scenario

4.1. Implementation

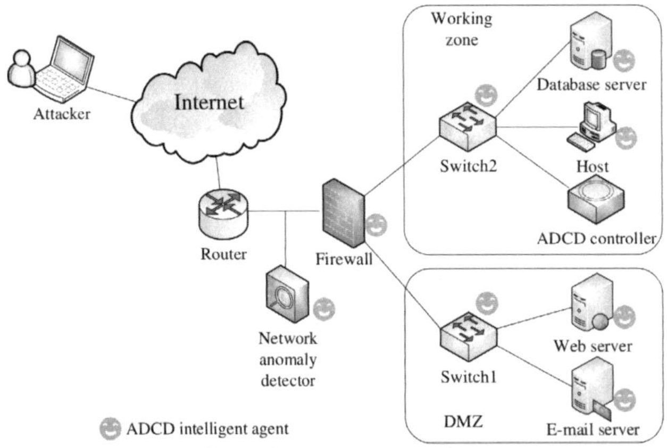

Figure 3 Implementation environment

As shown in Figure 3, it is a simulative enterprise network which is divided into demilitarized zone (DMZ) and working zone by the firewall security rules. There are 1 Web server and 1 E-mail server in DMZ, and 1 database server and 1 host in working zone.

According to the concept of ADCD, we make several implementations in the network as follows: 1) Install security softwares such as host firewall and host intrusion detectors on every device to detect anomalies and conduct defense actions; 2) Implement a network anomaly detector to detect abnormal traffics; 3) Deploy an ADCD controller in working zone; 4) Deploy ADCD intelligent agent on every device. Based on the implementations above, we create a multiagent system to realize ADCD. And the multiagent system is intelligent and ordered because of the rules defined in ADCD. What follows is the explanation for functions of ADCD controller and intelligent agent.

4.1.1. Functions of ADCD controller. The functions of ADCD controller are realizations of operations in ADCD, which including:

- Gathering all of the anomalies detected by devices, including abnormal file operations, abnormal network traffics, abnormal API callings and so on.
- Conducting misuse detection based on signatures of anomalies. This function is supported by some anti-virus softwares such as Kaspersky and Norton.
- Conducting anomaly detection based on anomalies and with the support of normal features. In this paper we choose the extended attack tree model[9] as the anomaly detection rules.
- Evaluating the effect of attacks to get the system security state. And we choose the multidimensional attack effect evaluation method[10] in this paper.
- Combining the attack type determined and system security state evaluated to conduct association analysis, and get the defense policy. For example, when the attack type is "Stuxnet virus", the policy will be "block communications and delete the malware". When the attack type is "unknown virus" and the anomaly is "abnormal network communications", the policy will be "mutate device's IP every 60 seconds in the B class address space".
- Distributing defense policy to relevant ADCD intelligent agent to conduct.

4.1.2. Functions of ADCD intelligent agent. The functions of ADCD intelligent agent focus on assistance to ADCD controller such as follows:

- Detecting or collecting anomalies detected in devices and sending them to ADCD controller.
- Receiving policies distributed by the ADCD controller, and resolving the policies into the standard instructions of a certain device and then send them to devices to conduct.

Note that all the functions of ADCD intelligent agent should be automatic without administrator's interference.

IOP Conf. Series: Materials Science and Engineering **435** (2018) 012064 doi:10.1088/1757-899X/435/1/012064

4.2. Defense process

We conduct a remote and multi-step attack as background to explain the defense process of ADCD. Firstly, we set a MS10_046 vulnerability on the working host, and a MS08_067 vulnerability on the database server. Then we use *Metasploit* (an attack testing platform) to attack as follow steps:

1) Use MS10_046 and DNS cheating to attack the working host and get a cmd shell, and then promote the privilege to system.
2) Take the working host as a springboard to conduct further attack to the database server using MS08_067 and get the administrator privilege.
3) Search classified files on the database server and send them back to attacker's computer, and also upload a Trojan to the database server as the backdoor.

In the above background, ADCD conducts following defense steps:

1) Firstly, the detector on working host captures abnormal communications launched by the device itself, and locates 2 processes, svchost.exe and rundll32.exe, by the process id. Then the detector on database server also detects abnormal communications and Windows register modification conducted by the device, and locates a process svchost.exe by the process id and a suspicious software Gh0st.exe by the record of Windows register. According to the FSM of ADCD, now the defense system is in state v_0. After the detection, the ADCD intelligent agents send these anomalies to ADCD controller.
2) ADCD controller conduct misuse detection on the anomalies received firstly, and find that the suspicious Gh0st.exe is a kind of Trojan. So the defense system goes into state v_1.
3) To determine what kind of types the attack causing the abnormal network communication is, the anomaly detection is conducted in ADCD controller. According to the detection rules proposed by [9], it can only be judged that some files are stolen by a remote computer, but the precise attack type is still unknown. So the defense system goes into state v_2.
4) Then, ADCD controller evaluate system's security state using the method of [10], and get a quantitative value. In the scenario of this paper, the quantitative value is 42.42. The defense system goes into state v_3.
5) Based on the result of attack detection and evaluation, ADCD controller conducts association analysis and outputs 3 defense policies, and the defense system goes into state v_4. The first policy is "delete Trojan":


```
<?xml version=”1.0” standalone=”yes”?>
<eventid>0010</eventid>
<policyid>0001</policyid>
<policyname>delete Trojan</policyname>
<policybody>
  <subject>databaseserver</subject>
  <target>Gh0st.exe</target>
  <action>delete</action>
</policybody>
<decision>true</decision>
```


The svchost.exe and rundll32.exe are system processes, so the attack can not be eliminated by deleting a malware. Therefore, the ADCD controller outputs the rest 2 policies: "IP block" and "IP mutation". The "IP block" policy is to cut off the current communications between attacker and victim host.


```
<?xml version=”1.0” standalone=”yes”?>
<eventid>0010</eventid>
<policyid>0002</policyid>
<policyname>IP block</policyname>
<policybody>
  <subject>firewall</subject>
  <target>10.100.200.11</target>
  <action>block</action>
</policybody>
```

```
<decision>true</decision>
```


The "IP mutation" policy is to decrease the probability of success for the attacks that may happen in the future. And the mutation parameters are determined according to the method of [7].


```
<?xml version="1.0" standalone="yes"?>
<eventid>0010</eventid>
<policyid>0003</policyid>
<policyname>IP mutation</policyname>
<policybody>
 <subject>Router</subject>
 <target>communicationparameter</target>
 <action>IP Random Mutation</action>
 <constraint>
  <mutationtarget>IP address</mutationtarget>
  <mutationperiod>60s</mutationperiod>
  <mutationspaee>65536</mutationspaee>
 </constraint>
</policybody>
<decision>true</decision>
```


6) ADCD controller sends the 3 policies above to the intelligent agents on Firewall and Switch2 to conduct. After these defense steps, anomalies are eliminated and the defense system goes into state v_5.

From the scenario above, we can find that there are two significant advantages of ADCD when attacks happen:

- ADCD can combine the traditional defense techniques such as anti-virus with the new moving target defense (MTD) techniques such as IP mutation cooperatively, to defense complex and multi-step attacks, and without changing the original network topology.
- Based on anomaly processing, ADCD can output targeted defense policies according to different attack types, no matter whether the attack is known or unknown. For example, ADCD can output a "delete Trojan" policy for known Trojans, or an "IP mutation" policy for unknown attacks that cause the communication anomalies.

5. Conclusion

A novel network proactive defense model called anomaly driven dynamic cooperative defense model (ADCD) is proposed and very detail explanations of ADCD's elements, operations, relations and rules are given. Furthermore, based on FSM, a formal analysis of ADCD is conducted to prove its security. Finally, A typical application scenario is displayed to verify the effectiveness of ADCD in defending complex and multistep attacks. We believe that ADCD can provide a new solution in defending network attacks.

Acknowledgments

This paper is supported by Henan Province Foundation and Frontier Technology Research Project (No. 142300413201), Fund for Zhengzhou Information Science & Technology Institute (No. 2016604703)

References

[1] Zhang T N, Yun X C, Zhang Y Z, Men C G, Sun J L. 2011 *J. Chinese Journal of Computers*, 34(2):216-228.

[2] Jajodia S, et al. 2011 *Moving Target Defense* (New York: Springer) 2011.

[3] Okhravi H, Rabe M A, Mayberry T J, Maybery T, Leonard W G, Hobson T, Bigelow D, Streilein,W W. 2013 *Survey of Cyber Moving Target Techniques (Electronic Materials).*

[4] Lee H C J, Thing V L L. 2004 *Proc. Int. Conf.Vehicular Technology* (New York: IEEE), 5:3291-3295.

[5] Dwivedi S, Angeri H R, Arora V J. 2008 *Architecture for Unified Threat Management (Electronic Materials).*

[6] CAI G L, WANG B S, WANG T Z, Luo Y B, Wang X F, Cui X W. 2016 *J. Journal of Computer Research and Development*, 53(5):968-987.
[7] Li N, Mitchell J C. 2003. *Lecture Notes in Computer Science*, 2562:58-73.
[8] Lei C, Ma D H, Zhang H Q, Yang Y J, Wang L M, 2017 *J. Chinese Journal of Computers*, Online Publishing 40(130).
[9] Wang L N, Tan C, Yu R W, Yin Z G. 2017 *J. Journal of Computer Research and Development*, 54(7):1537-1548.
[10] Zhan B F, Yin X C. 2011 *J. Computer Engineering and Design*, 32(8):2596-2599.

A Novel Steerable Single Layer Tubes of Surgcial Manipulator Based on Ni-Ti Alloy

Zhaoying Zhang[1,2], Fan Zhang [1,2*],Jing Sun [3],Minhua Zheng[3] and Ruijun Pan [4]

[1]College of Mechanical Engineering,Shanghai University of Engineering science ,Longteng Road No.333,Songjiang District ,Shanghai ,201620,P.R.China.
[2*]Intelligent Robotics Research Center of Shanghai of Engineering Science,Shanghai University of Engineering Science, Longteng Road No.333 Song Jiang District ,Shang hai ,201620,P.R.China.
[3]Department of General Surgery, Shanghai Minimally Invasive Surgery Center, Ruijin Hospital, Shanghai Jiao Tong University School of Medicine, Shanghai 200025, P.R.China
[4]Department of Digestive Surgery, Shanghai Minimally Invasive Surgery Center, Ruijin Hospital, Shanghai Jiao Tong University School of Medicine, Shanghai 200025, P. R. China.

E-mail:18019388213@163.com
*Corresponding Author:pdszhangfan@aliyun.com

Abstract. Objective Minimally Invasive Surgery (MIS) requires armamentarium with high positioning accuracy and maneuverability. At the same time ,the equipment need have more functions and smaller size, it can reduce patient's operative incision and pain perception when in operating can still keep steerability and stability.Method.Traditional design with multi-tubes assemabled, we designed a new stucture which just have single layer tube Results Finite element method can analysis this model which designed by another 3D drawing software in different size, varied exteral force can conclude that the angle of end deformation.3D model adopted the isometric atom with nickel titanium alloy we take advantage of Ni-Ti alloy's superelasticity which can recover to orignal shape in recoverable deformation interval , realise proximal end control distal end.

1. Introduction

In the early 19th century, philipp bozzini,a German, used candles and a series of lenses, by using the light of the candle crossed the lenses can image formation which internal features of human's bladder and urethra achieve exmine make sure these organ work like as usual whether or not.

In medical domain, Endoscopes are mostly used in neurosurgery [1]. Endoscope and it's appurtenances equipment can be penetrated into the cavity of human body through these natural passage or artificial opening of the human body. The internal tissue structure of human body can be imaged on the display device by imaging system, which can provide an intuitive reference object for doctors. Doctors can according these images which display internal tissue accurate cut off tumours or cysts.

In general, endoscope can be divided into hard tube endoscope and steerable endoscope by the means of control. When doctors employ endoscope into surgical operation select different types of endoscope on the base of different part which have different structure. Ordinarily, the hard tube endoscope is used for the examination of the pharynx, nasal cavity and brain, as shown in Figure 1B checking tympanic tumor via hard tube endoscope. Usually the head of a hard tube endoscope have sloping which can expanding the surgical field of vision can get more message about tissue. The head of hard tube endoscope has high stability and is easy to operate. It can be located accurately in the internal cavity of human body so as to avoid the injury of tissues and organs, reduce the probability of secondary injury. The pathological tissue which broken to pieces by some certain ways can be removed from the human tissue and transformed by the channel of the hard tube, it's channel can be use transport cure medicines, such as disinfected or eliminate the pernicious influcence medicine. It can also be provide a variety of working modes, without replacing the axis, it's head can be replaced, such as gripper or scissors can be installed[2,3,4].In this way, a variety of functions are relieved. The steerable endoscope's operational principle is that use rope traction, rope and rope between interference when in process of operation, as shown in Figure1C. Although the flexibility is larger, the angle range is wider than hard tube endoscope, but the precision of control is more difficult than hard tube endoscope. Endoscope technology has been continuously developed with the development of science and technology. Some high-tech equipment and technology are also applied with technology update. However, it's own shortcomings are becoming more and more obvious, precise positioning is not easy to achieve. The stability of terminal control still have improvment room.

Comprehensive consideration of two or more factors, the new equipment has the stability of the end of rigid endoscope and the large cavity. It is also has the large flexibility of steerable endoscope[5][6][7][8].Combine the advantages of the two design. The head of the endoscope is made into a flexible hinge, which makes the head bending, flexible and controllable in a certain range. The middle part of the endoscope is rigid enough to resist a certain amount of external load, such as the bending of the endoscope when the head is in contact with the internal tissue of the human body, which reduces the maneuverability of the endoscope and fails to remove the tumor tissue. This is not what we expect. There are many difficulties in the design of the device. The armamentarium which have multiple functions still have certain stiffness. A small size can reduce the area of the wound and make the patient recover quickly from illness. But at the same time, larger lumen still maintain with the decrease size, so that the removed lesion tissue fragments can be sucked out from the lumen of the tube, as shown Figure 1D.

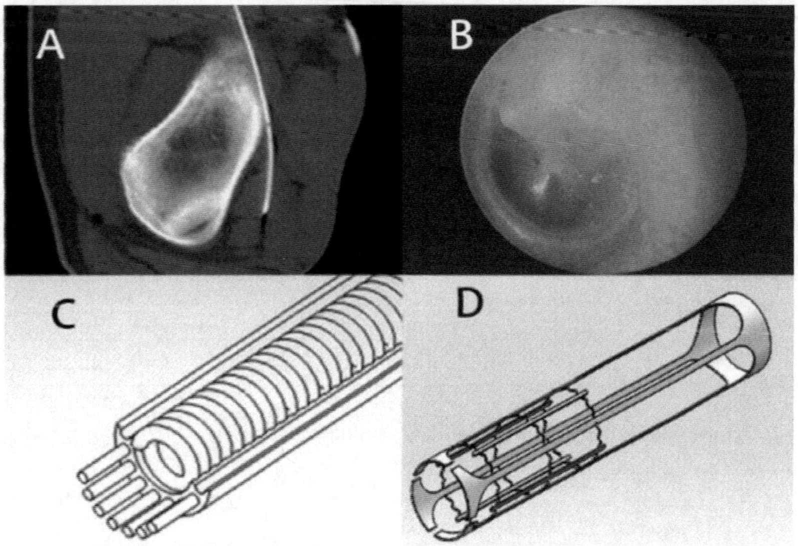

Figure.1. Application of laparoscopy in operation and real endoscope

Figure.1.(A)Products: Morrison Steerable Needle used for sampling, rounding the bone and carefully avoiding neurovascular bundle (*http://apriomed.com/products/morrison-steerable-needle/*)

[9]. (B)Using Hard endoscope check Tympanic tumor[10].(C)Traiditional steerable endoscope: outer cather has hole arround from circumference, in that metal wire in it which can be controled by these guide wires[11].(D)New technology designed in minivasive surgey which can imply lagre inner diameter and just have Single layer thin wall.

In the field of laparoscopic technology, Leonardo da Vinci robot has occupied an unshakable position[12].It's consists of three parts: the surgeon control's workbench, the bedside manipulator system, and the imaging system. It's patented wrist rotator can achieve 7 degrees of freedom, and the range of movement is even wider than the palm of the human hand. Greatly improve the flexibility of the instrument, can help doctors to complete the excision and suture more efficient. Traditional endoscopy has less flexibility, because the head of the endoscope does not have the same flexible mechanism as human wrist joint. Endoscopic surgery is more in the natural cavity or artificial opening of the single-hole surgery. Single-hole surgery requires more flexibility than the natural canal.

2. Present method

Many endoscope products use the structure of manipulable wire to realize the precise control of the end structure. The product structure of this design mostly adopts the structure with holes in the pipe wall, which metal wire inside the hole can realize movement along these holes. As shown as Figure 1C.Of course there are other structural options. For example, the company's products use it's own structure to achieve bidirectional bending. The product has a structural wall size of 0.82mm,an inner wall size of 0.68mm and a length of 170mm as shown Figure 1A[13].The structure of this design is compact, and it does not adopt the structure design of metal wire with holes in the pipe wall. It can control the far end bend through the control handle, so that it can accurately locate and remove the focus. The outer diameter and the inner diameter of the endoscope with wire are large and small, so that there can be room for the wire. The wire can pass through the holes in the tube wall to control the endoscope head. This greatly increases the cross section size of endoscope, reduce the rigidity of endoscope, reduce the stability of the end structure, reduce the ability to resist external interference, and is not conductive to the operation. Small internal diameter do not allow for other medical procedure, such as channeling broken tissue through channels to reduce the risk of direct exposure to air. A wire-controlled endoscope structure that connects the wire to the wall in the head by welding or otherwise. This operation reduce the service life of the endoscope and is difficult to maintain[14-17].

In view of the above problems, this paper adopts a new structure. By using the super-elasticity of Ni-Ti alloy material, the large deformation of the structure is realized, and the deformation of the terminal structure can be achieved by remote control. Nickel-titanium alloy materials were discovered in 1960 by u. S. Naval ordnance laboratory) at the National Navy Laboratory[18]. Nickel-titanium alloy materials are widely used in many fields, such as aerospace, medical domain and so on. For example, orthodontic nickel-titanium alloy arch wire, cardiac vascular stent and so on. Nickel-titanium alloy materials have two great properties: shape memory effect (SME) and superelasticity(SE)[19]. When the external load is removed, the sample can be restored to its original state without plastic deformation. We can see the property of Ni-Ti and Stainess Steel in Table 1.

Table 1. The property of Ni-Ti and Stainess Steel

property	Ni-Ti	Stainess Steel
Recoverable strain	8%	0.8%~1%
Biocompatibility	Excellent	Fair
Ultimate Tensile Strength(UTS)	~1240Mpa	~760Mpa
Density	6.45g/cm3	8.03g/cm3
Magnetic	No	Yes

3. Numerical analysis

We use another 3D designed model software design structure and we use the finite element software to analyze the model [20]. The deformation of the structure can be accurately calculated by applying external load to the structure as shown Figure 2A, we can change the load size, analysis different load caused deformation and the stress and strain of the structure can be seen from the deformation diagram. Before the stress-strain analysis, the structural model of the material is set up, and then the structure is meshed, and the whole structure is divided into discrete and finite meshes, it's vital to next step, the quality of meshes can influence the accuracy of results The stress and strain at any point can be analyzed by the finite element software when the structural load is applied. We can analyze it in finite element software by changing the size of the structure, such as thin-wall thickness, section shape or size of the connecting side rib, and the length of the deformed end, and by comparing the differences between them. Determine which parameters have the greatest effect on the end bending angle. Of course, through the finite element software analysis, we can analyze the maximum angle of the end structure deformation as shown Figure 2B.

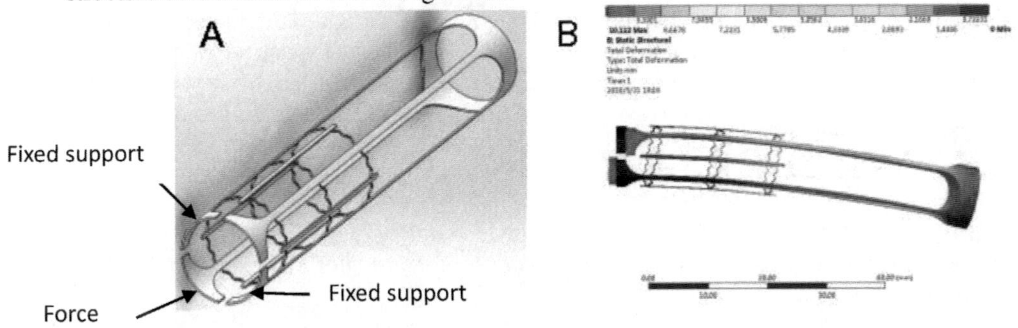

Figure.2.Apply external force and displacement diagram

Figure.2.(A)The direction of the arrow in the figure is the direction of the applied load.(B)The picture shows the deformation of applied 5N,we can see total deformation by analysis software.

4. Results and Disscusions

Use new model which have different size, such as length or demeter. Results shown as fellow as Figure.3.we use different model in order to make sure that indifferent enviorment which external force and self-size can have effect on deformation .

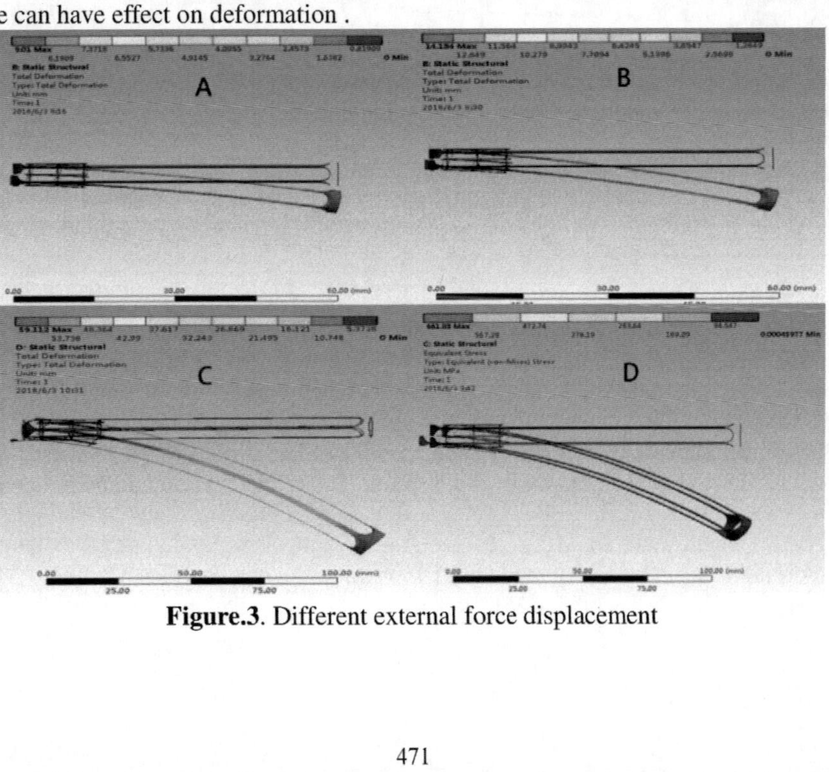

Figure.3. Different external force displacement

Figure.3.(A)The displacement cloud diagram of flexible mechanism with load of 1 N, inner diameter of 9.8 mm, wall thickness of 0.2 mm, length of 60 mm is represented.(B)The displacement cloud diagram of flexible mechanism with load of 3 N, inner diameter of 9.8 mm, wall thickness of 0.2 mm, length of 60 mm is represented.(C)The displacement cloud diagram of flexible mechanism with load of 5N,inner diameter of 9.8 mm, wall thickness of 0.2 mm, length of 120 mm is represented.(D)The displacement cloud diagram of flexible mechanism with load of 5N,inner diameter of 9.8 mm, wall thickness of 0.2 mm, length of 120mm is represented.

From the above displacement and strain cloud diagram, we can see that under the same external diameter the deformation of the distal mechanism can be increased by increasing the load. Under the condition that the external diameter remains constant, the end deformation of the flexible mechanism can be increased significantly by increasing the length of the flexible mechanism and increasing the external load. Compared with the traditional flexible mechanism controlled by flexible cable, the flexible mechanism of this kind of single-layer thin-walled pipe fittings will not occur crosstalk phenomenon. And there is a lot of space inside the monolayer to accommodate other devices, such as the optical fiber of a minimally invasive imaging system, or the splinter of a cyst or tumor, which can be sucked out by a suction device through the cavities. The near end structure is controlled and the end structure is deformed by long and thin strip transfer of tension and compression load. From the cloud figure.3.D,we can see that the end deformation angle of the monolayer thin-walled tubular structure can reach nearly 30° when the external load was applied by 5N.

Acknowledgements
The study is financially supported by Science and Technology Commission of Shanghai Municipality under grant No.17441901200.

References

[1] Frank Dewaele,Alain F. Kalmar,Frederic De Ryck,Nicolaas Lumen,Leonie Williams,Edward Baert a Hugo Vereeck,Jean Pierre Kalala Okito,Cyriel Mabilde Bart Blanckaert,Vincent Keereman, Luc Leybaert,Yves Van Nieuwenhove,Jacques Caemaert and Dirk Van Roost .A Novel Design for Steerable Instruments Based on Laser-Cut Nitinol.*Surgical Innovation* 2014, Vol. 21(3) pp 303-311.

[2] Semi Jeong,Hyunchul Choi,Gwangjun Go,Cheong Lee,Seong Young Ko,Jong-Oh Park,and Sukho
Park.International journal of congtrol,Automation and Systems .2017 vol 15(1) PP473-479.

[3] Kirsten R.Henken Peter R.Seevinck Jenny Dankelman John J van Dobbelsteen.Manually controlled steerable needle for MRI-guided percutaneous interventions.Med Biol Eng Comput (2017) 55:PP 235-244.

[4] Elif Ayvali,Mingyeen Ho,and Jaydev P.Desai. A Novel Discretely Actuated Steerable Probe for Percutaneous Procedures.Desai in Experimental Robotics (2014)

[5] G. Joerger A. Huang B. Bass B. Dunkin M. Garbey .Global Laparoscopy Positioning System with a Smart Trocar.2017 IEEE 17th International Conference on Bioinformatics and Bioengineering.

[6] Carlo A. Seneci,Gauthier Gras, Piyamate Wisanuvej, Jianzhong Shang,Guang-Zhong Yang.3DPrinting of Improved Needle Grasping Instrument for Flexible Robotic Surgery.2017 IEEE/RSJ International Conference on Intelligent Robots and Systems (IROS)September 24-28, 2017, Vancouver, BC, Canada.

[7] Joseph D. Greer,Tania K. Morimoto,Allison M.Okamura,and Elliot W Hawkes.Series Pneumatic Artificial Muscles (SPAMS) and Application to a Soft Continuum Robot.2017 IEEE International Conference on Robotics and Automation (ICRA)Singapore, May 29-June 3, 2017

[8] Sang Hong Qiang,Chang Li Min ,Liu Fen ,Yang Cheng Hao,Xu Li Ping.Design of a 6-DOF Cable-driven Minimally Invasive Surgical Instrument.Mechanical Science and Technology for Aerospace Engineering.February 2017 vol.36(2) .

[9] *http://apriomed.com/products/morrison-steerable-needle/.*
[10] Anne J. Moore,David W. Newell.Tumor Neurosurgery Principles and Practice. Springer-Verlag London Limited 2005; 1-297,489-532
[11] Breedveld P, Scheltes J, Blom E, Verheij J. A new, easily miniaturized steerable endoscope. IEEE Eng Med BiolMag. 2005:40-47
[12] Reduced-port robotic anterior resection for left-sided colon cancer using the da Vinci single-site platform. Bae S U,Jeong W K,Bae O S,et al. International Journal of Medical Robotics and Computer Assisted Surgery.2015.
[13] Chen Yuyu Liu Lei LI Bo Wei Chao Wang Shuxin LI Dichen.Design, Fabrication and Performance of a Flexible Minimally Invasive Surgery Manipulator Integrated with Soft Actuation and Variable Stiffness.*Journal of mechanical engineering*.PP1-9.
[14] Su He,Kong Kang.Modeling of Clamping Force Compensation of Cable-Driven Instrument for Minimally Invasive Surgical Robot .*Journal of Tianjin University (Natural Science and Engineering Technology Edition)*,2018,vol 51(05).PP453-459.
[15] Dariusz Patkowski,Konrad Rysiakiewicz,Jolanta Rysiakiewicz,Andrzej Kozie ††,Wojciech Apoznański,Jerzy Czernik. Minimally Invasive Surgery in Newborns and Infants - Own Experience. Polish Journal of Surgery,2009,vol 81(2).
[16] Florian Bösch,Michael Thomas,Jens Werner,Markus Guba,Martin Angele. Minimalinvasive Chirurgie des Ösophaguskarzinoms-Eine onkologisch sichere Methode mit signifikanten Vorteilen zur offenen Technik[J]. Klinikarzt,2014,vol 43(04).
[17] Mao Weiming,Liao Xiujun,Shao Shuxian,Wu Wenjing,Yu Yanyan,Yang Guangen. Comparative evaluation of colonoscopy-assisted transanal minimally invasive surgery via glove port and endoscopic submucosal dissection for early rectal tumor. *International journal of surgery* (London, England),2017,42.
[18] Gao Yunliang. Structure Design and Performance Study of Ni-Ti Alloy Vascular Stent .*General Research Institute for Nonferrous Metals*,2016.
[19] Tests on superelastic Ni–Ti SMA bars under cyclic tension and direct-shear: towards practical recentring connections.*Materials and Structures.*April 2015,vol 48(04), pp 1013-1030
[20] Wei Mingtang, Li Zhqiang,Zheng Qingli.Numerical analysis on longitudinal flexibility of a Ni-Ti coronary stent.*Journal of Medical Biomechanic*,2016,vol 31(01), pp 13-18.

Hybrid Harvester 3-RPS Robotic Parallel Manipulator

Zhongxing Yang[1] and Dan Zhang[1,*]

[1]Department of Mechanical Engineering, Lassonde School of Engineering, York University, Toronto, Canada

E-mail addresses: [1,]zxyang@yorku.ca; [1,*]dan.zhang@lassonde.yorku.ca

Abstract. This article presents the robotic structure and analysis of a hybrid energy harvester designed in the form of a reconfigurable 3-RPS parallel manipulator, which tracks and collects solar energy as a main function and also harvests energy from wind loads on the large solar panel. The reconfigurable design allows the harvester to lie flat on floor when destructive storms occur and then return to a working configuration after the storms. The workspace of the manipulator is determined by a minimum platform height algorithm that is developed to find all reachable positions that compose the workspace and also bring bending at root to least. The stiffness mapping of the harvester is significant in evaluating its capacity to take wind loads, which a transformation matrix is developed to convert to polar horizontal stiffness at end effector that is crucial to the intelligent control strategy in decision to enter lie-flat configuration.

1. Introduction

A solar tracker or harvester follows the sun to collect energy. Single-axis [1] and dual-axis [2] serial manipulators were designed as solar trackers. Since storms could damage solar tracking fields (North America [3] [4], and Pacific regions [5]), parallel manipulators which have higher stiffness and loading capacity [6] were designed with 2 [7] or 3 [8] DOF (mobility). The design of a general 3-RPS solar tracker [9] could be further improved to enable switch between a lie-flat [10] configuration for safety and a working configuration for enlarged workspace, by reconfiguration [11] methodology.

Loads on solar panels could be another energy source. Piezo-electric materials convert wind loads to electricity, as in applications [12], [13] and [14]. The wind loads on a flat panel could be generally divided to a lift load normal to the panel and a drag load parallel to the panel [15] [16].

This paper presents the design of a solar-piezo hybrid harvester with reconfiguration for dual modes. An algorithm is explained that seeks minimum platform height from a given orientation angle. A transformation matrix is developed that evaluates polar stiffness at end effector on a horizontal plane which is crucial to the intelligent control strategy in decision to enter lie-flat configuration.

2. Mechanical design

This tracking stand is a 3-RPS parallel manipulator with lie-flat and working configuration features.

The orientation angle and height of the solar panel could be adjusted by three actuated legs which are jointed to motor-driven reconfiguration bases. As to meet the dust-proof, load capacity and geometric requirements for out-door working conditions, the linear tables (type: Alpha 15-B-155) [17] and the linear actuators (type: DLA-12-40-311-200-IP65) [18] are selected to function as reconfigurable bases and telescopic legs respectively. The spherical joint with large motion range [19] and piezo-chips incorporated parallel structures [20] are considered. Figure 1 shows the design in two configurations.

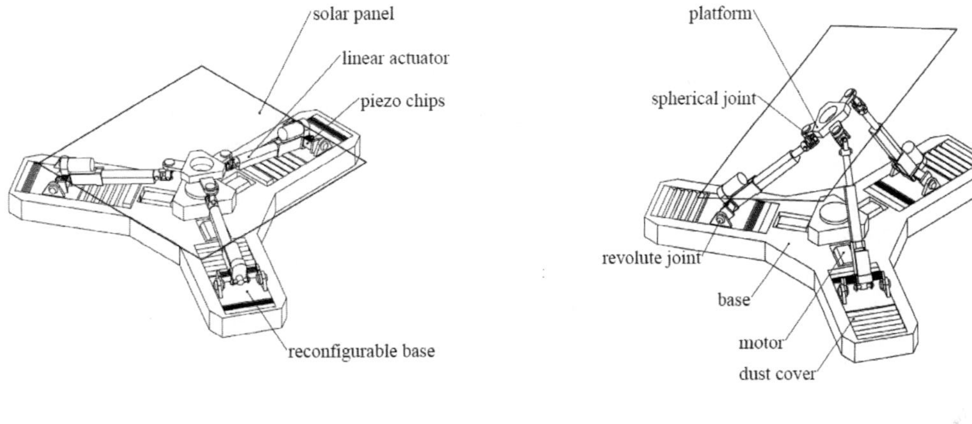

(a) (b)

Figure 1. The hybrid harvester (a) lie-flat configuration; (b) working configuration.

3. Inverse kinematics

The inverse kinematics solves for the actuation displacements from a given posture of the platform.

In figure 2. (a), the plane A represents the floor and the plane B represents a parallel plane above floor. The global coordinate system $A_o - X_oY_oZ_o$ is located at $\boldsymbol{A_o}$ underneath $\boldsymbol{B_o}$ by h_b.

$$\boldsymbol{X_o} = [1 \quad 0 \quad 0]^T; \boldsymbol{Y_o} = [0 \quad 1 \quad 0]^T; \boldsymbol{Z_o} = [0 \quad 0 \quad 1]^T \tag{1}$$

$$\boldsymbol{A_o} = [0 \quad 0 \quad 0]^T; \boldsymbol{B_o} = \boldsymbol{A_o} + h_b \cdot \boldsymbol{Z_o} \tag{2}$$

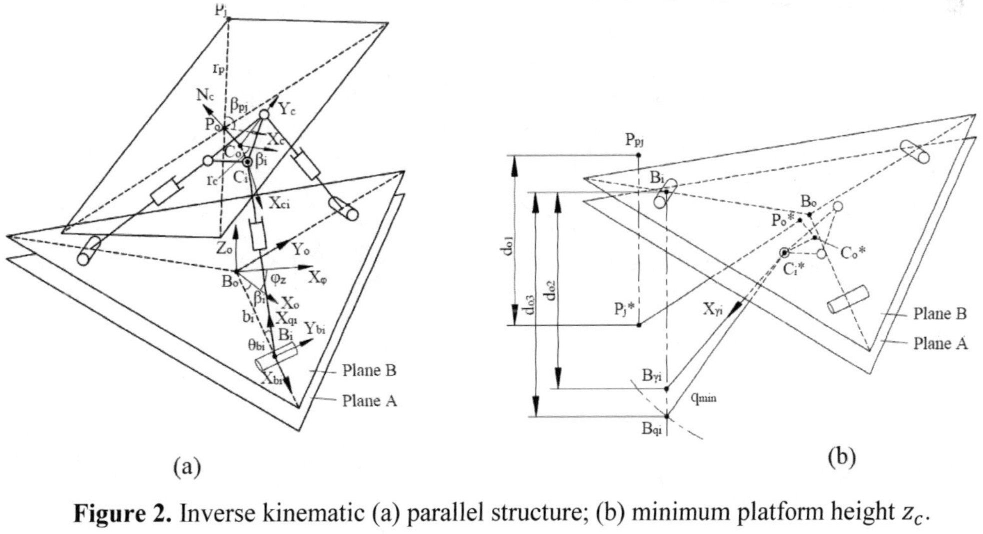

(a) (b)

Figure 2. Inverse kinematic (a) parallel structure; (b) minimum platform height z_c.

Three points $\boldsymbol{B_1}$, $\boldsymbol{B_2}$, and $\boldsymbol{B_3}$ are located on plane B where revolute joints are. One can connect lines $\boldsymbol{B_oB_i}$, and finds angle β_i between $\boldsymbol{X_o}$and $\boldsymbol{B_oB_i}$. The length of $\boldsymbol{B_oB_i}$ is b_i which can be adjusted by reconfiguration. Sub-coordinates $B_i - X_{bi}Y_{bi}Z_o$ are established at $\boldsymbol{B_1}$, $\boldsymbol{B_2}$, and $\boldsymbol{B_3}$.

$$\boldsymbol{X_{bi}} = \boldsymbol{Q_i} \cdot \boldsymbol{X_o}\,;\ \boldsymbol{Y_{bi}} = \boldsymbol{Q_i} \cdot \boldsymbol{Y_o};\text{ where } \boldsymbol{Q_i} = \begin{bmatrix} \cos\beta_i & -\sin\beta_i & 0 \\ \sin\beta_i & \cos\beta_i & 0 \\ 0 & 0 & 1 \end{bmatrix}\ i = 1,2,3 \tag{3}$$

The platform is a triangle with $\boldsymbol{C_o}$ at the center, while points $\boldsymbol{C_1}$, $\boldsymbol{C_2}$, and $\boldsymbol{C_3}$ are at the three vertices of the triangle where spherical joints are. One can connect lines $\boldsymbol{C_oC_i}$, the length of $\boldsymbol{C_oC_i}$ is a fixed value r_c. One can set a sub-coordinate $C_o - X_cY_cN_c$ at $\boldsymbol{C_o}$, where $\boldsymbol{Y_c}$ aligns with vector $\boldsymbol{C_oC_1}$, $\boldsymbol{X_c}$ lies on the triangle plane, and $\boldsymbol{N_c}$ be a unit vector normal to the platform. The angle β_i is between $\boldsymbol{X_c}$ and $\boldsymbol{C_oC_i}$.

A square solar panel is attached above the platform by h_p in $\boldsymbol{N_c}$ direction. The four points $\boldsymbol{P_1}$, $\boldsymbol{P_2}$, $\boldsymbol{P_3}$ and $\boldsymbol{P_4}$ are at the four points of the solar panel square. One can connect lines $\boldsymbol{P_oP_j}$, which forms an angle β_{pj} from $\boldsymbol{X_c}$, and the length of the lines is a fixed value r_p.

The vectors $\boldsymbol{C_oC_i}$ and $\boldsymbol{P_oP_j}$ from the perspective of coordinate $C_o - X_cY_cN_c$ are given below.

$$\boldsymbol{R_{ci}} = [r_c \cdot \cos\beta_i \quad r_c \cdot \sin\beta_i \quad 0]^T, i = 1,2,3 \tag{4.a}$$
$$\boldsymbol{R_{pj}} = [r_p \cdot \cos\beta_{pj} \quad r_p \cdot \sin\beta_{pj} \quad 0]^T, j = 1,2,3,4 \tag{4.b}$$

One can connect $\boldsymbol{B_iC_i}$ along which the prismatic joints are, the length of which is $q_i \in [q_{min}, q_{max}]$, and a unit vector $\boldsymbol{X_{qi}}$ is set along $\boldsymbol{B_iC_i}$. Vector $\boldsymbol{X_{qi}}$ forms an angle θ_{bi} above the plane B. In order to make sure that linear actuators operate under the moving platform, an angular constraint γ_{cm} is set which is the minimum angle the vector $\boldsymbol{X_{qi}}$ can form to platform plane ($C_o - C_1C_2C_3$).

Based on real values of the reconfigurable design, linear actuators and linear tables are selected from manufacturers, so that numerical analysis can be performed. Design parameters are shown in table 1.

Reconfiguration of the $\boldsymbol{B_i}$ along $\boldsymbol{X_{bi}}$ introduces an adjustment variable b_{i-adj}. When $b_{i-adj} = 0$, variable b_i is determined to allow minimum height of platform at the lie-flat posture.

$$\boldsymbol{B_i} = b_i \cdot \boldsymbol{Q_i} \cdot \boldsymbol{X_o} + \boldsymbol{B_o},\text{ where } b_i = q_{min} \cdot \cos\gamma_{cm} + r_c + b_{i-adj} \tag{5}$$

Table 1. Design parameters.

Parameters	Units	Values
h_b	m	0.092
r_p	m	0.5
r_c	m	0.1
h_p	m	0.05
$[q_{min}, q_{max}]$	m	[0.35, 0.55]
$\beta_{\mathrm{i},\ (\mathrm{i}=1,2,3)}$	rad	$\left[\frac{\pi}{2} \quad -\frac{\pi}{6} \quad \frac{7\pi}{6}\right]$
$\beta_{\mathrm{pj},\ (\mathrm{j}=1,2,3,4)}$	rad	$\left[\frac{\pi}{4} \quad -\frac{\pi}{4} \quad -\frac{3\pi}{4} \quad \frac{3\pi}{4}\right]$
γ_{cm}	rad	0.1745

Rotational matrix [21] needs platform rotation variables θ_x and θ_y around $\boldsymbol{X_o}$ and $\boldsymbol{Y_o}$ to convert vectors or matrix in $C_o - X_cY_cN_c$ to that in $B_o - X_oY_oZ_o$.

$$u_x = \frac{\theta_x}{\theta}\text{ and } u_y = \frac{\theta_y}{\theta},\text{ where } \theta = \sqrt{{\theta_x}^2 + {\theta_y}^2} \tag{6.a}$$

$$\boldsymbol{R} = \begin{bmatrix} c\,\theta + u_x{}^2(1 - c\,\theta) & u_x u_y(1 - c\,\theta) & u_y\, s\,\theta \\ u_x u_y(1 - c\,\theta) & c\,\theta + u_y{}^2(1 - c\,\theta) & -u_x\, s\,\theta \\ -u_y\, s\,\theta & u_x\, s\,\theta & c\,\theta \end{bmatrix} \tag{6.b}$$

One can set $\boldsymbol{C_o} = [x_c \quad y_c \quad z_c]^T$ as its position in global coordinate. As $\boldsymbol{B_i}$ is a revolute joint, then $\boldsymbol{B_o C_i}$ is perpendicular to $\boldsymbol{Y_{bi}}$, which is a relationship to find kinematic solutions for x_c and y_c.
So that,

$$\boldsymbol{C} = \begin{bmatrix} \boldsymbol{Y_{b2}} \cdot \boldsymbol{X_o} & \boldsymbol{Y_{b2}} \cdot \boldsymbol{Y_o} \\ \boldsymbol{Y_{b3}} \cdot \boldsymbol{X_o} & \boldsymbol{Y_{b3}} \cdot \boldsymbol{Y_o} \end{bmatrix}^{-1} \cdot \begin{bmatrix} \boldsymbol{R} \cdot \boldsymbol{R_{c2}} \cdot \boldsymbol{Y_{b2}} \\ \boldsymbol{R} \cdot \boldsymbol{R_{c3}} \cdot \boldsymbol{Y_{b3}} \end{bmatrix}, \text{ which satisfies } (\boldsymbol{R} \cdot \boldsymbol{R_{ci}} + \boldsymbol{C_o}) \cdot \boldsymbol{Y_{bi}} = 0 \tag{7.a}$$

$$x_c = \boldsymbol{C} \cdot \boldsymbol{X_o};\ y_c = \boldsymbol{C} \cdot \boldsymbol{Y_o} \tag{7.b}$$

The variable z_c is usually given by the motion planner, however improper selection of this value can result to invalid solution. In this article, an algorithm is explained that seeks the minimum height of platform center $\boldsymbol{C_o}$, z_c, if an eligible solution exists. This algorithm avoids missing valid solutions while keeping minimum solar panel height to mitigate the wind loads bending effects at structure root.
The minimum height algorithm starts by setting the initial value z_{co}.

$$\boldsymbol{C_o}^* = [x_c \quad y_c \quad z_{co}]^T, \text{ where } z_{co} = 0 \tag{8.a}$$

$$\boldsymbol{C_i}^* = \boldsymbol{R} \cdot \boldsymbol{R_{ci}} + \boldsymbol{C_o}^*;\ \boldsymbol{P_o}^* = h_p \cdot \boldsymbol{R} \cdot \boldsymbol{Z_o} + \boldsymbol{C_o}^*;\ \boldsymbol{P_j}^* = \boldsymbol{R} \cdot \boldsymbol{R_{pj}} + \boldsymbol{P_o}^* \tag{8.b}$$

In figure 2. (b), possible candidates of z_c are calculated by determining an offset as a supplement to z_{co}. The offsets are determined considering some criteria. Offset 1: none of $\boldsymbol{P_j}$ point should stay below plane B; Offset 2: an angular constraint γ_{cm} is met which is the minimum angle the leg can form to platform plane $(C_o - C_1C_2C_3)$ under this platform; Offset 3: none of q_i should be less than q_{min}.

Offset 1, d_{o1}:

$$d_{o1} = h_b - z_{pi-min}, \text{ where } z_{pi-min} = min\,(\boldsymbol{P_i} \cdot \boldsymbol{Z_o}) \tag{9.a}$$

$$z_{c-can1} = z_{co} + d_{o1} \tag{9.b}$$

Offset2, d_{o2}:
A unit vector $\boldsymbol{X_{ci}}$ is along the intersection line of planes $C_o - C_1C_2C_3$ and $X_{bi} - B_i - Z_o$.

$$\boldsymbol{X_{ci}} = \frac{\boldsymbol{Y_{bi}} \times \boldsymbol{N_c}}{|\boldsymbol{Y_{bi}} \times \boldsymbol{N_c}|}, \text{ where } \boldsymbol{N_c} = \boldsymbol{R} \cdot \boldsymbol{Z_o} \tag{10.a}$$

θ_{ci} is an acute angle between planes $C_o - C_1C_2C_3$ and $X_{bi} - B_i - Z_o$.

$$\theta_{ci} = \cos^{-1}|\boldsymbol{N_c} \cdot \boldsymbol{Y_{bi}}| \tag{10.b}$$

It is imagined that $\boldsymbol{X_{ci}}$ is rotated at $\boldsymbol{C_i}$ with an acute angle γ_{ci} about $\boldsymbol{Y_{bi}}$ to be altered to a new unit vector $\boldsymbol{X_{\gamma i}}$ which forms an angle γ_{cm} to the platform plane $C_o - C_1C_2C_3$ or to its projection on the platform in $\boldsymbol{N_c}$ direction. The arrowhead of $\boldsymbol{X_{\gamma i}}$ falls on a plane $(C_o - C_1C_2C_3)'$ that is parallel to platform plane and $\boldsymbol{X_{\gamma i}}$ is in plane $X_{bi} - B_i - Z_o$. The gap between planes $(C_o - C_1C_2C_3)'$ and $C_o - C_1C_2C_3$ is certain, and the intersection lines they make on plane $X_{bi} - B_i - Z_o$ are parallel and the gap between which can be determined with θ_{ci}.

IOP Conf. Series: Materials Science and Engineering **435** (2018) 012066 doi:10.1088/1757-899X/435/1/012066

$$\boldsymbol{X_{\gamma i}} = \boldsymbol{Q_i} \cdot \begin{bmatrix} \cos\gamma_{ci} & 0 & \sin\gamma_{ci} \\ 0 & 1 & 0 \\ -\sin\gamma_{ci} & 0 & \cos\gamma_{ci} \end{bmatrix} \cdot \boldsymbol{Q_i}^{-1} \cdot \boldsymbol{X_{ci}}\text{, where } \gamma_{ci} = \sin^{-1}\frac{\sin\gamma_{cm}}{\sin\theta_{ci}} \tag{10.c}$$

If $\boldsymbol{C_iB_i}$ is along $\boldsymbol{X_{\gamma i}}$, the length of $\boldsymbol{C_iB_i}$ for it to reach $\boldsymbol{B_i}$ is $q_{i-\gamma_{cm}}$.

With $\boldsymbol{X_{\gamma i}}$, $q_{i-\gamma_{cm}}$ and $\boldsymbol{C_i}^*$, one can find an imagined point $\boldsymbol{B_{\gamma i}}$ whose global coordinate in $\boldsymbol{Z_o}$ is z_{bi-c}.

$$z_{bi-c} = \left(q_{i-\gamma_{cm}} \cdot \boldsymbol{X_{qi}} + \boldsymbol{C_i}^*\right) \cdot \boldsymbol{Z_o}\text{, where } q_{i-\gamma_{cm}} = \frac{(\boldsymbol{B_i}-\boldsymbol{C_i}^*)\cdot \boldsymbol{X_{bi}}}{\boldsymbol{X_{\gamma i}}\cdot \boldsymbol{X_{bi}}} \tag{10.d}$$

Three legs each has its $\boldsymbol{B_{\gamma i}}$ height, z_{bi-c}, and the minimum among them is selected so that an offset is calculated based on the lowest $\boldsymbol{B_{\gamma i}}$, thus the other two legs will be satisfied with the offset lift.

$$d_{o2} = h_b - z_{bi-c-min}\text{, where } z_{bi-c-min} = \min(z_{bi-c}) \tag{10.e}$$

$$z_{c-can2} = z_{co} + d_{o2} \tag{10.f}$$

Offset 3, d_{o3}:

With $\boldsymbol{C_i}^*$ and q_{min}, one can find an imagined $\boldsymbol{B_{qi}}$ under $\boldsymbol{B_i}$ which ensures the minimum length of linear actuators be met in all legs. The global coordinate of $\boldsymbol{B_{qi}}$ in $\boldsymbol{Z_o}$ is z_{bi-q}.

$$z_{bi-q} = \boldsymbol{C_i}^* \cdot \boldsymbol{Z_o} - \sqrt{q_{min}^{\ 2} - \left((\boldsymbol{B_i} - \boldsymbol{C_i}^*) \cdot \boldsymbol{X_{bi}}\right)^2} \tag{11.a}$$

Three legs each has its $\boldsymbol{B_{qi}}$ height, z_{bi-q}, and the minimum among them is selected so that an offset is calculated based on the lowest $\boldsymbol{B_{qi}}$ and the other two legs will be satisfied with the offset lift.

$$d_{o3} = h_b - z_{bi-q-min}\text{, where } z_{bi-q-min} = \min(z_{bi-q}) \tag{11.b}$$

$$z_{c-can3} = z_{co} + d_{o3} \tag{11.c}$$

There are three z_c candidates, each is the minimum height requirement in its offset calculation that means it may have room to be enlarged. The maximum of the candidates is selected so that the other two offset requirements are included under it.

$$z_c = \max(z_{c-cank})\text{, k} = 1{,}23 \tag{12}$$

Once z_c is determined, the posture needs to be calculated again with updated values.

$$\boldsymbol{C_o} = [x_c \quad y_c \quad z_c]^T \tag{13.a}$$

$$\boldsymbol{C_i} = \boldsymbol{R} \cdot \boldsymbol{R_{ci}} + \boldsymbol{C_o}; \boldsymbol{P_o} = h_p \cdot \boldsymbol{R} \cdot \boldsymbol{Z_o} + \boldsymbol{C_o}; \boldsymbol{P_j} = \boldsymbol{R} \cdot \boldsymbol{R_{pj}} + \boldsymbol{P_o} \tag{13.b}$$

The motion of linear actuations are calculated based on the above platform position.

$$\boldsymbol{q_i} = |\boldsymbol{B_i} - \boldsymbol{C_i}| \text{ and } \theta_{bi} = \tan^{-1}\frac{(\boldsymbol{C_i}-\boldsymbol{B_i})\cdot \boldsymbol{Z_o}}{(\boldsymbol{B_i}-\boldsymbol{C_i})\cdot \boldsymbol{X_{bi}}} \tag{14}$$

The updated value of z_c will need to be verified. The posture with updated z_c need to meet a criteria that none of the linear actuators q_i extends q_{max}, fully extended size.

Within n_r pairs of (θ_x, θ_y), program loops calculate z_c at each (θ_x, θ_y) and then verify. The number of eligible solutions n_e is an index to evaluate the workspace.

4. Reconfiguration

The original $\boldsymbol{B_i}$ locations allows the lie-flat configuration but provides a limited workspace. Genetic algorithm is implemented to find another configuration that enlarges workspace, known as working configuration [22]. The reconfiguration design enable the switch between the two configurations.

The adjustments b_{i-adj} are the variables for the MATLAB optimization algorithm that seeks global minimum of fitness function f, while the ultimate aim is to enlarge the workspace index n_e. The optimization process and results are displayed in figure 3, table 2 and table 3.

$$\mathbf{X} = [b_{1-adj}\; b_{2-adj}\; b_{3-adj}] \tag{15.a}$$

$$f = -n_e \tag{15.b}$$

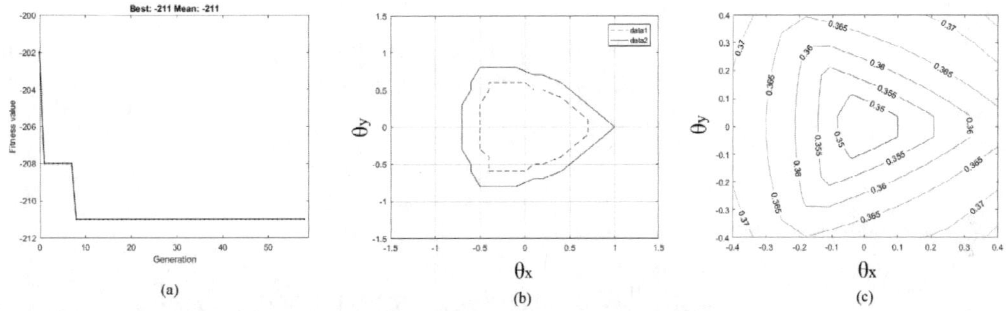

Figure 3. Optimization (a) genetic algorithm for larger workspace; (b) workspace boundary, dash: original; solid: optimized; (c) the z_c position mapping with optimized working configuration.

Table 2. Optimization parameters.

	Parameters	Units	Values
lower	X_{low}	m	$[-0.1 \quad -0.1 \quad -0.1]$
upper	X_{up}	m	$[0 \quad 0 \quad 0]$
search area	θ_x	rad	$-1.5{:}\,0.1{:}\,1.5$
search area	θ_y	rad	$-1.5{:}\,0.1{:}\,1.5$

Table 3. Optimization results.

	Variable **X**	Units	$f = -n_e$
original	[0 0 0]	$N.A.$	−126
optimized	[−0.098 − 0.098 − 0.097]	$N.A.$	−211
rounded	[−0.1 − 0.1 − 0.1]	$N.A.$	−211

5. Stiffness Mapping

The 3-RPS manipulator has 3 DOF, that means it has three independent motions, while the other motions in space are compliant to the independent motions. Article [23] has analyzed the inverse kinematics of 3-RRS parallel manipulator, of which the methods could be used to solve for other 3 DOF parallel manipulators.

5.1. Jacobian Matrix

In the global coordinate, the linear and angular velocities system of the end effector, center of the platform $\boldsymbol{C_o}$, is given below.

$$\boldsymbol{V_{co}} = [\partial x_c \quad \partial y_c \quad \partial z_c]^T \text{ and } \boldsymbol{\omega_{co}} = [\partial\theta_{cx} \quad \partial\theta_{cy} \quad \partial\theta_{cz}]^T \tag{16}$$

Only three of six space motion variables are independent. The variables have a relationship as below.

$$\boldsymbol{Y_{bi}} \cdot \boldsymbol{V_{ci}} = 0, \text{ where } \boldsymbol{V_{ci}} = \boldsymbol{V_{co}} + \boldsymbol{\omega_{co}} \times \boldsymbol{R} \cdot \boldsymbol{R_{ci}} \tag{17.a}$$

$$\boldsymbol{C_v} \cdot \boldsymbol{V_{co}} + \boldsymbol{C_w} \cdot \boldsymbol{\omega_{co}} = 0, \text{ where } \boldsymbol{C_v} = [\boldsymbol{Y_{b1}} \quad \boldsymbol{Y_{b2}} \quad \boldsymbol{Y_{b3}}]^T \text{ and } \boldsymbol{C_w} = \begin{bmatrix} \boldsymbol{R} \cdot \boldsymbol{R_{c1}} \times \boldsymbol{Y_{b1}} \\ \boldsymbol{R} \cdot \boldsymbol{R_{c2}} \times \boldsymbol{Y_{b2}} \\ \boldsymbol{R} \cdot \boldsymbol{R_{c3}} \times \boldsymbol{Y_{b3}} \end{bmatrix} \tag{17.b}$$

$$\boldsymbol{\omega_{co}} = -\boldsymbol{C_w}^{-1} \cdot \boldsymbol{C_v} \cdot \boldsymbol{V_{co}} \tag{17.c}$$

And the methods in [24] and [25] suggests to eliminate the compliant motion by dot multiplying a vector perpendicular to the compliant motions on both sides of the equation.

$$\boldsymbol{V_{qi}} + \omega_{qi} \cdot \boldsymbol{Y_{bi}} \times (\boldsymbol{C_i} - \boldsymbol{B_i}) = \boldsymbol{V_{ci}} \tag{18.a}$$

$$(\boldsymbol{C_i} - \boldsymbol{B_i}) \cdot \boldsymbol{V_{qi}} + 0 = (\boldsymbol{C_i} - \boldsymbol{B_i}) \cdot \boldsymbol{V_{ci}} \tag{18.b}$$

where,

$$\boldsymbol{V_{qi}} = \frac{(\boldsymbol{C_i} - \boldsymbol{B_i})}{|\boldsymbol{C_i} - \boldsymbol{B_i}|} \cdot \partial q_i = \boldsymbol{X_{qi}} \cdot \partial q_i \tag{18.c}$$

Rearrange the above, one has the following.

$$(\boldsymbol{C_i} - \boldsymbol{B_i}) \cdot \frac{(\boldsymbol{C_i} - \boldsymbol{B_i})}{|\boldsymbol{C_i} - \boldsymbol{B_i}|} \cdot \partial q_i = (\boldsymbol{C_i} - \boldsymbol{B_i}) \cdot (\boldsymbol{V_{co}} + \omega_{co} \times \boldsymbol{R} \cdot \boldsymbol{R_{ci}}) \tag{19.a}$$

$$q_i \cdot \boldsymbol{X_{qi}}^2 \cdot \partial q_i = q_i \cdot \boldsymbol{X_{qi}} \cdot (\boldsymbol{V_{co}} + \omega_{co} \times \boldsymbol{R} \cdot \boldsymbol{R_{ci}}) \tag{19.b}$$

$$\boldsymbol{X_{qi}}^2 = 1 \tag{19.c}$$

$$q_i \cdot \partial q_i = q_i \cdot \boldsymbol{X_{qi}} \cdot \left(\boldsymbol{V_{co}} - \boldsymbol{C_w}^{-1} \cdot \boldsymbol{C_v} \cdot \boldsymbol{V_{co}} \times \boldsymbol{R} \cdot \boldsymbol{R_{ci}}\right) \tag{19.d}$$

$$\partial q_i = \boldsymbol{X_{qi}} \cdot \left(\boldsymbol{V_{co}} - \boldsymbol{C_w}^{-1} \cdot \boldsymbol{C_v} \cdot \boldsymbol{V_{co}} \times \boldsymbol{R} \cdot \boldsymbol{R_{ci}}\right) \tag{19.e}$$

$$\partial q_i = \boldsymbol{X_{qi}} \cdot \boldsymbol{V_{co}} - \boldsymbol{X_{qi}} \cdot \left(\boldsymbol{C_w}^{-1} \cdot \boldsymbol{C_v} \cdot \boldsymbol{V_{co}}\right) \times (\boldsymbol{R} \cdot \boldsymbol{R_{ci}}) \tag{19.f}$$

$$\partial q_i = \boldsymbol{X_{qi}} \cdot \boldsymbol{V_{co}} + \boldsymbol{X_{qi}} \cdot (\boldsymbol{R} \cdot \boldsymbol{R_{ci}}) \times \left(\boldsymbol{C_w}^{-1} \cdot \boldsymbol{C_v} \cdot \boldsymbol{V_{co}}\right) \tag{19.g}$$

$$\partial q_i = \boldsymbol{X_{qi}} \cdot \boldsymbol{V_{co}} + \boldsymbol{X_{qi}} \times \boldsymbol{R} \cdot \boldsymbol{R_{ci}} \cdot \boldsymbol{C_w}^{-1} \cdot \boldsymbol{C_v} \cdot \boldsymbol{V_{co}} \tag{19.h}$$

$$\partial q_i = \boldsymbol{X_{qi}}^T \cdot \boldsymbol{V_{co}} + \left(\boldsymbol{X_{qi}} \times \boldsymbol{R} \cdot \boldsymbol{R_{ci}}\right)^T \cdot \left(\boldsymbol{C_w}^{-1} \cdot \boldsymbol{C_v}\right) \cdot \boldsymbol{V_{co}} \tag{19.i}$$

$$\partial q_i = \left(\boldsymbol{X_{qi}}^T + \left(\boldsymbol{X_{qi}} \times \boldsymbol{R} \cdot \boldsymbol{R_{ci}}\right)^T \cdot \left(\boldsymbol{C_w}^{-1} \cdot \boldsymbol{C_v}\right)\right) \cdot \boldsymbol{V_{co}} \tag{19.j}$$

$$\boldsymbol{J_{ri}} = \left(\boldsymbol{X_{qi}} + \boldsymbol{X_{qi}} \times \boldsymbol{R} \cdot \boldsymbol{R_{ci}}\right)^T \cdot \left(\boldsymbol{C_w}^{-1} \cdot \boldsymbol{C_v}\right) \tag{19.k}$$

$$\partial q_i = \boldsymbol{J_{ri}} \cdot \boldsymbol{V_{co}} \tag{19.l}$$

The Jacobian matrix is obtained as below.

$$\boldsymbol{J} = [\boldsymbol{J_{r1}} \quad \boldsymbol{J_{r2}} \quad \boldsymbol{J_{r3}}]^T \tag{20.a}$$

This reflects the relationship between actuation velocities and the end effector velocities.

$$[\partial q_1 \quad \partial q_2 \quad \partial q_3]^T = \boldsymbol{J} \cdot [\partial x_c \quad \partial y_c \quad \partial z_c]^T \tag{20.b}$$

5.2. Stiffness Matrix

Stiffness of end effector $\boldsymbol{K_c}$ is expressed as below, where $\boldsymbol{K_q}$ is the actuation stiffness.

IOP Conf. Series: Materials Science and Engineering **435** (2018) 012066 doi:10.1088/1757-899X/435/1/012066

$$\boldsymbol{K_c} = \boldsymbol{J}^T \cdot \boldsymbol{K_q} \cdot \boldsymbol{J}\text{ , where } \boldsymbol{K_q} = \begin{bmatrix} k_{q1} & 0 & 0 \\ 0 & k_{q2} & 0 \\ 0 & 0 & k_{q3} \end{bmatrix} \tag{21}$$

The stiffness along $\boldsymbol{X_o}$, $\boldsymbol{Y_o}$ and $\boldsymbol{Z_o}$ are given.

$$k_x = (\boldsymbol{K_c} \cdot \boldsymbol{X_o})^T \cdot \boldsymbol{X_o};\ k_y = (\boldsymbol{K_c} \cdot \boldsymbol{Y_o})^T \cdot \boldsymbol{Y_o};\ k_z = (\boldsymbol{K_c} \cdot \boldsymbol{Z_o})^T \cdot \boldsymbol{Z_o} \tag{22}$$

When $\boldsymbol{X_o}$ rotates around $\boldsymbol{Z_o}$ by φ_z, it forms a new vector $\boldsymbol{X_\varphi}$. The vector $\boldsymbol{X_o}$ expressed in coordinate with $\boldsymbol{X_\varphi}$ is given below.

$$(\boldsymbol{X_o})_\varphi = \boldsymbol{R_z} \cdot \boldsymbol{X_o}\text{ , where } \boldsymbol{R_z} = \begin{bmatrix} \cos(-\varphi_z) & -\sin(-\varphi_z) & 0 \\ \sin(-\varphi_z) & \cos(-\varphi_z) & 0 \\ 0 & 0 & 1 \end{bmatrix} \tag{23.a}$$

The velocity vector $(\boldsymbol{V_{co}})_\varphi$ represents the end effector velocity in coordinate with $\boldsymbol{X_\varphi}$. It replaces velocity $\boldsymbol{V_{co}}$ with transformation matrix $\boldsymbol{R_z}^{-1}$.

$$\text{As } \boldsymbol{V_{co}} = \boldsymbol{R_z}^{-1} \cdot (\boldsymbol{V_{co}})_\varphi,\ \partial q_i = \boldsymbol{J_{ri}} \cdot \boldsymbol{R_z}^{-1} \cdot (\boldsymbol{V_{co}})_\varphi\text{ , where } \boldsymbol{R_z}^{-1} = \begin{bmatrix} \cos\varphi_z & -\sin\varphi_z & 0 \\ \sin\varphi_z & \cos\varphi_z & 0 \\ 0 & 0 & 1 \end{bmatrix} \tag{23.b}$$

This provides a new Jacobian matrix with a rotation angle index of φ_z.

$$\boldsymbol{J}_{\varphi i} = \boldsymbol{J_{ri}} \cdot \boldsymbol{R_z}^{-1}\text{, so that } \boldsymbol{J}_\varphi = [\boldsymbol{J}_{\varphi 1} \quad \boldsymbol{J}_{\varphi 2} \quad \boldsymbol{J}_{\varphi 3}]^T \tag{23.c}$$

The stiffness $k_{x\varphi}$ along $\boldsymbol{X_\varphi}$ could be calculated.

$$\boldsymbol{K_\varphi} = \boldsymbol{J_\varphi}^T \cdot \boldsymbol{K_q} \cdot \boldsymbol{J_\varphi} \tag{23.d}$$

$$k_{x\varphi} = \left(\boldsymbol{K_\varphi} \cdot \boldsymbol{X_o}\right)^T \cdot \boldsymbol{X_o} \tag{23.e}$$

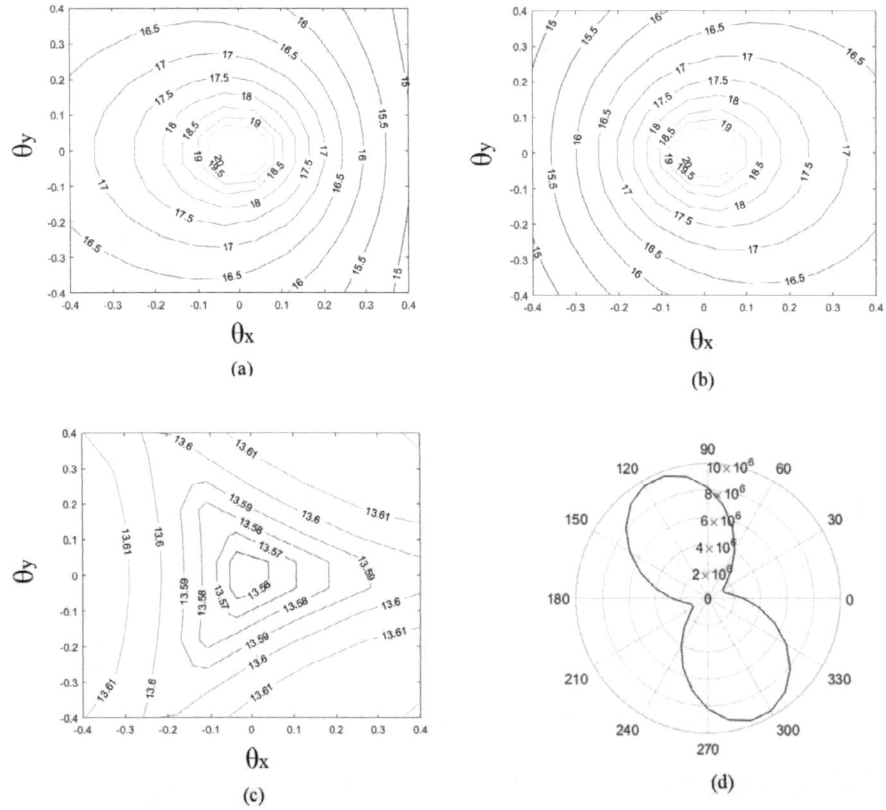

Figure 4. Stiffness mappings in prescribed motion range (a) log(k_x) N/m; (b) log(k_y) N/m; (c) log(k_z) N/m; (d) stiffness $k_{x\varphi}$ polar plot about φ_z $\theta_x = 0.4$ and $\theta_y = -0.5$ at working configuration $\boldsymbol{X} = [-0.1 - 0.1 \ -0.1]$.

The leg $\mathbf{B_iC_i}$ has the longitudinal stiffness $k_q = 4.9568 \times 10^5 \ N/m$, which has two components in series, linear actuator stiffness $k_a = 5 \times 10^5 \ N/m$ and eight parallel piezo chips stiffness each with k_p. The stiffness of k_p can be calculated with the design sizes [26]. The matireal Lead Zirconate Titanate (PZT) takes its Young's module $E_{pzt} = 63$ Gpa from [27].

$$\frac{1}{k_q} = \frac{1}{8 \cdot k_p} + \frac{1}{k_a}, i = 1,2,3 \tag{24}$$

Stiffness mapping visualize the directional stiffness within the selected range of workspace [28].

Figure 4. (a), (b) and (c) show the stiffness in directions along $\boldsymbol{X_o}$, $\boldsymbol{Y_o}$ and $\boldsymbol{Z_o}$. However the wind loads could be in any direction. As in figure 4. (d), for a certain configuration and posture of the manipulator, when φ_z varies within 0°~360°, the stiffness $k_{x\varphi}$ with respect to φ_z could be visualized with a polar plot, which helps to evaluate the manipulator loading capacity to horizontal winds in specific horizontal directions.

6. Conclusion

The design of a reconfigurable hybrid harvester enables 3-RPS parallel solar tracker to collect solar and wind energy at the same time. This design has a lie-flat feature that protects the manipulator from storms, and an enlarged workspace achieved by reconfiguration. Reconfiguration enables the manipulator to switch between the lie-flat configuration and the working configuration. Minimum platform height algorithm has been developed and demonstrated to find all eligible orientations of the platform. Stiffness mappings of the manipulator are plotted, after being converted to a polar plot, it assists to evaluate its capability to take wind loads in specific horizontal directions that is significant for lie-flat protection strategy against storms. The design adds favorable features to solar trackers that improves the energy collection efficiency and operational safety of the green industry.

Acknowledgement
The authors would like to thank the financial support from the Natural Sciences and Engineering Research Council of Canada (NSERC). The authors gratefully acknowledge the financial support from Kaneff Research Chairs program.

References

[1] T. Tudorache and L. Kreindler, 2010, "Design of a Solar Tracker System for PV Power Plants", Acta Polytechnica Hungarica, Vol. 7, No. 1, 2010
[2] D. Gnanarathinam.S, Sundaramurthy.S, and A. Wahi, 2015, "Design and Implementation of a Dual Axis Solar Tracking System", International Journal for Research in Applied Science & Engineering Technology, Vol. 3, Issue X, October 2015
[3] C. Thurston, 2015, "Ensuring Your Solar Array Doesn't Get Caught in the Wind", Renewable Energy World, Vol. 18, Issue 4, 2015
[4] "Solar Field in Ontario Sees Damage Following Winter Storm", from http://13wham.com/news/top-stories/solar-field-in-ontario-sees-damage-following-winter-storm
[5] "Tropical Storm Risk (TSR)" from http://www.tropicalstormrisk.com
[6] D. Zhang, Chapter 1 Introduction, Parallel Robotic Machine Tool, Springer 2010
[7] A. Cammarata, 2014, "Optimized Design of A Large-Workspace 2-DOF Parallel Robot for Solar Tracking Systems", Mechanism and Machine Theory Volume 83, January 2015, Pages 175–186
[8] "Design of Multiaxial Parallel Kinematic Machines with High Dynamic Capabilities" from http://www.ehu.eus/compmech/research-2/projects/design-of-multiaxial-parallel-kinematic-machines-with-high-dynamic-capabilities-dynamech
[9] Ashith Shyam R B and Ghosal A, 2015, "Three-Degree-of-Freedom Parallel Manipulator to Track the Sun for Concentrated Solar Power Systems", Chinese Journal of Mechanical Engineering, July 2015, Volume 28, pp 793–800, 2015
[10] Z. Yang and D. Zhang, "Reconfigurable 3-PRS Parallel Solar Tracking Stand", Proceedings of the ASME 2017 International Design Engineering Technical Conferences & Computers and Information in Engineering Conference, IDETC2017 August 6-9, 2017, Cleveland, Ohio, USA
[11] D. Zhang, Chapter 7 Reconfigurable Parallel Kinematic Machine Tools, Parallel Robotic Machine Tool, Springer 2010
[12] J. Sirohi and R. Mahadik, 2011, "Piezoelectric Wind Energy Harvester for Low-Power Sensors", Journal of Intelligent Material Systems and Structures 22(18) 2215–2228, 2011
[13] J.X. Tao, N.V. Viet, A. Carpinteri, and Q. Wang, 2017, "Energy harvesting from Wind by A Piezoelectric Harvester", Engineering Structures 133 (2017) 74–80
[14] J. Sirohi and R. Mahadik, 2012, "Harvesting Wind Energy Using a Galloping Piezoelectric Beam", Journal of Vibration and Acoustics Transactions of ASME, Vol. 134, FEBRUARY 2012
[15] L. M. Murphy, 1980, "Wind Loading on Tracking and Field Mounted Solar Collectors", Solar Energy Research Institute, United States. doi: 10.2172/6889663.
[16] S. Bhadurl and L. M. Murphy, 1985, "Wind Loading on Solar Collectors", Solar Energy Research Institute, United States, 1985
[17] Linear Table Alpha 15-B-155, HSB alpha Release: 12.12.2016
[18] Linear Actuators: DLA Series and Accessories, 50 mm to 300 mm stroke 1000 N, Transmotec 2018

IOP Conf. Series: Materials Science and Engineering **435** (2018) 012066 doi:10.1088/1757-899X/435/1/012066

[19] L. Schreiber and C. Gosselin, 2017, "Passively Driven Redundant Spherical Joint With Very Large Range of Motion", Journal of Mechanisms and Robotics Transactions of ASME, JUNE 2017, Vol. 9, 2017
[20] G. Yuan and D. H. Wang, 2017, "A Piezoelectric Six-DOF Vibration Energy Harvester Based on Parallel Mechanism: Dynamic Modeling, Simulation, and Experiment", Smart Materials and Structures, 26 (2017)
[21] "Rotational Matrix" from wikipedia.org
[22] "Genetic Algorithm: Find Global Minima for Highly Nonlinear Problems" from mathworks.com
[23] J. Li, J. Wang, W. Chou, Y. Zhang, T. Wang and Q. Zhang, "Inverse Kinematics and Dynamics of the 3-RRS Parallel Platform", Proceedings of the 2001 IEEE International Conference on Robotics and Automation, May 21-26 2001, Seoul Korea
[24] Y. Li and Q. Xu, "Kinematics and Stiffness Analysis for a General 3-PRS Spatial Parallel Mechanism", ROMANSY Conference 2004, June 14-18, 2004, Montreal, Canada
[25] S. R. Babu, V. R. Raju and K. Ramji, 2013, "Design for Optimal Performance of 3-RPS Parallel Manipulator Using evolutionary Algorithms", Transactions of the Canadian Society for Mechanical Engineering, Vol. 37, No. 2, 2013
[26] R. G. Budynas and J. K. Nisbett, Shigley's Mechanical Engineering Design, Ninth Edition, McGraw-Hill 2008, pp.1013
[27] "Material: Lead Zirconate Titanate (PZT)" from memsnet.org
[28] C. Gosselin, "Stiffness Mapping for Parallel Manipulators", IEEE Transactions on Robotics and Automation, Vol. 6, Issue 3, Jun 1990

AUTHOR INDEX

AUTHOR INDEX